ANIMATED NATURE.

THE

ILLUSTRATED

NATURAL HISTORY.

BY THE REV.

J. G. WOOD, M.A., F.L.S.,

AUTHOR OF "THE ILLUSTRATED NATURAL HISTORY OF MAN," "ANECDOTES OF ANIMAL LIFE,"
"MY FEATHERED FRIENDS," ETC. ETC.

WITH NEW DESIGNS
BY WOLF, ZWECKER, WEIR, COLEMAN, HARVEY, ETC. ETC.
ENGRAVED BY THE BROTHERS DALZIEL.

REPTILES, FISHES, MOLLUSCS,
&c. &c. &c

LONDON:
GEORGE ROUTLEDGE AND SONS, BROADWAY, LUDGATE HILL.
NEW YORK: 9, LAFAYETTE PLACE.

LONDON :
RICHARD CLAY AND SONS,
BREAD STREET HILL,
And at Bungay, Suffolk.

PREFACE.

——— —

As this Volume is conducted on precisely the same principles as those which characterised the two previous instalments of the work, no introductory preface is needful. I cannot, however, dismiss the Volume without returning my sincere thanks to the many friends who have aided me by their valuable assistance.

More especially are my acknowledgments due to those eminent naturalists, such as Mr. Blackwall and Mr. Busk, who liberally permitted the use of figures taken from unpublished volumes; and to the officers connected with the Zoological Department of the British Museum, without whose constant assistance I could not have carried out so laborious a task.

London, 1863.

AMERICAN RIVER TORTOISES AT HOME.

REPTILES.

THE remarkable beings which are classed together under the general title of REPTILES, or creeping animals, are spread over those portions of the globe where the climate is tolerably warm, and are found in the greatest profusion under the hotter latitudes. Impatient of cold, though capable of sustaining a temperature of such freezing chilliness

3.   B

that any of the higher animals would perish under its severity, and for the most part being lovers of wet and swampy situations, the Reptiles swarm within the regions near the equator, and in the rivers or vast morasses of the tropical countries the very soil appears to teem with their strange and varied forms. Indeed, the number of Reptiles to be found in any country is roughly indicated by the parallels of latitude, the lands near the equator being the most prolific in these creatures, and containing fewer as they recede towards the poles.

Some Reptiles inhabit the dry and burning deserts; but the generality of these creatures are semi-aquatic in their habits, are fitted by their structure for progression on land or in water, and are able to pass a considerable time below the surface without requiring to breathe.

This capacity is mostly the result of the manner in which the circulation and aeration of their blood is effected.

As has been shown in the two volumes on Mammalia and Birds, the heart in these animals is divided into a double set of compartments, technically termed auricles and ventricles, each set having no direct communication with the other. In the Reptiles, however, this structure is considerably modified, the arterial and venous blood finding a communication either within or just outside the two ventricles, so that the blood is never so perfectly aerated as in the higher animals. The blood is consequently much colder than in the creatures where the oxygen obtains a freer access to its particles.

In consequence of this organization the whole character of the Reptiles is widely different from that of the higher animals. Dull sluggishness seems to be the general character of a Reptile, for though there are some species which whisk about with lightning speed, and others, especially the larger lizards, can be lashed into a state of terrific frenzy by love, rage, or hunger, their ordinary movements are inert, their gestures express no feeling, and their eyes, though bright, are stony, cold, and passionless. Their mode of feeding accords with the general habits of their bodies, and the process of digestion is peculiarly slow.

Most of the Reptiles possess four legs, but are not supported wholly upon them, their bellies reaching the ground and being dragged along by the limbs. One or two species can support themselves in the air while passing from one tree to another, much after the fashion of the flying squirrels; and in former days. when Reptiles were apparently the highest race on the surface of the earth, certain species were furnished with wing-like developments of limb and skin, and could apparently flap their way along like the bats of the present time.

Excepting some of the tortoise tribe, the Reptiles are carnivorous beings, and many of them, such as the crocodiles and alligators, are among the most terrible of rapacious creatures. In this class of animals we find the first examples of structures which transmute Nature's harmless gifts into poison, a capacity which is very common in the later orders, such as the spiders and insects, and is developed to a terrible extent in some of the very lowest beings that possess animal life, rendering them most formidable even to man.

The skeleton of a true Reptile, from which class the *Batrachians, i.e.* the frogs, salamanders, and their kin are excluded, for reasons which will presently be given, is composed of well-ossified bones, and is peculiarly valuable to the physiologist. It is well known to all who have studied the rudiments of anatomy, that each bone is formed from several centres, so to speak, consisting of mere cartilaginous substance at its earliest formation, and becoming gradually ossified from several spots.

In the young of the higher animals these centres are only seen during their very earliest stages, and are by degrees so fused together that all trace of them is obliterated. But in the Reptiles it is found that many of the bones either remain in their separate parts, or leave so distinct a mark at the place where they unite, that their shape and dimensions are clearly shown. In the head of the adult crocodile, for example, the frontal bone is composed of five distinct pieces, the temporal of at least five pieces, and each side of the lower jaw-bone is composed of either five or six portions united by sutures.

With the exception of the tortoises, the Reptiles mostly possess a goodly array of teeth, set in the jaw or palate, and as a general fact, being sharp and more or less curved backward. Their bodies are covered with various modifications of the structure termed the dermal, *i.e.* skin skeleton, and are furnished with scales and plates of different forms. In some cases the scales lie overlapping each other like those of the fish, in others they are modified into knobby plates, and in some, of which the tortoises afford well-known examples, they form large flat plates on the back and breast, and scales upon the feet and legs.

The young of Reptiles are produced from eggs, mostly being hatched after they have been laid, but in some cases the young escape from the eggs before they make their appearance in the world. As a general fact, however, the eggs of Reptiles are placed in some convenient spot, where they are hatched by the heat of the sun. Some species are

SKELETON OF CROCODILE

very jealous about their eggs, keeping a strict watch over them, and several of the larger serpents have a curious fashion of laying the eggs in a heap, and then coiling themselves around them in a great hollow cone. The size of the eggs is extremely variable, for, although as a general fact those of the smaller Reptiles are large in proportion to the dimensions of the parent, those of the crocodiles and alligators are wonderfully small, not larger than those of our domestic geese, and in many cases much smaller. They are usually of a dull white colour, and in some instances are without a brittle shell, their covering being of a tough leathery consistence.

In form, and often in colour, the Reptiles exhibit an inexhaustible variety, and even each order displays a diversity of outward aspect unexampled in the two previous classes of Mammals and Birds. Strange, grotesque, and oftentimes most repulsive in appearance, though sometimes adorned with the brightest tints, the Reptiles excite an instinctive repugnance in the human breast; and whether it be a lizard, a snake, or a tortoise, the sudden and unsuspected contact of one of these beings will cause even the most habituated to recoil from its cold touch. This antipathy may, perhaps, have some connexion with the instinctive association of cold with death; but whatever may be the cause, the feeling is deep and universal.

TORTOISES.

THE very curious reptiles which are known by the general name of Tortoises, are remarkable for affording the first example of a skeleton brought to the exterior of the body, a formation which is frequent enough in the lower orders, the crustaceans and insects being familiar examples thereof. In these reptiles the bones of the chest are developed into a curious kind of box, more or less perfect, which contains within itself all the muscles and the viscera, and in most cases can receive into its cavity the head, neck, and limbs; in one genus so effectually, that when the animal has withdrawn its limbs and head, it is contained in a tightly closed case without any apparent opening.

The shell of the Tortoise is divided into two portions, the upper being termed the carapace, and the lower the plastron.

The carapace is formed by a remarkable development of the vertebræ and ribs, which throw out flat processes, and are joined together by sutures like the bones of the skull. The back is therefore incapable of movement, and from the arched shape of the bones is wonderfully strong when resting on the ground. In the Tortoises these bones are united throughout their entire length, but in the Turtles the ends of the ribs retain their original width.

The plastron is similarly formed of the breastbone, which is thought in these creatures to be developed to the greatest extent of which it is capable. It is composed of nine pieces, each being formed from one of the bony centres already mentioned. These bones are arranged in four pairs, and one in the centre of the front.

SKELETON OF TORTOISE.

As all the limbs have to be worked from the interior of the chest, amid the vital organs and muscles for moving them, they undergo considerable modification. The shoulder-blade, for example, is a curious three-branched bone, quite unique among vertebrate animals, the portion which represents the true shoulder-blade being almost cylindrical, one of the branches flattened, and the other cylindrical, but larger than the real blade-bone. This structure admits of the attachment of powerful muscles, and gives to the fore limbs the great strength which is needed for digging, swimming, climbing, and various modes of exertion. The strong curved bones of the fore limbs bear an evident analogy to the corresponding parts in the mole, with its powerful claws and feet, and its very long blade-bone.

The horny substance commonly termed "tortoiseshell," which is spread in flattened plates on the exterior of the bony case, is thought to be a modification of the scales found on lizards, serpents, &c., and which exist on the legs and other parts of the Tortoises themselves. The row of horny pieces which are found on the edge of the carapace also belong to the "dermal skeleton."

The Tortoises are quite devoid of teeth, the edge of the jaws being sharp and horny, so as to inflict a severe wound; and in many species one or both jaws are sharply hooked at the tip like a falcon's beak. The neck is always rather long, and in many species can be protruded to a considerable extent. Generally, the process of thrusting the neck from the shell is a slow one, but the withdrawal is accomplished with marvellous rapidity, on account of certain long muscles which tie the neck to the back of the carapace. Possibly these muscles, together with their tendons, would, when dried in the baking sunshine, produce musical sounds when touched, and thus give rise to the old poetical legend of the origin of the lyre.

The brain of the Tortoise is very small in proportion to the size of the animal, in the turtle weighing not quite one five-thousandth part of the whole body, and in the land Tortoise about one two-thousandth part. In man the brain is about one-fortieth the weight of the body.

The Tortoises produce their young from eggs, mostly soft and leathery in the texture of their covering, which are laid in some convenient spot, and left to be hatched by heat not derived from the parent. The circulation in the Tortoise is not very complete, but the arterial blood is redder and brighter than the venous.

IN the true TORTOISES the feet are club-shaped and the claws blunt, and the neck can be wholly withdrawn within the shell.

The first example of these creatures is the GOPHER, or MUNGOFA TORTOISE, a native of America. This is a rather pretty, though not brightly coloured species, its shell being mostly brownish yellow, boldly and variously clouded with rich dark brown. The lower jaw is yellow, and the whole of the plastron is yellow-brown. It is found plentifully in Georgia and Alabama, but according to Mr. Holbrook is not seen farther north than South Carolina. When full grown it is a moderately large species, from thirteen to more than fourteen inches in length, and very convex. The following interesting account of its habits is given by Mr. Holbrook in his valuable "North American Herpetology:"—

"They select dry and sandy places, are generally found in troops, and are very abundant in pine-barren countries. They are gentle in their habits, living entirely on vegetable substances. They are fond of the sweet potato (Convolvulus batatas), and at times do much injury to gardens by destroying melons, as well as bulbous roots, &c. &c. In the wild state they are represented as nocturnal animals, or as seeking their food by night: when domesticated—and I have kept many of them for years—they may be seen grazing at all hours of the day.

When first placed in confinement, they chose the lowest part of the garden, where they could most easily burrow. This spot being once overflowed by salt water in a high spring-tide, they migrated to the upper part, nearly eighty yards distant, and prepared anew their habitations. They seldom wandered far from their holes, and generally spent part of the day in their burrows. They delighted in the sun in mild weather, but could not support the intense heat of our summer noons; at those hours they retreated to their holes, or sought shelter from the scorching rays of the sun under the shade of broad-leaved plants. A tanyer (Arum esculentum) that grew near their holes was a favourite

GOPHER TORTOISE.—*Testudo Gopher.*

haunt. They could not endure rain, and retreated hastily to their burrows, or to other shelter, at the coming on of a shower.

As winter approached, they confined themselves to the immediate neighbourhood of their holes, and basked in the sunshine. As the cold increased, they retired to their burrows, where they became torpid; a few warm days, however, even in winter, would again restore them to life and activity.

The adults are remarkably strong, sustaining and moving with a weight of two hundred pounds or more. The female is generally larger than the male, with the sternum convex; the sternum of the male is concave, especially on its posterior part. The eggs are larger than those of a pigeon, round, with a hard calcareous shell; they are much esteemed as an article of food."

Perhaps the best known species of these creatures is the Common Land Tortoise, so frequently exposed for sale in our markets, and so favourite an inhabitant of gardens.

This appears to be the only species that inhabits Europe, and even in that continent it is by no means widely spread, being confined to those countries which border the Mediterranean.

It is one of the vegetable feeders, eating various plants, and being very fond of lettuce leaves, which it crops in a rather curious manner, biting them off sharply when fresh and crisp, but dragging them asunder when stringy, by putting the fore feet upon them, and pulling with the jaws. This Tortoise will drink milk, and does so by opening its mouth, scooping up the milk in its lower jaw, as if with a spoon, and then raising its head to let the liquid run down its throat.

One of these animals, which I kept for some time, displayed a remarkable capacity for climbing, and was very fond of mounting upon various articles of furniture, stools being its favourite resort. It revelled in warmth, and could not be kept away from the hearth-rug, especially delighting to climb upon a footstool that generally lay beside the fender. It used to clamber on the stool in a rather ingenious manner. First it got on its hind legs, rearing itself against the angle formed by the stool and fender. Then it would slowly raise one of its hind legs, hitch the claws into a hole in the fender, and raise itself

COMMON LAND TORTOISE.—*Testúdo Grœca.*

very gradually, until it could fix the claws of the other hind foot into the thick carpet-work of the stool. A few such steps would bring it to the top of the stool, when it would fall down flat, crawl close to the fender, and there lie motionless. If it were taken off twenty times a day, and carried to the other end of the room, it would always be found in its favourite resort in a few minutes.

This Tortoise had a curious kind of voice, not unlike the mewing of a little kitten. The Common Tortoise is known to live to a great age, and an interesting account of one of these animals may be found in White's "Selborne."

To this genus belongs a very large species, worthy of a passing description. This is the great INDIAN TORTOISE (*Testudo Indica*), a native of the Galapagos. This species is also known scientifically by the name of *Testudo planiceps*. Mr. Darwin writes as follows of this animal and its habits: "The Tortoise is very fond of water, drinking large quantities, and wallowing in the mud. The larger islands alone produce springs, and these are always situated toward the central parts, and at a considerable elevation. Hence broad and well-beaten paths radiate in every direction from the wells, even down to the sea-coast; and the Spaniards, by following them up, first discovered the watering-places.

When landed at Chatham Island, I could not imagine what animal travelled so methodically along the well-beaten tracks. Near the springs it was a curious spectacle to behold many of these great monsters, one set eagerly travelling onwards with outstretched necks, and another set returning, after having drunk their fill. When the Tortoise arrives at the spring, quite regardless of any spectator, it buries its head in the water above its eyes, and greedily swallows great mouthfuls, at the rate of about ten in a minute. The inhabitants say each animal stays three or four days in the neighbourhood of the water, and then returns to the lower country.

For some time after a visit to the springs the bladder is distended with fluid, which is said gradually to decrease in volume, and to become less pure. The inhabitants, when walking in the lower districts, and overcome with thirst, often take advantage of this circumstance by killing a Tortoise, and, if the bladder is full, drinking the contents. In one I saw killed, the fluid was quite limpid, and had only a very slightly bitter taste.

The inhabitants, however, always drink first the water in the pericardium, which is described as being best."

The flesh of these Tortoises is very good, and is largely eaten, both fresh and salted. A clear oil is also obtained from the fat. Those who catch these Tortoises do not choose to go through the trouble of cutting up and dressing an animal that is not quite fat, and, as the fitness of its condition cannot be ascertained by the ordinary process, a summary method is employed, viz. cutting a slit through the softer skin near the tail, so as to show the fat under the carapace. Should the Tortoise be in poor condition, it is allowed to go free, and, with the imperturbable temperament of the reptile race, seems to care little for the wound.

Dr. Livingstone mentions a species of Land Tortoise which is remarkable for its love of salt, and the extreme strength of the shell, which, as will be seen, baffles even the teeth of the hyæna, which can crush an ox-bone with ease.

"Occasionally we lighted upon Land Tortoises, which, with their unlaid eggs, make a very agreeable dish. We saw many of their trails leading to the salt fountains; they must have come great distances for this health-giving article. In lieu thereof they often devour wood-ashes. The young are taken for the sake of their shells, which, when filled with sweet-smelling roots, the women hang around their persons. When taken it is used as food, and the shell converted into a rude basin to hold food or water.

It owes its continuance neither to speed nor cunning. Its colour, yellow and dark-brown, is well adapted, by its similarity to the surrounding grass and brushwood, to render it undistinguishable; and though it makes an awkward attempt to run on the approach of man, its trust is in its bony covering, from which even the teeth of a hyæna glance off foiled.

When this long-lived creature is about to deposit her eggs, she lets herself into the ground by throwing the earth up around her shell until only the top is visible; then, covering up the eggs, she leaves them until the rains begin to fall, and the fresh herbage appears; the young ones then come out, their shells still quite soft, and unattended by their dam, begin the world for themselves. Their food is tender grass, and a plant named 'thotona,' and they frequently resort to heaps of ashes, and places containing efflorescence of the nitrates for the salts these contain."

THE curious Tortoise which is known only by the comparatively scientific name of PYXIS inhabits several parts of the world, and is not uncommon in some portions of India and Madagascar, from which latter localities the specimens in the British Museum have been brought.

In common with one or two other species, hereafter to be described, the Pyxis has the power of drawing its head, neck, and limbs within the shell and then shutting itself down by means of a lid, formed by the movable front of the sternum. In most of this tribe of reptiles, the sternum is hard and immovable, but in the Pyxis, it moves on a leathery kind of hinge, so as to open when the creature wishes to thrust out its head and limbs, and to close firmly when it withdraws within the shelter of its bony armour.

In order to permit of this total withdrawal into the shell, the carapace is oval and more convex than is usually the case, so as to afford a sufficient space for the reception of the head and limbs. These, too, are rather diminutive in proportion to the size of the animal, and so formed as to be packed into a small compass. The Tortoise employs this curious mode of guarding its vulnerable points whenever it fears danger, and is then so securely locked up in its armour-plates that it is safe from almost every enemy except man. The word Pyxis is Greek, and is very appropriately given to this species, its signification being a box.

The Pyxis is a pretty, but not a large species. The colour is extremely variable, scarcely any two individuals being precisely alike, but the general colours are yellow and black. On the carapace the plates are marked with a number of radiating triangular spots, and on the plates which edge the shell there are lines of black. Below, the yellow generally takes a more orange tint, and is diversified with black marks round its edge.

THE PYXIS.—*Pyxis arachnoïdes.*   PYXIS.   Under side with closed shell.)

WE now come to a group of Tortoises called TERRAPINS.

These creatures are inhabitants of the water, and are mostly found in rivers. They are carnivorous in their diet and take their food while in the water. They may be known by their flattened heads, covered with skin, sometimes hard but often of a soft consistency, and their broad feet with the toes webbed as far as the claws.

The LETTERED TORTOISE is, together with its companion, an American species of the large genus Emys, examples of which are found in various portions of the world, and of which nearly fifty species are known to zoologists. All these creatures have their heads covered with a thin but hard skin.

The Lettered Terrapin is very common in Northern America, and is found in the rivers, ponds, lakes, or even the marshy grounds, where it can obtain an abundant supply of food. It is fond of reptiles, and causes great destruction among the frogs in their earlier stages of existence. It also has a great liking for worms, and, like the green crab of our own coasts, is very apt to take the fisherman's bait, and exasperates him greatly by making him pull up nothing but a little Tortoise when he thought he had caught a fine fish. Regular anglers therefore bear an intense hatred to this Tortoise.

It is easily kept in captivity, and will then feed on many substances, preferring those of an animal nature, and being very fond of various reptiles. It will also eat vegetable substances, and one of these Tortoises was fond of purslain (*Portulacea oleracea*).

In colour it is very pretty, though rather variable. Generally, it is dark brown above, and the edges are boldly scribbled with broad scarlet marks, something like the letters of some strange language. Below it is yellow and the head is yellow and black.

The CHICKEN TORTOISE is also found in North America.

It is very common in the ponds, lakes, or marshy grounds, and though very plentiful, and by no means quick in its movements, is not easily caught, owing to its extreme wariness. Hundreds of these Tortoises may be seen reposing on logs, stones, or the branches of fallen trees, where they are apparently an easy prey. But they are very

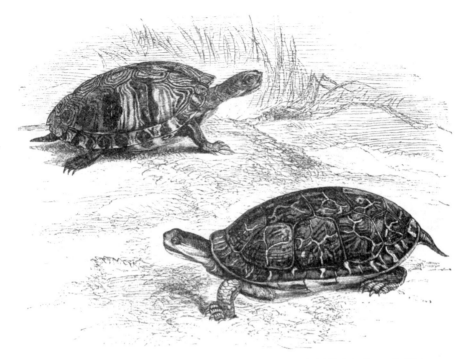

LETTERED TORTOISE.—*Emys scripta.*　　　CHICKEN TORTOISE.—*Emys reticularia.*

sensitive to the approach of an enemy, and the first that perceives the coming danger tumbles off its perch and falls into the water with a great splash that arouses the fears of all its companions, who go tumbling and splashing into the water in all directions, and in a few seconds not a Tortoise is to be seen where they were so plentiful before they took alarm.

The Chicken Tortoise swims well, but not rapidly, and as it passes along with its head and neck elevated above the surface, it looks so like the dark water-snake of the same country, that at a little distance it might readily be mistaken for that reptile.

It is rather a small species, seldom exceeding ten inches in length. Its flesh is remarkably excellent, very tender and delicately flavoured, something like that of a young chicken, so that this Tortoise is in great request as an article of food, and is largely sold in the markets, though not so plentifully as the common salt-water terrapin. Its colour is dark brown above, and the plates are scribbled with yellow lines, and wrinkled longitudinally. The neck is long in proportion to the size of the animal, so long indeed that the head and neck together are almost as long as the shell. The lower jaw is hooked in front.

AN allied species, popularly called the QUAKER TORTOISE and scientifically *Emys olivacea*, is remarkable for the extreme length of the claws of the fore feet, the three middle claws being elongated in a manner that irresistibly reminds the observer of the nails belonging to a Chinese mandarin of very high rank.

The SALT-WATER TERRAPIN is a well-known species, living in North and South America, where it is in great request for the table.

The generic name of Malaclemys, or Soft Terrapin, has been given to this species on account of the formation of the head, which is covered with soft spongy skin. The head is large in proportion to the size of the animal and flattened above.

This Terrapin lives in the salt-water marshes, where it is very plentiful, and from which it never travels to any great distance. During the warm months of the year, it is lively and constantly searching after prey, but when the cold weather comes on, it burrows a hole in the muddy banks of its native marsh and there lies buried until the warm sunbeams of spring break its slumbers and induce it once more to seek the upper earth and resume its former active existence.

It is more active in its movements than is the case with the Tortoises in general, and can not only swim rapidly, but walk with tolerable speed. It is very shy, and discovers approaching peril with a keenness of perception that could scarcely be expected from one of these shielded reptiles, whose dulness and torpidity have long been proverbial.

SALT-WATER TERRAPIN.—*Malaclémys concéntrica.*

Mr. Holbrook, in his valuable "North American Herpetology," writes as follows concerning this Terrapin.

"They are very abundant in the salt marshes around Charleston, and are easily taken when the female is about to deposit her eggs in the spring and early summer months. They are then brought in immense numbers to market; yet, notwithstanding this great destruction, they are so prolific that their number appears undiminished. Their flesh is excellent at all times, but in the northern cities it is most esteemed when the animal has been dug out of the mud in its state of hibernation. The males are smaller than the females, and have the concentric striæ more deeply impressed."

The colour of this Salt-water Terrapin is rather variable, but is usually dark greenish brown on the upper surface, and yellow on the plates which surround the edge of the shell. Below it is yellow, and in many specimens it is marked with variously shaped spots of dark grey. The lower jaw is furnished with a hook, and the sides of the head are dusty white sprinkled with many small black spots.

VERY many species of Tortoise are extremely variable in their colour, but there are few which are so remarkable in this respect as the creature which is appropriately named the BOX TORTOISE.

This species belongs to America, and is found spread over the whole of the Northern States. It is very plentiful in the localities which it favours, and although so small a creature, is able by means of its wonderful organization to protect itself against almost every foe. Many of the Tortoises can withdraw their limbs and head into their shell, leaving open, however, the apertures through which this movement is achieved, so that the animal might be killed or hooked out by a persevering foe, such as the jaguar, which is known to attack turtles, insinuate its lithe paw within the shell, and scoop out the inhabitant with its sharp curved claws.

But in those instances where the animal has the power of closing the openings through which the legs, tail, and head protrude, there is hardly any mode of getting at the flesh without breaking the shell, a feat beyond the power of any animal, except perhaps an elephant, to perform. Certain birds, it is said, are clever enough to soar to a great height with the Tortoise, and break the shell by letting it fall upon a convenient rock, but this

BOX TORTOISE.—Cistuda Carolina.

story does not seem to be very strongly attested. Several species possess this valuable capability, but none to so perfect a degree as the Box Tortoise, which, according to the Rev. Sydney Smith's felicitous summary, need fear no enemy except man and the boa constrictor, the former taking him home and roasting him, and the latter swallowing him entire and consuming him slowly in its interior, as the Court of Chancery does a large estate.

With regard to this curious propensity, it is evident that there is some analogy between these Tortoises and certain mammalia, which are also able to withdraw themselves within the protection of certain armour with which they are furnished. In the case of the hedgehog the animal assumes more of an offensive than a defensive character, and relies, not on an impenetrable covering, for the skin is soft, and a pointed weapon can find an easy entrance between the spines, but on the bristly array of bayonet-like spikes that protrude their threatening points in every direction, and bid a tacit defiance to the foe.

The scale-covered manis again, although guarded with successive layers of broad, horny plates, is, in point of fact, less protected when rolled up than when walking quietly along; for when at rest, the scales overlap each other like the tiles of a house, so that any weapon would glance aside, but when curled up the scales are erected and leave a passage for the arrow or the spear between them.

The real defence of the hedgehog lies in the points of its quills, and of the manis in the razor-like edges of its scales, but the defence of the Tortoise is wholly inaggressive, and is more allied to that of the armadillo or perhaps the singular pichiciago (*Chlamydophorus truncatus*), a most remarkable little creature with a curious shelly covering spread over nearly the whole upper surface and down the hind quarters. A description of this animal may be found in the volume on the Mammalia, page 770. There are again many of the lower animals which have a similar mode of defence, a very familiar example being the well-known pill-woodlouse so common in our gardens, which rolls itself into a round ball when alarmed, and permits itself to be handled and even rolled along the ground without displaying any signs of life.

The Box Tortoise is a terrestrial species, and always keeps to the dry forest-lands, detesting the vicinity of water. It is commonly found in the pine forests, because they are always on thoroughly dry soil, and on account of its fondness for such localities is

MUD TORTOISE. *Kinosternon Pennsylvanicum.*

sometimes known by the popular name of the Pine Terrapin. The negroes call it by the name of Cooter. In the wild state it mostly feeds on insects, and is peculiarly fond of the cricket tribe, but in captivity it will eat almost any food that is offered, taking insects, meat, apples, or even bread.

It is a very little creature, being when adult a very little more than six inches in length. In colour it is extremely variable, but is generally yellowish brown, striped with a brighter hue, and sometimes mottled with black. Of a number of specimens in the British Museum, no two are exactly alike, some being yellow, spotted with black, while others exactly reverse these tints, and are black, spotted with yellow. Others again are yellow with black rays, and others olive with yellow rays and streaks. The carapace has a very slight keel along its upper edge.

The upper jaw of this species is furnished with a rather broad hook, and the lower jaw is also hooked, but not so boldly.

THE common MUD TORTOISE, so called from its mud-haunting propensities, is an example of rather a curious genus of Tortoises, inhabiting America.

It is an odd little creature, being when adult not quite four inches in length, and moving with moderate speed. It is mostly found in ponds and muddy pools, where it feeds upon fish, aquatic insects, and similar diet, catching even the active fish without much difficulty. I lately saw some aquatic Tortoises, which I think belonged to this genus, which had to be ejected from a large basin of a fountain because they killed the newts which inhabited the same locality. Their movements in the water were so deliberate that it was not until they were detected in the very act of biting the newts that their delinquencies were discovered. Their mode of attack was simply to creep under their victim as it balanced itself in the water or swam gently within reach, and then to secure it with a quiet snap of its beak.

Like the lettered terrapin, already mentioned, it has a vexatious habit of taking the angler's bait, and causes many a fisherman to lose his temper when pulling up a useless little Mud Tortoise instead of the fish on which he had set his heart. It seizes the worm just as it catches the newts, taking it so quietly into its mouth that the float is hardly shaken by the touch. But when the fisherman pulls his line, the Tortoise kicks, pulls and flounces about in so energetic a style that it often deludes the angler into the idea that he has hooked quite a fine fish.

This species has a decided smell of musk, a peculiarity which is found in others of the same genus, one of which (*Kinosternon odoratum*) goes by the appropriate, though not very refined, name of Stink-pot, in consequence of the powerful musky odour which it exudes.

The colour of the Mud Tortoise is mostly dusky brown above, and chestnut below, though this colouring is liable to some variation in different individuals. The tail is thick and pointed, and horny at the tip. The head is large, and there are four large warty appendages on the chin.

THE last example of the Terrapin is that singular animal which is appropriately called the ALLIGATOR TERRAPIN, from the great resemblance which it bears to that reptile.

It is also an American species, and lives mostly in the water. When adult it reaches a large size, often exceeding three feet in length, and as it is very fierce of disposition, lithe of neck, and strong of jaw, it is somewhat dreaded by those who have had a practical acquaintance with its powers. The jaws of this animal are sharp edged, and remarkably strong, cutting like the blades of steel shears. Mr. Bell remarks that he has seen one of these creatures bite asunder a stick of half an inch in diameter. When caught, therefore, the captors always cut off these dangerous heads at once.

Mr. Holbrook gives the following interesting account of the Alligator Terrapin and its habits :—

"It is found in stagnant pools or in streams where the waters are of sluggish motion. Generally they prefer deep water, and live at the bottom of rivers ; at times, however, they approach the surface, above which they elevate the tip of their pointed snout, all other parts being concealed, and in this way they float slowly along with the current, but if disturbed, they descend speedily to the bottom.

They are extremely voracious, feeding on fish, reptiles, or any animal substance that falls in their way. They take the hook readily, whatever may be the bait, though most attracted by pieces of fish ; in this way many are caught for market. It is, however, necessary to have strong hooks and tackle, otherwise they would be broken, for the animal puts forth great strength in its struggles to escape, both with its firm jaws and by bringing its anterior extremities across the line. When caught, they always give out an odour of musk, more or less distinct ; sometimes in very old animals it is so strong as to be disagreeable.

Occasionally it leaves the water, and is seen on the banks of rivers or in meadows, even at a distance from its accustomed element. On land, his motions are awkward : he walks slowly, with his head, neck, and long tail extended, elevating himself on his legs like the alligator, which at that time he greatly resembles in his motions. Like the alligator also, after having walked a short distance, he falls on his sternum to rest for a few moments, and then proceeds on his journey.

In captivity they prefer dark places, and are exceedingly ferocious ; they will seize upon and bite severely anything that is offered them, and their grasp upon the object with their strong jaws is so tenacious, that they may even be raised from the ground without loosing their hold.

ALLIGATOR TERRAPIN.—*Chelydra serpentina.*

In many of the northern cities they are brought in numbers to market and are esteemed excellent food, though I think that they are far inferior to the green turtle, the soft-shelled, or even several of the emydes. They are kept for months in tubs of fresh water, and feed on such offal as may be given them, though they never become fat or increase much in weight."

Though a very valuable and curious reptile, the Alligator Terrapin is far from beautiful, with its little dusky shell, its long knob-covered tail, its singular legs and feet, and its great sharply toothed jaws. On account of its habit of snapping fiercely at its opponents, it is often called by the name of Snapping Turtle, a title, however, which rightly belongs to a species which will shortly be described.

Its head is large and covered with a hard wrinkled skin; the neck is long, thick, and furnished with a number of projecting tubercles. Under the chin are two distinct barbels. When adult, the shell is so formed that a depression runs along the centre, leaving a kind of keel at each side of the central line; but when young, the shell forms three distinct keels. It is rather flat, oblong, and at the hinder portion is deeply cleft, so as to form a row of blunt teeth, but while young the teeth are sharp. The tail is stout, long, and is furnished with a series of large blunt tubercles along its central line.

WE now arrive at another family of Tortoises, termed Chelydes, an example of which is the remarkable MATAMATA, the acknowledged type of its family.

All the Chelydes have broad flattened heads, long broad contractile necks, and when in repose have a curious custom of bending their necks under the side of the carapace. Their feet are webbed in order to enable them to pass rapidly through the water, and there is always a lobe between the claws. They are aquatic Tortoises, carnivorous, and voracious, and only feed while in the water. When swimming, the whole of the shell is kept below the surface.

The Matamata is certainly the most remarkable of aspect among all the Tortoises, and perhaps may lay claim to be considered one of the oddest-looking animals in the world, far exceeding in its grotesque ungainliness even the wild and weird creations of the middle-age painters.

MATAMATA.—*Chelys Matamata*

This Tortoise inhabits Southern America, and is most plentiful in Cayenne. Formerly it was very common, but on account of the excellence of its flesh, it has been subject to such persecution, that its numbers have been considerably diminished. It haunts the lakes and rivers, where it swims well and with some speed. As is the case with most aquatic tortoises, it is carnivorous, and feeds on fish, reptiles, and other creatures, which it captures by a sudden snap of its sharp beak. In general, it appears not to care for chasing the intended prey, but conceals itself among the reeds and herbage of the river-side, and from its hiding-place thrusts out its neck suddenly upon its victims as they pass unsuspectingly within reach of their destroyer. On occasion, however, it will issue from its concealment, dart rapidly through the water and seize a fish, reptile, or even a water-fowl, and then retire with its prey to its former hiding-place.

It is a large and formidable creature, attaining, when adult, to a length of three feet.

The head of the Matamata is most singular in shape, and remarkable for the strange appendages which are placed upon it. The head itself is much flattened, and rather broad, and the snout is prolonged in a most extraordinary manner, so as to form an elongated and flexible double tube.

On the top of the head are two membranous prolongations of the skin, standing boldly from the head, and having much the appearance of ears. From the chin hang two curiously fringed membranes, and the throat is decorated with four similar membranes, but of larger size and more deeply fringed. The neck is long, and bears upon its upper surface two rows of small membranous tufts, deeply fringed, and greatly resembling, in every point but that of size, the tufts on the chin and throat. The limbs are powerful, and the tail is short.

The shell of the Matamata is rather convex, broader before than behind, and rather flattened in the middle of the back. The shields are elevated, rather sharp at their tips, and are arranged so as to form three regular keels along the back.

A NEARLY allied species of river Tortoise is worthy of a passing notice. It is the NEW HOLLAND CHELODINE, sometimes called the YELLOW CHELODINE, from the olive-yellow colour of the plastron.

SNAPPING TURTLE.—*Triónyx feróx.*

This remarkable reptile may almost deserve the name of the Snake Tortoise, its long flexible neck, and flat, narrow, and pointed head, having a very serpentine aspect. As its name imports, it is an inhabitant of Australia, and is found most commonly in New Holland. It is a water-loving creature, not caring much for rivers and running streams, but haunting the pools, marshes, and stagnant waters, where it lives in the midst of abundance, finding ample food among the fishes and aquatic reptiles which generally swarm in such localities. It is an active animal, traversing the water with considerable speed, and capturing its prey by means of its sharp jaws.

The gape is very large, and the jaws are comparatively slender. The shell is broad, rather flattened, and the shields are thin and smooth, not being elevated as in the preceding species. The general colour of the shell is brown above and yellow below, each shield having a black line round its edge.

WE now arrive at another family of the Tortoises, known popularly as Soft Turtles— a rather inaccurate title, inasmuch as they are not turtles but Tortoises—and scientifically as *Trionycidœ.* The latter title is of Greek origin, signifying three-clawed, in allusion to the fact that, although the species belonging to the family have five toes on each foot, only the three inner toes of each foot are armed with claws.

These Tortoises are rather interesting to the careful observer, because the peculiar structure of the external covering permits the formation of the skeleton to be seen without the necessity for separating the shells. In particular, the method in which the breast-bone is developed into the broad flattened plate which forms the plastron, can clearly be

3.

C

seen through the skin, and even the position of the sutures can be made out without much difficulty.

The head of these creatures is rather oval and flattened, the jaws are horny, but covered with hanging fleshy lips, and the mouth is lengthened into a cylindrical trunk. The neck is long and can be contracted, the feet are short, very wide, and the toes are connected together by strong webs. They all live in warm climates, and are found in rivers and lakes.

The typical species is the celebrated FIERCE TRIONYX, or SNAPPING TURTLE, a reptile which derives its former title from the exceeding ferocity of its disposition, and the latter from the method in which it secures its prey or attacks its foes. It is found spread over many parts of North America.

This fierce and determined marauder of the waters is even more formidable than the two previous species, and not only causes terror among the smaller creatures which inhabit the same localities, but is even dreaded by man, whose limbs have often been severely wounded by the bite of these ferocious reptiles. Like the aquatic Tortoises, it is carnivorous in its habits, and is terribly destructive among the fish, smaller quadrupeds, birds, and reptiles. Lurking on the banks it snatches away many an unfortunate animal as it comes to drink, or seizes the water-fowl that have ventured too close to their terrible neighbour. So fiercely carnivorous is this Tortoise, and so voracious is its appetite, that it will even catch young alligators, and devour them in spite of their teeth and struggles.

The flesh of this species is very delicate, tender, and richly flavoured, so that it often meets the doom which it has inflicted on so many other animals. As it is so voracious, it will take almost any kind of bait, provided that it be composed of animal substance, but it prefers fish, and cannot resist a hook so baited.

Its captor's work, however, is not confined to hooking and drawing it ashore, as the Snapping Turtle, when it finds itself with a hook firmly fixed in its jaws, and itself being irresistibly dragged from the water, seems possessed with tenfold ferocity, writhing its long flexible neck, darting its head furiously at its foes with the rapidity of a serpent's stroke, and snapping sharply with its formidable jaws, one bite of which would shred away the fingers from the hand, or the toes from the feet, as easily as the gardener's scissors sever the twigs and leaves. Such a misfortune has indeed been known to occur. Mr. Bell records an instance where a Snapping Turtle, that was being conveyed to England, contrived to reach the hand of one of the sailors in its fierce struggles, and bit off one of his fingers.

The eggs of the Snapping Turtle are very spherical in form, and brittle of substance. The female lays a large number of these eggs, from fifty to sixty being the usual average, and always deposits them in some dry situation. In order to find a suitable spot for the deposition of her eggs, the female leaves the water, and is often forced to traverse a considerable distance before she can find a spot sufficiently dry for her purpose. Sometimes she will even ascend a very steep acclivity in her anxiety to find a locality that is quite dry, covered with sandy soil, and exposed to the full rays of the sun. She begins her task about May, and the little Tortoises are hatched in July.

The following curious account of the tenacity of life possessed by these creatures has been kindly forwarded to me :—

" As regards the tenacity of life of the Snapping Turtle, and the sympathy (*rapport*) which seems to exist between its severed limbs and main trunk, for some time after the separation has taken place, I witnessed a very curious incident when staying at a farm in Massachusetts, U.S. America.

When I had brought the animal home, suspended by its tail, I killed it by chopping its head off, yet the head would open and shut the mouth, and roll its eyes. When I held a stick between the opened jaws it closed them with violence, and kept hold of it. Meanwhile the headless body was crawling on the ground.

About a quarter of an hour after having severed the head from the body, my mother had got boiling water, which I threw over the body, placed in a tub, in order to make the horny matter separate from the flesh ; the moment this was done the back heaved and the sides were puffed out as if wind were blown between skin and flesh, and instantaneously the head, which lay about three or four feet from the tub, on the ground, opened its mouth with a slight hissing sound, let go its hold on the stick, and the part of the neck adhering to the head expanded, as if also wind was blown into it, and both body and head lay motionless and dead.   After having taken out thirty-four eggs, I took out the heart, which, strange to say, was still throbbing with life, contracting and expanding.   I put it upon a plate, where it kept on beating until about noon the following day."

The shape and general appearance of this creature may be learned from the engraving better than by a page of description, and it is only necessary to point out that, in this species, the front edge of the carapace is furnished with a great number of tooth-like points, all radiating from the shell.   These teeth, or tubercles, distinguish it from another American species, appropriately termed the Unarmed Trionyx (*Triónyx múticus*).

BEFORE taking leave of the Soft Turtles, we must cast a casual glance at two rather curious species.   The one is the TYRSE (*Tyrse*, or *Trionyx Nilóticus*), a native of Africa, as its name imports.   This animal is found in the Nile, and other African rivers, and is a good representation of the American reptile, being very fierce, strong, and voracious, and said to devour the young crocodiles, just as the snapping turtle eats young alligators. The shell of the Tyrse is rather convex, but often is flattened along the line of the vertebræ, and its back is olive-green spotted with yellow or white.

The other species is the DOGANIA (*Dogania subplanus*, or *Trionyx subplanus*).   This curious-looking reptile is an Asiatic species, and is found in India.   Its neck seems preternaturally long, and supports a very large head, broad behind, and produced into a conical muzzle in front.   The shell is rather oval, much flattened, and quite conceals the conical tail.   Its colour is brown, mottled largely with yellow ; the head is also yellow and brown.   The ribs are not fully united together until the animal has attained a rather advanced age.

WE now arrive at the TURTLES, a group that can be distinguished by many unmistake-able marks.   Their feet are very long, those of the fore limbs being longest, flat, expanded at the end, and often furnished with flattened claws.   In fact the feet are modified into fins or paddles, in order to suit the habits of these reptiles, which only feel themselves at home in the water, and are often met at sea some hundreds of miles from the nearest land. The ribs of the Turtles, instead of being united throughout their length, as in the tortoises, are only wide, flat, and united for part of their length, the remaining portions being free, and radiating like the spokes of a wheel.

These reptiles inhabit the seas of the torrid and the temperate zones, and their food is mostly of a vegetable nature, consisting of various seaweeds, but there are a few species which are animal feeders, and eat creatures such as molluscs, star-fish, and other marine inhabitants.   Several species are remarkably excellent for food, and caught in great numbers for the table, while others are equally useful in supplying the beautiful translucent substance known by the name of tortoiseshell.   Their head is rather globular, and their jaws are naked and horny, and are capable of inflicting a severe wound.

THE first example of the Turtles is the LUTH, or LEATHERY TURTLE, so called from the soft leather-like substance with which its shell is covered.

This species is found in the Atlantic, Pacific, and Indian Oceans, where it grows to a very large size, often weighing more than sixteen hundred pounds, and measuring eight feet in length.   Being a very good swimmer, owing to the great development of the limbs, especially the fore legs, it ventures far out to sea, and is occasionally driven to strange countries.   Specimens of this reptile have been taken on the coast of France, and one or two on our own shores.   These individuals were rather large, weighing about seven or eight hundred pounds.

The Leathery Turtle feeds on fish, crustacea, molluscs, radiates, and other animals, and its flesh seems to be hurtful, causing many symptoms of poisoning in those who eat it. It resorts in numbers to the Tortugas, or Turtle Islands, of Florida, for the purpose of laying its eggs, which are generally about three hundred and fifty in number, and are laid in two sets.

LUTH, OR LEATHERY TURTLE.—*Sphargis coriacea*

This species is remarkable for having no horny plates, the bones of the carapace and plastron being covered with a strong leathery skin, smooth in the adult animal, but covered with tubercles in the young. Along the back run seven ridges, sharp, and slightly toothed in the full-grown Turtle, but bluntly tubercled in the young. The eye is very curious, as the lids are set vertically instead of horizontally, and when the creature opens and shuts its eyes, have a very singular effect. The jaws are very formidable, being sharply edged, deeply scooped with three rounded notches in the front of the upper jaw, so as to form two curved sharply pointed teeth, and the extremity of the lower jaw is strongly hooked.

The legs of the Leathery Turtle are very long, especially the two fore limbs, which, in a specimen measuring eight feet in total length, were nearly three feet long, and more than nine inches wide. The feet are not furnished with claws, but the toes have a little horny scale at their tips, which take the place of the claws. The general colour of this animal is dark brown, with pale yellow spots, but sometimes the skin is irregularly pied with black and white.

ANOTHER well-known species of Turtle deserves a passing notice. This is the LOGGERHEAD TURTLE, or CAOUANE (*Caouana caretta*), sometimes called the RHINOCEROS TURTLE.

This fine species has a wide range of locality, being found in the most warm seas. It is extremely powerful, fierce, and voracious, biting with great force, and cutting hard substances without much difficulty. According to Catesby, "the Loggerhead Turtles are the boldest and most voracious of all other Turtles. Their flesh is rank and little sought for, which occasions them to be more numerous than any other kind. They range the ocean over, an instance of which, among many others that I have known, happened the 20th of April, 1725, in latitude 30° north, when our boat was hoisted out, and a Loggerhead Turtle struck, as it was sleeping on the surface of the water.

This, by our reckoning, appeared to be midway between the Azores and the Bahama Islands, either of which places being the nearest land it could have come from, or that

HAWKSBILL TURTLE.—*Caretta imbricata.*

they are known to frequent, there being none on the north continent of America farther north than Florida. It being amphibious, and yet at so great a distance from land in the breeding-time, makes it the more remarkable. They feed mostly on shell-fish, the great strength of their beaks enabling them to break very large shells." Several other species belong to the same genus.

In general appearance this species is not unlike the common Green Turtle, which will presently be described, but the shell is broader, deeper coloured, and has two more plates on the back. The plates along the upper part of the back are six-sided, rather square, and keeled. There are two claws on each foot.

THE well-known CARET, or HAWKSBILL TURTLE, so called from the formation of the mouth, is a native of the warm American and Indian seas, and is common in many of the islands of those oceans. One or two specimens have been taken on our coasts.

The Hawksbill Turtle is the animal which furnishes the valuable "tortoiseshell" of commerce, and is therefore a creature of great importance. The scales of the back are thirteen in number, and as they overlap each other for about one-third of their length, they are larger than in any other species where the edges only meet. In this species, too, the scales are thicker, stronger, and more beautifully clouded than in any other Turtle. The removal of the plates is a very cruel process, the poor reptiles being exposed to a strong heat which causes the plates to come easily off the back. In many cases the natives are very rough in their mode of conducting this process, and get the plates away by lighting a fire on the back of the animal. This mode of management, however, is injurious to the quality of the tortoiseshell. After the plates have been removed, the Turtle is permitted to go free, as its flesh is not eaten, and after a time it is furnished with a second set of plates. These, however, are of inferior quality and not so thick as the first set.

When first removed, they are rather crumpled, dirty, opaque, brittle, and quite useless for the purposes of manufacture, and have to undergo certain processes in order that these defects may be corrected. Boiling water and steam are the two principal agents in this part of the manufacture, the plates being boiled and steamed until they are soft and clean, and then pressed between wooden blocks until they are flat. The tortoiseshell possesses the valuable property of uniting together perfectly, if two pieces are thoroughly softened, heated, and then subjected to the action of a powerful press. By this mode of treatment, the tortoiseshell can be formed into pieces of any size or thickness, and can even be forced into moulds, retaining, when cold, a perfect impression of the mould. Even the chippings and scrapings of this valuable substance are collected, and being heated and pressed, are formed into solid cakes fit for the purposes of manufacture.

The uses to which this costly and beautiful substance are put, are innumerable. The most familiar form in which the tortoiseshell is presented to us is the comb, but it is also employed for knife-handles, boxes, and many other articles of ornament or use.

This species is not nearly so large as the green Turtle, and its flesh is not used for food. The eggs, however, are thought to be a great delicacy. It is remarkable that when these eggs are boiled, the albumen, or "white" as it is popularly called, does not become firm. The external membrane is white, flexible, and the eggs are nearly spherical in their form. Their number is very great, and the animal usually lays them in sets at intervals of about three weeks.

The young are generally hatched in about three weeks after the eggs are laid in the sand, the hot rays of the sun being the only means by which they attain their development. When first excluded from the shell, the young Turtles are very small and soft, not obtaining their hard scaly covering until they have reached a more advanced age. Numberless animals, fish, and birds feed on these little helpless creatures, and multitudes of them are snapped up before they have breathed for more than a few minutes. The rudiments of the scales are perceptible upon the backs of these little creatures, but the only hard portion is the little spot in the centre of each plate, which is technically called the areola, the layers of tortoiseshell being added by degrees from the edges of the plates.

Many birds are always hovering about the islands where Turtles lay their eggs, and as soon as the little things make their appearance from the sand and hurry instinctively towards the sea, they are seized by the many foes that are watching for their prey. Even when they reach the water, their perils are not at an end, for there are marine as well as aerial and terrestrial foes, and as many fall victims in the water as on land. So terrible is the destruction among these reptiles in their early days of life, that were it not for the great number of eggs laid, they would soon be extirpated from the earth.

Three specimens of the Hawksbill Turtle have been found on our shores, and one of them, which was taken alive in 1774, was conveyed ashore and placed in a fish-pond, where it lived until winter. This specimen was caught in the Severn.

The shell of the Hawksbill Turtle is rather flat, and heart-shaped. When young, the centre of each plate is rather pointed, but in the adult animal the points are worn away and never restored. The plates surrounding the edges of the shell are arranged so as to form strong teeth pointing towards the tail. In the younger specimens, there are two keels running the length of the plastron, but in the older individuals these are worn away like the projections on the back. The jaws are strongly hooked at their tips, and the under jaw shuts within the upper. The tail is very short. The colour of this species is yellow richly marbled with deep brown above. The under parts are yellowish white, splashed with black on the areola in the half-grown and younger individuals, and the head is brown, the plates being often edged with yellow.

The best known of all the Turtles is the celebrated GREEN TURTLE, so called from the green colour of its fat.

This useful animal is found in the seas and on the shores of both continents, and is most plentiful about the Island of Ascension and the Antilles, where it is subject to incessant persecution for the sake of its flesh. The shell of this reptile is of very little use, and of small value, but the flesh is remarkably rich and well-flavoured, and the green fat has long enjoyed a world-wide and fully deserved reputation.

In Europe the flesh of the Green Turtle is little but an object of luxury, attainable only at great cost and dressed with sundry accompaniments that increase rather than diminish its natural richness. But in many instances, more especially on board ship,

GREEN TURTLE.—*Chelónia víridis*

when the sailors have been forced to eat salt provisions until the system becomes deteriorated, and the fearful scourge of scurvy is impending over crew and officers, the Turtle becomes an absolute necessity, and is the means of saving many a noble vessel from destruction, by giving the crew a healthful change of diet, and purifying the blood from the baneful effects of a course of salted provisions.

Landsmen have little notion of the real texture and flavour of "salt junk," their ideas being generally confined to the delicately corned and pinky beef or pork that is served up to table, with the accompaniments of sundry fresh and well-dressed vegetables. Whereas, salt junk is something like rough mahogany in look and hardness, and salted to such a degree as almost to blister the tongue of a landsman. It may easily be imagined how any one who has been condemned to a course of this diet for a lengthened time would welcome fresh meat of any kind whatever, and we need not wonder at the extraordinary relish with which sailors will eat sharks, sea-birds, and various other strangely flavoured creatures.

Even in such favoured countries as England, the flesh and fat of the Turtle are valuable in a medicinal point of view, and will supply in a more agreeable, though more costly manner, the various remedies for consumptive tendencies, decline, and similar diseases, of which cod-liver oil is the most familiar and one of the most nauseous examples.

Formerly, before steam power was applied to vessels, the Turtle was extremely scarce and very expensive, but it can now be obtained on much more reasonable terms. Many vessels are now in the habit of bringing over Turtles as part of their cargo, and it is found that these valuable reptiles are easily managed when on board, requiring hardly any attention. The following short account of some captive Turtles has been kindly presented to me by a partaker of their voyage and their flesh.

" The Island of Ascension is a great resort of Turtle, which are there captured and retained prisoners in some large ponds from which they are occasionally transferred to H.M. ships for ' rations' for the crew. These Turtle may be seen in the ponds, lazily moving along, one above another, sometimes three or four deep. They occasionally come to the surface to take breath, and will splash about at times quite merrily, as though ignorant that their destiny tended towards conversion into soup and cutlets. At the best, however, they are lethargic, awkward creatures.

About half a dozen fine Turtle were conveyed on board our ship during my stay at the Island of Ascension; they were unwieldy monsters, measuring rather more than four feet six inches in length, and about three feet in breadth. They were allowed to lie either in the boats, or on the after-part of the poop, and seldom disturbed themselves unless the vessel gave an extra roll, or they were stirred up by a pail of water being thrown over them or a wet swab rubbed over their hooked beaks.

Their tenacity of life was remarkable; they remained on board ship during upwards of three weeks without any food, and their only refresher was a cold bath, derived from the before-mentioned pail of water, which they usually received with a dreamy lengthy sort of hiss. Even after their three weeks' starvation, they died very hard. One, whose throat was cut in the morning, and from whose body numerous eggs had been extracted, was giving an occasional flap with her fins late in the afternoon; the fact of her throat having been cut and her body otherwise mutilated appeared merely to produce the effect of ultimately damaging her constitution, and I have grave doubts whether the fact of her ceasing to move was not as much due to the destruction of the various membranes as to the extinction of her reptilian life."

As these animals are large and very powerful, it is not a very easy task to secure and bring them on board. The usual plan is to intercept them as they are traversing the sands, and to turn them over on their backs, where they lie until they can be removed. Many of the tortoise tribe can recover their position when thus overturned, but the Green Turtle is quite unable to restore itself to its proper attitude, and lies helplessly sprawling until it is lifted into the boat and taken on board. In many cases the creature is so enormously heavy that the united strength of the pursuers is inadequate to the task, and they are consequently forced to employ levers and so to tilt it over.

Sometimes the Turtle is fairly chased in the water and struck with a curious kind of harpoon, consisting of an iron head about ten inches in length, and a staff nearly twelve feet long. The head is only loosely slipped into a socket on the staff, and the two are connected with a cord. Two men generally unite in this chase, one paddling the canoe and the other wielding the harpoon. They start towards the most likely spots, and look carefully at the bottom of the sea, where it is about six or ten feet in depth, to see whether the expected prey is lying at its ease and does not perceive them.

Sometimes they are forced to give chase to a Turtle on the surface, and sometimes the individual on which they had fixed, takes the alarm, and swims away. In either case they continually pursue the single swimming reptile, until it is fatigued with constant irritation, and sinks to the bottom to rest. No sooner has the Turtle assumed this position than the harpooner lowers his weapon into the water, takes an accurate aim, and then drives the steel spike deep into the shell. Off dashes the Turtle, carrying with it the harpoon. Were it not for the peculiar construction of the harpoon, the weapon would soon be shaken off, and the Turtle escape, but as the shaft slips readily off the head, there is no leverage and the steel head remains fixed, towing after it the long wooden shaft, which soon tires out the poor victim. When thoroughly fatigued, it is drawn to the surface, a rope put round it, and either taken into the boat or hauled ashore.

The food of this Turtle consists of vegetable substances, mostly algæ, which is found in great abundance in those warm climates. This animal grows to a very great size, as may be imagined from the fact that it often requires the united aid of three men to turn it over. A very pure limpid oil is obtained from these species, useful for burning in lamps and other similar purposes. A fat full-grown specimen will sometimes furnish thirty pints of this substance.

The eggs of the Turtle are thought as great delicacies as its flesh, and it is rather a remarkable fact, that although the flesh of the hawksbill Turtle is distasteful to all palates and hurtful to many constitutions, the eggs are both agreeable in flavour and perfectly harmless. It is while the female Turtle is visiting shore for the purpose of depositing her eggs that she is usually captured, as these sea-loving reptiles care little for the shore except for this purpose. So admirable an account of the manner in which the Turtle behaves when laying her eggs is written by Audubon, that the description must be given in his own words.

" On nearing the shore, and mostly on fine calm moonlight nights, the Turtle raises her head above the water, being still distant thirty or forty yards from the beach, looks around her, and attentively examines the objects on shore. Should she observe nothing likely to disturb her intended operations, she emits a loud hissing sound, by which such of her enemies as are unaccustomed to it are startled, and apt to remove to another place, although unseen by her.

Should she hear any more noise, or perceive any indication of danger, she instantly sinks and goes off to a distance ; but should everything be quiet, she advances slowly towards the beach, crawls over it, her head raised to the full stretch of her neck, and when she has reached a place fitted for her purpose, she gazes all around in silence. Finding all well, she proceeds to form a hole in the sand, which she effects by removing it from under her body with her hind flappers, scooping it out with so much dexterity, that the sides seldom, if ever, fall in. The sand is raised alternately with each flapper as with a ladle, until it has accumulated behind her, when, supporting herself with her head and fore-part on the ground, she, with a spring from each flapper, sends the sand around her, scattering it to the distance of several feet.

In this manner the hole is dug to the depth of eighteen inches or sometimes more than two feet. This labour I have seen performed in the short space of nine minutes. The eggs are then dropped one by one and disposed in regular layers to the number of one hundred and fifty, or sometimes nearly two hundred. The whole time spent in this operation may be about twenty minutes. She now scrapes the loose sand back over the eggs, and so levels and smooths the surface, that few persons on seeing the spot would imagine that anything had been done to it. This accomplished to her mind, she retreats to the water with all possible despatch, leaving the hatching of the eggs to the heat of the sand.

When a Turtle, a loggerhead for example, is in the act of dropping her eggs, she will not move, although one should go up to her, or even seat himself on her back ; but the moment it is finished, off she starts, nor would it be possible for one, unless he were as strong as Hercules, to turn her over and secure her."

CROCODILES AND ALLIGATORS.

ACCORDING to the arrangement of the national collection in the British Museum, the link next to the tortoise tribe is formed of an important group of reptiles, containing the largest of the reptilian order, larger indeed than most present inhabitants of the earth, if we except one or two African and Indian animals, and some members of the cetaceous tribe. As is the case with nearly all reptiles, they are carnivorous, and owing to their great size, strength of muscle, voracity of appetite and the terrible armature of sharp teeth with which their jaws are supplied, they are the dread of the countries which they inhabit, ruling the rivers with a sway as despotic as is exercised by the lion and tiger on land, the eagle in the air, or the shark in the seas.

On account of the peculiar manner in which their bodies are covered with square, keeled, bony plates embedded in the skin, and protecting the body with an armour that effectually guards its upper and more exposed portions from any ordinary weapon, they are separated from the true lizards and scientifically termed EMYDOSAURI, or Tortoise-lizards, the bony plates being considered to have a certain analogy with those of the shielded reptiles. By some zoological authors these animals are termed LORICATA or Mailed Reptiles, from the Latin word *lorica* which signifies a coat of mail or cuirass.

Although these creatures are capable of walking upon land, for which purpose they are furnished with four legs, they are more fitted for the water than its shores, and are swift and graceful in the one, as they are stiff, awkward, and clumsy on the other. Through the water they urge their course with extraordinary speed, their long, flattened, flexible tail answering the double purpose of an oar and a rudder, but on land their bodies are so heavy and their legs are so weak, that they can hardly be said to walk, a term which seems to imply that the body is wholly supported by the legs, but to push or drag themselves along the ground, on which rests a considerable portion of their weight.

The head of these creatures is always rather elongated, and in some species is lengthened into a narrow and prolonged snout. Each jaw is furnished with a row of sharply pointed and rather conical teeth. These teeth are hollow, mostly grooved on the surface, and are replaced when they fall by new teeth that grow behind them, and in process of time push the old ones out of their sockets.

The nostrils are placed at the very extremity of the skull and upon a slightly raised prominence, so that the animal is able to breathe by merely exposing an inch or so above the water, and thus can conceal itself from almost any foe, or make an unsuspected approach upon its prey. There is yet another more important use for the position of the nostrils. The Crocodiles feed on fishes and various water-loving creatures, but also are in the habit of lurking by the river-bank, and suddenly seizing upon any unfortunate animal that may come to drink. Suppose, for example, that a calf or a dog is thus dragged into the water, the reptile grasps it across the body, and sinks below the surface so as to keep the head of the victim below water while itself can breathe by means of the elevated nostrils.

But as during this process the mouth is held widely open, it might be rationally presumed that considerable inconvenience would be caused by the water running down the throat. Such would indeed be the case, were not this difficulty provided for by a simple yet very wonderful contrivance. At the back of the throat, a pair of thin cartilaginous plates are so arranged, that when the animal opens its mouth the pressure of the water rushing into the mouth immediately closes one upon the other, and effectually prevents the passage of a single drop, the closure being in exact proportion to the volume of water. The structure indeed is very like that of the valves of the heart. The channels which lead from the nostrils run very far back through the skull, and open behind the throat valves, so that respiration is in no way impeded. They cannot, however, swallow their prey while under water, but are obliged to bring it on shore for that purpose. The tongue is small, and fastened down to the lower jaw throughout its length, so that it was formerly thought that the Crocodiles were destitute of that organ.

There is rather a curious structure in the vertebræ of the neck. These bones are furnished with short transverse processes like false ribs, which have the effect of preventing the animal from turning its head from side to side. On land therefore, where its feeble limbs are so inadequate to the support of the long and heavy body, it can easily be avoided by any one of ordinary agility. The eyes are large, and set rather far back upon the head. The ears are carefully guarded from the ingress of water by a pair of tightly closing valves. Below the throat are a pair of glands which secrete a substance having a strong musky scent which is very disagreeable, and in old individuals taint the whole flesh with its rank odour and render it uneatable to ordinary palates.

The young of these reptiles are hatched from eggs, which are strangely small in proportion to the large dimensions of the adult animal, the newly hatched offspring being so small as hardly to be recognised as belonging to the same species as their parents, especially as there are certain differences of shape hereafter to be mentioned.

These great reptiles are divided, or rather fall naturally, into two families, namely, the Crocodiles and the Alligators. All the members of these families can be easily distinguished by the shape of their jaws and teeth, the lower canine teeth of the Crocodiles fitting into a *notch* in the edge of the upper jaw, and those of the Alligators fitting into a *pit* in the upper jaw. This peculiarity causes an obvious difference in the outline of the head, the muzzle of the Crocodiles being narrowed behind the nostrils, while that of the Alligators forms an unbroken line to the extremity. A glance therefore at the head will suffice to

AFRICAN CROCODILES AT HOME

settle the family to which any species belongs. In the Crocodiles, moreover, the hind legs are fringed behind with a series of compressed scales.

OUR first example of the Crocodiles is the very remarkable GAVIAL, or GANGETIC CROCODILE, sometimes known by the name of NAKOO.

GAVIAL, OR GANGETIC CROCODILE.—*Gaviális Gangéticus.*

This curious reptile is one of the largest, if not the very largest of its order, sometimes reaching a length of twenty-five feet. As its popular name imports, it is a native of India, and swarms in many of the Indian rivers, the Ganges being greatly infested with its presence. It is a striking animal, the extraordinary length of its muzzle giving it a most singular and rather grotesque aspect.

This prolongation of the head varies considerably according to the age and sex of the individual. In the young Gavial, for example, just hatched from the egg, the head is short and blunt, and only attains its full development when the creature has reached adult age. The males can be distinguished from the other sex by the shape of the muzzle, which is much smaller at the extremity. There are many teeth, the full complement being about one hundred and twenty. They are similar in appearance, and about equal in length.

The colour of this species is dark olive-brown, spotted with black. Several species of African Gavials are known to zoologists, besides the Asiatic animal, but on account of the different formation of the head, such as the absence of a swollen muzzle in the male, and some important variations in the plates of the neck and back, they are placed in another genus, and termed False Gavials. In the British Museum examples may be found, among which may be named BENNETT'S GAVIAL (*Mecistops Bennettii*), an inhabitant of Western Africa, and the False Gavial (*Mecistops cataphractus*). Some naturalists, however, think that these animals are only varieties of the same species.

WE now arrive at the true Crocodiles, in which the jaws are moderately lengthened, wide, flat, tapering, and rather dilated at the extremities. The most peculiar of these reptiles is the long-celebrated CROCODILE of Northern Africa.

This terrible creature is found chiefly in the Nile, where it absolutely swarms, and though a most destructive and greatly dreaded animal, is without doubt as valuable in the water as the hyæna and vulture upon the land. Living exclusively on animal food, and rather preferring tainted or even putrefying to fresh meat, it is of great service in devouring the dead animals that would otherwise pollute the waters and surrounding atmosphere.

EGYPTIAN CROCODILE.—*Crocodilus vulgáris.*

It also feeds on fish, which it can catch by means of its great swiftness in the water, and is a dangerous foe to cattle or other beasts that come to the river-side for drink. Some persons relate that when its intended victim does not come sufficiently near to be snapped up, the Crocodile crawls to the banks, and with a sweep of its long and powerful tail strikes the poor creature into the water, where it is immediately seized in the Crocodile's ready jaws.

Human beings have a great dread of this terrible reptile. Many instances are known where men have been surprised near the water's edge, or captured when they have fallen into the river. There is, it is said, only one way of escape from the jaws of the Crocodile, and that is to turn boldly upon the scaly foe, and press the thumbs into his eyes, so as to force him to relax his hold, or relinquish the pursuit. Mr. Petherick relates a curious instance, where a man was drawing water, and was chased by a Crocodile into the recess in the earth in which he was standing while working the lever of the "shadoof." The man crouched as far back as he could squeeze himself, and the Crocodile tried to follow him, but got itself so firmly wedged in the narrow channel through which it was endeavouring to force its way, that it could neither reach the man, whose trembling breast was within a span of the reptile's terrible teeth, nor retreat from the strange position into which it had forced itself. After spending some time in terror, the poor man contrived to give the alarm to his comrades, who came running to his assistance, and despatched the Crocodile as it lay helplessly fixed in the crevice.

The plates which cover the skin of the Crocodile are of exceeding hardness, so hard, indeed, that they are employed as armour by some ingenious warriors. A coat of natural scale armour formed from the Crocodile skin may be seen in the British Museum. Even a rifle ball may be turned by these horny plates, provided that it strikes rather obliquely; and they are impervious to ordinary steel weapons. Modern rifles, however, especially if the ball is hardened with solder or tin, make little account of the plates, but cut their way through them without difficulty.

As this reptile is so dangerous and costly a neighbour to the inhabitant of the river banks, many means have been adopted for its destruction. One such method, where a kind of harpoon is employed, is described by Dr. Rüppell: "The most favourable season is either the winter, when the animal usually sleeps on sand-banks, luxuriating in the rays of the sun, or the spring, after the pairing time, when the female regularly watches the sand islands where she has buried her eggs. The native finds out the place, and on the south side of it, that is, to the leeward, he digs a hole in the sand, throwing up the earth to the side which he expects the animal to take. Then he conceals himself, and the Crocodile, should it fail to observe him, comes to the accustomed spot and soon falls asleep.

The huntsman then darts his harpoon with all his force at the animal, for in order that its stroke may be successful, the iron ought to penetrate to the depth of at least four inches, in order that the barb may be fixed firmly in the flesh. The Crocodile, on being wounded, rushes into the water, and the huntsman retreats into a canoe, with which a companion hastens to his assistance. A piece of wood, attached to the harpoon by a long cord, swims on the water, and shows the direction in which the Crocodile is moving. The huntsmen, pulling at this rope, drag the beast to the surface of the water, where it is again pierced by a second harpoon. . . .

When the animal is struck, it by no means remains inactive; on the contrary, it lashes instantly with its tail, and endeavours to bite the rope asunder. To prevent this, the rope is made of about thirty separate slender lines, not twisted together, but merely placed in juxtaposition, and bound round at intervals of every two feet. The thin lines get between the teeth or become entangled about them."

In spite of the great strength of the reptile, two men can drag a tolerably large one out of the water, tie up his mouth, twist his legs over his back, and kill him by driving a sharp steel spike into the spinal cord just at the back of the skull.

There are many other modes of capturing and killing the Crocodile, such as a hook baited with meat, to which the voracious reptiles are attracted by the cries of a pig, which is pulled by the tail or otherwise maltreated, for the purpose of eliciting those ear-piercing yells which aggrieved swine always produce. The yelping of a dog answers the same purpose, and is used in the same manner. In some cases the negroes are bold enough to engage the Crocodile in its own element, and to attack it with a long knife, which they plunge into the belly.

The eggs of the Crocodile are about as large as those of the goose, and many in number, so that these terrible reptiles would overrun the country, were they not persecuted in the earliest stages by many creatures, who discover and eat the eggs almost as soon as they are laid. It is curious that the Crocodile is attended by a bird which warns it of danger, just as the rhinoceros has its winged attendant, and the shark its pilot fish. The Crocodile bird is popularly called the ziczac, from its peculiar cry.

SEVERAL other species of Crocodiles are known, among which two species are deserving of a short notice, namely, the INDIAN CROCODILE (*Crocodílus porósus*), and the AMERICAN CROCODILE (*Crocodilus Americánus*). As the name of alligator is popularly given to these and other reptiles, there is great confusion respecting the precise animal which is under discussion.

The Indian Crocodile, as its name imports, is an Asiatic species, and is found largely in India. It is sometimes called the DOUBLE-CRESTED CROCODILE, because the head is furnished with two long ridges extending from the front of the eye over the upper jaw. This species is common in Ceylon, and literally swarms in the still waters and tanks, though it is but rarely found in rapid streams, and never except in the low lands, the hill marshes being free from these pests. Respecting this animal, Sir E. Tennent writes as follows :—

" The species which inhabit the fresh water is essentially cowardly in its instinct, and hastens to conceal itself on the approach of man. A gentleman who told me the circumstance, when riding in the jungle, overtook a Crocodile evidently roaming in search of water. It fled to a shallow pool almost dried by the sun, and thrusting its head into the mud till it covered up its eyes, it remained unmoved in profound confidence of perfect concealment.

In 1833, during the progress of the pearl fishery, Sir Robert Wilmot Horton employed men to drag for Crocodiles in a pond which was infested with them in the immediate vicinity of Aripo. The pool was about fifty yards in length by ten or twelve wide, shallowing gradually to the edge, and not exceeding four or five feet in the deepest part.

As the party approached the pond, from twenty to thirty reptiles, which had been basking in the sun, rose and fled to the water. A net, specially weighted so as to sink its lower edge to the bottom, was then stretched from bank to bank, and swept to the farther end of the pond, followed by a line of men with poles to drive the Crocodiles forward. So complete was the arrangement, that no individual could avoid the net; yet, to the astonishment of the Governor's party, not one was to be found when it was drawn on shore, and no means of escape was apparent or possible except dashing into the mud at the bottom of the pond."

The extreme tenacity of life possessed by these reptiles is well exemplified, though in a rather painful manner, by an incident which occurred in Ceylon. A fine specimen had been caught by a hook, to all appearance killed, the viscera removed, and the aperture kept open by a stick placed across it. A few hours afterwards the men came to their victim with the intention of cutting off the head, but were much surprised to find the spot vacant. On examination of the locality it was evident that the creature had recovered itself in some strange manner, crawled away for some distance, and made its escape into the water.

The same author also describes the habits of another species, the MARSH CROCODILE (*Crocodilus palustris*), sometimes known by the names of MUGGER, or GOA; an animal which has a large range of locality, being found in Asia and Australia. Sometimes this species grows to a great length; there is in the British Museum a skull twenty-six inches in length, denoting a total length of thirty-three feet.

This animal is in the habit of traversing considerable distances in search of water, but, according to the Singhalese, its feet are sadly cut in passing over the hard stony ground. If it is baffled in its search, it returns to the exhausted pool, burrows beneath the mud, and there waits until released by the rains. Sir E. Tennent mentions one instance where he saw the recent impress of a Crocodile in the mud from which it had just emerged, and he was told of a curious incident which befell an officer attached to the surveying department. Having pitched his tent, he had retired to rest as usual, but during the night he was disturbed by a movement of the earth below his bed. On the following morning the mystery was solved by the appearance of a Crocodile, which made its way from under the bed.

As is the case with the common Crocodile of Egypt, the young of this reptile are very small when hatched, but so fierce even in their early days, that they can be caught by pushing a stick towards them, letting them bite it, and pulling them out before they loosen their hold. A gentleman who has resided for eight years in Ceylon told me that one of his friends was so taken with the appearance of these little reptiles that he captured one, packed it carefully, and took it home. On arriving in his house he put the Crocodile, then about nine or ten inches long, into a basin of water and left it. Shortly afterwards a little boy, one of his children, peeped into the basin, and seeing the Crocodile, gave it a push with his finger. The fierce little creature at once snapped at the offending finger, and held it so tightly that the poor child could not shake it off, and ran screaming about the house with the young Crocodile dangling at the end of his finger, until it was removed by an attendant.

ANOTHER well-known species is the AMERICAN CROCODILE, so often and so wrongly termed the alligator. This reptile is found in the tropical and hotter parts of America, and is very common in some localities. When first hatched, the young seem to feed only on living insects, and according to the experiments of M. Bosc, they would not even touch the insects with which they were supplied until their intended prey began to crawl. In Carolina these creatures pass their winter under the mud. During the summer they become lively at night, and make such a hideous bellowing that a person unaccustomed to it has no chance of sleeping. They also have a habit of clattering their jaws together with a loud noise.

ANOTHER species, the MARGINED CROCODILE (*Crocodilus marginátus*), resides in the rivers of Southern Africa. It may be distinguished from the Egyptian species by the great concavity of the forehead, and the strong keels of the dorsal or back plates. I am indebted to Captain Drayson, R.A., author of "Sporting Scenes among the Kaffirs," for the following account of the Margined Crocodile and its habits, from which it appears that the reptile is formidable not only to the creatures on which it usually feeds, but to man himself :—

"About two or three miles from the Bay of Natal there is a river called the Umganie; into this river a lake called the Sea-Cow Lake empties itself. The lake was, during my residence at Natal, the retreat of several hippopotami and Crocodiles, both of which were in the habit of *treking* into the Umganie river. Often when riding round the banks of this lake, I have disturbed two or three Crocodiles, which were stealing amongst the reeds and long grass, in hope of stalking a fat toad or a sleepy guana. Sometimes a scaly reptile might be awakened from his doze by the sound of my horse's feet, and would rush through the long reeds towards his retreat. Their movement is much more rapid than would be supposed from their appearance, and they care nothing for a fall head over tail, but almost fling themselves down the steep banks when alarmed.

On the banks of the Umganie were several Kaffir kraals, in one of which resided a man who had been roughly treated by a Crocodile. This man, seeing me pass his residence, called to me and asked as a favour that I would watch at a particular part of the river until I shot a rascally Crocodile that had nearly killed him. The Crocodile, he informed me, always made its appearance about sundown, and he hinted that a position might be selected so that the sun would dazzle the Crocodile and prevent him from seeing me. Finding that I was willing to gratify his revenge, he limped out of the inclosure surrounding his huts, and offering me his snuff-gourd, he at my request gave me the following account of his escape.

He had so frequently crossed the stream below his huts at all times of day, and had seen Crocodiles of small dimensions, that he had become as it were familiarized to them, and did not imagine that there was any danger to be expected from them. One evening, at about sundown, he was wading across the river, the water of which reached above his waist. Suddenly he felt himself seized by the under part of his thigh, whilst he was at the same instant dragged under water. His wife was following him, and seeing him fall she scrambled forward to the place where he had disappeared, and thus caused considerable noise and splashing, which (or something else, perhaps the toughness and bad flavour of the Kaffir) had the effect of making the Crocodile quit his hold on the Kaffir, not however without tearing off great portion of the under-part of his thigh. The man with difficulty escaped to the shore, but he remained a cripple for life, unable to do more than put the toes of his foot on the ground."

WE now come to the ALLIGATORS, the second family of those huge reptiles which may be known, as has already been mentioned, by the lower canine teeth fitting into pits in the upper jaw. They are divided into three genera, all of which are inhabitants of the New World. They are indiscriminately called Alligators, Crocodiles, or Caymans, by the natives or the non-zoological traveller, and there is consequently much difficulty in identifying the particular species. The genus Alligator may be known by the partly webbed toes, the outer toe being free.

The COMMON ALLIGATOR inhabits Northern America, and is plentifully found in the Mississippi, the lakes and rivers of Louisiana and Carolina and similar localities. It is a fierce and dangerous reptile, in many of its habits bearing a close resemblance to the crocodiles and the other members of the family.

Unlike the crocodile, however, it avoids the salt water, and is but seldom seen even near the mouths of rivers, where the tide gives a brackish taste to their waters. It is mostly a fish-eater, haunting those portions of the rivers where its prey most abounds, and catching them by diving under a passing shoal, snapping up one or two victims as it passes through them, tossing them in the air for the purpose of ejecting the water which has necessarily filled its mouth, catching them adroitly as they fall, and then swallowing

ALLIGATOR.—*Alligátor Mississipensis*

them. Though timid, as are most reptiles as long as their passions are not touched, the Alligator has within it a very mine of furious rage, which, when aroused, knows no fear. Urged by a blind instinct that sees no obstacles and hardly deserves so intellectual a name as anger, it flings itself upon the assailants, and only ceases its attack as its last breath is drawn.

No easy matter is it to drive the breath out of an Alligator, for its life seems to take a separate hold of every fibre in the creature's body, and though pierced through and through with bullets, crushed by heavy blows, and its body converted into a very pin-cushion, spears taking the place of the pins, it writhes and twists, and struggles with wondrous strength, snapping direfully with its huge jaws, and lashing its muscular tail from side to side with such vigour that it takes a bold man to venture within range of that terrible weapon.

It is fortunate for the assailant that its head is not gifted with mobility equal to that of the tail. The Alligator can only turn its head very slightly indeed, on account of two bony projections, one on each side of the head, which are efficient obstacles to any but the smallest lateral motion. The antagonist may therefore easily escape if on land, by springing aside before the reptile can turn. He must, however, beware of its tail, for the Alligator when angry, sweeps right and left with that powerful member, and deals the most destructive blows with wonderful rapidity. Still, the creature would rather avoid than seek a combat, and does not act in this fashion until driven to despair.

In some parts of America they catch the Alligator in a very ingenious manner. An ordinary hook is said to be of little service against such a quarry, and the natives employ a kind of mixture between a hook and grapnel which very effectually answers their purpose. This so-called hook is made of four sticks of hard tough wood barbed at each end, slightly curving and bound together at one end so as to cause all the upper barbs to radiate from each other. This apparatus is baited with the flesh of some animal, and suspended just about a foot from the water, the other end of the rope being made fast to a tree or strong stake.

As soon as the Alligator takes this bait and begins to pull at the cord, the barbs begin to make their way into its throat, and it is evident from the construction of the hook that

3.

the more the animal pulls, the firmer are the barbs struck into its throat. When thus hooked, its struggles are terrific, and Mr. Waterton, who succeeded in capturing a fine specimen more than ten feet in length, had the greatest difficulty in securing it without damaging its appearance.

The eggs of the Alligator are small and numerous. The parent deposits them in the sand of the river side, scratching a hole with her paws, and placing the eggs in a regular layer therein. She then scrapes some sand, dry leaves, grass and mud over them, smoothes it and deposits a second layer upon them. These eggs are then covered in a similar manner and another layer deposited until the mother reptile has laid from fifty to sixty eggs. Although they are hatched by the heat of the sun and the decaying vegetable matter, the mother does not desert her young, but leads them to the water and takes care of them until their limbs are sufficiently strong and their scales sufficiently firm to permit them to roam the waters without assistance.

As is the case with the crocodiles, the young Alligators are terribly persecuted by birds and beasts, and are even in danger of being eaten by the old males of their own species. During the winter months the Alligator buries itself in the mud, but a very little warmth is sufficient to make it quit its retreat and come into the open air again. While lively, especially at night, it is a most noisy animal, bellowing in so loud a tone and in so singular a cadence that even the nightly concert of jaguars and monkeys is hardly heard when the Alligators are roaring.

It sometimes attains to a great size, and is then formidable to man. Mr. Waterton mentions a case when one of these creatures was seen to rush out of the water, seize a man and carry him away in spite of his cries and struggles. The beast plunged into the river with his prey, and neither Alligator nor man were afterwards seen.

The JACARE, or YACARE (*Jacare sclerops*), also belongs to this family. It inhabits Brazil and is not uncommon. It may be known by the ridge across the face between the eyes, the scarcely-webbed hind feet and the fleshy eyelids. On account of the aspect of its eyes it is sometimes called the Spectacled Cayman. It is said that although this reptile attains a very large size, it will not attack a man even in the water, provided that he always keeps in motion. They pass the night in the water and the day on the shore, where they lie sleeping on the sand, dashing into the water if alarmed.

WE now leave the crocodiles and alligators, and proceed to another order of reptiles. These creatures are termed Amphisbænidæ, from two Greek words signifying to go both ways, in allusion to the shape of the animal, which looks as if it had a head at each extremity. In former times indeed, it was thought that not only could these reptile creep backward and forward with equal ease, but that they absolutely possessed two veritable heads. None of these reptiles are of great size. They are divided into four families, three of which are without external feet, and the members of the other family only possess the front pair of legs very slightly developed. Their eyes are very minute and entirely covered with skin, so that their sight must be of the most limited character. As in the case of the mole, however, this deprivation of sight does not interfere with the welfare of the animal, for it lives mostly beneath the earth, where eyes would be useless.

The SOOTY AMPHISBÆNA is a native of Southern America, being found most plentifully in Brazil and Cayenne. It lives almost wholly underground, boring its way through the soft earth like the common worm, and traversing the soil with considerable address. It feeds upon animal substances, and is very fond of ants, termites and their young. Indeed, it is no extraordinary occurrence on breaking down a termite's nest, to find an Amphisbæna within, luxuriously curled up in the midst of plenty. Ant's nests below the ground are often penetrated and ransacked by this reptile.

Being too small to injure man by sheer force, and being devoid of poisonous teeth, this creature is quite harmless except to the insects on which it feeds. It is able to crawl in either direction with nearly equal ease and rapidity, and on account of the bluntness of its tail and the almost imperceptible eyes, affords some reason for the popular idea of its possessing two heads.

SOOTY AMPHISBÆNA.—*Amphisbæna Americana.*

In speaking of this reptile, Stedman has the following remarks. "This is the snake which, supposed blind, and vulgarly said to be fed by the large ants, is in this country honoured with the name of King of the Emmets. The flesh of the Amphisbæna, dried and reduced to a fine powder, is confidently administered as a sovereign and infallible remedy in all cases of dislocation and broken bones, it being very naturally inferred that an animal which has the power of healing an entire amputation in its own case, should at least be able to cure a simple fracture in the case of another."

This process of reasoning alludes to a curious popular error respecting the Amphisbæna. The people of the countries which it inhabits believe that if one of these reptiles is cut in two, each half, being furnished with a separate head, hastens to its fellow part, and neatly fitting the severed surfaces, repairs the breach and is soon restored to its original condition.

It is rather a dull and sluggish animal when exposed to light, crawling slowly upon the ground, twisting itself lazily about, and opening its mouth in a purposeless kind of fashion without any definite intention of biting or escaping.

The colour of the Sooty Amphisbæna is rather variable, but consists of black and white. Its length is about three feet. There are several species of this genus in the British Museum, one of which, the White Amphisbæna (*Amphisbæna alba*) is of a white colour, and remarkable for a little pellucid dot in the front edge of each scale.

CLOSELY allied to this creature is another reptile, very appropriately called the CHEIROTES, or HAND-EARED LIZARD (*Cheirótes lumbricóïdes*). This is a native of Brazil, and as far as is known, is of subterranean habits, like the amphisbæna.

The Cheirotes is the only example of all the amphisbænas that possesses external limbs, and even in this instance, they are small and but slightly developed. There are no hind legs, but the two fore legs are set just behind the head; nearly in the place where the ears might be expected to be seen. They are very short, rather flat and strong, and are terminated with five toes, four of which are armed with a tolerably strong claw. The fifth toe is very small and without a claw.

The head of this creature is no larger than the body, the teeth are conical, moderately strong and slightly curved backwards, the muzzle is arched, the tongue horny at the tip, the tail is short, and there is a row of small pores on the under side of the abdomen. Its length is about eight or ten inches, and its colour is yellow, spotted with brown above, and whitish below. This species is the sole representative of its family. The other two

WHITE-THROATED REGENIA.—*Regénia albogulóris.*

families, namely the Trigonophidæ and the Lepidosternidæ, may easily be distinguished oy the fact that in the former the teeth are set in the margin of the jaws instead of on their inner side as in the other families, and that in the latter, the scales on the chest are larger and of different shapes, whereas in the other two families they are all squared. Moreover the pores under the abdomen are absent.

WE now leave the shielded reptiles and proceed to the Scaled Lizards. These creatures for ma very large and important group, and may be distinguished from the previous section by the covering of the body, which is formed of scales either granular or overlapping each other, instead of the straight-edged plates which cover the bodies of the tortoise and crocodiles. The tongue of these animals is rather long, nicked at the tip, and often capable of extension. The young are produced from eggs, sometimes hatched before being deposited, but generally after they have been laid in some suitable spot. The eggs are covered with a rather soft, leathery shell.

The true LIZARDS have four limbs, generally visible, but in a few instances hidden under the skin. Their body is long and rounded, and the tail is tapering and mostly covered with scales set in regular circles or "whorls." The mouth cannot be dilated as in the snakes, because the under jawbones are firmly united in front instead of being separable as in the serpents. The ear has a very singular appearance, the drum or "tympanum" being mostly distinct and exposed.

There are twenty-four families of true Lizards, and passing by several anatomical and structural distinctions, which will be found at the end of the volume, we will proceed at once to the first family, called the MONITORS. In all these creatures the head is covered with very little, many-sided scales; the tongue is long, slender, and capable of being withdrawn into a sheath at its base; the scales are small, rounded, and arranged in cross rings, those of the side resembling those of the back; the legs are four in number, and each foot has five toes. They are all inhabitants of the Old World, and are seldom if ever found far from water.

OUR first example is the WHITE-THROATED REGENIA, or WHITE-THROATED VARAN, a remarkably fine and powerful species of Lizard, inhabiting Southern Africa. A rather full and accurate description of this Lizard is given by Dr. Smith.

NILOTIC MONITOR.—*Mónitor Nilóticus*.

"It is usually discovered in rocky precipices or on low stony hills, and when surprised seeks concealment in the chinks of the former or the irregular cavities of the latter, and where any irregularities exist on the surface of the stones or rocks, it clasps them so firmly with its toes that it becomes a task of no small difficulty to dislodge it, even though it be easily reached. Under such circumstances the strength of no one man is able to withdraw a full-grown individual, and I have seen two persons required to pull a specimen out of a position it had attained, even with the assistance of a rope tied in front of its hinder legs. The moment it was dislodged it flew with fury at its enemies, who by flight only saved themselves from being bitten. After it was killed, it was discovered that the points of all the nails had been previously broken or at the moment it lost its hold.

It feeds upon crabs, frogs, and small quadrupeds, and from its partiality to the two former, it is often found among rocks near running streams, which fact having been observed by the natives, has led them to regard it as sacred, and not to be injured without danger of drought."

This fine Lizard has large, oblique nostrils, a shortish tail with a double keel on its upper surface, and the scales are oblong and have a blunt ridge or keel. The head is short and the scales of the body are large, convex, and surrounded with granulations. The length of the full-grown Regenia is nearly five feet, and its colour is dark brown, above variegated with large white spots, and paler beneath, especially under the throat.

The NILOTIC MONITOR, or VARAN OF THE NILE, as it is sometimes called, is, as its name imports, a native of those parts of Africa through which the Nile, its favourite river flows.

The natives have a curious idea that this reptile is hatched from crocodile's eggs that have been laid in hot elevated spots, and that in process of time it becomes a crocodile. This odd belief is analogous to the notion so firmly implanted in the minds of our own sea-side population, that the little hermit crab, which is found so plentifully in periwinkle shells, is the young of the lobster before it is big and hard enough to have a shell of its own.

It is almost always found in the water, though it sometimes makes excursions on land in search of prey. To the natives it is a most useful creature, being one of the appointed means for keeping the numbers of the crocodile within due bounds. It not only searches on land for the eggs of the crocodile, and thus destroys great numbers before they are hatched, but chases the young in the water, through which it swims with great speed and agility, and devours them unless they can take refuge under the adult of their own species, from whose protection the Monitor will not venture to take them.

When full grown, the Nilotic Monitor attains a length of five or six feet. The colour of this species is olive-grey above, with blackish mottlings. The head is grey, and, in the young animal, is marked with concentric rows of white spots. Upon the back of the neck is a series of whitish yellow bands, of a horse-shoe, or semilunar shape, set cross-wise, which, together with the equal-sized scales over the eyes, serve as marks which readily distinguish it from many other species. The under parts are grey, with cross bands of black, and marked with white spots when young.

Specimens belonging to this genus are scattered over the greater part of the world. For example, the INDIAN MONITOR (*Monitor dracœna*) is found in the country from which it takes its name. It is rather a prettily marked animal, being brown with black spots when old, and yellow eye-like marks when young. Another species, GOULD'S MONITOR (*Monitor Gouldii*), inhabits Australia, being most commonly found on the western side of that land.

WE now arrive at another family of Lizards, called, from the typical species, the Teguexins. In these reptiles, the head is covered with large, regular, many-sided shields, the sides are flat, and the throat has a double collar.

Our first example is the TEGUEXIN, or VARIEGATED LIZARD, so called on account of the contrasting colours with which it is decorated. It is also known by the name of SAFEGUARD, a title which has been given to it because it is thought to give notice, by hissing, of the approach of the alligator. The monitors derive their name from a similar belief, they being thought to warn human beings of the approach of poisonous serpents.

Several species of Teguexin are known, all inhabiting the warmer portions of America, and possessing similar habits. It is said that, although strong and agile, they do not ascend trees, but range at will the hot sandy plains or the dense damp underwood on the margins of lakes and rivers, into which they plunge if alarmed, and remain below the surface until the danger has passed away, their capacious lungs and imperfect circulation permitting them to endure a very long immersion without inconvenience.

The Teguexin is a large and powerful Lizard, exceeding five feet in length when full grown, and extremely active. It feeds mostly, if not entirely, upon animal food, and makes great havoc among snakes, frogs, toads, and other semi-aquatic creatures. It often indulges in diet of a higher nature, and when it can find an opportunity, devours poultry, or breaks and eats their eggs. Sometimes it has been known to eat Lizards of a closely allied species, a fact which has been proved by finding some bones, and other portions of the Ameiva lizard within the stomach of a Teguexin that had been killed. Together with these relics were found the shelly wing-cases of beetles, and the skins of sundry caterpillars.

The teeth of this species are strong, and the reptile can bite with great force. It is a bold and determined combatant when attacked, and if it succeeds in grasping a foe, retains its hold with the pertinacity of the bulldog. The flesh of the Teguexin is eaten, and thought to be excellent. According to Azara, the skin of its tail, when separated into rings, is considered to be a safeguard against paralysis, and worn for that purpose, as well as to remove tumours, another healing power which it is supposed to possess.

TEGUEXIN.—*Tenus Teguexim.*

The general colouring of the Teguexin is as follows. The upper parts are deep black, with bold mottlings of yellow or green. On the upper part of each side there are two series of white spots, and the under parts are mostly yellow, with black bands. The colouring is, however, extremely variable.

THE curious little AMEIVA, which has just been mentioned as falling a victim to the previous species, is closely allied to the Teguexin. It is rather a pretty Lizard, with a very long whip-like tail, and peculiarly elongated toes on the hinder feet. The long tail is covered with a series of scales, arranged in rings, of which about one hundred and twenty have been counted in a perfect specimen. The colour of the Ameiva is dark olive, speckled with black on the nape of the neck and front of the back. On the sides are rows or bands of white spots edged with black, from which peculiarity it is sometimes called the Spotted Lizard. There are many species of Ameiva, inhabiting either Central America, or the West Indian Islands.

The SIX-LINED TARAGUIRA also belongs to the Teguexins. This pretty little Lizard, with its dark green body, and yellow streaks, inhabits North America. Mr. Holbrook makes the following remarks respecting its habits. "This is a very lively, active animal, choosing dry and sandy places for its residence, and is frequently met with in the neighbourhood of plantations, or near fences and hedges. Most usually it is seen on the ground in search of insects; its motions are remarkably quick, and it runs with great speed. It is very timid. It feeds on insects, and generally seeks its food towards the close of the day, when they may be seen in corn fields, far from their usual retreat; and not unfrequently I have met male and female in company."

SIX-LINED TARAGUINA.—*Cnemidóphorus sex-lincátus.*

The colour of this little reptile is dark brown, with a perceptible purple gloss on the back, diversified with six narrow streaks of yellow, one line on each side reaching from the eye to the middle of the tail. The abdomen is bluish white, with a silvery lustre, and the throat is silvery white. The length of this species is from nine to eleven inches.

A brief notice must also be given of two curious species, also belonging to the same family. The first is the SPURRED CENTROPYX, or SPURRED LIZARD (*Céntropyx calcátus*), so called from two pair of small, sharp, horny spikes, which are set at each side of the base of the tail. The colour of this species is olive-green above, with three streaks of a paler hue, and a double series of black spots on the back. Below it is greenish white.

The other species is the GREAT DRAGON (*Ada Guianensis*), a native of tropical America. This fine Lizard is generally from four to nearly six feet in length, and is strong and nimble. It does not appear to be so good a swimmer as some of the preceding species, but runs fast, and can climb trees with great agility. It is generally found among the marshy and low-lying lands, though it spends more time on the land than in the water.

It is a desperate fighter when attacked, and as it has a habit of hiding itself in a deep burrow, and bites fiercely at the hand that is thrust forward to seize it, is not easily captured. It is, however, much sought after, as its flesh is very good, and the eggs are thought to be great delicacies. There are usually from thirty to forty eggs. The general colour of this reptile is olive, yellow beneath, and mottled with brown.

THE true LIZARDS, or Lacertinidæ, now come before our notice. The tongue of these reptiles is long, flat, can be thrust out to some distance, and very deeply forked. The teeth are hollow at their roots, the scales are keeled, and the sides are flat. They are scattered over the greater part of the globe.

England possesses at least two examples of this family, one of which, the SCALY LIZARD, is extremely common.

This pretty little reptile is extremely plentiful upon heaths, banks, and commons, where it may be seen darting about in its own quick, lively manner, flitting among the grass stalks with a series of sharp, twisting springs, snapping up the unsuspecting flies as they rest on the grass blades, and ever and anon slipping under shelter of a gorse bush, or heather tuft, only to emerge in another moment brisk and lively as ever.

These little creatures are so quick and sharp sighted, that it is not very easy to catch them, especially if they are among gorse bushes, for they twist about so adroitly, that a very smart movement of the hand is required to follow them, and the prickly points of the gorse are always lurking among the grass, to the detriment of a tender skin. They can swim tolerably if thrown into the water, but do not seem to seek that element voluntarily. I have generally found that when flung into water, they lie for a short time quite motionless, with their limbs extended, and tail straight, as if bewildered with the sudden change. They soon, however, get their head towards shore, and then, with a serpentine movement of the tail, scull themselves to land.

This is one of the reptiles that produces living young, the eggs being hatched just before the young Lizards are born. With reptiles, the general plan is to place the eggs in some spot where they are exposed to the heat of the sunbeams; but this Lizard, together with the viper, is in the habit of lying on a sunny bank before her young ones are born, apparently for the purpose of gaining sufficient heat to hatch the eggs. This process is aided by the thinness of the membrane covering the eggs.

SCALY LIZARD.—*Zoótoca vivipara.*

The colour of this little Lizard is extremely variable, but in general, the upper parts are olive-brown, with a dark brown line along the middle of the back, this line being often broken here and there. Along each side runs a broader band, and between these bands are sundry black spots and splashes. The under parts are orange, spotted with black in the male, and olive-grey in the female. The total length of the Scaly Lizard is about six inches.

THE beautiful EYED LIZARD, or GREAT SPOTTED GREEN LIZARD, as it is sometimes called, from the colours with which it is decorated, is a native of Southern Europe, and various other warm portions of the world, being found in Algiers, Senegal, and parts of America.

This creature inhabits dry spots, where the sun has most power, and may be seen among hedges, underwood, or loose stones, running about in search of food,

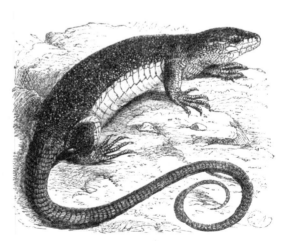

EYED LIZARD.—*Lacerta ocelláta.*

and displaying the gem-like brilliancy of its clothing, as it darts from spot to spot with the agility which characterizes all the species of this genus.

It is of rather a fierce nature, having little fear, and boldly attacking any antagonist

GREEN LIZARD.—*Lacerta viridis.*

that may assail it. If it be irritated with a stick, it will turn sharply upon the offending weapon, and bite it smartly, and if a dog attempts to seize it, the courageous little creature will spring upon its muzzle, and maintain its hold with such pertinacity, that it will suffer itself to be killed rather than relinquish its grasp. In consequence of this combative character, it is greatly respected by the inhabitants of the country where it dwells, and being thought to be poisonous as well as ferocious, is dreaded with a fear quite as keen, though not so reasonable, as would be inspired by a rattlesnake or cobra.

The home of this species is generally made under the roots of trees if the soil be sufficiently dry and sandy to suit its habits. Otherwise it will excavate a tunnel in the side of a bank or under a hedge, always choosing a southern aspect so as to ensure the warmth which its nature seems to demand. Sometimes it settles upon a soft sandstone rock for its domicile, and hollows out a deep burrow in the softest part of the rock, mostly choosing the loose sandy layers that often occur between two tolerably broad strata of rock. Like the rest of the Lizards it feeds on insects and similar creatures, darting on them with great speed and certainty of aim.

The colour of this Lizard is very beautiful, rendering it one of the most lovely of its tribe. The ground colour of the body is bright glittering green as if covered with an armour of emeralds, upon which are set along the sides some rather large eye-like spots of rich azure. A kind of network of black is also spread over the body, sometimes running in well-defined lines, and sometimes composed of rows of black dots. The temples of the Eyed Lizard are covered with unequal many-sided scales, rather convex in their form. Its length when full grown is about fifteen or sixteen inches, but it is very variable in size as well as in colour.

A VERY beautiful species of this genus is common in Jersey and many parts of Europe, Asia, and Africa. This is the GREEN LIZARD, sometimes called the JERSEY LIZARD. As its name imports, this reptile is of a green colour, and with the exception of the preceding species, is as beautiful a creature as can be seen.

Like the eyed Lizard, it haunts sunny spots, and may be found in orchards, gardens, shrubberies, copses and similar localities, where it can find plenty of food and obtain concealment when alarmed. Old ruins, too, are greatly haunted by this beautiful Lizard, which flits among the moss-covered stones with singular activity, lying at one moment as if asleep in the sunbeams, or crawling slowly as if unable to proceed at any smarter pace, and then, when the hand is thrust towards it, disappearing with a rapidity that looks like magic.

Since the great demand for ferneries and vivaria of different descriptions has arisen, this Lizard has been brought over to England in great numbers, as it is a beautiful ornament to a glass fern-case, and is sufficiently hardy to be kept alive with a very little care. It seems to revel in the sunshine, and there are few objects more beautiful than the emerald green hues of this Lizard as the sunbeams flash and glitter on its resplendent surface.

It is susceptible of kindness, and can soon be tamed by those who choose to take the trouble of familiarizing themselves with their bright and lively favourite. Although sufficiently bold and apt to bite if it fancies itself aggrieved, it can be so thoroughly tamed that it will come and take flies out of the hand. In France and other countries this pretty harmless little creature is greatly dreaded, the popular belief attributing to it sundry destructive powers of the same nature as those which our rustic population believe to be exercised by the common newt.

The colour of this beautiful creature is rich shining green above, a little blue sometimes appearing upon the head, and the quality of the green being rather variable in different individuals. A multitude of little golden spots are also perceptible on the back, and similar dots of black are not unfrequently sprinkled over the surface. Underneath, the green fades into a yellower hue.

UNTIL comparatively later years, the SAND LIZARD was confounded with the scaly Lizard, which has recently been described.

This reptile is extremely variable in size and colouring, so variable, indeed, that it has often been separated into several species. Two varieties seem to be tolerably permanent, the brown and the green; the former, as it is believed, being found upon sandy heaths where the brown hues of the ground assimilate with those of the reptile, and the green variety on grass and more verdant situations, where the colours of the vegetation agree with those of the body.

Though quick and lively in its movements, it is not so dashingly active as the scaly Lizard, having a touch of deliberation as it runs from one spot to another, while the scaly Lizard seems almost to be acted upon by hidden springs. It does not bear confinement well, and in spite of its diminutive size and feeble powers, will attempt to bite the hand which disturbs it in a place whence it cannot escape. When it finds itself hopelessly imprisoned, it loses all appetite for its food, hides itself in the darkest corner of its strange domicile, and before many days have passed, is generally found lying dead on the ground.

Unlike the scaly Lizard, this species lays its eggs in a convenient spot and then leaves them to be hatched by the warm sunbeams. Sandy banks with a southern aspect are the favoured resorts of this reptile, which scoops out certain shallow pits in the sand, deposits her eggs, covers them up, and then leaves them to their fate. Mr. Bell, who

SAND LIZARD.—*Lacerta agilis.*

has paid great attention to this subject, has remarked that the eggs are probably laid for a considerable period before the young are hatched from them.

As has been already remarked, the colouring of this creature is exceedingly variable in different individuals. Generally it is sandy brown above, with some faint bands of a darker brown with rows of black spots, which sometimes have a whitish dot in their centre.

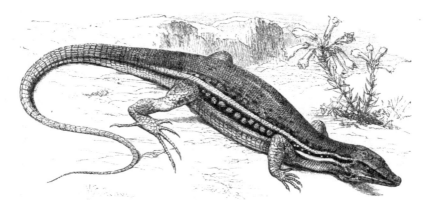

CAPE SPINE-FOOT.—*Acanthodáctylus.*

The sides have a tinge of green more or less distinct, and the under surface is white. In some individuals the green is very distinct, and Mr. Bell thinks with some reason that the many written accounts of discovering the green or Jersey Lizard in England, may be referred to the green variety of this species, which shone with peculiar lustre in consequence of the sunbeams being reflected from the shining sides. I know of one instance where the true *Lacerta viridis* was captured and killed near Oxford, but I believe that it must have been a wanderer from one of the numerous fern-cases that are to be seen in so many houses. The average length of the Sand Lizard is about seven inches or a little more.

PASSING by a series of genera affording but few interesting points, we come to the curious animal called the CAPE SPINE-FOOT. The generic name Acanthodáctylus, signifies Thorn, or Spine-toed, and is very appropriately given to this animal and the other species of the same genus. All the Spine-foot Lizards are inhabitants of Africa, and most of them are found towards the northern portion of that continent.

According to Dr. Smith, "this Lizard is found on the sandy districts of Great Namaqualand, and where the surface of the country is irregular it is generally met on the highest spots. Where small sand-hills occur, it resorts to them in preference to the other localities, and from the peculiar assistance it derives from the serrated fringes which edge its toes, it runs over the loose sand on the steep surfaces of those slopes with great activity. It feeds on insects."

The colour of this Lizard is a very peculiar brown above, changing from yellow-brown to a much warmer hue, partaking of the orange. The top of the head is mottled with dark brown, and the back is freckled with the same hue. From the eyes run two whitish bands on each side, the lower terminating at the hind leg and the upper reaching some distance along the tail. Between and about these bands are bold brown mottlings in the male, and an orange wash in the female. The upper part of the legs are also mottled with dark brown. The toes are very long, especially those of the hind foot, and are edged with a fringe composed of sharply pointed scales. The female is larger and more clumsily made than the male.

ANOTHER pretty species of Lizard, termed the NAMAQUA EREMIAS, is found in the portion of Africa from which it derives its name. The name Eremias signifies a dweller in a wilderness, and is given to this and several other species because it is always found in hot and arid situations, the sandy flats between Cape Town and Little Namaqua-land being its most favoured localities.

NAMAQUA EREMIAS.—*Erèmias Namaquensis.*

It is chiefly remarkable for the great length and slenderness of its tail, which measures five and a half inches in length, although the head and body together are only two inches long. The colour of the back and upper parts is delicate brown mottled with a deeper hue, and along the back are drawn four narrow lines of light reddish orange. The sides are cream-yellow, the upper portions of the legs are olive-brown, and the under surface of the animal is yellowish white. There is a trifling variation in the colouring according to the age of the individual. Thirteen or fourteen species of this genus are known to zoologists, most of them being natives of Africa.

OUR last example of the true Lizards or Lacertinidæ is the curious little creature termed the ELEGANT OPHIOPS. Two species are known as belonging to this genus, and they can at once be separated from the true Lizards by the character of the eyelids, which are only rudimentary and hardly visible, so as to have gained for their owners the generic title of Ophiops, or Serpent-eyed Lizards.

The Elegant Ophiops inhabits the south-eastern portions of Europe, and the neighbouring parts of Asia. The shores of the Mediterranean appear to be favourite localities of the Ophiops, and in those places it is not at all uncommon. It is lively and active in character, and, like the rest of the same family, feeds on insects, which it catches by suddenly springing on them as they repose from their aërial excursions or crawl along the ground. Like most Lizards, it is rather variable in colouring, but the general tints are as follows. The back and upper parts are olive, sometimes deepening into bronze. Along each side run two bands of pale yellow, and between the bands are sundry black spots, also arranged in lines, but varying in form, size, and number according to the age of the individual. The under parts are white.

Quitting the true Lizards, we come to another family of reptiles, called the Zonuridæ, or Band-tailed Lizards, because the scales of the tail are arranged in regular series or rings, and by their overlapping cause the edges to stand out boldly in whorls. Along the sides of these reptiles runs a distinct longitudinal fold, covered with little granular scales, and the eyes are furnished with two valvular lids.

The COMMON ZONÚRUS, or ROUGH-SCALED CORDYLE, is a native of Southern Africa, and very plentiful at the Cape, where it may be seen among the rocks or in sunny localities flitting from spot to spot with some speed, though not exhibiting the singular activity which is possessed by many of the smaller Lizards. It is chiefly remarkable for the curious aspect of the tail, with its whorls of spike-tipped scales, which looks as if

ELEGANT OPHIOPS.—*Ophiops elegans.*

a number of thimbles had been deeply notched round their edges and then thrust into one another.

There is a somewhat similar reptile called the COMMON CORDYLE (*Cordylus polygonus*), but it may be distinguished by a peculiarity of structure which has caused it to be placed in a different genus. In the members of the genus Zonurus, the eyelids are opaque, as is generally the case, but in the genus Cordylus there is a smooth transparent spot in the centre of the lower eyelid.

The form of the Rough-scaled Cordyle is rather stout and flattened, as accords with the comparative slowness of its movements. In colour it is variable, but the usual tints are orange-yellow on the back, sides, and tail, fading into yellow on the head and white on the under parts. This species may be distinguished from the other Cordyles by the smooth shields of the head and the rhomboidal-shaped scales of the back, which are larger in the centre than on the sides and decidedly keeled. On the flanks the keels are so long as to become spines, and the sides of the neck are covered with sharp spine-like scales.

THE second species in the illustration is the FALSE CORDYLE, which is placed in a separate genus on account of the shape and size of the scales upon the back and sides. Instead of being large and tolerably even in size, as in the preceding genus, they are very small and granular, alternating with bands of larger scales, which are three-sided, convex, and slightly keeled. These scales are largest on the sides of the back. The generic name Microlepidotus signifies small-scaled, and is given to these creatures in allusion to the minute scales of the back and sides.

The habits of this reptile are much like those of the previous species. Dr. A. Smith writes as follows respecting this creature, after describing the singular variations of colour to which it is subject :—

"Each of the varieties appeared to be restricted to its own localities, and, so far as my observations extend, no specimens of two varieties are ever found in the same localities. All the varieties inhabit rocky situations, and when they have a choice they invariably prefer precipices and the stony walls of difficultly accessible ravines. In this situation they wander carelessly in search of food or warmth, unless alarmed by what they may regard as enemies. On being closely approached in their retreats they are with difficulty captured, as by aid of the prominences on the hinder edge of each temple, they hold on with a tenacity which is quite surprising, and by them they occasionally offer such an

ROUGH-SCALED CORDYLE.—*Zonurus Cordylus.*   FALSE CORDYLE.—*Pseudocordylus microlepidotus.*

effectual resistance to the force applied from behind, that the tail breaks off from the body before the reptile is secured."

As, in Dr. Smith's work, the description of the different varieties occupy nearly five quarto pages of letterpress, it is evidently impossible to give more than a general description in this volume. Suffice it to say, that in one variety, found on the Table Mountain and about Cape Town, the colour is ochry yellow above banded with dark brown; in another, which inhabits the rocks about Algoa Bay, it is yellow, with bold black bars along the back; another, which lives on the banks of the Orange River, is brown above, warming into bright chestnut in the male and olive-green mottled with dusky black in the female; and a fourth variety, which is found in the high mountainous regions about Natal, is bright green, with an olive-green stripe and short bars of the same tint across the back. The tail is also banded with two shades of green, one a deep olive, and the other having a much yellower hue. The female of this variety is without the bands, and is only mottled with dark olive and spotted with the same hue along the sides. The length of the False Cordyle is about eighteen inches.

A SMALL group of reptiles is collected under the generic title of Gerrhosauri, or Basket-Lizards, because the arrangement of their scales and colouring has an effect as if the body had been covered with delicate wicker-work, such as is employed to protect glass flasks from injury.

These Lizards are natives of Southern Africa, where they are far from uncommon. They are all rather pretty in form and colouring, but the most pleasing in general appearance is BIBRON'S GERRHOSAURUS (*Gerrhosaurus Bibróni*). This animal is found near the Orange River, and may be seen slipping about among the rocky sides of the dark ravines that are so plentiful in that neighbourhood. It is a very shy and timid creature, and if it fancies itself watched by an unfriendly eye, or suspects the least shadow of danger, it quietly glides under the heap of dead wood and dried leaves which collect in abundance in such localities, and will not venture out again until it is tolerably sure that the danger has passed away.

SAUROPHIS.—*Saurophis tetradáctylus.*

As is the case with most of these Lizards, there is considerable variation of colouring, but in general the upper surface is dark brown, and the sides of the head, the throat and front of the fore limbs are bright scarlet. Along the back run four yellow lines, of which the two central only extend as far as the hind legs, whereas the two outer streaks are continued to the extremity of the tail. It is not a large species, being about ten or eleven inches in length.

THE generic name SAUROPHIS, which is given to the reptile next in order, is of Greek origin and signifies Lizard-Snake, in allusion to the very serpentine aspect of its body.

This singular creature inhabits Southern Africa, and at first sight might be easily mistaken for a serpent as it crawls about the ground, its four tiny limbs being far too weak to render it any great assistance in progression, which is achieved, as in the serpents, by continual movement of the projecting edges of the scales. Very little is known of its habits.

The head of this reptile is of a somewhat pyramidal shape and covered with shields, as are both temples. The scales of the back are slightly grooved and a small keel runs across their length ; they are regularly arranged in fourteen series. On the abdomen, the shields are in six rows. There are four very small and feeble limbs, each of which is furnished with four little short and compressed toes, with rather long claws at their extremities. The body is long and cylindrical, and a decided groove runs along each side. Its colour is tawny brown, each scale being of a deeper hue at its edge, so as to give a slightly mottled appearance to the creature. The legs and lower edge of the temple are white, spotted with little dots of black.

ON account of the great rapidity of its movements, the reptile represented in the accompanying illustration has received the appropriate title of TACHYDROME, a name derived from the Greek, and signifying a swift runner.

This pretty little Lizard is an Asiatic animal, being mostly found in China, Cochin China, and Java. Although its limbs are much larger and more powerful in proportion to the size of the body than those of the preceding species, its tail is of such great comparative length and so slender in its proportions, that, quick as is the creature in all its movements, it has much of a serpentine aspect. The tail indeed is longer in proportion to the body than is the case with any other of the order, being three times the length of the body and head, and tapers from the body like the thong of a whip from its handle.

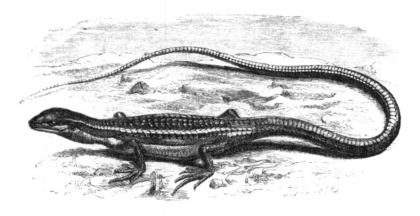

TACHYDROME.—*Tachydromus sexlineatus.*

The collar of this creature is covered with scales and decidedly toothed. The scales of the back are nearly square in form, slightly overlap each other, and are arranged in four longitudinal series. Each scale has a decided keel along its length. The scales of the sides are small and granular, and those of the abdomen and throat are larger, strongly keeled, and boldly overlap each other, a provision which is evidently intended for the purpose of aiding the creature in progression, and enabling it to hold itself firmly in any cleft into which it may have retreated. The scales of the common snake answer the same purpose, as any one may prove by taking a snake by the tail and drawing it backwards over a carpet, or by allowing itself to insinuate half of its body into a crevice in a rock or old wall, and then endeavouring to draw it out again by pulling at its tail.

The colour of this pretty Lizard is dark olive above. On each side a bold white streak, edged on either side with black, runs from the base of the head to the insertion of the tail. On the sides of the body and neck are a multitude of little black dots, each having a white centre, and between these dots the colour is blue, glossed with golden yellow. The abdomen and under parts are pure shining white, and the tail is generally olive, though in some specimens it has something of a metallic or iridescent lustre, and gleams with golden or coppery reflections. Between the nostril and the eye runs a short black line, and on the temples are two similar lines, with a white streak between them. The total length of the Tachydrome is about one foot.

In the curious snake-like Lizard called the SCHELTOPUSIC, or PSEUDOPUS, the limbs are almost entirely absent, the front pair being altogether wanting, and not even exhibiting a trace of their locality, while the hind pair of legs are only indicated by two slight scale-like appendages at the junction of the tail with the body. It is often the case that with reptiles in which the limbs are externally wanting, their bones, although very small and delicate, are found beneath the skin. But in the Scheltopusic, the only indication of legs is found in a pair of very tiny bones attached to the pelvis, and exhibiting the merest rudiment of the missing limb.

Moreover, the pelvis itself is very small and slight, and is itself scarcely more than rudimentary in its form, though affording one of the needful transition links between the quadrupedal Lizards and the footless snakes, some of which, indeed, possess the rudiments of limbs even in a more doubtful state than is found in the Scheltopusic. In consequence of the absence of limbs, the movements of this reptile are completely those of a serpent, and so snake-like is it in all its gestures, that in the countries where it resides, it is popularly considered as a serpent, as is the case with the blindworm of England.

3.   E

SCHELTOPUSIC.— *Pseudopus Pallasii*

The Scheltopusic is a native of the coast of Northern Africa, and is also found in Dalmatia, the Morea, and parts of Siberia, where it is called by the title under which it is now generally known. It seems to be rather a timid creature, and very mistrustful of strange sights or sounds, always remaining within the vicinity of some familiar spot, whither it seeks an immediate retreat if disturbed.

Thickly wooded valleys, where the underwood is dark and dense, and the vegetation is rank and heavy, are favourite localities of this harmless and weaponless reptile, which has no mode of defence if attacked, and can only retreat from the approach of danger by gliding silently under the brushwood and insinuating itself in some dark crevice, where it lies secure. So watchful is this creature, that although its movements are rather slow, it is not very easily captured, mostly gliding away in so silent a manner that it has reached its haven of safety before its presence is even suspected.

Even if it be seen and followed, it is not readily captured after once it has succeeded in burying itself among the brushwood, for its colour is sufficiently sombre to harmonize so well with the dark soil and dead sticks and leaves among which it resides, that its outline can with difficulty be discerned, even by a practised eye. As is the case with most reptiles, it loves to emerge from its retreat and crawl to some spot where the sunbeams have thoroughly warmed the ground, and there to lie basking in the genial heat. While thus occupied, it is not so wary as at other times, and may be approached and secured before it can make good its retreat.

The whole aspect of this reptile is so serpentine that it has been attacked and killed under the impression that it was a poisonous snake, and great has been the surprise of its slayers to find that they had destroyed, not a venomous serpent, but a harmless Lizard. This creature has been often captured alive and kept in confinement. In its wild state it feeds mostly on insects, the smaller reptiles, and similar creatures, sometimes gliding into a nest of newly hatched birds and swallowing them. This propensity was once exhibited by a captive Scheltopusic; it had fed very contentedly on hard-boiled eggs, until one day it contrived to gain access to a nest full of very young birds, and swallowed the whole brood.

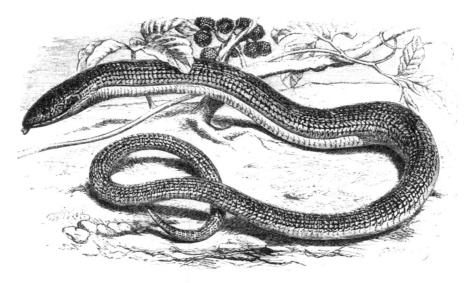

GLASS SNAKE. *Ophisaurus ventralis.*

The jaw-teeth of this reptile, although not of a venomous character, are strong, and those of the palate, although small, are probably useful in aiding the creature to secure and swallow its prey. The tongue is thin and covered with little papillæ of various sizes. Along each side runs a rather deep groove or furrow, which, on a closer inspection, is found to be double. The scales of the back are rather shining and closely set, and there is a slight keel running along the centre of each scale, which is shown more distinctly on the tail than on the body. The keel is shown more distinctly in the young than in the adult.

The colour of this reptile is rather variable, but in general the ground colour of the body is chestnut, profusely dotted with blackish spots, caused by the dark edges or spots of each scale. These scales are arranged in a regular series of thirteen longitudinal rows. The eye is bright golden green, and has a very beautiful appearance, as it contrasts well with the chestnut and black of the body and head. The young Scheltopusic is very different from its parent in the colouring, being grey above, with rather obscure bands of greyish brown, and the under surface is grey, with a whitish lustre. The length of the Scheltopusic is about eighteen inches, the tail occupying about three-fifths of the whole measurement.

The curious reptile which is appropriately called the GLASS SNAKE is a native of North America.

In this creature there is not even a vestige of limbs, so that it is even more snake-like than the preceding species. The generic title of Ophisaurus is of Greek origin, signifying Snake-lizard, and is given to the reptile on account of its serpentine aspect. The reader may remember that on page 48 there is an account of the saurophis, a name which is exactly the same as that of the present species, except that the one is called the lizard-snake, and the other the snake-lizard, a distinction which, in the present case, is without a difference, so that the two reptiles might exchange titles and yet be appropriately named.

The Glass Snake is indeed so singularly like a serpent that it can only be distinguished from those reptiles by certain anatomical marks, such as the presence of eyelids, which are wanting in true serpents, the tongue not sheathed at the base, and the solid jawbones, which in the serpents are so loosely put together that the parts become widely separated when the mouth of the creature is dilated in the act of swallowing its prey.

The Glass Snake is one of the earliest of the reptile tribe to make its appearance in the spring, shaking off its lethargy and coming out of its home to bask in the sunbeams and look after the early insects, long before the true snakes show themselves. It is generally found in spots where vegetation is abundant, probably because in such localities it finds a plentiful supply of the insects, small reptiles, and other creatures on which it feeds.

It is fond of frequenting the plantations of sweet potato (*Convolvulus batatas*), and during harvest-time is often dug up together with that vegetable. The home of this reptile is made in some very dry locality, and it generally chooses some spot where it can be sheltered by the roots of an old tree, or a crevice in a convenient bank. It moves with tolerable rapidity, and its pursuer must exercise considerable quickness before he can secure it.

To catch a perfect specimen of the Glass Snake is a very difficult business, for when alarmed, it has a remarkable habit of contracting the muscles of its tail with such exceeding force that the member snaps off from the body at a slight touch, and sometimes will break into two or more pieces if struck slightly with a switch, thus earning for itself the appropriate title of Glass Snake. Our common blindworm, which will be described in a future page, possesses a similar capacity, and often uses it in a rather perplexing fashion. Catesby remarks that this separation of the tail into fragments is caused by the construction of the joints, "the muscles being articulated in a singular manner quite through the vertebræ." The tail is more than twice the length of the body, from which it can only be distinguished by a rather close inspection.

The head of the Glass Snake is small in proportion to the body, rather pyramidal in shape. Along each side of the body runs a rather deep d uble groove. The colouring of this creature is extremely variable, but is generally as follows : The head is mottled above and at the sides with black and green, and the jaws are edged with yellow. The upper part of the body is marked with multitudinous lines of black, green, and yellow, and the abdomen is bright yellow along its length. In the tail there are about one hundred and forty rings of scales. Sometimes the upper surface is black on the sides and neck, and brown on the back, the head being marbled with yellow and black ; another variety is chestnut above, with white spots edged with black, and the under parts pale orange ; while a third variety is grey mottled with black. The total length of this reptile is from two to three feet.

FOUR small families now follow, containing but very few individuals. The first of these is called the CHALCIDÆ. These reptiles have long cylindrical bodies, with a slight granular groove on the front of each side, and four very short rudimentary limbs. The typical species of this family is the CHALCIS (*Chalcis flavescens*), a native of tropical America, Guiana, and the neighbouring parts. The fore feet have three toes, but the hind feet are undivided, so as to form a single toe. The scales are squared, and arranged in twenty longitudinal series on the back, and six series on the abdomen.

THE next family, the ANADIADÆ, contains, as far as is known, only one species, the EYED ANADIA (*Anadia ocellata*), thought to inhabit tropical America. In this creature the lower eyelids are pellucid, the scales of the back and sides six-sided and not over-lapping each other, while those of the abdomen are squared. The limbs are four in number, and there are five unequal and rather flattened toes on each foot. The colour of this species is pale brown, with a bronze gloss, deepening on the sides and having some white spots edged with black towards the front. Beneath it is shining white.

IN the family of the CHIROCOLIDÆ there is likewise only one species, called the CHIROCOLE (*Heterodactylus imbricatus*), a native of Brazil. This creature has a double collar, and the ears are hidden beneath the skin. The scales of the back, the sides and the tail, are six-sided, rather sharp, arranged in regular rings, and furnished with keels. Those of the abdomen are squared and arranged longitudinally in six rows. There are four short legs, with five toes on each foot, the thumb of the fore limbs being only rudimentary. The colour of the Chirocole is brown, with a pale streak on each side.

THE fourth family is the CERCOSAURIDÆ, containing two genera. These animals have the ears distinct, the throat with a double series of shields, and the collar distinct. On the back and upper part of the tail the scales are large, boldly keeled, and arranged into a regular longitudinal series. The scales of the under portions are squared and flat. There are four limbs, each with five unequal toes. A good type of this family is afforded by the EYED CERCOSAURUS (*Cercosaura ocellata*). The body of this creature is long and rather cylindrical. Its colour is black with four white streaks, the head and the under parts are yellowish, and the sides are sprinkled with green, and variegated with eight or nine white spots edged with black.

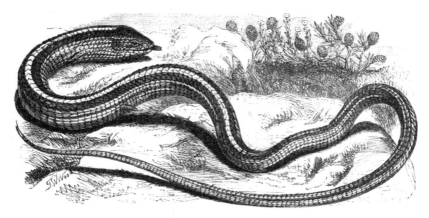

ANGUINE LIZARD.—*Chamæsaura anguina*

OUR last example of the Cyclosaurian reptiles is the ANGUINE LIZARD, or CHAMÆSAURA, the only representative of its family.

The Anguine Lizard is a native of Southern Africa, and is obtained from the Cape of Good Hope; of its habits there is but little known. It is a curious-looking creature, exceedingly snake-like in general appearance, its four limbs being of the most rudimentary character, small, delicate, feeble, not even separated into toes at the extremity, but ending in a single claw as if the whole limb were only composed of one small joint. These imperfect limbs are wholly useless for progression, those of the anterior extremity being hardly larger than the long narrow scales with which the body is covered, and the hinder pair exhibiting but very little more development.

So perfectly serpentine is the form of this creature that the mark of separation between the tail and body is so slightly defined that the precise line of junction is almost invisible, whereas in the common blindworm, itself a most snake-like reptile, the line of demarcation is plainly shown by a decided diminution in the diameter. The tail is very long and slender, measuring more than twice the length of the body.

The head of the Anguine Lizard is covered with regular many-sided shields, and the temples and the whole of the body and tail are clothed with scales, their edges projecting boldly, and arranged in a series of regular rings or "whorls." Along the back there are six rows of broad scales, and on the sides and abdomen the scales are long, narrow, and with a decided keel running along their central line. There is no groove along the sides, which are rounded. Upon the head the plates are rather long, keeled, and project very slightly over each other. The ears are distinct. The colour of the Anguine Lizard is brown, and along each side runs a long yellow streak.

A SECOND tribe of Lizards now comes before our notice. These are the GEISSOSAURI, a title derived from two Greek words, the former signifying the eaves of a house, and the latter a Lizard. As in this tribe there are many families, and more than eighty genera, it will be impossible to give more than a very slight account of these reptiles, or even to mention more than a small number selected as types of the large or small groups which they represent.

Indeed, the lower we descend in the scale of creation, the more numerous the species seem to become, and the more perplexing is the task of selecting those species which are worthy of mention on account of their scientific characteristics, and yet possess sufficient individuality to interest the general reader.

To watch the greater number of reptiles in their wild state, is a task simply impossible for any human being to achieve. Many reptiles live in dry and thirsty lands, where no creatures but the white ant and the Lizard seem to acquire moisture, and through which the traveller can only pass with hasty steps, dreading the delay of each minute lest his precious store of water should fail, and leave him to perish by the most terrible of deaths.

Others reside on the sides of precipitous rocks, over which the enterprising traveller can only pass at hazard of life and limb, and in any case would not be able to watch the proceedings of the shy and timid Lizards that find their home among these craggy recesses, and retreat into them on the slightest alarm. But the chief residence of the reptile race is to be found in hot climates, and in low, swampy ground, where the morasses are ever filled with decaying vegetable matter, and exhale a soft, thick miasma, as deadly to the white man as the fumes of arsenic, and injurious even to the dark-skinned native, who can breathe unharmed a fetid atmosphere that would smite down his white master as quickly and surely as if he were struck with a bullet, and who only attains his fullest development under these conditions.

In these dread regions, their seething putridity concealed by all the luxuriant vegetation of tropical climes, like a royal mantle flung over a festering corpse, the reptile race abound, the poisoned air being to these creatures the very breath of life, and the surrounding decay the sustaining power of their existence. Indeed, the object of their lives seems to be, by individual transmutation of poisons into living flesh, to destroy by slow but certain degrees the mass of decaying vegetation, and so to prepare an abiding place for beings of a higher order than themselves.

On placing ourselves even in imagination amid such scenes, we seem to be transported back into the former ages of our earth, when man could find no resting-place for his foot, and no atmosphere which he could breathe and live ; when the greater part of the soil was little more than soft mud, the air thick, dank, heavy, and overcharged with decomposition, and the multitude of strange reptiles that bored their slimy way through the deep ooze, crawled lazily upon the slowly hardening banks, or urged their devious course through the turbid waters, were the physically ruling though morally subservient powers of the world.

Little is wanting to complete the illusion except to give to every object an increase of dimensions, for the vegetation of those days was rank and luxuriant to a degree that is now well indicated, though on a smaller scale, by the foliage of the tropics, and the huge forms of the ancient and now extinct reptile race are closely reproduced by the more familiar inhabitants of the swamp before us.

As the expanse of putrefaction was greater in those epochs, so the miasma destroyers were larger. Frogs and toads as big as calves, reptilian quadrupeds as large as elephants, and reptilian bats expanding leatherywings as wide as those of the pelican, were fit inhabitants of the atmosphere which they breathed, and in which their mission was consummated. Now that the marshy districts are smaller and less poisonous, the reptile race that inhabits them is of smaller dimensions.

The earth has now been so far purified by successive generations and regenerations of

PYGOPUS.—*Pygopus lepidopodus*

life and death, added to human ingenuity and industry, that its harmful districts occupy but a comparatively small portion of its surface, the greater part of the world being suitable for human habitations, the black man settling as a pioneer, a hewer of wood and drawer of water, where the white man cannot yet abide. But in all those localities where the miasmatic exhalations impall the land with their pestilential mantle, and scatter the seeds of death on every breeze, the reptiles may be found luxuriating amid the deadly elements, and thriving in spots where the foot of man dares not tread, and his inquiring eye ventures not to penetrate.

THE first family of this tribe is distinguished by the apparent absence of eyelids, those organs being only rudimentary and scarcely visible, so as to give to the eyes a superficial resemblance to those of the serpents. On account of this peculiarity, the reptiles belonging to this family are termed the Gape-eyed Skinks. Their bodies are spindle-shaped, their tongues are scaly, nicked at the tip, their teeth are conical, and their limbs are four in number, and very feeble.

These creatures are found in various parts of the globe, but Australia seems to be their favourite home. The PETE or AUSTRALIAN TILIQUA (*Crytoblépharus Boutonii*) is a good example of the Gape-eyed Skinks or GYMNOPHTHALMIDÆ, a long name derived from two Greek words signifying naked-eyed. As its name imports, this reptile is a native of Western Australia, but it is also found in other parts of the world, specimens having been taken in Timor and the Mauritius. The colour of the Pete is olive, sometimes with a wash of bronze, mottled with brown, and variegated with little black streaks. Sometimes there is a bright yellow streak on each side. Its eyelid is circular and scaly, and the three upper scales are the largest.

THE next family is well represented by the PYGOPUS, or NEW HOLLAND SCHELTOPUSIC, a curious reptile that inhabits Australia.

This creature might easily be mistaken for the snake-like Lizard called the Schelto-pusic, which has already been described on page 49, as the two fore legs are entirely absent, and the hinder pair are very small, rudimentary, and set so closely against the body that they would escape a casual glance. They are flattish, covered with scales, and are not even divided into joints or toes, so that they are wholly useless for progression, the Pygopus creeping along after the ordinary fashion of snakes.

If the creature be turned on its back, a curious arrangement of scales is seen. Between the bases of the lower limbs, several large shield-like scales are seen, and just above them is a row of rather long and arched scales, extending in a semicircular form from one limb to the other, and looking much like the stones that are set upon the summit of an arched doorway. Each of these scales is pierced with a circular pore, so that the general effect is very striking. The whole body of this reptile is very long in proportion to its width, and it has altogether a very serpentine aspect.

The head of the Pygopus is rather short, and is covered above with some rather large shields, that upon the top of the head being equal to any two others in size. The scales of the back are keeled, and its colour is coppery grey, with five rows of rather oblong white spots with black centres, and a few black streaks drawn obliquely upon the sides of the neck.

The DELMA (*Delma Fraseri*) is very like the pygopus, but may be distinguished from it by the scales of the back, which are smooth and without keels, by the shorter hinder limbs, the absence of the pores, and the elliptical shape of the pupil of the eye, that of the pygopus being circular.

Two more small families of reptiles are worthy of a passing notice. The first is that which is represented by a single species, the APRASIA (*Aprasia pulchella*), and remarkable for being destitute of limbs, and having none of the pores which have just been mentioned. The body is lengthened, and covered with six-sided scales on the upper surface and flanks. The scales are quite smooth, and their colour is pale brown, with a dot of dark hue in the centre of each scale, giving a sort of variegated aspect. Along the flanks these dots become longer, so that they almost join each other, and form imperfect streaks on the sides. The lips are yellow. This reptile inhabits Western Australia.

THE next family only contains one genus, which, like the preceding creature, inhabits Australia. In these reptiles the head is long and flattened, the pupil of the eye elliptical and upright, the scales are oval, smooth, and overlap each other, and the curious pores are present, each set in the front edge of a scale. BURTON'S LIALIS (*Lialis Burtoni*) may be taken as an example of this family. The colour is olive above, with five imperfect brown streaks, and grey below, with large whitish spots.

THE large and important family of the Skinks contains between forty and fifty genera, nearly each of which possesses one or more species, concerning which there is something worthy of notice. In these reptiles the head is rather squared than rounded, and covered regularly with horny shields. The body is mostly spindle-shaped, though sometimes of a cylindrical form, and very much elongated, in which case the legs are generally rudimentary, and sometimes altogether wanting externally. Our common blindworm is a familiar example of this structure. The tail suits the form of the body, being cylindrical in the long-bodied species, and tapering in those of a more spindle-like shape.

The genus in which the COMMON, or OFFICINAL SKINK is placed, is now so restricted, that it only contains a single species; but in the earlier times of zoological science, its rules were so greatly relaxed, that many species were admitted within its limits.

In this genus the muzzle is wedge-shaped, the scales are thin and smooth, and the tail conical and pointed. The toes are rather flattened, and fringed on the side. The eyes are guarded by distinct eyelids, the lower of which is covered with scales. The palate is furnished with teeth, and has a longitudinal groove, and the ears are small, and toothed in front. There are four short and rather stout limbs, tolerably strong, and enabling the creature to make its escape from its enemies by rapidly sinking below the sandy soil on which it is usually found.

The Skink is a native of Northern Africa, and is very common in some localities.

COMMON SKINK.—*Scincus officinalis.*

Specimens are said to have been found in some portions of Asia, and it seems to be clearly proved to inhabit Syria and several parts of India.

It is a tolerably active little Lizard, not running fast or far, but contenting itself with hanging about the same locality, and feeling itself more secure on the sandy soil of its native districts, than if wandering at large on the plains. Indeed, unless it is alarmed, or except when it is aroused to short exertions by the presence of its prey, the Skink seldom troubles itself to hurry its pace beyond a slow crawl; and not even when most startled, does it attempt to seek safety in flight. No sooner does it perceive the approach of danger, than it slips below the sand with such singular speed and adroitness, that those who have witnessed this performance, say that it seems rather to be gliding into some hole already excavated, than to be engaged in the labour of sinking a tunnel for the purpose of aiding its escape. Several travellers have seen the Skink thus bury itself, and have all carried away the same opinion of its powers.

If quietly approached, it may often be detected sleeping in the hot sunbeams, lying stretched at length upon the stones or rocks, and so far steeped in slumber, that it may be approached quite closely without taking alarm.

The name of Officinal Skink has been given to this reptile on account of the high place which it formerly held among the medical profession, and the extreme value which it was thought to possess when dried, pounded, made up neatly into draughts or boluses, and used as a medicine. There is hardly a disease to which the human race is liable, which was not thought curable by the prepared body of this reptile, certainly not the least repulsive of all the disgusting substances which the early physicians delighted to choose from the animal, vegetable, and mineral kingdom, to fill their multitudinous boxes and bottles, and to inflict upon their patients. Sometimes a physician would even evince his belief in the efficacy of his medicine by taking it himself, and would swallow, with full belief in its healing powers, the burnt liver of a hyæna, the moss from a dead man's skull, the grated flesh of a mummy, or the remains of a pounded lizard, together with many other substances too revolting to mention.

Did a warrior receive a wound from a poisoned arrow, or was a woodman bitten by a venomous snake, there was nothing so effectual for the cure as the dried flesh of the Skink, sometimes called El Adda, and sometimes known by the name of Dhab. He who provided himself with this all-powerful medicine was secure against fits of all kinds, which never attacked the system fortified by a dose of powdered Skink, or were speedily driven away if the sufferer had not previously partaken of this panacea. All skin diseases were cured by the Skink, and even the fearful elephantiasis yielded to its potent sway.

Were the system too inexcitable and lethargic, and did the blood course too slowly through the veins, a little Skink powder would restore the natural powers to their full vigour. Or, on the contrary, if the patient happened to be feverish, restless, with a burning forehead, a parched skin, and a hurried pulse, a dose of the same useful medicine would cool the system, cure the headache, and bring the pulse to its normal state. It is an infallible remedy for worms, eradicates cancer, and removes cataract. In fine, a satisfactory estimate of its valuable properties may be gained by perusing, in the daily journals, any advertisement of any patent medicine, together with the list of maladies for which it is a certain remedy.

Even in the present day, this medicine is in great vogue among the sages of the East. Should the reader happen to travel into eastern lands, and fall sick of a fever, be afflicted with a sunstroke, find himself suddenly smarting with a nettle-rash, catch a cold, or suffer from sand-blindness, the remedy which will, in all probability, be offered to him, will consist of this universal panacea. In the time of the ancients, the Skink was in much favour as a medicine, and was imported largely to Rome, ready prepared in white wine. The heads and feet were considered the most efficient portions of the animal, and were relied upon as infallible renovators of a constitution broken by age, or shattered by excess.

Wherever modern civilization has most penetrated, the Skink has, happily for itself, fallen greatly in medical estimation, and in some places is entirely rejected from the pharmacopeia ; though there are not wanting some European physicians who assert that the creature really does possess some valuable properties, but that it has fallen into disrepute through the over-estimate which had been formed of its powers, and which naturally created a reaction in the opposite direction.

In Southern Egypt it still commands the firm belief of the people, and is hunted down with the greatest zeal, as it not only can be applied to the personal ailings of the captors, but can be quickly dried in the burning sunbeams, and sent to Cairo and Alexandria, where it commands a ready sale.

In its habits, this Skink much resembles the generality of terrestrial Lizards of its size and locality. As it seeks for safety below the sand, it is generally to be seen upon the hillocks of fine loose sand which are collected by the south wind, at the foot of any tree which may manage to survive in so ungenial a soil, or are blown against the hedges of the more cultivated land. It generally lies quietly upon the sand, but occasionally starts into vigorous action when it perceives an insect passing within easy reach, makes a sudden rush, captures its prey, and subsides again into its former inactive repose. Beetles are its favourite food, and of these insects it will eat a considerable quantity, but can preserve life for a lengthened period without taking any food at all.

Should it be disturbed, it instantly sinks below the sand, with almost magical quickness ; and according to M. Lefebvre, who collected a great number of these Lizards in the year 1828, a few seconds suffice it for constructing and retiring into a burrow several feet in depth. Although armed with tolerably strong teeth and claws, it does not attempt to bite when captured, and any scratch inflicted on the hand of the captor is merely caused by its struggles while endeavouring to effect its escape.

The general colour of the Officinal Skink is reddish dun, crossed with bands of a darker hue above. Below and upon both the flanks, it is of a silvery whiteness. It is, however, liable to considerable variations, of which the most important may be briefly denoted as follows : In one variety, the upper parts are yellow, or silver-grey, with seven or eight large brown spots on the sides. In another, the head is yellow, the upper parts are chestnut-brown, profusely sprinkled with little white spots, each scale having two, or even

BROAD-HEADED PLESTIODON. — *Plestiodon laticeps.*

three, white dots upon the surface. The back is marked with a series of broad white bands, generally five or six in number, and having a black patch at either extremity of each band. In another variety, the upper parts are silvery grey, splashed with pure white, and variegated with irregular brown spots. But however great may be the variations, they are all confined to the upper surface, the abdomen, flanks, and under surface retaining their beautiful silvery whiteness. The banded variety is the most common. The Officinal Skink is by no means a large reptile, seldom exceeding eight inches in length, and being generally about six or seven inches long.

PASSING by one or two genera of considerable extent, such as Hinulia and Mocoa, the members of which are mostly found in Australia, though there are species which inhabit China, Java, the Philippines and New Zealand, we come to a reptile very well known by the popular title of the SCORPION LIZARD, and called more scientifically, as well as more correctly, the BROAD-HEADED PLESTIODON.

In spite of the rather alarming name which the terrors of the ignorant have caused them to bestow upon it, the Scorpion Lizard is one of the most harmless, as well as one of the most useful little creatures that inhabit the earth.

It is a native of Northern America, and is spread over a very large tract of country. This curious Lizard is one of the species that delights in trees, and of which we shall see more in a future page. It generally resides in some tree buried in the depths of the forest, and remains at a considerable elevation above the ground, never liking to make its home less than thirty or forty feet above the earth, and often placing itself at a much greater height.

The domicile in which this reptile most delights is the deserted home of a woodpecker, which has brought up her little family, and forsaken the burrow which had taken such time and trouble to hollow from the decaying wood. Here the Scorpion Lizard takes up its residence, and here it remains snugly concealed unless it is alarmed by an enemy at the gate of its wooden fortress, when it runs nimbly to the entrance, and pokes out its red head with so threatening a gesture, that its intending assailant, thinking it must possess a store of poison to assume so resolute an aspect, retreats from the spot and leaves the reptile in quiet possession of its abode.

Happily for the Lizard, the belief in its venomous propensities is widely diffused and deeply engrained in the popular mind, so that without having a single dangerous property except that of undaunted courage when driven to bay, it has established a reputation for ability to avenge itself when injured, which is of no less service to reptiles than men. Not that it is wholly destitute of offensive weapons, for its teeth are strong and sharp, its feet powerful, and its claws are sufficiently pointed to scratch rather deeply.

The Scorpion Lizard is naturally a very timid and retiring creature, and on the approach of danger slips quietly out of the way, wisely preferring flight to combat. But if seized, the captor will have no small struggle before he can fairly secure his small but determined quarry, for the creature bites so fiercely with its sharp teeth, retains its hold with such bull-dog tenacity, and kicks and scratches with such hearty goodwill, that the non-zoological populace may well be excused for thinking it to be venomous in tooth as well as in temper. The bite, indeed, is so severe, and the creature has such power of jaw, that the wounds inflicted are always exceedingly painful for an hour or two, and might give rise to the idea that the teeth were poisonous like those of the rattlesnake.

The Scorpion Lizard is seldom seen except upon trees, where it can mostly find a sufficiency of food among the insects that always haunt the branches of trees, and of drink in the dew-drops that collect at morning and evening. When, however, it needs a more abundant diet, it descends to the ground for a short visit, but after satisfying its wants, it returns to its tree, runs easily up the trunk, and again establishes itself in its burrow.

The head of the Scorpion Lizard is very broad at the base, and narrows rather suddenly to the snout, which is slightly elongated. The upper part of the head is of a bright red colour. The body is olive-brown above, and the throat, abdomen, and whole of the under parts, are yellowish white. Just in front of the ear is a series of oblong tubercles, and the temples are smooth and covered with rather large shields. The feet are large in proportion to the size of the body, and the toes are rather compressed and exceedingly delicate, in fact almost thread-like in form. The length of the Scorpion Lizard is generally about eleven or twelve inches.

THERE is a closely allied species, also common in North America, popularly called the BLUE-TAIL and scientifically the FIVE-LINED PLESTIODON (*Plestiodon quinquelineátum*).

Like the preceding species, the Blue-Tail inhabits the deepest forests, but is not one of the arboreal reptiles, being always found upon the earth, usually remaining within a short distance of its home, which is made in one of the numerous decaying tree-stumps which are found in these vast forests. Its food consists of insects, which it catches principally upon the ground.

The head of this Lizard is red like that of the scorpion, but of a much paler quality. The body is olive, with five longitudinal white streaks, the central stripe being forked in front, and with two black bands. The tail is brownish, with a decided wash of blue during the life of the animal, a colouring which has earned for it the popular name of Blue-Tail. It is, however, subject to slight variations in the colour and shape of the markings. There are several little lobes in front of the ears. The length of the Blue-Tail is about eight or ten inches.

A REPTILE which bears some resemblance to the scorpion Lizard is found in Jamaica and the West Indian islands, where it seems to take the place of that creature, and to enjoy a reputation almost as bad, with as little cause. The negroes call it by the name of MABOUYA (*Mabouya ágilis*), but as they apply that term to anything which is, or which they consider to be venomous, and as there are very many really poisonous creatures in those countries, and many more which are falsely thought to be so, the word is rather vague in its application.

The Mabouya is a good climber, running up trees with perfect facility, and having a tendency to traverse the huts of the negroes, much to the consternation of the inmates.

Its usual habitation, however, is made in the holes of old decaying trees, and except during the very hot weather, it mostly remains at home. There is another reptile, inhabiting the same country and to which the same title is applied, and which will be mentioned in a future page.

The lower eyelid of the Mabouya is remarkable for a little transparent disc in the centre, the palate is without teeth, and the scales are smooth. Along the back run four black streaks, the two central stripes extending only to the middle of the body, while the two external lines are prolonged nearly to the insertion of the hinder limbs.

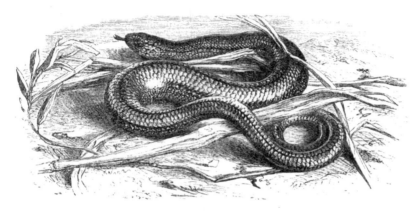

BLINDWORM.—*Anguis fragilis.*

THE great family of the Skinks finds a familiar representative in the common BLIND-WORM or SLOW-WORM of England, which from its snake-like form and extreme fragility, might well deserve the title of the English glass snake. In this reptile there is no external trace of limbs, the body being uniformly smooth as that of a serpent, and even more so than in some of the snakes, where the presence of the hinder pair of limbs is indicated by a couple of little hook-like appendages. Under the skin, however, the traces of limbs may be discovered, but the bones of the shoulders, the breast, and the pelvis are very small and quite rudimentary.

This elegant little reptile is very common throughout England, and is spread over the greater part of Europe and portions of Asia, not, however, being found in the extreme north of Europe. In this country it is plentiful along hedgerows, heaths, forest lands, and similar situations, where it can find immediate shelter from its few enemies, and be abundantly supplied with food. It may often be seen crawling leisurely over a beaten footpath, and I have once captured it while crossing a wide turnpike road near Oxford.

Why the name of the Blindworm should have been given to this creature I cannot even conjecture, for it has a pair of conspicuous though not very large eyes, which shine as brightly as those of any animal, and are capable of good service. Indeed, all animals which prey upon insects, and similar moving things, must of necessity possess well-developed eyes, unless they are gifted with the means of attracting their prey within reach, as is the case with some well-known fishes, or chase it by the senses of hearing and touch, as is done by the mole. Moreover, the chief food of the Blindworm consists of slugs, which glide so noiselessly that the creature needs the use of its eyes to detect the soft mollusc as it slides over the ground on its slimy course. Speed is not needful for such a chase, and the Blindworm accordingly is slow and deliberate in all its movements except when very young, when it twists and wriggles about in a singular fashion as often as it is touched.

The great fragility of the Blindworm is well known. By a rather curious structure of the muscles and bones of the spine, the reptile is able to stiffen itself to such a degree, that on a slight pressure, or trifling blow, or even by the voluntary contraction of the body, the tail is snapped away from the body, and on account of its proportionate length, looks just as if the creature had been broken in half. The object of this curious property seems to be to insure the safety of the animal. The severed tail retains, or rather acquires, an extraordinary amount of irritability, and for several minutes after its amputation, leaps and twists about with such violence, that the attention of the foe is drawn to its singular vagaries, and the Blindworm itself creeps quietly away to some place of shelter.

Even after the movements have ceased, they may be again excited by touching the tail with a stick, or even with the finger, when it will jump about with a vigour apparently undiminished. On frequently repeating the process, however, the movements become perceptibly less active, and after a while the only sign of movement will be a slight convulsive shiver. Half an hour is, as far as my own experience goes, the limit to which this irritability endures.

I well remember meeting with an incident of this nature near Dover, where I came suddenly upon a reptile among the rank grass and underwood, that I at first took for a viper, and at which I aimed a thrust with a little twig of decaying wood, which broke at once. Immediately after the thrust, something began to hop and plunge about most violently just by my feet, and having a very wholesome dread of a viper's fangs, I jumped back a step or two, to the great indignation of a swarm of bees, which had settled themselves in the ruins of an old wooden hut close to the spot. They at once intimated their displeasure in that wing-language so expressive to all bee-owners, so, hastily tossing the writhing object to a distance with the shattered remnant of the stick, I got away from the bees, and experimented for some time on the tail of the Blindworm, as it proved to be. Even the flight through the air, and the heavy fall, seemed to have little or no effect upon the irritability of the severed member, and when I reached it after its fall, I found it hopping about quite merrily.

When the tail of the Blindworm is thus snapped off, the scales of the body project all round the fractured portion, forming a kind of hollow into which the broken end of the tail can be slipped.

According to popular notions, the Blindworm is a terribly poisonous creature, and by many persons is thought to be even more venomous than the viper, whereas it is perfectly harmless, having neither the will nor the ability to bite, its temper being as quiet as its movements, and its teeth as innocuous as its jaws are meek. I fancy that the origin of this opinion may be found in the habit of constantly thrusting out its broad, black, flat tongue with its slightly forked tip ; for the popular mind considers the tongue to be the sting, imagining it to be both the source of the venom, and the weapon by which it is injected into the body, and so logically classes all creatures with forked tongues under the common denomination of poisonous animals.

It is said that this reptile will bite when handled, but that its minute teeth and feeble jaws can make no impression upon the skin ; and also that when it has thus fastened on the hand of its captor, it will not release its hold unless its jaws be forced open. For my own part, and I have handled very many of these reptiles, I never knew them attempt to bite, or even to assume a threatening attitude. They will suddenly curl themselves up tightly, and snap off their tails, but to use their jaws in self-defence is an idea that seldom appears to occur to them.

The pertinacity with which the notion of the Blindworm's venomous properties is implanted in the rustic mind is really absurd. During the summer of this year, I passed some little time in the New Forest, and having gone round to the farms in the neighbourhood, as distances of several miles are euphuistically called, begged to have all reptiles brought to me that were discovered during the haymaking. In consequence, the supply of vipers and snakes was very large, and on one occasion a labourer came to the house, bare-headed, his red face beaming with delight, and his manner evincing a proud consciousness of deserving valour. Between his hands he held his felt hat tightly crumpled together, and within the hat was discovered, after much careful manœuvring, the head of a Blindworm emerging from one of its folds.

As I put out my hand to remove the creature, the man fairly screamed with horror, and even when I took it in my hand, and allowed it to play its tongue over the fingers, he could not believe that it was not poisonous. No argument could persuade that worthy man that the reptile was harmless, and nothing could induce him to lay a finger upon it; the prominent idea in his mind being evidently, not that the Blindworm had no poison, but that I was poison-proof. To add to his alarm, the creature had snapped off its tail during the rough handling to which it had been subjected, a proceeding which, by his peculiar process of reasoning, only corroborated its venomous properties.

In its wild state the Blindworm feeds mostly on slugs, but will also eat worms and various insects. Some persons assert that it devours mice and reptiles, but that it should do so is a physical impossibility, owing to the very small dimensions of the mouth, and the structure of the jaw, the bones of which are firmly knitted together, and cannot be separated while the prey is being swallowed, as is the case with the snakes.

In captivity it seems to reject almost any food except slugs, but these molluscs it will eat quite freely. The specimen from which the illustration was drawn, has been in my possession for about four months, and has proved a very interesting creature. After keeping it for a fortnight, I procured six or seven white garden slugs, and placed them in the glass vessel together with the Blindworm.

The reptile instantly saw its prey, but did not move from its place, merely following with a slow movement of the head the course of one of the slugs that crawled within an inch or two of its nose. Presently it raised its head very deliberately, and hovered over the slug as it glided along, and after following it for an inch or two, quickly opened its mouth to the full extent, lowered its head, and grasped the slug just behind the head, squeezing it with some force, and causing a great commotion among the muscles of the foot.

Presently it relaxed its hold a little, again opened its mouth and took a fresh grasp, and after three or four of these movements, it contrived—how I cannot comprehend, though I have watched the creature over and over again—to get the head of the slug down its throat. The process of swallowing was then very easy, and after a few more efforts, the whole of the mollusc had disappeared. After resting for a few minutes, it attacked another slug precisely in the same manner; but I have seldom seen it eat more than two or three at one meal. By degrees it caught and ate all the slugs, and will finish a dozen in a week or ten days.

Upon the 12th of September my Blindworm unexpectedly became the mother of a numerous progeny, nine little Blindworms having made their appearance in the world during the night. They are remarkably pretty little creatures, and so unlike their parent that few persons would attribute them to the same species. They are much more serpentine in their general aspect, their heads being considerably wider than their necks, whereas in the adult the head and neck are as nearly as possible of the same width.

Their colour is shining creamy yellow above, and jetty black below, the line of demarcation running along the flanks, and being very sharply defined. Along the back runs a narrow black line, which upon the head is expanded, and then divides so as to form a letter Y. Just above the nose is another forked black mark, looking like an inverted V, and both these letters have a notable circular enlargement at the angle. As the creature grows, the V mark becomes gradually uncertain, and finally disappears, but the black line down the back, and its Y-like termination, retain their position through life, though they are not so conspicuous as in the young, owing to the darker colouring of the surface.

How these little things feed I cannot make out. They were born on the 12th of September, as has already been mentioned, and though they have now lived for about five weeks, have grown considerably, and have always been very lively, they have taken no food so far as I can discover. For the first three weeks of their life, they lived in a glass jar closed at the top, and with an inch or so of dry earth at the bottom, in which there could be no nourishment. A little milk was poured on the mould now and then, and they perhaps may have licked the moistened earth, and so have obtained some little nourishment, though they were never seen to do so, and indeed appeared perfectly indifferent to the milk.

When I introduced the slugs, the odd little reptiles acted just as their mother was doing, followed the slugs about with their heads, hovered over them, made believe to eat them, and then were quietly walked over by their intended prey, which being nearly twice as big as themselves, proceeded on its course without paying the least regard to the tiny reptiles, whose bodies were not larger than ordinary knitting needles, and easily glided over them, or put them to ignominious flight.

After they had been in the jar for some time, I fitted up an old aquarium in a manner intended to imitate as far as possible their natural home, building a bank of earth and stones at either end, laying turf in the middle, and planting ferns upon the banks, with moss round their roots. They enjoyed the change very greatly, immediately proceeded to burrow in all directions through the earth and among the stones, until they established a whole series of tunnels through which they can glide at will, and seem to take great pleasure in permeating their establishment at all hours, especially delighting in pushing their way through the moss and then retreating into their burrows.

On a cold day they bury themselves below the mould, but the first gleam of sunshine that plays among the green fern-leaves brings them from their recesses, and causes them to glide about the moss and turf most merrily. Sometimes, when they are coiled asleep within their home, their bodies are pressed against the glass, and it is curious to see how immovable they will lie in spite of tapping the glass, but how soon they wake up and brisk they become when the glass is warmed. Even a few warm breaths upon the glass suffice to awake them.

I think that I have discovered another kind of subsistence for the young, but that has only been possible since they have been placed in the aquarium, or rather, the fernery, as it is now. Sundry very minute insects of the dipterous order may be seen flitting about within the glass, probably having been introduced with the turf and ferns, and it is possible that the young Blindworms may contrive to catch and eat these creatures, and derive some nutriment from them in spite of their diminutive size.

When wild, the Blindworm generally retires to its winter quarters towards the end of August, or even sooner, should the weather be chilly. The localities which it chooses for this purpose are generally dry and warm spots, where the dried leaves and dead twigs of decayed branches have congregated into heaps, so as to afford it a safe refuge. Sometimes it bores its way into masses of rotten wood; and on heathery soils, where the ground slopes considerably, it selects a spot where it will be well sheltered from the winter's rains and snows, and burrows deeply into the dry loose soil.

It is singular to see the creature emerging from the ground when the least touch will soil the fingers, and to see how totally free from earth-stains is the bright glittering skin of the reptile, upon which not a particle of mud can cling. I once detected upon the head of my specimen a projection which I thought was a little lump of mud, I having just watered the ferns and turf, greatly to the discomfiture of the Blindworms, both old and young; but upon close examination I found it was only a little pebble which had lodged upon the head, as the reptile came hastily out of its burrow to avoid the water. So quietly did the Blindworm move, that the stone retained its place upon the head for several minutes, and did not fall off until I startled the creature, and caused it to turn its head rather sharply.

The Blindworm would be a most useful inhabitant of a garden, not at all repulsive, and, indeed, very seldom seen, its instinct teaching it to remain within some dark recess during the day, and only to come out at night when the slugs leave their earthy hiding-places, and commence feeding. Moreover, it is very prolific, and needs no special appliances, as is the case with the frog and toad, which require the presence of water to produce and hatch their young, and for the little reptiles to come to maturity. Sometimes the number of young is twelve or thirteen, and sometimes there are only seven or eight. The usual average is, however, nine or ten; and they are very hardy little things, requiring no care whatever.

Being one of the earliest to retire into its winter quarters, the Blindworm is one of the first reptiles to leave them, appearing before either the snake or the viper. The reason for this early appearance is simple enough. Neither creature can venture into action when it can find no food, the active powers of the body causing a waste which must be restored with nutriment. The snake feeds upon frogs, and therefore cannot leave its winter's home until it finds the frogs ready for it. The frogs, again, which feed upon insects, must wait until the vegetation has attained sufficient luxuriance to afford food for their insect prey ; but the Blindworm, which finds its nourishment among the molluscs which devour the earliest leaves, is able to leave its winter quarters as soon as the vegetation begins fairly to sprout, and the slugs to devour it.

Even during the winter, a warmer sunbeam than usual will tempt the Blindworm to come to the mouth of its burrow, poke out its head, and enjoy the temporary, but cheering warmth. My own specimens have not yet made any preparations towards retiring to winter quarters, though the usual time has passed away nearly two months ago, a circumstance which is probably due to the warmth of their home, and the occasional supply of slugs which I now and then put into the case.

Like the snakes, the Blindworm casts its skin at regular intervals, seeming to effect its object in various modes, sometimes pulling it off in pieces, but usually stripping it away, like the snakes, by turning it inside out, just as an eel is skinned. Some persons, who have witnessed the process, state that this eversion is only extended to the base of the tail, and that the entire tail is drawn out of the skin like a hand out of a glove. Mr. G. Daniel mentions, that a Blindworm in his possession cast its skin in so many pieces, that the largest portion was only two inches in length. The process began by a split along the abdomen, and the head was the last part extricated from the rejected integument. This mode of shedding the skin was, however, owing, in all probability, to some weakness in the individual, or to the want of the usual aids, such as the stems of grass, heather, and other vegetation, against which the reptile contrives to rub itself, so as to assist its efforts in peeling off the cuticle. The colour of the Blindworm is rather variable. In my own specimen, now crawling over the paper on which I write, and blotting it sadly, the colour is dark olive-brown above, with a shining silvery lustre, and diversified with a narrow black line along the back, and a broader black line down each side. The flanks are greyish white, mottled with black, and the under parts are nearly black, variegated with a little grey. The Y-like mark on the head is still apparent, but there is no trace of the inverted V. On the sides of the head, the mottlings of grey and black are very bold, and round the neck runs a collar of black. This mark, however, may have been caused by the stupidity of the captor, who was so frightened at the contortions of the reptile, that he tied a string round its neck to form a safe handle with which to carry it.

Mr. Bell, in his volume on the British reptiles, states that the tail is obtuse, and in the illustration gives it much too short an aspect, though he states that it rather varies in length, in some cases being not more than half the length of the body, while in others it nearly equals the head and body together. In my own specimen, the tail is by no means obtuse, but very slender and well pointed, and can be so tightly curled at its extremity as not to be removable without damage to the creature. While held in the hand, it generally twists the tip of the tail firmly round one of the fingers, not in a spiral position, but so as to make one complete circle, the extremity of the tail just touching the spot where the circle commences. The total length of this specimen now lying flat against a two-foot rule, towards which I have just succeeded in coaxing it by a judicious arrangement of light and shade, and an occasional touch with the finger, is thirteen inches and a half. The body and head occupy precisely six inches, and the remaining seven inches and a half are given to the tail. The spot where the body ends and the tail begins is very evident, the diameter of the body diminishing slightly but suddenly.

THE family of the Skinks contains so many interesting creatures, that it is difficult to make a satisfactory selection, and impossible to avoid a feeling of regret at the necessity for passing so many species without even a cursory notice. Before, however, proceeding to the next family, we must give a short notice of one or two rather conspicuous species.

The first is the SPINE-BACKED LIZARD of New Guinea (*Tribolonótus Novæ Guineæ*), a very remarkable creature, notable for the singular formation of the scales which cover the back, and in allusion to which the creature has been placed under the generic name Tribolonotus. This long word is of Greek origin, signifying calthrop-backed: calthrops being certain horrible instruments thrown on the ground to check the advance of cavalry, and consisting of four iron spikes, set round a ball in such a manner, that when flung on the ground, three points rest on the earth, and the other projects perpendicularly into the air.

Though really harmless, the Spine-backed Lizard is a most formidable looking creature, the whole of the back being covered with long and sharply pointed spikes, formed by a modification of the scales, that project boldly in all directions, and fully justify the generic name. Even on the tail the scales, which are arranged in whorls, are long, pointed, and project over each other, so as to give a very formidable aspect to this member. Even the head is armed with these pointed scales, which become larger and larger as they approach the neck. The colour of this Lizard is brown above, and greyish white below.

ANOTHER notable member of this family is the well-known GALLIWASP (*Celestus occíduus*).

This reptile is a native of the West Indian Islands, and is very common in Jamaica, where it is held in great, but groundless dread, by the inhabitants, and especially by the negroes. It generally haunts damp situations, and is mostly found in marshy lands, near water, or hidden under rocks where moisture is retained by the nature of the ground. It is thought that when the Galliwasp is irritated, its bite is as venomous as that of a poisonous snake, and causes immediate death. On account of the dread in which it is held, the negroes call it by the name of Mabouya, in common with the reptile which has already been described on page 60.

The colour of the Galliwasp is brown of various tones, diversified with cross bands of blackish brown. It is about one foot in length. There are several species belonging to this genus, all being found in Jamaica.

The last example of the Skinks which can be mentioned in these pages is SAGRA'S DIPLOGLOSSUS, or DOUBLE-TONGUED LIZARD (*Diploglossus Sagra*).

This reptile is a native of Cuba, and is found in localities where the air is cool, and the soil light and moist. It is an active little creature, and moves from place to place with much agility. In this reptile the tongue is rather large, covered with little scale-like papillæ in front, becoming more thread-like behind. The colour is grey, with a bronzy lustre, and a black streak runs along each side.

THE next family of Lizards contains only one species, the OPHIOMORE (*Ophiomorus miliáris*), and is separated from the skinks and the sepsidæ on account of a formation of the scales of the head, which seems to place it in an intermediate position between those two families. There are no external limbs, and the whole body and tail are long cylindrical, tapering, and serpentine in aspect. The colour of the Ophiomore is brown above, covered with numerous tiny black dots arranged in regular lines along the body, and being larger upon the sides. The under parts are white, and the sides are grey. It is a native of Northern Africa, and has been brought from Algiers.

IN the SEPSIDÆ, a family which contains seven genera, there are always external limbs, mostly four in number, but in one genus, Scelotes, the front pair of legs are wanting, and the hinder pair are small and divided at the extremity into two toes only.

The typical species of this family is the common SEPS, or CICIGNA, a curious snake-like Lizard, found in various parts of the world, and not uncommon in many portions of Europe. Specimens have been taken in the south of France, in Italy, Sardinia, Syria, and the north of Africa. The name of Seps is of Greek origin, and signifies corruption   From

SEPS, OR CICIGNA.—*Seps triddctylus.*

ancient times to the present day, this harmless little reptile has been held in great dread by the natives of the country wherein it dwells, being considered as a deadly enemy to cattle, biting them at night during their sleep, and filling their veins with corruption. Horses, and especially mares, were thought to be the most frequent sufferers from the bite of this reptile.

The legs of the Seps are very weak, and set far apart, so that the creature trusts but little to the limbs for its powers of locomotion, and wriggles itself along after the fashion of the snakes. The food of the Seps consists of worms, small snails, slugs, insects, spiders, and similar creatures, its general habits seeming to resemble those of the blindworm. Like the lizard, when the winter approaches, it burrows deeply in the loose soil, and remains hidden until the succeeding spring.

The teeth of the Seps are small, conical, and simple, and there are no teeth on the palate, which is grooved longitudinally. The eyelids are scaly, and the lower has a transparent disc. The toes are three in number on each foot, and very feeble; the tail is conical and pointed. The colour of the common Seps is grey, with four longitudinal brown streaks, which, on a closer inspection, are found to consist of a succession of brown dots.

Two members of the Sepsidæ deserve a passing notice before we pass to the next family. The first is the CAPISTRATED SPHÆNOPS (*Sphænops sepsoides*).

This reptile is a native of Northern Africa, but seems to have a rather restricted range, being seldom if ever found out of Egypt. In some parts of that country it is very plentiful, being found in the rice grounds, under hedges, and on the roads where the wheels of passing vehicles have worn deep ruts. Indeed, it appears to have a predilection for ridged ground, over which it passes with considerable speed, and is not to be captured without the exercise of some agility. It is quite harmless, and even when caught, struggles with all its might to escape, but does not attempt to bite the hand that holds it. Like many other reptiles of similar form, it burrows in the ground, but makes its tunnel so near the surface of the ground, and in so horizontal a direction, that the foot of a traveller will often lay open the superficial retreat and render its inmate homeless for a time.

ACONTIAS.—*Acontias meleagris*.

The ancient Egyptians seem to have held this little reptile in religious veneration, as there are several known instances where it has been honoured with the ceremony of embalming, and placed in the sacred tombs, together wtth other creatures formerly reverenced as types of divinity.

The Sphænops has four legs, moderately well shaped, but rather weakly formed, and the feet are divided into four toes, each of which is furnished at the extremity with a claw. The head is wedge-shaped, rounded in front, the palate is without teeth, and the lower eyelid is transparent. The general colour of the Sphænops is pale brown, with a longitudinal series of black dots, and a black streak on each side of the muzzle.

Our last example of this family is the Tiligugu (*Gongylus ocellátus*), or Eyed Tiliqua, another of the numerous reptiles classed under the common title of Mabouya by the ignorant and fearful.

It inhabits the countries bordering the Mediterranean, and is found in Sardinia, Malta, Egypt, and even in Teneriffe. Like the preceding species, it is quick and active in its movements, and when seized does not attempt to bite. It is a lover of dry and elevated spots, where the sand is loose, and there are plenty of stones under which it may hide itself. The food of this reptile consists of insects. Beside the names which have already been mentioned, it is also called Lacepede's Galliwasp and the Ocellated Skink.

In colouring it is one of the most variable of reptiles, but the general tints are grey with a bronze gloss, diversified by a number of white spots edged with black. It has four legs, the toes are five on each foot, the head is conical, with a rounded muzzle, and the lower eyelid has a transparent disc.

The Acontiadæ form the next family, which contains three genera. The head is small, the upper eyelid is either very small or altogether absent, the body is cylindrical, and the limbs, when present, are very weak and small. In two of the genera, Nessia and Evesia, there are four limbs, in the former with three toes, and in the latter with the feet small, imperfect, and not divided into toes. The upper eyelid is distinct though small.

In the ACONTIAS or JAVELIN SNAKE the limbs are absent externally, and the upper eyelid is rudimentary. The body of this reptile is elongated and cylindrical, not unlike that of the common blindworm of England. The name Acontias is derived from a Greek word signifying a javelin, and has been given to this creature on account of the shape of the head, which bears some resemblance to the point of a spear. Some writers think that the name is given in allusion to its quick movements when seizing its prey. As in shape, so in habits it resembles the blindworm, and like that reptile is very common in the fields and under hedges. It is a South African reptile, and is found abundantly at the Cape of Good Hope.

In its colouring this is a very handsome little creature, being sometimes called the Painted Acontias (by the French writers *La Peintade*), in allusion to the variegated tints with which it is bedecked. Like many other reptiles, especially those which are lightly coloured, it is susceptible of much variation. Generally, however, it is rich chestnut-brown above, profusely dotted with bright yellow, the spots being arranged in series of varying number, one specimen having eight rows of spots, while another has only six. The scales are smooth, the teeth are conical and rather blunt, the eyes are very small, and the tail is short and rather rounded at the tip.

ANOTHER curious family of reptiles possesses only two limbs at the most, the front pair being always, and the hinder pair sometimes, wanting. There is a curious cup-like shield on the chin, the body and tail are cylindrical, and both eyes and ears are apparently absent, but may be found hidden under the skin, where the greater part, if not the whole, of their functions must be in abeyance. In consequence of this remarkable privation they are classed together under the very appropriate name of Typhlinidæ, a term derived from the Greek, and which signifies blindness.

In the typical species, the TYPHLINE, or BLIND ACONTIAS (*Typhline Cuvieri*), as it is sometimes, but rather erroneously, called, the limbs are entirely absent, and the creature looks about as helpless a being as can well be imagined, having no apparent legs, feet, eyes nor ears. The Typhline inhabits Southern Africa, and is found at the Cape of Good Hope. In its colouring it is rather variable, being generally of a brownish hue with spots of purple upon the hinder part of the scales of the back, and sometimes of a yellowish tint with violet spots.

THERE are so many reptiles scattered over the world, and they are divided by modern systematic zoologists into so many families, that it is only possible to give a short description of one or two examples of each family, while to supply illustrations would be wholly impracticable without nearly doubling the amount of space that can be allotted to them.

The next family in the catalogue of the British Museum is called by the name of Typhlopsidæ, or Blind Reptiles, a title which has been given to them because their eyes are either very small or altogether wanting externally. In all these animals the head is broad, rather flattened, and has a large erect plate near the muzzle. The mouth is small, semilunar in shape, and placed under the muzzle in a manner somewhat resembling that of the sharks. The tail is cylindrical, and has a large shield or plate at the tip, sometimes conical and sometimes spine-shaped.

In the TYPHLOPS, the typical species, the head is nearly covered by a single, very large shield, which is rather bent downwards in front. The tail is very short and tapers suddenly, and the scales of the body are small and uniform. It inhabits India, where it is not uncommon, though, in consequence of its earth-loving habits, it is not very often seen except by those who know its localities, and search purposely for the hidden reptile. It moves over the ground with some rapidity, and burrows easily, penetrating to a depth of three or four feet during the rainy season. At other times it is mostly content with the shelter of large stones and similar places of refuge.

Owing to the small size and the rather remarkable position of the mouth, the Typhlops is unable to act on the offensive, and when captured, although it attempts to glide through the fingers, does not even offer to bite. It is wonderfully tenacious of life, and according to Dr. Russell, will live for some time even when immersed in spirits of wine. The general colour of the Typhlops is yellowish white.

This family contains also the Clawed-snake (*Onychophis*)—so called because the bony shield on the muzzle is erect, keeled, and bent over into a claw-like shape—and the Silver Snakes (*Argyrophis*), a small group of reptiles, deriving their popular name from the silvery lustre of their scales.

THE last family of the sub-order Leptoglossæ, or Slender-tongued Lizards, is the group of reptiles termed the Rough-tailed Lizards, or UROPELTIDÆ. In these Lizards the head is rather compressed, flat above and sharp towards the muzzle. The eyes are of moderate size, and without eyelids, a bony scale answering the purpose. The body is cylindrical and covered with regular six-sided scales, sometimes ridged, but mostly smooth. The tail is also cylindrical, and abruptly terminated, as if cut off obliquely. There are no external limbs, and by most systematic naturalists the Rough-tails have been placed among the serpents, which they very closely resemble, except in the arrangement of certain scales, and the short, abruptly truncated tail.

According the elaborate catalogue of the British Museum, the Rough-tailed Lizards are divided into three genera, separated from each other by the formation of the scales that cover the tail. While moving, the Rough-tails aid themselves by pressing the truncated tail against the ground. As a typical species, we may select the PHILIPPINE SHIELD-TAIL (*Uropeltis Philippinus*), a reptile which, as its name imports, inhabits the Philippine Islands. In this creature the tail is rather flattened, and covered above with a curious "flat, roundish, radiating, granular shield." On the lower side of the tail the scales are arranged in six rows. The colour of the Philippine Shield-tail is brown above and white beneath, the line of demarcation being very distinct, and regularly waved.

A NEW sub-order now comes before our notice, the members of which are distinguished by the formation of their tongues, which, instead of being flat and comparatively slender, as in the preceding Lizards, are thick, convex, and have a slight nick at the end. On account of this structure, the species of this sub-order are termed PACHYGLOSSÆ, or Thick-tongued Lizards.

These reptiles are divided into sundry groups, the first of which is termed the NYCTISAURA, or Nocturnal Lizards. These creatures have eyes formed for seeing in the dusk, circular eyelids which, however, cannot meet over the eye-ball, and in almost every case the pupil is a long narrow slit like that of the cat. The body is always flattened. The limbs are four in number, tolerably powerful, and are used in progression.

Of these Lizards, the first family is the GECKOTIDÆ, or Geckos, a very curious group of reptiles, common in many hot countries, and looked upon with dread or adoration by the natives, sometimes with both, where the genius of the nation leads them to reverence the object of their fears, and to form no other conception of supreme power than the capability of doing harm.

The FAN-FOOT, or HOUSE GECKO, is a native of Northern Africa, and is very common in Egypt, and is found, as its name imports, in houses, traversing the floor and walls with astonishing address, in search of its food, which consists of worms, insects, and similar creatures. The natives have a very great dread of this creature, asserting that it is extremely poisonous—the poison not being injected by the teeth, but exuding from the lobules of the toes. The generic title Ptyodactylus, or Toe-spitter, is given to the reptile in allusion to this idea. It is said by Hasselquist, that if a Gecko is taken in the hand, the poisonous matter which is immediately shed over the skin from the feet of the captive, causes an instantaneous eruption, similar to that produced by the sting of a nettle. The same traveller proceeds to relate an incident which is hardly so much in accordance with

COMMON GECKO.—*Gecko verus.*   FAN-FOOT.—*Ptyodáctylus Geoku.*

probability, namely, that two women and a girl were lying at the point of death from having eaten some cheese over which one of these reptiles had walked.

So great is the dread inspired by this creature, that in Cairo it is popularly termed Abou-burs, or father of the leprosy. The people fancy that it purposely poisons their provisions, and that it is especially fond of communicating the venom to salted meat of all kinds. In former times the Fan-foot was endowed with even greater powers of offence, its teeth being added to its weapons, and asserted to be capable of leaving their impression even on steel, though in point of fact, the jaws of the Geckos are rather feeble, and their teeth very small, and hardly able to pierce even the human skin.

The Geckos are indebted for their power of traversing perpendicular walls to the formation of their feet, which, although greatly varied in the different genera, have the same essential qualities in all. In this genus the toes are expanded at their extremities, into a round disc, and furnished with claws which are sheathed in a notch cut in the front of the disc. The colour of the Fan-foot is reddish brown spotted with white.

THE COMMON GECKO, or RINGED GECKO, is an Asiatic species, being as common in India as the preceding species in North Africa. It may be easily known from the fan-foot by the large tubercles upon the back.

This reptile has much the same habits as the fan-foot, and possesses equally the ability to run over a perpendicular wall. During the day-time it conceals itself in some chink or dark crevice, but in the evening it leaves its retreat, moving rapidly and with such perfectly silent tread that the ignorant natives may well be excused for classing it among supernatural beings. The Gecko occasionally utters a curious cry, which has been compared to that peculiar clucking sound employed by riders to stimulate their horses, and in some species the cry is very distinct and said to resemble the word Geck-o, the last syllable being given smartly and sharply.   On account of this cry, the Geckos are variously called Spitters, Postilions, and Claqueurs.

During the cold months of the year the Geckos retire to winter quarters, and are thought to retain their condition during this foodless season by means of two fatty masses at the base of the abdomen, which are supposed to nourish them as the camel is nourished by the hump.   The male is smaller than the female, and the eggs are very spherical, and covered with a brittle chalky shell.   The colour of the Gecko is reddish grey with white spots.   The scales of the back are flat and smooth, and there is also a series of rather large tubercular projections arranged in twelve rather distinct rows.

CLOSELY allied to these two reptiles is the SPOTTED GECKO, or SPOTTED HEMIDACTYLE, a rather pretty species of Gecko found in various parts of Asia, and tolerably common in India, China, and Ceylon.   Sir Emerson Tennent, in his valuable work on Ceylon, gives a very interesting account of this little creature, and relates two curious anecdotes, exhibiting the readiness with which even a Gecko can be tamed by kind treatment.

" In a boudoir where the ladies of my family spent their evenings, one of these familiar and amusing little creatures had its hiding place behind a gilt picture-frame, and punctually as the candles were lighted, it made its appearance on the wall to be fed with its accustomed crumb ; and if neglected, it reiterated its sharp quick call of *chic-chic-chit,* till attended to.   It was of a delicate grey colour, tinged with pink, and having by accident fallen on a work-table, it fled, leaving its tail behind it, which, however, it reproduced within less than a month.   This faculty of reproduction is doubtless designed to enable the creature to escape from its assailants; the detaching of the limb is evidently its own act.

In an officer's quarters in the fort of Colombo, a Gecko had been taught to come daily to the dinner-table, and always made its appearance along with the dessert.   The family were absent for some months, during which the house underwent extensive repairs, the roof having been raised, the walls stuccoed, and ceilings whitened.   It was naturally surmised that so long a suspension of its accustomed habits would have led to the disappearance of the little Lizard, but on the return of its old friends, at their first dinner it made its entrance as usual the instant the cloth had been removed."

ANOTHER rather curious species is the TURNIP-TAILED GECKO (*Thecadáctylus rapi-caudus*), so called from the odd shape of its tail, which, when reproduced, is very much swollen at the base, and, with its little conical extremity, has an almost absurd resemblance to a young turnip.   It is worthy of mention, that all the Geckos possess the faculty of reproducing their tails when those members have been lost by some accident, and that the second tail is mostly very unlike the original.   Before the creature has suffered (if it does suffer) this mutilation, the tail is covered with scales of the same structure and form as those of the back ; but when the tail is reproduced, it is generally supplied with little squared scales arranged in cross series.   In examining a Gecko therefore, it is necessary to ascertain whether the tail be in its normal condition or only a second and altered edition of that member.

The colour of the Turnip-tailed Gecko is brown, mottled boldly with a darker tint, and speckled with tiny dots of dark brown.   The scales of the back are six-sided, and on each side of the base of the tail there is a prominent conical tubercle.   This species inhabits Tropical America.

THE very remarkable reptile which is figured in the accompanying illustration, is a native of Java.

The FRINGED TREE GECKO, or SMOOTH-HEADED GECKO, is especially worthy of notice on account of the broad membranous expansions which fringe the sides of the head, back, limbs, and tail.   On the body this membrane is covered with scales, and waved on its edges, but on the tail the waves become suddenly deepened, so as to form bold

FRINGED TREE GECKO.—*Ptychozoön homalocéphela*

scollops. The toes are webbed to the tips, and, with the exception of the thumb-joint, are furnished with claws at the swollen extremity. The scales of the back are smooth and flat, and even the membranous fringes are covered with scales.

Formerly this creature was thought to be aquatic in its habits, but it is now known to live on trees, and to employ the membranous expansions in aiding it in its passage from branch to branch, much after the well-known fashion of the flying squirrels. The generic title, Ptychozöön, is composed of two Greek words, the former signifying a fold of a garment, and the latter a living being. The general colour of the Fringed Tree Gecko is brown above, with a slight yellowish tinge along the spine, and crossed with small dark brown lines, very narrow and deeply waved. A line of similar appearance and of a bold zig-zag form encircles the top of the head, looking as if a dark brown string had been tied at the ends, formed into a rude circle and then pinched at intervals so as to cause deep indentations. Below it is of a whitish grey colour.

THE curious and rather interesting little Lizard called the CAPE TARENTOLA, is an inhabitant, as its name signifies, of the Cape of Good Hope, and is found spread over a considerable portion of Southern Africa.

This reptile is of slower habits than the generality of the Geckos, and moves along with deliberate and apparently purposeless steps. It is almost invariably seen upon or near decayed wood, and is frequently found under the bark of dead trees, clinging tightly to the trunk, and shielded by the bark from the unwelcome glare of daylight. In all probability, it finds abundance of food in the same locality, for the space between the bark and wood of a decaying or dead tree, is generally filled with insects of various kinds and in their different states of existence, beside being the chosen home of millipedes, spiders, and similar creatures.

Although a slow mover, the Cape Tarentola can, after the manner of its kin, ascend smooth and perpendicular objects with perfect ease and noiseless motions, and can even traverse and cling to a ceiling or a cross-beam without difficulty, and there remain motionless for hours. Like the generality of the Gecko family, it detests the daylight.

and the bright beams of the sun are a torture to this dweller in darkness, which, if over-taken by daylight while out of its refuge, crawls away to the nearest cranny and there buries itself until the evening hours bring with them the desired shades, and restore the animal to its wonted activity. It is extremely shy, and even in the dusk it will avoid the dangerous approach of an intruder by silently slipping under the cover of the loose bark, or hiding itself among the decaying wood.

CAPE TARENTOLA.—*Taréntola Capensis*

It is quite a little creature, rarely measuring more than four inches in length, and often not reaching even those moderate dimensions. As is the case with many Lizards, it is liable to certain variations in colouring, but its general tints are as follows: The back and upper portions of the body are yellowish brown with a decided yellow wash, and banded with several dark brown bars, rather curved. Scattered over the body are certain protuberant scales of a lighter hue. The tail is a pale brownish purple with a reddish gloss, and speckled with warm chestnut-brown. The abdomen, and the under portions of the body and limbs are ochry yellow and the eyes are, although devoid of expression and of a passionless brightness like polished stone, very shining and of a bright orange-brown. The whole form of this Lizard is rather thick and clumsy.

As this family contains at least forty genera, it is manifestly impossible to mention more than a few species, which can be accepted as types of the family, and serve as links to render the chain of nature complete. Passing, therefore, several series of genera, we will give a short time to one or two species of Gecko before proceeding to the next family.

The WOODSLAVE, as the reptile is popularly termed by the natives of the country where it resides, or the BANDED SPHÆRODACTYLE (*Sphærodáctylus sputátor*), as it is more scientifically called by British zoologists, is a small species of Gecko found in most of the American islands, and is spread over many portions of South America; and is held in great dread by the white and dark population. It is generally supposed to possess a store of venomous saliva, causing the part of the body on which it falls to swell grievously, and to eject this poisonous substance from some distance upon those who chance to vex its irascible temper. The specific term sputator signifies a spitter, and has been given to the reptile on account of this supposed propensity. The poisonous saliva is said to be black.

The Woodslave has no claws on its toes, the pupil of the eye is round, and the eyelid circular. The back and tail are covered with small scales. The colour is generally black and yellow, arranged in cross bands, and there is a white streak on each side of the head. There are several species belonging to this genus, all inhabiting similar localities.

THE reader will remember that in the turnip-tailed Gecko, mentioned on page 72, the tail is curiously swollen at the base after its reproduction. In the LEAF-TAILED GECKO, otherwise called WHITE'S PHYLLURE, (*Phyllúrus platúrus*), the tail is always rather long, flattened considerably, very broad, with a deep notch at its junction with the body, and a shallower double notch in the centre. Along the middle there also runs a shallow groove, and the entire aspect is so quaint, not to say ludicrous, that on seeing a specimen of this odd-looking Lizard, the first impression on the mind is that the tail has been cleverly manufactured and attached to the body by artificial means. This Gecko is a native of New Holland.

Both the scientific names of the Leaf-tailed Gecko refer to the singular formation of its tail, the one signifying Leaf-tail, and the other Broad-tail. The head of this reptile is very broad at the base, very sharp at the snout, and the skin adheres so closely to the bone as to exhibit the form of the skull through its substance. The toes are long, slender, and rather compressed. Along the sides runs a fold of skin, very slight, but sufficiently conspicuous. The tail is very thin and leaf-like; along the edge runs a series of spiny scales, and its surface is covered with rather long conical tubercles arranged in cross rows. The colour is brown, and a number of little spiny tubercles are scattered over the back.

In taking leave of the Geckos, we must cast a hasty glance at their feet. In many of their movements the Geckos bear a curious likeness to the common fly, and when one of these reptiles is seen gliding along a perpendicular wall with noiseless step, or clinging with perfect ease to an overhanging beam, quite regardless of the fact that it is hanging with its back downwards, the resemblance is irresistible. And on inspecting the foot and its structure, the resemblance which this member bears in many species to the well-known foot of the fly, is remarkably close and worthy of attention.

WE now arrive at an important tribe of Lizards, called by the name of Strobilosaura, a title derived from two Greek words, one signifying a fir-cone and the other a lizard, and given to these creatures because the scales that cover their tails are set in regular whorls, and bear some resemblance to the projecting scales of the fir-cone. In all these reptiles the tongue is thick, short, and very slightly nicked at the tip. The eyes have circular pupils, and are formed for·day use.

THE first family of these Lizards consists of those creatures which are grouped together under the general title of IGUANA. This word is employed extremely loosely, as the name of Iguana is applied to many species of Lizards, such as the monitors and the varans, which in reality have little in common with the true Iguanas. These reptiles can mostly be distinguished from the rest of their tribe by the formation of their teeth, which are round at the roots, swollen and rather compressed at the tip, and notched on the edge. There are generally some teeth on the palate. All the true Iguanas inhabit the New World. As the family of Iguanas is extremely large, and contains more than fifty genera, we can only examine a few of the most interesting species, the first of which is the COMMON IGUANA.

This conspicuous, and in spite of its rather repulsive shape, really handsome Lizard, is a native of Brazil, Cayenne, the Bahamas, and neighbouring localities, and was at one time very common in Jamaica, from which, however, it seems to be in process of gradual extirpation.

In common with those members of the family which have their body rather compressed, and covered with squared scales, the Iguana is a percher on trees, living almost wholly among the branches, to which it clings with its powerful feet, and on which it finds the greater part of its food. It is almost always to be found on the trees that are in the vicinity of water, and especially favours those that grow upon the banks of a river, where the branches overhang the stream.

Though not one of the aquatic Lizards, the Iguana is quite at home in the water, and if alarmed, will often plunge into the stream, and either dive or swim rapidly away. While swimming, it lays its fore legs against the sides, so as to afford the smallest possible resistance to the water, stretches out the hinder legs, and by a rapid serpentine movement of its long and flexible tail, passes swiftly through the waves. It has considerable power of enduring immersion, as indeed is the case with nearly all reptiles, and has been known to remain under water for an entire hour, and at the end of that time to emerge in perfect vigour.

From the aspect of this long-tailed, dewlapped, scaly, spiny Lizard, most persons would rather recoil than feel attracted, and the idea of eating the flesh of so repulsive a creature would not be likely to occur to them. Yet in truth, the flesh of the Iguana is justly reckoned among one of the delicacies of the country where it resides, being tender, and of a peculiarly delicate flavour, not unlike the breast of a spring chicken. There are various modes of cooking the Iguana, roasting and boiling being the most common. Making it into a fricassee, however, is the mode which has met the largest general approval, and a dish of Iguana cutlets, when properly dressed, takes a very high place among the delicacies of a well-spread table.

The eggs too, of which the female Iguana lays from four to six dozen, are very well flavoured and in high repute. It is rather curious that they contain very little albumen, the yellow filling almost the entire shell. As is the case with the eggs of the turtle, they never harden by boiling, and only assume a little thicker consistence. Some persons of peculiar constitutions cannot eat either the flesh or the eggs of the Iguana, and it is said that this diet is very injurious to some diseases. The eggs are hid by the female Iguana in sandy soil near rivers, lakes, or the sea-coast, and after covering them with sand, she leaves them to be hatched by the heat of the sun.

In consequence of the excellence of the flesh and eggs, the Iguana is greatly persecuted by mankind, and its numbers considerably thinned. Those who hunt the animal for sport or merely to supply their own homes, generally employ a noose for the purpose, which they cast dexterously round the neck of the reptile as it sits on a branch, and then by a sudden and sharp jerk loosen its hold, and secure it. The creature is very bold, having but little idea of running away, and in general is so confident of its capability of frightening away its antagonist by puffing up its long dewlap, and looking ferocious, that it is captured before it discovers its mistake. Even when caught, it has no notion of yielding without a struggle, but bites so fiercely with its sharp leaf-like teeth, and lashes so vigorously with its long whip-like tail, that it is not secured without some trouble and risk. It is also very tenacious of life, and does not readily die even from repeated blows with heavy sticks, so that the spear or the pistol are often employed to kill it.

Those, however, who hunt the Iguana for sale, are obliged to have recourse to other expedients, such as nets, and dogs, the latter being trained to secure the Iguana without killing it. Many persons set out on regular expeditions of this sort, embarking in a little vessel and visiting numbers of different islands and inlets in chase of the Iguana. Those which they can succeed in taking alive, have their mouths carefully secured to prevent them from biting, and are then stowed away in the hold, where they will live for a considerable time without requiring any nourishment. Those which are killed, they either eat on the spot, or salt them down in barrels for winter consumption. Were the Iguanas quick of foot, they would seldom be captured, but, fortunately for the hunters, they cannot run fast, and according to the quaint language of Catesby, who visited the Bahamas about 1740, "their holes are a greater security to them than their heels."

The food of the Iguana seems to consist almost entirely of fruits, fungi, and other

IGUANA.—*Iguana tuberculáta.*   NAKED-NECKED IGUANA.—*Iguana delicatíssima.*

vegetable substances, and it is known that in captivity it feeds upon various leaves and flowers. Yet it has been said by some persons who have seen the Iguana in its native state, that it eats eggs, insects, and various animal substances. Perhaps these creatures were not the true Iguanas, but belonged to the monitors, varans, or similar carnivorous Lizards.

The Iguana is capable of domestication, and can be tamed without much difficulty by those who are kind to it and accustom it to their presence. It will even permit itself to be carried about in its owner's arms, though it will not permit a stranger to approach.

The general aspect of the Iguana is most remarkable, and can perhaps be better understood by reference to the illustration than by any lengthened description. Suffice it to say that the head is rather large, and covered above with large scales. The mouth is enormously wide, and studded around the edge with those singularly shaped teeth which have already been described. About the angles of the jaw there are generally some large, solitary, rounded scales. The chin is furnished with a kind of dewlap, large, baggy, and capable of being inflated at the will of the animal, scaly, and edged in front with a row of bold, tooth-like projections. The sides of the neck are covered with tubercles. The tail is extremely long, and very thin and tapering. The usual colour of the Iguana is dark olive-green, but is rather variable even in the same individual, being affected by change of weather, or locality, or temper. On the sides a few brown bands are generally seen, and the tail is marked with brown and green of various tones, the two colours being arranged in alternate rings. The average length of the Iguana is about four feet, but it often attains a much greater size, reaching a length of six feet or a little more.

The NAKED-NECKED IGUANA was long confounded with the preceding species, bearing a great resemblance to that reptile in colour, form, and habit, and being found in the same localities. It can, however, be readily distinguished from the common Iguana by the absence of tubercles upon the sides of the neck. Along each side of the lower jaw runs a series of large strong scales. The general colour of this species is bluish green, darker on the back than on the abdomen. Its flesh is esteemed equally with that of the preceding species.

BESIDES these Iguanas, there are one or two which deserve a short notice, although our limited space does not permit of a lengthened description. One of these animals is the MARBLED IGUANA or CAMALEAO (*Pólychrus marmorátus*), also a native of Brazil and Central America. This species has the throat compressed into a small dewlap, and the scales of the back and sides equal. There is no crest upon the back and tail. Its colour is brown, mottled with bold marblings and diverging lines of a darker hue, and sometimes having a slight purple gloss.

The APLONOTE (*Aloponótus Ricardi*) is another species of Iguana, having its head covered with small equal many-sided plates, and its throat dilated into a small pouch without the toothed projections in front. A shallow crest runs along the back and tail, and the back is without scales, but covered with multitudinous granular tubercles of a very small size. The tail is compressed. The colour of this species is blackish brown, variegated with many spots of tawny brown.

Another curious species is appropriately called the HORNED IGUANA (*Metopóceros cornútus*), deriving its name from the horn-like projections upon its head. Upon the forehead there is a large horn-like tubercle, and two pairs of large horny plates between the nostrils. There is a crest upon the back, but it is very low between the shoulders, and upon the loins it is not continuous. It inhabits St. Domingo.

IT has already been mentioned that the Iguana possesses the power of swimming to a large extent, and that it is capable of sustaining a long submersion without suffering any injury.

There is a curious species of Iguana, the MARINE OREOCEPHALE, which exists upon the seashore, and passes a considerable portion of its time in the water. This creature was first made known to science by Mr. Darwin, who found it on the coasts of the Galapagos islands, and describes its habits in the following words:—

"It is a hideous-looking creature, of a dirty black colour, stupid and sluggish in its movements. The usual length of a full-grown one is about a yard, but there are some even four feet long. I have seen a large one which weighed twenty pounds. These lizards are occasionally seen some hundred yards from the shore swimming about, and

MARINE OREOCEPHALE.—*Oreocéphalus cristátus*

Captain Collnett in his voyage says that they go out to sea in shoals to catch fish. With respect to the object I believe he is mistaken, but the facts stated on such good authority cannot be doubted.

When in the water, the animal swims with perfect ease and quickness by a serpentine movement of its body and flattened tail, the legs during this time being perfectly motionless and closely collapsed on its sides. A seaman on board sunk one with a heavy weight attached to it, thinking thus to kill it directly, but when, an hour afterwards, he drew up the line, the lizard was quite active. Their limbs and strong claws are admirably adapted for crawling over the rugged and fissured masses of lava which everywhere form the coast. In such situations, a group of six or seven of these hideous reptiles may oftentimes be seen on the black rocks, a few feet above the surf, basking in the sun with outstretched legs."

In this reptile the throat is not formed into a pendent pouch, but the skin is much crumpled, so that the animal can dilate it at will. The whole body is covered with sharp, rough, tubercular scales, and a crest of longer scales runs along the back. The teeth are sharp and three-lobed, and although, when the wide mouth is opened, they present a very formidable array of weapons, the creature is quite harmless, and feeds on vegetable diet, seaweeds forming the chief part of its subsistence. The middle toes are united by a strong web, and the claws are large. There is some difference in the aspect of the young and adult, this distinction being most obvious in the head, where the scales are rather convex in the young, but in the adult are enlarged into unequal and rather high tubercular shields.

BASILISC.—*Basiliscus Americanus.*

IN the earlier ages of science, when a few facts were struggling their way through the superincumbent mass of fiction that had so long caused Natural History to be little more than a collection of moral fables, the BASILISC was a creature upon whose wondrous properties the inventive pens of successive narrators were never tired of dilating. Crowned with a royal diadem, emblematical of its sovereign rule, the Basilisc held supreme sway over the reptile race, and derives its name of Basilisc, or kinglike, "because he seemeth to be the King of Serpents, not for his magnitude or greatnesse. For there are many serpents bigger than he, as there be many four-footed beasts bigger than the lyon, but because of his stately face and magnanimous minde."

The Basilisc was thought to be an occasional *lusus naturæ*, having during his life no companion of his own kind, and to derive his existence from an egg laid by a cock when he was very old, and sat upon by a snake. Some scientific writers, however, better informed than the more popular zoologists, said that the egg was not incubated by a snake, but by a toad.

Before the Basilisc all living creatures but one were forced to fly, and even man would fall dead from the glance of the kingly reptile's eye. "This poyson," says Topsel, "infecteth the air, and the air so infected killeth all living things, and likewise all green things, fruits and plants of the earth: it burneth up the grasse whereupon it goeth or creepeth, and the fowls of the air fall down dead when they come near his den or lodging. Sometimes he biteth a man or beast, and by that wound the blood turneth into choler, and so the whole body becometh yellow or gold, presently killing all that touch it or come near it." Even a horseman who had taken into his hand a spear which had been thrust through a Basilisc, "did not only draw the poyson of it into his own body and so dyed, but also killed his horse thereby."

The only creature that could stand before the Basilisc and live, was said to be the cock, whose shrill clarion the bird-reptile held in such terror, that on hearing the sound, it fled into the depths of the desert and there concealed itself. Travellers, therefore, who were forced to pass through the sandy deserts of Libya, were advised always to carry with them a supply of strong lively loud-voiced cocks, by whose vigorous crowings they would be protected from the Basiliscs haunting those parts.

There is an old proverb, "No smoke without fire," and this saying is verified in the present case. In some parts of Tropical America there is a perfectly harmless Lizard of no great dimensions, belonging to the family of the Iguanas, and having a bold crest on the back of its head. It is probable that one of these reptiles was imported into the Old World at some time now forgotten, and that its rather odd shape and the crest on its head were seized upon by the first describers, and reported with continually increasing exaggerations by succeeding writers.

Like the rest of the Iguanas, this animal is a good climber of trees, it can swim well, and its food consists apparently of insects and the various little creatures which frequent the water and the foliage of its banks.

Although quite innocuous, it certainly is rather forbidding, and when it obtains its greatest length of three feet, presents a sufficiently formidable appearance to warrant in some degree the wild and fabulous tales which were deduced from its strange shape. Along the back, instead of the row of pointed spines which generally cross the back of the Iguanas, runs a broad crest-like membrane, another broad membrane occupying the upper surface of the tail. These curious appendages are supported by a series of slender bones, formed by elongations of the vertebræ of the back and tail, so that the animal looks exactly as if the fins of a fish had been grafted on the body of a reptile. There is a slight pouch on the throat, and the palate is toothed.

CRESTED ANOLIS.—*Xiphosurus velifer*

MANY species of the Lizard tribe are called by the name of Anolis, but are divided by systematic zoologists of the present day into several distinct genera. The CRESTED ANOLIS inhabits some of the hotter portions of America and the neighbouring islands.

The chief point of interest in this Lizard is the curiously expansile throat, which, in common with others of the same genus, it is able to expand at will. When terrified, it tries to escape, but if it finds itself deprived of all means of eluding its antagonist, it turns to bay, and by puffing out the throat until it assumes a very great size, endeavours thereby to intimidate the foe. While thus engaged, the creature has the faculty of continually altering its colour; the hues of the body to a certain degree, but more especially those of the throat, changing with a rapidity that is said even to surpass the famed powers of the chameleon.

It is an active little creature, traversing perpendicular objects with nearly as much ease as the Gecko, and to aid it in these movements the last joint but one of the toes is

3. G

swollen, so as to form a pad, and is covered below with cross ridges, so as to enable the creature to take a firm hold of the object to which it is clinging.   The food of the Anolis consists chiefly of insects, which are captured by means of singular address on the part of the Lizard.   The Anolis can run up and down trees, walls, or rocks, with such rapidity, and leap so boldly from one spot to and her, that at a little distance its movements might easily be mistaken for those of a bird.

Though not aquatic in its habits, and apparently not taking willingly to the water, the Anolis is mostly to be found in the woods and thickets that are in the close neigh-bourhood of a stream or lake.   It is a timid, yet a restlessly inquisitive animal ; for although it hides itself with instinctive caution on hearing the approach of a footstep, it is of so curious a nature that it must needs poke its head out of its hiding-place, and so betray itself in spite of its timidity.   So absorbed, indeed, is the Anolis in gratifying its curiosity that it will allow itself to be captured in a noose, and often falls a victim to the rude and inartificial snares made by children.   Its voice is a little sharp chirruping sound ; and by imitating these notes, the children decoy it within reach of the fatal noose.

The usual resting-place of the Crested Anolis is within the hollow of some decaying tree, where also the female deposits her eggs.

The colour of the Crested Anolis is dark ashen blue, a blackish spot being apparent on each side.   Along the nape of the neck and the back runs a series of long compressed scales, forming a toothed crest, and on the basal half of the tail is a fin-like crest, strengthened by bony rays.   The throat-pouch is extremely large, and when inflated gives to the reptile quite an ungainly appearance.   The greatest known length of the Crested Anolis is about eighteen inches, but the other species are generally of much smaller dimensions.   The name Xiphosurus is of Greek origin, and signifies Sword-tail.

OF the restricted genus Anolius, we take two examples.   In this genus the back and nape of the neck are either smooth, or have a low crest formed by two series of short scales.   The scaly plate at end of the muzzle is erect.   All these Lizards are very active, inhabiting trees, and jumping about from branch to branch with wonderful skill, and clinging even to the pendent leaves by means of their curiously formed feet.

This GREEN CAROLINA ANOLIS is, as its name imports, a native of North America, where it is tolerably common.   It is a pretty lively little creature, specially brisk and active in its movements.

This Lizard is, according to Holbrook, " a bold and daring animal, haunting outhouses and garden fences, and in new settlements it even enters the houses, walking over the tables and other articles of furniture in search of flies.   It is very active, climbing trees with great rapidity, and leaping with ease from branch to branch and from tree to tree, securing itself even on the leaves by means of the oval disks of the fingers and toes, which enable it also to walk easily on glass, and on the sides and ceilings of rooms.   It feeds on insects, and destroys great numbers, seizing them suddenly and devouring them, unre-strained even by the presence of man."

Towards the spring, the Green Anolis becomes quarrelsome, and is so exceedingly pugnacious, that the adult males hardly ever meet without a fight, the vanquished usually coming off with the loss of his tail—a misfortune, however, that sometimes occurs to both the combatants.   This Lizard is seldom seen in all its beauty except when engaging in battle, for at the sight of its antagonist it remains stationary for a moment, nods its head up and down two or three times, as if to work itself into a proper state of fury, puffs out its dewlap, which then becomes of a light scarlet, and having gone through all these preliminaries, it leaps on its foe and the struggle begins.   As the summer draws on, the irascibility of its temper diminishes, and during the whole summer and early autumn these pretty Lizards may be seen amicably associating together.   They are fond of basking in the sun, and will then dilate their dewlaps, at the same time assuming the most brilliant emerald hues.

The colour of this reptile is extremely variable, altering even in the same individual according to the season of the year, the temperature, the health, or even the present state of the creature's temper.   Generally the whole upper surface is beautiful golden green, and

RED-THROATED ANOLIS.—*Anólius bulláris.*   GREEN CAROLINA ANOLIS.—*Anólius principális.*

the abdomen white, with a tinge of green. The dewlap, or throat-pouch, is white, with a few little spots and five bars of red, which colour, when the pouch is inflated, spreads over its whole surface. The total length of this reptile is nearly seven inches.

THE second species, the RED-THROATED ANOLIS, is a native of America and the neighbouring isles.

It is a brisk and lively little creature, darting about the ground, over rocks, among the branches, or upon the leaves, with equal address. It is, perhaps, a little too fond of fighting, and terribly apt to quarrel with others of its own kind. Those who have witnessed a combat between two of these Lizards say that it is remarkable for ferocity, courage and endurance. They face each other with expanded throats and glaring eyes, their skin changing its lustrous colouring, and their whole being instinct with fury.

As during each combat one or two females are generally spectators of the fight. it is probable they may be the cause of war, and that the victor may receive his reward from one of the female witnesses of his prowess. So furious do they become, that the conqueror is said to devour the vanquished, who, however, sometimes runs away as fast as he can, and escapes with the loss of his tail, which is left writhing in the victor's mouth and soon swallowed. Those who have thus lost their tails seem to be greatly affected by the mutilation, and are timid and languishing afterwards.

The inflated throat part of the angry animal has a very curious effect, as it becomes of a bright cherry-red, due probably to the excited state of the creature.

Mr. Bell, in his work on British reptiles, mentions a curious anecdote of one of these Lizards which was worsted in combat with a common garden-spider. "The activity of the smaller insectivorous Lizards, when in pursuit of their food, is exceedingly curious and interesting. They watch with all the caution of a cat, and dart upon their prey with the quickness of lightning.

In the act of seizing their food, however, they must necessarily be exposed to some danger from the noxious qualities of the insects which they indiscriminately attack. The following fact would seem to indicate that even in our own temperate climate, an insect not generally recognised as poisonous may inflict a fatal injury on its saurian enemy.

Some years since, I had in my possession two living specimens of the beautiful little green Anolis of the West Indies, a Lizard about the size of our smallest species. I was in the habit of feeding them with flies and other insects, and having one day placed in the cage with them a very large garden-spider (*Epeira diadema*), one of the Lizards darted at it, but seized it only by the leg. The spider instantly ran round and round the creature's mouth, weaving a very thick web round both jaws, and then gave it a severe bite on the lip, just as this species of spider usually does with any large insect which it has taken. The Lizard was greatly distressed, and I removed the spider and rubbed off the web, the confinement of which appeared to give it great annoyance, but in a few days it died, though previously in as perfect health as its companion, which lived for a long time afterwards."

With regard to the injury produced by the bite of the spider, I can say from personal experience that even to human beings, especially those who are tender-skinned, the bite of the common garden-spider is extremely painful. I have suffered for some hours from the bite of one of these creatures, and I have seen the arm of a young lady flushed and swollen, because a garden-spider had bitten the back of her hand. The pain is something like that produced by the sting of a wasp, but more dull, and seeming to throb with the pulse.

The colour of the Red-throated Anolis is greenish blue, excepting on the throat when the creature is excited. There is no crest on the nape and back, but the tail is slightly toothed above. When full grown it is about the size of our sand Lizard.

CROWNED TAPAYAXIN.—*Phrynosóma Blainvillii*

OUR last example of this large and interesting family is the CROWNED TAPAYAXIN, one of the singular North American reptiles which are popularly known by the name of Horned Toads, their general form and mode of sitting being extremely toad-like.

This animal is not at all uncommon in California, and is said when at liberty in its wild state to move with much rapidity over the ground in search of its insect prey. Its habits in confinement, however, do not carry out this statement, as it is then sluggish to a degree, remaining for many consecutive hours in precisely the same attitude, heedless of the falling rain or the burning rays of the sun, and scarcely changing its position even when pushed with the finger. It is quite harmless, in spite of its very formidable looks, and does not attempt to avenge itself upon its captor, however roughly it may be handled.

FRINGED DRAGON.—*Draco fimbriátus.*　　　　　FLYING DRAGON.—*Draco volans*

After a while it can be made to know its owner, and will even take flies and other insects out of his hand. Little red ants seem to be its favourite food, but it lives on beetles, and insects of various kinds.

The head of this curious reptile is armed with long, pointed, conical spines, set around its edge and directed backward. Shorter and stouter spines, but of a triangular shape, are scattered over the back, and extend even over the odd, short, and pointed tail. Each edge of the tail is armed with a strong row of spines, giving it a regularly toothed appearance. The general colour of the Crowned Tapayaxin is grey, variegated with several irregular bands of rich chestnut-brown. The head is light brown blotched with a darker hue, and the under parts are ochry yellow, marked with sundry blotches of dark grey.

The family which comes next in order is that in which are included the AGAMAS, a group of Lizards which have been appropriately termed the Iguanas of the Old World. In the members of this family the teeth are set upon the edge of the jaws, and not upon their inner side, as in the true Iguanas of the New World. Between thirty and forty genera are contained in this family, and some of the species are interesting as well as peculiar beings.

Perhaps the most curious of all this family, if not, indeed, the most curious of all the reptiles, is the little Lizard which is well known under the title of the FLYING DRAGON.

This singular reptile is a native of Java, Borneo, the Philippines and neighbouring islands, and is tolerably common. Some writers believe that this creature was the original source from which the many fables respecting the formidable dragon of ancient and modern mythology were derived. Perhaps, however, the real clue to the various fables that were once so common respecting the formidable dragon may be found in one of the huge saurians of the ancient days, which had survived its comrades, and preserved its existence upon the earth after man had been placed upon this planet.

The most conspicuous characteristic of this reptile is the singularly developed membranous lobes on either side, which are strengthened by certain slender processes from the first six false ribs, and serve to support the animal during its bold leaps from branch to branch. Many of the previously mentioned Lizards are admirable leapers, but they are all outdone by the Dragon, which is able, by means of the membranous parachute with which it is furnished, to sweep through distances of thirty paces, the so-called flight being almost identical with that of the flying squirrels and flying fish.

When the Dragon is at rest, or even when traversing the branches of trees, the parachute lies in folds along the sides, but when it prepares to leap from one bough to another, it spreads its winged sides, launches boldly into the air, and sails easily, with a slight fluttering of the wings, towards the point on which it had fixed, looking almost like a stray leaf blown by the breeze. As if in order to make itself still more buoyant, it inflates the three membranous sacs that depend from its throat, suffering them to collapse again when it has settled upon the branch. It is a perfectly harmless creature, and can be handled with impunity. The food of the Flying Dragon consists of insects.

The colour of this reptile is variable, but is usually as follows: The upper surface is grey with a tinge of olive, and daubed or mottled with brown. Several stripes of greyish white are sometimes seen upon the wings, which are also ornamented with an angular network of dark blackish brown. Sometimes the black is rather plentiful upon the wings, forming four or five oblique bands near the edge. It is a small creature, measuring only a few inches in length.

The FRINGED DRAGON is mostly found in Sumatra, where it seems to be tolerably common. In habits, and in general appearance, this reptile bears a great resemblance to the preceding species, from which, however, it may be known by the conspicuous black spots on its wings, each spot being surrounded with a ring of white. The head is greyish white, covered with an irregular network of dark brown, and on the throat are a number of circular specks covered with granular scales. Upon the under parts of the male, the scales are rather large and keeled, and upon the wing are a number of rather short white dashes of a partly triangular shape. Along the sides runs a series of small triangular keeled scales.

Besides these species there are several other flying Dragons, all inhabiting similar localities. They are divided into genera on account of the different structure of the ear and the position of the nostrils. The tail of all the Dragon Lizards is extremely long, and very slenderly formed.

A VERY curious reptile of this family deserves a passing notice. This is the TIGER LIZARD, or GONYOCEPHALE (*Gonyocéphalus chameleontína*), a native of Java. This creature is remarkable for the high and deeply toothed crest which runs along the nape of the neck, like the crest of an ancient helmet, and far overtops the head, although the upper part of the skull is much raised by an enlargement of the orbits. A large but compressed pouch hangs from the lower jaw and throat, and is prolonged so as to form an angular fold just before the shoulder. A toothed crest runs along the back, but is barely one quarter the height of that which passes over the nape, and the tail is long and compressed. The colour of this Lizard is green, with variable streaks and scribblings of black, and the legs are deeply banded. The Tiger Lizard sometimes attains a length of three feet.

The Lizards of this family are remarkable for the extraordinary modifications of form which they exhibit. In one species, such as the tiger Lizard, a row of long spike-like scales is raised upon the neck, in the dragons the skin of the sides is dilated to an enormous extent, and even the ribs are drawn out like wire and turned out of their usual course to support the membranous expansion, and in the FRILLED LIZARD the neck is furnished with a large plaited membrane on each side, forming a most remarkable appendage to the animal without any apparent object.

FRILLED LIZARD.—*Chlamydosaurus Kingii.*

The Frilled Lizard is a native of Australia, and, like most of the family, is generally found on trees, which it can traverse with great address. It seems to be a bold and courageous animal, trusting to its formidable teeth and generally ferocious aspect as a means of defence. "As we were pursuing our walk in the afternoon," writes Captain Grey, "we fell in with a specimen of the remarkable Frilled Lizard. It lives principally in trees, though it can run very swiftly along the ground. When not provoked or disturbed, it moves quietly about, with its frill lying back in plaits upon the body ; but it is very irascible, and directly it is frightened, it elevates the frill or ruff, and makes for a tree, where, if overtaken, it throws itself upon its stern, raising its head and chest as high as it can upon the fore legs ; then, doubling its tail underneath the body, and displaying a very formidable set of teeth from the concavity of its large frill, it boldly faces an opponent, biting furiously whatever is presented to it, and even venturing so far in its rage as to fairly make a charge at its enemy.

We repeatedly tried the courage of this Lizard, and it certainly fought bravely whenever attacked. From the animal making so much use of its frill as a covering and means of defence for its body, this is probably one of the uses to which nature intended the appendage should be applied."

This remarkable Lizard was discovered by Mr. Allan Cunningham, who caught the first specimen as it was perching on the stem of a small decayed tree.

The general colour of the Frilled Lizard is yellow-brown mottled with black, and it is remarkable that the tongue and the inside of the mouth are also yellow. The frill, which forms so conspicuous an ornament to this creature, is covered with scales, and toothed on the edge. It does not come to its full size until the animal has attained maturity, and increases in regular proportion to the age of its owner. In the young, the frill does not even reach the base of the fore limbs, while in the adult it extends well beyond them. The head is somewhat pyramidal in shape, and four-sided. There is no pouch on the throat. A small crest runs along the nape of the neck, but does not extend to the back. The tail is long and tapering, and like the back, is devoid of a crest. The eyes are rather prominent during the life of the reptile, and the tongue is thick, short, and nicked at the end. It is rather a large species, measuring when full grown nearly a yard in total length.

In the genus Grammatophora, the head is three-sided, and rather flattened, with a sharpish muzzle. There is no throat-pouch, but the skin of the chest is folded into a kind of cross plait. The tail is long, conical, rather flattened at the base, and covered with overlapping keeled scales. All the members of this genus inhabit Australia.

The MURICATED LIZARD, or GRAMMATOPHORE, is a native of New Holland. It is almost entirely arboreal in its habits, being seldom if ever seen except on trees, which it traverses with remarkable agility, being quick, sharp, and dashing in its movements. It feeds on insects, and is enabled to catch them as they settle on the leaves or branches.

MURICATED LIZARD, OR GRAMMATOPHORE.—*Grammatóphora muricáta.*

It also eats the caterpillars, grubs, and other larvæ, which it can find in profusion among the boughs.

The colouring of this Lizard is rather variable. Generally the back is brownish grey, traversed by sundry brownish bars, running longitudinally on the body and transversely upon the legs and tail. Upon the nape of the neck and the back runs a crest composed of triangular compressed scales, having two or three similar rows of pointed scales at each side. Upon the sides of the nape are rows of triangular keeled scales, and the sides are covered with little compressed scales intermixed with large keeled shields. The toes are long, and the two central ones are much longer than the others. This is a small Lizard, only measuring when full grown about fourteen inches.

The STELLIO, sometimes called the HARDIM by the Arabs, is a well-known Lizard inhabiting Northern Africa, Syria, and Greece.

It is a very active little creature, haunting the ruins of ancient dwellings, heaps of stones, rocks, and similar localities, among which it flits from spot to spot with ceaseless activity. It has a curious habit of bending or nodding its head downwards, a movement which is greatly resented by the stricter Mahometans, who are pleased to consider the Lizard as offering an insult to their religion by imitating them in their peculiar actions of prayer. The more religious among them, therefore, take every opportunity of killing the Stellio, blending amusement, piety, and destructiveness with a happy appreciation of their several merits, earning a good position in Paradise on easy terms, and consoling

EGYPTIAN MASTIGURE.—*Uromastix spinipes.* STELLIO OR HARDIM.—*Stellio Cordylina.*

themselves for the present dearth of infidel heads by slicing off those of the unbelieving Lizards.

The Stellio lives almost entirely on the various insects that flit about the sand, and its quick rapid movements are needed to secure its prey. A kind of cosmetic was anciently made from this reptile, and even at the present day, the Turks employ it in the offices of the toilet.

The colour of the Stellio is olive-green above, clouded with black, and the under parts are yellow, sometimes tinged with green. There is no crest upon the nape of the neck, and the scales of the tail are rather large, and arranged in distinct whorls. There is no decided throat-pouch, but the skin of the throat is loose and plaited into a single cross fold towards its base. The body is rather flattened, and there is a longitudinal plait on each side. The tail is round and conical.

IN the restricted genus AGAMA—a word, by the way, which is not derived from any classical source, but is simply the popular name among the natives of Jamaica—the scales of the back are flat and keeled, and the third and fourth toes are nearly equal in length. The throat is marked with one longitudinal fold, and one or sometimes two transverse folds towards its base. Upon the sides of the neck and near the ears are curious groups of spiny scales. There is a slight crest along the back, the body is rather flattened, and the tail is long, tapering, and is covered with whorls of boldly projecting scales.

In a very old work on natural history, it is stated that the Lizards which have their tails thus armed with sharp spiny scales, make use of them in a rather singular fashion. They feed, according to these old writers, on cattle and other animals, and judging that from their small size they cannot bring an ox or a cow home after they have killed it, they jump on its back, cling tightly there with their feet, and by judicious lashing of the sharp tail, guide the animal to their home, where they give the fatal bite.

The SPINOSE AGAMA (*Agama colonorum*) is a well-known example of this genus, residing in Northern Africa, and plentiful in Egypt. The colour of this reptile is brown :

MOLOCH.—*Moloch horridus.*

the scales on the sides of the neck are very long and sharp, and those of the back are broad, boldly keeled, and sharply pointed, so that the creature presents rather a formidable appearance. The tail is long and powerful.

THERE is a very remarkable Lizard belonging to this family, called the EARED MEGA-LOCHILE, or sometimes, though wrongly, the EARED AGAMA.

This curious creature is found in Russia, several specimens from that country being in the collection of the British Museum. In this genus, containing, as far as is at present known, only one species, the head is flat and round, the eyes large, and the ears sunken and concealed under the skin. On the angle of the mouth at each side is placed a large membranous fold of skin, curved so as to bear a close resemblance to a large external ear, and boldly toothed on its edge. The neck is rather contracted, as if pinched, and has a cross fold below. The back has no crest, the tail is much flattened throughout its length, and the toes are long and very strongly toothed on the edge. The colour of this reptile is grey and brown, with a slight green wash upon the top of the head.

The EGYPTIAN MASTIGURE, or SPINE-FOOTED STELLIO, is a native of Northern Africa, and was said, though wrongly, to be the reptile spoken of by the ancients as the land-crocodile. A figure of this creature may be seen on page 89.

This species attains a rather large size, a full-grown specimen sometimes measuring a yard in length. It is an inhabitant of desert spots, preferring old ruins, rocky ground, and similar localities, where it can obtain instant refuge in case of alarm. The colour of this reptile is bright grass-green during life, but, as is generally the case with all these animals, the brilliant colours fade soon after death, and change to dingy blackish brown if the skin be stuffed, or to mottled greys, browns, and blacks, if preserved in spirits. The head of this creature is rounded, the back without a crest, the skin of the throat so folded as partly to cover the ears, and the ears themselves are oblong, and toothed in front. The tail is rather flattened, and furnished with transverse rows of large scales, boldly keeled, and sharply pointed. A few conical spines are scattered upon the upper part of the thigh, the sides, and loins.

THE last example of the Agamidæ which can be figured in these pages is the most ferocious-looking of the whole family, and were its dimensions much enlarged, would be universally allowed to be the most terrible-looking creature on the face of the earth. Many reptiles are spiny in different parts of their bodies, but this creature, appropriately termed the MOLOCH, bristles like a hedgehog with sharp spikes, which project both above and below in such profusion, that this Lizard almost seems to have been formed for the purpose of testing the number of effective spikes that can be planted on a given space. The creature is all spikes, and thorns, and projections. Upon the top of the head two very large spikes are seen, projecting from each eyebrow, and on the back of the neck is a large rounded protuberance, covered with little spiny scales, and having one long projecting spine on each side. On the back, the arrangement is very curious. A number of long spines are scattered at intervals over the surface, each of which is surrounded by a circle of lesser spines. It is worthy of notice that these large spines are hollow, and fit upon protuberances of the skin much in the same way that a cow's horn is sheathed on its core. The whole head and limbs are covered with spines similar in formation, but smaller in size. The tail is covered with long, sharp, spiny scales, arranged in whorls, and boldly radiating from their centre; and even the toes are covered as far as the long, sharp claws, with boldly keeled scales. The general colour of this reptile is palish yellow, spotted regularly with brown above and below, with dark red blotches edged with black. The Moloch is a native of Australia.

THE last tribe of the Lizards contains but one genus and very few species. From their habit of constantly living on trees, these creatures are called DENDROSAURA, or TREE LIZARDS. In these, the scales of the whole body are small and granular, and arranged in circular bands. The tongue is very curious, being cylindrical and greatly extensile, reminding the observer of a common earthworm, and swollen at the tip. The eyes are as peculiar as the tongue, being very large, globular, and projecting, and the ball is closely covered with a circular lid, through which a little round hole is pierced, much like the wooden snow-spectacles of the Esquimaux. The body is rather compressed, the ears are concealed under the skin, and the toes are separated into two opposable groups, so that the creature can hold very firmly upon the boughs. All the Dendrosaura are inhabitants of the Old World. The tail is very long and prehensile, and is almost invariably seen coiled round the bough on which the reptile is standing.

The most familiar example of the Dendrosaura is the common CHAMELEON, a reptile which is found both in Africa and Asia.

This singular reptile has long been famous for its power of changing colour, a property, however, which has been greatly exaggerated. as will be presently seen. Nearly all the Lizards are constitutionally torpid, though some of them are gifted with great rapidity of movement during certain seasons of the year. The Chameleon, however, carries this sluggishness to an extreme, its only change being from total immobility to the slightest imaginable degree of activity. No one ever saw a Chameleon even walk, as we understand that word, while running is a feat that no Chameleon ever dreamed of.

When it moves along the branch upon which it is clinging, the reptile first raises one foot very slowly indeed, and will sometimes remain foot in air for a considerable time, as if it had gone to sleep in the interim. It then puts the foot as slowly forward, and takes a good grasp of the branch. Having satisfied itself that it is firmly secured, it leisurely unwinds its tail, which has been tightly twisted round the branch, shifts it a little forward, coils it round again, and then rests for a while. With the same elaborate precaution, each foot is successively lifted and advanced, so that the forward movements seem but little faster than the hour-hand of a watch.

The extreme slowness and general habits of this animal are well depicted in an account of a tame Chameleon, kindly presented to me by Captain Drayson, R.A.

" I once owned a Chameleon, which was a very quaint creature. He had been captured by some Kaffir boys, whom I found laughing immoderately at the animal, a practice which I found very common amongst these people whenever they saw one of these reptiles. For a trifle the creature became my property, and I carried him to a little wattle and daub house in which I then resided. Being anxious to watch the private habits of my visitor, I drove a stick into the wall, and placed him upon it. The stick was about four feet in length and half an inch in diameter, so that the locomotion of the Chameleon was rather limited.

The first peculiarity I remarked about him was the very slow methodical way in which he moved. To turn to the right-about would occupy him several minutes, whilst to move from one end of the stick to the other was a recreation of which he was sparing, a whole day being devoted to this performance. There was something rather antique in his general appearance, both as regards his form and movements ; the long independent-moving, swivel eyes, giving him the characteristics of an Egyptian production, whilst the habit of puffing himself out occasionally, and of hissing, made him seem old-fashioned in the extreme.

I was disappointed when I found how slight was the variation in his colour. I had been led to believe that if placed on a scarlet, blue, or black groundwork he would soon assume the same hue ; this I found was a delusion. His usual colour was a light yellowish green, and this he could alter to a dark blue or brown-green, and he could make several dark brown spots become very prominent on his skin.

The method I used to adopt to make him show off, was to rub his side with my finger. He objected to this treatment, and used to puff away pompously, and vary his tints, as it appeared to me, by means of contracting or expanding his muscles under the skin. He looked very lantern-like, as though he were merely skin and ribs, and he was never found guilty of eating anything. Sometimes I saw flies settle upon him, a liberty which he did not resent. He merely turned one of his swivel eyes towards the delinquent and squinted calmly at it. Occasionally I put a fly in his mouth, and forced him to keep it there ; he took the affront very coolly, and the fly was seen no more. So hollow did he appear, that I frequently listened to hear if the flies were buzzing about inside him, but all was quiet. He stayed on the stick during two months. I then gave him a run out of doors, but having left him a few minutes, he took advantage of my absence and levanted, after which I saw him no more."

The food of the Chameleon consists of insects, mostly flies, but, like many other reptiles, the Chameleon is able to live for some months without taking food at all. This capacity for fasting, together with the singular manner in which the reptile takes its prey, gave rise to the absurd fable that the Chameleon lived only upon air. To judge by external appearance, there never was an animal less fitted than the Chameleon for capturing the winged and active flies. But when we come to examine its structure, we find that it is even better fitted for this purpose than many of the more active insect-eating Lizards.

The tongue is the instrument by which the fly is captured, being darted out with such singular velocity that it is hardly perceptible, and a fly seems to leap into the mouth of the reptile as if attracted by magnetism. This member is very muscular, and is furnished at the tip with a kind of viscid secretion which causes the fly to adhere to it. A lady who kept a Chameleon for some time, told me that her pet died, and when they came to examine it, they found that its tongue had in some strange way got down its throat, an accident which they took to be the cause of its death. Its mouth is well furnished with teeth, which are set firmly into its jaw, and enable it to bruise the insects after getting them into its mouth by means of the tongue.

The eyes have a most singular appearance, and are worked quite independently of each other, one rolling backwards while the other is directed forwards or upwards. In connexion with this subject some very curious and valuable remarks will be found on page 94. There is not the least spark of expression in the eye of the Chameleon, which looks about as intellectual as a green pea with a dot of ink upon it.

CHAMELEON.—*Chaméleo vulgáris*.　　　　LARGE-NAPED CHAMELEON.—*Chaméleo bifurcus*.

Owing to the exceeding slowness of its movements, it has no way of escaping when once discovered, and as a French writer well says, "un Caméléon aperçu est un Caméléon perdu." Great numbers of these creatures fall victims to enemies of every kind, and were it not that their colour assimilates so well with the foliage on which they dwell, and their movements are so slow as to give no aid to the searching eye of their foes, the race would soon be extinct. The Chameleon has an odd habit of puffing out its body for some unexplained reason, and inflating itself until it swells to nearly twice its usual size. In this curious state it will remain for several hours, sometimes allowing itself to collapse a little, and then reinflating its skin until it becomes as tense as a drum and looks as hollow as a balloon.

The Chameleon is readily tamed, if such a word can be applied to the imperturbable nonchalance with which it behaves under every change of circumstance. It can be handled without danger, and although its teeth are strong, will not attempt to bite the hand that holds it. It is, however, rather quarrelsome with its own kind, and the only excitement under which it has been seen to labour is when it takes to fighting with a neighbour. Not that even then it hurries itself particularly or does much harm to its opponent, the combatants contenting themselves with knocking their tails together in a grave and systematic manner.

A few words on the change of colour will not be out of place. The usual colour of the Chameleon when in its wild state is green, from which it passes through the shades of violet, blue and yellow, of which the green consists. In this country, however, it rarely retains the bright green hue, the colour fading into yellowish grey, or the kind of tint which is known as *feuille-morte*. One of the best and most philosophical disquisitions on this phenomenon is that of Dr. Weissenbaum, published in the "Magazine of Natural History" for 1838. The writer had a living Chameleon for some time, and gives the result of his observations in the following words :—

"The remote cause of the difference of colour in the two lateral folds of the body, may be distinctly referred to the manner in which the light acts upon the animal. The statement of Murray that the side turned towards the light is always of a darker colour, is perfectly true ; this rule holds good with reference to the direct and diffused light of the sun and moon as to artificial light. Even when the animal was moving in the walks of my garden, and happened to come near enough to the border to be shaded by the box edging, that side so shaded would instantly become less darkly coloured than the other.

Now the light in this way seldom illumines *exactly* one half of the animal in a more powerful manner than the other, and as the middle line is constantly the line of demarcation between the two different shades of colour, we must evidently refer the different effects to two different centres, from which the nervous currents can only radiate, under such circumstances, towards the organs respectively situated on each side of the mesial line. Over these centres, without doubt, the organ of vision immediately presides ; and, indeed, we ought not to wonder that the action of light has such powerful effects on the highly irritable organization of the Chameleon, considering that the eye is most highly developed. The lungs are but secondarily affected, but they are likewise more strongly excited on the *darker side, which is constantly more convex than the other.*

Many other circumstances may be brought forward in favour of the opinion that the nervous currents in one half of the Chameleon are going on independently of those in the other ; and that the animal has two lateral centres of perception, sensation and motion, besides the common one in which must reside the faculty of concentration.

Notwithstanding the strictly symmetrical construction of the Chameleon as to its two halves, the eyes move independently of each other, and convey different impressions to their different centres of perception : the consequence is, that when the animal is agitated, its movements appear like those of two animals glued together. Each half wishes to move its own way, and there is no concordance of action. The Chameleon, therefore, is not able to swim like other animals ; it is so frightened if put into water, that the faculty of concentration is lost, and it tumbles about as if in a state of intoxication.

On the other hand, when the creature is undisturbed, the eye which receives the strongest impression propagates it to the common centre, and prevails on the other eye to follow that impression, and direct itself to the same object. The Chameleon, moreover, may be asleep on one side and awake on the other. When cautiously approaching my specimen at night with a candle, so as not to awake the whole animal by the shaking of the room, the eye turned toward the flame would open and begin to move, and the corresponding side to change colour, whereas the other side would remain for several seconds longer in its torpid and changeable state, with its eye shut."

It seems probable that the change of colour may be directly owing to the greater or less rapidity of the circulation, which may turn the Chameleon from green to yellow, just as in ourselves an emotion of the mind can tinge the cheek with scarlet, or leave it pallid and death-like. Mr. Milne Edwards thinks that it is due to two layers of pigment cells in the skin, arranged so as to be movable upon each other, and so produce the different effects.

The young of the Chameleon are produced from eggs, which are very spherical, white in colour, and covered with a chalky and very porous shell. They are placed on the ground under leaves, and there left to hatch by the heat of the sun, and the warmth produced by the decomposition of the leaves. The two sexes can be distinguished from each other by the shape of the tail, which in the male is thick and swollen at the base.

THERE are nearly twenty species of Chameleons known to zoologists at the present day, all presenting some peculiarity of form or structure. One of the most remarkable species is the LARGE-NAPED CHAMELEON or Fork-nosed Chameleon, as it is sometimes called.

This creature inhabits Madagascar, that land which nourishes so many strange forms of animal life. It is also found in India, the Moluccas, and Australia. When full grown, the muzzle of the male is very deeply cleft, or forked, the two branches diverging from each other, and presenting the singular appearance shown in the illustration. The female has no horns, and in the male they are short and blunt while the creature is young, not obtaining their full length and sharpness until it has attained full age. These curious forked projections belong to the skull, and are not merely a pair of prolonged scales or tubercles.

THE large and important order at which we now arrive, consists of reptiles which are popularly known as SNAKES, or more scientifically as OPHIDIA, and to which all the true serpents are to be referred.

Almost every order is bordered, so to speak, with creatures so equally balanced between the characteristics of the orders that precede and follow it, that they can be with difficulty referred to their right position. Such, indeed, is the case with the Ophidia, from which are excluded, by the most recent systematic zoologists, the amphisbænians and many other footless reptiles, now classed among the lizards. The greater number of the Snakes are without any vestige of limbs, but in one or two species, such as the pythons, the hinder pair of limbs are represented by a pair of little horny spurs placed just at the base of the tail, and are supported by tiny bones that are the undeveloped commencements of hinder limbs. Indeed, several of the true lizards, the common blindworm, for example, are not so well supplied with limbs as these true Snakes.

The movements of the serpent tribe are, in consequence, performed without the aid of limbs, and are, as a general rule, achieved by means of the ribs and the large cross scales that cover the lower surface. Each of these scales overlaps its successor, leaving a bold horny ridge whenever it is partially erected by the action of the muscles. The reader will easily see that a reptile so constructed can move with some rapidity by successively thrusting each scale a little forward, hitching the projecting edge on any rough substance, and drawing itself forward until it can repeat the process with the next scale. These movements are consequently very quiet and gliding, and the creature is able to pursue its way under circumstances of considerable difficulty.

Oftentimes the Snake uses these scales in self-defence, offering a passive resistance to its foe when it is incapable of acting on the offensive. Any one may easily try this experiment by taking a common field Snake, letting it glide among the stubble or into the interstices of rocky ground, and then trying to pull it out by the tail. He will find that even if the reptile be only half concealed, it cannot be dragged backward without doing it considerable damage, for on feeling the grasp, it erects all the scales and opposes their edges so effectually to the pull that it mostly succeeds in gliding through from the hand that holds it. I have often lost Snakes by allowing them to insinuate themselves into crevices, and have been fain to let them escape rather than subject them to the pain, if not absolute damage, which they must have suffered in being dragged back by main force.

The tongue of the Snakes is long, black, and deeply forked at its extremity, and when at rest is drawn into a sheath in the lower jaw. In these days it is perhaps hardly necessary to state that the tongue is perfectly harmless, even in a poisonous serpent, and that the popular idea of the "sting" is entirely erroneous. The Snakes all seem to employ the tongue largely as a feeler, and may be seen to touch gently with the forked extremities the objects over which they are about to crawl or which they desire to examine. The external organs of hearing are absent.

The vertebral column is most wonderfully formed, and is constructed with a special view to the peculiar movements of the serpent tribe. Each vertebra is rather elongated, and is furnished at one end with a ball and at the other with a corresponding socket, into which the ball of the succeeding vertebra exactly fits, thus enabling the creature to writhe and twine in all directions without danger of dislocating its spine. This ball-and-socket principle extends even to the ribs, which are jointed to certain rounded projections of the vertebræ in a manner almost identical with the articulation of the vertebræ upon each other, and as they are moved by very powerful muscles, perform most important functions in the economy of the creature to which they belong.

Sometimes the Snakes advance by a series of undulations, either vertical or horizontal, according to the species, and when they proceed through water, where the scales of the abdomen would have no hold of the yielding element, their movements are always of this undulatory description. The number of vertebræ, and consequently of ribs, varies much in different species, in some Snakes being about three hundred.

The jaws of the serpents are very wonderful examples of animal mechanics, and may be cited among the innumerable instances where the existing construction of living beings has long preceded the inventions of man. We have already seen the invaluable mechanic invention of the ball-and-socket joint exhibited in the vertebræ of the Snakes, and it may be mentioned that in the spot where the limbs of almost all animals, man included, are joined to the trunk, the ball-and-socket principle is employed, though in a less perfect manner than in the Snakes. It is by means of this beautiful form of joint that posture-masters and mountebanks are able to contort their bodies and limbs into so many wonderful shapes, the muscles and tendons yielding by constant use and enabling the bones to work in their sockets without hindrance. Indeed, a master of the art of posturing is really an useful member of society, at all events to the eye of the physiologist, as showing the perfection of the human form, and the wonderful capabilities of man, even when considered from the mere animal point of view.

In the jaw of the serpents, we shall find more than one curious example of the manner in which human inventions have succeeded, if, indeed, they have not been borrowed from some animal structure.

All the Snakes are well supplied with teeth; but their number, form, and structure differ considerably in the various species. Those Snakes that are not possessed of venomous fangs have the bones of the palate as well as the jaws furnished with teeth, which are of moderate size, simple in form, and all point backward, so as to prevent any animal from escaping which has ever been grasped, and acting as valves which permit of motion in one direction only.

The bones of the jaws are, as has already been mentioned, very loosely constructed, their different portions being separable, and giving way when the creature exerts its wonderful powers of swallowing. The great python Snakes are well known to swallow animals of great proportionate size, and any one may witness the singular process by taking a common field Snake, keeping it without food for a month or so, and then giving it a large frog. As it seizes its prey, the idea of getting so stout an animal down that slender neck and through those little jaws appears too absurd to be entertained for a moment, and even the leg which it has grasped appears to be several times too large to be passed through the throat. But by slow degrees the frog disappears, the mouth of the Snake gradually widening, until the bones separate from each other to some distance, and are only held by the ligaments, and the whole jaw becoming dislocated, until the head and neck of the Snake look as if the skin had been stripped from the reptile, spread thin and flat, and drawn like a glove over the frog.

No sooner, however, has the frog fairly descended into the stomach, than the head begins to assume its former appearance; the elastic ligaments contract and draw the bones into their places, the scales, which had been far separated from each other, resume their ordinary position, and no one would imagine, from looking at the reptile, to what extent the jaws and neck have recently been distended. As many of the Snakes swallow their prey alive—the frog for example, having been heard to squeak while in the stomach of its destroyer—the struggles of the internal victim would often cause its escape, were it not for the array of recurved teeth, which act so effectually, that even if the Snake wished to disgorge its prey it could not do so. Mr. Bell had in his collection a small Snake which had tried to swallow a mouse too large even for the expansile powers of a Snake's throat, and which had literally burst through the skin and muscles of the neck.

The lower jaw, moreover, is not jointed directly to the skull, but to a most singular development of the temporal bone, which throws out two elongated processes at right angles with each other, like the letter L laid horizontally ⌐, so that a curious double lever is obtained, precisely after the fashion of the well-known " throwing-stick" of the aboriginal Australians, which enables those savages to fling their spears with deadly effect to a distance of a hundred yards.

The teeth of the venomous Serpents will be described in connexion with one of the species.

The Serpents, in common with other reptiles, have their bodies covered by a delicate epidermis, popularly called the skin, which lies over the scales, and is renewed at tolerably regular intervals. Towards the time of changing its skin, the Snake becomes dull and sluggish, the eyes look white and blind, owing to the thickening of the epidermis that covers them, and the bright colours become dim and ill-defined. Presently, however, the skin splits upon the back, mostly near the head, and the Snake contrives to wriggle itself out of the old integument, usually turning it inside out in the process. This shed skin is transparent, having the shape of each scale impressed upon it, being fine and delicate as goldbeater's-skin, and being applicable to many of the same uses, such as shielding a small wound from the external air. In two very fine specimens of cast skins, formerly belonging to a viper and boa-constrictor, now lying before me, the structure of each scale is so well shown, that the characteristics of the two reptiles can be distinguished as readily as if the creatures were present from whose bodies they were shed. Even the transparent scale that covers the eyes is drawn off entire, and the large elongated hexagonal scales that are arranged along the abdomen, and aid the animal in its progress, are exhibited so boldly that they will resist the movement of a finger drawn over them from tail to head.

THE first sub-order of Snakes consists of those Serpents which are classed under the name of VIPERINA. All these reptiles are devoid of teeth in the upper jaw except two long, poison-bearing fangs, set one at each side, and near the muzzle. The lower jaw is well furnished with teeth, and both jaws are feeble. The scales of the abdomen are bold, broad, and arranged like overlapping bands. The head is large in proportion to the neck, and very wide behind, so that the head of these Snakes has been well compared to an ace of spades. The hinder limbs are not seen.

In the first family of the Viperine Snakes, called the CROTALIDÆ, the face is marked with a large pit or depression on each side, between the eye and the nostril. The celebrated and dreaded FER-DE-LANCE belongs to this family.

This terrible reptile is a native of Brazil, and in some parts is very common, owing to its exceeding fecundity and its habit of constant concealment. It has an especial liking for the sugar plantations, and a field of canes is seldom cut without the discovery of seventy or eighty of these venomous creatures. Martinique and St. Lucia are terribly haunted by this Snake, which is held in great dread by the natives and settlers. In general, the Serpents, even those of a poisonous character, avoid the presence of man, but the Fer-de-Lance frequently takes the initiative, and leaping from its concealment, fastens upon the passenger whose presence has disturbed its irritable temper, and inflicts a wound that is almost invariably fatal within a few hours.

Even in those cases where the sufferer recovers for the time, the system is terribly injured, and the latent virulence of the poison can hardly be eliminated from the frame, even at the cost of painful boils and ulcerations which last for many years. The nervous system is also much affected, as giddiness and paralysis are among the usual consequences of the strong venom which this reptile extracts, by some inexplicable chemistry, from perfectly harmless food. Convulsions, severe pain at the heart, together with distressing nausea, are among the many symptoms produced by this poison.

To escape this creature in its chosen haunts is a matter of very great difficulty, as it is either concealed under dead leaves, among the heavy foliage of parasitic plants, or

FER-DE-LANCE.—*Craspedocephalus lanceolatus.*

coiled up in the nest of some poor bird whose eggs or young it has devoured, and from this spot of vantage makes its stroke, swift and straight as a fencer's thrust, and without the least warning by hiss or rattle to indicate its purpose.

All animals dread the Fer-de-Lance; the horse prances and snorts in terror on approaching its hiding-place, his whole frame trembles with fear, and he cannot be induced by spur or whip to pass within striking distance of this formidable reptile. Birds of all kinds have a horror of its presence, and will pursue it from place to place, or hover near the spot on which it is resting, fluttering their wings, stretching their necks, and uttering hoarse cries of mingled rage and terror. The honey guide is especially fearful of this Serpent, and has often guided a man, not as he supposed, to the vicinity of a hive of wild bees, but to the resting-place of this venomous Snake. The pig, when in good condition, is said to be the only animal that can resist the poison, the thick coating of fat which covers the body preventing the venom from mingling with the blood. It is said, indeed, that a fat hog cares nothing for Fer-de-Lance or rattlesnake, but receives their stroke with contemptuous indifference, charges at them fearlessly, tramples upon them until they are disabled, and then quietly eats them.

Against the effects of this poison, there seems to be no certain remedy; but the copious use of spirits has lately appeared to neutralize in some measure the full virulence of a Snake bite. The amount of strong spirits which can be drunk under such circumstances is almost incredible, its whole force seeming to be employed in arming the nerves against the enfeebling power of the poison. Some recent and valuable experiments have shown, that if a man, bitten by a venomous Serpent, can be kept in a state of semi-intoxication through the use of spirituous liquors, this rather strange process will give him almost his only hope of escape.

Yet nothing is made in vain, and terrible as is this creature to man, it is of no small use to him even in the localities where it is most dreaded. But for the presence of the Fer-de-Lance and one or two other Serpents closely allied to it, the sugar plantations would be devastated by the rats which crowd to such fertile spots, and on which this Snake chiefly feeds.

As is the case with many Serpents, the colour of the Fer-de-Lance is rather variable. Its usual tints are olive above with dark cross bands, and whitish grey below, covered with very minute dark dots. The head is brown. This reptile attains a considerable size, being generally five or six feet long, and occasionally reaching a length of seven or eight feet. The tail ends in a horny spine which scrapes harshly against rough objects, but does not rattle.

CLOSELY allied to the fer-de-lance is another poisonous Serpent of Southern America, remarkable for the very large size to which it attains, and the glowing radiance of its fearful beauty. This is the CURUCUCU, more familiarly known by the popular title of BUSHMASTER, (*Láchesis mutus*).

Mr. Waterton, who has incidentally mentioned this Snake in his "Wanderings," has kindly sent me the following information about this terrible creature. "The Bushmaster will sometimes reach fourteen feet in length. The Dutch gave it the name of Bushmaster on account of its powers of destruction, and being the largest poisonous Snake discovered. It still continues to have the same name among the colonists of British Guiana. Its Indian name is COUANACOUCHI. It is a beautiful Serpent, displaying all the prismatic colours when alive, but they disappear after death. All these three species (the Bushmaster, Labarri, and Coulacanara) inhabit the trees as well as the ground, but as far as I could perceive, they never mount the trees with a full stomach."

THE name of WATER VIPER is appropriately given to the creature now before us, in consequence of its water-loving habits.

It is a native of many parts of America, and is never seen at any great distance from water, being found plentifully in the neighbourhood of rivers, marshes, and in swampy lands. It is a good climber of trees, and may be seen entwined in great numbers on the branches that overhang the water. On the least alarm, the reptile glides from the branch, drops into the water and wriggles its way into a place of safety. The object of climbing the trees seems to be that the creature delights to bask in the sun, and takes that method of gratifying its inclination where the whole of the soil is wet and marshy. But in those localities where it can find dry banks and rising grounds, the Water Viper contents itself with ascending them and lying upon the dry surface enjoying the genial warmth.

It is a most poisonous reptile, and is even more dreaded by the negroes than the rattlesnake, as like the fer-de-lance, it will make the first attack, erecting itself boldly, opening its mouth for a second or two, and then darting forward with a rapid spring. At all times it seems to be of an aggressive character, and has been known to chase and bite other Snakes put into the same cage, the poor creatures fleeing before it and endeavouring to escape by clinging to the sides of the cage. But when several other individuals of the same species were admitted, the very Snake that had before been so ferocious became quite calm, and a box containing four or five specimens has been sent on a journey of many miles without any quarrels ensuing among the inmates.

The food of the Water Viper consists of fishes, which it can procure by its great rapidity of movement and excellent swimming powers, of reptiles and even of birds. Mr. T. W. Wood has favoured me with an account of the manner in which a Water Viper, in the collection of the Zoological Society, devoured the prey that was put before it.

"A short time ago I had the good fortune to be at the Zoological Gardens when the reptiles were fed. Some sparrows and titlarks were put into the apartment containing several specimens of the Water Viper. The sparrows seemed very much terrified, and soon huddled together in a corner, afraid, as I suppose, of the spectators.

One of the titlarks, however, bolder than the rest, ran about as if at home. One of the Water Vipers perceiving it quiet for a moment, seemed to fix its eye upon the poor little creature. The reptile commenced moving towards the bird slowly but surely, their eyes being intently fixed upon each other. When the Serpent had approached within about half an inch, it opened its mouth and seized the bird by the side, its left wing being grasped in the Snake's mouth. The ill-fated bird instantly gave two or three convulsive struggles, the head then dropped, the eyes closed, and all was over ; a drop of

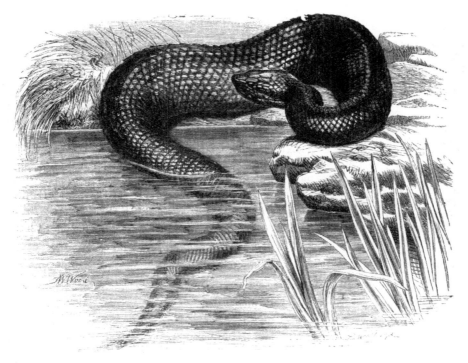

WATER VIPER.—*Cenchris piscivorus.*

blood oozed slowly out of the bird's bill. The reptile did not release the bird after it was bitten, but began to swallow it almost immediately.

Another titlark was then introduced by the keeper. This bird was, when I approached, lying on its side as if dead. Another Water Viper seized its head and commenced swallowing it, the bird struggling violently; at each effort of deglutition the venomous fangs were seen to move forward. In this case the poison did not take such rapid effect, as the bird was evidently alive when it disappeared down the reptile's throat."

The colour of the Water Viper is greenish brown, taking a yellowish tone along the sides, and banded with blackish brown. It seldom exceeds two feet in length. This serpent is also known by the popular names of COTTON-MOUTH and WATER MOCASSIN SNAKE. The COPPER-HEAD SNAKE of the same country, (*Cenchris contortrix*) is closely allied to it.

THE well-known and terrible RATTLESNAKE now comes before us. Two species of Serpents are popularly called by this name, both of which will be described.

This dreaded reptile is a native of North America, and is remarkable for the singular termination to the tail, from which it derives its popular name. It has already been mentioned that the fer-de-lance has a long horny scale at the tip of its tail, and in the Rattlesnake this appendage is developed into a rather complicated apparatus of sound.

At the extremity of the tail are a number of curious loose horny structures, formed of the same substance as the scales, and varying greatly in number according to the size of the individual. It is now generally considered that the number of joints on the "rattle" is an indication of the reptile's age, a fresh joint being gained each year

RATTLESNAKE.—*Uropsophus durissus.*

immediately after it changes its skin and before it goes into winter quarters. There is, however, another opinion prevalent among the less educated, which gives to the Rattlesnake the vindictive spirit of the North American Indian, and asserts that it adds a new joint to its rattle whenever it has slain a human being, thus bearing on its tail the fearful trophies of its prowess, just as the Indians wear the scalps of their slain foes.

The joints of this remarkable apparatus are arranged in a very curious manner, each being of a somewhat pyramidal shape, but rounded at the edges, and being slipped within its predecessor as far as a protuberant ring which runs round the edge. In fact, a very good idea of the structure of the rattle may be formed by slipping a number of thimbles loosely into each other. The last joint is smaller than the rest, and rounded. As was lately mentioned, the number of these joints is variable, but the average number is from five or six to fourteen or fifteen. There are occasional specimens found that possess more than twenty joints in the rattle, but such examples are very rare.

When in repose, the Rattlesnake usually lies coiled in some suitable spot, with its head lying flat, and the tip of its tail elevated in the middle of the coil. Should it be irritated by a passenger, or feel annoyed or alarmed, it instantly communicates a quivering movement to the tail, which causes the joints of the rattle to shake against each other, with a peculiar skirring ruffle, not easily described but never to be forgotten when once heard. All animals, even those which have never seen a Rattlesnake, tremble at this sound, and try to get out of the way. Even a horse newly brought from Europe is just as frightened as the animal that has been bred in the same country with this dread Serpent, and at the sound of the rattle will prance, plunge, and snort in deadly fear, and cannot be induced to pass within striking distance of the angry Snake.

It has already been mentioned that swine are comparatively indifferent to the Rattlesnake, and will trample it to death and eat it afterwards. It is certain that they will eat a dead Rattlesnake, though almost any other animal will flee from the lifeless carcase nearly as swiftly as from the living reptile. Perhaps the thick coating of fat that clothes the body of the well-fed swine may neutralize the poison of the venomed teeth, and so enable the hog to receive the stroke with comparative impunity. The peccary is also said to kill and devour the Rattlesnake without injury, and deer are reported to jump upon it and kick its life out with their sharp hoofs.

Fortunately for the human inhabitants of the same land, the Rattlesnake is slow and torpid in its movements, and seldom attempts to bite unless it is provoked, even suffering itself to be handled without avenging itself. Mr. Waterton tells me in connexion with these reptiles : " I never feared the bite of a Snake, relying entirely on my own movements. Thus, in presence of several professional gentlemen, I once transferred twenty-seven Rattlesnakes from one apartment to another, with my hand alone. They hissed and rattled when I meddled with them, but they did not offer to bite me." The fer-de-lance Snake is, as has already been mentioned, most fierce and irritable in character, taking the initiative, and attacking without reason. But the Rattlesnake always gives notice of its deadly intentions, and never strikes without going through the usual preliminaries. When about to inflict the fatal blow, the reptile seems to swell with anger, its throat dilating, and its whole body rising and sinking as if inflated by bellows. The tail is agitated with increasing vehemence, the rattle sounds its threatening war-note with sharper ruffle, the head becomes flattened as it is drawn back ready for the stroke, and the whole creature seems a very incarnation of deadly rage. Yet, even in such moments, if the intruder withdraw, the reptile will gradually lay aside its angry aspect, the coils settle down in their place, the flashing eyes lose their lustre, the rattle becomes stationary, and the Serpent sinks back into its previous state of lethargy.

It is rather curious that the Rattlesnake varies much in its powers of venom and its irritability of temper according to the season of the year. During the months of spring it will seldom attempt to bite, and if it does strike a foe, the poison is comparatively mild in its effects. But after August, and before it seeks its winter quarters, the Rattlesnake is not only more fierce than at any other time of the year, but the venom seems to be of more fearful intensity, inflicting wounds from which nothing escapes with life.

The rapidity of the effects depends necessarily on the part which is bitten. Should the points of the teeth wound a moderately large vein or an artery, the venom courses swiftly through the blood, and the victim dies in a few minutes. But if, perchance, the tooth should pierce some fleshy and muscular part of the body, the poison does not have such rapid effect, and the injured person may be saved by the timely administration of powerful remedies. There seems, indeed, to be no one specific for the bite of this reptile, as the effects vary according to the individual who happens to be bitten, and the state of health in which the sufferer may be at the time. Immediate suction, however, and the unsparing use of the knife appear to be the most efficacious means of neutralizing the poison, and strong ammonia and oil have been employed with good results. Catesby, in writing about this reptile, remarks that he has known instances where death has occurred within two minutes after the infliction of the bite.

The food of the Rattlesnake consists of rats, mice, reptiles, and small birds, the latter of which creatures it is said to obtain by the exercise of a mysterious power termed fascination, the victim being held, as it were, by the gaze of its destroyer, and compelled to remain in the same spot until the Serpent can approach sufficiently near to seize it. It is even said that the Rattlesnake can coil itself at the foot of a tree, and by the mere power of its gaze force a squirrel or bird to descend and fling itself into the open mouth waiting to receive it.

These phenomena have been strongly asserted by persons who say that they have seen them, and are violently denied by other persons who have never witnessed the process, and therefore believe that the circumstances could not have happened. For my own part

I certainly incline to the theory of fascination, thinking that the power exists, and is occasionally employed, but under peculiar conditions. That any creature may be suddenly paralysed by fear at the sight of a deadly foe is too well known to require argument, and it is therefore highly probable that a bird or squirrel, which could easily escape from the Serpent's jaws by its superior agility, might be so struck with sudden dread on seeing its worst enemy that it would be unable to move until the reptile had seized it.

Birds, especially, are most sensitive in their nature, and can be fascinated in a manner by any one who chooses to try the experiment. Let any bird be taken, laid on its back, and the finger pointed at its eyes. The whole frame of the creature will begin to stiffen, the legs will be drawn up, and if the hand be gently removed, the bird will lie motionless on its back for any length of time. I always employ this method of managing my canaries when I give them their periodical dressing of insect-destroying powder. I shake the powder well into their feathers, pour a small heap of it on a sheet of paper, lay the bird in the powder, hold my finger over its eyes for a moment, and leave it lying there while I catch and prepare another bird for the same process. There is another way of fascinating the bird, equally simple. Put it on a slate or dark board, draw a white chalk line on the board, set the bird longitudinally upon the line, put its beak on the white mark, and you may go away for hours, and when you return the bird will be found fixed in the same position, there held by some subtle and mysterious influence which is as yet unexplained.

Thus far there is no difficulty in accepting the theory of fascination, but the idea of a moral compulsion on the part of the Snake, and a perforced obedience on the part of its victim, is so strange that it has met with very great incredulity. Still, although strange, it is not quite incredible. We all know how the immediate presence of danger causes a reckless desire to see and do the worst, regardless of the consequences, and heeding only the overpowering impulse that seems to move the body without the volition of the mind. There are many persons who cannot stand on any elevated spot without feeling so irresistible a desire of flinging themselves into the depths below, that they dare not even stand near an open window or walk near the edge of a cliff. It may be that the squirrel or bird, seeing its deadly enemy below, is so mentally overbalanced that it is forced to approach the foe against its own will, and is drawn nearer to those deadly fangs by the very same impulse that would urge a human being to jump over the edge of a precipice or from the top of a lofty building.

Every squirrel or every bird may not succumb to the same influence, just as every human being does not yield to the insane desire of jumping from heights, and it is probable that a Rattlesnake may coil itself under a tree and look all day at the squirrels sporting upon the branches, or the birds flitting among the boughs, without inducing one of them to become an involuntary victim. Yet it is possible that out of the many hundreds that could see the Serpent, one would be weak-minded enough to yield to the subtle influence, and, instead of running away, find itself forced to approach nearer and nearer the fearful reptile.

Some persons acknowledge the fact that the bird approaches the Snake, and is then snapped up, but explain it in a different manner. They say that the bird is engaged in mobbing or threatening the Snake, just as it might follow and buffet a hawk, an owl, or a raven, and in its eagerness approaches so closely that the Snake is able to secure it by a sudden dart. Such is very likely to be the case in many instances, as the little birds will often hover about a poisonous Snake, and, by their fluttering wings and shrieking cries, call attention to the venomous reptile. But the many descriptions of the fascinating process are too precise to allow of such a supposition in the particular instances which are mentioned.

Even the common Snake of England can exercise a similar power. I have seen one of these Snakes in chase of a frog, and the intended victim, although a large and powerful specimen of its race, fully able to escape by a succession of leaps such as it would employ if chased by a human being, was only crawling slowly and painfully like a toad, its actions reminding one of those horrid visions of the night when the dreamer finds himself running or fighting for his life, and cannot move faster than a walk, or strike a blow that would break a cobweb. In such cases, the victim may be taken from the pursuer, but unless it is carried to a considerable distance, it will soon be in the jaws of the Serpent a second time.

It is worthy of notice that in all such instances, a sudden sound will seem to break the spell and snap the invisible chain that binds the victim to its destroyer. If birds are spell-bound by finger or chalk line, as has already been described, a quick movement or a heavy footstep will release them from their bonds, and a sudden shout will in a similar manner enable a bird to break away from the Serpent into whose jaws it was on the point of falling. One of my friends when in Canada saw a little bird lying on the ground, fluttering about as if dusting itself, but in a rather strange manner, and on his nearer approach, a Snake glided from the spot, and the bird gathered its wings together and flew away. The Snake was one of the harmless kind, and being taken to the house of the person who had interrupted it in its meal, served to keep the premises clear of rats and mice. The Serpent is not the only creature to which this singular power is attributed, for the natives of Northern Africa assert that the lion is also gifted with this influence, and can induce certain hapless men and women to leave their homes and follow him into the woods. This, however, is only a popular tradition among the natives, and has met with no corroboration.

The Rattlesnake retires to its winter quarters as soon as the increasing coldness of the weather gives it warning to seek a home where it can find protection against the frosts. Sometimes the Snake chooses a convenient hole or crevice for this purpose, but in general it prefers the neighbourhood of marshy ground, and harbours under the heavy masses of a certain long-stemmed moss (*sphagnum palustre*) which grows plentifully in such situations. In such localities the Rattlesnake may be found during the winter, either coiled up in masses containing six or seven individuals, or creeping slowly about beneath the protecting moss. Many of these fearful Snakes are killed during the cold months by persons who are acquainted with their habits, and surprise them in their winter quarters.

The general colour of the Rattlesnake is pale brown. A dark streak runs along the temples from the back of the eye, and expands at the corner of the mouth into a large spot. A series of irregular dark brown bands are drawn across the back, a number of round spots of the same hue are scattered along the sides, upon the nape of the neck and back of the head.

ANOTHER species of Rattlesnake which is known by the same title, is also common in North America. This is the CASCAVEL, or BANDED RATTLESNAKE (*Crótalus hórridus*), a reptile very similar in rattle and general habits to the preceding species, but distinguishable by the colouring of the body. Like the common Rattlesnake, the ground colour of the body is brown, but there is a broad dark streak on each side of the neck, and the body is covered with oblique cross-bars which leave diamond-shaped spots.

A THIRD species, the SMALL, or MILIARY RATTLESNAKE, (*Crotalóphorus miliáris*), is also well known in North America.

This serpent is thought to be even more dangerous than either of the preceding reptiles, because its dimensions are so small that a passenger is liable to disturb it before he sees the deadly creature in his path, and the sound of the rattle is so feeble that it is inaudible at the distance of two or three paces, and can only be heard when special attention is paid to it. It is a prolific species, and still maintains its numbers, in spite of the constant persecution to which it is subjected.

The food of the Miliary Rattlesnake consists of mice, frogs, insects, and similar creatures, which it mostly obtains by darting suddenly upon them as they pass near the spot where the reptile is lying. This serpent is fond of coiling itself on the fallen trunks of trees, decaying stumps, or similar situations. Fortunately, it is very easily killed, a smart blow dealing instant death even from a very small stick. The colour of this reptile is brownish olive, darker upon the cheeks, which are diversified by a narrow white streak from the back of the eye. A series of brown spots runs along the centre of the back, and the sides are ornamented with two rows of brown spots, each spot corresponding with a space in the other row. The abdomen is sooty black, marbled with a darker and rather more polished hue. An irregular dark brown band runs along each side of the nape and the crown of the head.

WE now come to the second great family of poisonous Serpents, namely the Vipers, or Viperidæ. All the members of this family may be distinguished by the absence of the pit between the eyes and the nostrils. There are no teeth in the upper jaw except the two poison-fangs.

A rather celebrated species of these Snakes is the TIC-POLONGA or KATUKA (*Daboia elegans*), a native of Asia, and perhaps of Brazil. This Serpent is much dreaded, its poison being of a very deadly character. A chicken that was bitten by a Tic-polonga died in thirty-six seconds, and a dog bitten by the same creature was dead in twenty-six minutes after receiving the injury. It is tolerably common in India and Ceylon, but is not so familiarly known as the cobra and other species, because it is not employed for public exhibition as is the case with those Serpents.

Sir Emerson Tennent, in his well-known " Natural History of Ceylon," writes thus of the Tic-polonga. "These formidable Serpents so infested the official residence of the District Judge of Trincomalie, in 1858, as to compel his family to abandon it. In another instance, a friend of mine, going hastily to take a supply of wafers from an open tin case which stood in his office, drew back his hand on finding the box occupied by a Tic-polonga coiled within it."

The word Tic-polonga signifies Spotted-polonga, the latter word being a kind of generic title given by the natives to many Serpents, no less than eight species being classed under this common title. It is said that the Tic-polonga and the cobra bear a mortal hatred towards each other, and to say that two people hate each other like the Tic-polonga and cobra is equivalent to our proverb respecting the cat and dog. The Tic-polonga is said always to be the aggressor, to find the cobra in its hiding place, and to provoke it to fight. There are many native legends in Ceylon respecting the ferocity of this Snake.

Its general colour is brown; there are two dark brown spots on each side of the back of the head, and a yellow streak runs between them. Upon the body are three rows of oblong brown spots, edged with white.

THE terrible PUFF ADDER is closely allied to the preceding species.

This reptile is a native of Southern Africa, and is one of the commonest, as well as one of the most deadly, of poisonous Snakes. It is slow and apparently torpid in all its movements, except when it is going to strike, and the colonists say that it is able to leap backwards so as to bite a person who is standing by its tail. Captain Drayson, R.A., who has seen much of this reptile and its habits, has kindly forwarded to me the following short account of this creature :—

"This formidable looking reptile is more dreaded than any other of the numerous poisonous Snakes in Africa, a fact which mainly results from its indolent nature. Whilst other and more active Snakes will move rapidly away upon the approach of man, the Puff Adder will frequently lie still, either too lazy to move, or dozing beneath the warm sun of the south. This reptile attains a length of four feet, or four feet six inches, and some specimens may be found even longer; its circumference is as much as that of a man's arm. Its whole appearance is decidedly indicative of venom. Its broad ace-of-clubs-shaped head, its thick body, and suddenly tapered tail, and its chequered back, are all evidences of its poisonous nature. It derives its popular name from its practice of puffing out or swelling the body when irritated.

In a country so infested with poisonous snakes as are some portions of South Africa, it is surprising that there are not more instances of lives having been lost by this means. It is, however, as rare to hear of a person having been bitten and dying from the bite of a poisonous Snake in South Africa as it is to hear of a death in England from the bite of a mad dog. The fact, however, is that all Snakes will, if possible, make their escape when man approaches them, and it is merely when they are trodden upon, or are oppressed by their own superabundant poison, that they are disposed to bite an animal unsuited for their food.

PUFF ADDER.—*Clotho arietans.*

An infuriated Puff Adder presents a very unprepossessing appearance. I once saw a female of this species in a most excited state. She had been disturbed in her retreat under an old stump by some Kaffirs, who were widening the highroad through the Berea bush at Natal. She had several young ones with her, and showed fight immediately she was discovered. The Kaffirs were determined to kill the whole family, but were fearful of approaching her. Happening to pass at the time of the discovery, I organized a ring, and, procuring some large stones, directed the Kaffirs to open fire. After a few minutes the excited lady was killed, and she and her young were carefully buried in a retired locality, lest some bare-footed Kaffir might tread upon her head, and thus meet his death."

There is certainly in nature no more fearful an object than a full-grown Puff Adder The creature grovels on the sand, winding its body so as to bury itself almost wholly in the tawny soil, and just leaving its flat, cruel-looking head lying on the ground and free from sand. The steady, malignant, stony glare of those eyes is absolutely freezing as the creature lies motionless, confident in its deadly powers, and when roused by the approach of a passenger, merely exhibiting its annoyance by raising its head an inch or two, and uttering a sharp angry hiss. Even horses have been bitten by this reptile, and died within a few hours after the injury was inflicted. The peculiar attitude which is exhibited in the illustration is taken from life, one of the Puff Adders in the collection of the Zoological Society having been purposely irritated. In the background is seen another individual of the same species, as it usually lies, half-buried in the sandy soil.

It is rather curious that the juice of tobacco is an instant poison to these creatures, even more suddenly deadly to them than their poison to the human beings who can absorb the tobacco juice with impunity. The Hottentots will often kill the Puff Adder by spitting in its face the juice of chewed tobacco, or making it bite the end of a stick which has been rubbed in the tobacco oil found in all pipes that have been long used without being cleaned.

The Bushmen are in the habit of procuring from the teeth of this serpent the poison with which they arm their tiny but most fearful arrows. In the capture of the Puff Adder they display very great courage and address. Taking advantage of the reptile's sluggish habits, they plant their bare feet upon its neck before it has quite made up its reptilian mind to action, and, holding it firmly down, cut off its head and extract the poison at their leisure. In order to make it adhesive to the arrow point, it is mixed with the glutinous juice of the amaryllis.

There seems to be no certain remedies for the bite of the Puff Adder. Ammonia appears to be the least inefficacious substance for that purpose, and the natives occasionally attempt to heal the injury by splitting a living fowl across the breast, and applying the still-palpitating halves to the wound. There is a kind of seed called the "gentleman bean," which is said to have a beneficial effect. If one of these beans be placed on the recently inflicted wound, it adheres with great firmness, and is said to absorb the poison from the system, and to fall off as soon as this object is achieved. The Bushmen are in the habit of swallowing the poison whenever they kill a Puff Adder and do not need its venomous store for their arrows, hoping thereby to render themselves proof against its effects. When examined under the microscope, the poison resolves itself into minute crystalline spiculæ, not unlike those of Epsom salts, which must be kept perfectly dry or they will soon vanish from the glass on which they are placed.

The colour of the Puff Adder is brown, chequered with dark brown and white, and with a reddish band between the eyes. The under parts are paler than the upper.

SEVERAL other deadly serpents of the same country are closely allied to the puff adder. The first is the DAS ADDER or RIVER JACK (*Clotho nasicornis*) of the colonists, remarkable for the long curved horn or spine upon the nose, formed by the peculiar development of the scales over the nostril. This curious structure is only found in the male. In colour it is much darker than the puff adder, being black, marbled with a paler hue, and decorated with sundry lozenge-shaped spots along the back.

The BERG ADDER (*Clotho Atropos*) is another of these fearful reptiles. As its name denotes, it is found more among the hills and stony ranges than on the plains, but is not unfrequently found upon the flats, and will sometimes intrude into very awkward positions, such as the floor of a hut, or even the bed upon which some wearied man is about to cast himself. It is not quite so poisonous as the puff adder, though its looks are quite as unprepossessing, and it never bites unless purposely irritated, or trodden upon.

It is an ugly, thick-bodied, slow crawling creature, with a suddenly tapering tail and a most evil looking head. It is not a large reptile, its average length being about eighteen inches. Its colour is olive-grey, marbled on the sides, and decorated along the back with four rows of dark squared spots.

YET one more species of this genus deserves a passing notice. This is the HORNED ADDER (*Clotho cornúta*), sometimes, but erroneously, called the Cerastes, a term that is rightly applied to another Serpent shortly to be described. It sometimes goes by the popular name of HORNSMAN. It derives its name of Horned Adder from the groups of little thread-like horns that are seen on the head, one group appearing above each eye. In some works of Natural History, it is called the PLUMED VIPER, in allusion to these curious groups. It is not very graceful in form, being decidedly short, squat, and puffy in shape, but is very prettily marked, its body being richly marbled with chestnut, covered with a multitude of minute dots, and variegated with four rows of dark spots along the back, two rows running on each side of the vertebral line.

THE true CERASTES or HORNED VIPER is a native of Northern Africa, and divides with the cobra of the same country the questionable honour of being the "worm of Nile" to whose venomous tooth Cleopatra's death was due.

The bite of this most ungainly looking Serpent is extremely dangerous, though, perhaps, not quite so deadly as that of the cobra, and the creature is therefore not quite so much dreaded as might be imagined. The Cerastes has a most curious appearance, owing to a rather large horn-like scale which projects over each eye, and which, according to the natives, is possessed of wonderful virtues. They fancy that one of the so-called horns contains the supply of poison for the teeth, and that the other, if pounded and the powder rubbed over the eyelids, will enable the fortunate experimenter to see all the wealth of the earth—a privilege which, according to the peculiar cast of the Oriental mind, is of nearly as much value as the actual possession. The reader may remember a tale in the "Arabian Nights" in which a similar story is narrated.

The Cerastes has, according to Bruce, an awkward habit of crawling until it is alongside of the creature whom it is about to attack, and then making a sidelong leap at its victim. He relates an instance where he saw a Cerastes perform a certainly curious feat. "I saw one of them at Cairo crawl up the side of a box in which there were many, and there lie still as if hiding himself, till one of the people who brought them to us came near him, and though in a very disadvantageous position, sticking, as it were, perpendicularly to the side of the box, he leaped near the distance of three feet, and fastened between the man's forefinger and thumb, so as to bring the blood."

CERASTES, OR HORNED VIPER.—*Cerastes Hasselquistii.*

The man who was thus bitten happened to be one of the men who profess Serpent charming, and avow themselves to be proof against the bite of any poisonous Snake. In this instance no ill effects followed the hurt, although Bruce proved that the poison-fangs had not been extracted, by making the reptile bite a pelican, which died in about thirteen minutes. Some persons have suggested that in this, as well as in other similar instances, the man was a clever juggler, who substituted a really venomous specimen for a Snake whose poison-fangs had been extracted. But in any case it would be necessary to handle the really poisonous reptile for the purpose of effecting the exchange, and, in my opinion, the necessarily rough handling of the creature would be a matter of no small danger. Bruce enters into this subject at some length, and records the result of a long series of experiments in a form which, though very interesting, is now so familiar as to need no quotation.

That in many instances the poison-teeth of venomous Serpents have been extracted, in order to allow the performer to play his tricks with them without harm, is very well known, but the fact of acknowledged and detected imposture does not invalidate the reality which is clumsily imitated by pretenders, any more than a forgery disproves the existence of a genuine document. More will be said on this subject when we come to the different species of cobra.

The Cerastes usually lives in the driest and hottest parts of Northern Africa, and lies half-buried in the sand until its prey should come within reach. Like many Serpents, it can endure a very prolonged frost without appearing to suffer any inconvenience ; those kept by Bruce lived for two years in a glass jar without partaking of food, and seemed perfectly brisk and lively, casting their skins as usual, and not even becoming torpid during the winter.

The colour of the Cerastes is pale brownish white, covered irregularly with brown spots. Its length is about two feet.

PASSING to another genus of venomous Snakes, we come to a rather pretty little Serpent, an inhabitant of India, and called by the natives HORATTA PAM (*Echis carináta*). It is said to be very dangerous in spite of its small dimensions, and to require a double dose of Serpent medicine in order to counteract the effects of its poison. Its colour is greyish brown, darkening into rather deep brown on the head, and variegated with angular white streaks on the body, and large oblong spots on the head, edged with a deeper hue. Its length is about fifteen or sixteen inches.

THE common ASP or CHERSÆA (*Vipera aspis*) is nearly allied to the preceding species.

This Snake is common in many parts of Europe, and is plentiful in Sweden and the neighbouring countries, besides being distributed over nearly the whole continent. It is much dreaded, and with reason, for its bite is very severe, and in some cases will cause death. As is the case with other venomous reptiles, the Asp is most dangerous during the hottest months of the year, and it has well been remarked that there is probably some connexion between the electrical state of the atmosphere, and the venom of Serpents, as the poison is always most deadly and the creatures most fierce when the electrical condi tions of the atmosphere are disturbed, and the thunder-clouds are flying quickly through the air. When a person is bitten in one of his limbs, he quickly digs a hole and buries the injured part below the surface of the earth, as the fresh mould is thought to be very efficacious in alleviating the ill effects of the poison. Should the injury be in a toe or a finger, the rougher but more effectual remedy of instant amputation is generally employed.

The colour of this reptile is olive above, with four rows of black spots. The two middle rows are often placed so closely together, that they coalesce and form a continuous chain of black spots along the spine, very like the well-known markings of the common viper.

ANOTHER venomous Snake, the AMMODYTE or SAND-NATTER (*Vipera Ammodýtes*). belongs to the same genus as the asp.

This reptile inhabits southern Europe, and is generally found in rocky localities. The bite of this creature is very dangerous, and the remedies employed are generally of little efficacy. Enlarging the wound with a thorn, and squeezing a garlic upon the part bitten, is the general mode of alleviating the pain, but is of little use to the injured person. Its colour is olive above, with a broad oblique dark streak on each temple, two similar streaks on each side of the head, and a wavy dark line along the crown of the spine.

THE common VIPER, or ADDER, is very well known in many parts of England, but in some localities is very plentiful, while in others it is never seen from one year's end to another.

Many persons mistake the common grass Snake for the Viper, and dread it accordingly. They may, however, always distinguish the poisonous reptile from the innocuous, by the chain of dark spots that runs along the spine, and forms an unfailing guide to its identifi- cation. Fortunately for ourselves, it is the only poisonous reptile inhabiting England, the variously-coloured specimens being nothing more than varieties of the same species.

Like most reptiles, whether poisonous or not, the Viper is a very timid creature, always preferring to glide away from a foe rather than to attack, and only biting when driven to do so under great provocation.

The following interesting account of a Viper's bite and its consequences, has been kindly forwarded to me by Mr. W. C. Coleman :—

"Several years ago, in my school-boy days, I had an experience with a Viper, which may possibly interest such of your readers as have not enjoyed a similar intimacy with the creature, especially as it places the Viper character in a somewhat more amiable light than it is usually represented.

One cold damp day in the beginning of May, I was out in the country on a foraging expedition; birds' nests and objects of natural history in general being the objects of search. Entering, in the course of exploration, a likely coppice, I descried a blackbird's nest perched among some tangled stems of underwood three or four feet from the ground. A glance at the interior, however, soon showed that some other marauder had forestalled me, as the sole occupants of the nest were some crushed and empty egg-shells, and scanty remains of the fluid contents spilt about. "A weasel," thought I, but wrongfully, as it happened, for on turning away in dudgeon, a rustling movement among the herbage on the ground a couple of yards off, attracted my eyes and ears; and there I saw the undoubted spoiler of the nest, a large Viper, moving away briskly with his tail in the direction of the nest.

VIPER, OR ADDER.—*Pelias Berus.*

A little knowledge is a dangerous thing, and my slight natural history reading, assisted by bad engravings, had helped me to fancy that I knew the Viper from the common Snake well enough; and so, deciding that this was only a common harmless Snake, I made a plunge at the creature and apprehended him with my unprotected hand. Receiving no bite, I was now confirmed in my idea of the beast's perfect innocence (except in the bird's-nest matter), and decided on adopting him as a pet. So presently set off home, a distance of more than two miles, taking my serpentine friend in my hand. Not always in my hand, however, for to beguile the homeward journey I proceeded to try sundry experiments on the supple backbone and easy temper of the animal, occasionally tying him round my neck, and so wearing him for a considerable distance; then twining him round my wrist into a fancy bracelet, and weaving him into various knots and devices according to taste, all this with perfect impunity on my part, and the utmost apparent good humour on his.

On the road, a kind farmer of my acquaintance, whose natural history lore was more practical than my own, endeavoured to convince me that I was 'harbouring a Viper in my bosom,' but I was not going to hear my good-tempered playmate called bad names; put my finger into the Adder's very mouth to prove he had no idea of biting, and so passed on, in much conceit with myself as an accomplished herpetologist.

We thus reached home in perfect safety and amity. My brothers and sisters greeted the stranger with some little instinctive horror at first, but got over that feeling when they heard of his innocent nature and amusing capabilities, in proof of which I repeated the necktie experiment, &c. About this stage, however, I must mention that he exhibited a somewhat unpleasant phenomenon common to the Snake tribe in general, who can relieve themselves of the torpor consequent on a heavy meal, by disgorging the same when irritated and requiring restoration of their usual activity. The rejectamenta in this case consisted of portions of unhatched young birds, thus confirming the nest robbery.

Being thus lightened, and perhaps stimulated by the warmth of a fire in the room, he was now lively enough, unhappily for me, for on essaying to continue my experiments, by tying him into a double knot, his endurance was at an end ; one dart at my finger and a sharp puncture told me that the thing was done. Then, too late, I recollected that the ' Adder is distinguished by a zigzag chain of dark markings down the back,' and sure enough the vile creature before me had those very marks. In a rage, I battered his life out with a stick, lest he should do more damage, and then settled down to watch the progress of the poison within my system.

It was not slow to take effect ; first the wound looked and felt like a nettle sting, then like a wasp sting, and in the course of a few minutes the whole joint was swollen, with much pain. At this juncture my father, a medical man, arrived from a country journey, and set the approved antidotes to work, ammonia, oil, and lunar caustic, to the wound, having previously made incisions about the punctured spot, and with paternal affection attempted to draw out the poison by suction ; but nothing availed, and all sorts of horrid symptoms set in, fainting, sickness, delirium, and fever ; the hand and whole arm to the shoulder greatly swollen and discoloured, with most intense pain. This state of things lasted for several days. I forget the exact time, but I was not fully restored for more than a fortnight after the bite.

Since that day I have taken care to put my acquaintance with Serpents on such a footing as to be able at a glance to tell the species of any of our three English Snakes ; a piece of useful knowledge most easily gained, and well worth the acquirement."

It was a most providential circumstance that the reptile did not bite him immediately after its capture, and that the wound was inflicted on the finger and not on the neck, as in the one case he could hardly have reached his home, and in the other, the great swelling might have caused suffocation, as is known to be the case with persons bitten in the neck by other poisonous Serpents.

A FEW words will not be out of place respecting the alleged capability of the Viper of receiving its progeny into its mouth when in danger.

A long-standing controversy on this subject has elicited a vast amount of correspondence, the whole of which seems to resolve itself into two divisions, namely, communications from a great number of persons who assert that they have seen the young Vipers crawl into their parent's open mouth, and letters from two or three persons who say that they did not do so, because such a proceeding is impossible, and contrary to the laws of nature.

One of the most learned of the objectors remarks, that no amount of testimony can prevail against reason, and that the persons who assert that they have seen the young Vipers crawl into their mother's mouth, have fallen into the dangerous fallacy of believing what they saw. Now this argument, novel though it may be to the scientific world in general, is perfectly familiar to theologians as being the sheet-anchor of a certain school of controversialists, who deny the credibility of the miraculous events narrated in the Scriptures. It has been repeatedly exploded in polemical controversy, and long abandoned by impartial thinkers, inasmuch as it assumes a knowledge of all the laws of nature, and contracts the power of the Divine Creator of the Universe within the narrow limits of the individual idiosyncracy and mental capacities of the disputant.

It has ever been conceded that, in all ages, the testimony of credible witnesses has been the surest mode of confuting false reasoning and thereby eliciting truth ; so that when any unprejudiced reasoner finds that a favourite theory is contradicted by the testimony of even one trustworthy observer, much more when the united accounts of many competent judges all tend to the same point, he feels that it is time for him to reflect whether, however perfect may be the form of his syllogism, there may not be something wrong with his premises. Reasoning is more liable to falsity than the senses to deception. It is easy enough to talk of a flagrant violation of the laws of nature, but before we venture to do so it is as well to be quite certain that we are sure of the full extent of those laws. Who is there, even among the most learned, that can define the full working of even a single known law and its ever-varying action under different circumstances? And who can venture to say that some hitherto unrecognised law may not be in existence, which, if known and acknowledged, would account for the circumstances which at present seem so unaccountable?

In the second place, if we are not to depend upon the testimony of our acknowledged senses, on what are we to depend for the whole of natural philosophy, astronomy, or, indeed, any other established science? It is simply on the testimony of our senses that all existing sciences are founded, and even analogous reasoning is not admitted as valid proof of an asserted fact. There is hardly any new discovery which does not destroy some old and respectable theory, and give entirely a new idea of the law of nature on which it depends.

The operation of the senses is in itself one of the known laws of nature, by which we discover facts and through which we are enabled to exercise our reasoning faculties. A human being without the senses of sight, hearing, and touch, would be the dullest animal on the face of the earth, and as long as the privation lasted, would hold a lower place than a sponge or a medusa. If we once acknowledge that the evidence of the senses is not to be believed, we must reject the whole of the physical sciences. Astronomical observations, chemical experiments, geological surveys, anatomical researches, and the whole of natural history, must be at once thrown aside if such a theory is to be consistently carried out; and for the same reason, the courts of law must be abolished, depending as they do on the personal observations of human beings, mostly illiterate, and often ignorant to a degree. Repeated observations are the only method of ascertaining the laws of nature, and if they show that certain events, however strange they may appear, have really occurred, they surely prove, not that the senses of the witnesses were deceived, but that another law of nature has been discovered.

Were the Viper the only creature of whom such an act is related, the phenomenon would be less worthy of belief; but there is hardly a poisonous Snake of any country by whom the same act is not said to be performed, the narrators not being professed naturalists with a theory, but travellers, hunters, and settlers, casually noting the result of their personal experience. I cannot but think that the accumulated testimony of many trustworthy persons, acting independently of each other, accustomed to observation, and mostly unaware of the importance that would be afterwards attached to their words, is entitled to some respect, and affords legitimate grounds to the truth-seeker, not for contemptuous denial, but for further investigation.

Several observant inhabitants of the Forest of Dean assert that both sexes assume this protective habit, the male as well as the female receiving the young into the mouth in cases of sudden danger. In those localities, the head of the Viper is always chopped off as soon as the reptile is killed, and the Viper-catchers say that in such cases the young Vipers frequently are seen crawling out of the severed neck.

I certainly never saw the Viper act in this manner, but I have had very few opportunities of watching this reptile in a wild state and noting its habits; whereas those who spend their lives in the forests, and especially those men who add to their income by catching or killing these reptiles, speak of the reception of the young into the mouth of the parent, as a fact too well known to be disputed.

It has been objected that the young would be consumed by the gastric juice of the parent—one of the most sensible objections that has been made. But this assertion has been invalidated by the researches of able anatomists and experimentalists, such as Mr. F. T. Buckland, &c., who have discovered, by careful dissection, two facts; the one, that the young may be concealed within the expansile body of the parent without entering the true stomach at all, the œsophagus or gullet forming a highly expansile antechamber between the throat and the actual stomach ; and the other, that if they should happen to do so, the gastric juice would not hurt them. Incredible, therefore, as the possibility of such an act may seem, it can but be acknowledged that the weight of practical testimony is wholly in its favour. Moreover, the various suggestions offered to account for the deception practised by the Viper upon the eyes of observers, just as if it had been a professed conjurer performing before an audience, are really puerile in the extreme, and if they happen to affect the written testimony of one person, they are contradicted by the written testimony of another. It is to be hoped that if the Viper really does act in the manner stated, a specimen may be obtained with the young still within her body, and attested in such a manner that no objector may invalidate the proof by saying that the old one had been captured and the young pushed down her throat by force.

The head of the Viper affords a very good example of the venomous apparatus of the poisonous Serpents, and is well worthy of dissection, which is better accomplished under water than in air. The poison-fangs lie on the sides of the upper jaw, folded back and almost undistinguishable until lifted with a needle. They are singularly fine and delicate, hardly larger than a lady's needle, and are covered almost to their tips with a muscular envelope through which the points just peer. The poison-secreting glands and the reservoir in which the venom is stored, are found at the back and sides of the head, and give to the venomous Serpents that peculiar width of head which is so unfailing a characteristic. The colour of the poison is a very pale yellow, and its consistence is very like that of salad oil, which, indeed, it much resembles both in look and taste. There is but little in each individual, and it is possible that the superior power of the larger venomous Snakes of other lands, especially those under the tropics, may be due as much to its quantity as its absolute intensity. In a full-grown rattlesnake, for example, there are six or eight drops of this poison, whereas the Viper has hardly a twentieth part of that amount.

On examining carefully the poison-fangs of a Viper, the structure by which the venom is injected into the wound will be easily understood. On removing the lower jaw, the two fangs are seen in the upper jaw, folded down in a kind of groove between the teeth of the palate and the skin of the head, so as to allow any food to slide over them without being pierced by their points. The ends of the teeth reach about halfway from the nose to the angle of the jaw, just behind the corner of the eye.

Only the tips of the fangs are seen, and they glisten bright, smooth and translucent, as if they were curved needles made from isinglass, and almost as fine as a bee's sting. On raising them with a needle or the point of the forceps, a large mass of muscular tissue comes into view, enveloping the tooth for the greater part of its length, and being, in fact, the means by which the fang is elevated or depressed. When the creature draws back its head and opens its mouth to strike, the depressing muscles are relaxed, the opposite series are contracted, and the two deadly fangs spring up with their points ready for action. It is needful while dissecting the head to be exceedingly careful, as the fangs are so sharp that they penetrate the skin with a very slight touch, and their poisonous distilment does not lose its potency even after the lapse of time.

The next process is to remove one of the teeth, place it under a tolerably good magnifier and examine its structure, when it will be seen to be hollow, and, as it were, perforated by a channel. This channel is however seen, on closer examination, to be formed by a groove along the tooth, which is closed except at the one end whence the poison exudes, and the other at which it enters the tooth. If the tooth be carefully removed, and the fleshy substance pushed away from its root, the entrance can be seen quite plainly by the aid of a pocket lens. The external aperture is in the form of a very narrow slit upon the concave side of the fang, so very narrow, indeed, that it seems too small for the passage of any liquid.

There are generally several of the fangs in each jaw, lying one below the other in regular succession. From the specimen which has just been described, I removed four teeth on each side, varying in length from half to one-eighth the dimensions of the poison-fangs.

3.

DEATH ADDER.—*Acánthophis antárctica.*

The Viper seems to be well aware of the power of its fangs, and to discriminate between animate and inanimate antagonists. I have tried in vain to make a Viper bite a stick with which I was irritating it, but no sooner did a kitten approach than the reptile drew back its head, and made its lightning-like dart at the little creature with such rapidity, that it would have gained its point had not its back been so much injured as to deprive it of its natural powers.

The ordinary food of the Viper is much the same as that of the common Snake, and consists of mice, birds, frogs, and similar creatures. It is, however, less partial to frogs than the common Snake, and seems to prefer the smaller mammalia to any other prey. The young of the Viper enter the world in a living state, having been hatched just before they are born. The fat of the Viper was once in high estimation as a drug, and the older apothecaries were accustomed to purchase these reptiles in considerable numbers. Even now this substance is in some repute in many agricultural districts, being employed as a remedy for cuts, sprains or bruises, and especially as a means of alleviating the painful symptoms of a Viper's bite.

The colour of the Viper is rather variable, but the series of very dark marks down the back is an unfailing sign of the species, and is permanent in all the varieties. Generally the ground colour is greyish olive, brown, or brownish yellow; along the back runs a chain of zigzag blackish markings, and a series of little triangular spots is found upon each side. The largest specimen I have yet seen in a wild state was one of the yellow varieties. Sometimes the ground is brick-red, and now and then a nearly black specimen is found. Mr. Bell mentions an example where the ground colour was greyish white, and the markings jetty black.

THE reptile that is called by the significant title of DEATH ADDER, or DEATH VIPER, is a native of Australia, where its poisonous fangs render it an object of much fear. A very excellent, though short description of this Snake, is given by Mr. Bennett in his "Wanderings in New South Wales."

"The most deadly Snake in appearance, and I believe also in effect, is one of hideous aspect, called by the colonists the Death Adder, and by the Yas natives 'Tammin,' from having a small curved process at the extremity of the tail; or, more correctly, the tail terminating suddenly in a small curved extremity, bearing some resemblance to a sting; it is considered by popular rumour to inflict a deadly sting with it.

This hideous reptile is thick in proportion to its length; the eye is vivid yellow, with a black longitudinal pupil. The colour of the body is difficult to be described, being a complication of dull colours, with narrow, blackish bands shaded off into the colours which compose the back; abdomen slightly tinged with red; head broad, thick, and flattened. The specimen I examined measured two feet two inches in length, and five inches in circumference. A dog that was bitten by one died in less than an hour. The specimen I examined was found coiled up near the banks of the Murrumbidgee river; and being of a torpid disposition, did not move when approached, but quietly reposed in the pathway, with its head turned beneath its belly."

The generic title of Acanthophis, or Thorny-Snake, is given to this species on account of the structure of the tail, which is furnished at its extremity with a recurved horny spine.

We now arrive at a very remarkable family of Snakes, which pass their lives in water, either fresh or salt, and are river or sea Serpents as the case may be. In order to enable them to pass through the waters without injury to the organs of respiration, the nostrils are furnished with a valve so as to prevent the ingress of water while the creature is below the surface.

A good example of these marine Serpents is the BLACK-BACKED PELAMIS (*Pélamis bicolor*), the Nalla Whallagee Pam of the Indian fishermen. This Snake is found only at sea, and is said seldom if ever to approach the shore, except for the purpose of depositing its eggs, which are laid on the beach sufficiently near high-water mark for the young Snakes to seek their congenial element as soon as they are hatched. The Black-backed Pelamis is frequently found sleeping on the surface of the sea, and is then caught without much difficulty, as it is forced to throw itself on its back before it can dive. It has been suggested that this movement is intended to expel the air in the ample lungs. Sometimes it is unwillingly captured by the fishermen in their nets, and is an object of considerable dread to them on account of the formidable character of its teeth. In these Serpents the fangs are but little larger than the other teeth of the jaw, but can be distinguished by their slightly superior size and the groove that runs along their front edge.

In this, as in many other species of water Snake, the tail is flattened, or rather compressed, for a considerable portion of its length, and forms a powerful organ of propulsion. The colour of this species is black above and yellow below and on the flanks. The upper lip is also yellow, and the tail is of the same hue, variegated with black on the edges and spotted with black on the sides. Its average length is rather less than one yard.

The SHOOTER SUN (*Hydrophis obscúra*) is another of the sea Serpents. This reptile is also one of the Indian species and inhabits the sea or the saline waters of the river-mouths, not being able to exist in fresh water. It is an admirable swimmer, but is very awkward on dry land, and cannot survive for any length of time unless it has access to salt water. The outline of this Serpent is most remarkable. The head and neck are almost absurdly minute in proportion to the wide thick body, bearing about the same proportion as the tip of the little finger does to the wrist. The tail is also very wide, extremely blunt, and compressed.

The markings of this reptile are rather curious. The ground colour is black. There is a large yellow spot on each side of the head, a series of pale grey-brown spots runs on each side of the neck, and a row of large rounded white marks is arranged along the back so as to form a richly variegated pattern of boldly contrasted colours.

The CHITTUL (*Hydrophis sublœvis)* is another of these marine Snakes, and is found in India and Ceylon. It is rather a large species, sometimes exceeding five feet in length

ACROCHORDE.—Acrochordus Javanicus.

and is handsomely coloured. It is extremely venomous, a fowl that had been bitten by a Chittul dying within five minutes after receiving the injury. The ground colour of this Snake is yellow, and the body is covered with an irregular row of black rings. Some black bands also cross the neck.

In the ACROCHORDE, sometimes called the Oular Carron, the tail, instead of being flattened, is rounded, conical, and very short, diminishing in diameter in a very sudden manner. It is a native of Java, and is said to be wholly vegetarian in its diet, the stomach having been found to contain nothing but half-digested fruit. The flesh of the Acrochorde is said to be excellent.

Upon the head are a number of little scales, each of which is divided into three ridges. The creature is in the habit of distending its body with air to a very great extent, and when it so acts, the scales separate from each other and make the head and body look as if they were covered with tubercles. The general colour is brown in the adult, and brown banded and streaked with a darker hue in the young.

The CHERSYDRUS (*Chersydrus granulátus*) is a rather curious aquatic Serpent, found in Asia and most common in Java. It is sometimes called the Banded Acrochorde, but wrongly so, as its tail, instead of being round and conical, is flat, compressed, and sword-like in shape. It inhabits the bottoms of marine creeks and the mouths of rivers. The Javanese call it Oular Limpè. The body of this reptile is covered with small scales, each boldly keeled in the centre, and its colour is black and white arranged in alternate rings.

The ERPETON, or HERPETON as the name is sometimes written, is a truly curious reptile, of no great size, but bearing a pair of appendages on the head that seem to serve no recognised purpose save to bewilder zoologists. The muzzle of this creature is covered with scales, and on each side of it rises the curious appendage which is seen in the illustration. This remarkable organ is soft, but completely covered with scales and defended by them. Of its habits nothing appears to be known, and even its country is dubious. Its colour is pale brown streaked with white.

ERPETON. — *Erpeton tentaculatum*

THE sombre and rather unsightly CERBERUS, better known by its native name of KAROO BOKADAM, is an Asiatic reptile, being found in India, the Philippines, Ceylon, Borneo, and similar countries. It is an ugly looking Serpent, but is not much dreaded, and is thought to be practically non-venomous. It is a stout, thick-bodied Snake, with a very large head in proportion to the size of its neck, though small in comparison with the body. The mouth is not large, and the teeth are small, regular, and set rather closely together. The nostrils of this Serpent are very small, and placed close to each other almost on the very tip of the muzzle. The eyes are small, round, and projecting as if squeezed out of the head, and are surrounded by a curious circle of nearly triangular scales, much as a circular window in a brick wall is edged with wedge-shaped bricks.

The general colour of this Serpent is greyish brown above, covered with narrow bands of black set rather closely together. The abdomen is black mottled with yellow, the sides are white with spots of pale brown, and the lips and throat are of the same tint but spotted with black. The tail is nearly black. The usual length of this Serpent is about three feet six inches.

WE now arrive at a very important family of serpents, including the largest species found in the order. These Snakes are known by the popular title of Boas, and scientifically as Boidæ, and are all remarkable, not only for their great size and curious mode of taking their prey, but for the partial development of the hinder limbs, which are externally visible as a pair of horny spurs, set one on each side at the base of the tail, and moderately well developed under the skin, consisting of several bones jointed together. In most of the species the tail is rather short and strongly prehensile. The peculiar habits of these enormous Snakes will be mentioned in connexion with the various species. The first of these creatures is the DIAMOND SNAKE of Australia, (*Morélia spilótes*), a very handsome species and tolerably common. It is called the Diamond Snake on account of the pattern of the colours, which are generally blue, black, and yellow, arranged so as to produce a series of diamonds along the back. The CARPET SNAKE (*Morelia variegata*), of the same country is closely allied to it. Both these reptiles are variable in their colouring.

KAROO BOKADAM.—*Cérberus cinéreus.*

THE members of the restricted genus PYTHON are remarkable for their habit of depositing the eggs together and coiling their bodies round them, so as to form a large conical heap. The common grass Snake of England is said to perform the same feat. The true Pythons are inhabitants of Asia, and are generally found in India. The common ROCK SNAKE of India (*Python molúrus*), is a good example of this genus. The natives believe that the little spurs are useful in fighting, and therefore cut them off whenever they capture the reptile. It is the Pedda-Poda of the Hindoos. It is not one of the largest of its kind, usually attaining a length of ten or eleven feet, and not being held in much dread. A fowl that was inclosed in a cage with one of these serpents, soon obtained the mastery over her terrible companion, and was seen quietly pecking at its head.

One of these reptiles that was kept at the gardens of the Zoological Society, once made a curious mistake while being fed, and had well-nigh sacrificed the life of its keeper. The man had approached the reptile with a fowl in his hand and presented it as usual to the Snake. The Serpent darted at the bird, but as it was just then shedding its skin and nearly blind, it missed its aim, and instead of seizing the bird, grasped the keeper's left thumb, and instinctively flung its coils around his arms and neck, as is customary when the animal seized is of considerable size.

The keeper tried to force the Snake's head from its hold, but could not reach it as he was bound in the folds of the Snake. He then cast himself on the ground in order to battle to the greatest advantage, but would probably have succumbed to the fearful pressure, had not two keepers providentially entered the room, and by breaking away the Serpent's teeth released the man from his terrible assailant. Except the fright and a few wounds from the serpent's teeth, no evil results ensued.

ANOTHER species of Indian Rock Snake, called by the natives ULAR SAWA (*Hypsirhína aër*), is tolerably common, and in its habits resembles the preceding species. It often attains to a very considerable size, and is said when full-grown to be about thirty feet in length. This terrible Snake has been known to kill mankind, crushing the body in its

NATAL ROCK SNAKE.—*Hortulia Natalensis.*

numerous folds until nearly every bone was broken. In one such instance, the man had been caught by the right wrist, as was seen by the marks of the Serpent's teeth.

The handsome NATAL ROCK SNAKE, or PORT NATAL PYTHON, as it is sometimes called, now comes under our notice. It is a fine, handsome species, sometimes attaining a great length, and being most beautifully coloured. During life and when in full health and in the enjoyment of liberty, this, in common with many other Snakes, has a beautiful rich bloom upon its scales, not unlike the purple bloom of a plum or grape. Should, however, the Snake be in ill health, this bloom fades away, and in consequence, we seldom if ever see it on the scales of the Serpents which have been brought to England, and are kept in glass-fronted cases in lieu of the wide desert, and only a blanket to creep into instead of the rocky crevices of their native country.

The dimensions of this reptile are often very great. Dr. A. Smith has seen a specimen measuring twenty-five feet in length, exclusive of a portion of the tail which was missing. Flat skins of this creature are, however, very deceptive, and cannot be relied upon, as they stretch almost as readily as India-rubber, and during the process of drying are often extended several feet beyond the length which they occupied while surrounding the body of their quondam owner.

The teeth of this Serpent are tolerably large, but not venomous, and although of no insignificant size, are really of small dimensions when compared with the size and weight of their owner. Few persons have any idea of the exceeding heaviness of a large Snake, and unless the reptile has been fairly lifted and carried about, its easy gliding movements have the effect of making it appear as if it were as light as it is graceful.

Both jaws are thickly studded with these teeth, and their use is to seize the prey and hold it while the huge folds of the body are flung round the victim, and its life crushed out of its frame by the contracting coils. In order to secure its prey, the Rock Snake acts after the manner of all this family. It waits in some spot where it knows that its victim will pass, coils its tail round some object, such as a tree or a stone, so as to give it a firm hold, and then, rapidly darting at the prey, it draws back its head, carrying the poor victim into the fatal grasp of its folds. It usually seizes by the throat, and retains its hold until the crushed animal is quite dead.

The following interesting account of the Rock Snake of Natal has been kindly forwarded to me by Captain Drayson, R.A.:—

"The Rock Snake is somewhat rare, even in the least populous districts, and, in consequence of its retired habits and silent method of moving, it is not frequently seen. Although on an average I traversed the forests and plains near my various stations at least five times a week, I saw but seven Rock Snakes during a period of nearly three years. This Snake retreats into rocky crevices, or amongst the most tangled brushwood, after it has devoured its prey, which consists of toads, frogs, lizards, such as guanas, &c., birds of any size, and even small bucks. Its bite is quite harmless compared to that of the poisonous Snakes, and it destroys its victims by pressure.

So cautious is this Snake to remain quite quiet if it think itself unseen, that on one occasion I nearly rode over a rather large Boa, which lay on a small path along which I was riding. On each side of this path there was a dense jungle, and there was merely room for one animal to travel along it. I happened to 'pull up' my pony to examine the surrounding bush, when I noticed that his erected ears indicated that he had seen game, he being a most accomplished shooting pony. Upon looking on the path before me I observed a very large Snake, lying perfectly still, and looking at me in a very suspicious manner. The reptile being partly concealed by the long grass I could not see whether or not it was a poisonous snake, so I quietly 'reined back' about a yard, and shot the creature through the body. The coils and contortions were something terrific to see, as the monster fought hard for his life; but even the bone and muscle of a Boa has but a poor chance against gunpowder and lead. A charge of buck shot in the head settled the business, and cleared the path of a very disagreeable vis-à-vis. This Snake measured about sixteen feet in length, and was in very fair condition, having a fine bloom on his skin. He had resided about a hundred yards from a long vlei (lagoon), in which frogs and lizards abounded.

A much larger Rock Snake was shot by me some time after this, and measured upwards of seven yards. I once had an opportunity which rarely occurs to Englishmen, viz. that of trying my speed with a young Boa constrictor. Upon returning from shooting one afternoon I crossed the Umbilo river near Natal, and shortly after observed a coran flying up and down in a very singular manner. This bird being very good eating, I dismounted, and commenced stalking him, and approached within a few yards of him without being discovered. I then noticed a snake creeping towards the coran, which merely flew on a few feet and then settled again. The Snake again approached the bird, which, however, seeing me, became disenchanted, and was making its escape when I shot it, and then turned my attention to the Snake, which remained quite still. I soon saw that the animal was a young Rock Snake about twelve feet long, and, being desirous to obtain a live specimen of this reptile, I ran to my pony, where on the saddle I had a long leather strap, with which I hoped to noose the young Boa.

Upon returning to the scene of the coran's death, I found the Snake making off as fast as he could towards a clump of thick bush. Immediately starting after him, I headed him after a race of about sixty yards, when he turned and tried another direction. I failed in noosing him, and, finding that he would probably escape into the bush, I was compelled to knock him on the head with a dead branch which happened to be near me. I believed him to have been killed outright; but upon conducting a naturalist to the scene on the following morning the Snake had vanished, a fact which, combined with subsequent experience of the Snake nature, induces me to believe that he was merely stunned by the blow, and became refreshed during the cool of the evening, after which he retreated to his stronghold."

ABOMA. — *Epicrates Cénchria.*

The colour of the Natal Rock Snake is olive, variegated with yellow cross-bands and spots, edged with deep black. The head is marked with an arrow-headed spot, and a dark streak runs from the back of the eye. The under parts and the sides of the face are yellow.

There are several other species inhabiting Africa, resembling the preceding creature in general habits and appearance.

THE splendid RINGED BOA of America, sometimes called the ABOMA, has been celebrated for its destructive powers, and in ancient times was worshipped by the Mexicans and propitiated with human sacrifices. Naturally the people of the country would feel disposed to awe in the presence of the mighty Snake whose prowess was so well known by many fatal experiences, and this disposition was fostered by the priests of the Serpent deity, who had succeeded in taming several of these giant Snakes, and teaching them to glide over and around them, as if extending their protection to men endowed with such supernatural powers.

This Serpent destroys its prey after the fashion of its family, merely by squeezing it to death between its folds. While thus engaged, the reptile does not coil itself spirally round the victim, but wraps fold over fold, to increase its power, just as we aid the grasping strength of one hand by placing the other over it. It is said that the Snake can be removed from its prey by seizing it by the tail, and thus unwinding it. Moreover, a heavy blow on the tail, or cutting off a few feet of the extremity, is the best way of disabling the monster for the time.

This creature is rather variable in its colouring, the locality having probably some influence in this respect. Generally it is rich chocolate-brown, with five dark streaks on the top and sides of the head, a series of large and rather narrow dark rings along the back, and two rows of dark spots on the sides. Sometimes a number of large spots are seen on the back, and white streaks on the sides. In all the members of this genus, the hinder limbs or "spurs" of the male are larger and stronger than in the female.

ANOTHER American species, the DOG-HEADED BOA or BOJOBI (*Xiphosôma caninum*), is notable for the formidable armament of teeth which line the mouth, and the beautiful green colour of its skin. As is the case with all the Boidæ, this species is only found in the hottest parts of the country, and is most plentiful in Brazil. It may be known from the other species partly by its green colour, partly by the deep pits on the plates that edge the lips, and partly by the regular ring of scales that surrounds the eye. This Snake is sometimes called the ARARAMBOYA.

WE now come to the BOIGUACU or true BOA CONSTRICTOR, a title which is indifferently applied to all the family, and with some degree of appropriateness, inasmuch as they all kill their prey by pressure or constriction.

This magnificent reptile is a native of Southern and Tropical America, and is one of those Serpents that were formerly held sacred and worshipped with divine honours. It attains a very large size, often exceeding twenty feet in length, and being said to reach thirty feet in some cases. It is worthy of mention, that before swallowing their prey, the Boas do not cover it with saliva, as has been asserted. Indeed, the very narrow and slender forked tongue of the Serpent is about the worst possible implement for such a purpose. A very large amount of this substance is certainly secreted by the reptile while in the act of swallowing, and is of great use in lubricating the prey, so as to aid it in its passage down the throat and into the body, but it is only poured upon the victim during the act of swallowing, and is not prepared and applied beforehand.

The dilating powers of the Boa are wonderful. The skin stretches to a degree which seems absolutely impossible, and the comparison between the diameter of the prey and that of the mouth through which it has to pass, and the throat down which it has to glide, is almost ludicrous in its apparent impracticability, and unless proved by frequent experience, would seem more like the prelude to a juggler's trick than an event of every day occurrence. To such an extent is the body dilatable, that the shape of the animal swallowed can often be traced through the skin, and the very fur is visible through the translucent eyes, as the dead victim passes through the jaws and down the throat.

There is a popular idea among the inhabitants of the country in which the Boa lives, that if it attacks a man in a forest, he may possibly escape by slipping round a tree in such a manner that the Serpent may squeeze the trunk of the tree, mistaking it for the body of the man, and so burst itself asunder by the violence of its efforts. Whether any one has escaped by this rather transparent device is not mentioned.

The colour of the Boa Constrictor is rich brown, and along its back runs a broad chain of large blackish spots of a somewhat hexagonal shape, and of pale white spots scooped at each end. These dark and pale spots are arranged alternately and form a really pretty pattern, and should the colours be faded, as is always the case when the skin has been renewed, the species may be recognised by the arrangement of the scales round the eyes, which are set in a circle, are thirty in number, and are separated from the scales of the lips by two rows of smaller scales.

AN equally celebrated Snake, the ANACONDO, is also figured in the illustration on page 123.

This gigantic serpent is a native of Tropical America, where it is known under several names, La Culebra de Agua, or Water Serpent, and El Traga Venado, or Deer-Swallower being the most familiar. The flesh of this Serpent, although firm and white, is seldom if ever eaten by the natives, although the flesh of Serpents is considered a delicacy by many nations. Within the body is a large amount of fat from which can be obtained a very considerable quantity of oil. This oil is thought to be a specific for many complaints,

BOIGUACU.—*Boa Constrictor.*　　　　ANACONDO.—*Eunectes marinus.*

especially for rheumatism, strains, and bruises. Seven or eight gallons of fine oil can be extracted from one of these reptiles, but the process of draining off the oil is generally performed in so careless a manner, that half of the amount is usually wasted.

Sir R. Ker Porter has some curious remarks on the Anacondo. "This serpent is not venomous nor known to injure men (at least not in this part of the New World);

however, the natives stand in great fear of it, never bathing in waters where it is known to exist. Its common haunt, or rather domicile, is invariably near lakes, swamps, and rivers; likewise close to wet ravines produced by inundations of the periodical rains; hence, from its aquatic habits, its first appellation (*i.e.* Water Serpent). Fish, and those animals which repair there to drink, are the objects of its prey. The creature lurks watchfully under cover of the water, and while the unsuspecting animal is drinking, suddenly makes a dart at the nose, and with a grip of its back-reclining double range of teeth, never fails to secure the terrified beast beyond the power of escape."

Compression is the only method employed by the Anacondo for killing its prey, and the pestilent breath which has been attributed to this reptile is wholly fabulous. Indeed, it is doubtful whether any Snake whatever possesses a fetid breath, and Mr. Waterton, who has handled Snakes, both poisonous and inoffensive, as much as most living persons, utterly denies the existence of any perceptible odour in the Snake's breath. It is very possible that the pestilent and most horrible odour which can be emitted by many Snakes when they are irritated, may have been mistaken for the scent of the breath. This evil odour, however, is produced from a substance secreted in certain glands near the tail, and has no connexion with the breath.

The colour of the Anacondo is rich brown; two rows of large round black spots run along the back, and each side is decorated with a series of light golden yellow rings edged with deep black.

ONE or two members of this family are worthy of a passing notice. The well-known YELLOW SNAKE of Jamaica (*Chilabothrus inornàtus*) is allied rather closely to the boa and the anacondo. It is a rather handsome reptile, being of an olive-green upon the head and front part of the body, covered with a multitude of little black lines, drawn obliquely across the body. The hinder part of the body is black, spotted with yellowish olive.

ANOTHER member of this family, the CORAL SNAKE (*Tortrix Scytale*) is a well-known inhabitant of Tropical America, and is feared or petted by the natives, according to the locality in which it happens to reside. In some parts of the country, the native women, knowing it to be perfectly harmless, and being pleased with the bold contrast of black and pale gold which decorate its surface, are in the habit of taming it and of placing it round their necks in lieu of a necklace. In other parts of the country, however, the natives believe it to be terribly poisonous, and flee from its presence with terror.

It lives chiefly on insects, worms, and caterpillars, and is very timid. This creature does not taper so gradually from the middle of the body to the tail as is usual in most Serpents, but is nearly of the same cylindrical form throughout its length. The ground colour of this Serpent is pale yellow, decorated with jetty black rings, about sixty in number, that are drawn irregularly over its surface. The Coral Snake never grows to any great size, and seldom reaches two feet and a half in length.

WE now come to another section of the Serpents, termed COLUBRINÆ, the members of which are known by the broad, band-like plates of the abdomen, the shielded head, the conical tail, and the teeth of both jaws. Some of them are harmless and unfurnished with fangs, whereas some are extremely venomous and are furnished with poison-fangs in the upper jaw. These, however, do not fold down like those of the viper and rattlesnake, but remain perfectly erect. The formation of the fangs again differs in the various species. In some the fang is grooved for the introduction of poison into the wound whereas in others it is perforated nearly throughout its length.

SCHAAP-STICKER.—*Psammophýlax rhombeátus*

As an example of the first family of these Serpents, we may take the common BROWN SNAKE of America (*Conocéphalus striátus*).

This reptile is quite harmless, and is plentiful in many portions of America, having rather a wide range of locality. Although common it is not conspicuous, for its small dimensions, its sombre hue, and its retiring habits serve to conceal it from the general gaze. It is usually found hiding under the bark of trees, in stone heaps, or among the crevices of rocky ground, choosing those localities because it feeds principally on insect prey, and can find abundance of food in such places. Its colour is greyish brown above and white below. It is a small species, rarely reaching eleven inches in length.

THE large family of the Coronellidæ contains many curious Serpents, among which may be mentioned the well-known SCHAAP-STICKER of Southern Africa.

This Snake has a rather wide range of country, being spread over nearly the whole of Southern Africa, and very common at the Cape of Good Hope. It is a handsome little reptile, prettily marked, and brisk and lively in its movements, as is required for the purpose of catching the agile prey on which it feeds. The Schaap-sticker lives mostly on insects and small lizards, and darts upon them with great swiftness of movement. It is generally found crawling among heaps of dead leaves, or trailing its variegated form over grassy banks, where it finds the prey on which it subsists.

The colour of this Serpent is extremely variable, and decidedly different in the old and young. In the young specimen, the spots that ornament the back are darker than in the adult, and there is generally a little wash of green over the surface. The general colour of this Snake is brown, with a greyish or golden tint according to the individual. Along the back run several rows, usually three or four in number, of dusky spots, generally of a somewhat oval or rhombic form, and edged with deep black. In one specimen in the British Museum, the spots have coalesced so as to form three continuous bands running along the body. The length of the Schaap-sticker is about two feet.

ANOTHER species belonging to this family (*Coronella Austriaca*) has been once or twice captured in England, but its occurrence is extremely rare. In two cases where this creature has been caught, the locality was also inhabited by the sand lizard (*Lacerta stirpium*), and it is rather remarkable, that upon the Continent, where the Snake is tolerably common, these two species are generally found in the same locality. In general appearance, this Snake is not unlike the viper, and is about the same size, attaining a length of two feet when adult. It may, however, easily be known from the viper, by the absence of the chain of dark lozenge-shaped marks upon the back, for which is substituted a double series of small dark spots, one row at each side of the spine. There is a dark patch upon the shoulder and head, and under the eyes runs a blackish streak. The body is generally brown, but the depth and tone of the ground colour and the markings are extremely variable, but are almost always darker towards the head. Below, the colour is light brown, often marbled with black. The neck is large, being scarcely smaller than the body.

The BLACK SNAKE or ZWARTE SLANG (*Coronella cana*) of Southern Africa, belongs to the same genus.

This reptile is common throughout Southern Africa, but is not very often seen, on account of its timid habit of hiding itself in some crevice, except when in search of food, or when coiled up in repose enjoying the hot beams of the sun. When young, it frequents little hillocks covered with stones, but when it reaches adult age, it takes to the plains, preferring those that are of a sandy nature, interspersed with little shrubs. It is a shy reptile, and mostly runs away when alarmed. Sometimes, however, it will turn upon the pursuer, and if grasped, will coil itself round the arm and squeeze so tightly, that the hand becomes numbed and unable to retain its hold.

Many Snakes are variable in their colouring, but the Black Snake is, perhaps, the most remarkable among them for this peculiarity. Usually, as its name imports, it is black, but sometimes it is bright chestnut. Many specimens are grey, mottled with black, while others are chestnut, marbled with deep rich brown. When full grown, it attains a length of seven feet.

OUR common GRASS SNAKE, or RINGED SNAKE, is a good example of the Natricidæ.

It is extremely plentiful throughout England, being found in almost every wood, copse, or hedgerow, where it may be seen during the warm months of the year, sunning itself on the banks, or gently gliding along in search of prey, always, however, betraying itself to the initiated ear by a peculiar rustling among the herbage. Sometimes it may be witnessed while in the act of creeping up a perpendicular trunk or stem, a feat which it accomplishes, not by a spiral movement, as is generally represented by artists, but by pressing itself firmly against the object, so as to render its body flatter and wider, and crawling up by the movement of the large banded scales of the belly, the body being straight and rigid as a stick, and ascending in a manner that seems almost inexplicable.

The Ringed Snake is perfectly harmless, having no venomous fangs, and all its teeth being of so small a size that even if the creature were to snap at the hand, the skin would not be injured. Harmless though the Serpent be, it will occasionally assume so defiant an air, and put on so threatening an aspect, that it would terrify those who were not well acquainted with its habits. I have kept numbers of these Snakes, and have often known them, when irritated, draw back their heads and strike at the hand in true viperine fashion. Indeed, the venomous look of the attitude is so strong, that I never could resist the instinctive movement of withdrawing the hand when the Snake made its stroke, although I knew full well that no injury could ensue.

The food of the Ringed Snake consists mostly of insects and reptiles, frogs being the favourite prey. I have known Snakes to eat the common newt, and in such cases the victim was invariably swallowed head first, whereas the frog is eaten in just the opposite direction. Usually, the frog, when pursued by the Serpent, seems to lose all its energy, and instead of jumping away, as it would do if chased by a human being, crawls slowly like a toad, dragging itself painfully along as if paralysed. The Snake, on coming up with its prey, stretches out its neck and quietly grasps one hind foot of the frog, which thenceforward delivers itself up to its destroyer an unresisting victim.

The whole process of swallowing a frog is very curious, as the creature is greatly wider than the mouth of the Snake, and in many cases, when the frog is very large and the Snake rather small, the neck of the Serpent is hardly as wide as a single hind leg of the frog, while the body is so utterly disproportioned, that its reception seems wholly impossible. Moreover, the Snake generally swallows one leg first, the other leg kicking freely in the air. However, the Serpent contrives to catch either the knee or the foot in its mouth during these convulsive struggles, and by slow degrees swallows both legs. The limbs seem to act as a kind of wedge, making the body follow easily, and in half an hour or so the frog has disappeared from sight, but its exact position in the body of the Snake is accurately defined by the swollen abdomen. Should the frog be small, it is snapped up by the side and swallowed without more ado.

In captivity, this Snake will eat bread and milk, and insects of various kinds, such as the cockroach, mealworm, or any beetle that may be found running about under stones and leaves. It always, however, prefers frogs to any other food, and seems to thrive best on such a diet.

The skin or slough of the Ringed Snake is often found in the hedgerows or on waste grounds, entangled among the grass stems and furze through which the creature had

RINGED SNAKE, OR GRASS SNAKE.—*Tropidonótus natrix.*

crawled with the intention of rubbing off the slough against such objects. In some parts of the country, the rejected slough is thought to be a specific against the headache, and is tied tightly round the forehead when employed for alleviating pain.

The Ringed Snake is fond of water, and is a good swimmer, sometimes diving with great ease and remaining below the surface for a considerable length of time, and sometimes swimming boldly for a distance that seems very great for a terrestrial creature to undertake. This reptile will even take to the sea, and has been noticed swimming between Wales and Anglesea.

I have often seen tame Snakes taken to an old deserted stone-quarry for a bath in the clear water which had collected there. Generally the Snake would swim quietly from one side to another, and might then be recaptured, but on sundry occasions it preferred diving to the very bottom, and there lay among the stones, heedless of all the pelting to which it was subjected, and impassive as if perfectly acquainted with the harmless nature of stones projected into water. Nothing would induce the Snake to move but a push with a stick, and as the water was rather deep and the quarry wide, a stick of sufficient length was not readily found. The motions of the Snake while in the water are peculiarly graceful, and the rapid progress is achieved by a beautifully serpentine movement of the body and tail.

This Snake is susceptible of kindness, and if properly treated, soon learns to know its owner, and to suffer him to handle it without displaying any mark of irritation. Though harmless and incapable of doing any hurt by its bite, the Snake is not without other means of defence, its surest weapon being a most abominable and penetrating odour, which it is capable of discharging when irritated, and which, like that of the skunk, adheres so closely to the skin or the clothes, that it can hardly be removed even by repeated washings. Moreover, it is of so penetrating a nature, that it cannot be hidden under artificial essences, being obtrusively perceptible through the most powerful perfumes, and rather increasing than diminishing in offensiveness by the mixture. The reptile will, however, soon learn to distinguish those who behave kindly to it, and will suffer itself to be handled without ejecting this horrible odour.

The young of the Ringed Snake are hatched from eggs, which are laid in strings in some warm spot and left to be hatched by the heat of the weather or other natural means. Dunghills are favourite localities for these eggs, as the heat evolved from the decaying vegetable matter is most useful in aiding their development, and it often happens that a female Snake obtains access into a hothouse and there deposits her eggs. Some persons say that the mother is sometimes known to remain near the eggs, and to coil herself round them as has already been related of the boa. The eggs are soft, as if made of parchment, and whitish. They are found in chains containing fifteen or twenty, and are cemented together by a kind of glutinous substance.

During the winter the Snake retires to some sheltered spot, where it remains until the warm days of spring call it again to action. The localities which it chooses for its winter quarters are always in some well sheltered spot, generally under the gnarled roots of ancient trees, under heaps of dry brushwood, or deep crevices. In these places the Snakes will congregate in great numbers, more than a hundred having been taken from one hollow. A few years ago I saw a hole from which a great number of Ringed Snakes had been taken ; it was situated in a bank, at some depth. The colour of the Ringed Snake is greyish green above and blue black below, often mottled with deep black. Behind the head is a collar of golden yellow, often broken in the middle so as to look like two patches of yellow. Behind the yellow collar is another of black, sometimes broken in the middle also. Along the back run two rows of small dark spots, and a row of large, oblong spots is arranged down each side. Both the colour and the shape of the spots are very variable.

The length of this reptile is generally about a yard, but it sometimes attains a length of four feet. The female is always larger than the male. The generic title Tropidonotus is formed from two Greek words signifying keel-backed, and is given to these Serpents because the scales of the back are keeled.

THE two Snakes represented in the accompanying illustration are well-known inhabitants of Northern America, and are dreaded from their fierce aspect, but without the least reason.

The HOG-NOSE SNAKE is so called from the odd formation of the muzzle, which is rather blunt, and slightly turned up at the tip, something like the snout of a hog. It generally frequents moist and marshy localities, as the edges of rivers and ponds, where it finds a plentiful subsistence among the toads, frogs, lizards, and insects which swarm in such spots.

Although as harmless as our ringed Snake, and of similar dimensions, so that it need not be feared on account of its bodily strength, the Hog-nose Snake is rather feared by those who are not acquainted with its structure and habits. If it be irritated in any way, it assumes a most threatening attitude, coils itself like a rattlesnake, flattening its head after the fashion of venomous Serpents, utters a furious hiss, and strikes at the foe with the rapidity of lightning. Yet all this flourish of defiance is without the least foundation, and although it might serve to intimidate the ignorant, only raises the mirth of the better

HOG-NOSE SNAKE.—*Heterodon platyrhinus.*   BLACK VIPER.—*Heterodon niger.*

instructed. For the Serpent does not even open its mouth when it strikes, but darts its closed jaws at the foe, without even inflicting the trifling wounds which might be caused by its small but needle-like teeth. Even if pushed about with a stick, and handled in the roughest manner, it never bites, but contents itself with its impotent personation of the venomous Snakes.

Sometimes it tries other arts, and instead of simulating envenomed rage, pretends to be dead and lies motionless, hoping to escape as soon as the enemy has gone away. So perfectly does it assume the semblance of death, suffering itself to be tossed about without displaying the least sign of life, the muscles relaxed and the body hanging loosely and heavily in the hand, that experienced naturalists have been repeatedly deceived, and only discovered the deception by seeing the reptile make its escape after they had left it lying apparently dead upon the ground.

The colour of the Hog-nose Snake is rather variable, but is generally of a darker or lighter brown above, with a row of large blotches of a different shade of brown running along the sides. Sometimes these blotches are so large, that they unite across the back and form broad bands. There is a dark band between the eyes. The average length of this reptile is about three feet.

THE sombre BLACK VIPER belongs to the same genus as the preceding species, and is very similar to that reptile in many of its habits. It is a very ugly and ungraceful-looking Snake, with a neck of great width, and a head very narrow in front and very wide behind, and is by no means a pleasing object to the eye. It does not frequent the marshy

3.   K

CORN-SNAKE.—*Cóluber guttátus.*   THUNDER SNAKE.—*Cóluber Getúlus.*

localities so constantly as the hog-nose, but prefers the more elevated and drier situations, having a great fondness for the pine-barren districts where the soil is dry and the fallen leaves afford it a shelter and a hunting-ground. It feeds mostly on little mammalia, certain reptiles, and insects. Like the hog-nose Snake, it hisses and strikes with fangless jaws when irritated, and on account of its thick body, flat, wide head, and little glittering eyes, has so venomous an aspect, that it terrifies almost any antagonist for the moment, and then glides away before he has recovered from the instinctive shock to the nerves.

The colour of the Black Viper is wholly black above, without any spots, though on the living Snake there are indications here and there of a deeper tint. The under parts are blackish slate, and the throat takes a whiter hue. It is but a little Serpent, in spite of all its airs, being seldom more than twenty inches in length.

THE family of the Colubrinæ is represented in most parts of the world, North America possessing a large number of examples.

The CORN-SNAKE of America may be reckoned among the most handsome of its tribe. This pretty reptile is extremely common in many parts of America, although it is not very frequently seen, owing to its dislike of daylight. As long as the sun is above the horizon, the Corn-Snake conceals itself in some hiding-place, and issues from its home as soon as the shades of evening begin to approach. It is fearless after its fashion, and has an instinctive liking for the habitations of mankind, haunting farms and houses, where it does considerable service by devouring rats and mice. Occasionally it takes toll in the form of a chicken, but its services most certainly outbalance its little perquisites. It will even enter houses, and can be tamed and made quite familiar. Sometimes it takes a fancy to frequent the roadside, and may be seen quietly coiled and at rest, or trailing its beautiful scales out of the reach of wheels or hoofs.

The colours of this Serpent are brilliant, and arranged in a bold and striking manner. The general colour is rich chestnut-red, and along each side runs a series of large patches of a brighter, but deeper red, each patch being edged with jetty black. There is also a row on each side of much smaller spots of an oval shape, just outside the larger row, and arranged alternately with them. These spots are golden yellow, and are also edged with black. There are some similar spots on the head, and a streak is generally found over each temple. The under parts are silvery white, boldly chequered with black. The length of the Corn-Snake varies from five to six feet.

ANOTHER example of this genus is the THUNDER SNAKE, so called from the threatening black and white of its body, which seems to have a lowering aspect, and to menace poison as the thunder-cloud augurs lightning. Sometimes it is known by the name of KING SNAKE, or CHAIN SNAKE, the latter title being given because the black and white markings of the body are arranged alternately in a chain-like fashion.

The Thunder Snake is mostly found in moist and shady places, where it feeds upon small quadrupeds, reptiles, and birds if it can catch them. The portentous aspect of this Snake is fully carried out by its character, which is fierce, quarrelsome, and aggressive to a degree seldom found even in poisonous Serpents, and in a fangless Snake not at all to be expected. If put in a box with other Serpents, it always quarrels and fights with them; and in one instance, when a Thunder Snake had been introduced into a cage where a miliary rattlesnake was residing, it attacked the venomous reptile in spite of its poisonous weapons, overpowered, killed, and ate it. Some persons think that a deadly feud always rages between the Thunder Snake and rattlesnake, but the truth of this supposition is somewhat dubious. In the instance just mentioned, the creature would probably have treated a Serpent of any species in precisely the same manner.

The Thunder Snake is coloured after a very peculiar fashion. All along the body run alternate bands of jetty black and pure white, the black being very broad and the white very narrow, and not reaching completely across the body. The head is also mottled and scribbled with black upon white after a curious and most complicated fashion. The full length of this Serpent is about four feet.

Two other examples of this genus require a short notice, as they are frequently mentioned in local works and books of travel.

The CHICKEN SNAKE (*Coluber quadrivittatus*) derives its name from its habit of entering farms and houses and stealing chickens from the roost. As, however, it feeds largely on rats and mice, its services in this respect may in all probability counterbalance the loss caused by its thefts. Like the corn-Snake, it is soon tamed, and will become very familiar. In colour it is a very delicate looking reptile, being of a soft bright golden brown, and having four narrow stripes upon a rich dark brown running the whole length of the body. In length it is usually about four feet six inches, though a few specimens attain the length of six and even seven feet. This is also a North American reptile.

The MILK SNAKE, or HOUSE SNAKE (*Coluber eximius*) is common in many parts of North America, and has derived its popular names from its habit of entering houses and its fondness for milk, which some persons fancy it obtains from the cows. Its general food consists of mice and insects, and, like the preceding species, it is probably of some use to the farm where it takes up its residence, and worthy of the encouragement which it sometimes receives.

In the general arrangements of the markings, it is not unlike the corn-Snake, with which it has often been confounded, especially after the fresh beauty of its colours has been dimmed by death, or extracted and changed by spirits. There are similar rows of patches along the sides, but in this species the spots are much broader, often coalescing over the back and forming bands, and the general hue of the body is a beautiful blue

BLACK SNAKE.—*Coryphodon constrictor.*

tinge. The under parts are silver-white, boldly tesselated with oblong and sharply defined marks of black. The length of the Milk Snake is generally about four feet.

Other species belonging to the genus Coluber are found in Australia, India, Japan, China, and Europe, the latter (*Coluber Æsculapii*) being the Serpent which is represented by the ancients as twined round the staff of Æsculapius and the caduceus of Mercury.

The BLACK SNÀKE of America is perhaps the best known of the numerous Serpents, which, happening to be black or dark brown, have been called by the same title.

This Snake is common in Northern America, where it is sometimes known under the name of RACER, on account of its great speed. It is a perfectly harmless, but highly irascible reptile, especially during the breeding season, when it seems to become endowed with an unreasoning ferocity, which, happily for the world, is seldom found in reptiles better provided with offensive weapons. It has a curious habit of rustling its tail among the herbage in such a manner as to resemble the whirr of the dreaded rattlesnake, and then darts at the object of its rage and inflicts a tolerably severe bite, thereby inducing great terror on the part of the sufferer, who, in the hurry of the moment, naturally believes that he has been bitten by the rattlesnake itself.

It is fond of climbing trees in search of young birds, eggs, and similar dainties, and even in that position, is of so tetchy a disposition, that when irritated, it will descend in order to attack its foe. Even if confined with other Snakes, it becomes quarrelsome, fights with them, and if possible will kill them.

The haunts of the Black Snake are usually to be found along the edges of streams and ponds or lakes, and the reptile is mostly to be seen in shady spots, well sheltered by brushwood. Sometimes, however, it goes further a-field, and wanders over the free country, traverses rocky soil, or glides along the roadside. It is a most useful reptile, being very fond of rats, and able from its great agility to climb over walls or buildings in search of its prey, and to insinuate its black length into their holes. It also feeds much on birds, especially when they are young, and is consequently an object of detestation to the feathered tribes. It often happens that the locality of the Black Snake is indicated by the proceedings of the little birds, which collect above their hated enemy, scold with

COACH-WHIP SNAKE.—*Herpetodryas flagelliformis.*
(Lower figure.)

GREEN SNAKE.—*Cyclophis æstivus.*
(Upper figure.)

harsh cries, flutter their wings noisily, and by dint of continual annoyance will often drive the reptile away from the locality. It has been thought that this Serpent was in the habit of killing its prey by pressure, after the fashion of the boas, but this statement has not been satisfactorily confirmed.

The colour of this Snake is blue-black above, and ashen slate below, becoming rather whiter upon the throat. In some specimens a number of spots are observed upon the back of a deeper and duller hue than the general tint. In length the Black Snake generally reaches from five to six feet.

THE small, but interesting family of the Dryadidæ contains a number of Serpents remarkable for the slender elegance of their form, the delicate beauty of their colouring, and the singular swiftness of their movements.

The well-known COACH-WHIP SNAKE of North America is an useful example of this family. This remarkable reptile has not earned its popular name without good reason, for the resemblance between one of these Serpents and a leather whip-thong is almost incredibly close. The creature is very long in proportion to its width, the neck and head are very small, the body gradually swells towards the middle and then as gradually diminishes to the tail, which ends in a small point. The large smooth scales are arranged in such a manner that they just resemble the plaited leather of the whip, and the polished brown-black of the surface is exactly like that of a well-worn thong.

The movements of this Snake are wonderfully quick, and when chasing its prey, it seems to fly over the ground. The mode of attack is very remarkable. Seizing the doomed creature in its mouth, it leaps forward, flings itself over the victim, envelopes it with coil upon coil of its lithe body, so as to entangle the limbs and bind them to the body, and, in fact, makes itself into a living lasso. One of these Snakes was seen engaged in battle with a hawk, and would apparently have conquered in the seemingly unequal combat had not the foes been separated. It had grasped the hawk by one wing, had dragged it to the ground, and had succeeded in disabling the terrible claws from striking, when the sudden approach of the narrator alarmed the Snake, which released its hold, darted into the bushes, and permitted the rescued hawk to fly away in peace.

The colour of this Serpent is rather variable. Generally it is shining black above and lighter beneath, with splashes of purple-brown. Sometimes, however, it is cream or clay-coloured, and occasionally has been seen almost white. But, whatever colour may be the body, the portion near the head is always raven-black. The length of this Snake is about five or six feet.

ANOTHER very slender Snake, also a native of America, is closely allied to the preceding species. This is the GREEN SNAKE, well known for its grass-green colour and its singular activity.

The Green Snake is fond of climbing trees, traversing the boughs in search of food with marvellous celerity, and darting at its insect prey through considerable distances. So slender is this Serpent, that a specimen which measures three feet in length, will barely reach one-third of an inch in thickness at its widest part. Partly owing to this extreme delicacy of form, and partly on account of the leaf-green colour of its body, the Green Snake is not easily seen among the foliage, and in many cases would be undiscovered but for its rapid and energetic movements. The food of this Snake consists mostly of insects. It is very readily tamed, and many persons are fond of carrying the beautiful creature about them, tying it round their throats as a necklace, or as a bracelet on the wrist. The eye corresponds in beauty to the rest of the person, being very large and of a beautiful topaz-yellow.

The colour of the Green Snake is delicate grass-green above, and silvery white below. Its average length is about three feet.

BRAZIL possesses a most lovely example of these Serpents, the EMERALD WHIP SNAKE (*Philodryas viridissimus*).

Dr. Wucherer, of Bahia, writes as follows concerning this pretty species in a letter quoted by Sir J. E. Tennent, in his "Natural History of Ceylon:"—"I am always delighted when I find that another tree-Snake has settled in my garden. You look for a bird's nest: the young ones have gone, but you find their bed occupied by one of these beautiful creatures, which will coil up its body of two feet in length within a space not larger than the hollow of your hand.

They appear to be always watchful, for at the instant you discover one, the quick playing of the long, black, forked tongue, will show you that you too are observed. On perceiving the slightest sign of your intention to disturb it, the Snake will dart upwards through the branches and over the leaves, which scarcely seem to bend beneath the weight. A moment more, and you have lost sight of it. Whenever I return to Europe, you may be sure that in my hothouse these harmless lovely creatures shall not be missing."

The green colour of this species is paler below than above.

The GREY SNAKE of Jamaica (*Dromicus ater*) is another instance of this family. It is often called the BLACK SNAKE, but as that title has already been employed, it is better to use the popular name which is first mentioned.

This reptile is extremely plentiful in Jamaica, where it is mostly found haunting heaps of dead leaves, rocks, and buildings. It is especially fond of the crevices found in old walls, and will lie for hours with its head and neck hanging out of some cranny, partially awaiting the approach of any miserable lizard which may come within reach while searching after flies. It is rather a savage ophidian, darting fiercely at its adversary if irritated, and inflicting a wound which, though not dangerous, is very unpleasant, and causes the limb to swell and ache for some time. It is said, that if it is attacked by a dog, it strikes at the eyes, and can blind the poor creature. While preparing to strike, it dilates its neck, and flattens its head, so as to look as like a venomous Serpent as its limited means will permit.

The colour of the Grey Snake is exceedingly variable. Mostly it is uniformly black, with a tinge of brown, but it often happens that the former tint is subservient to the latter, and in many cases the colour is grey, sometimes of a uniform tint, and sometimes variegated with large dark spots. The length of this Snake is rather more than three feet.

THE little family of the Dasypeltidæ possesses but one genus, but is remarkable for the formation of the teeth, and their use. The teeth of the jaws are very minute and scanty, being at the most only six or seven in number; but some sharp and strong processes issue from the hinder vertebræ of the neck, through holes in the membranes, and form a series of tooth-like projections in the gullet.

The most familiar example of this family is the ROUGH ANODON of Southern Africa. The name Anodon is of Greek origin, and signifies toothless. This reptile lives almost

ROUGH ANODON. —Dasypeltis scabra.

wholly upon eggs, which it eats after a curious fashion. When it finds a nest, it takes the eggs into its mouth, where they lie unharmed, on account of the absence of teeth, so that the shell is not broken, and the liquid contents are preserved. When, however, the reptile swallows the egg, it passes into the throat, and meets the saw-like row of vertebral teeth which have just been mentioned. In its passage, the shell is cut open by these teeth, and the muscular contraction of the gullet then crushes the eggs, and enables the contents to flow down the Snake's throat. These bony processes are tipped with enamel like real teeth.

The colour of this remarkable Serpent is brown, with a row of black marks along the back, sometimes coalescing into a continuous chain, a series of smaller spots upon each side, and some arrow-head marks upon the head of a jetty black.

THE next family is composed of the Tree-Serpents, or Dendrophidæ, so called from the habit of residing among the branches of trees.

Our first example of this family is the well-known BOOMSLANGE of Southern Africa. In pronouncing this word, which is of Dutch or German origin, and signifies Tree-Snake, the reader must remember that it is a word of three syllables. The Boomslange is a native of Southern Africa, and is among the most variable of Serpents in colouring, being green, olive, or brown; of such different colours, that it has often been separated into several distinct species.

Dr. A. Smith has given the following valuable description of the Boomslange and its habits:—

"The natives of South Africa regard the Boomslange as poisonous, but in their opinion we cannot concur, as we have not been able to discover the existence of any gland manifestly organised for the secretion of poison. The fangs are inclosed in a soft pulpy sheath, the inner surface of which is commonly coated with a thin glairy secretion. This secretion possibly may have something acrid and irritating in its quality, which may, when it enters a wound, occasion pain and swelling, but nothing of greater importance.

BOOMSLANGE.—*Buoëphalus Capensis.*

The Boomslange is generally found on trees, to which it resorts for the purpose of catching birds, upon which it delights to feed. The presence of a specimen in a tree is generally soon discovered by the birds of the neighbourhood, who collect around it, and fly to and fro, uttering the most piercing cries, until some one, more terror-struck than the rest, actually scans its lips, and almost without resistance, becomes a meal for its enemy. During such a proceeding, the Snake is generally observed with its head raised about ten or twelve inches above the branch, round which its body and tail are entwined, with its mouth open, and its neck inflated, as if anxiously endeavouring to increase the terror which it would almost appear it was aware would sooner or later bring within its grasp some one of the feathered group.

Whatever may be said in ridicule of fascination, it is nevertheless true, that birds, and even quadrupeds also, are, under certain circumstances, unable to retire from the presence of certain of their enemies; and what is even more extraordinary, unable to resist the propensity to advance from a situation of actual safety into one of the most imminent danger. This I have often seen exemplified in the case of birds and Snakes; and I have heard of instances equally curious, in which antelopes and other quadrupeds have been so bewildered by the sudden appearance of crocodiles, and by the grimaces and contortions they practised, as to be unable to fly or even move from the spot towards which they were approaching to seize them."

THE beautiful BOIGA, sometimes called the AHÆTULLA, also belongs to the family of Tree-Serpents. This pretty and graceful creature inhabits Borneo, and on account of the extreme gentleness of its disposition, and the ease with which it is tamed, the children are in the habit of considering it as a kind of living toy, and allow it to twine around their bodies, or carry it about in their little hands without the least alarm. It is a most active Serpent, living in trees, and darting its lithe form from branch to branch with arrow-like celerity, leaping, as it were, from the coiled folds in which it prepares itself for the spring, and passing through the boughs as if shot from a bow, its glittering scales flashing an emerald or sapphirine radiance, as it glances through the sunbeams.

The head of the Boiga is long and slender, as beseems the delicate body ; the eye is large, full, and round, and the gape very wide. The upper part of its body is rich shining

BOIGA.—*Ahætulla liocercus.*

blue, shot with sparkling green ; and three bright golden stripes run along the body, one traversing the spinal line, and another passing along each side. Behind each eye is a bold jetty black streak, and immediately below the black line runs a stripe of pure white.

The specific name ought properly to be spelled leiocercus. It is of Greek origin, and signifies smooth-tail, in allusion to the smooth-surfaced scales of the back and tail.

THE family of the Wood-Snakes, or Dryiophidæ, as they are learnedly called, contains some interesting and rather curious reptiles. The upper figure in the illustration represents the GOLDEN TREE-SNAKE, which is a native of Mexico. It is a most lovely species, and of a most singular length, looking more like the thong of a "gig whip" than a living reptile. It lives in trees, and in many respects resembles the preceding species. It is not so gorgeously decorated as the boiga, but its colours are beautifully soft and delicate. The general tint of this Serpent is grey, tinged with yellow, and having a golden reflection in certain lights, and being decidedly iridescent in others. The body is profusely covered with minute dottings of black.

The lower figure represents the LANGAHA, one of the Serpents of Madagascar, remarkable for the singular appendage to the head. The muzzle is extremely elongated, and is furnished with a fleshy projection, about one-third as long as the head, and covered with small scales. There is another species, the COCK'S-COMB LANGAHA (*Langaha crista-galli*), also a native of Madagascar, which is known from the ordinary species by the form of the appendage, which is toothed something like the comb upon a cock's head. The colour of the Langaha is reddish brown.

GOLDEN TREE-SNAKE.—*Dryiophis acuminate.* LANGAHA.—*Lángaha nasuta.*

A VERY beautiful example of the Wood-Snakes is found in Ceylon. This is the BROWN WOOD-SNAKE (*Passerita mycterizans*). Like the langaha, the snout of this Serpent is furnished with an appendage, which is pointed, and covered with scales, and is about one-fourth as long as the head. This appendage is conspicuous, but its use is not very plain. It lives almost wholly in trees, and is nocturnal in its habits, traversing the boughs at night for the purpose of catching the small birds as they sleep, taking their young out of the nest, and preying upon the lizards and geckos which also prowl about the trees by night in search of their insect food. There are two varieties of this beautiful Serpent, one being bright green above, with a yellow stripe down each side, and paler below; while the other is brown, glossed with purple, and without the yellow stripe. This variety is rare. The length of these Snakes rarely exceeds three feet.

THE DIPSAS and its congeners may be known from the preceding Snakes, which they much resemble in general form, by the large size of the head compared with the extremely delicate and slender neck. The body, too, is much wider in the centre, causing the neck and tail to appear disproportionately small. This Snake is a native of many parts of Asia.

DIPSAS.—*Eudipsas cynodon.*   BANDED BUNGARUS.—*Bûngarus fasciâtus.*

and is found in the Philippines. The name Dipsas is derived from a Greek word, signifying thirst, and is given to this snake because the ancients believed that it was eternally drinking water and eternally thirsty, and that to allay in some degree the raging drought, it lay coiled in the scanty springs that rendered the deserts passable. As they considered almost all Serpents to be venomous, and, according to the custom of human nature, feared most the creatures of which they knew least, they fancied that the waters were poisoned by the presence of this dreaded Snake. Lucan, in the Pharsalia, alludes to this idea :—

> " And now with fiercer heat the desert glows,
> And mid-day gleamings aggravate their woes;
> When lo ! a spring amid the sandy plain
> Shows its clear mouth to cheer the fainting train.
> But round the guarded brink, in thick array
> Dire aspics rolled their congregated way,
> And thirsting in the midst the horrid Dipsas lay.
> Blank horror seized their veins, and at the view,
> Back from the fount the troops recoiling flew."

The ancient writers also averred that the bite of the Dipsas inoculated the sufferer with its own insatiate thirst, so that the victim either died miserably from drought, or killed himself by continually drinking water.

The colours of the Dipsas are not brilliant, but are soft and pleasing. The general tint is grey, banded with brown of different shades, sometimes deepening into black. The top of the head is variegated with brown, and a dark streak runs from the eye to the corner of the mouth.

THE BANDED BUNGARUS is a native of India, where, from its habits, it is sometimes called the Rock-Serpent. The name Bungarus is a most barbarous Latinization of the native word Bungarum-Pamma, which, though not euphonious, has at all events the advantage of being indigenous, and might have been spared the further distortion of being wrested into a sham classical form. In this reptile the head is rather flat and short, and the muzzle is rounded. The upper jaws are furnished with grooved fangs.

The colour of the Banded Bungarus is very variable, but always consists of some light hue, relieved by bands or rings of jetty black along its length.

AN allied species, the SERPENT-EATING HAMADRYAS (*Hamadryas elaps*), is notable for the peculiarity from which it derives its name. It feeds almost wholly on reptiles, devouring the lizards that inhabit the same country, and also living largely on Snakes. Dr. Cantor says of 'this Serpent that it cannot bear starvation nearly so well as most reptiles, requiring to be fed at least once a month. "Two specimens in my possession were regularly fed by giving them a Serpent, no matter whether venomous or not, every fortnight. As soon as this food is brought near, the Serpent begins to hiss loudly, and expanding its hood, rises two or three feet, and retaining this attitude as if to take a sure aim, watching the movements of the prey, darts upon it in the same manner as the naja tripudians (*i.e.* the cobra) does. When the victim is killed by poison, and by degrees swallowed, the act is followed by a lethargic state, lasting for about twelve hours."

The Hamadryas is fond of water, will drink, and likes to pass the tongue rapidly through water as if to moisten that member. It is a fierce and dangerous reptile, not only resisting when attacked, but even pursuing the foe should he retreat, a proceeding contrary to the general rule among Serpents. The poison of this creature is virulent and active, a fowl dying in fourteen minutes, and a dog in less than three hours, after receiving the fatal bite, although the experiments were made in the cold season, when the poison of venomous Snakes is always rather inactive. The poisonous secretion reddens litmus paper very slightly, and, as is the case with most Serpent poisons, loses its efficacy by being exposed to the air. The native name of the Hamadryas is Sunkr Choar.

The colour of this Snake is generally of an olive hue, auburn, and pale below, but there is a variety marked with cross-bands of white. It is a large species, varying from four to six feet in length, while some specimens are said to reach ten feet.

WE now come to some of the most deadly of the Serpent tribe, the first of which is the well-known COBRA DI CAPELLO, or HOODED COBRA of India.

This celebrated Serpent has long been famous, not only for the deadly power of its venom, but for the singular performances in which it takes part. The Cobra inhabits many parts of Asia, and in almost every place where it is found, certain daring men take upon themselves the profession of Serpent-charmers, and handle these fearful reptiles with impunity, cause them to move in time to certain musical sounds, and assert that they bear a life charmed against the bite of these reptilian playmates. One of these men will take a Cobra in his bare hands, toss it about with perfect nonchalance, allow it to twine about his naked breast, tie it round his neck, and treat it with as little ceremony as if it were an earthworm. He will then take the same Serpent—er apparently the same—make it bite a fowl, which soon dies from the poison, and will then renew his performances.

Some persons say that the whole affair is but an exhibition of that jugglery in which the Indians are such wondrous adepts; that the Serpents with which the man plays are harmless, having been deprived of their fangs, and that a really venomous specimen is adroitly substituted for the purpose of killing the fowl. It is moreover said, and truly, that a Snake, thought to have been rendered innocuous by the deprivation of its fangs, has bitten one of its masters and killed him, thus proving the imposture.

Still, neither of these explanations will entirely disprove the mastery of man over a venomous Serpent. In the first instance, it is surely as perilous an action to substitute a venomous Serpent as to play with it. Where was it hidden, why did it not bite the man instead of the fowl, and how did the juggler prevent it from using its teeth while he was conveying it away? And, in the second instance, the detection of an impostor is by no means a proof that all who pretend to the same powers are likewise impostors.

The following narrative of Mr. H. E. Reyne, quoted by Sir J. E. Tennent in his "Natural History of Ceylon," seems to be a sufficient proof that the man did possess sufficient power to induce a truly poisonous Serpent to leave its hole and to perform certain antics at his command. "A Snake-charmer came to my bungalow in 1854, requesting me to allow him to show me his Snakes dancing. As·I had frequently seen them, I told him I would give him a rupee if he would accompany me to the jungle and catch a Cobra that I knew frequented the place.

He was willing, and as I was anxious to test the truth of the charm, I counted his tame Snakes, and put a watch over them until I returned with him. Before going, I examined the man, and satisfied myself he had no Snake about his person. When we arrived at the spot, he played upon a small pipe, and after persevering for some time, out came a large Cobra from an ant-hill which I knew it occupied. On seeing the man, it tried to escape, but he caught it by the tail and kept swinging it round until we reached the bungalow. He then made it dance, but before long it bit him above the knee. He immediately bandaged the leg above the bite, and applied a Snake-stone to the wound to extract the poison. He was in great pain for a few minutes, but after that it gradually went away, the stone falling off just before he was relieved.

When he recovered, he held up a cloth, at which the Snake flew, and caught its fangs in it. While in that position, the man passed his hand up its back, and, having seized it by the throat, he extracted the fangs in my presence and gave them to me. He then squeezed out the poison on to a leaf. It was a clear oily substance, and when rubbed on the hand, produced a fine lather. I carefully watched the whole operation, which was also witnessed by my clerk·and two or three other persons."

COBRA DI CAPELLO.—*Naja tripudians.*

With regard to the so-called charming of Serpents, there is no need of imagining these men to be possessed of any superhuman powers; for these, and most of the venomous Serpents, are peculiarly indolent, and averse to using the terrible weapons which they wield; in proof of which assertion, the reader may recollect that Mr. Waterton, though not pretending to be a Snake-charmer, carried a number of rattlesnakes in his bare hand without being bitten for his meddling. Not that I would positively assert that the Snake charmers do not possess some means of rendering themselves comparatively proof against the Serpent's bite; for it is reasonable to conclude that, just as a secretion of a cow will, when it has been suffered to pervade the system, render it proof against the poison of the small-pox, there may be some substance which, by a kind of inoculation, can guard the recipient against the poison of the Cobra. In the last century, the one was quite as irremediable as the other.

Another fact is yet to be mentioned. In almost every instance where a poison, vegetable or animal, is likely to gain access to human beings, Nature supplies a remedy at no great distance, just as, to take a familiar instance, the dock is always to be found near the nettle. There certainly are many poisons for which no sure remedy has been discovered, and, until lately, the venom of the Cobra ranked among that number. Recently, however, some important discoveries have been made, which seem to prove that the bite of the Cobra may be cured in two methods, viz. the external application of certain substances to the wound, and the internal administration of others. As the general character of the Cobra is almost precisely the same as that of many other venomous Serpents, and has long been familiar to the public, I shall devote the greater portion of the space, not to the creature itself, but to the remedies for its bite.

The first of these remedies is a plant belonging to the group of birth-worts, and known to botanists by the name of *Aristolochia Indica*.

This plant has long been considered as a valuable remedy for the bite of the Cobra, but the accounts of its use and mode of operation have mostly been vague and scarcely trustworthy. I have, however, been fortunate enough to obtain much valuable information on this subject from R. Lowther, Esq., formerly Commissioner in India, who was accustomed to employ this plant very largely in cases of Cobra-bites, and has kindly forwarded the following communication on the subject:—

" According to your request I have the pleasure of inclosing a statement of one out of at least twenty cases of Snake-bites, in which the exhibition of the *Aristolochia Indica* was attended with complete success, on patients who were brought to my house on a litter, in a perfect state of coma from the bites of venomous Snakes.

The *Aristolochia Indica* is noticed by medical writers as a powerful stimulant, much extolled as a remedy for Snake-bites, in support of which I need only refer you to my detailed statement, as also to the circumstances under which the plant was transferred to my garden at Allahabad. The gentleman from whom I received it (Mr. Breton, Deputy Collector of Customs) gave me the following account of it.

A Cobra, to the great alarm of his servants, had taken up its abode in a mound of earth, formed by white ants, in the vicinity of his house. A party of Snake-catchers having one day made their appearance in the village, Mr. Breton was afforded the opportunity of getting rid of the reptile by having it dug out of its lodgment. After having reached a considerable depth, the man at work used his finger for the purpose of ascertaining the direction of the hole. This seemed to have been its termination, or nearly so, as the Snake caught hold of his finger. His companion immediately ran off to the bank of a stream near at hand, and brought back some leaves, which, having bruised with a stone, he administered to his friend's relief. Mr. Breton requested the man to take him to the plant, which he forthwith removed to his own garden. The Snake-catcher informed him the plant was a specific, and that they usually carried the dried root about with them in case of need.

Mr. Breton, having been subsequently appointed to Allahabad, brought the plant away, and was successful in the treatment of numerous cases. On being removed to a distant station, he transferred the plant to me. The plant is a creeper, and sheds its leaves at that season when Snakes, for the most part, are lying inert in their holes. I should have mentioned, that the Cobra above referred to was killed in the hole.

There are several species of *Aristolochia*, all of them I believe stimulant; but the *Indica* is that which I refer to, it is intensely bitter and strongly aromatic.

In one bad case which came under my treatment, in which large doses had been exhibited, I gave an additional leaf to the patient to take home, but to be used only in case of relapse. Her husband informed me that, although quite recovered, she took the extra dose at 1 o'clock in the morning, and became so giddy that in attempting to move she reeled about like a drunken creature.

A young Hindoo woman was brought to my door in a 'charpoy,' or litter, in a state so apparently lifeless from a Snake-bite, that I had no hesitation in refusing to prescribe. An officer, who was on a visit at my house at the time, considered the woman beyond the power of human relief, and advised me to send her away, as my failure would bring discredit on a remedy which was attracting public notice. In this instance the patient was as cold as marble; there was no pulsation; countenance death-like.

The woman's husband manifested great distress at my refusal, at the same time urging that as the remedy had been prepared, I might, at any rate, give his wife the chance of recovery. I explained to him my motives, and my firm belief that his wife was dead long before he had reached my door. However, rather than add to his distress by persisting in my refusal, I forced her jaws open, and poured down her throat three medium-sized leaves of the *Aristolochia Indica*, reduced to a pulp, with ten black peppercorns, diluted with a graduated ounce of water. The remedy having flowed into her stomach, I directed her body to be raised and supported in a sitting posture, and with some anxiety, though without the slightest prospect of success.

I attentively watched her features, and in the course of eight or ten minutes I observed a slight pulsation on her under lip. I instantly directed her husband, with the aid of my own servants, to drag her about for the purpose, if possible, of increasing the circulation. Supported by two men, holding her up by the waist and arms, she was moved about, her feet helplessly dragging after her. After the lapse of a few minutes, I perceived an attempt on the part of the patient to use her feet. I accordingly directed them to raise her body sufficiently high to admit of the soles of her feet being placed on a level with the ground. In a few minutes she gave a deep inspiration, accompanied with a kind of shriek, manifesting the return of consciousness. This was followed by an exclamation, "A fire is consuming my vitals!" At this time her chest and arms were *deadly cold.* I immediately gave her the pulp of one leaf in an ounce of water, which greatly alleviated the burning sensation in the stomach.

She was then enabled to explain the position of the wound on her instep, which had the appearance of a small speck of ink, surrounded by a light-coloured circle. I had the part well rubbed with the *Aristolochia*, after which she was able to walk without assistance. I kept her walking up and down for at least a couple of hours. Having expressed herself entirely recovered, I allowed her to depart. She called on the following morning to show herself.

The Snake unfortunately escaped, but the woman described it as a 'Kala Samp,' which is the term ordinarily used for the Kobra Kapelle.

I have written the above entirely from memory, the case having occurred eight or nine years ago.

A middle-aged woman was brought to my door in the early part of the rainy season, who had been bitten by a Snake at daybreak, while stooping down for the purpose of sweeping the floor. She called out to the people of the house that a rat had bitten her, and nothing more was thought of it, as her attention was directed to her infant who became fractious for the breast. She accordingly went to bed to give the child sustenance, and not long afterwards complained of giddiness. It was suggested to her that a Snake might have bitten her, but she referred to a hole in the mud-wall from which the rat must have darted out.

Nothing further transpired until the household were alarmed on finding her in a state of insensibility, foaming at the mouth, and the infant at her breast. They were then convinced that a Snake must have done the mischief, and immediately carried her off to the charmer! After detaining the woman for a full hour, the fellow coolly told her friends to take her off to the Commissioner, who would prescribe for her. The poor woman had been dead for some time before the incantations were finished. On arriving at my house, I found the deceased in a state of incipient decomposition, and, having heard the statement of her friends, directed them to take the body away for the performance of funeral rites, and to lose no time in bringing her infant, who was said to be suffering from the effects of the poison.

The poor thing reached my house in a state of insensibility, though not dead. Its head was hanging on its shoulder, and when raised beyond the perpendicular would fall on the opposite shoulder. The body was not cold, and that was the only indication that death had not supervened. I selected one of the smallest of the leaves of the *Aristolochia*, and pounded one-third of it, and, with a small table-spoonful of water, poured the solution into the stomach. After the lapse of four or five minutes the child heaved a deep sigh, opened its eyes wildly, gave a loud scream, and afterwards became quite composed. The child was brought to me on the following morning quite well."

As this plant is so valuable, and seems likely to become an acknowledged remedy, a few lines may be spared for a short description of the species, and the mode of its action.

The *Aristolochia Indica* is one species of a rather large genus, inhabiting many parts of the world, but being most plentiful in the hotter regions. It is a creeping plant, and the specimens grown by Mr. Lowther were trained upon a trellis-work, which they clothed with their narrow, abruptly pointed leaves. Another species of this group of plants, the *Aristolochia serpentina*, is not uncommon in parts of North America, where it is known under the title of the Virginian Snake-root. An infusion of this plant is used as a specific against ague and liver affections.

The fresh leaf of the *Aristolochia Indica* is, when tasted, very bitter and aromatic, bearing some resemblance to quinine in the clear searching quality of the bitter. It is remarkable that when persons are suffering from the poison of the Cobra they describe it as being sweet. There is certainly a kind of sweetness in the leaf, for on chewing a dried leaf of this plant, kindly sent me by Sir W. Hooker, from the collection in the botanical gardens at Kew, I find it to be rather, but not very bitter, with a pungent aroma, something like that of the common ivy, and a faint, though decided sweetness as an after-flavour.

It is not a universal specific, for when experiments were tried by getting some dogs bitten by the Cobra, and treating them with this leaf, they died to all appearance sooner than if they had been entirely neglected. Mr. Lowther has made rather a curious series of experiments on the Cobra's poison and the mode of its action, and has found that while human beings become cold as marble under the influence of the venom, dogs are affected in precisely an opposite manner, being thrown into a high fever, from which they die. The body of a dog killed by a Cobra's bite, will remain quite hot for some ten hours. The *Aristolochia*, therefore, which is a powerful stimulant, rather aids than counteracts the operation of the poison.

In the case of a human being, however, the effect of this remedy seems to be infallible, and Mr. Lowther informs me that he always kept a mortar and pestle by the plant, so that no time should be lost in bruising the leaf, and mixing it thoroughly with water, before pouring it down the throat of the sufferer. The admixture of water was necessary, because, in most instances, the patient was insensible, and the jaws stiffened, so that the mouth needed to be opened forcibly, and the preparation poured down the throat.

THE second mode of cure employed by the natives of India, Ceylon, and even of some parts of Africa, is the now celebrated Snake-stone, so carefully described by Sir J. E. Tennent in his "Natural History of Ceylon." On being bitten by a Cobra, the sufferer applies one of these remarkable objects to each puncture, where they adhere strongly for a variable space of time, five or six minutes appearing to be the usual average. They seem to absorb the blood as it flows from the wound, and at last fall off without being touched, when the danger is considered to be over. This mode of application is general throughout all parts of the world where the Snake-stone is known.

Through the kindness of Sir J. E. Tennent, I have been enabled to make a careful inspection of these objects, and to peruse the original letters relating to their use. They are flattish, shaped something like the half of an almond with squared ends, rather light, bearing a very high polish, and of an intense black—in fact, looking much as if they were rudely cut from common jet. The value of these singular objects is placed beyond doubt by the carefully accredited narratives lately published.

In one case, a native was seen to dart into the wood, and return, bearing a Cobra, about six feet in length, grasping it by the neck with the right hand and by the tail with the left. The Serpent was powerful, and struggled so hard, that its captor was forced to call for assistance. As, however, he held the reptile awkwardly, it contrived to get its head round, and to the horror of the spectators, fastened on his hand, retaining its hold for several seconds. The white bystanders at once gave up the man for lost, but his companion speedily produced from his waistband two Snake-stones, one of which he applied to each puncture. They clung firmly, seemed to absorb the flowing blood, and in a minute or two relieved the extreme pain which the man was already suffering. Presently both Snake-stones dropped simultaneously, and the man declared that the danger had then passed away.

Another native then took from his stores a little piece of white wood, passed it over the head of the Cobra, grasped it by the neck and put it into his basket, averring that when armed with this weapon, a man could handle any kind of Snake without being bitten.

A similar instance is related by Mr. Lavalliere, formerly District Judge of Kandy, and forwarded to Sir J. E. Tennent by the writer, together with the materials employed. The woody substances will presently be described; at present our business is with the Snake-stone, or Pamboo-Kaloo as the natives call it.

The formation of these objects has long been a mystery, and they have been made into a very profitable article of commerce by those who possess the secret. The monks of Manilla are said to be the chief makers of Snake-stones, and to supply the merchants, by whom they are distributed throughout so many countries.

One of these stones was sent for analysis to Mr. Faraday, who pronounced it to be made of charred bone, and in all probability to have been filled with blood, and again charred. "Evidence of this is afforded, as well by the apertures of cells or tubes on its surface, as by the fact that it yields and breaks under pressure, and exhibits an organic structure within. When heated slightly, water rises from it and also a little ammonia, and if heated still more highly in the air, carbon burns away, and a bulky white ash is left, retaining the shape and size of the stone." This ash is composed of phosphate of lime, and Sir J. E. Tennent remarks, with much judgment, that the blood discovered by Mr. Faraday was probably that of the native to whom the Snake-stone was applied.

Another light has been thrown on the subject by Mr. R. W. H. Hardy, who states that the Snake-stone is in use in Mexico, and that it is formed by cutting a piece of stag's-horn into the proper shape, wrapping it tightly in grass or hay, folding it in sheet-copper so as to exclude the air, and calcining it in a charcoal fire.

Being desirous of testing the truth of this recipe, I procured a piece of stag's-horn, cut it into proper shape, and exposed it to the heat of a fierce charcoal fire for an hour and a half. On removing it from the copper, the hay had been fused into a black mass, easily broken, and forming a complete cast of the inclosed horn, which fell out like an almond from its shell.

On comparing the charred horn with the veritable Snake-stones, I find them to be identical except in the polish. The fracture of both is the same, and when exposed to a white heat in the air, my own specimen burned away, leaving a white ash precisely as related of the real specimen, and the ashes of both are exactly alike, saving that my own is of a purer white than that specimen calcined by Mr. Faraday, which has a slight tinge of pink, possibly from the absorbed blood. On throwing it into water, it gave out a vast amount of air from its pores, making the water look for a few seconds as if it were newly opened champagne, a peculiarity which agrees with Thunberg's description of the Snake-stone used at the Cape, and imported at a high price from Malabar. The rather high polish of the Cingalese Snake-stone I could not rightly impart to my own specimen, probably for want of patience. I found, however, that by rendering the surface very smooth with a file, and afterwards with emery paper, *before* exposing it to the fire, it could be burnished afterwards by rubbing it with polished steel. Even in the original objects, the polish is not universal, the plane side being much rougher than the convex.

We will now pass to the little pieces of woody substance, by which the natives assert

3.

that they hold dominion over the Serpent tribe. It has already been mentioned that the native who produced the Snake-stones, employed a small piece of wood as a charm to render the Snake harmless while he handled it. Mr. Lavalliere, in the course of his narration, remarks that the man who was bitten proceeded to bandage his leg above the wound, and to stroke it downwards with a piece of some root. I have also inspected the identical substances used in the two cases just narrated, and have come to the conclusion that no virtue resides in the particular plant from which the charm is taken, but the whole of its value lies in the confidence with which the possessor is inspired.

There are three specimens of charmed woods, all belonging to different plants. One is apparently a part of an aristolochia, another is so small and shrivelled that it cannot be identified, while the third, on being cut and tasted, proves to be nothing more or less than a piece of common ginger. This fact serves to establish the theory of Mr. Waterton, that there is no particular secret in Snake-charming, except the possession of confidence and unhesitating resolution.

ONE notable peculiarity in the Cobra is the expansion of the neck, popularly called the hood. This phenomenon is attributable, not only to the skin and muscles, but to the skeleton. About twenty pairs of the ribs of the neck and fore part of the back are flat instead of curved, and increase gradually from the head to the eleventh or twelfth pair, from which they decrease until they are merged into the ordinary curved ribs of the body. When the Snake is excited, it brings these ribs forward so as to spread the skin, and then displays the oval hood to best advantage. In this species, the back of the hood is ornamented with two large eye-like spots, united by a curved black stripe, so formed that the whole mark bears a singular resemblance to a pair of spectacles.

The native Indians have a curious legend respecting the origin of this mark, and their reverence for the reptile. One day when Buddha was lying asleep in the sun, a Cobra came and raised its body between him and the burning beams, spreading its hood so as to shade his face. The grateful deity promised to repay the favour, but forgot to do so. In those days the Brahminny kite used to prey largely on the Cobras, and worked such devastation among them, that the individual who had done Buddha the forgotten service ventured to remind him of his promise, and to beg relief from the attacks of the kite. Buddha immediately granted the request by placing the spectacles on the Snake's hood, thereby frightening the kite so much that it has never since ventured to attack a Cobra.

It is rather curious that many persons fancy that the Cobra loses a joint of its tail every time that it sheds its poison, this belief being exactly opposite to the popular notion that the rattlesnake gains a new joint to its rattle for every being which it has killed.

The colour of this Serpent is singularly uncertain, and the British Museum possesses several specimens of each variety. In some cases the body is brownish olive, and the spectacles are white, edged with black. Another variety is also brownish olive, but covered with irregular cross-bands of black. The spectacles are remarkably bold, white, edged with black. Other specimens are olive, marbled richly with brown below. The spectacles are like those of the last variety. Sometimes a few specimens are found of a uniform brownish olive without any spectacles; others are black with white spectacles, and others, again, black without spectacles. Even the number of rows in which the scales are disposed is as variable as the colour. The specimens without spectacles seem to come from Borneo, Java, the Philippines, and other islands. The length of the Cobra di Capello is usually between three and four feet.

The AFRICAN COBRA or HAJE is equally poisonous with its Asiatic relative. It is sometimes called SPUUGH-SLANGE, or Spitting-Snake, on account of its power of projecting the poisonous secretion to a distance. It effects this object by a sudden and violent expiration of the breath, and, if aided by the wind, will strike an object at the distance of several feet. Gordon Cumming mentions an instance of his suffering from the poison of this Serpent. "A horrid Snake, which Kleinberg had tried to kill with his loading-rod, flew up at my eye and spat poison in it. I endured great pain all night; the next day the eye came all right again." This short narrative was much ridiculed when the work first

HAJE, OR SPUUGH-SLANGE.—*Naja haje.*

appeared, familiar as the existence of the Spitting-Snake has been to naturalists for many years.

The Haje is one of the fiercest among poison-bearing Snakes, seldom running from an adversary, but generally turning to fight, and not unfrequently beginning the attack. Generally, it moves slowly, but when angry, it darts at its foe, and strikes and spits with such rapid energy, that the antagonist stands in need of a quick hand and eye to conquer the furious reptile. It is a good climber, and is in the habit of ascending trees in search of prey. It is fond of water, and will enter that element voluntarily. While immersed, it swims well, but slowly, scarcely elevating its head above the surface.

In colouring it is one of the most variable of Snakes. Sometimes it is light yellow-brown, either of a uniform tint, or covered with irregular blotches. Other specimens are black when adult, having, when young, a series of broad yellow bands on the fore part of the body. Another variety is black, with a greyish white spectacle-like mark on the neck, and the fore part of the abdomen yellow, with some broad cross-bands. This is the variety shown in the illustration. It is rather curious that the hood of the black specimens is not so wide as in the yellow and brown varieties. The length of the Haje is about five or six feet.

One of the brightest and loveliest of Serpents is the Bead Snake of North America.

This beautiful little reptile inhabits the cultivated grounds, especially frequenting the sweet-potato plantations, and burrowing in the earth, close to the roots of the plants, so that it is often dug up by the negroes while getting in the harvest. It possesses poison-fangs, but is apparently never known to use them, permitting itself to be handled in the roughest manner, without attempting to bite the hand that holds it.

The colours of this Snake are bright, pure, and arranged in a manner so as to contrast boldly with each other. The muzzle and part of the head are black, the remainder of the head is golden yellow, and the front of the neck jetty black. A narrow band of golden yellow with undulating edges comes next the black, and is followed by a broad band of the lightest carmine. From this point the whole of the body and tail are covered with narrow rings of golden yellow, alternating with broad bands of carmine and jetty black.

BEAD SNAKE.—*Elaps fulvius*

Towards the tail the carmine bands become paler and more of a vermilion hue, and for the last four inches there are no red bands, the black and yellow alternating equally. The extreme tip of the tail is yellow. The Bead Snake never attains any great size, seldom exceeding two feet in length.

IT is very remarkable that the terrible LABARRI Snake of South America (*Elaps lemniscátus*) should be closely allied to and belong to the same genus as the bead Snake of the Northern States. Mr. Waterton states that this Serpent is fond of lying coiled on a stump of a tree or some bare spot of ground, where it can hardly be distinguished from the object on which it is reposing. The same writer remarks in a letter to me, that "the Labarri Snake has fangs, and is mortally poisonous when adult. It exhibits the colours of the rainbow when alive, but these colours fade in death. I have killed Labarri Snakes eight feet long."

WE now arrive at a most curious family, known by the possession of very long poison-fangs, perforated, and permanently erect. They only include one genus, of which the best known species is the NARROW-HEADED DENDRASPIS (*Dendraspis angústiceps*).

This Serpent is very long, slender, and unusually active and a good climber, exceeding the haje in this accomplishment. It is found in Southern Africa, and is tolerably common at Natal. Its colour is olive-brown washed with green above, and a paler green below. It is rather a large though very slender Snake, sometimes reaching the length of six feet.

THE last example of the Serpent tribe is the ATRACTASPIS of Southern Africa (*Atractaspis irreguláris*). The fangs of this Snake are longer in proportion than those of any other known Serpent, reaching nearly to the angle of the mouth. They are so long, indeed, that Dr. Smith is of opinion that the creature cannot open its mouth sufficiently wide to erect the fangs fully, so that the poison-teeth are always directed backwards. They still, however, serve an important purpose; for when the Atractaspis seizes its prey, the poison-fangs necessarily pierce the skin, so as to inject the venom into the body of the victim, and from their shape act as grapnels, by which all attempts at escape are foiled.

Very little is known of the habits of this Snake, but it is thought to burrow in loose ground.

The colour of the Atractaspis is blackish green above, shaded with orange-brown, and orange-buff below. It is a small Serpent, rarely measuring more than two feet in length.

THE BATRACHIANS are separated from the true reptiles on account of their peculiar development, which gives them a strong likeness to the fishes, and affords a good ground for considering these animals to form a distinct order. On their extrusion from the egg, they bear no resemblance to their parents, but are in a kind of intermediate existence, closely analogous to the caterpillar or larval state of insects, and called by the same name. Like the fish, they exist wholly in the water, and breathe through gills instead of lungs, obtaining the needful oxygen from the water which washes the delicate gill-membranes. At this early period they have no external limbs, moving by the rapid vibration of the flat and fan-like tail with which they are supplied. While in this state they are popularly called tadpoles, those of the frog sometimes bearing the provincial name of pollywogs. The skin of the Batrachians is not scaly, and in most instances is smooth and soft. Further peculiarities will be mentioned in connexion with the different species.

These creatures fall naturally into two sub-orders, the leaping or tail-less Batrachians, and the crawling Batrachians. The leaping Batrachians, comprising the frogs and toads, are familiar in almost all lands, and in England are well known on account of their British representatives.

The tongue plays an important part in separating the frogs and toads into groups, and in the first group the tongue is altogether absent, these creatures being in consequence called Aglossa, or tongueless Batrachians.

THE first of these creatures, the XÉNOPUS of Western and Southern Africa (*Dactylethra lœvis*), is remarkable for possessing nails on its feet, the first three toes being tipped with a sharply pointed claw or nail. The family is very small, comprising only one genus, and, as far as is known, two species. The colour of the Xenopus is ashy brown, veined with blackish brown. It is rather a large species.

THE celebrated SURINAM TOAD has long attracted attention, not for its beauty, as it is one of the most unprepossessing of beings, but for the extraordinary way in which the development of the young is conducted.

When the eggs are laid, the male takes them in his broad paws, and contrives to place them on the back of his mate, where they adhere by means of a certain glutinous secretion, and by degrees become embedded in a series of curious cells formed for them in the skin. When the process is completed, the cells are closed by a kind of membrane, and the back of the female Toad bears a strong resemblance to a piece of dark honey-comb, when the cells are filled and closed. Here the eggs are hatched, and in these strange receptacles the young pass through their first stages of life, not emerging until they have attained their limbs, and can move about on the ground.

The skin of this, as well as of other batrachians, is separated from the muscles of the back, and allows room for the formation of the cells, being nearly half an inch thick. The full-sized cells are much deeper than long, and each would about hold a common horse-bean, thrust into it endways. The mouths of the cells assume an irregularly hexagonal form, probably because their original shape would be cylindrical, were they not squeezed against each other.

When the young have attained their perfect state, they break their way through the cover of the cells, and present a most singular aspect as they struggle from the skin, their heads and paws projecting in all directions. In the museum of the College of Surgeons may be seen some very good specimens of the Surinam Toad, some being entire, and others

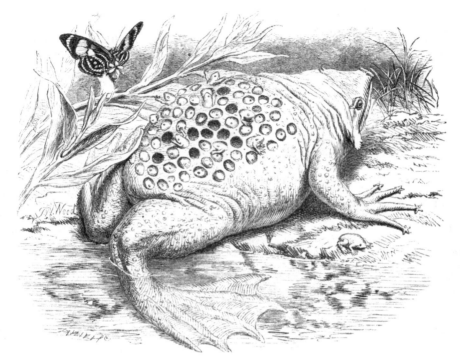

SURINAM TOAD.—*Pipa Americana.*

dissected so as to show the cells and their structure. After the whole brood have left their mother's back, the cells begin to fill up again, closing from below as well as from above, and becoming irregularly puckered on the floors. The cells in the middle of the back are the first developed; the whole process occupies rather more than eighty days.

As its name implies, this singular creature inhabits Surinam, but is also found in various parts of Central America. In spite of its repulsive aspect, the negroes are said to eat its flesh.

The colour of the Surinam Toad is brownish olive above, and whitish below. The skin is covered with a large number of tiny and very hard granules, among which are interspersed some horny tubercular projections. The snout is of a very curious shape, the nostrils being lengthened into a kind of leathery tube. The throat of the male is furnished with a very large bony apparatus, of a triangular box-like shape, and within are two movable pieces by which the voice is modulated.

WE now come to the Batrachians with tongues. In the greater number of these creatures, the tongue is fastened to the front of the mouth and free behind, the tip pointing down the throat. The prey is taken by the rapid throwing forward of this tongue, and its equally rapid withdrawal into the mouth, carrying the doomed creature on its tip, with such celerity that the eye can hardly follow the movement.

The skeleton of the adult Frog is worthy of a short notice before we proceed to the farther investigation of these remarkable creatures. The first point which strikes the observer is the shape of the head, and the enormous size of the orbits of the eyes, which are so large, that when the skull is placed flat upon an open book, several words can be read through the orifices. Very little room is left for the brain, and, in consequence, the intellectual powers of the Frog are but slender.

The vertebra are furnished with projections at each side, but the ribs are totally wanting. On account of this deficiency, the process of respiration cannot be maintained as is usual among the better developed beings, but is similar to that which is employed by the tortoises. The needful movements are made not by the sides but by the throat, so that if a quiescent Frog be watched, it appears to be continually gulping something down its throat, as is indeed the case, the material being air, which is thus forced into the beautifully formed lungs. The formation of the pelvis is rather peculiar, and can be comprehended by reference to the illustration.

The hind-legs are extremely long, and the toes so much lengthened, that in the common English Frog the middle toe occupies about three-fifths of the length of the entire body, and in some species is even more produced. Owing to the peculiar shape of the limbs, the Frog when reposing sits almost upright, and is at once ready for the extra-ordinarily long leaps which it can take when alarmed. The usual mode of progression is by a series of jumps, though of short range, but the creature will often crawl after the fashion of the toad—the presence of a snake seeming almost always to have the effect of causing the change of action.

The skin of the Frog is very porous, and is capable of absorbing and exuding water with wonderful rapidity. If a Frog, for example, be kept for some time in a perfectly dry spot, it loses its fine, sleek condition, becomes thin and apparently emaciated, and assumes a very pitiable appearance. But if it be then placed merely on wet blotting paper, its thirsty skin drinks the needful moisture, and it soon becomes quite plump and fresh. A familiar proof of the extreme porosity of the skin is afforded by the dead Frogs which are often

SKELETON OF FROG.

found on the high road or dry paths in the middle of summer, and which are dried into a shrivelled, horny mass, which would be shapeless but for the bones of the skeleton around which the skin and muscles contract.

The whole of these creatures are most tenacious of life, suffering the severest wounds without appearing to be much injured at the time, and bearing the extremes of cold and hunger with singular endurance. Heat, however, is always distasteful to the Frog, and when carried to any extreme becomes fatal. In the hot countries, where Frogs of various species exist, they all unite in the one habit of avoiding the hot beams of the sun by hiding in burrows or crevices during the day, and only emerging from their refuge in the night-time, or during rainy weather. Many species even dive below the muddy soil of pools as soon as the water has nearly disappeared, and there remain moist, torpid, and content until the next rains refill their home with the needful waters.

Most of the Frogs have a power of changing the colour of the skin, which is often found to lose its brightest tints and become dark brown or nearly black in a very short space of time. Any sudden alarm will often produce this change, the presence of a snake being an almost unfailing means of effecting this object; and it is known that the colour of the Frogs is greatly affected by the locality in which they are at the time placed. The Tree-Frogs are more subject to this change of colour than the ordinary species; but even our common English Frog is well known to alter from yellow to brownish black in a very short space of time. This change is produced by some mental emotion acting upon certain masses of pigment or colouring matter in the skin; and for a farther elucidation of the subject, I must refer the reader to my "Common Objects for the Microscope," where the pigment masses are drawn as seen through the microscope, and their peculiar action explained.

ONE of the most singular members of this group of animals is the PARADOXICAL FROG (*Pseudis paradoxa*).

This curious creature is a native of Surinam and South America, and is remarkable for the enormous size of the larva, or tadpole. As a general rule, and indeed, as might be expected, the generality of the batrachians are smaller in their larval than in their adult state; the tadpole of the common Frog being a good example. But the Paradoxical Frog exhibits a phenomenon which is perhaps found in none of the higher animals, though common enough among the non-vertebrated beings, and is less in its adult state than in its preliminary form of tadpole.

The tail of this tadpole is exceedingly voluminous, and the body has other envelopes or appendages, which, when thrown off as it proceeds to its perfect state, reduce the bulk so greatly that the earlier observers thought that the creature reversed the usual order of nature, and from a Frog became a tadpole. Some persons went even farther, and said that it was changed from a Frog into a fish. The appropriate title of Paradoxical was given to it in allusion to this opinion.

Strange, however, as this phenomenon may appear, and remarkable as it undoubtedly is, it finds abundant parallels in the insects, where the larva is often of greater bulk than the perfect insect, or imago, as it is technically called. We may take for example the common silkworm, where the caterpillar is extremely large when compared with the moth into which it afterwards changes; or that great, fat, bulky, subterranean grub, which eats continually for three years, becomes so obese that it is forced to lie on its side, and afterwards turns into the neat, compact, and active little cockchaffer.

The colour of the Paradoxical Frog is greenish, spotted with brown, and streaked irregularly with brown along its legs and thighs. The snout is tapering, and rather pointed in front.

OUR next example of the Ranidæ is the AFRICAN BULL-FROG.

This fine species is spread over the whole of Southern Africa, but is found most plentifully towards the eastern coast, where it always frequents springs, pools, or the vicinity of fresh water. It is most impatient of drought, and when a more than usually dry season has parched the ground and rendered the hot soil uncomfortable for the delicate skin of the creature's feet and abdomen, these Frogs are said to congregate in the pools in great numbers, and just before the water has quite dried up, to burrow deeply into the soft mud and there lie until the next rains bring the welcome moisture.

Fifty of these large Frogs have been seen gathered together in one little pool, far from any other water. It is, moreover, evident that they must have some place of concealment, for they are sure to appear in great numbers after a few heavy rains, and it is quite consistent with probability that they should possess a simple and obvious method of preserving their lives during the frequent droughts of the climate in which they reside.

Dr. Livingstone mentions this fine species in his well-known work on Southern Africa. "Another article of which our children partook with eagerness was a very large Frog, called 'Matlamétlo.'

These enormous Frogs, which, when cooked, look like chickens, are supposed by the natives to fall down from thunder-clouds, because after a heavy shower the pools which are filled, and retain water a few days, become instantly alive with this loud croaking pugnacious game. This phenomenon takes place in the driest parts of the desert, and in places where to an ordinary observer there is not a sign of life.

AFRICAN BULL-FROG.—*Tomopterna adspersa.*

Having been once benighted in a district of the Kalahari, where there was no prospect of getting water for our cattle for a day or two, I was surprised to hear in the fine still evening the croaking of Frogs. Walking out until I was certain that the musicians were between me and our fire, I found that they could be merry on nothing else but a prospect of rain.

From the bushmen I afterwards learned that the Matlamétlo makes a hole at the root of certain bushes, and there ensconces himself during the months of drought. As he seldom emerges, a large variety of spider takes advantage of the hole, and makes its web across the orifice. He is thus furnished with a window and screen gratis, and no one but a bushman would think of searching beneath a spider's web for a Frog. They completely eluded any search on the occasion referred to; and as they rush forth into the hollows filled by the thunder-showers when the rain is actually falling, and the Bechuanas are cowering under their skin garments, the sudden chorus struck up simultaneously from all sides seems to indicate a descent from the clouds.

The presence of these Matlamétlo in the desert in a time of drought was rather a disappointment, for I had been accustomed to suppose that the note was always emitted by them when they were chin-deep in water. Their music was always regarded in other spots as the most pleasant sound that met the ear after crossing portions of the sandy desert; and I could fully appreciate the sympathy for these animals shown by Æsop, himself an African."

It is a large and handsome species, but becomes duller in colour as it increases in age. The young, however, are very lightly tinted. The general colour is greenish brown above, with a decided rusty wash, variegated with mottlings of reddish brown, and streaked and spotted with yellow. The green takes a brighter and purer hue along the sides of the head and legs. The abdomen is yellow, mottled with orange, and the chin is striped and splashed with brown. The eyes are very curious and beautiful, being of a rich chestnut hue, covered with a profusion of little golden white dots, which shine with a metallic lustre.

When young, the yellow lines on the body are edged with jetty black, and the legs are covered with bold black bars. The head is stout and rather flat, and the skin of the body

PICKEREL-FROG.—*Rana palustris.* SHAD-FROG.—*Rana halecina.*

is puckered into longitudinal folds. The lower jaw is remarkable for two large, bony, tooth-like projections in front. The ordinary length of a full-grown specimen is about six inches.

WE now come to the very large genus of which our common English Frog is so familiar an example, and which finds representatives in all except cold latitudes. The very handsome SHAD-FROG derives its popular name from its habit of making its appearance on land at the same time that the shads visit the shore. The specific title *halecina* also alludes to this circumstance, as the Indian word for a shad is halec. As the herrings appear about the same season, the Frog is called by the Swedes SILL-HOPPETOSSE, or Herring-hopper, the latter name being much in use in Pennsylvania.

This Frog requires much moisture, and is seldom seen at any distance from the banks of rivers or pools of fresh water. Sometimes, however, when the dew lies very heavily on the grass, the Shad-Frog makes its way over the fields to spots far from the water-side, but takes care to return before the hot sunbeams have dried up the grateful moisture of the herbage. The food of this reptile consists chiefly of insects. It is a very active creature, and ever lively, making leaps of eight or ten feet in length.

It is thought by many persons to rank among the handsomest of the froggish tribe. The general colour is light golden green, variegated with four rows of olive spots, edged with rich gold. One regular row of these spots runs along each side of the spine, and the others are scattered rather vaguely along the sides. The throat is white with a silvery lustre, and the abdomen whitish yellow. The aural vesicles are brown, with a circular centre of azure-blue, and look like two little targets on the side of the head. The eyes are very large, of a beautiful golden lustre, and with a bold black streak drawn horizontally through their centre. The legs are exceedingly long in proportion to the size of the body, being five inches in length, whereas the body measures scarcely three inches. This length of limb and lightness of body adds greatly to the leaping powers, for which this creature is so celebrated.

ANOTHER very common and very pretty Frog is found in Northern America. This is the PICKEREL-FROG, so called because it enjoys a sad pre-eminence among anglers as a bait for pike, too fortunate if it can be snapped up at once by the voracious fish, instead of dangling for a season in mid-water, with a hook delicately inserted under its skin so as to keep it lively as long as possible, and prevent it from losing by death its attractive appearance.

It is mostly found in or near the salt marshes, and is remarkable for possessing a powerful and extremely disagreeable odour. In spite, however, of this seeming drawback, its flesh is said to be very delicate, and to be quite as good as that of the edible Frog of Europe.

The colouring of this species is very striking, on account of its irregularly squared aspect. The ground tint is pale brown above, covered with moderately large square spots of dark brown arranged like the stones of a tesselated pavement, and producing a somewhat regular pattern. A bright yellow line, not raised above the general surface, runs from behind each eye, and the under parts are yellowish white. It is quite a little Frog, being under three inches in length.

UPON the accompanying illustration is presented a figure of the celebrated BULL-FROG of America, one of the largest and most conspicuous of its kind.

This enormous batrachian is perhaps the best swimmer among the Frog race, having been known to live for several years in water without any support for its feet. It leads a solitary life for the greater part of its existence, living in a hole near the water, and seldom leaving its domicile by day unless when suddenly alarmed. If frightened by an unknown sound or sight, the Bull-Frog leaps at once into the water, and instead of diving to the bottom immediately, skims along the surface for a few yards before it disappears.

During the breeding-season, these huge Frogs assemble together in great multitudes, congregating to the amount of four or five hundred in some pool or marsh, sitting with their bodies half submerged, and making night hideous with their horrid bellowing cries. Few persons, except those who have had personal experience, and who have lost night after night of needful sleep by the ceaseless noise, can imagine the loudness of voice and variety of tone possessed by the different species of Frogs. And travellers who lie awake at night, unwilling hearers of the nocturnal concerts, are disposed to envy the happy ignorance of those whose calmer lot is cast in countries where the drummings, bellowings, chatterings, and pipings of the Frog race are practically unknown. Among these nightly musicians the Bull-Frog is the loudest and most pertinacious; mostly remaining quiet by day, but sometimes exulting in a black cloud or a heavy shower, and raising its horrid din even in the hours of daylight.

It is a most voracious creature, feeding mostly on snails and similar prey, which it catches on its nocturnal excursions from its domicile, but often devouring animals of a larger size, such as crayfish, two of which crustaceans have been found in the stomach of a single Bull-Frog, and even gobbling down an occasional chicken or duckling. Taking advantage of its voracity, the inhabitants of the country are in the habit of catching it by means of a rod and line. The hook is generally baited with an insect, and gently drawn along the ground near the Frog, which leaps upon it, seizes it, and is hooked without difficulty. It is rather curious that the Frog will not touch the insect as long as it is allowed to rest quietly on the ground, but as soon as the line is pulled, so as to make the insect move, it is at once pounced upon. Our English Frogs and toads have the same custom.

The flesh of the Bull-Frog is very delicately flavoured, and in some places the creature is kept in captivity and fed for table.

BULL-FROG.—*Rana mugiens*.

This species is exceedingly active, making leaps of eight or ten feet in length and five feet in height. There is a well-known story of a race between a Bull-Frog and an Indian, the former to have three jumps in advance, and the distance about forty yards to a pond from which the Frog had been taken. When the parties were ready to start, the glowing tip of a burning stick was applied to the Bull-Frog, which set off at such a rate, and made such astonishing leaps to get into the welcome water, that its human opponent was vanquished in the race.

In some places this creature is never disturbed, as it is supposed, perhaps with some justice, to aid in keeping the water pure. The popular name of Bull-Frog is derived from its cry, which is said to resemble the bellowing of the animal whose name it bears. Several species of Frog have been classed under the same popular name.

The colour of the Bull-Frog is brown, mottled with black above, and taking a greener hue upon the head. The abdomen is greyish white, and the throat is white dotted with green. The length of the head and body of the large species is rather more than six inches, and a fine specimen will sometimes measure nineteen or twenty inches from the nose to the extremity of its feet. The skin of the back is smooth, and without any longitudinal fold.

THERE is another tolerably common species inhabiting the same country, which is also popularly called the Bull-Frog, but is known among men of science as *Rana clámitans*. It may be readily distinguished from the bull-Frog, which it otherwise greatly resembles, by the presence of a glandular fold on each side of the back. It is a very noisy creature, with a sharper and more yelping cry than the preceding species. When disturbed, it shoots at once into the water, and there sets up its peculiar cry. It is more active than the common bull-Frog, and if once released, is almost certain to escape, from the great length and rapidity of its leaps, the creature never seeming to pause between two jumps, but springing off the earth with an instantaneous rebound not unlike the flying leaps of the jerboa or kangaroo. It is a moisture-loving species, and is never found far from water.

WE now come to the best known of all the batrachians, the COMMON FROG of Europe.

COMMON FROG.—*Rana tempordria.*

The general form and appearance of this creature are too well known to need much description. It is found plentifully in all parts of England, wandering to considerable distances from water, and sometimes getting into pits, cellars, and similar localities, where it lives for years without ever seeing water. The food of the adult Frog is wholly of an animal character, and consists of slugs, possibly worms, and insects of nearly every kind, the wire-worm being a favourite article of diet. A little colony of Frogs is most useful in a garden, as they will do more to keep down the various insect vermin that injure the garden, than can be achieved by the constant labour of a human being.

The chief interest of the Frog lies in the curious changes which it undergoes before it attains its perfect condition. Every one is familiar with the huge masses of transparent jelly-like substance, profusely and regularly dotted with black spots, which lie in the shallows of a river or the ordinary ditches that intersect the fields. Each of these little black spots is the egg of a Frog, and is surrounded with a globular gelatinous envelope about a quarter of an inch in diameter. According to gipsy lore, rheumatism may be cured by plunging into a bath filled with Frog spawn.

On comparing these huge masses with the dimensions of the parent Frog, the observer is disposed to think that so bulky a substance must be the aggregated work of a host of Frogs. Such, however, is not the case, although the mass of spawn is forty or fifty times larger than the creature which laid it. The process is as follows: The eggs are always laid under water, and when first deposited, are covered with a very slight but firm membranous envelope, so as to take up very little space. No sooner, however, are they left to develop, than the envelope begins to absorb water with astonishing rapidity, and in a short time the eggs are inclosed in the centre of their jelly-like globes, and thus kept well apart from each other.

In process of time, certain various changes take place in the egg, and at the proper period the form of the young Frog begins to become apparent. In this state it is a black grub-like creature, with a large head and a flattened tail. By degrees it gains strength, and at last fairly breaks its way through the egg and is launched upon a world of dangers, under the various names of tadpole, pollywog, toe-biter or horsenail.

As it is intended for the present to lead an aquatic life, its breathing apparatus is formed on the same principle as the gills of a fish, but is visible externally, and when fully developed consists of a double tuft of finger-like appendages on each side of the head. The tadpole, with the fully developed branchiæ, is shown at fig. *a* on the accompanying illustration. No sooner, however, have these organs attained their size than they begin again to diminish, the shape of the body and head being at the same time much altered, as is seen in fig. *b*. In a short time they entirely disappear, being drawn into the cavity of the chest and guarded externally by a kind of gill cover.

Other changes are taking place meanwhile. Just behind the head two little pro-. jections appear through the skin, which soon develop into legs, which, however, are not

at all employed for progression, as the tadpole wriggles its way through the water with that quick undulation of the flat tail which is so familiar to us all. The creature then bears the appearance represented in fig. *c*.

Presently another pair of legs make their appearance in front, the tail is gradually absorbed into the body—not falling off, according to the popular belief—the branchiæ vanish, and the lungs are developed. Fig. *d* represents a young Frog just before the tail is fully absorbed.

The internal changes are as marvellous as the external. When first hatched, the young tadpole is to all intents and purposes a fish, has fish-like bones, fish-like gills, and a heart composed of only two chambers, one auricle and one ventricle. But in proportion to its age, these organs receive corresponding modifications, a third chamber for the heart being formed by the expansion of one of the large arteries,

TADPOLES.

the vessels of the branchiæ becoming gradually suppressed and their place supplied by beautifully cellular lungs, formed by a development of certain membranous sacs that appear to be analogous to the air-bladders of the fishes.

The Frog, contracted as are its intellectual powers, is yet susceptible to human influence, and can be tamed by kind treatment. Mr. Bell mentions a curious instance where one of these creatures became so completely domesticated, that it used to come nightly from a hole in the skirting-boards where it had established itself, partake of food offered to it by the members of the family, and even jump upon the hearth-rug in winter in order to enjoy the warmth of the fire. A favourite cat, which inhabited the same house, took a strange fancy to the Frog, and these seemingly incongruous companions were to be constantly seen sitting together on the hearth-rug, the Frog nestling under the soft warm fur of the cat. The Frog was, however, more than a year an inmate of the house before it became domesticated, and for many months would retreat to its stronghold when approached.

Stories of so-called "showers of Frogs" are often seen in the papers, and as a general rule are little to be credited, the solution of the supposed phenomenon being merely that a shower of rain has induced the creatures to come simultaneously from their retreats. There are, however, instances where credible spectators have seen them fall, and in such cases the little creatures were probably sucked up by a waterspout, or even by a brisk whirlwind, together with the water in which they were disporting, carried away for some distance, and at last dropped on the ground, as is sometimes the case with sticks, stones, and leaves, picked up by a passing whirlwind.

The general colour of the common Frog is greenish yellow, or brown, the same individual often passing through all these colours in a few days. A long patch of blackish brown or warm brown is placed behind each of the eyes, and it is yellowish white below. There are no teeth in the lower jaw, and only a single row of very tiny teeth in the upper jaw and on the palate. The ordinary length of the Frog is rather less than three inches. and the total length of the hinder leg is about four inches.

BANDED FROG.—*Rana fasciata.*

A VERY pretty species of this genus is found in Southern Africa. This is the BANDED FROG, remarkable for the beautiful stripes which adorn its body, and the inordinate length of the second toe of the hind foot.

This pretty creature is not very plentiful in any one locality, but is spread widely throughout the Cape district and the whole of Southern Africa. It is very active, being a good leaper, and brisk in all its movements. The second toe of the hind foot is truly remarkable. The whole of the toes are but slightly webbed, and project boldly beyond the connecting membrane; but the second toe is nearly as long as the whole body, which is longer than in the generality of Frogs. The object of this exceeding development is not very clear.

The general colour of this species is wood-brown, upon which are drawn six dark streaks, the two centre stripes running nearly the entire length of the body. The hinder part of the thigh is orange-brown, and the under parts are yellowish white. The length of the head and body is nearly two inches.

THE celebrated EDIBLE FROG, or GREEN FROG of Europe (*Rana esculenta*), also belongs to this large genus. This handsome species is common in all the warmer parts of the Continent, but in the vicinity of large cities is seldom seen, except in the ponds where it is preserved, and whence issues a horrid nocturnal concert in the breeding time. The proprietors of these froggeries supply the market regularly, and draw out the Frogs with large wooden rakes as they are wanted. In Paris these creatures are sold at a rather high price for the table, and as only the hind legs are eaten, a dish of Frogs is rather an expensive article of diet.

It is needful to make a very early visit to the market, four or five A.M. being about the best time, to see the manner in which the Frogs are brought to market. They are generally sold by women, each of whom has by her side two tubs or barrels, one containing living Frogs, and the other having a leather band nailed to the side, in which is stuck a sharp, broad-bladed knife. When the purchaser has bargained for a certain number, the seller plunges her left hand into the one barrel, brings out a Frog by its legs, lays it across the edge of the second barrel, and with a single cut of the knife, severs the hind legs just above the pelvis, leaving the whole of the body and fore-quarters to fall into the tub. The hind legs are then carefully skinned, and dressed in various ways, that with white sauce seeming to be the best, at all events according to my own taste. They require considerable cooking, but when properly dressed have a most delicate and

HORNED FROG.—*Cerátophrys cornuta.*

peculiar flavour, which has been compared, but not very happily, to the wing of a chicken. I would suggest that a mixture of the smelt and the breast of the spring chicken would convey a good idea of the Edible Frog when cooked.

Poachers are very apt to invade the froggeries, and without entering the boundaries often contrive to kidnap a goodly number of the inmates by a very curious mode of angling, something like "bobbing" for eels. They get a very long fishing-rod, tie a line of sufficient length to the tip, and at the end of the line they fasten, in place of a hook and bait, a simple piece of scarlet cloth. Thus prepared, they push the rod over the fence, let the scarlet rag just touch the surface of the water, and shake the rod so as to make the rag quiver and jump about. The Frog, thinking that it has found a very savoury morsel, leaps at the rag, closes its mouth firmly upon it, and is neatly tossed out of the water and over the hedge before it can make up its mind to loosen its hold.

The colour of this species is bright green spotted with black, and having three bold yellow stripes along the back. The under parts are yellowish. In size it is rather larger than the common species.

THE remarkable HORNED FROG is one of the quaintest species among the Frog tribe.

There are several species belonging to this genus, all inhabiting Southern America, and all notable for the singular development of the upper eyelids, which are prolonged into hard, horn-like points. In the present species the back is furnished with a bony shield, and the prominences over the eyes are bold and well defined. The body is short, stout, and squat, the skin covered with tubercles and folds, and the opening of the mouth enormous. It is a large and voracious species, one specimen when opened being found to have swallowed a full-grown land-Frog (*Cystignathus fuscus*), belonging to the same genus as our next example. The toes are long, powerful, and with hardly a vestige of web except just at the base.

THE little ORNATE LAND-FROG affords a remarkable contrast to the last-mentioned species on account of its small dimensions, the activity of its movements, and the beauty of its colouring.

It is found in Georgia and South Carolina, and is always seen on land and dry spots, its thirsty frame being amply supplied by the dews and casual rains without needing immersion in water. Indeed, this Frog is so little conversant with the element usually so familiar to all its tribe, that if thrown into water, it makes no attempt to swim, but lies helplessly sprawling on the surface. On land, however, it displays wonderful activity, being of an extremely lively nature, and making long and bold leaps in rapid succession, so that it is not to be captured without considerable difficulty.

The colour of this species is rather variable, but is generally of a soft dove tint, on which are placed several oblong marks of deep rich brown edged with golden-yellow. Below it is silvery white granulated with grey. It is a very little species, measuring only one inch and a quarter when full-grown.

ANOTHER species of this genus, the SENEGAL LAND-FROG (*Cystignathus Senegalensis*), inhabits Southern Africa.

It resides in burrows in the ground, and is tolerably quiet, except before rain or on a dull day, when it begins to pipe, and continues its curious cry for several hours together. The voice of this Frog is a sharp piping whistle several times re-

ORNATE LAND-FROG.—*Cystignathus ornatus.*

peated. Dr. A. Smith relates that he was greatly puzzled on hearing this strange whistling sound, and made many a fruitless search after the utterer. At last one of the Hottentots showed him the animal in its burrow, and after that time he was able to procure as many as were desired.

The head and body of this species are short, puffy, and smooth, and the colour is yellowish grey, with three longitudinal bands. Below, it is yellowish white without any mottlings. Its length is about two inches.

THE pretty PAINTED FROG is a European species, being found in Greece, Sicily, and Sardinia. It has a rather wide range of locality, as it is not uncommon in Northern Africa, along the banks of the Nile, and is tolerably plentiful along the shores of the Mediterranean.

It is fond of water, but seems careless whether it be salt or fresh, and is found indifferently in rivers, streams, lakes, and the saline morasses. The common esculent Frog possesses similar habits, and the two species are often seen in company. The food of the Painted Frog consists of insects, spiders, slugs, and snails, both terrestrial and aquatic. There is a difference in the web of the toes in the sexes, those of the female being scarcely webbed at all, while in the male the membrane extends to half their

PAINTED FROG.—*Discoglossus pictus.*

length.   The thumb is quite rudimentary, and its place is indicated by a small tubercular projection.

The colour and general aspect of the skin are extremely variable, the difference seeming to be quite capricious, and not depending on sex or locality.   The ground colour is usually yellowish green or olive, decorated with spots and having several white longitudinal streaks.   In some specimens the skin is smooth, while in others it is covered with tubercles, and the spots are seldom alike in two individuals, sometimes running together so as to form continuous bands.   The white lines too are often partially, and sometimes wholly absent.   In this species the male does not possess any vocal sacs.

THE reader will remember that in the description of the Surinam Toad, on page 149, mention was made of the curious manner in which the female carries her eggs upon her back until they have passed through their preliminary stages of existence.   A noteworthy analogy, close in some respects, but failing singularly in others, is to be found in the NURSE FROG of Europe (*Alytes obstétricans*).

In this species it is the male that undergoes the anxieties of watching over the young offspring, his mate being comparatively free from that duty.

When the eggs, about sixty in number, are laid, he takes possession of them, and fastens them to his legs by means of a glutinous substance, and carries them about with him wherever he goes.   In process of time, the eggs swell, and become so transparent that the black eyes of the future young are seen through their envelopes.   Their careful parent then proceeds to some spot where he can find still water, deposits them, and departs,

SOLITARY FROG.—*Scaphiopus solitarius.*

rejoicing in his freedom. The young soon burst their way through the envelopes in which they had been surrounded, and swim off merrily.

Except at such times, the Nurse Frog is seldom seen in the vicinity of water, and even at that season, the creature does not care to swim about or even to enter the water. The colour of this species is olive-brown with small dark spots. Several specimens, with their eggs, may be seen in the collection of the British Museum.

THE very odd-looking species which is popularly and appropriately termed the SOLITARY FROG is a native of North America, and is remarkable for several peculiarities of form, the eye and the foot being chiefly notable.

It is a land-loving species, never seen in or near water except during the breeding season. During the greater part of the year it resides in holes which it scoops in the sandy soil, and at the bottom of which it sits watching for prey, much like a gigantic ant-lion. In order to assist it in digging, the animal is furnished with a flat, sharp-edged spur, with which it scoops out the loose soil. Sometimes, however, it wedges itself into the sand, tail foremost, and shovels its way downwards much after the fashion of the crab. The hole is about six inches in depth.

Quick though it is in this labour, it is but a sluggish and inactive creature when compared with most of its kin, being a very poor leaper, and slow in most of its movements. It is generally to be seen in the month of March, just after the spring rains, and is a very hardy species, caring little for cold, and traversing the snow without apparent inconvenience.

The eye of the Solitary Frog is very beautiful, and at the same time most remarkable. It is large, full, and of a rich topaz hue, and across its centre run two bold black lines at right angles to each other, so as to form a cross very like that which is seen upon starch grains when viewed by polarized light.

Altogether, the aspect of this species is very unique. It looks much more like a toad than a frog, and has a remarkably blunt snout. Its general colour is olive, mottled with brown above, and covered with tubercles. Along each side of the spine runs a line of "king's yellow," and the under parts are yellowish white. The average length of the Solitary Frog rather exceeds two inches.

THE last of the true Frogs which can be mentioned in this work is the BOMBARDIER (*Bombinator igneus*), a native of many parts of Europe, and common in France.

It is fond of water, and seldom found in very dry localities. When disturbed, it has the power of emitting a strong and very unpleasant odour of garlic, which serves it as a means of defence, like the penetrating scent of the common ringed snake. It is active, and can both swim and leap well. The eggs are laid in long strings, and the tadpole is of a very large size when compared with the earliest state of its perfect existence, and, like the paradoxical Frog already described, is larger in the tadpole state than after it has assumed its perfect form.

The colour of the Bombardier is greyish brown above, and orange below, marbled or spotted with blue black.

WE now arrive at another section of Batrachians, including those creatures which are known under the title of Toads, and of which the COMMON TOAD of Europe is so familiar an example. The members of this section may be known by the absence of teeth in the jaws and the well-developed ears.

The general aspect and habits of this creature are too well known to require more than a cursory notice. Few creatures, perhaps, have been more reviled and maligned than the Toad, and none with less reason. In the olden days, the Toad was held to be the very compendium of poison, and to have so deadly an effect upon human beings, that two persons were related to have died from eating the leaf of a sage bush under which a Toad had burrowed. Still, even in those times, it was held to possess two virtues, the one being the celebrated jewel supposed to be found in its head, and the other the power of curing bleeding at the nose.

This jewel could not be procured by dissection, but must be obtained by causing the owner to eject it. " But the art," says one of the quaint old writers, "is in taking of it out, for they say it must be taken out of the head alive before the Toad be dead, with a piece of cloth of the colour of red Scarlet, wherewithal they are much delighted, so that while they stretch out themselves as it were in sport upon that cloth, they cast out the stone of their head, but instantly they sup it up again, unless it be taken from them through some secret hole in the said cloth, whereby it falleth into a cistern or vessel of water into which the Toad dareth not enter, by reason of the coldnesse of the water. . . . The probation of this Stone is by laying of it to a live Toad, and if she lift up her head against it, it is good, but if she run away from it, it is a counterfeit."

The same writer gives, in his own racy language, an account of the use to which even so venomous an animal as a Toad may be put by those who know how to employ the worst things for the best purposes. " Frederic, the Duke of Saxony, was wont to practis in this manner. He had ever a Toad pierced through with a piece of wood, which Toad was dryed in the smoak or shadow, this he rowled in a linnen cloth; and when he came to a man bleeding at the nose, he caused him to hold it fast in his hand until it waxed hot, and then would the bloud be stayed. Whereof the Physitians could never give any reason, except horrour and fear constrained the bloud to run into his proper place, through fear of a Beast so contrary to humane nature. The powder also of a Toad is said to have the same vertue."

For these, and other similar opinions too numerous for mention, there is some little foundation. The skin of the Toad's back is covered thickly with little glands, and some larger glands are gathered into two sets, one at each side of the back of the head, and secrete a liquid substance, with sufficient acridity to make the eyes smart should they be touched with this fluid, and to force a dog to loose his hold, if he should pick up a Toad in his mouth, and run away with open jaws and foaming mouth. The glands at the back of the head secrete a large quantity of liquid, and if pressed, will eject it in little streams to the distance of a few inches.

In France, this poor creature is shamefully persecuted, the idea of its venomous and spiteful nature being widely disseminated and deeply rooted. The popular notion is that

TOAD.—*Bufo vulgáris.*

the Toad is poisonous throughout its life, but that after the age of fifty years it acquires venomous fangs like those of the serpents. I once succeeded, but with great difficulty, in saving the life of a fine fat Toad that was leisurely strolling in the Forest of Meudon and had got into a rut too deep for escape. I had stooped down to remove the poor creature from danger, but was dragged away by the bystanders, who quite expected to see me mortally bitten, and who proceeded to slaughter the Toad on the spot. "Every one kills Toads in France," said they.

Hearing from them, however, that tobacco was instantaneously fatal to Toads, I made a compromise that they might kill it by putting tobacco on it, but in no other way. The experiment was accordingly tried, and I had the pleasure of seeing the creature walk away with the tobacco on its back, quite unconscious that it ought to have been dead. One of the spectators not only insisted upon the quinquegenarian fangs, but averred that he had a pair at home in a box. However, I never could induce him to show them to me.

In point of fact, the Toad is a most useful animal, devouring all kinds of insect vermin, and making its rounds by night when the slugs, caterpillars, earwigs, and other creatures are abroad on their destructive mission. Many of the market-gardeners are so well aware of the extreme value of the Toad's services, that they purchase Toads at a certain sum per dozen, and turn them out in their grounds.

Dull and apathetic as the Toad may seem, it has in it an affectionate and observant nature, being tamed with wonderful ease, and soon learning to know its benefactors and to come at their call. Mr. Bell had one of these creatures, which was accustomed to sit on one hand and take its food out of the other. Many persons have possessed tame Toads, which would leave their hiding place at the sound of a whistle or a call, and come hastily up to receive a fly, spider, or beetle. Toads can be rendered useful even in a house, for they will wage unceasing war against cockroaches, crickets, moths, flies, and other insect pests.

It is worthy of notice, that the Toad will never catch an insect or any other prey as long as it is stationary, but on the slightest movement, the wonderful tongue is flung forward, picks up the fly on the tip, and returns to the throat, placing the morsel just in the spot where it can be seized by the muscles of the neck, and passed into the stomach. So rapidly is the act performed, that Mr. Bell has seen the sides of a Toad twitching convulsively from the struggles of a beetle just swallowed, and kicking vigorously in the stomach.

Entomologists sometimes make a curious use of the Toad. Going into the fields soon after daybreak, they catch all the Toads they can find, kill them, and turn the contents of their stomachs into water. On examining the mass of insects that are found in the stomach, and which are floated apart in the water, there are almost always some specimens of valuable insects, generally beetles, which from their nocturnal habits, small dimensions, and sober colouring, cannot readily be detected by human eyes.

The Toad will also eat worms, and in swallowing them it finds its fore-feet of great use. The worm is seized by the middle, and writhes itself frantically into such contortions that the Toad would not be able to swallow it but by the aid of the fore-feet, which it uses as if they were hands. Sitting quietly down with the worm in its mouth, the Toad pushes it further between the jaws first with one paw and then with another, until it succeeds by alternate gulps and pushes to force the worm fairly down its throat.

These paws are also useful in aiding it to rid itself of its cuticle, which is shed at intervals, as is the case with many reptiles and Batrachians. The process is so singular, and so admirably described by Mr. Bell, that it must be given in his own words :—

" I one day observed a large Toad, the skin of which was particularly dry and dull in its colour, with a light streak down the mesial line of its back ; and on examining further, I discovered a corresponding line along its belly. This proved to arise from an entire slit in the old cuticle, which exposed to view the new and brighter skin underneath. Finding, therefore, what was going to happen, I watched the whole detail of this curious process.

I soon observed that the two halves of the skin, thus completely divided, continued to recede further and further from the centre, and became folded and rugose ; and after a short space, by means of the continued twitching of the animal's body, it was brought down in folds on the sides. The hinder leg, first on one side and then on the other, was brought forward under the arm, which was pressed down upon it, and on the hinder limb being withdrawn, its cuticle was left inserted under the arm, and that of the anterior extremity was now loosened, and at length drawn off by the assistance of the mouth. The whole cuticle was thus detached, and was now pushed by the two hands into the mouth in a little ball, and swallowed at a single gulp. I afterwards had repeated opportunities of watching this curious process, which did not materially vary in any instance."

Though apparently unfit for food, the Toad is eaten by some nations, and certainly is not more unprepossessing than the iguana. The Chinese, however, are in the habit of eating a species of Toad for the purpose of increasing their bodily powers, thinking that the flesh of this creature has the property of strengthening bone and sinew.

This animal is extremely tenacious of life, and is said to possess the power of retaining life for an unlimited period if shut up in a completely air-tight cell. Many accounts are in existence of Toads which have been discovered in blocks of stone when split open, and the inference has been drawn that they were inclosed in the stone while it was still in the liquid state, some hundreds of thousands of years ago, according to the particular geological period, and had remained without food or air until the stroke of the pick brought them once more to the light of day.

Such an account appears at once to be so opposed to all probability as to challenge a doubt ; but if there had been sufficient testimony, even to one such fact, an unprejudiced thinker would be justified in placing it among the wonderful but veritable occurrences that occasionally startle mankind. But there really seems to be no account which is sufficiently accurate to permit of such a conclusion. In more than one case, the whole story has proved to be nothing more than an imposition ; and in others, there is hardly sufficient evidence to show that some crevice did not exist, which would supply the inclosed animal with sufficient air for its narrow wants, and permit many minute insects to crawl into the cavity which held the imprisoned Toad.

NATTERJACK.—*Bufo calamita.*

There is no doubt that in many cases a little Toad has crept into a rocky crevice after prey or in search of a hiding place, and by reason of its rapid increase in size been unable to make its exit. As, moreover, the creature is very long lived, it would, by frequent movements, give a polish to the walls of its cell in a few years; a circumstance that has been employed as a proof of the antiquity of the Toad and its residence. Similar instances are known where the animal has been found enclosed in timber. Here, however, is less difficulty in accounting for the fact, because the growth of wood over a wounded part is often extremely rapid, and has been known to cause the enclosure of nails, tools, and even birds' nests with their eggs. Even in such a case, there is not sufficient evidence to prove that the closure was absolutely perfect, and that the Toad was hermetically sealed in the wooden walls of its cell.

Dr. Buckland made some experiments on this supposed property of the Toad, and inclosed a number of these creatures in artificial chambers, made to represent as nearly as possible the rock and wood in which the imprisoned Toads have been found. None of these experiments met with success; and in those cases where the Toads lived longest, the plaster was found imperfect. Some of the Toads whose cells were really air-tight died in a month or two.

It may, however, be reasonably urged that such experiments do not fairly represent the original conditions under which an animal could survive for so long a period, and that in order to carry out the experiment in a consistent manner, the Toads ought to have been procured when very young, inclosed in a chamber with a moderate aperture, and that aperture lessened gradually, so as to prepare the creature by degrees for its long fast and deprivation of air. For a good summary of this subject and a collection of almost every narrative, I may refer the reader to Mr. Gosse's "Romance of Natural History," second series.

The development of the Toad is much like that of the Frog, except that the eggs are not laid in masses but in long strings, containing a double series of eggs placed alternately. These chains are about three or four feet in length, and one-eighth of an inch in diameter. They are deposited rather later than those of the Frog, and the reptiles, which are smaller and blacker than the Frog larvæ, do not assume their perfect form until August or September. The general colour of the Toad is blackish-grey with an olive tinge, and the tubercles which stud the surface are brown. Beneath, it is yellowish white, tinged with grey, and in some specimens spotted with black. The full size of the Toad is not well ascertained, as it seems to have almost unlimited capacities for increasing in size together with years. The length of a very large specimen is about three inches and a half.

ANOTHER species of Toad, the NATTERJACK, is found in many parts of England. It may be known from the common species by the short hind-legs, the more prominent eyes, the less webbed feet, the yellow line along the middle of the back, and the black bands on the legs. It is not so aquatic as the common Toad, haunting dry places, and seldom approaching water except during the breeding season. Its ordinary length is about three inches.

The GREEN or VARIABLE TOAD (*Bufo víridis*) is rather a handsome species, and is found plentifully in the south of France. It derives its popular names from the large spots of deep green with which its upper surface is adorned. Many of the Batrachians possess the capability of changing their hues according to locality or through mental emotion, and the Green Toad is extremely conspicuous in this respect, wearing different colours in light and shade, sleep and wakefulness.

The WARTY TOAD of Fernando Po (*Bufo tuberósus*) is a singular looking species, remarkable for the extreme development of the hard tubercles on the back, and being among Batrachians analogous to the moloch among lizards, or the porcupine among mammalia. The whole upper surface of the body is thickly covered with large tubercles, each having a horny spine in the centre. The glands on the back of the head are large and very conspicuous. Even the under parts are covered with tubercles, but without the spine in the centre. Above each eyelid is a group of horny tubercles, so that the creature presents a most remarkable appearance. Its length is about three inches.

OUR last example of these creatures is the large AGUA TOAD of America (*Bufo agua*).

This large species digs holes in the ground, and resides therein. It is one of the noisiest of its tribe, uttering a loud snoring kind of bellow by night and sometimes by day, and being so fond of its own voice that even if taken captive it begins its croak as soon as it is placed on the ground. It is very voracious, and as it is thought to devour rats, has been imported in large numbers from Barbadoes into Jamaica, in order to keep down the swarms of rats that devastate the plantations. When these creatures were first set loose in their new home, they began to croak with such unanimous good-will that they frightened the inhabitants sadly, and caused many anxious householders to sit up all night.

This Toad grows to a great size, often obtaining a length of seven inches, and nearly the same measurement in breadth. It may be recognised by the great enlargement of the bone over the eyes, and the enormous dimensions of the glands behind the head. Its colour is extremely variable.

WE now come to the Tree-Frogs, or Tree-Toads, so called from their habits of climbing trees, and attaching themselves to the branches or leaves by means of certain discs on the toes, like those of the geckos. In the first family the toes are webbed, and the processes of the vertebræ are cylindrical. A good example will be found in the SAVANNAH CRICKET FROG of America.

This species is very common in its own country, and is found throughout a very large range of territories, specimens in the collection of the British Museum having been taken from several Northern and Southern States of America. It is a light, merry little animal, uttering its cricket-like chirp with continual reiteration, even in captivity. Should it be silent, an event sometimes greatly to be wished, it can at any time be roused to utterance by sprinkling it with water. It is easily tamed, learns to know its owner, and will take flies from his hand.

This species frequents the borders of stagnant pools, and is frequently found on the leaves of aquatic plants and of shrubs that overhang the water. It is not, however, possessed of such strongly adhesive powers as the true Tree-Frogs, and is unable to sustain itself on the under side of a leaf. It is very active, as may be surmised from the slender body and very long hind-legs, and, when frightened, can take considerable leaps for the purpose of avoiding the object of its terror.

The colour of this species is greenish brown above, diversified by several large oblong spots edged with white, and a streak of green, or sometimes chestnut, which runs along the spine and divides at the back of the head, sending off a branch to each eye. The legs are banded with darkbrown, and the under surface is yellowish grey with a slight tinge of pink. It is but a little creature, measuring only an inch and a half in length.

SAVANNAH CRICKET FROG.—*Acris gryllus.*

Another species (*Hyla Carolinensis*) is sometimes called by the same popular title, because its voice, like that of the preceding species, bears some resemblance to that of a cricket. Being one of the true Tree-Frogs, it is not a frequenter of the water, but proceeds to the topmost branches of trees, and there chirps during the night.

ANOTHER family, containing the well-known Tree-Frog of Europe, has the toes webbed, and the processes of the vertebræ flattened. The best-known species is the common GREEN TREE-FROG of Europe, now so familiar from its frequent introduction into fern-cases and terrestrial vivaria.

This pretty creature is mostly found upon trees, clinging either to their branches or leaves, and being generally in the habit of attaching itself to the under side of the leaves, which it resembles so strongly in colour, that it is almost invisible even when its situation is pointed out. When kept in a fern-case, it is fond of ascending the perpendicular glass sides, and there sticking firmly and motionless, its legs drawn closely to the body, and its abdomen flattened against the glass.

The food of the Tree-Frog consists almost entirely of insects, worms, and similar creatures, which are captured as they pass near the leaf whereto their green foe is adhering. It is seldom seen on the ground except during the breeding season, when it seeks the water, and there deposits its eggs much in the same manner as the common Frog. The tadpole is hatched rather late in the season, and does not attain its perfect form until two full months have elapsed. Like the Toad, the Tree-Frog swallows its skin after the change. The common Tree-Frog is wonderfully tenacious of life, suffering the severest wounds without seeming to be much distressed, and having even been frozen quite stiff in a mass of ice without perishing.

The following interesting account of a young Tree-Frog is by Mr. G. S. Ullathorne :—

" My acquaintance with this interesting reptile (which had already passed through all the stages of the tadpole state) began in the following manner :—

I was at school in Hanover at the time, and used frequently to take walks in the neighbouring woods, with a companion. During one of these walks we came across three Green Frogs (or rather they came across our path). Guessing at once they were Tree-Frogs, and thinking that they were just the things to keep, we were 'down upon them,' and tied them up in our handkerchiefs. I contented myself with one, and let my companion have the others. When I arrived safely at my journey's end with my Frog, I procured for him a good-sized glass jar, put a little water in the bottom, a branched stick for him to climb up (though he generally preferred the sides of the glass), covered the top of the jar with a piece of muslin, and installed him on a shelf with a salamander (*Salamandra maculosa*), a ring snake (*Natrix torquata*), and various other 'pets.'

My great amusement was to watch the little creature eat. When I put a fly into his jar, as long as the fly remained quiet, the Frog took no notice of it, but directly the fly began buzzing about, the Frog would wake up from his lethargic state, and on a suitable opportunity would make a leap at the poor fly, adroitly catch it in his mouth (though he sometimes missed his mark), and, I need hardly add, swallow it. On one occasion, I gave my little favourite a very large 'blue-bottle,' almost as large as himself, but nothing daunted, he caught it in his mouth and endeavoured to swallow it, though in vain, for had I not been there I verily believe he would have been choked.

Before he changed his skin, which he did now and then, his colour became much darker and looked more dirty, and he went into quite a torpid state, but when the event was over, he appeared greener and livelier than ever. One day, after I had had him some time, I was playing upon the pianoforte, when I was astonished by an extraordinary sound, but on looking round I discovered the cause of the great noise, for there was my Frog swollen to an immense extent under the chin, and croaking in a very excited manner, making quite a loud noise. I mention this circumstance because it has been imagined that a *solitary* Tree-Frog will not croak, but mine certainly proved to the contrary, for though the first croaking was evidently the effect of the piano, yet he would frequently croak after that time without being excited by any apparent noise whatever. I may here mention that the noise of a quantity of Frogs croaking and nightingales singing, has frequently kept me awake for a considerable time during a spring night.

And now comes the most melancholy part of my story. Leaving my Frog carelessly on the window-sill, I went to school ; when I came back there was the glass certainly, and the Frog also, but oh ! distressingly melancholy to relate, the water was quite hot from the intense heat of the sun, and the poor Frog was scorched, or rather boiled to death—he was quite discoloured, being instead of green a sort of yellow. And thus ends my tale."

The colour of this species is green above, sometimes spotted with olive, and a greyish yellow streak runs through each eye towards the sides, where it becomes gradually fainter, and is at last lost in the green colour of the skin. In some specimens there is a greyish spot on the loins. Below, it is of a paler hue, and a black streak runs along the side, dividing the vivid green of the back from the white hue of the abdomen.

The CHANGEABLE TREE-TOAD is a native of many parts of America, being found as far north as Canada, and as far south as Mexico. It is a common species, but owing to its faculty of assimilating its colour to the tints of the object on which it happens to be sitting, it escapes observation, and is often passed unnoticed in spots where it exists in great numbers.

This is a curious and noteworthy species, as it possesses the capability of changing its tints to so great an extent that its true colours cannot be described. It is usually found on the trunks of trees and old moss-grown stones, which it so nearly resembles in colour, that it can hardly be detected, even when specially sought. The skin of this creature will in a short time pass from white through every intermediate shade to dark-brown, and it is not an uncommon event to find a cross-shaped mark of dark-brown between the shoulders. Old and decaying plum-trees seem to be its favourite resting-places, probably because the insects congregate on such trees.

GREEN TREE-FROG.—*Hyla arbórea.* CHANGEABLE TREE-TOAD.—*Hyla versicolor.*

It is a noisy creature, especially before rain, and has a curious liquid note, like the letter *l* frequently repeated, and then ending with a sharp, short monosyllable. During the breeding season, this Frog leaves the trees and retires to the pools, where it may be heard late in the evening. In the winter it burrows beneath the damp soil, and there remains until the spring. The contour of this species is very toad-like in shape and general appearance, and this resemblance is increased by the skin glands, which secrete a peculiarly acrid fluid.

The upper surface of this creature is, as has already been remarked, too variable for description. There is always, however, a little bright yellow on the flanks, and the under surface is yellowish white, covered with large granulations. The length of this species is about two inches.

In the Pouched Frog we find a most singular example of structure, the female being furnished with a pouch on her back, in which the eggs are placed when hatched, and carried about for a considerable period.

This pouch is clearly analogous to the living cradle of the marsupial animals. It is not merely developed when wanted, as is the case with the cells on the back of the Surinam Toad, but is permanent, and lined with skin like that of the back. The pouch does not attain its full development until the creature is of mature age, and the male does not possess it at all. When filled with eggs the pouch is much dilated, and extends over the whole back nearly as far as the back of the head. The opening is not easily seen without careful examination, being very narrow, and hidden in folds of the skin.

Its colour is very variable, but green has the predominance. It is found in Mexico, but many specimens in the British Museum have been brought from the Andes of Ecuador.

A very curious species, called the Lichened Tree-Toad (*Trachycéphalus lichenátus*), inhabits Jamaica, and is described by Mr. Gosse in his " Naturalist's Sojourn " in that island.

POUCHED FROG.—*Nototrema marsupiatum.*

It derives its name from the aspect of the head, which looks as if it was overgrown with lichens. It is generally found among the wild pine trees, and is very active, being able to take considerable leaps. Sometimes it puffs out its body, and causes a kind of frothy moisture to exhale from the skin. This moisture adheres to the fingers like gum, and causes the Frog to leave a trail behind it like that of a snail or slug.

The colour of the Lichened Tree-Toad is pale red mottled with brown, and having a large patch of the same colour between the shoulders. The muzzle and sides are pale green, spotted with dark reddish brown, and below it is whitish grey, the chin being speckled with reddish brown. The head is flattened, sharply pointed at the muzzle, and studded with sharp bony ridges. Its ordinary length is about four inches.

ANOTHER species of the same genus, the MARBLED TREE-TOAD (*Trachycéphalus marmorátus*), is described by the same writer :—

" One of them was taken in a bedroom at Savannah-le-Mar, one night in October, having probably hopped in at the open window from the branches of a mango tree only a few feet distant. I was surprised at its change of colour, in this respect resembling the chameleon and anoles, or still nearer, the geckos.

When I obtained it, the whole upper parts were of a rich deep amber brown, with indistinct black bands. On looking at it at night, to my surprise I saw a great alteration of hue. It was paler on the head and back, though least altered there; on the rump and on the fore and hind legs it was become a sort of semi-pellucid drab, marked with minute close-set dark specks. When disturbed, it presently became slightly paler still, but in a few minutes it had recovered its original depth of tint. In the course of half an hour it displayed again the speckled dark hue, and now uniformly so, save a black irregular patch or two on the head, and a dark patch between the mouth and each eye. The belly, which was very regularly shagreened, was of a dull buff, not susceptible of change. Its eyes retained their proverbial beauty, for the irides were of a golden-brown tint, like sun-rays shining through tortoiseshell.

This specimen was about as large as a middling English Frog, being two inches and a quarter in length.

BLUE FROG.—*Pelodryas cœruleus.*

While in captivity, if unmolested, it spent a good deal of time motionless, squatting flat and close, with shut eyes as if sleeping, but sometimes it was active. I kept it in a basin covered with a pane of glass, for facility of observation. It would keep its face opposite the window, altering its position pertinaciously if the basin were turned, though ever so gently. It took no notice of cockroaches, nor of a large flesh-fly, which buzzed about it, and even crawled over its nose. If taken in the hand, it struggled vigorously, so as to be with difficulty held; once or twice, while thus struggling, it uttered a feeble squeak; but if still retained, it would at length inflate the abdomen with air, apparently a sign of anger. It leaped, but not far."

THE very odd-looking species which is shown in the accompanying illustration is the sole representative of a family, remarkable for having webbed toes, flattened processes of the vertebræ, and glands at the back of the head.

The BLUE FROG, as it is called from its hue, inhabits Australia, and is not uncommon at Port Essington, whence several specimens have been brought to this country. The head of this species is broader than long, the muzzle short and rounded, and the gape very large. The secreting glands at the back of the head are large, and extend in a curve over the ear as far as the shoulder. They are pierced with a large number of pores, and by their shape and dimensions give to the creature a very singular aspect. The discs of the fore-feet are extremely large, and the toes of the hind-feet are about three-quarters webbed. The colour of the Blue Frog is light uniform blue above, and below silvery white. Its length is about three inches and a half.

THE large and handsome BICOLOURED TREE-FROG is the only species at present known as belonging to the family.

In this creature the toes are not webbed, but in other respects the form resembles that of the preceding family, except perhaps that the processes of the vertebræ are wider in proportion to their volume. The Bicoloured Tree-Frog inhabits South America, Brazil, and Guiana, and seems to be tolerably common. Possibly its bright and boldly contrasting colours render it more conspicuous than its green and olive relatives. The popular name of this creature is very appropriate, as the whole of the upper parts are intense azure, and the under parts pure white, or white tinged with rose. The thighs and sides are spotted with the same hue as the abdomen.

BICOLOURED TREE-FROG.—*Phyllomedusa bicolor.*

PASSING over the small section of Frogs (*Micrhylina*) distinguishable by their toothed jaws and imperfect ears, and represented by a single species, we come to the third section of these animals (*Hylaplesura*), known by their toothless jaws and perfectly developed ears. Of this section, the Two-Striped Frog affords a good example.

This species is a native of Southern Africa, and is chiefly found in the eastern and northeastern parts of the colony of Cape Town. It lives almost entirely upon or in trees, and may be seen either in the cavities of a decaying trunk, or clinging to the bark in close proximity to one of these holes.

In Dr. A. Smith's "Illustrations of the Zoology of Southern Africa," there is so curious and important an account of the imprisonment of this species in the bole of a tree, that it must be given in his own words :—

" On the banks of the Limpopo river, close to the tropic of Capricorn, a massive tree was cut down to obtain wood to repair a waggon. The workman, while sawing the trunk longitudinally, nearly along its centre, remarked on reaching a certain point—' It is hollow, and will not answer the purpose for which it is wanted.'

He persevered, however, and when a division into equal halves was effected, it was discovered that the saw in its course had crossed a large hole, in which were five specimens of the species just described, each about an inch in length. Every exertion was made to discover a means of communication between the external air and the cavity, but without success. Every point of the latter was probed with the utmost care, and water was left in each half for a considerable time, without any passing into the wood. The inner surface of the cavity was black, as if charred, and so was likewise the adjoining wood for half an inch from the cavity.

The tree, at the part where the latter existed, was nineteen inches in diameter, the length of the trunk was eighteen feet ; the age, which was observed at the time, I regret

TWO-STRIPED FROG.—*Brachymerus bifasciatus.*

to say, does not appear to have been noted. When the Batrachia above mentioned were discovered, they appeared inanimate, but the influence of a warm sun, to which they were subjected, soon imparted to them a moderate degree of vigour. In a few hours from the time they were liberated, they were tolerably active, and able to move from place to place, apparently with great ease."

The colour of this species is deep liver-brown above, with two longitudinal yellow stripes, beginning at the eyes and extending as far as the base of the hind legs. A forked yellow mark appears between these stripes just where they end, and the limbs are liver-brown, spotted with yellow. The under parts are very pale brownish red, profusely variegated with pale yellow spots. In length it is nearly two inches. The generic name Brachymerus is derived from two Greek words, signifying short-thighed.

The TINGEING FROG of Southern America (*Hylaplésia tinctória*) is worthy of a casual notice.

This creature is so called because the Indians are said to employ it for imparting a different tinge to the plumage of the green parrot. They pluck out the feathers on the spots where they desire to give the bird a different coloured robe, and then rub the wounded skin with the blood of this Frog. The new feathers that supply the places of those that have been removed, are said to be of a fine red or yellow hue.

It is found in various parts of Southern America, and is common in Surinam, where it mostly inhabits the woods, traversing the branches and leaves by day, and at night concealing itself under the loose bark. Like the common tree-Frog of Europe, it seldom visits the water except during the breeding-season for the purpose of depositing its eggs.

In colour it is extremely variable. Some specimens are black, with a white spot on the top of the head, and two stripes of the same colour running from the head along each side. In certain individuals there are cross bands of white between the stripes. Other examples are grey above and black below; some are wholly black, spotted with large round white marks; others are black; others are grey, spotted with black; while a few specimens are brown, with a large white spot on each side, and two white bands on the fore limbs.

The RHINOPHRYNE is remarkable as being the only known example among the Frogs where the tongue has its free end pointing forward, instead of being directed towards the throat.

RHINOPHRYNE.—*Rhinophrynus dorsalis.*

This curious species inhabits Mexico, and can easily be recognised by the peculiar form of its head, which is rounded, merged into the body, and has the muzzle abruptly truncated, so as to form a small circular disc in front. The gape is extremely small, and the head would, if separated, be hardly recognisable as having belonged to a Frog. There are two glands by the ears, but although they are of considerable dimensions, they are scarcely apparent externally, being concealed under the skin. The legs are very short and thick, and the feet are half-webbed. Each hind foot is furnished with a flat, oval, horny spur formed by the development of one of the bones. There are no teeth in the jaws, and the ear is imperfect. The colour of the Rhinophryne is slate-grey, with yellow spots on the sides and a row of similar spots along the back. Sometimes these latter spots unite so as to form a jagged line down the back.

WE now arrive at the Crawling Batrachians, technically called Amphibia Gradientia. All these creatures have a much elongated body, a tail which is never thrown off as in the frogs and toads, and limbs nearly equal in development, but never very powerful. Like the preceding sub-order, the young are hatched from eggs, pass through the preliminary or tadpole state, and, except in a very few instances, the gills are lost when the animal attains its perfect form. Both jaws are furnished with teeth, and the palate is toothed in some species. The skin is without scales, and either smooth or covered with wart-like excrescences. There is no true breast-bone, but some species have ribs.

The development of the young from the egg is not quite the same as that of the tailless Batrachians. Instead of being deposited in masses or long strings, the eggs are laid singly, and are hatched in succession. When the young are first hatched they bear some resemblance to the tadpole of the frog, the gills being very conspicuous. In these creatures, however, the fore-legs make their appearance first, and are soon followed by the hinder pair, whereas in the frogs the hind-legs are seen for some time before the fore-limbs are visible externally. Further remarks will be made on this subject when we come to our well-known representative of this sub-order, the common newt or eft.

SALAMANDER.—*Salamandra maculosa.*

The celebrated Salamander, the subject of so many strange fables, is a species found in many parts of the continent of Europe.

This creature was formerly thought to be able to withstand the action of fire, and to quench even the most glowing furnace with its icy body. It is singular how such ideas should have been so long promulgated, for although Aristotle repeated the tale on hearsay, Pliny tried the experiment, by putting a Salamander into the fire, and remarks with evident surprise, that it was burned to a powder. A piece of cloth dipped in the blood of a Salamander was said to be unhurt by fire, and certain persons had in their possession a fire-proof fabric made, as they stated, of Salamander's wool, but which proved to be asbestos.

Another fable related of this creature still holds its ground, though perhaps with little reason. I have already mentioned one or two instances of the prejudices which are so deeply ingrained in the rustic mind, and given a short account of the superstitions prevalent in France regarding toads. The Salamander there suffers an equally evil reputation with the toad, as may be seen by the following graphic and spirited letter addressed to the *Field* newspaper :—

" Returning homeward a few evenings ago from a country walk in the environs of D——, I discovered in my path a strange-looking reptile, which, after regarding me stedfastly for a few moments, walked slowly to the side of the road, and commenced very deliberately clambering up the wall. Never having seen a similar animal, I was rather doubtful as to its properties ; but, reassured by its tranquil demeanour, I put my pocket-handkerchief over it, and it suffered itself to be taken up without resistance, and was thus carried to my domicile. On arriving *chez moi*, I opened the basket to show my captive to the servants (French), when, to my surprise and consternation, they set up such a screaming and hullabaloo, that I thought they would have gone into fits.

' *Oh! la, la, la, la, la !—Oh! la, la, la, la, la !*' and then a succession of screams in altissimo, which woke up the children, and brought out the neighbours to see what could be the matter.

' *Oh, monsieur a rapporté un sourd !*'

' *Un sourd !*' cried one.

' Un sourd !' echoed another.

' UN S-O-U-R-D ! ! !' cried they all in chorus ; and then followed a succession of shrieks.

3.

When they calmed down into a mild sample of hysterics, they began to explain that I had brought home the most venomous animal in creation.

' *Oh ! le vilain bête !* ' cried Phyllis.

' *Oh ! le méchant !* ' chimed in Abigail; ' he kills everybody that comes near him ; I have known fifty people die of his bite, and no remedy in the world can save them. As soon as they are bitten they *gonflent, gonflent,* and keep on swelling till they burst, and are dead in a quarter of an hour.

Here I transferred my curiosity from the basket to a glass jar, and put a saucer on the top to keep it safe.

' *Oh ! Monsieur,* don't leave him so; if he puts himself in a rage, nothing can hold him. He has got such force that he can jump up to the ceiling ; and wherever he fastens himself he sticks like death.'

' Ah ! it's all true,' cried my landlady, joining the circle of gapers ; ' Oh ! *la, la ! Ça me fait peur ; ça me fait tr-r-r-embler !* '

' Once I saw a man in a haycart try to kill one, and the *bête* jumped right off the ground at a bound and fastened itself on the man's face, when he stood on the haycart, and nothing could detach it till the man fell dead.'

' *Ah ! c'est bien vrai,*' cried Abigail ; ' they ought to have fetched a mirror and held it up to the *bête,* and then it would have left the man and jumped at its *image.*'

The end of all this commotion was that, while I went to inquire of a scientific friend whether there was any truth in these tissue of *bêtises,* the whole household was in an uproar, *tout en émoi,* and they sent for a *commissionnaire* and an ostler with a spade and mattock, and threw out my poor *bête* into the road, and foully murdered it, chopping it into a dozen pieces by the light of a stable lantern ; and then they declared that they could sleep in piece !—*les misérables !*

But there were sundry misgivings as to my fate, and as with the Apostle, ' they looked when I should have swollen or fallen down dead suddenly ; ' and next morning the maids came stealthily and peeped into my room to see whether I was alive or dead, and were not a little surprised that I was not even *gonflé,* or any the worse for my *rencontre* with a *sourd.*

And so it turned out that my poor little *bête* that had caused such a disturbance was nothing more nor less than a Salamander—a poor, inoffensive, harmless reptile, declared on competent authority to be noways venomous ; but whose unfortunate appearance and somewhat Satanic livery have exposed it to obloquy and persecution."

This notion of the poisonous character of the Salamander is of very old date, as the reader may see by referring to any ancient work on Natural History. One of the old writers advises any one who is bitten by a Salamander to betake himself to the coffin and winding-sheet, and remarks that a sufferer from the bite of this animal needs as many physicians as the Salamander has spots. If the Salamander crawled upon the stem of an apple-tree, all the crop of fruit was supposed to be withered by its deadly presence, and if the heel of a man should come in contact with the liquid that exudes from the skin, all the hair of his head and face would fall off.

There is certainly an infinitesimally minute atom of truth in all this mass of absurdities, for the Salamander does secrete a liquid from certain pores in its surface, which, for the moment, would enable it to pass through a moderate fire, and this secretion is sufficiently acrid to affect the eyes painfully, and to injure small animals if taken into the mouth.

The Salamander is a terrestrial species, only frequenting the water for the purpose of depositing its young, which leave the egg before they enter into independent existence. It is a slow and timid animal, generally hiding itself in some convenient crevice during the day, and seldom venturing out except at night or in rainy weather. It feeds on slugs, insects, and similar creatures. During the cold months it retires into winter quarters, generally the hollow of some decaying tree, or beneath mossy stones, and does not reappear until the spring

The ground colour of this species is black, and the spots are light yellow. Along the sides are scattered numerous small tubercles.

CRESTED NEWT.—*Triton cristátus.*　　　　SMOOTH NEWT.—*Lophínus punctátus.*

THE common NEWT, ASKER, EFFET, EFT, or EVAT, as it is indifferently termed, is well known throughout England. At least two species of Newt inhabit England, and some authors consider that the number of species is still greater. We shall, however, according to the system employed in this work, follow the arrangement of the British Museum, which accepts only two species, the others being merely noted as varieties.

The CRESTED NEWT derives its popular name from the membranous crest which appears on the back and upper edge of the tail during the breeding season, and which adds so much to the beauty of the adult male.

This creature is found plentifully in ponds and ditches during the warm months of the year, and may be captured without difficulty. It is tolerably hardy in confinement, being easily reared even from a very tender age, so that its habits can be carefully noted.

At Oxford we had some of these animals in a large slate tank through which water was constantly running, and which was paved with pebbles, and furnished with vallisneria and other aquatic plants, for the purpose of imitating as nearly as possible the natural condition of the water from which the creatures had been taken. Here they lived for some time, and here the eggs were hatched and the young developed.

It was a very curious sight to watch the clever manner in which the female Newts secured their eggs; for which purpose they used chiefly to employ the vallisneria, its long slender blades being exactly the leaves best suited for that purpose. They deposited an egg on one of the leaves, and then, by dexterous management of the feet, twisted the leaf round the egg, so as to conceal it, and contrived to fasten it so firmly that the twist **always** retained its form. The apparent shape of the egg is oval, and semi-transparent, but on looking more closely, it is seen to be nearly spherical, of a very pale yellow-brown, and inclosed within an oval envelope of gelatinous substance.

When the young Newt is hatched it much resembles the common tadpole, but is of a lighter colour, and its gills are more developed. It rapidly increases in size, until it has attained a length of nearly two inches, the fore-legs being then tolerably strong, and the hinder pair very small and weak. The gills are at this time most beautiful objects, and if the young creature be properly arranged under the microscope, the circulation of the blood, as seen through their transparent walls, is one of the most exquisite sights that the microscope can afford.

The legs now attain greater strength, the gills become gradually more opaque and slowly lessen in size, being at last entirely absorbed into the body. In exact proportion to the diminution of the gills the lungs increase in size, and the animal undergoes exactly the same metamorphosis as has already been related of the frog, being changed, in point of fact, from a fish into a batrachian. The tail, however, remains, and is made the principal, if, indeed, not the only means by which the Newt propels itself through the water.

When it has passed through its changes, the Newt is no longer able to lead a sub-aquatic life, but is forced to breathe atmospheric air. For this purpose it rises to the surface at tolerably regular intervals, puts its snout just out of the water, and, with a peculiar little popping sound, ejects the used air from its lungs and takes in a fresh supply.

Towards the breeding season the male changes sensibly in appearance; his colours are brighter, and his movements more brisk. The beautiful waving crest now begins to show itself, and grows with great rapidity, until it assumes an appearance not unlike that of a very thin cock's-comb, extending from the head to the insertion of the hinder limbs, and being deeply toothed at the edge. The tail is also furnished with a crest, but with smooth edges. When the animal leaves the water, this crest is hardly visible, because it is so delicate that it folds upon the body and is confounded with the skin; but when supported by the water, it waves with every movement of its owner, and has a most graceful aspect.

After the breeding season, the crest diminishes as rapidly as it arose, and in a short time is almost wholly absorbed. Some remnants of it, however, always remain, so that the male may be known, even in the winter, by the line of irregular excrescences along the back. The use of this crest is not known, but it evidently bears a close analogy to the gorgeous nuptial plumage of many birds, which at other times are dressed in quite sober garments.

The Newt feeds upon small worms, insects, and similar creatures, and may be captured by the simple process of tying a worm on a thread by the middle, so as to allow both ends to hang down, and then angling as if for fish. The Newt is a ravenous creature, and when it catches a worm, closes its mouth so firmly that it may be neatly landed before it looses its hold. Some writers recommend a hook, but I can assert, from much practical experience, that the hook is quite needless, and that the Newt may be captured by the simple worm and thread, not even a rod being required.

It is curious to see the Newt eat a worm. It seizes it by the middle with a sudden snap, as if the jaws were moved by springs, and remains quiet for a few seconds, when it makes another snap, which causes the worm to pass farther into its mouth. Six or seven such bites are usually required before the worm finally disappears.

The skin or epidermis of the Newt is very delicate, and is frequently changed, coming off in the water in flakes. I found that my own specimens always changed their skin as often as I changed the water, and it was very curious to see them swimming about with the flakes of transparent membrane clinging to their sides. The skin of the paws is drawn off just like a glove, every finger being perfect, and even the little wrinkles in the palms being marked. These gloves look very pretty as they float in the water, but if removed they collapse into a shapeless lump.

The food of the Newt consists of worms, insects, and even the young of aquatic reptiles. I have seen a large male Crested Newt make a savage dart at a younger individual of the same species, but it did not succeed in eating the intended victim.

This creature is very tenacious of life, and the muscular irritability of the body seems to endure for a long time after the creature is dead. One of these animals, that had been dead for some time, whose heart and lungs had been removed, and whose limbs had been pinned out ready for dissection, was so retentive of this singular irritability, that when the tail was touched with the point of a scalpel, the body and limbs writhed so actively as to free the limbs from their attachments. On repeating the experiment, it was found that this susceptibility gradually departed, lingering longest towards the body. The eel possesses an even greater degree of this muscular irritability, as is well known by all who have made an eel-pie or seen it prepared. The tail of the blindworm, too, which has already been described, is equally irritable when separated from the body.

The colour of the Crested Newt is blackish or olive-brown, with darker circular spots, and the under parts are rich orange-red, sprinkled with black spots. Along the sides are a number of white dots, and the sides of the tail are pearly white, becoming brighter in the spring. The length of a large specimen is nearly six inches, of which the tail occupies rather more than two inches and a half.

The STRAIGHT-LIPPED NEWT of Mr. Bell (*Triton Bibronii*) is only ranked as a variety of this species in the official catalogue of the British Museum. In this variety the upper lip does not overhang the lower, and the skin is more tubercular than in the ordinary examples.

The MARBLED NEWT (*Triton Marmorátus*) is a continental species, and is found plentifully in the southern parts of France.

It is a much larger species than the preceding, often attaining the length of eight or nine inches. It mostly lives in the water, but will leave that element voluntarily when the weather is stormy, or even if the hot sunbeams are too powerful to please its constitution. A rather powerful and not very pleasant odour is exhaled from this creature. During the winter it leaves the water, seeks for some hole in a decaying tree, and there remains until the following spring. The colour of the Marbled Newt is olive-brown above, marbled with grey and dotted with white on the back. The head is grey, with black dots and spots. Along the centre of the back runs a streak of white and orange, and the under parts are dotted with white.

The SMOOTH NEWT is more terrestrial in its habits than the crested species, and is often seen at considerable distances from water.

By the rustics this most harmless creature is dreaded as much as the salamander in France, and the tales related of its venom and spite are almost equal to those already mentioned. During a residence of some years in a small village in Wiltshire, I was told some very odd stories about this Newt, and my own powers of handling these terrible creatures without injury was evidently thought rather supernatural. Poison was the least of its crimes, for it was a general opinion among the rustics in charge of the farm-yard that my poor Newts killed a calf at one end of a farmyard, through the mediumship of its mother, who saw them in a water-trough at the other end; and that one of these creatures bit a man on his thumb as he was cutting grass in the churchyard, and inflicted great damage on that member.

The worst charge, however, was one which I heard from the same person. A woman, he told me, had gone to the brook to draw water, when an Effert, as he called it, jumped out of the water, fastened on her arm, bit out a piece of flesh, and spat fire into the wound, so that she afterwards lost her arm.

All the Newts possess singular powers of reproducing lost or injured members, this faculty proving them to hold a rather low place in the scale of creation. The Smooth Newt has been known to reproduce the tail, and even the limbs; and in one case an eye was removed entirely, and reproduced in a perfect state by the end of the year.

This species may be known by its smooth and non-tubercular skin, and its small size. During the breeding season the male wears a crest, which runs continuously from the head to the end of the tail, and is not so deeply cleft as that of the crested species.

This ornament is very delicate and beautiful, and at the height of the season is often edged with beautiful carmine or violet. The colour is brownish grey above and bright orange below, covered with round spots of black. In the autumn and during the winter the abdomen becomes much paler. The length of this species is about three inches and a half.

The PALMATED WATER NEWT of Mr. Bell (*Lissotríton pálmipes*) is held by the authorities of the British Museum to be merely a variety of this species.

WE now arrive at another family, known by the curious manner in which the teeth of the palate form a broken cross-series.

The first example is the JAPANESE SALAMANDER (*Onychodáctylus Japónicus*), remarkable for having, during the larval state and in the breeding season, claws upon the toes. Its colour is purplish black, variegated irregularly with white, and the claws are black. It is thought by the natives to possess medical properties, and they employ its flesh in sundry ailments, killing, and drying it in the sun for better preservation.

AMBYSTOME, OR SPOTTED EFT.—*Ambystoma Carolina.*

ANOTHER example of this family is the AMBYSTOME, or SPOTTED EFT, of North America.

This species is not uncommon in the countries which it inhabits, and is found in some numbers in Pennsylvania. The eggs of this creature are not deposited singly and in the water, as is the case with the newts, but are laid in small packets, and placed beneath damp stones. The head of the Ambystome is thick, convex, and with the muzzle rounded. Its colour is deep violet-black above, and purple-black below, with a row of circular or oval yellow spots along the sides. These spots are large in proportion to the dimensions of the individual, and have a very bold effect. The genus is rather large, containing about eleven acknowledged species. One of them, *Ambystoma talpoideum*, or Mole-like Ambystome, derives its name from its habit of burrowing in the ground after the fashion of the mole. It lives in South Carolina, and is found on the sea-islands. The fore-limbs are peculiarly short and stout, and the body is rather thick and clumsily made.

WE now come to a very remarkable creature, the AXOLOTL, which is presumed to be but the larva or tadpole state of some very large batrachian. Like many other enigmatical animals, it has been bandied about considerably in the course of investigation, and according to the latest observations, the original opinion seems to be correct, namely, that it is not

AXOLOTL.—*Axolotles guttatus.*

an adult crawling batrachian with perpetual gills, but that it is in its preliminary or tad-pole stage of existence. Mr. Baird makes the following sensible remarks on this subject:—

"It so much resembles the larva of *Ambystoma punctata*, in both external form and internal structure, that I cannot but believe it to be the larva of some gigantic species of this genus. It differs from all other perennibranchiates in possessing the larval character of the gular or opercular flap, this being unattached to the adjacent integuments, and free to the extremity of the chin. The non-discovery of the adult is no argument against its existence. I had caught hundreds of the very remarkable larva of *Pseudotriton salmoneus* near Carlisle before I found an adult. Until then I knew nowhere to refer the animal, supposing this species to exist no nearer than the mountains of New York and Vermont.

As may be seen from the illustration, the gills or branchiæ are quite as large in pro-portion as those of the newt in its larval state. They are three in number on each side, and furnished with fringes.

The Axolotl inhabits Mexico, where it is tolerably plentiful, and in some places is found in such numbers that it is sold in the markets for the table. It frequents the lake surrounding the city of Mexico, and, according to Humboldt, is also found in the cold waters of certain mountain lakes at a considerable elevation above the sea.

The colour of this remarkable creature is rather dark greyish brown, covered thickly with black spots. The length varies from eight to ten inches.

ANOTHER small order now comes before us, containing a few species, and only two very small families. In all these creatures the body is long and lizard-like, the legs four and feeble, and the gills internal, but permanent throughout life.

OUR first example of this family is the now celebrated GIGANTIC SALAMANDER.

This is undoubtedly one of the least attractive of the vertebrate animals, being dull in habits, sombre in colour, with a sort of half-finished look about it, and not possessing even that savage ugliness which makes many a hideous creature attractive in spite of its uncomeliness. It is a native of Japan, and even in that country seems to be rare, a large sum being asked for it by the seller. It lives in the lakes and pools that exist in the basaltic mountain ranges of Japan.

GIGANTIC SALAMANDER.—*Siebóldia maxima*

Dr. Von Siebold brought the first living specimen to Europe, and placed it in a tank at Leyden, where it was living when the last accounts were heard, having thus passed a period of many years in captivity. Its length is about a yard. Two specimens were brought over at the same time, being of different sexes, but on the passage, the male unfortunately killed and ate his intended bride, leaving himself to pass the remainder of his life in celibacy. It fed chiefly on fish, but would eat other animal substances.

A fine specimen is now living in the Zoological Gardens, and has attracted much notice in spite of its ugliness and almost total want of observable habits. It is very sluggish and retiring, hating the light, and always squeezing itself into the darkest corner of its tank, where it so closely resembles in colour the rock-work near which it shelters itself, that many persons look at the tank without even discovering its presence. The length of this specimen is about thirty-three inches, and if it survives, it may possibly attain even a larger size.

The head of this creature is large, flattened, and very toad-like in general aspect, except that it is not furnished with the beautiful eyes which redeem the otherwise repulsive expression of the toad. The head is about four inches wide at the broadest part, and is covered with innumerable warty excrescences. The eyes are extremely small, placed on the fore part of the head, and without the least approach to expression, looking more like small glass beads than eyes.

The whole upper part of the body is covered thickly with excrescences, and even the under part of the rounded toes are studded with little tubercles, which can be plainly seen with a magnifying lens as the creature presses its feet against the glass wall of its tank. Despite of its sluggish nature, it is quite able to obtain its own subsistence by catching the

MENOPOME.—*Protonopsis horrida.*

fish on which it feeds, and the keeper told me that even in captivity it easily catches the fish that are put into its tank. On the journey, it was mostly fed upon eels, and at the present time it eats eels as well as other fish, provided they are rather small.

It is well to mention casually in this place that the human-looking skeleton, discovered at Œningen in 1726, and long supposed to be the fossil skeleton of a man who had perished in the deluge, is nothing more than the bones of a huge Salamander, closely allied to the present species. The colour of the Gigantic Salamander is a very dark brown, with a tinge of chocolate, and taking a lighter and more yellowish hue upon the under surface of the feet.

THE great MENOPOME of America has been honoured with a large array of names, among which are TWEEG, HELLBENDER, MUD DEVIL, and GROUND PUPPY, the first being an Indian name, and the others given to the creature in allusion to its mud-loving habits or the ferocity of its disposition.

The Menopome inhabits the Ohio and Alleghany rivers, and it is a fierce and voracious animal, so dangerous a foe to fish and other living beings that it is in some places known by the name of Young Alligator. It is very ugly, and rather revolting in appearance, so that the fishermen stand in great awe of the fierce, active beast, and think it to be venomous as well as voracious. The teeth, however, are very small in proportion to the size of the creature. Its colour is slaty grey, with dark spots, and a dark streak runs through the eye. Its length is about two feet.

THE second family of this order is represented by its typical species, the CONGO SNAKE. This curious creature is a native of America, and is found rather plentifully near New Orleans, in Florida, Georgia, and South Carolina. It is fond of burrowing in mud, and will often descend to a depth of three feet below the surface of the soil, acting indeed more like an earthworm than a vertebrate animal. Many of these creatures have been accidentally dug out while deepening or clearing ditches. The negroes are much afraid of the Congo Snake, and think it to be poisonous, a belief which has its only foundation in fear, generated by ignorance.

CONGO SNAKE.—*Amphiuma means.*

The legs are extremely small and feeble, and there are only two toes on each foot. Its colour is dark blackish grey above, and lighter beneath. Another species, the THREE-TOED CONGO SNAKE (*Muranopsis tridactyla*), is much like the common Congo Snake, from which it may be distinguished by possessing three toes on each foot instead of two. The length of both these creatures is from two to three feet. These two species constitute the whole of the family to which they belong.

AMONG these remarkable animals, the orders multiply themselves rapidly. The Pseudophidia, or False Serpents, include some very curious species, whose position remained long unsettled. There is but one family, and all its members have very long and cylindrical bodies, no limbs, a very short tail, and a smooth, wrinkled skin, in which are embedded a multitude of minute scales. The two worm-like creatures in the illustration on page 187 afford good examples of this very remarkable family.

THE left-hand figure represents the WHITE-BELLIED CÆCILIA. The name Cæcilia is derived from a Latin word signifying blindness, and is given to the creature because the eyes are always minute, and in some species are hidden under the skin. This species inhabits Southern America, and, like the rest of its kin, burrows deeply under the ground after the fashion of the earthworm, to which it bears so strong an external resemblance, preferring wet and marshy ground to dry soil. Its body is rather thick and cylindrical, and is surrounded by about one hundred and fifty incomplete rings. The muzzle is rounded and so is the tail. There are teeth in the jaws and on the palate, all of which are short, strong, and conical; the tongue has a curiously velvety feel to the touch. Below each nostril there is a small pit, sometimes taken for a second nostril.

The colour of the White-bellied Cæcilia is blackish, marbled with white along the under surface.

THE right-hand figure represents the SLENDER CÆCILIA, so called on account of its slight form. In this species the body is smooth throughout the greater part of its length, but towards the tail the skin is gathered into fifteen circular folds pressed closely together.

WHITE-BELLIED CÆCILIA.—*Cæcilia tentaculata.*   SLENDER CÆCILIA.—*Cæcilia gracilis*

The muzzle is rather broad and rounded. The body of the Slender Cæcilia is extremely elongated, being about two feet in length, and not thicker than an ordinary goosequill. Its colour is almost wholly black.

THE small but very remarkable order of animals which stands next in our list, has proved an insoluble enigma to the systematic zoologists, who not only are unable to decide upon any order to which it may belong, or in what precise relation it stands to other reptiles, but are not even able to announce positively its class, or to say whether it is a reptile or a fish. The three species which comprise this order—if indeed they do not form a separate class—are so fish-like in most parts of their anatomy and their general habits, that they might be regarded as belonging to the fishes, were not they allied to the reptiles by one or two peculiarities of their structure. Some accurate and experienced anatomists accordingly place these creatures among the fishes, while others, equally experienced, consider them as belonging to the reptiles. In fact, the position in which these creatures are placed depends wholly on the amount of importance given to the reptilian or piscine characters.

THE species represented in the engraving is that which is found in Africa, inhabiting the beds of muddy rivers, and is known by the name of LEPIDOSIREN or MUD-FISH.

The habits of this creature are very remarkable. Living in localities where the sun attains a heat so terrific during a long period of the year that the waters are dried and even their muddy beds baked into a hard and stony flooring, these animals would be soon extirpated unless they had some means of securing themselves against this periodical infliction, and obtaining throughout the year some proportion of that moisture for lack of which they would soon die. The mode of self-preservation during the hot season is very like that which has already been mentioned in the case of certain frogs and other similar creatures, but is marked by several curious modifications.

When the hot season has fairly commenced, and the waters have begun to lessen in volume, the Lepidosiren wriggles its way deeply into the mud, its eyes being so constructed that the wet soil cannot injure them, and the external nostrils being merely two shallow blind sacs.   After it has arrived at a suitable depth, it curls itself round, with its tail wrapped partly over the head, not unlike the peculiar attitude assumed by fried whitings, except that its flexible spine enables it to squeeze the two sides closer together than can be accomplished in that fish, and in that position awaits the coming rains.   It will lie in a torpid condition for a very considerable space of time, depending entirely on the advent of rain for the re-assumption of vitality.

After it has curled itself up and resigned itself to the exigencies of its condition, a large amount of a slimy substance is secreted from the body, which has the effect of making the walls of its cell very smooth, and probably aids in binding the muddy particles together.   When the rains fall, the moisture penetrates rapidly through the fissures of the earth, cracked in all directions by the constant heat, reaches the cell of the Lepidosiren, dissolves its walls, and restores the inhabitant to life and energy.

Several specimens have been brought to Europe, most of which I have had opportunities of seeing while alive, as well as of examining parts of their structure after death.

LEPIDOSIREN.—*Protopterus annectens.*

While retained in an ordinary aquarium, it passes much of its time in an apparently semi-torpid condition at the bottom of the tank, generally seeking the darkest corner and squeezing itself along one of the perpendicular angles of the case.   It was found, however, that whenever the surface of the water was disturbed, the creature woke up, as it were, and rose to see what was the matter.   In this way it could be induced to come at a signal to take the food on which it lived.

Farther investigations and experiments on a larger scale, such as were carried out at the Crystal Palace, afforded a considerable insight into the habits of this singular creature. I have much pleasure in acknowledging the kind assistance given to me by the directors of that institution, and the facilities which they have afforded me in my inquiries, and especially in returning thanks to Mr. F. Wilson, of the Zoological department, who was in charge of the various specimens of this creature, and who took a most kindly interest in furnishing information respecting the habits of his former charges.

Several batches of these animals have been kept alive at the Crystal Palace, all of which have died, some after a life of only a few weeks, and others after surviving for three years. It will, however, be useless to follow the fortunes of each separate individual, and we will therefore only examine the general habits which seem to be common to all.

The Lepidosirens, or Mud-fish as they are popularly called, were sent while still in their muddy nests, or "cocoons," according to the technical term, and, in one instance, three specimens were inclosed in a single lump of hard mud, weighing when dry about twenty pounds.

One of the cocoons is now lying before me, together with the dried and shrivelled body of its former inhabitant, still curled up in the singular fashion already mentioned. The walls of the cocoon are composed of a thick greyish clay, quite hard and dry, and intermixed here and there with remnants of vegetable matter. The hollow in which the Lepidosiren resided is quite smooth in the interior, but gives no idea of the real shape of the inhabitant, the cell seeming to be somewhat large, most probably on account of the coat of mucous substance with which it was lined, and part of which is to be seen still adhering, like flakes of dry membrane, to the sides of the cell.

By rapidly tearing this membranous substance with an oblique bearing, it can be in some places split like a scrap of paper under similar circumstances, but when placed under the microscope, it shows no signs of organization, being of a light brown colour, irregularly mottled with black. When burned, it rapidly takes fire and bursts into flame, giving out a very nauseous odour, like that which is perceived on burning the wing-case of a beetle, and leaves a firm black ash, of nearly the same shape and form as before the light was applied to it.

The remainder of this substance is found loosely adhering to the body of the former inhabitant, and can be easily stripped off.

On being immersed in water, the earthy cocoons fell to pieces as if they had been made of sugar, and the imprisoned creatures were thus released. At first they were exceedingly sluggish, and hardly stirred, but after the lapse of an hour or two they became tolerably alert.

One of these specimens died after it had been kept about six weeks, and a good plaster-cast of it is now before me. Its length is ten inches, and the circumference of the head, just in front of the fore pair of limbs, is exactly three inches. The scales are tolerably well marked, and are shown even in the plaster-cast, though in the living animal there is hardly a trace of them. They are also very evident after the creature has been immersed in spirits for some time. In taking a cast of the Lepidosiren, the mucous secretion with which the body is covered affords a serious obstacle to the correctness of the image, as it is apt to adhere to the plaster, and pull away with it some portions of the skin.

A fellow-specimen, that floated dead from its cocoon, is also before me, bent on itself in the manner usual among these creatures, and with its mouth widely open, showing the peculiar teeth.

Finding, as has already been mentioned, that the Lepidosiren would rise to the surface of the water when a splashing was made, the attendants used to feed it by paddling about with the finger, and then holding a piece of raw beef in the spot where the disturbance had been made. The creature used to rise deliberately, snatch the meat away, and, with a peculiarly graceful turn of the body, descend to its former resting-place for the purpose of eating its food.

The mode of eating was very remarkable. Taking the extreme tip of the meat between its sharp and strongly formed teeth, it would bite very severely, the whole of the head seeming to participate in the movement, just as the temporal muscles of the human face move when we bite anything hard or tough. It then seemed to suck the meat a very little farther into its mouth and gave another bite, proceeding in this fashion until it had subjected the entire morsel to the same treatment. It then suddenly shot out the meat, caught it as before by the tip, and repeated the same process. After a third such manœuvre it swallowed the morsel with a quick jerk. The animal always went through this curious series of operations, never swallowing the meat until after the third time of masticating.

After a while, it was thought that the water in which it lived was not sufficiently warm to represent the tepid streams of its native land, and its tank was consequently sunk in the north basin of the building, where the water is kept at a tepid heat for the purpose of nourishing the tropical plants which grow in it. Here the creature remained for some time, but at last contrived to wriggle itself over the side of its tank, and roam about in the large basin quite at liberty.

It remained here for some time, and being deprived of its ordinary supply of raw beef, took to foraging for itself. The gold-fish with which the basin is stocked became its victims, and it was quite as destructive as an otter would have been. It had quite a fancy for attacking the largest fish; and though apparently slow in its movements, could catch any fish on which it had set its wishes. As the fish was quietly swimming about, suspecting no evil, the Lepidosiren would rise very quietly beneath it until quite close to its victim, just as the terrible ground-shark rises to take its prey. It then made a quick dart with open mouth, seized the luckless fish just by the pectoral fins, and with a single effort bit entirely through skin, scales, flesh, and bone, taking out a piece exactly the shape of its mouth, and then sinking to the bed of the basin with its plunder. The poor fish was never chased, but was suffered to float about in a half-dead state, and numbers of mutilated gold-fish were taken out of the basin.

I have several times seen the creature while swimming about in search of a dinner, and have been much struck with the exceeding grace of its movements, which indeed very strongly resemble those of the otter.

At last its depredations were checked, for when the basin was cleansed, according to custom, a portion was fenced off, so that the Lepidosiren could not get out, and the gold-fish could not get in.

Not choosing to supply a succession of gold-fish, out of each of which the fastidious creature would only take one bite, the superintendent bethought himself of frogs, and fed the animal regularly with these batrachians. But having been warned, by the effects on the gold-fish, not to trust his fingers within reach of the teeth that could inflict such very effective bites, he got a long stick, cleft one end of it, put one hind-foot of the frog into the cleft, and held it on the surface of the water, so that the struggles of the intended victim should agitate the surface, and warn the Lepidosiren that its dinner was ready. No sooner did the frog begin to splash, than the Lepidosiren rose rapidly beneath it, seized it in its mouth, dragged it off the stick like a pike striking at a roach, and sunk to the bottom with its prey. Not a vestige of the frog was ever seen afterwards; and Mr. Wilson naturally conjectures that the poor victim was gradually chewed up, like the beef with which the creature was formerly fed.

Under this regimen the Lepidosiren grew apace, and in three years had increased from ten inches in length and a few ounces in weight, to thirty inches long, and weighing six pounds and a quarter. The rapidity of its growth may be accounted for by the fact, that it had fed throughout the entire year, instead of lying dormant for want of water during half its existence, and its size was apparently larger than it would be likely to attain in its native state.

Thinking that perhaps the creature might need its accustomed season of repose—happily called æstivation, in opposition to the term hibernation—it was well supplied with clay similar to that from which its cocoon had been formed, but without any result, the animal evincing no disposition to avail itself of the stores so thoughtfully collected in its behalf. This is, I think, a very interesting example of the manner in which nature accommodates herself to circumstances, and is paralleled by many other instances in the several departments of Natural History. Bees, for example, on finding themselves within easy distance of a sugar plantation, have been known to decline honey making; and the same result has occurred when they were transported to fertile localities where the honey-bearing flowers are in blossom throughout the year.

As an example of a similar phenomenon occurring in the vegetable kingdom, I may instance some Australian flowers brought over by Mr. Howitt, and planted in his garden in the suburbs of London. These plants were at first sadly puzzled by the seasons, wanting to blossom just as our winter had set in, but in the course of a few years they grew gradually later in blossoming, until they had found the proper season, and then were content to put forth their leaves and flowers at the same time as the indigenous plants.

The cause of this specimen's regretted death was rather curious. In the winter time, when the basins were cleaned, the animal was removed from the north to the south basin while the former was being emptied. Unfortunately, the fires which warmed the water were suffered to expire during the night, and in the morning the poor Lepidosiren was found chilled to death.

The history of this creature is not only interesting, but is valuable as it shows the comparative advantages of watching the habits of animals in large and small habitations. Had, for example, the creature lived from the first in the large basin, its remarkable mode of eating its food could not have been observed, as it always seeks the bottom of its prison for that purpose; while, had it been always kept in the glass tank, its graceful movements and fish-eating propensities would never have been discovered.

The bones of the Lepidosiren are, when first taken from the body, of a bright green colour, and so gelatinous in structure, that if left in the water they would probably dissolve. After a time, however, the green colour fades, though traces of it can still be discerned. The bones of the head are, however, of a firmer character, as is needful for the management of the sharp and powerful teeth; and in the skull of the above-mentioned specimen, now lying before me, and from which this illustration has been carefully drawn, the green tint still lingers on several of the bones. The drawing is of the natural size.

SKULL OF LEPIDOSIREN.

The teeth are most remarkable, looking as if they were made from a ribbon of enamel-covered bone, plaited in a series of very deep undulations in front, and sweeping off at each side with a bold curve. Those of the palate and lower jaw are so made that they lock into each other, the folds exactly corresponding, and fitting into each other with such exactness, that no creature when seized could hope to escape without much detriment. The edges of this continuous tooth-ribbon, if I may so call it, are very sharp, and armed with small saw-like teeth, rather worn away in front, but very perceptible on the sides. In the very front of the upper jaw are two little pointed teeth, set apparently loosely in the soft parts of the nose, and quite useless for biting. When, however, the skull is removed from the body, and cleared of muscle and other soft parts, these teeth retain their place, and by the hardening of their attachments become tightly fixed in the skull.

During life the points of these teeth project very slightly through those two little holes just inside the upper lip, which are considered as the internal nostrils. While the creature is alive, the teeth cannot be seen even when the mouth is open, being covered by a very soft and yielding substance, through which they seem to cut when in use.

The external aspect of this creature is very singular, the chief characteristics being its eel-like form, and the four long slender projections which stand in the place of limbs, and are analogous to similar structures in certain reptiles already described and figured. These are not true limbs, and the cartilaginous ray by which they are supported has no joint. They are quite soft and flexible, as if they were made of leather, and are of very trifling use in locomotion. The two fore-limbs are set at the shoulders, just behind the head, and widely separated from each other, while the hinder pair are quite close together at their bases. In the species just described, two short tubercular appendages, about an inch in length, accompany the larger limb-like projections, and, except in dimensions, bear a close resemblance to those organs. I may take this opportunity of remarking that the creature is not known to leave the water and to crawl on land, and that the figure is represented as lying partly on the bank of the stream, not to intimate that it is capable of terrestrial locomotion, but merely in order to show the peculiar shape to better advantage.

The specimen now (January, 1862) living in the Zoological Gardens has not attained to any great size, being scarcely half as large as the individual just described, though it has lived in England for three years. The tank in which it resides is small, and may probably account for the slight increase in dimensions. It is interesting to watch this creature move about its prison, as the peculiar screw-like or spiral movement of the limbs is well exhibited.

The whole body is covered with rather large scales, embedded deeply in the skin, and not easily to be seen in living specimens. The name of Lepidosiren, or Scaly Siren, is given to this creature on account of its scaly covering. At about one-third of the distance from the head to the tip of the tail a rather narrow and fin-like membrane arises, which runs completely round the tail until it is terminated close to the bases of the hind pair of limbs. It is strengthened throughout by a series of soft jointed rays.

The flesh of the Lepidosiren is very soft and white, and is thought to be excellent for the table, so that in its native country it is dug up from its muddy bed and used for food. It usually burrows to a depth of eighteen inches. This creature possesses both lungs and gills, the latter organs being twofold, the external gills being tufted on the under side, and the internal gills being placed on the edge of the divisions between the gill openings on the side of the neck. The heart is more reptilian than piscine, having three compartments, two auricles and one ventricle, and affords one of the strongest reasons for ranking the creature among the former class.

There are several species of Lepidosiren, divided, in the catalogue in the British Museum, into two genera, distinguished from each other by the number of ribs. The species which is found in Southern America, and is there known under the popular name of CARAMURU (*Lepidosiren paradoxa*), has fifty-five pairs of ribs, whereas the African species has only thirty-six pairs. The colour of the Lepidosiren is darkish brown with a wash of grey.

THE next order of Crawling Batrachians is called by the name of Meantia, and contains a very few but very remarkable species. In all these creatures the body is long and smooth, without scales, and the gills are very conspicuous, retaining their position throughout the life of the animal. There are always two or four limbs, furnished with toes, but these members are very weak, and indeed rudimentary, and both the palate and the lower jaw are toothed.

The first example of this order is the celebrated PROTEUS, discovered by the Baron de Zois, in the extraordinary locality in which it dwells.

At Adelsberg, in the duchy of Carniola, is a most wonderful cavern, called the Grotto of the Maddalena, extending many hundred feet below the surface of the earth, and consequently buried in the profoundest darkness. In this cavern exists a little lake, roofed with stalactites, surrounded with masses of rock, and floored with a bed of soft mud, upon which the Proteus may be seen crawling uneasily, as if endeavouring to avoid the unwelcome light by which its presence is known. These creatures are not always to be found in the lake, though after heavy rains they are tolerably abundant, and the road by which they gain admission is at present a mystery.

The theory of Sir H. Davy is, " that their natural residence is a deep subterraneous lake, from which in great floods they are sometimes forced through the crevices of the rocks into the places where they are found; and it does not appear to me impossible, when the peculiar nature of the country is considered, that the same great cavity may furnish the individuals which have been found at Adelsberg and at Sittich."

Whatever may be the solution of the problem, the discovery of this animal is extremely valuable, not only as an aid to the science of comparative anatomy, but as affording another instance of the strange and wondrous forms of animal life which still survive in hidden and unsuspected nooks of the earth.

Many of these animals have been brought in a living state to this country, and have survived for a considerable time when their owners have taken pains to accommodate their condition as nearly as possible to that of their native waters. I have had many opportunities of seeing some fine specimens, brought by Dr. Lionel Beale from the cave at Adelsberg. They could hardly be said to have any habits, and their only custom seemed to be the systematic avoidance of light. Dr. Beale has kindly forwarded to me the following account of these curious creatures :—

PROTEUS.—*Proteus anguinus.*

" One of the Proteuses I brought over from Adelsberg lived for five years, and, what is very interesting, passed four years of his life in the same water, a little fresh being added from time to time to make up for the loss by evaporation. He lived in about a quart of water, which was placed in a large globe, this being kept dark by an outer covering of green baize. Perhaps half a pint of water may have been added during two years.

He was not once fed while he was in confinement, and one of his companions died soon after taking a worm before he had been two years in this country.

The one I kept was very active, and his movements were as rapid as those of an eel. He was thinner just before death than when he was brought from the cave, but the loss of substance was so very slow as not to be perceptible from year to year, and to the last he retained the power of performing very active muscular movements.

His external gills always contracted when a strong light was thrown upon them. The circulation of the blood in the vessels of these organs was very often exhibited ; the animal being placed in a long tube with a flat extremity, provided with an arrangement for the constant supply of water, and on several occasions some of the large blood corpuscules were removed for the purpose of microscopical examination, so that the animal was not placed under the most favourable circumstances for living without food.

There are probably very few more striking examples of very slow death from starvation than this, and it is probable that the ultimately fatal results were as much caused by

NECTURUS.—*Nectúrus laterólis.*

confinement, change of air and temperature, and occasional exposure to light for some hours, as from mere starvation. It is well-known, for example, that, as a general rule, the Batrachia endure starvation most remarkably."

The gills of the Proteus are very apparent, and of a reddish colour, on account of the blood that circulates through them. I have often witnessed this phenomena by means of the ingenious arrangement invented by Dr. Beale, by which the creature was held firmly in its place while a stream of water was kept constantly flowing through the tube in which it was confined. The blood discs of this animal are of extraordinary size; so large, indeed, that they can be distinguished with a common pocket magnifier, even while passing through the vessels. Some of the blood corpuscules of the specimen described above are now in my possession, and, together with those of the lepidosiren, form a singular contrast to the blood corpuscules of man, the former exceeding the latter in dimensions as an ostrich egg exceeds that of a pigeon.

The colour of the Proteus is pale faded flesh tint, with a wash of grey. The eyes are quite useless, and are hidden beneath the skin, those organs being needless in the dark recesses where the Proteus lives. Its length is about a foot. What are the natural habits of this strange animal, what is its food, of what nature is its development, and what is its use, are a series of problems at present unanswered. By some writers it has been thought to be merely the larval state of some large Batrachian at present unknown; but the anatomical investigations that have been made into its structure seem to confirm the idea that it is a perfect being, and one of those species which carry the gills throughout their whole existence.

In the NECTURUS, the head is much broader and flatter and the tail shorter than in the preceding species. This animal belongs to the same family as the proteus, but is a native of America, being found in the Mississippi and several of the lakes. It is rather a large animal, attaining, when adult, a length of two or three feet, and being of a thick and sturdy make. The gills of this creature are large and well tufted, and the limbs are furnished with four toes on each foot, but without claws.

The general colour of this creature is olive-brown above, dotted with black, and with a black streak from the nostril through the eye, and along each side to the tail. Below it is blackish brown with olive spots.

OUR last example of the Batrachians is the curious SIREN, or MUD-EEL, as it is sometimes called, on account of its elongated eel-like form and its mud-loving habits.

It is a native of several parts of America, and is found most plentifully in Carolina, where it haunts the low-lying and marshy situations. The rice-grounds seem to be its most favoured localities, the muddy soil being the substance best adapted for its means of

SIREN.—*Siren lacertina.*

progression. Its food seems to consist almost entirely of worms and various insects, of which it will consume a considerable quantity every day. A fine specimen that lived for some time in the Zoological Gardens used to feed upon earthworms, of which it would devour about eighteen or twenty every two days. This individual passed the greater part of its time beneath the thick stratum of soft mud with which the bed of the basin was profusely covered. This was a very long specimen, and by an uninitiated observer would probably have been taken for an eel.

The head of the Siren is small in proportion to the size of the animal, the eye is very small, and the gill tufts are three in number on each side, and beautifully plumed. It has only one pair of legs, the hinder set being wanting, and the front pair are extremely small, and of no practical use in progression. It has only three toes on each foot. The colour is dark blackish brown, and the length of a fine specimen is about three feet.

FISHES.

In the Fishes, the last class of vertebrated animals, the chief and most obvious distinction lies in their adaptation to a sub-aqueous existence, and their unfitness for life upon dry land.

There are many vertebrate animals which pass the whole of their lives in the water, and would die if transferred to the land, such as the whales and the whole of the cetacean tribe, an account of which may be found in vol. i. page 521. But these creatures are generally incapable of passing their life beneath the waters, as their lungs are formed like those of the mammalia, and they are forced to breathe atmospheric air at the surface of the waves. And though they would die if left upon land, their death would occur from hunger and inability to move about in search of food, and in almost every case a submersion of two continuous hours would drown the longest breathed whale that swims the seas.

The Fishes, on the contrary, are expressly formed for aquatic existence; and the beautiful respiratory organs, which we know by the popular term of "gills," are so constructed that they can supply sufficient oxygen for the aeration of the blood. They have not the power, as is sometimes imagined, of separating the oxygen, which, in its combination with certain proportions of hydrogen, composes the element in which they live, but are able to take advantage of the atmospheric air which is contained in the water.

Any reader who happens to possess a globe with gold-Fish can prove, and doubtlessly has proved, the truth of this assertion. It often happens that when the supply of water is insufficient, or the mouth of the vessel too small to permit the air to be absorbed by the water in sufficient volume, the Fish come gasping to the surface, and there swim with gaping mouths, sucking in the air with audible gulps. But if a little water be taken up in a cup or spoon, and dashed back from a little height, so as to cause a sharp splash, or, better still, if a syringe be employed for the same purpose, so as to drive a quantity of atmospheric air into the water; the Fish soon become contented, their anxious restlessness abates, and they quietly swim backward and forward, without displaying any more signs of uneasiness.

The reason that Fishes die when removed from the water, is not because the air is poisonous to them, as some seem to fancy, but because the delicate gill membranes become dry and collapse against each other, so that the circulation of the blood is stopped, and the oxygen of the atmosphere can no longer act upon it. It necessarily follows, that those Fish whose gills can longest retain moisture will live longest on dry land, and that those whose gills dry most rapidly will die the soonest. The herring for example, where the delicate membranes are not sufficiently guarded from the effects of heat and evaporation, dies almost immediately it is taken out of the water; whereas the carp, a fish whose gill-covers can retain much moisture, will survive for an astonishingly long time upon dry land, and the anabas, or climbing perch, is actually able to travel from one pool to another, ascending the banks, and even traversing hot and dusty roads.

The entire shape of these creatures, subjected though it be to manifold variations, is always subservient to the great object of passing rapidly through the ponderous liquid in which they swim, so as to enable them to secure their prey or avoid their enemies. Even in creatures of such different shapes as the sharks, the eels, the salmon tribe, and the flat fish, the capacity for speed is really wonderful, and is in all effected by simple and beautiful modifications of one mechanical principle, that of the inclined plane or screw.

In all Fishes, the power of progression lies in the wonderfully muscular tail with its appended fin, and the creature drives itself forward by repeated strokes of this organ in exactly the same manner that a sailor urges a boat through the water by the backward and forward movements of a single oar in the stern.

To show the power of this principle, I will mention that being on one occasion left with a party of friends on board a fishing barge in a small lake, and deserted by an ill-conditioned boatman, who refused either to put us ashore or take us to a better fishing-ground, and so went misanthropically home to his dinner, I called to mind the progression of the Fishes, and straightway became independent of the boatman. After hauling up the anchor, I inserted the butt end of the largest fishing-rod into the head of the rudder so as to form an extempore tiller, and by moving the rudder gently to and fro I was able to propel the barge in any direction and to any distance. We thus traversed the lake at our pleasure, drove the barge ashore at its further extremity, and left the boatman to find it and take it back as he could.

Even the eels and the flat Fishes, with their gracefully serpentine movements, adopt this mode of progression, though it is not so apparent as in the Fish whose bodies are less flexible and accordingly employ more force in the tail itself.

The fins are scarcely employed at all in progression, but are usually used as balancers and occasionally to check an onward movement. Before proceeding further, I may mention that all the fins of a Fish are distinguished by appropriate names. As they are extremely important in determining the species and even the genus of the individual, and as these members will be repeatedly mentioned in the following pages, I will briefly describe them.

Beginning at the head and following the line of the back, we come upon a fin, called from its position the "dorsal" fin. In very many species there are two such fins, called, from their relative positions, the first and the second dorsal fins. The extremity of the body is furnished with another fin, popularly called the tail, but more correctly the caudal fin. The fins which are set on that part of the body which corresponds to the shoulders are termed the "pectoral" fins; that which is found on the under surface and in front of the vent is called the abdominal fin, and that which is also on the lower surface, and between the vent and the tail, is known by the name of the "anal" fin. All these fins vary extremely in shape, size, and position. The figure on page 198 exhibits all these fins.

The gill-cover, or operculum as it is technically called, is separated into four portions, and is so extensively used in determining the genus and species that a brief description must be given. The front portion, which starts immediately below the eye, is called the "præ-operculum," and immediately behind it comes the "operculum." Below the latter is another piece, termed from its position, the "sub-operculum," and the lowest piece, which touches all the three above it, is called the "inter-operculum." Below the chin and reaching to the sub-operculum, are the slender bones, termed the "branchiostegous rays," which differ in shape and number according to the kind of Fish.

The scales with which most of the Fish are covered are very beautiful in structure, and are formed by successive laminæ, increasing therefore in size according to the age of the Fish. They are attached to the skin by one edge, and they overlap each other in such a manner as to allow the creature to pass through the water with the least possible resistance. The precise mode of overlapping varies materially in different genera. Along each side of the Fish runs a series of pores, through which passes a mucous secretion formed in some glands beneath. In order to permit this secretion to reach the outer surface of the body, each scale upon the row which comes upon the pores is pierced with a little tubular aperture, which is very perceptible on the exterior, and constitutes the "lateral line." The shape and position of this line are also used in determining the precise position held by any species. In comparing the scales taken from different Fishes, it is always better to take those from the lateral line.

The heart of the Fish is very simple, consisting of two chambers only, one auricle and one ventricle. The blood is in consequence cold.

The hearing of Fishes appears in most cases to be dull, and some persons have asserted that they are totally destitute of this faculty. It is now, however, known that many species have been proved capable of hearing sounds, and that carp and other fish can be taught to come for their food at the sound of a bell or whistle. The internal structure of the ear is moderately developed, and there are some curious little bones found within the cavity, technically called otoliths.

The sense of touch seems to have its chief residence in the mouth and surrounding parts, the scaly covering rendering the surface of the body necessarily obtuse to sensation. The smell seems to be strongly developed, if it be possible to pronounce an opinion from the size and distribution of the nasal nerves. The brain is very small in these creatures, and from its shape, as well as its dimensions, denotes a low degree of intelligence.

SKELETON OF PERCH.

The arrangement of the bones is very curious, and is so complicated that a better idea can be formed by examining the accompanying illustration than by reading many pages of laboured description. The skeleton is that of the common perch.

In the anatomy of the Fishes there are many other interesting structures, which will be described when treating of the particular species in which they are best developed.

THE fishes comprised in the first order, are called by the rather harshly sounding title of Chondropterýgii, a term derived from two Greek words, the former signifying cartilage and the latter a fin, and given to these creatures because their bones contain a very large amount of cartilaginous substance, and are consequently soft and flexible. The bones of the head are rather harder than those of the body and fins.

It is necessary, before entering into any description of the different species, to premise that the arrangement of the fishes is a most difficult and complicated subject, in which no two systematic naturalists seem to agree entirely. I have, therefore, followed the course which has been adopted throughout the whole of this work, and accepted the arrangement given in the catalogues of the British Museum.

The cartilaginous fishes are again subdivided into groups, in the first of which the gills are quite free, and the members of this group are accordingly called by the name of Eleutheropómi, or free-gilled fishes. What quality in the fishes should give birth to such polysyllablic and harsh-sounding names, is not easy to say; but the fact is patent that not even in botany is the scientific terminology so repulsive as in the fishes. I shall endeavour, as far as possible, to avoid this technical language, and to throw the scientific descriptions to the end of the work, as in the two former volumes; and the reader may feel sure when his attention is struck by a long and difficult name, that it is only used in consequence of the exigencies of the occasion.

The first family, of which the common STURGEON is a good and familiar example, are

STURGEON.—*Acipenser áttilus.*

at once known by the cartilaginous or bony shields with which the head and body are at intervals covered.

In this remarkable fish the mouth is placed well under the head, and in fact seems to be set almost in the throat, the long snout appearing to be entirely a superfluous ornament. The mouth projects downwards like a short and wide tube, much wider than long, and on looking into this tube no teeth are to be seen. Between the mouth and the extremity of the snout is a row of fleshy finger-like appendages, four in number, and apparently organs of touch.

One or two species of Sturgeon are important in commerce, as two valuable articles, namely isinglass and caviare, are made from them. The former substance is too well known to need a description, and the mode of preparing it for use is briefly as follows. The air-bladder is removed from the fish, washed carefully in fresh water, and then hung up in the air for a day or two so as to stiffen. The outer coat or membrane is then peeled off, and the remainder is cut up into strips of greater or lesser length, technically called staples, the long staples being the most valuable. This substance affords so large a quantity of gelatinous matter, that one part of isinglass dissolved in a hundred parts of boiling water will form a stiff jelly when cold.

Caviare is made from the roe of this fish, and as nearly three millions of eggs have been taken from a single fish, the amount of caviare that one Sturgeon can afford is rather large. It is made by removing all the membranes, and then washing the roe carefully with vinegar or white wine. It is next dried thoroughly in the air, well salted, subjected to strong pressure in order to force out all moisture caused by the wet-absorbing properties of the salt, and is lastly packed in little barrels for sale. The caviare made on the Caspian is considered the best. In England it is not much eaten, but in Russia it forms a large item in the national consumption, probably on account of the great number of fasts observed by the Greek Church. The roes of several other fish are employed in the same manner, and in Italy, a substance called "botargo" is prepared from the roe of a species of mullet.

The common Sturgeon has sometimes, but not very often, been found in our rivers, and whenever it is captured in the Thames within the jurisdiction of the Lord Mayor, it is termed a royal fish, and becomes the property of the Crown. It is not unfrequently taken near our shores, more especially on the eastern coast, and most persons are familiar with the occasional appearance of one of these fine fish on a fishmonger's stall. The flesh of the Sturgeon is held in some estimation, and in the olden English days, it was always reserved for the table of the king. Some very fine specimens have sometimes been caught in our rivers, the largest on record having weighed four hundred and sixty pounds. The size of this specimen may be imagined from its weight, as another individual which weighed only one hundred and ninety pounds measured eight feet in length.

The body of the Sturgeon is elongated, and slightly five-sided from the head to the tail. Along the body run five rows of flattened bony plates, each plate being marked with slight grooves in a radiating fashion, and having a pointed and partly conical spine on each plate, the points being directed towards the tail. The plates along the summit of the back are the largest.

THERE are many species of Sturgeons, and among the most remarkable are those which are shown in the accompanying illustration, both being natives of the rivers of North America.

The two smaller figures represent the SHOVEL-FISH, so-called from the curious form of its head, which is flattened, rounded, and really not unlike the implement from which it derives its popular title. The two figures are given in order to show the different aspect of the adult and young. The reader will doubtlessly remark the large size and conspicuous arrangement of the bony scales along the body.

THE central figure represents the SPOONBILL STURGEON, sometimes called, in allusion to the singular shape of the head, the PADDLE-FISH. This creature is remarkable for several reasons. In the first place, the uncommonly elongated and flattened snout is sufficiently conspicuous to arrest the attention of even the most casual observer, and in the second place, the body is quite smooth, and wants those bony plates which generally form so characteristic an adornment of the Sturgeon. Several specimens of this remarkable fish are now in the British Museum, and were taken from the Ohio and Mississippi.

THE very singular family of the Chimæridæ contains a few but remarkable species.

Both these creatures are sufficiently quaint and ungainly in aspect; but as the NORTHERN CHIMÆRA has sometimes been found on our coasts, that species has been chosen for the illustration. It is also known by the title of RABBIT-FISH, probably on account of its general aspect, and KING OF THE HERRINGS, because it follows the shoals of those fishes during their wonderful migrations, and makes great havoc among their numbers. The appendage to the top of the head is also looked upon by the Norwegians in the light of a kingly crown, and has contributed towards its royal title. It is known in some localities under the name of SEA CAT.

This species is mostly found in the Northern seas, and is, when living, a most beautiful creature, its body glowing with golden brown variegations upon a white ground. The title of Gold and Silver Fish is sometimes given to the Northern Chimæra in consequence of this gorgeous colouring. The pupil of the eye is green, and the iris is white. It feeds mostly upon the smaller fish, but finds much of its subsistence among the various molluscs, crustaceans, and other inhabitants of the ocean. The flesh is not considered good, being hard and coarse.

SHOVEL-FISH.—*Scaphiorhynchus cataphractus.*　　　　SPOONBILL STURGEON.—*Polyodon spátula.*

The form of this fish is very peculiar, the body being to.erably large and rounded towards the point, and the tail tapering rapidly until it ends in an elongated thong, almost like the lash of a whip. The second dorsal fin commences immediately behind the first, and extends along the tail nearly to the extremity of its lengthened filamentary termination. The sexes may readily be distinguished from each other, both by the

NORTHERN CHIMÆRA.—*Chimæra monstrosa.*

shape of the head and first dorsal fin, and by a pair of bony appendages close to the ventral fins. It is not a large species, seldom exceeding a yard in length.

In the seas of the southern hemisphere, there is another species of Chimæra, called from its locality, the SOUTHERN CHIMÆRA (*Callorhynchus antárctica*) or ELEPHANT-FISH, the latter title being given to it on account of the extraordinary prolongation of the snout. The Araucanian name for this species is CHALGUA ACHAGUAL. The snout of this fish is developed into a strange cartilaginous prolongation, which is bent backwards in a hook-like form, and is thought by some persons to bear a resemblance to a common hoe.

The tail of this species does not correspond in oddity with its head, being without the long filament that gives so strange an aspect to its Northern relative. The colour is satiny white mottled with brown, and the size is about the same as that of the Northern Chimæra.

THE fishes belonging to the next sub-order have their gills fixed by their outer edge to the divisions in the gill-openings at the side of the neck. This sub-order includes the Sharks and the Rays, many representatives of which creatures are found on the British coasts.

The first family of this large and important group is known by the name of Scyllidæ, and its members can be recognised by several distinguishing characteristics. They have spout-holes on the head, and the gill-openings are five in number on each side. Sometimes there only seem to be four openings, but on closer examination the fourth and fifth are found set closely together, the opening of the fifth appearing within that of the fourth. The teeth are sharp and pointed, and the tail is long, notched on the outer side, and is not furnished with a fin.

One of the commonest British species is the LITTLE DOG-FISH, called by several other names, as is usual with a familiar species that is found in many localities. Among such names are SMALL SPOTTED DOG-FISH, LESSER SPOTTED SHARK, MORGAY, and ROBIN HUSS.

LITTLE DOG-FISH.—*Scyllium canicula.*   ROCK DOG-FISH.—*Scyllium catulus.*

This fish is plentiful on our coasts, especially in the southern extremity of England, and is often thought a great nuisance by fishermen, whose bait it takes instead of the more valuable fish for which the hook was set. It generally remains near the bottom of the water, and is a voracious creature, feeding upon crustaceans and small fish. It often follows the shoals of migrating fish, and on account of that custom is called the Dog-fish. Generally its flesh is neglected, but when properly dressed, it is by no means unpalatable, and is said to be sometimes trimmed and d essed in fraudulent imitation of more valuable fish.

The skin of this and other similar species is rough and file-like, and is employed for many purposes. The handles of swords, where a firm hold is required, are sometimes bound with this substance; and joiners use it in polishing the surface of fine woods so as to bring out the grain. It is also employed instead of sand-paper upon match boxes.

The egg of this species is very curious in form and structure, and is often found on the sea-shore, flung up by the waves, especially after a storm. These objects are familiar to all observant wanderers by the sea-shore, under the name of mermaid's purses, sailor's purses, or sea purses. Their form is oblong with curved sides, and at each angle there is a long tendril-like appendage, having a strong curl, and in form not unlike the tendrils of the vine. The use of these appendages is to enable the egg to cling to the growing seaweed at the bottom of the ocean, and is to prevent it from being washed away by the tide. After a storm, however, when the agitated waves have torn up the beds of marine wrack and other seaweeds that usually lie in still calmness beneath their sheltering waters, and especially during the time of low tide, these objects may be found lying upon the uncovered and dripping shore, their strong but delicate tendrils entwined in almost inextricable complexity among the salt-loving vegetation of the ocean, and their tiny inmates as yet imperfect and unborn.

Water, which to these creatures contains the breath of life, gains access to the imprisoned sharkling, through two slight longitudinal apertures, one towards each end of the egg; and it is a very remarkable fact that in these waters the undeveloped young are furnished with small external gills, which are afterwards absorbed into the system—a phenomenon curiously analogous to the structure of the tadpole. The substance of the egg-shell, if such a term can be applied to the envelope which contains the young, is of a moderately stiff horny character, becoming harder when dry, and of a semi-transparent yellowish hue, not very unlike, though not so clear as the yellow portions of tortoise-shell.

For the escape of the young Shark, when strong enough to make its own way in the wider world of waters, an outlet is provided in the opened end of the envelope, which opens when pushed from within and permits the little creature to make its way out, though it effectually bars the entrance against any external foe. When it first leaves its horny home, the neophyte Shark bears with it a capsule, containing a portion of the nutrimental principle of the egg, as is seen in the chicken of the common fowl, and is enabled to exist upon this substance until it has attained the power of foraging for itself, when the small remainder of the capsule is absorbed into the abdomen.

The head of the Little Dog-fish is rather flat upon the top, there is a little spiracle or blow-hole behind each eye, and the shape of the mouth is somewhat like a horse-shoe. The general colour of the body is pale reddish on the upper parts, covered with many little spots of dark reddish brown; below it is yellowish white. The length of this species is about eighteen inches.

THE second species shown in the engraving is called the Rock Dog-fish, because it is often found on rocky coasts. From its superior size, it is also known by the name of Large Spotted Dog-fish, and on several coasts it goes by the curt and not euphonious name of Bounce.

The habits of this fish are so like those of the preceding species, that they need no description.

It may readily be distinguished from the little dog-fish by the large size and fewer number of the spots, as well as by the shape of the ventral fins, which in this species are nearly squared at the end, whereas in the former they are of a diamond-like form. The colour of the Rock Dog-fish is brownish grey above, without the red tinge of the little dog-fish, and covered rather sparingly with large patches of blackish brown. Below it is whitish. The length of a fine specimen will sometimes be nearly a yard.

ANOTHER species of Dog-fish, namely the Black-mouthed Dog-fish, or the Eyed Dog-fish (*Pristidúrus melanóstomus*), is mentioned by Mr. Yarrell among the list of British fishes. It may be at once distinguished from either of the preceding species by its large snout, and a row of small flat and sharp-edged prickles, arranged in saw-like fashion on the upper rim of the tail fin. The generic title Pristidurus, or Saw-tail, is given to the fish in allusion to this peculiarity. Its colour is light brown on the upper surface, sprinkled with spots, the smaller of which are scattered irregularly, and the larger arranged in four rows, two on each side. Its length is between two and three feet.

The BLUE SHARK, so called from the fine slaty blue colour of its skin, is a not unfrequent visitor of our shores, and is the object of the deadliest hatred to the fishermen, who are sometimes doomed to see their fish stolen, their nets cut to pieces, and their lines hopelessly ruined by this fish, without the least power of checking its depredations.

About the month of June, according to Mr. Couch's observations, this Shark makes its appearance on our coasts, and has sometimes been so plentiful that nine or ten have been taken by our fishing boats in a single day. As the fishermen are hauling up their

BLUE SHARK.—*Squalus glaucus.*

lines with the fish upon the hooks, the Blue Shark will follow the fish as it is drawn upwards, seize upon it, and hook itself for its trouble. Exasperated by the unsuspected check upon its maraudings, it tries to bite the line asunder, a feat easily performed by its lancet-like teeth with their notched edges.

Sometimes, however, it takes to another stratagem, and as soon as it feels the hook, rolls itself round so rapidly on its axis, that it winds the line round its body into a mass of inextricable entanglement. So effectually is this feat achieved, that in spite of the value of the line, the fishermen have been known to give up any attempt to unravel its knotty convolutions. This fish has another fashion of biting the line asunder without any apparent reason.

Perhaps, however, it never is so thoroughly destructive as in the pilchard season, when it follows the vast shoals of these fish to our shores and devours them whole-sale. Even when they are inclosed in the net, the Blue Shark is not to be baffled or deprived of its expected banquet, for, swimming along the whole length of the net, it bites at the inclosed fish, caring nothing for the meshes, and taking out large mouthfuls of mingled net and pilchards, swallows them together.

There is hardly a season passes when the capture of a fine specimen of the Blue Shark on our coasts is not recorded in the local papers. The sailors have an idea that this voracious fish is able to succour her young when in danger, by opening her mouth and letting them swim down her throat. It is undoubtedly true, that living young have been found in the stomach of large sharks; but whether they had been swallowed as a means of protection, is by no means proved. The reader will doubtlessly remember the similar stories that have been told of the viper and other poisonous snakes.

In the foreground of the engraving is introduced the skull of a large Shark, for the purpose of showing the terrible teeth with which it is armed, and which lie in several rows, ready to take the place of those which are broken or cast off when their work is done. From these teeth, which cut like broken glass, the natives of many savage lands make tools and weapons of war, by ingeniously fixing them into wooden handles.

The voracity and dulness of nerve belonging to the Shark is really wonderful. One of my friends was fishing after a large Shark that was following the vessel, and after a little time, succeeded in inducing the fish to take the great hook that had been nicely baited with pork to suit his palate. Too sudden a jerk, however, having been given to the line, the hook tore its way through the side of the cheek, setting the Shark free. The wound was a terrible one, and bled profusely, but the Shark seemed to care little or nothing about it, still hovered about the bait, as if unable to resist its attractions, and after a little while, was hooked a second time and hauled safely on board.

The capture of a Shark is always an event on board ship, especially if she be a sailing vessel and the wind has fallen. A hook made for the purpose, is secured to a fathom or so of iron chain, the Shark being capable of biting through a rope in an instant, and in no way so particular in its diet as to need fine tackle. Indeed, as in the last-mentioned instance, the creature seems to be perfectly aware of the danger, but to be incapable of resisting the tempting morsel. The other end of the chain is firmly lashed to a stout rope, and the latter secured to the vessel, as one rush of a powerful Shark would pull half a dozen men overboard.

All things being ready, a good large piece of pork is fixed tightly on the hook, and allowed to tow overboard. The Shark, being to the full as inquisitive as the cat, comes up with true feline curiosity, and sniffs at the bait with an air of deliberate scrutiny. Sometimes, it having perhaps lately partaken of a good meal, it is very coy about taking the bait, and keeps the anxious anglers above in a state of tantalized impatience for an hour or more. Generally, however, it dashes at the bait at once, and has even been known to leap from the water and hook itself before the bait had even reached the surface.

Now begins a mighty struggle, and all is eager excitement. The Shark knows no wiles, but uses all its great strength to tear away from the hook by sheer force, having apparently but slight sense of pain, and in many cases would do so were not a check put upon its efforts by a rope knotted into a bowline and dexterously slipped over its tail. Being now held by both extremities, it is shorn of its strength like Samson without his locks, and lifted on deck by both lines. Sometimes a trident-like harpoon, technically called a "grains," the handle of which is heavily loaded with lead to make it fall with greater force, is dropped upon the struggling fish.

Being brought on deck, however, the struggles of the creature recommence with tenfold violence. Twisting with marvellous agility, snapping right and left with its murderous teeth, and dealing heavy blows with its terrible tail, it makes the deck tremble under its strokes, until some experienced sailor runs in with an axe, and with a blow across the tail, reduces the creature to malignant impotence. The muscles of the Shark are endowed with astonishing irritability, and long after the body has been cut to pieces and parts of it cooked and eaten, the flesh will quiver if pricked with a knife-point; the separated heart will beat steadily while lying on the bare boards, and the jaws of the severed head will snap with frightful vehemence if any object be put between the teeth.

Sailors generally make high festival at the dismemberment of a Shark, and have great delight in opening the creature for the purpose of finding out the articles which it had swallowed. For a Shark, when following a vessel, will eat anything that falls overboard. The contents of a lady's workbox, a cow's hide entire, knives, hats, boots, and all kind of miscellanea have been found in the interior of a Shark; while on one occasion were discovered the papers of a slaver, which had been flung overboard when the vessel was overhauled, and by means of which papers so strangely recovered, the vessel was identified and condemned.

The colour of this species is beautiful slate-blue above, and white below.

HAMMER-HEADED SHARK.—*Sphyrnias zygæna*.

THE remarkable fish depicted in the accompanying illustration affords a striking instance of the wild and wondrous modifications of form assumed by certain creatures, without any ascertained purpose being gained thereby. We know by analogous reasoning that some wise and beautiful purpose is served by this astonishing variation in form; but as far as is yet known, there is nothing in the habits of this species that accounts for the necessity of this strange shape.

The shape of the body is not unlike that of the generality of Sharks, but it is upon the head that the attention is at once rivetted. As may be seen from the figure, the head is expanded laterally in a most singular manner, bearing, indeed, no small resemblance to the head of a hammer. The eyes are placed at either end of the projecting extremities, and the mouth is set quite below, its corners just coinciding with a line drawn through the two projecting lobes of the head. It is worthy of notice, that several of our commonest British insects—those beautiful dragon-flies belonging to the genus Agrion—have heads modelled on a very similar principle, and there are some exotic insects where this singular shape is even more exaggerated, the eyes being set quite at the end of long lateral footstalks.

This species attains to a considerable size, seven or eight feet being a common measurement, and specimens of eleven or twelve feet having been known. Its flesh is said to be almost uneatable, being hard, coarse, and ill-flavoured. The Hammer-headed Shark produces living young, and from the interior of a very fine specimen captured near Tenby in 1839, and measuring more than ten feet in length, were taken no less than thirty-nine young, all perfectly formed, and averaging nineteen inches in length.

TOPE.—*Galeus canis.*　　　　SMOOTH HOUND.—*Mustelus vulgaris.*

Several species of Hammer-headed Sharks are known, among which the Heart-headed Shark (*Sphyrnias Tiburo*), has the best developed head, and the Broad-headed Shark (*Sphyrnias láticeps*), the most so. Another species, the Tudes (*Sphyrnias Tudes*), thought to inhabit the Mediterranean, and the shores of Southern America, is inter-mediate between the two extremes.

The general colour of this species is greyish brown above, and greyish white below.

THE destructive and voracious fish, which is indiscriminately known by the names of TOPE, PENNY DOG, or MILLER'S DOG, according to the particular coast near which it is found, is another familiar British representative of that great shark family, from whose larger developments we are in this favoured country happily free.

The Tope is commoner towards the southern than the northern coasts, but wherever it is found, it is an intolerable nuisance, behaving itself much after the example set by the blue shark, and being, in proportion to its dimensions, quite as injurious to the fishing interest. Like the last-mentioned species, it produces living young, the number of a single family being about thirty. They are born in May and June, and mostly remain on our coasts through their first winter, not retiring into deep water till they have entered their second year.

Like the blue shark, the Tope is fond of robbing the fishermen's hooks, and will in like manner endeavour to free itself when hooked, biting through the line, or rolling round with such rapidity that it winds the long cord about its body into tangled knots. The upper surface of the Tope is slaty grey, becoming lighter towards the abdomen, which is nearly white.

THE prettily marked and curiously toothed SMOOTH HOUND is also known under the titles of SKATE-TOOTHED SHARK and RAY-TOOTHED DOG, the two latter titles being appropriately given it on account of its curious and beautifully formed teeth, which resemble in form the cylinders of a crushing mill, and are used for a similar purpose.

The jaws, instead of being studded with rows of sharp and knife-like teeth, are supplied with two rounded projections on which the flat-topped teeth are set closely together like the stones of a mosaic, and which are so formed that they roll over each other as the jaws are closed, producing a crushing effect of enormous power. These curious teeth are rendered needful by the food on which the Smooth Hound lives, namely the hard-shelled crustaceans, whose armour of proof is nevertheless soon comminuted under the bony rollers.

As may be inferred from the character of its food, the Smooth Hound is not destructive to the fisheries, and may be allowed to live in harmless security. Its flesh is said to be tolerably well-flavoured, and even moderately tender. It produces its young in a living state, but is not very prolific, the number at a birth rarely exceeding ten or twelve. Almost as soon as born they retire into deep water, so that though a tolerably plentiful species, it is not seen so often as those which live in shallow waters.

The colour of the Smooth Hound is pearly grey, and above the lateral line, which in this species is very strongly marked, the body is decorated with small round white spots, very conspicuous while the creature is young, but becoming fainter when it attains maturity. The under parts are whitish yellow.

BEFORE noticing some of the larger and more terrible species, we must not omit the PORBEAGLE, sometimes called the BEAUMARIS SHARK (*Isúrus cornúbicus*), a fish of a wonderfully mild aspect for a Shark, and notable for a very porpoise-like aspect. The name of Porbeagle is in fact owing to this resemblance. This species feeds on fish of various kinds, three full-grown hakes having been found in the stomach of one individual, and derives some of its subsistence from the larger molluscs. It attains a rather large size, five or six feet being a common length. Its colour is uniform greyish black above, and white below.

THE dreadful WHITE SHARK, the finny pirate of the ocean, is happily almost a stranger to our shores, though a stray specimen may now and then visit the British Islands, there to find but scant hospitality.

This is one of the large species that range the ocean, and in some seas are so numerous that they are the terror of sailors and natives. One individual, whose jaws are still preserved, was said to have measured thirty-seven feet in length, and when we take into consideration the many instances where the leg of a man has been bitten off through flesh and bone as easily as if it had been a carrot, and even the body of a boy or woman severed at a single bite, this great length will not seem to be exaggerated.

Many portions of this fish are used in commerce. The sailors are fond of cleaning and preparing the skull, which, when brought to England, is sure of a ready sale, either for a public museum, or to private individuals who are struck with its strange form and terrible armature. The spine, too, is frequently taken from this fish, and when dried, it passes into the hands of walking-stick makers, who polish it neatly, fit it with a gold handle, and sell it at a very high price. One of these sticks will sometimes fetch six or seven pounds. There is also a large amount of oil in the Shark, which is thought rather valuable, so that in Ceylon and other places a regular trade in this commodity is carried on.

The fins are very rich in gelatine, and in China are, as is said, employed largely in the manufacture of that gelatinous soup in which the soul of a Chinese epicure delights, and of which the turtle soup of our metropolis is thought by Chinese judges to be a faint penumbra or distant imitation. The flesh is eaten by the natives of many Pacific islands; and in some places the liver is looked upon as a royal luxury, being hung on boards in the sun until all the contained oil has drained away, and then carefully wrapped up in leaves and reserved as a delicacy.

3.

WHITE SHARK, OR LAMIA.—*Carcharodon Rondelétii*

These islanders have a very quaint method of catching the Shark—absurdly impotent in theory, but strangely efficacious in practice. They cut a large log of wood into the rude resemblance of a canoe, tie a rope round the middle, form the end of the rope into a noose, and then set it afloat, leaving the noose to dangle in the water. Whether induced by curiosity, or by what strange impulse urged, is not very clear, but the fact is patent that before the noose has been floating very long, a Shark is sure to push its head through it, and on backing as soon as it feels the obstruction, is caught by the tightening of the noose. The natives then go off in their canoes, chasing the bewildered Shark, who is unable to dive on account of the floating log, and who is so lustily battered about the head with the heavy clubs so admirably made by those ingenious natives, that it is soon killed and hauled ashore in triumph.

The colour of the White Shark is ashen brown above, and white below.

THE upper figure in the accompanying illustration represents the BASKING SHARK, otherwise known by the name of SAIL-FISH and SUN-FISH. The first and last names are derived from its habit of lying motionless on the surface of the water, evidently enjoying the rays of the sun; and the intermediate term refers to the sail-like aspect of its first dorsal fin, which projects high out of the water when the fish is swimming near the surface, as is shown in the figure. In the Orkneys it is called the HOMER. This word has no reference to the Greek poet, with whom the rough fishermen are not likely to have much acquaintance, but is a contraction of Hoe-mother, the fish being thought to be the parent of the hoe, or picked dog-fish, a species which will presently be described.

THRESHER, OR FOX SHARK.—*Alópias vulpes.*   BASKING SHARK.—*Cetorhinus máximus.*

It is a magnificent fish, often attaining to a length of thirty-five or thirty-six feet. It does not appear, however, to be dangerous in proportion to its size, its teeth being very small in proportion to its dimensions, and the only food found in its stomach being the remains of crustaceans and probably of echini.

The Basking Shark is not very uncommon on our shores, especially if westerly winds have been prevailing. It seems to be of a rather dull and listless character, allowing itself to be approached quite closely by a boat, without giving any signs of alarm until the bow of the boat actually touches its person. Owing to this sluggard mode of life, it can easily be harpooned, but then bursts into furious energy with startling quickness, dives like lightning to a great depth, so as to require a very considerable length of rope, and putting forth the vast powers that have been lying dormant in the warm embraces of the sunbeams, like the might of Hercules in Omphale's arms, dashes away with such speed, and plunges with such wrathful violence, that its capture is an achievement of great difficulty and no small danger.

The gill apertures of the Basking Shark are extremely long, reaching almost across the neck. The head is conical, the muzzle short, and the eyes near the snout. The skin is very rough to the touch, whether the hand be passed from head to tail or *vice versâ,* and the colour is blackish brown, glossed with a bluish tint.

THE lower figure is that of a well-known species, familiar under the names of THRESHER, FOX SHARK, SEA FOX, and SEA APE. It is at once to be recognised by the peculiar form of the head and the wonderfully long upper lobe of the tail, which equals in length the body from the tip of the snout to the base of the tail. The lower lobe is quite short, and in no way conspicuous.

This fish is appropriately called the Thresher on account of its habit of using its long and flexible tail after the fashion of a quarter-staff, and dealing the most tremendous blows on or near any object that may excite its ire. Sometimes it seems to employ its tail in playing off a practical joke or frightening away dolphins or other creatures that are disporting themselves in apparent security. The following short account by Captain Crow will give a good idea of the powers of this tremendous weapon when wielded by the iron muscles of the Thresher :—

" One morning during a calm, when near the Hebrides, all hands were called up at three A.M. to witness a battle between several of the fish called Threshers or Fox Sharks and some swordfish on the one side and an enormous whale on the other. It was in the middle of summer, and the weather being clear and the fish close to the vessel, we had a fine opportunity of witnessing the contest. As soon as the whale's back appeared above the water, the Threshers, springing several yards into the air, descended with great violence upon the object of their rancour, and inflicted upon him the most severe slaps with their long tails, the sounds of which resembled the reports of muskets fired at a distance.

The swordfish in their turn attacked the distressed whale, striking from below, and thus beset on all sides, and wounded, where the poor creature appeared, the water around him was dyed with blood. In this manner they continued tormenting and wounding him for many hours, until we lost sight of him, and I have no doubt that they in the end completed his destruction." This strange alliance of two different fish against a marine mammal is a truly curious circumstance, and may have a deeper meaning than appears on the surface.

The food of the Thresher consists mostly of fish, and in the stomach of one of these creatures taken off the coast of Cornwall were found a quantity of young herrings. The colour of the Thresher is dark slaty blue above, and the same colour, but mottled with white, below.

AMONG the British Sharks, the PICKED DOG-FISH deserves notice, on account of the curious weapons from which it derives its name.

In front of each dorsal fin is placed a strong and sharply pointed spine, or pike, which has caused the fish to receive its popular name in most parts of the coast. The word is a dissyllable, and pronounced Pick-ed. On some of our shores it is called the BONE DOG, and in the Orkneys it is known by the name of the HOE.

These spines form aggressive weapons of a rather formidable character, the fish having the capability of directing a blow with wonderful accuracy. Mr. Couch says, that he has known the Picked Dog-fish able to pierce a finger if laid on its head, and never to miss its aim. When about to strike, it bends its body like a bow, and suddenly lashes out in the intended direction. It is a very common species, especially during the herring season, as it follows the shoals of those fish for the purpose of feeding on them. Even the tiny, quarter-grown young, not half the size of their intended prey, instinctively follow the herrings, though it is manifestly impossible that they should be able to eat them.

The Picked Dog-fish is destructive to the fishing trade, not only on account of its large appetite and the number of fish it consumes, but because it cuts the hooks away from the lines with its sharp teeth. As, moreover, it is extremely plentiful, some twenty thousand having been captured at one haul of a seine net, the destruction which it causes can be readily imagined. Sometimes this fish assembles in large shoals, and then the fishermen avenge themselves of their injuries, by shooting their nets around them, and capturing them by boats' loads at a time. Their flesh is tolerably good, a useful oil is obtained plentifully from the liver, while the refuse portions are most valuable as manure, and are strewed in unfragrant richness over the fields, warning the nostrils at a considerable distance that the next year's crop is likely to be successful, and that a nearer approach is undesirable except to the farmer and the entomologist.

PICKED DOG-FISH.—*Acánthias vulgaris.*

The colour of the Picked Dog-fish is slaty grey above, diversified, when young, with a few white spots, and the under parts are yellowish white. The skin is rough if stroked from the tail to the head, and smooth when rubbed in the reverse direction. The average length of this species is about eighteen inches.

The GREENLAND or NORTHERN SHARK (*Dalátias boreális*) must receive a brief notice, as it is frequently mentioned in accounts of whaling voyages.

This species is remarkable for the very small proportionate size of the fins, and for the manner in which the points of the teeth diverge from the centre of the jaw. It is a great foe to the whale and whalers, and is so heedless of danger when intent on satisfying its hunger, that it will follow a dead whale to the ship, mix boldly with the men who are engaged in cutting the blubber, thrust its head boldly among them, and at every bite scoop out lumps as large as a man's head.

So deeply engaged is the creature in this interesting occupation, that even if a man should slip into the water from the smooth oily skin of the whale, the Greenland Sharks take no notice of him, but continue their depredations on the whale. Even after the long whaling knife has been thrust through its body, it will dart off for the moment on feeling the wound, but will soon return to the same spot and continue its banquet. It also feeds on crustaceans and small fishes. Many specimens are nearly if not wholly blinded by a parasitic animal technically called *Lernœa elongata*, some three inches in length, which fastens upon the corner of the eye and lives upon its fluids.

The colour of this species is brown with a shade of deep blue. Its length, when full grown, is about fourteen feet.

ANOTHER curious species of Shark, called appropriately the SPINOUS SHARK (*Echinorhínus spinósus*), is notable for the spine-topped bony tubercles which are scattered over the surface of the body. The greater number of these spinous projections are boldly hooked, in a manner not unlike the thorns of the common bramble, and the points are directed backwards; others, however, are quite straight and stand upright. The object of these curious spines is not clearly known. They are very small in proportion to the size of the fish, and it is said that the males are more thickly studded with them than the females.

ANGEL-FISH.—*Squatina vulgaris.*

The colour is dark leaden grey on the head and back as far as the first dorsal fin, the remainder being reddish yellow with mottlings and cloudings of purple and brown. On the abdomen are irregular spots of vermilion. The chin and sides of the mouth are white. The average length of a full-grown specimen seems to be about seven or eight feet. In most, if not in all, of these creatures, the female is larger than the male, as is the case with the birds of prey.

THE dark-skinned, wide-mouthed, leather-finned and thorn-backed fish which is shown in the illustration, is popularly known throughout many parts of England, France, and Italy by the name of the ANGEL-FISH, a term singularly inappropriate except on the well-known principle "lucus a non lucendo," or perchance as leaving the spectator the option of choosing the kind of angel which the creature is thought to resemble.

Sooth to say, it is as hideous a fish as is to be found in the waters, and from all accounts is as unprepossessing to the inhabitants of the sea as to those of the land, being voracious to a degree, and attaining a size that causes it to be a most formidable foe to the many fishes on which it feeds. It is also known by the name of MONK-FISH, in allusion to the rounded head, which was thought to bear some resemblance to the shaven crown of a monk ; and in some places is called the SHARK RAY because it seems to be one of the connecting links between the sharks and the rays, and has many of the characteristics of both. On some parts of the British coasts it is known as the KINGSTON.

It has many of the habits of the flat-fishes, keeping near the bottom, and even wriggling its way into the muddy sand of the sea-bed so as to conceal its entire body. As in the course of these movements it disturbs many soles, plaice, flounders, and other flat-fishes that inhabit the same localities, it snaps them up as they endeavour to escape, and devours great quantities of them, so that it is really a destructive fish upon a coast.

It is most common upon the southern shores, and has there been taken of considerable size, attaining a weight of a hundred pounds. Unfortunately the flesh is now thought to be too coarse for the table, though it was formerly in some estimation, so that the creature is useless to the fisherman, who can only avenge himself for his losses by killing the destructive creature, but cannot repay himself by eating or selling it. The skin, however, being rough, is of some small use in the arts, being dried and employed, like

SAW-FISH.—*Pristis antiquorum.*

that of the dog-fish, for polishing joiner's work, and it is in some places manufactured into a sort of shagreen.

As may be seen by the illustration, the eyes are set rather far back on the upper part of the head, and a little behind each eye is the temporal orifice, very large in proportion to the dimensions of the fish, very long, and set transversely on the head. The wide mouth, which opens in front of the head and not below as in the sharks is furnished with rather long and sharply pointed teeth. The colour of the upper parts is dark chocolate-brown mottled with a darker hue, and very rough. Along the back runs a row of short sharp spines, their points directed backwards, and the under parts are smooth and of a dull brownish white. The length of an adult specimen is seven or eight feet.

WE now arrive at the Rays. The first family of these fishes is evidently intermediate between the sharks and the skates, and is in many respects a very interesting and remarkable group of fishes. The common SAW-FISH, so well known from the singular development of the snout, is a good example of this family.

It has a very wide range of locality, being found in almost all the warmer seas, and even in the cold regions near the pole. In the illustration, a view of the head and saw is given in the foreground, and the general shape of the fish is shown in the partly submerged figure above.

The snout of this fish is greatly prolonged, and flattened like a sword-blade. On either edge it bears a row of tooth-like projections, firmly imbedded in the bone, few, short, and wide apart at the base of the beak, but becoming larger and set closer together towards the point. The form of the sockets into which the teeth are received, and their rather enlarged termination, are conspicuously indicated on the surface of the saw-blade. The tip of the saw is covered with hard granular scales. The number of teeth is not the same in every individual; in a specimen in my possession there are twenty-eight on each side of the saw.

It is said that, like the sword-fish, this creature will attack the whale, thrusting its armed beak into the soft blubber-covered body of the huge cetacean, and avoiding, by its superior agility, the strokes of the tortured animal's tail, any blow of which, if it succeeded in its aim, would crush the assailant to death. The Saw-fish does certainly use this weapon for the destruction of fish. Captain Drayson has informed me that when lying off the Cape, he has more than once seen a Saw-fish come charging among a shoal of fishes, striking right and left with the serrated edges of the saw, and killing or disabling numbers of the fish by this process.

In all the Saw-fishes the skin is covered with minute rounded or hexagonal scales, arranged like the stones of a mosaic. The blow-holes are very large, and are set some distance behind the eyes. The mouth is on the under surface of the head, and is furnished with a crushing apparatus, not unlike that which has already been described as belonging to the smooth hound dog-fish.

The colour of the Saw-fish is dark grey above, nearly black in some individuals, the sides are ashen, and the abdomen white. It oftens attains a great size, measuring fifteen or eighteen feet in length, including the saw.

The TENTACULATED SAW-FISH (*Pristióphorus cirrátus*) is worthy of notice as forming a transition link between the sharks and the true Saw-fish. In this creature, the snout is lengthened; and armed with spines, but these structures are of different lengths, hooked, and only attached to the skin, and not implanted in the bone, as is the case with the true Saw-fish.

IN the true Rays or Raidæ, the fore part of the body is flattened and formed into a disc-like shape, by the conjunction of the breast fins with the snout.

Our first example of the Rays is the TORPEDO, a fish long celebrated for its power of emitting at will electrical shocks of considerable intensity. In consequence of this property, it is sometimes called the CRAMP-FISH, CRAMP RAY, ELECTRIC RAY, or NUMB-FISH.

The object of this strange power seems to be twofold, namely, to defend itself from the attacks of foes, and to benumb the swift and active fish on which it feeds, and which its slow movements would not permit it to catch in fair chase. It does not always deliver the electric shock when touched, though it is generally rather prodigal of exercising its potent though invisible arms, but will allow itself to be touched, and even handled, without inflicting a shock. But if the creature be continually annoyed, the shock is sure to come at last, and in such cases with double violence. It has been observed, moreover, that the fish depresses its eyes just before giving its shock.

The power of the shock varies greatly in different individuals, with some being so strong as to cause the recipient to fall to the ground as if shot, and with others, so feeble that it is hardly perceived. According to M. de Quatrefages, the fishermen are sometimes unpleasantly made aware that they have captured a Torpedo in their meshes, by the sudden shock through their arms and breast as they are hauling in their net. Anglers, too, are sometimes struck by means of the line which they are holding, and I presume that in either case the line must be wet, or it would not act as a conductor of the electrical fluid.

One of these fishes was placed in a vessel of water, and a duck was forced to swim about in the same vessel. The Torpedo soon became excited, and in a few hours the duck was dead. Fish also of different kinds are killed by this remarkable influence, and it is plausibly suggested by one writer, that this mode of destruction would render them liable to rapid decomposition, and would aid the organs of digestion in a creature like the Torpedo, where they are but imperfectly developed.

The shocks of this fish were once used as remedies for gout and fevers. In the first case, the patient had to lay his foot on the Torpedo, and bravely hold it in its place.

EYED TORPEDO.—*Torpédo ocúlata.*

despite of all the shocks sent by the angry fish through the sensitive limb of the **aggressor**; and in the latter case the Torpedo was used, as it were, to frighten the fever out of the system. The patient was stripped, and the Torpedo placed successively to the joints, trunk, and extremities, so that the whole of the body and limbs were permeated in their turn by the electric shock.

That the stroke of the Torpedo is veritable electricity is a fact which was once much disputed, but is now conclusively proved by a host of experiments. Needles have been magnetised by it just as if the shock had been that of a galvanic battery, the electrometer showed decided proofs of the nature of the fluid that had been sent through it, and even the electric spark has been obtained from the Torpedo—very small, it is true, but still recognisably apparent. It is rather curious, that in the course of the experiments it was discovered that the upper surface of the Torpedo corresponded with the copper plate of a battery, and the lower surface with the zinc plate.

The structure of the electrical organ is far too complex to be fully described in this work, as it would require at least forty or fifty pages, and a large number of illustrations. I will, however, give a brief summary of the strange organ by which such wonderful results are obtained, and any of my readers who would like to examine it more in detail, will find ample information in an article on the subject by Dr. Coldstream, in the "Cyclopædia of Anatomy and Physiology," and from a valuable series of wax models in the museum of the College of Surgeons.

Briefly, then, this organ is duplex, and consists of a great number of columns, placed closely against each other, each inclosed in a very thin membrane. These columns are again built up, as it were, of flat discs, separated by a delicate membrane, which seems to contain fluid. This structure may be roughly imitated by piling a number of coins upon each other, with a bladder between each coin and its successor—in fact, a kind of voltaic pile. The length of the columns, and consequently the number of discs, varies according to their position in the body. The columns extend quite through the creature, from the skin of the back to that of the abdomen, and are clearly visible on both sides, so that those of the middle are necessarily the longest, and those at either end become gradually shorter. In many large specimens, more than eleven hundred columns were counted, **and the** number of discs is on an average a hundred to the inch. It seems, from the **best** researches, that the growth of this organ is produced, not by the increase of each column but by a continual addition to their number. A vast amount of blood-vessels pass through the electric organ, and it is permeated with nerves in every direction.

How the electrical effect is produced is a very deep mystery. In fact, we know scarcely aught of this marvellous power, save the knowledge that it pervades all nature, and even in its external manifestations is one of the most ethereal and most potent of the second means through which the will of the Creator guides His universe. That the same electrical principle exists in all animals is familiarly known, and also that it is far more intense in some individuals than in others of the same species. It is known that the contact of two different kinds of flesh, such as the muscle of a fish and an ox, both newly killed, will produce similar effects ; and that it exists so largely in human beings, that no two individuals can place themselves on isolated stools, and join their hands, without emitting so much electricity by that slight contact, that the instrument will record its presence.   But the origin of this wonderful power eludes our mental grasp like the receding waters of the mirage, and the increase of our knowledge serves but to betray the extent of our ignorance.

I cannot but think that this subtle and potent emanation, which is able to strike the victim through an intervening space of the fluid common to both aggressor and sufferer, has some affinity with the still more subtle and equally mysterious influence by which certain of the serpent race are enabled to paralyze or attract the creatures which they could not secure by actual contact. It may possibly be that the electric powers of the Torpedo, which need water or some other conducting substance for their exercise, are, after all, but a more concentrated and palpable manifestation of that force, which enables the rattle-snake to arrest an animal not in physical contact with itself, the pointed finger to lay a bird motionless on its back until released by a sudden sound or touch, and one human being to influence his fellow without the use of words, and to attract or repel him by an irresistible though invisible agency.

It is rather remarkable that even the Torpedo, gifted with such puissant arms, dealing pain and death around at will, should find at all events one foe insensible to the electric stroke, and perhaps even needing its exciting influence to preserve it in health.   This is a parasitic creature, termed scientifically the Branchellion, which clings to the Torpedo and feeds upon its juices, quite indifferent to all the shocks which its victim dispenses.   It generally measures from an inch to an inch and a half in length.

This fish is found in the Mediterranean, and the Indian and Pacific Oceans, and occasionally off the Cape, and has now and then been captured on our coasts.   Happily, the Torpedo does not attain a very great size, one of the largest specimens being about four feet long, and weighing sixty or seventy pounds.

The Rays are well represented in England by several large and curious species.   One of the commonest examples is the THORNBACK SKATE or RAY, so called from the large number of thorny projections which are scattered over its back and especially along the spine.   This species is represented by the upper figure in the illustration.

The Thornback is one of our common Rays, and is taken plentifully on the shores of England, Scotland, and Ireland.   As is the case with many of the same genus, the flesh is considered rather good, and is eaten both when fresh and when salted for consumption during stormy weather.   Autumn and winter are the best seasons for procuring this fish, as the flesh is then firm and white, while during the rest of the year it is rather liable to become flabby.   Thornbacks taken in November are thought to be the best.

This species, like the rest of the Rays, feeds on crustacea, flat-fish, and molluscs, and as many of these creatures possess very hard shells, the Rays are furnished with a crushing mill of teeth, which roll on each other in such a way that even the stony shell of a crab is broken up under the pressure.   It is notable that the teeth differ in the two sexes when adult.   Those of the female are flat on the top, but those of the male throw out a strong angular projection, which is so arranged that the projections of one jaw exactly fit into the interstices of the other, and the roller-like arrays of teeth bear a wonderful resemblance to the well-known clod-crushing machine.

THORNBACK SKATE.—*Raia clavata.*　　COMMON SKATE.—*Raia batis*

The young of this and other Skates are produced from eggs, whose form is familiar to every visitor to the seashore, where they go by the popular name of Skate-barrows. Their colour is black, their texture leathery, thin, and tough, and their form wonderfully like a common hand-barrow, the body of the barrow being represented by the middle of the egg, and the handles by the four projections at the angles. The empty cases are continually thrown on the beach, but it is seldom that the young are found inclosed, except after a violent storm, or when obtained by means of the dredge.

This species is notable for certain thorny appendages to the skin, which are profusely sown over the back and whole upper surface, and among which stand out conspicuously a few very large tubercular spines, with broad, oval, bony bases, and curved, sharp-pointed projections. Fifteen or sixteen of these bony thorns are found on the back. Along the spine runs a single row of similar spines, and at the commencement of the tail it is accompanied by another row on either side, making that member a very formidable instrument of offence. In point of fact, the tail is as formidable a weapon as can be met with, and the manner in which this living quarter-staff is wielded adds in no slight degree to its power. When angered, the Skate bends its body into a bow-like form, so that the

STING RAY.—*Trygon vastinaca*

tail nearly touches the snout, and then, with a sudden fling, lashes out with the tail in the direction of the offender, never failing to inflict a most painful stroke if the blow should happen to take effect.

The colour of the Thornback Skate is brown, diversified with many spots of brownish grey, and the under parts are pure white.

The COMMON SKATE, sometimes called the TINKER, is so well known that only a very short description is needed.

This fish is found on all our coasts in great plenty, and sometimes attains to a really large size, a fine specimen having been known to weigh two hundred pounds. The fishermen have a custom of calling the female Skate a Maid, and the male, in consequence of the two elongated appendages at the base of the tail, is called the Three-Tailed Skate. It is a very voracious creature, eating various kinds of fish, crustaceans, and other inhabitants of the deep.

The colour of this species is greyish brown on the upper surface, and a little reddish brown and black-brown are found on the edges of the broad fins. Below, it is greyish white, over which divers darker lines are drawn, and upon which are scattered a great number of bluish spots with small sharp points among them.

TERRIBLE as is the armed tail of the thornback skate, and severe as are the wounds that can be inflicted by it, the STING RAY is furnished with a weapon even more to be dreaded, and capable of causing a still more serious injury.

The tail itself of this species is long, flexible, whip-like and smooth, so that were it unaided by any additional armature, it could only inflict a sharp and stinging blow, which, however painful, would do no more damage than the cut of a horsewhip. As, however, may be seen in the illustration, the tail is further armed with a projecting bony spine, very sharp at the point, and furnished along both edges with sharp cutting teeth. When attacked or irritated, the Sting Ray suddenly strikes its whip-like tail around the offender in lasso fashion, and holding him tightly against the barbed spine, wields the latter with such strength and rapidity that it lacerates the flesh to a frightful and dangerous extent, in some cases even causing the death of the victim.

EAGLE RAY. – *Myliobatis áquila*.

Along the coast, where the offensive powers of this fish are familiarly and practically known, an opinion prevails that the bony spine is supplied with poison. This notion, however, is one of the many popular errors on similar subjects, having been founded on the aggravated inflammation that sometimes follows the wounds caused by the Sting Ray. There is no poison whatever in this bone, and any such symptoms are due, not to the inherent venom of the weapon, but to the unsound constitution of the sufferer.

The reader will at once perceive the exact resemblance between the spine of the Sting Ray and the many-barbed spears used by the savage inhabitants of the Pacific islands. In fact, this spine not only furnished them with the original idea of those cruel weapons, but is constantly taken from the fish and affixed to the shaft of a lance. In their eyes, its great merit—and one which they imitate in their manufactured weapons—is that when the spear is struck into the body of a foe, the jagged blade is sure to snap asunder at the point where it enters the body, leaving several barbs fixed in the wound without any handle by which they may be withdrawn.

It is found that in the Sting Ray, a second spine exists below the first, which is provided in order to supply the place of the first in case it should be broken off or dragged out.

The Sting Ray is in some places called the FIRE FLAIRE, probably on account of the very red colour of the flesh when cut open. This fish is not approved for the table, being rank and disagreeable in flavour.

The colour of the Sting Ray is greyish yellow above, taking a slaty blue tint towards the middle of the body, and spotted with brown when the creature is young. Below, it is white. The eyes are golden colour, the temporal orifice behind each eye is extremely large, and the tail is very thick and muscular at the base. The spine is set about one-third of its length from the base. The mouth and teeth are small.

IN some respects, such as the long tail and double-barbed spine with which it is armed, the EAGLE RAY bears some resemblance to the preceding species, but may be readily distinguished from that fish by the projecting head, the bluntness of the snout, the very great length and comparative tenuity of the tail, the shortness of the spine, and the diminutive size of the temporal apertures. In some places this fish is called the Whip Ray, in allusion to the extreme length of the slender tail.

The flesh of the Eagle Ray is not eaten, being hard, rank, and disagreeable, but the liver is thought to be eatable, and a large quantity of good oil is obtained from it. It has been found on our coasts, a specimen having been taken at Berwick in 1839, but it is mostly found in the Mediterranean and more southern seas. It sometimes attains to a very large size, weighing as much as eight hundred pounds. Its colour is dark brown above, deepening towards the edges, and greyish white below.

BEFORE quitting these fish entirely, a short notice must be given of several interesting species, of which figures cannot be inserted for want of space.

The first is the HORNED RAY (*Cephaloptera Johnii*), sometimes called, from its huge dimensions, horned head, dark body, and lowering aspect, the SEA DEVIL. There are, however, several species which are popularly called by the latter title.

This enormous creature is found in the Mediterranean and the warmer seas in general, and has been taken in the nets together with the tunny. The flesh is not eaten except by the very poor, but the supply of oil from the liver is abundant and valuable. There seem to be hardly any bounds to the size which this creature will attain. M. Le Vaillant saw three of these huge fish sporting round the ship in lat. 10° 15' N. long. 350° W. and, after some persuasion, induced the crew to attempt their capture. They secured the smallest of the three, and when it was brought on board, it was found to measure twenty-eight feet in width, twenty feet in length, to weigh a full ton, and to have a mouth large enough to swallow a man.

This gigantic Ray feeds almost wholly on fishes and molluscs. On account of their horned heads, the Italian fishermen call the old ones cows and the young calves. A strong attachment seems to exist between the male and female, for it has more than once happened that when one fish has been harpooned or otherwise captured, its mate has hung about the boat until it shared the same fate with its deceased partner ; and in one instance, where the female had been caught in a tunny net, the male was seen wandering about the net for several days, and at last was found dead in the same partition where his mate had been captured. So, in common justice, the name of Sea Devil ought not to be applied to so loving and faithful a creature.

The colour of the Horned Ray is very dark black-blue above, and grey-white beneath. The jaws and mouth are proportionately greater than is generally the case with these fishes. The tail is long, thin, and smooth for the first quarter of its length, after which it is furnished with tubercles. At its base there is a sharp, flattened spine, armed, like that of the preceding species, with a double row of barbs.

THERE are several other British Rays, among which may be briefly mentioned the LONG-NOSED SKATE (*Raia Salviani*), remarkable for the great length of the snout ; the FLIPPER SKATE (*Raia intermedia*), notable for the olive-green colour of the upper surface, and the numerous white spots with which it is covered ; the BORDERED RAY (*Raia marginata*), which may be known by the dark edge to the side fins, or wings as they are generally called, and the three rows of sharp spines on the tail ; and lastly, the HOMELYN RAY (*Raia miraletus*), which may be distinguished by the large size of the eyes and temporal orifices, and the bold dark spots on the sides.

WE now arrive at the vast order of the SPINE-FINNED FISHES, known scientifically as the ACANTHOPTERYGII. In all these fishes, the skeleton is entirely bony, and part of the rays of the dorsal, anal, and ventral fins are formed into spines, in some species very short, and in others of extraordinary length.

Without devoting more time or space to the purely scientific and anatomical characteristics, which will be separately described at the end of the volume, we will proceed at once to the various species of this vast and important order. I may here mention, that, whenever possible, I have selected British fish as examples of the various genera, employing only those foreign species that are needful to fill up the links of the chain, or that are worthy of notice from some remarkable points in their form or their habits.

THE first family is well represented in England by many pretty and interesting species, of which the two creatures figured in the engraving are familiar examples.

The THREE-SPINED STICKLEBACK is one of our commonest British fishes, and is known in different parts of England under the names of TITTLEBAT, PRICKLEFISH, and SHARPLIN.

It is a most bold and lively little fish, hardly knowing fear, pugnacious to an absurd degree, and remarkably interesting in its habits. Even more voracious than the perch, it renders great service to mankind in keeping within due bounds the many aquatic and terrestrial insects, which, although performing their indispensable duties in the world, are so extremely prolific, that they would render the country uninhabitable were they allowed to increase without some check.

So voracious and fearless indeed is this little creature that it always forms the earliest game of the juvenile angler, who need not trouble himself in the least about the temper of his hooks, the fineness of his tackle, or the delicate balance of his float. Any one can catch a Stickleback without rod, float, or even hook. All that is needful is to repair to

TEN-SPINED STICKLEBACK.—*Gasterósteus pungítius.*   THREE-SPINED STICKLEBACK. —*Gasterósteus aculeátus*

the nearest streamlet, armed with a yard or two of thread and a walking-stick. Thin twine will answer very well instead of the thread, and even the stick is not absolutely needed. Having proceeded thus equipped to the bank of the stream, a worm may be picked out of the ground, tied by the middle to the thread and thrown quite at random into the water.

The Sticklebacks will not be in the least frightened by the splash, but rather rejoice in it as calling their attention to food. In a moment the worm will be the centre of a contending mass of little fishes, rolling over and over, struggling to the utmost of their power, and entirely hiding the worm from sight. Now let the angler quickly lift the bait out of the water, swing it on shore, and he will almost certainly find that he has captured two Sticklebacks, one hanging to each end of the worm, and retaining its hold so perseveringly that it can hardly be induced to relinquish its gripe. This process may be repeated at pleasure, and as the Sticklebacks never seem to learn wisdom, a large store may soon be accumulated. This is a good way of stocking an aquarium, as the strongest and liveliest fish are sure to be caught first.

I have caught them by hundreds in a common butterfly-net, by the simple stratagem of lowering the net into the water, dangling the worm over the ring, and by degrees lowering the worm and raising the net until I had the whole flock within the meshes.

Should the reader be disposed to place his newly captured specimens in an aquarium, he must make up his mind that they will fight desperately at first, and until they have satisfactorily settled the championship of the tank their intercourse will be of the most aggressive character. Never were such creatures to fight as the Sticklebacks, for they will even go out of their way to attack anything which they think may possibly offend them, and they have no more hesitation in charging at a human being than at one of their own species. I have known one of these belligerent fish make repeated dashes at my walking stick, knocking his nose so hard against his inanimate antagonist that he inflicted a perceptible jar upon it, and in spite of the blows which his nose must have suffered, returning to the combat time after time with undiminished spirit.

These combats are, however, most common about the breeding season, when every adult Stickleback challenges every other of his own sex, and they do little but fight from morning to evening. They are as jealous as they are courageous, and will not allow another fish to pass within a certain distance of their home without darting out and offering battle.

Any one may see these spirited little combats by quietly watching the inhabitants of a clear streamlet on a summer day. The two antagonists dart at each other with spears in rest, snap at each other's gills or head, and retain their grasp with the tenacity of a bull-dog. They whirl round and round in the water, they drop, feint, attack, and retreat, with astonishing quickness, until one confesses itself beaten, and makes off for shelter, the conqueror snapping at its tail, and inflicting a parting bite.

Then is the time to see the triumphant little creature in all the glory of his radiant apparel ; for with his conquest he assumes the victor's crown : his back glows with shining green, his sides and head are glorious with gold and scarlet, and his belly is silvery white. It is a little creature certainly, but even among the brilliant inhabitants of the southern seas, a more gorgeously coloured fish can hardly be found. If the conqueror Stickleback could only be enlarged to the size of a full-grown perch or roach, it would excite the greatest admiration. It is curious, that the vanquished antagonist loses in brilliance as much as the conqueror has gained ; he sneaks off ignominiously after his defeat, and hides himself, dull and sombre, until the time comes when he, too, may conquer in fight, and proudly wear the gold and scarlet insignia of victory.

These struggles are not only for mastery, but are in so far praiseworthy, that they are waged in defence of home and family.

The Stickleback is one of the very few fish who build houses for their young, as a defence against the many foes which are ever lying in wait for the destruction of the eggs or the newly hatched young. These nests are built of various vegetable substances, and their structure is admirably described in the following passage extracted from an educational magazine of 1834, and quoted by Mr. Couch in his valuable history of the British fish :—

" In a large dock for shipping on the river Thames, thousands of Pricklefish were bred some years ago, and I have often amused myself for hours by observing them. While multitudes have been enjoying themselves near the shore in the warm sunshine, others have been busily engaged making their nests, if a nest it can be called. It consisted of the very minutest pieces of straw or sticks, the exact colour of the ground at the bottom of the water on which it was laid, so that it was next to an impossibility for any one to discover the nest, unless they saw the fish at work, or observed the eggs.

The nest is somewhat larger than a shilling, and has a top or cover, with a hole in the centre, about the size of a very small nut, in which are deposited the eggs or spawn. This opening is frequently concealed by drawing small fragments over it, but this is not always the case. Many times have I taken up the nest, and thrown the eggs to the multitude around, which they instantly devoured with the greatest voracity. These eggs are about the size of poppy seeds, and of a bright yellow colour ; but I have at times seen them almost black, which, I suppose, is an indication that they are approaching to life.

In making the nest, I observed that the fish used an unusual degree of force when conveying the material to its destination. When the fish was about an inch from the nest, it suddenly darted at the spot, and left the tiny fragment in its place, after which it would be engaged for half a minute in adjusting it. The nest, when taken up, did not separate, but hung together like a piece of wool."

This interesting little account is, as Mr. Couch remarks, doubly valuable, as not being the work of a professed naturalist, but of an observant lover of nature, who saw some curious phenomena, and recorded them in simple and unpretending language. The fifteen-spined Stickleback, a marine species, also makes a nest, though hardly of so careful a construction.

The Three-spined Stickleback is very fond of inhabiting the mouths of rivers where they empty themselves into the sea, the brackish water appearing to suit its constitution. It can therefore be easily acclimatized to new conditions, and a specimen that has been taken from an inland stream can soon be brought to inhabiting the water of a marine aquarium, though such water is usually, in consequence of evaporation, more salt than that of the sea.

As a general fact, the flesh of the Stickleback is despised as an article of food, and in my opinion wrongly so. I have often partaken of these little fish fried, or even baked, and think them decidedly palatable—delicate, crisp, and well-flavoured, with the slightest possible dash of bitter that gives a unique piquancy to the dish. At all events, the young of the Stickleback and the minnow frequently do duty as whitebait, and the guests never discover the deception. Yet there is hardly any place in England where even the starving poor will condescend to eat this delicate and nutritious little fish, which can be scooped by thousands out of any streamlet, and does not require more trouble in cooking than the red herring. The only use that at present seems to be made of this fish is to spread it over the ground as manure, an office which it certainly fulfils admirably, but might, in all probability, be better employed in feeding man than manuring his fields. An oil is sometimes expressed from them, and the refuse carted off to the fields, but the value of the oil seems hardly to repay the trouble of procuring it.

Mr. Yarrell mentions a considerable number of British Sticklebacks; but Dr. Günther, in his elaborate catalogue of Acanthopterygian fishes in the British Museum, comprises several species together, as only varieties and not different species. For example, the QUARTER-ARMED STICKLEBACK (*Gasterosteus gymnūrus*) or SMOOTH-TAILED STICKLEBACK, known by its four or five scaly plates above the pectoral fin; the HALF-ARMED STICKLEBACK (*Gasterosteus semiarmatus*), where the plates extend throughout half the length of the body; the HALF-MAILED STICKLEBACK (*Gasterosteus semiloricatus*), where they extend still farther; and the NEW YORK STICKLEBACK (*Gasterosteus Noveboracencis*), are all considered as being only varieties of the species which has just been denoted.

THE left-hand figure on page 223 represents the TEN-SPINED STICKLEBACK. This fish is nearly, if not quite, as plentiful as the three-spined species, and is perhaps the smallest of our river-fish.

It may be readily distinguished by the nine or ten spines upon the back, all in front of the dorsal fin, and by the absence of plates upon the sides. All the Stickle-backs are voracious little creatures, and I am told by an angler friend that they destroy quantities of the spawn of other fish, and seize upon the young as soon as they are hatched. He also informs me that they are extremely capricious in their choice of locality. For example, at the head of a mill-stream they may be found by thousands, while at the tail of the same stream not a single Stickleback can be found. There is a still part of the New River, where they are so plentiful that the roach fisher is entirely baffled in his sport by these little creatures eating his bait before it sinks to its full depth, and yet the middle of the stream is quite free from them.

The Ten-spined Stickleback does not like salt water, and cannot be acclimatized to the marine aquarium like its three-spined relative. All the Sticklebacks are remarkable for the comparative nakedness of the skin, which for the most part bears no scales, as in the generality of fish, and in the Ten-spined species is wholly naked. The place of the scales is supplied by certain bony or scaly plates upon the side, and it is the nakedness of the skin which permits the colours of these little fish to glow with such bright and changeful hues.

3.

The colour of this species is green upon the back, and on the abdomen and sides silvery white spotted minutely with black. The fins are very slightly tinged with yellow. The length of the Ten-spined Stickleback is variable, but rarely exceeds two inches.

The FIFTEEN-SPINED STICKLEBACK, SEA-ADDER, or BISMORE, is wholly a marine species, and is common on all our coasts.

It is remarkably elongated in proportion to its width, and this formation, together with its armature of sharp tooth-like spines, has gained it the name of Sea-Adder. It is a voracious creature, feeding on all sorts of marine animals, molluscs, worms, eggs, and fry, and minute crustaceans. Mr. Yarrell advises the collector of marine crustaceans to examine carefully the stomachs of the shore-frequenting fishes, and especially of this species, as he will be likely to discover some curious species of those animals, too active or too small to lodge in his net, but unable to avoid the quick eye and ready jaws

FIFTEEN-SPINED STICKLEBACK.—*Gasterósteus spináchia.*

of the Stickleback. The same writer mentions that on one occasion, when a Fifteen-spined Stickleback had been caught with a net and placed in water together with a small eel, three inches in length, the voracious creature seized on the eel in a very short time, and contrived to swallow it. The eel, however, was too long to be wholly accommodated in the stomach of the Stickleback, and after a while was disgorged, only partly digested.

This, as well as the other species, is of very changeful colouring, its tints altering according to the circumstances of the moment. As in the case of the frog, already alluded to when treating of that creature, the colour of the Stickleback varies with singular rapidity, being dull or bright according to the mental emotions of the individual. The specimen above mentioned was so alarmed when captured, that it changed from its former brilliant tints to pale yellow and brown, remained in that state for eighteen hours, and then suddenly regained its former brightness.

The Fifteen-spined Stickleback makes a nest for its eggs, and watches it as faithfully as the preceding species. The materials are composed of bits of the delicate green or purple seaweeds, woven among the branches of growing corallines, and bound together with a kind of thread of animal matter, so as to form a pear-shaped mass, about as large as the closed fist. The eggs, which are very large, and of a light amber colour, are not lodged in a hollow within this nest, but distributed in little packets throughout the mass. Mr. Couch gives the following curious account of a nest of the

JAPANESE SINGLETHORN.—*Monocentris Japónicus.*

Fifteen-spined Stickleback, made in the end of an old rope: "A singular instance of constructive skill and patience in the formation of its nest, which occurred within my knowledge, is deserving of remembrance. The situation selected was the loose end of a rope, from which the separated strands hung out about a yard from the surface, over a depth of four or five fathoms, and to which the materials could only have been brought, of course, in the mouth of the fish, from the distance of about thirty feet.

They were formed of the usual aggregation of the finer sorts of green and red seaweed, but they were so matted together in the hollow formed by the untwisted strands of the rope, that the mass constituted an oblong ball of nearly the size of the fist, in which had been deposited the scattered assemblages of spawn, and which was bound into shape with the thread of animal substance already described, and which was passed through and through in various directions, while the rope itself formed an outside covering to the whole.

We can scarcely suppose that such a nest can have been the work of more than a couple of fishes, but the grains of spawn had grown to almost the size of radish-seeds, and in collective bulk seemed greatly disproportionate to the size of the parent, and only to be explained by the well-known fact, that the ova of fishes generally obtains an increase of bulk by the absorption of water after exclusion; which fluid may be supposed to exert considerable influence on the further development of the young. The embryo of this fish, as is believed to be the case with many others, is not found to bear a close resemblance to the parent, and in fact may be said to pass through a decided metamorphosis in the course of its final development."

The jaws of this species are much elongated, and the under jaw protrudes well beyond the upper. Along the back runs a row of fifteen spines, short, sharp, slightly hooked backwards, and each with a very slight membrane. Along the lateral line runs a series of keeled scales, and on the abdomen are two bony plates, bearing on their inner edges two sharp spines of unequal length. If the body be severed at the centre, the line of section will be nearly a pentagon. The upper part of the head, body, and tail is olive-green, taking a golden tint on the sides. The rest of the body is silvery white. The length of this fish varies from five to seven inches.

MEDITERRANEAN HOPLOSTETHUS —*Hoplostethus Mediterraneus.*

We now arrive at another family of fishes, in which the body is rather compressed—*i.e.* flattened sideways—the eyes are large, and the mouth oblique. It is scientifically known by the name of Berycidæ, and all its members are inhabitants of the tropical and temperate seas.

Our first example of this family is the JAPANESE SINGLETHORN.

In all the fishes of this genus, the scales are rather large, very strong, and so closely compacted together that they form a strong mailed covering to the body. The name of Monocentris or Singlethorn is derived from the curious modification of the ventral fins, which are devoid of membrane, and are reduced to a single very strong and rather lengthened spine, and a few very short rays. In the place of the dorsal fin are four or five thick spines, and the shield-like scales of the body are rough, projecting, and keeled.

The Japanese Singlethorn is an inhabitant, as its name imports, of the seas of Japan, and is almost, if not quite, the only species of its genus. It is chiefly remarkable for the size of its head, the strong thorn-like spines, and the mailed suit of hard and projecting scales. It is of a tolerably uniform colour, its whole body being silvery white, and its length is about six or seven inches.

The large-eyed and deep-bodied fish which is shown in the illustration, derives its name of HOPLOSTETHUS, or Armed-breast, from the strong and sharp spines which are placed on the scapular bone and the angle of the præoperculum. Like the last-mentioned species, it seems to be the only member of its genus.

This fish is found in the Mediterranean, and is not uncommon on the coast of Madeira. It is remarkable not only for the offensive weapons with which it is armed, but for the large, full eye, the saw-like series of notches on the abdomen, and the beautiful rosy hue of its scales. The dorsal fin of this fish is single, but is composed of two distinct portions, the one being supported by strong spinous rays, and the other by soft and flexible rays. The muzzle is very short, rounded, and does not protrude; the tail is deeply forked, and the serrated portion of the abdomen consists of eleven, twelve, or thirteen scales. The body is very deep in proportion to its length.

BEFORE proceeding to the next family, we must casually notice two large genera belonging to this family. Of the first genus, the MURDJAN PERCH (*Myripristis murdjan*) is a good example.

This handsome fish is found off the coasts of India and in the Red Sea, and can be easily recognised by its beautiful colouring, its large scales, short muzzle, and prominent chin. The general colour of this splendid fish is bright rose-pink, beautifully mottled by a rich violet edge to each scale. The soft portions of the dorsal, ventral, and anal fins are boldly margined with white, and the front rays have a cross band of violet-brown. The tail fin is edged with white, and a longitudinal stripe of violet-brown traverses each lobe. About fourteen or fifteen species of this genus are known.

OF the next genus, the SCARLET PERCH (*Holocentrum rubrum*) is rather a striking example.

This fine fish inhabits the Asiatic seas, and there are specimens in the British Museum from the Red Sea, Amboyna, Louisiade Archipelago, the Philippines, Japan, and China. On the operculum are two strong spines, the upper being the larger. The colour of this fish is shining red, diversified with eight bands of greyish white. The outer edges of the tail fin are black, and there is a patch of the same colour on the ventral fins. This genus contains many very handsome species, and in almost every case the prevailing colours are red and violet.

WE now come to the large and important family of the Perches, which comprises many of the handsomest and most valuable fishes. The members of this family are found in all parts of the globe.

PERCH.—*Perca fluviatilis.*

The COMMON PERCH is well known as one of our handsomest river-fish, and, on account of its boldness and the voracious manner in which it takes the bait, and the active strength with which it struggles against its captor, is a great favourite with many anglers.

Moreover, when captured, and placed in an aquarium, it very soon learns to distinguish the hand that feeds it, and will come to the surface and take food from the fingers. It has a fashion of seizing its food with a rather sharp jerk, and then snatches it away with such violence, that when it takes the hook, it will drag a stout cork float several inches below the surface, and, by the force of its own stroke, will mostly hook itself without any exertion on the part of the angler. Bold-biting, however, as is its reputation, there are some seasons of the year when it is almost impossible to catch a Perch, and even the shy and gently nibbling roach is an easier prey.

The Perch is a very hardy fish, living for a long time when removed from the water, in consequence of the structure of the gill-cover, which prevents the delicate membranes of the branchiæ from becoming dry. It will, in consequence, endure being transported over considerable distances, if it be only watered occasionally, a capacity which enables the proprietor of a fish-pond to stock it without difficulty. In some countries, where fish is a common article of food, and enforced on certain days by ecclesiastical law, these fish are kept in ponds, caught in nets, put into baskets with grass which is always kept wet, and taken to the markets, where they remain through the day, and if not sold, are carried back to their pond in the evening and replaced.

It does not seem to be a cold-enduring fish, however hardy it may be in other respects ; and though it is plentiful in almost every lake and river of England and Wales, it is hardly to be found in the waters north of the Tweed. All the temperate parts of Europe possess this well-known species.

The Perch is a truly voracious fish, feeding upon all kinds of aquatic worms, insects, and fishes, preferring the latter diet as it becomes older and larger. The smaller fish, such as minnows, young roach, dace, and gudgeons, are terribly persecuted by the Perch, and a bait formed of either of these fish, or a good imitation of them, will generally allure the finest Perches to the hook. Although generally inhabiting mid or deep water, it will sometimes come to the surface to snap up a casual fly that has fallen into the water, and on several occasions has been captured by anglers when fishing with a fly for trout.

Practical fishermen say that the Perch is almost the only fish which the pike does not venture to attack, and that if a pike should make one of its rushing onslaughts on a Perch, the intended prey boldly faces the enemy, erects the dorsal fin with its array of formidable spines, and thus baffles the ever-hungry aggressor. Still, it is an article of faith with some anglers, that a young Perch from which the dorsal fin has been removed is one of the surest baits for pike. Perhaps they think that the pike is so delighted to find a Perch unarmed, that it seizes the opportunity to feed upon a luxury which it can seldom obtain.

The Perch is not often seen in the middle of a stream, preferring to haunt the banks, and from under their shadow to watch the little fish and other creatures on which it feeds. This habit is common to many, if not to most of the carnivorous river-fishes, the pike and trout being also bank lovers, and having special retreats whither they betake themselves, and which they will not suffer any other fish to approach. Deep holes by the bank are favourite resorts of the Perch, and on a fine day, when the water is clear, it is often possible to see them in their home, swimming gently to and fro, and never stirring from the narrow limits of the hole in which they reside.

By careful management, it will sometimes be possible to capture every member of the party, for if the bait be quietly let down among them, and each Perch when hooked drawn smartly out of the water, the survivors seem to care nothing about the sudden exit of their companion, and successively fall victims to the same fate. It is, however, as a rule, unsafe to let one fish get off the hook, as these creatures, though mute to human ears, have their own silent speech, and are able to communicate ideas among themselves.

The eye of a Perch is said to be an almost irresistible bait for these fish, and the sufferer has even been known to break away from the hook, to which one of its eyes clung, and to be again captured by biting at its own eye.

The flesh of the Perch is white, firm, well flavoured, and is thought to be both delicate and nutritious.

The Perch is not a large fish, from two to three pounds being considered rather a heavy weight. Individuals, however, of much greater dimensions have been, though rarely, captured. One of the finest Perches ever taken in England, was captured in the river Avon, in Wiltshire, by a night-line baited with a roach ; its weight was eight pounds. Specimens of five or six pounds are occasionally taken, but are thought so valuable, that the captor generally sends the account of his success to some journal.

The colour of the Perch is rich greenish brown above, passing gradually into golden white below. Upon the sides is a row of dark transverse bands, generally from five to seven in number. The first dorsal fin is brown, with a little black between two or three of the first and last rays; the second dorsal and the pectoral are pale brown, and the tail and other fins are bright red.

THE fine fish so well known under the name of BASSE, or SEA-DACE, or SEA-PERCH, is common on many of our coasts, and is considered by anglers as affording good sport.

It seems, from the accounts of practical sportsmen, to bite with readiness at a bait, but to be a difficult fish to secure, on account of its tender mouth, its ingenious stratagems, and its great strength. When hooked, it leaps, plunges, and swims with such force and swiftness, that the captor is forced to exercise the greatest skill in preventing it from breaking away. One of its favourite ruses is to double back under the boat, in hopes of cutting the line against the keel, or gaining a fixed point by which it may be able to drag the hook from its mouth.

Even when fairly tired out, and drawn to the edge of the boat, it is by no means secured, for its scales are so hard that a very sharp blow of the gaff is needed to fix the hook in its side, and its gills and fins are so formidably armed, that it cannot be grasped

BASSE. *Labrax lupus.*

with impunity. The spines of the dorsal fin, in particular, are strong and sharp as packing-needles, and the various portions of the operculum are edged with projecting teeth that cut like lancets. Many are the wounds that have been inflicted by the sudden twist and wriggle of the Basse, when grasped in a careless manner. When lifted into the boat, the hook is not to be taken from the mouth without some risk, as may be imagined on reference to the illustration.

It is a voracious fish, and derives its name of "lupus," or wolf, in consequence of its insatiate appetite. It feeds upon other fish and various inhabitants of the sea. Mr. Couch states that it is very fond of woodlice, and is bold enough to venture among rocks in a tempest for the sake of snapping up these creatures, as they are washed by the waves and beaten by the winds from their places of concealment among the stones.

The flesh of the Basse is very excellent, and is thought to be in best condition when the fish is small, measuring about eighteen inches in length. The colour of this fish is dark dusky blue on the back, and silvery white on the abdomen; the fins are brown. It sometimes attains a very large size, having been known to weigh upwards of twenty pounds. It seldom, however, reaches such extreme dimensions, and a specimen of fifteen pounds' weight is thought to be a remarkably fine one.

GIANT PERCH.—*Lucioperca Sandra*

THE pretty little RUFFE (*Acerina cérnua*) is common in many English rivers, where it is sometimes known under the name of POPE, the reason for the latter title not being very clear.

In general appearance the Ruffe bears some resemblance to the perch, the shape of its body and the thorny fins being not unlike those of that handsome fish. It may, however, be immediately distinguished from the perch by its spotted fins, and the absence of the dark band over the sides. Moreover, the dorsal fin is single. It is a tolerably bold biter, and takes a hook readily when baited with a little bright-red worm.

The colour of the Ruffe is light olive-brown above, and silver-white on the abdomen; the flanks are yellowish brown. The back, dorsal fin, and tail, are covered with little brown spots, set so closely in the tail as to resemble bars, and upon the gill-covers there is a little pearly green. The length of this fish seldom exceeds six or seven inches.

A REMARKABLY fine fish, called the GIANT PERCH, is found in many of the rivers and lakes of Germany and Eastern Europe.

This handsome species derives its name of Lucioperca, or Pike-Perch, from the resemblance which it bears to both these fishes, having the lengthened body of the one, and the spine-armed fins of the other. It has, however, nothing to do with the pike, and is closely allied to the perch, belonging, indeed, to the same family. The teeth are rather large, and are thought to resemble those of the pike in length and sharpness.

The colour of the Giant Perch is greenish olive above, banded with brown. Below, it is white. It is a very fine fish, attaining, when full-grown, to a length of three or four feet. There are several species belonging to the same genus.

A VERY handsome fish, that is popularly but erroneously called the AMERICAN PIKE, has derived its name from the elongated and somewhat pike-like form of its body. The teeth, however, are even, and bear no resemblance to those of the real pike.

The flesh of this fish is thought to be good for the table, and as the dimensions to which the creature attains are often considerable, it is really one of the valuable inhabitants of the American waters. It is one of the sea-loving species, and is mostly found on the Atlantic shores of tropical America. Many specimens are in the British Museum, some of which were taken in the West Indies, others off the coast of Guiana, some from Bahia, and others from Surinam. The general colour of the American Pike is silvery white, tinged on the back with green, and becoming a pure shining white on

AMERICAN PIKE. — *Centropomus undecimalis.*

the abdomen. The dorsal fins are two in number, the first being shortish, and having eight very strong and sharp spines. The second spine of the anal fin is very long and sharp, and the præoperculum is armed with two sharply toothed edges.

THE well-known BLACK BASSE of America (*Centropristis atrarius*) inhabits the rivers and lakes of North America.

This fine fish is a really valuable species, on account of its large dimensions and the excellence of its flesh, and the attention of scientific men has lately been turned towards its preservation. In the Patent Reports of 1859 upon some Black Basse that were transferred to Waramang Lake, Connecticut, in 1853, it is said that they multiplied very rapidly, grew at the average rate of one pound per annum, and ordinarily attained a weight of five pounds or a little more. They are very hardy, and can be taken from one locality to another if placed in a tub of water covered with a wet canvas. So rapid, indeed, is its increase, that although less than a hundred were originally placed in the lake, they have probably increased to several millions in a space of seven years.

It is a marvellously bold-biting fish, and affords good sport to all anglers, whether they only fish for the sake of the amusement, using a fly or other delicate bait, or whether they merely seek to take their prey as a matter of business, and employ small fish as a bait, or the obstruction "spoon," whose treacherous glitter the Black Basse is seldom able to withstand. It is an active and powerful fish, and when hooked struggles so long and so fiercely, that it tests all the angler's skill before it can be safely landed.

The colour of the Black Basse is brown, washed with golden green, and mottled with dark spots on the centre of each scale, darker on the back, and becoming nearly white on the abdomen. When newly caught, the body is traversed with several dark bands. It is a very fine fish, specimens having been known to weigh nearly twenty pounds.

CLOSELY allied to these fish is an enormous genus, containing about one hundred and forty known species, from which the OUATALIBI, or RUDDY SERRANUS (*Serránus ouatalibi*), is selected as an example.

STONE BASSE.—*Polyprion cérnium.*

This beautiful fish inhabits the warm Caribbean sea, and is plentiful upon the West Indian coasts. Its colour is bright red, and the head, body, and sometimes the dorsal fin, are profusely powdered with small blue spots, edged with black. Just by the joint of the lower jaw there is a pair of largish black spots, and on the back of the tail, immediately behind the dorsal fin, is another black spot. Of its habits nothing interesting is told.

The STONE BASSE is an inhabitant of the British seas. It is otherwise known as COUCH'S POLYPRION, in honour of the eminent naturalist who first made it known as a British species, and as JEW-FISH and WRECK-FISH—the last title being given to it on account of its habit of frequenting drifting timbers, apparently for the purpose of feeding upon the various marine creatures that swarm about such localities. In Madeira it is called CHERNE, when full grown, and CHERNOTTE when young.

Barnacle-laden timber seems to have great attractions for the Stone Basse, and it is mentioned by Mr. Yarrell that a becalmed vessel was surrounded for a fortnight with these fish, probably on account of the trailing barnacles with which her planking was covered. Their presence was most valuable, as they were caught in great numbers, and the men fed almost wholly upon them for twelve or fourteen days.

From examination of the stomach, the Stone Basse seems to feed mostly on small fish of various kinds, sardines having been found in its interior in large quantities. Molluscs also form part of its food. It lives mostly in the deeper waters, preferring a rocky bottom, and generally remaining deeply immersed, unless attracted to the surface by the presence of its food.

When following floating timbers, it is a remarkably bold fish. Mr. Couch remarks thus upon its habits : " When a piece of timber, covered with barnacles, is brought by the currents from the more southern regions which these fishes inhabit, considerable numbers of them sometimes accompany it. In the alacrity of their exertions, they pass over the wreck in pursuit of each other, and sometimes for a short space are left dry on the top, until a succeeding wave bears them off again. From the circumstance of their being usually found near floating wood covered with barnacles, it might be supposed that this shell-fish forms their food ; but this does not appear to be the case, since, in many that were opened, nothing was found but small fishes. Perhaps the young fishes follow the floating wood for the sake of the insects that accompany it, and thus draw the Stone Basse after them."

The colour of the Stone Basse is dark purple-brown above, and silvery white below. The fin-membranes are brown, and the tail is tipped with white. When young, it is mottled with darker and lighter brown. The lower jaw is larger than the upper, and over the operculum runs horizontally a bold bony ridge, ending in a sharp point directed backwards. There is also a row of short sharp spines over the eye, and the first ray of the ventral fins and the first three rays of the anal fin are furnished with strong thorny spines, so that the fish is armed at all points, and when struggling violently is likely to inflict rather severe wounds on the hand that grasps it incautiously.

PASSING by many large genera, which cannot be noticed for lack of space, we come to a very odd-looking fish, called perforce, for want of a popular title, the OREOSOMA, a name framed from two Greek words, and literally signifying hilly-bodied. As the reader may see by reference to the engraving, the name is very appropriate. The upper figure shows its aspect from below.

This remarkable little fish was captured in the Atlantic by Peron, and has ever been esteemed as one of the curiosities of the animal kingdom. Upon the body there are no true scales, but their place is supplied by a number of bony or horny protuberances, of a conical shape, and serving no ascertained purpose. These cones may be divided into two distinct sets, the larger set being arranged in two ranks, four on the back and ten on the abdomen, and among them are placed the smaller set. The body of this fish is very deep in proportion to its length, and the operculum has two ridges, terminating in flattened angles. There are two dorsal fins, the first armed with five spines.

OREOSOMA.—*Oreosóma Atlánticum.*

BEFORE leaving this family we must briefly examine another very large genus, here represented by the BANDED MULLET (*Apógon fasciátus*).

This fish is found off the Feejee Islands, upon the coast of Mozambique, and in the Australian and Moluccan seas. The genus to which it belongs comprises about sixty species, all inhabiting the warmer waters, and some entering the mouths of rivers. They are most plentiful in the Indian and Australian seas, but are never seen in the colder waters of the northern and southern regions. The scales of these fish are large, and fall off almost at a touch. The gill-cover is rather formidably armed, the operculum bearing spines, and the præoperculum having a double-notched ridge.

The colouring of the Banded Mullet is bold and striking. The general tint of the body is a glowing rose, and a series of broad dark bands are drawn along the body, four or five on each side, and one on the back. At the base of the tail fin is a large round black spot, and a black band runs across the root of the second dorsal and anal fins.

THE next family, termed the Pristipomidæ, after the typical genus, forms a large and somewhat important group of fishes. They are all carnivorous, *i.e.* they feed upon fish in preference to other diet; they have no molar or cutting teeth, and all inhabit the waters of the warm and temperate regions of the globe. The greater number of the species are marine, but a few are found in the rivers.

As an example of the typical genus, we will take the KAKAAN (*Pristipóma hasta*), a species found in the "Red Sea, along the east coast of Africa, through all the Indian seas to the northern shores of Australia."

In this prettily marked species, the dorsal fins are separated by a notch, rather variable in depth, and the fourth dorsal spine is much elongated, being indeed equal to half the length of the head. The second spine of the anal fin is also long and sharp. The colouring of the Kakaan is seldom precisely the same in any two individuals, but the body is always covered with a great number of brown spots, arranged with some degree of regularity. Sometimes these spots fall into horizontal lines, so as to look at a little distance like a series of brown bars drawn along the body, while in other specimens the spots are gathered into vertical bands. There are also several series of circular brown spots on both the dorsal fins.

CAPEUNA.—*Hæmulon quadrilineatum.*

The CAPEUNA, or FOUR-STREAKED RED-THROAT, is a remarkably pretty fish, and a good example of the genus to which it belongs.

The generic title of Hæmulon is given to these fishes on account of the bright ruddy colour of that part of the lower jaw which is concealed when the jaws are shut. The French call this genus Rougegueule. The profile of their rather elongated head is thought to bear some resemblance to that of a pig. The Capeuna is most beautifully coloured, as will be seen when the description is compared with the figure. The spines of the dorsal fins are tolerably firm, but cannot be termed strong or formidable, and the same may be said of the lengthened second spine of the anal fin. The eye is large and full, and the tail is deeply forked. A rich brown band runs along the whole of the body just above the dorsal line, and a corresponding band is drawn immediately below it. Between the upper band and the spinous portion of the dorsal fin, a short brown streak is drawn, looking as if dashed hastily with one sweep of a brush, and a still shorter stripe of the same colour runs along each side of the head just above the eye. From the eyes are drawn two wider stripes of rich golden yellow, which pass beneath the lateral line, and run to a considerable distance, the lower streak being continued as far as the tail fin, and the upper reaching to the middle of the soft portion of the dorsal fin, where it turns slightly upwards.

IN the fish represented in the accompanying illustration, the reader may see one of those remarkably coloured species for which the warmer seas are so famous, and whose vivid colouring and striking forms put to shame the comparatively sober inhabitants of the northern waters.

What connexion there may be between colours and caloric is one of the unsolved enigmas of creation, and though it is most evident that such a connexion exists, its principles and even its results are at present shrouded in mystery.

CUVIER'S BODIAN.—*Diagramma lineátum.*

The tints which decorate the finny inhabitants of these tepid waters are brilliant beyond all power of description, and the most glowing colours of the artist, though painted on a ground of burnished gold, fail to convey more than a dim idea of the wondrous chromatic effects produced by the living creatures. Even the patterns in which these colours are arranged are as unexpected as they are effective, and the art student would gain no slight knowledge of that most difficult science of colour, were he to visit the tropical seas, and study the fishes as they swim calmly in the crystalline water, amid the forests of waving seaweeds or branching corals.

The harmony of the tints is not less remarkable than their brilliancy, for the brightest and most glowing colours are flung boldly together in kaleidoscopic profusion, and in defiance of all the conventional rules by which artists like to govern themselves and others, are so exquisitely harmonious that not a tint could be altered or removed without destroying the entire chromatic effect. Examples of some of these fish will be given in the course of the succeeding pages, and the reader will see that, even when labouring under the disadvantage of substituting plain black and white for their natural colours, they must be truly the humming-birds of the ocean.

The CUVIER'S BODIAN is a species spread over the greater part of the Indian seas, and caught, though it appears but rarely, on the coasts of Ceylon being most **frequently** captured on the southern shores and upon rocky ground. The Cingalese name is Deweeboraloowah. In colour it is a remarkably handsome fish, though not of such pure primary tints as others which will presently be mentioned. The colour of this fish is yellowish brown on the back, changing gradually to reddish grey on the sides, and fading to simple grey on the abdomen. The head, tail, and fins are bright golden yellow, and the bars and patches of darker colour are deep chocolate-brown. Its average length is from eighteen to twenty inches.

THE next family, the Mullidæ, finds a well-known representative in the common SURMULLET of the British seas, sometimes called the STRIPED RED MULLET, on account of the yellow longitudinal stripes that are drawn along the body.

This fish is celebrated for the excellence of its flesh, and in the time of the ancients was one of the most costly luxuries that the wealthy epicure could place upon his table, from forty to sixty pounds being paid for a fish weighing six or seven pounds. These

SURMULLET.—*Mullus Surmuletus.*

dimensions are but rarely reached, and never, as it is believed, on the comparatively cold shores of England. The liver is held to be the best part of this fish, but the whole of its flesh is firm, white, and delicately flavoured. Its value in the market is extremely variable, owing to its migratory habits, being at one time caught by hundreds in the trawl or mackerel nets, while at other times there is not a single individual to be found. There seems, however, to be one definite rule in its migrations, namely, that it approaches the shore in the summer time, and in the winter retires into deep water, whence it can only be taken in the trawl net.

The colour of this fish is extremely beautiful, but, as Mr. Yarrell remarks, the changing tints of red and purple are due, not to the natural colouring of the scales, but to the effects of violence. "If closely examined, it will be observed that where the scales happen not to have been removed, the natural colour is little more than a pale pink, passing into white on the belly, the lower part of the sides having three or four longitudinal stripes; but that the mixture of purple and bright red which ornaments every part of this fish is the consequence of violence: every scale removed by force—and but little is necessary—increases this colour; it is produced by extravasated blood lying under the transparent cuticle, but above the true skin."

The long barbules with which the lower jaw is furnished are supposed to be organs of

THREE-BANDED MULLET.—*Upenéus trifasciátus.*

touch, as they are well furnished with nerves, extremely sensitive, and may aid the fish in distinguishing one substance from another under dark overhanging rocks, where the eyes would be of no service. They are composed of long muscular fibres, covered with skin, and strengthened by a single cartilaginous ray that passes along the centre.

The average weight of the Surmullet is about two pounds, and its ordinary length eighteen inches.

ANOTHER species of this genus, the PLAIN RED-MULLET (*Mullus barbátus*), has occasionally been taken on the British coasts, where, however, it seems to be of very rare occurrence.

In general habits it closely resembles the preceding species, but may be distinguished from that fish by the almost vertical line of the head, which rises abruptly from the muzzle to the eyes, and by the different colouring. In the Plain Red-Mullet the back is light pink, the sides and part of the abdomen dark red, and there is a single yellow streak below the lateral line.

A RATHER extensive genus belonging to the present family cannot be passed over without some notice, as it contains many fish which are remarkable for their form and colouring, if not for their habits or utilities.

The THREE-BANDED MULLET is a native of the Indian and Polynesian Seas, and has been taken off the coasts of China, Amboyna, Celebes, Ceylon, and India. It is an extremely variable species, so much so indeed that it has been indiscriminately called the Three-banded or the Two-banded Mullet, according to the number of stripes worn by the particular individual. The common variety of this fish is marked as follows : Behind the eye is a large black spot of an oblong form, a broad cross-band runs over the tail, and another from the front portion of the soft dorsal fin. Sometimes a third similar band rises from the spinous portion of the dorsal fin. Between these bands the scales are either yellow or white, according to the peculiarity of the individual. The upper half of the dorsal fin is mostly streaked with white and black longitudinal bands, and the remainder is wholly black. The anal fin is also marked with similar black bands.

BRAIZE.—*Pagrus vulgaris.*

Another variety of this fish is marked in a very singular manner. The front portion of the body is black, diversified with two white longitudinal bands, one drawn from above the eye to the beginning of the soft dorsal fin, and the other running obliquely from the cheek just below the eye to the lateral line. On each side of the tail there is a large black spot. About twenty species of this genus are known, all belonging to the warmer seas, and for the most part marked with bold stripes or spots.

The family of the SPARIDÆ is represented by the BRAIZE, otherwise known as the BECKER, PANDORA, and KING OF THE SEA-BREAM.

This is a common fish in the Mediterranean, and has occasionally, though very rarely, been taken on the British coasts.

On leaving the precincts of its native sea, it seems to take a north or north-westerly course. It is on some occasions a rather sociable fish, swimming in little shoals; approaching the shore in the spring and retiring into deep water towards the middle of summer. In habits it resembles the common sea-bream, of which Mr. Couch writes as follows : " In its general habits it might be considered a solitary fish, as where they most abound, the assemblage is formed commonly for no other purpose than the pursuit of food. Yet there are exceptions to this, and fishermen inform me of instances in which multitudes are seen congregated at the surface, moving slowly along, as if engaged in some important expedition. This happens most frequently over rocky ground in deep water."

There are several varieties of this fish, the teeth and number of fin rays differing in certain individuals. Its colour is uniform red. According to Dr. Günther, Mr. Yarrell has mistaken the Spanish Sea-Bream (*Pagellus erythrinus*) for the present species, giving the figure and description of one and the vignette of the other.

The COMMON SEA-BREAM (*Pagellus centrodontus*), so well known on our coasts, is a handsome fish, notable for its large round eyes, and the reddish grey hue of its body. It is sometimes called the GILT-HEAD, because part of the head looks as if it were silvered, and when young, it goes by the name of CHAD.

It is tolerably common, especially on the southern coasts, in the summer and beginning of autumn, but seems to be unable to endure the cold, and passes into deeper water at the first indications of winter. In the early summer, the young Sea-Breams, or Chads as they are then called, haunt the rocks in great numbers, and give good sport to anglers, biting freely at a baited hook, and struggling with some violence in spite of their modest six inches of length. The Chads do not possess the dark patch above the pectoral fin, and this mark is not obtained until the fish has reached its first autumn and is about half-grown. In their earlier stage, when they are only an inch or two long, they are devoured in great numbers by the larger fish.

The food of the Sea-Bream consists of various animal and vegetable substances; and the strong array of teeth which line its jaws are admirably adapted to the use for which they were made, namely to nibble the green seaweeds from the face of the rocks and to bruise them when taken into the mouth.

The flesh of the Sea-Bream is not ordinarily thought of much value for the table, but Mr. Yarrell mentions a plan by which it can be rendered palatable. "When thoroughly cleaned, the fish should be wiped dry, but none of the scales should be taken off. In this state it should be broiled, turning it often, and if the skin cracks, flour it a little to keep the outer case entire. When on table, the whole skin and scales turn off without difficulty, and the muscle beneath, saturated with its own natural juices, which the outside covering has retained, will be found of good flavour."

The eye of this fish is very large and of a beautiful golden yellow. The cheeks and part of the gill-cover are scaly, and a portion of the surface in front of and under the eyes has a metallic lustre. The general colour is reddish with a tinge of grey, becoming lighter on the sides, and fading into white below. Above the base of the pectoral fin is a rather large dark patch, which on a closer examination is seen to consist of a number of smaller spots. A few very faint bands are drawn along the sides. The dorsal and anal fins are brown, the ventrals grey, and the pectorals and tail fin red.

As allusion has been made to the term GILT-HEAD as one of the popular names of the sea-bream, it is as well to mention that the title rightly belongs to a closely allied species *Chrysophrys aurata*, a fish that properly inhabits the Mediterranean, but has occasionally been taken on the British coasts.

This fish derives its name from the semilunar golden spot over the eye. At the upper part of the edge of the operculum there is a violet patch. The back is blue, fading delicately into silver-grey, and the sides are longitudinally banded with golden streaks. The fins are greyish blue, and at the bases of the dorsal and anal fins the scales are so raised at each side, that the fin looks as if it were set in a groove. This arrangement is seen in many of the fish belonging to this family.

WE now arrive at a large family, containing a series of fishes remarkable for their extraordinary shape, their bold and eccentric colouring, and their curious habits. In Dr. Günther's elaborate arrangement of the Acanthopterygiian fishes, this family is called by the name of Squamipinnes, or scaly-finned fishes, because "the vertical fins are more or less densely covered with small scales;" the spinous portions sometimes not scaly. They are nearly all carnivorous fishes, and for the most part are exclusively inhabitants of the tropical seas or rivers. Their bodies are very much compressed and extremely deep in proportion to their length, and the mouth is usually small and placed in front of the snout.

THE first group of this family, or sub-family as it might be called, is termed Chæto-dontína, from the large typical genus of the group. Their mouths are small, and furnished with several rows of very tiny, slender, and bristle-like teeth, a peculiarity of structure that has gained for them their scientific name Chætodontína, a term composed of two Greek words, the former signifying hair, and the latter a tooth. The colours of the species belonging to this group are brilliant in tint, and are generally arranged in bold stripes or spots. Black and yellow are the prevailing hues, but blue and green are found in some species.

3.   R

WANDERING CHÆTODON.—*Chætodon pictus.*
Lower figure.

BEAKED CHÆTODON.—*Chelmo rostrátus.*
Upper figure.

The WANDERING CHÆTODON is an example of a very large genus, comprising about seventy species, all of which are striking from their shape and colour. Some of them are almost circular or disc-like in the general contour of their figure, and the arrangement of the markings is very conspicuous. The muzzle is moderate in length, and the scales are rather large in proportion to the dimensions of the body.

The Wandering Chætodon is a native of the waters extending from the Red Sea to Polynesia, and is one of the common fishes of the Ceylonese coasts.

The colours of this fish are very beautiful, and are arranged after a very curious fashion. The ground colour of the body is golden yellow, on which a number of purplish brown lines are drawn in a manner that can readily be understood by reference to the illustration. Those that start-from the upper edge of the gill-cover, are drawn obliquely towards the centre of the dorsal fin, and from the last of these lines a number of streaks issue nearly at right angles, take a slight sweep downwards and then converge towards the tail. From the upper part of the head a broad black band descends to the angle of the interoperculum, and envelops the eye in its progress. The dorsal fin has a narrow black edge, and a black band extends along the soft portion of the same fin, crosses the tail, and is continued on the anal fin, which has a black and white edge. Two bold black bands are drawn across the tail. It is not a large species, rarely exceeding one foot in length.

THE second figure on the same engraving represents a most remarkable species, called, from the form of its mouth, the BEAKED CHÆTODON.

The curiously elongated muzzle is employed by this fish in a rather unexpected manner, being used as a gun or bow, a drop of water taking the place of the arrow or bullet. Perhaps the closest analogy is with the celebrated "sumpitan," or blow-gun, of the Macoushi Indians, a tube through which an arrow is driven by the force of the breath. The Beaked Chætodon feeds largely on flies and other insects, but is not forced to depend, as is the case with nearly every other fish, on the accidental fall of its prey into the water. If it sees a fly or other insect resting on a twig or grass-blade that over-hangs the water, the Chætodon approaches very quietly, the greater part of its body submerged, and its nose just showing itself above the surface, the point directed towards the victim. Suddenly, it shoots a drop of water at the fly with such accuracy of aim, that the unsuspecting insect is knocked off its perch, and is snapped up by the fish as soon as it touches the surface of the water.

This habit it continues even in captivity, and is in consequence in great estimation as a household pet by the Japanese. They keep the fish in a large bowl of water, and amuse themselves by holding towards it a fly upon the end of a slender rod, and seeing the finny archer strike its prey into the water. Another fish, which will be figured and described in the following pages, possesses the same faculty, but is not so remarkable for its eccentric form and the bold beauty of its tints.

The Beaked Chætodon inhabits the Indian and Polynesian seas, and has been taken off the west coast of Australia, where it is usually found in or near the mouths of rivers. Over the head and body of this species are drawn five brownish cross-bands edged with darker brown and white, and in the middle of the soft dorsal fin there is a rather large circular black spot edged with white.

Several other species of this genus are recognised, one of which, the LONG-BEAKED CHÆTODON (*Chelmo longirostris*), is truly remarkable for the exceeding development of the snout, which considerably exceeds half the length of the head. This species is also notable for a large triangular patch of jetty black, which covers the upper surface of the head, the neck, and the side of the head as far as the lower edge of the eye. There is also a circular spot of the same hue on the anal fin. This species is a native of Amboyna.

THE very remarkable fish which is depicted on the accompanying illustration adds to the singular shape of all the group the peculiarly elongated dorsal spine from which it has received its name of LONG-SPINED CHÆTODON, or CHARIOTEER. It also well exhibits the scale-covered fins, a structure which is indicative of the large family to which it belongs. Both scientific names are of Greek origin, the former signifying a charioteer, the long slender spine representing the whip ; and the latter signifies "single-horned," in allusion to the same peculiarity.

The fourth dorsal spine of this species is enormously elongated and whip-like ; its use not being as yet ascertained or even conjectured with any show of reason. Over each eye

LONG-SPINED CHÆTODON.—*Heniochus monóceros.*

is a conical projection, not easily distinguished, on account of the deep black hue with which it is coloured, and a similar protuberance arises on that part of the fish which is by courtesy termed the nape of the neck. It may be seen in the figure, about half-way between the snout and the beginning of the dorsal fin. Three very broad black bands are drawn across the body; their edges are sharply defined, as if a painter had drawn them with black varnish. The foremost band commences at the first dorsal spine, and sweeps over the neck, upper part of the head, snout, and chin, the eye being imbedded, as it were, in the black ground, and shining with great vividness on account of the contrast. The second band passes from the fifth to the seventh dorsal spines to the abdomen, being rather narrow at the top, and widening as it passes downwards below, but not comprising the pectoral fin. The third band starts a little below the central streak, and is drawn rather obliquely over the body, through the hinder portion of the anal fin. A remarkably fine specimen in the British Museum was captured off the Mauritius.

The members of the curious genus to which the SEMILUNAR HOLOCANTHUS belongs are remarkable for a very strong, sharp-pointed, thorny spine with which the præoperculum is armed. These curious fish are found in almost all tropical seas.

SEMILUNAR HOLOCANTHUS.—*Holocanthus semicirculátus.*

All these fish are notable for the strange fashion of their colouring, and the present species affords a good example of the infinite variety with which the inhabitants of the warmer seas are tinted. There are no brilliant colours in this fish, the striking effects being produced by the bold contrasts of black and white, toned down with a little blue. The ground colour of the body is jetty black, upon which are drawn a number of curved concentric lines, alternately pure white and pale blue, the blue lines being narrower than the white. The tail fin is also black, and is marked with cross lines of white. The scales of this species are small, and the spine is smooth.

Nearly forty species of this genus are now known, all of which possess some remarkable peculiarity in colouring. There is, for example, the RINGED HOLOCANTHUS (*Holocanthus annularis*), where the shoulder is decorated with a blue ring, and the body is marked with six or seven arched blue stripes, all radiating from the base of the pectoral fin. The SPOTTED HOLOCANTHUS (*Holocanthus maculôsus*) has a number of black semilunar spots on the fore part of the body; the CILIATED HOLOCANTHUS (*Holocanthus ciliaris*) is marked with an azure ring on the nape of the neck, and a number of blue spots and streaks about the head; the EMPEROR HOLOCANTHUS (*Holocanthus Imperator*) has a number of blue lines upon the head, chest, and anal fin, a large black spot on the shoulder, and the body decorated with many waved orange-coloured streaks; and lastly the ARCHED HOLOCANTHUS (*Holocanthus arcuatus*), though not so brightly clad, is quite as striking a species as any that has been mentioned, simply on account of the single arched stripe that is drawn along the body, from the eye to the end of the dorsal fin, taking a slight upward curve like a bent bow.

It is said of one of the species LAMARCK'S HOLOCANTHUS (*Holocanthus Lamarckii*), that the attachment between the sexes is very strongly developed, and that if one individual be captured, its mate will haunt the fatal spot and even fling itself ashore or into the net in the eagerness of its search.

OF another group or sub-family of the scale-finned fishes, the ARCHER FISH is a good example.

This curious species is a native of the East Indian and Polynesian seas, and possesses the power of shooting water at its prey with even more force than the beaked chætodon

ARCHER FISH.—*Toxōtes 'aculātor*

So powerful indeed is the projectile force, and so marvellously accurate is the aim, that it will strike a fly with certainty at a distance of three or even four feet. In general appearance there is little to attract attention about this fish, the only remarkable point in its form being the greatly elongated lower jaw, which may possibly aid it in directing the liquid missile on which it partially depends for its subsistence as does a hunter on the accuracy of his rifle. The general colour of the Archer Fish is greenish, and the short wide bands across the back are dark brown with a shade of green. Two species of this genus are known.

BANDED CHILODACTYLE.—*Chilodáctylus zonatus*

As an example of the next family, the Cirrhitidæ, we take the BANDED CHILODACTYLE. The family to which this fish belongs is a very small one, containing only eight genera,

RED SCORPION-FISH.—*Scorpæna scrofa.*

none of which comprise many species. Altogether, this family is not larger than many single genera. The members of which it is composed are all exotic species, inhabiting the "seas of the tropical regions and the southern temperate parts of the Pacific," and never, so far as is known, making their appearance on the British coasts.

It would, however, be unwise to assert that such fish never do come within reach of our coasts, for it is manifestly impossible to decide the precise range of any active inhabitant of the ocean. Perplexing as is the task of ascertaining the habitation of migrating birds, the difficulty of fixing the range of fishes is far less easy to overcome, as the transition from the tropical to the temperate, and from them to the colder seas, is so extremely gentle, that a fish of errant disposition, or one that has been caught in a long lasting storm might be, and has been often, driven into strange waters which it does not know, and from which it can find no retreat.

The Banded Chilodactyle is a native of the Chinese and Japanese seas, and there are specimens in the British Museum from both these localities. The dorsal fin of this fish has the fourth spine much elongated, and the membrane of the spinous portion deeply notched. The pectoral fins have also one spine, the last but four, considerably lengthened and pointing backwards. The colour of the fish is light brown, with several bands of a darker brown and spots of the same hue. A rather broad blackish band runs along the soft portion of the dorsal fin, and on the tail fin are a number of round grey spots edged with brown. When this fish attains to a considerable age, a pair of elevated tubercles make their appearance, one on the forehead and another on the snout.

THE large and important family of the Triglidæ, or Gurnards, is represented by several British fishes. This family contains a great number of species, many of which are most remarkable, not only for their beautiful colours, which alone are sufficient to attract attention, but also for the strange and wild shape, and large development of the fins. They are carnivorous fish, mostly inhabiting the seas, a very few species being able to exist in fresh water. They are not swift or strong swimmers, and therefore remain, for the most part, in deep water. Some, however, are able, by means of their largely developed pectoral fins, to raise themselves into the air, and for a brief space to sustain themselves in the thinner element. The mouth is mostly large, and in some cases the gape is so wide, and the head and jaws so strangely shaped, that the general aspect is most repulsive.

ON account of its fiery colour and ungainly aspect, the RED SCORPION-FISH has long been supposed to possess qualities as dangerous as its appearance is repulsive, and has been termed the SEA SCORPION and SEA DEVIL from the supposed venom of its spines and frowardness of its temper. It is, however, a harmless fish enough, not capable of inflicting such severe injuries as several species that have already been described. When captured, it certainly plunges and struggles violently in its endeavours to escape, and if handled incautiously it will probably inflict some painful injuries with its bony spears. This result, however, is attributable to the carelessness of the captor and to the natural desire for liberty, and not to any malevolent propensities innate in its being.

The flesh of the Red Scorpion-fish is dry and flavourless, and is seldom brought to table, being eaten only by the poor, who cannot afford to throw away even so tasteless an article of food. In some parts of the world a useful oil is obtained from its liver.

The general colour of this species is red, marbled with brown upon the body and fins. There is a rather conspicuous blotch of blackish brown on the dorsal fin between the sixth and ninth dorsal spines. It is not at all a large fish, the average length of a full-grown individual being about eighteen inches, which in a few very fine specimens is extended to two feet.

ANOTHER species, the SPOTTED SCORPION-FISH (*Scorpæna porcus*), inhabits the same localities, and has similar habits. Both these fish are extremely voracious, as may be inferred from their wide mouth and general aspect, feeding on the smaller fish and similar creatures. They have a habit of lying in ambush under overhanging tufts of seaweed, and thence issuing in chase of any unfortunate little fish that may happen to pass near the fatal spot. All the fish of this genus are remarkable for their large head, with its armature of spines and odd skinny flaps, and the curious naked groove that runs along its summit. The pectoral fins are always large and rounded, and the body is mostly decorated with sundry skinny appendages. Examples of this genus are found in all the tropical seas, extending as far north as the Mediterranean, and to the Atlantic shores of Northern America.

The general colour of the Spotted Scorpion-fish is brownish red, marbled with dark brown and dotted with black. In some individuals the dots are arranged in lines round the dark marblings. A few half-grown specimens have been seen with a black blotch on the dorsal fin, and a number of black dots on the tail fin, arranged so as to form cross-bands. This is not quite so large a species as the preceding.

THE extraordinary creature depicted in the engraving, which is known to British residents by the name of RED FIRE-FISH, and to the natives of Ceylon by the title of GINI-MAHA, inhabits the greater part of the tropical seas, from Eastern Africa, through the Indian seas, to Australia. In the British Museum are many specimens of this wild and weird-looking being, some having been taken from Northern, and others from Southern Africa, several from Amboyna, one or two from China, and others from Australia.

This fish is remarkable for the singular development of the dorsal and pectoral fins, the latter being of such vast proportionate size that they were formerly supposed to act like the corresponding organs in the flying fish, and to raise the creature out of the water into the air. Such, however, is not the case, for the rays which carry the connecting membrane are not supported by a corresponding strength of bone as in the true flying fishes, and are far too weak to serve that purpose. Indeed, the object of this remarkable development is one of the many mysteries with which the inquiring zoologist is surrounded, and which make his task so exhaustlessly fascinating.

The structure of the entire skeleton is very interesting to comparative anatomists, but is too complicated, and requires too many technical terms to be described in these pages.

The Red Fire-fish is common off the Ceylonese coast, and is said to be rather valuable as an article of food, its flesh being very white, firm, and nutritious. The native fishermen hold this species in some dread, thinking that it can inflict an incurable wound with the sharp spines which arm its person and stand out so boldly in every direction. This idea, however, is without any foundation; for although the thorny spines may prick the hand deeply and painfully, they carry no poison, and inflict no venomed hurt.

RED FIRE-FISH.—*Ptéröis völitans.*

The general colour of the Red Fire-fish is pinky brown, barred with darker brown, and the head is redder than the body. The huge pectoral and dorsal fins are reddish brown, crossed with bold bars of black; the ventral fin is black, dotted with white spots, and the rest of the fins, including that of the tail, are light brown, spotted with black. It is by no means a large fish, being generally about seven or eight inches in length. There are nine or ten species of this genus.

ONE or two notable fishes require a cursory notice.

The SEA LOCUST (*Apistos Israelitórum*) is a native of the Red Sea, and is remarkable as being the only flying fish of those strange waters. It is particularly plentiful on that part of the coast near which the Israelites were forced to wander for a space of forty years, and on that account has received its specific title. Ehrenberg has noticed that it is very abundant near Tor, and that several specimens fell into his boat almost every time that the sea was agitated. He further throws out a suggestion, that the quails to which allusion is made in the sacred volume are really the Sea Locusts, but this conjecture seems to be entirely gratuitous, and is unsupported by facts.

There is, in truth, no particular reason why the Hebrew word, which is translated as "quails," should not signify the bird in question; and at all events, it certainly seems to be a feathered being of some kind, and not a fish, even though that fish does occasionally raise itself into the air for a brief space. The Arab name for this fish is Gherad-el-bahr, signifying literally, locust of the sea. The generic name, Apistos, signifies faithless or treacherous, and is given to this fish on account of the sharp spines which jut from the head, and which can inflict a painful wound.

YELLOW SCORPÆNA.--*Hemitripterus Americānus*

ANOTHER curious fish is the SEEPAARD of the Dutch (*Agriopus torvus*), a native of the seas around the Cape of Good Hope.

It is a rather powerfully armed species, on account of the strong, sharp, and recurved spines of the dorsal fin, but its head is not supplied with the thorny projections that render the preceding fish so perilous to handle. The dorsal fin of the Seepaard is single, and the spinous portion is greatly developed, rising in a bold curve over the shoulders and back like the crest of an ancient helmet, and being continued almost as far the tail. Very little is known of this fish, though it is far from uncommon, and is eaten by the Dutch colonists of the Cape.

Its colour is brown, mostly marbled with black, and the skin is smooth

THE strange and quaintly decorated fish which is represented in the accompanying illustration is, as it name imports, an inhabitant of the American coast, being found on the Atlantic shores of Northern America.

This odd-looking species frequents the same localities as the cod, and is often taken at the same time as that fish. The skin of the YELLOW SCORPÆNA is devoid of scales, and the ventral and pectoral fins are enveloped in thick skin. The head is depressed, naked, and is covered with a series of loose skinny appendages, that flap and wave about in the water without any apparent purpose. It is also armed with a number of rather sharp spines. There are two dorsal fins, the first being so deeply scooped that at one time the fish was described as possessing three dorsals. The first four spines of the dorsal fin are very long, and the membrane is deeply scooped between the fourth and fifth spines. The general colour of this fish is yellow, tinged more or less with red, and in some specimens marbled with brown. The length of a very fine specimen is about two feet, but the ordinary average is from fourteen to eighteen inches.

THERE is a very ugly fish, found throughout the warmer oceans, from the Indian seas to Polynesia, called by the natives of the Isle of France the FI-FI, a very appropriate name, signifying hideous. Its scientific title is *Synanceia verrucosa*.

It is not easy to imagine any living creature more frightfully repulsive than this species, which looks as if it had been originally but an undeveloped idea of a fish only

FILAMENTOUS GURNARD.—*Pelor filamentosum.*

partially carried out, with a body covered with tubercles, deceased from confluent small·pox, and its surface in an advanced stage of decomposition. The head of this creature is enormous in proportion to the size of its body, and the skin, which hangs loosely about it, is soft, spongy, wrinkled, warty, and when touched adheres to the fingers as if covered with glue. In most places where it is known, it is held in much dread, and is thought to be capable of inflicting poisoned wounds by means of the sharp and slender spines with which it is armed. Its colour is blackish, mottled at random with white, grey, and brown. Sometimes it is almost wholly black.

I HAVE already mentioned that the present family is rich in strange and eccentric forms, the head being apparently crushed out of all shape, hung about with scraps of depending skin, and armed with sharp projecting spines; the body oddly coloured, and the fins developed into the most extraordinary shapes, as if intended to show the infinite variety of Nature, and the contracted powers of human conjecture. There seems to be no reason whatever for the singular development of the fins in several of these species, for the odd shape of the head, or for the flaps of loose skin that depend therefrom like casual tatters on a mendicant's professional costume.

The FILAMENTOUS GURNARD affords another example of this apparent capriciousness of grotesque formation, the shape of this very remarkable fish being better understood by a figure than by verbal description alone. It is found on the coasts of the Isle of France. It appears to feed mostly upon crustaceans and molluscs, and the bony remnants of certain cuttle-fish have been found in its stomach. Its colour is greyish brown, marbled with a deeper hue of the same tint, and covered with minute spots of white.

There is another species of this genus which is coloured in a rather bold and pleasing manner. This is the SPOTTED PELOR (*Pelor maculatum*), which derives its name from the manner in which the black hue of the skin is variegated with white. In this species there are three large white patches on the back, and three more on the dorsal fin. Some circular white spots are scattered on the head, and a white ring encircles the eyes. The pectoral fins are decorated with a bold white band, and the tail fin is marked with two white bands alternating with the same number of black stripes.

THE odd-looking fish, which is known by the name of the THREE-LOBED BLEPSIAS, is one of those species to which the ancient naturalists had affixed certain names without any apparent motive for so doing.   There is no particular meaning in the word, and the sum of information obtainable from lexicons is, that it signifies a certain fish.

The members of this genus are found on the coasts of Kamschatka, and some fine specimens in the British Museum were obtained from the New Orcas Islands, in the Gulf

THREE-LOBED BLEPSIAS.—*Blepsias cirrhosus*

of Georgia.   This species is not very common, but may easily be known from its congener, the TWO-LOBED BLEPSIAS (*Blepsias bilobus*), by the peculiar manner in which the spiny portion of the dorsal fin is notched so as to form the whole fin into three distinct lobes. In the second species this structure is not seen.   In both, the soft portion of the dorsal fin is greatly developed, and the body and fins are boldly marked with dark streaks upon a lighter surface.   The body is entirely covered with prickles.

WE now come to a very familiar and not very prepossessing fish ; the well-known BULL-HEAD, or MILLER'S THUMB, sometimes called by the name of TOMMY LOGGE.

This large-headed and odd-looking fish is very common in our brooks and streams, where it is generally found under loose stones, and affords great sport to the juvenile fisherman.   In my younger days, the chase of the Bull-head was rather an exciting one, and was carried out without hook or line, or indeed any aid but the hands.   This fish has a habit of hiding itself under loose stones, and on account of its flat, though wide head, is enabled to push itself into crevices which are apparently much too small to contain it.

By practice, the stones which seemed most likely to shelter a Bull-head were soon noted, and an experienced eye was not very long in detecting the presence of the fish.   The Bull-head has an inveterate habit of wriggling its tail, thus creating a current of water which betrays its whereabouts.   The mode of catching the fish was, to wade very quietly to the stone, put both hands into the water, raise the stone smartly with the left hand, and make a rapid grasp with the right on the place where the fish was supposed to be lying. If the stone were tolerably large, the Bull-head could generally be picked out of its retreat by the fingers of one hand, while those of the other hand stopped up the entrance of the cranny, and prevented the intended prey from making its escape.   The Bull-head is rather slow of movement, and when the sheltering stone is suddenly removed, it seems perplexed and bewildered, and merely flounces about in the same spot for a second or so before darting off to secure itself by flight.

The name of Miller's Thumb is derived from the peculiarly wide and flattened head, which is thought to bear some resemblance to the object whence its name is taken. A miller judges of the quality of the meal by rubbing it with his thumb over his fingers as it is shot from the spout, and by the continual use of this custom, the thumb becomes gradually widened and flattened at its extremity. The name of Bull-head also alludes to the same width and flatness of the skull.

BULL-HEAD.—*Cottus góbio.*

The Bull-head is a voracious little fish, feeding on various water insects, worms, larvæ, and the young fry of other fish. It is a representative of a rather large genus, comprising about twenty-six or twenty-seven known species, which are spread over all the northern and temperate parts of the world. In Russia the Bull-head is believed by the general public to possess the same quality as is attributed to the kingfisher by our own rustic population, and to indicate the direction of the wind by always keeping its head turned to windward when it is dried and suspended horizontally by a thread.

The mouth of this little fish is very wide, and contains numerous minute teeth. There is one spine on the præoperculum, and the operculum ends in a flattened point. The general colour of the smooth skin is very dark brown on the back, white on the abdomen, and greyish white on the flanks. The rays of the fins are spotted with dark blackish brown and white, rather variable in different individuals, and the fins are marked with dark brown dots. The eyes are yellow, and the pupil very dark blue. It is but a small fish, averaging four, and seldom exceeding five, inches in length.

Several other species of this genus inhabit England. There is the SHORT-SPINED COTTUS, or SEA SCORPION (*Cottus scorpius*), which, as its name denotes, is one of the marine species. It is a very common fish, being found plentifully under heavy seaweeds and stones in the pools that are left above low-water mark by the retreating tide. The name of Scorpion is given to it on account of the sharp spines with which its head is armed, no less than eight sharp and four rather blunt prickles being found on the head. The rays of the dorsal and pectoral fins are also sharply pointed, so that it must be cautiously handled by those who wish to escape wounded fingers. This is a much prettier species than the preceding fish, its body being rich purple-brown, mottled with a warm red hue, and in the adult male there are some stripes of red on the pectoral fins, and the abdomen is brightly decorated with some snowy white circular spots on glowing scarlet. In this country, its extreme length seldom exceeds eight inches.

ANOTHER and much more formidable species is the well-known FATHER LASHER, LONG-SPINED COTTUS, or LUCKY PROACH (*Cottus bubalis*). In colour this species is very like the sea scorpion, but it may readily be distinguished from that fish by the array of long and sharply pointed spines with which its head is armed.

There are two spines proceeding from the back of the head and pointing towards the tail, four spines on the præoperculum, and three on the operculum, besides those on the snout and other parts of the head and front of the body. The fish is quite aware of the formidable nature of these bony spears, and whenever it is threatened or touched, it spreads out the gill-covers, so as to present the sharp points in all directions, like the quills of an angry porcupine. It is a rather large species, measuring ten inches in length. It is common on our coasts, and like the preceding species may be taken in the rock pools at low water.

The FOUR-HORNED COTTUS (*Cottus quadricornis*) is sometimes, but rarely, found on the shores of England. It may be easily known by the four bony protuberances on the crown of the head. There are four spines on the præoperculum. Its general colour is brown above, and greyish white below, the sides being yellow. The lateral line is marked with rough points.

ARMED PLATYCEPHALUS.—*Platycéphalus grandispinis*

THE generic name of Platycephalus, which is appropriately given to this and the other fish placed in the same group, is of Greek origin, and signifies Broad-head.

In the illustration, the great width of the head is not seen, but its very considerable flatness is well shown. On looking down on the head, it is seen to be even wider in proportion than that of the bull-head, but is narrower towards the snout and not so rounded. The body is also flattened in front, but assumes a more cylindrical form towards the tail.

The ARMED PLATYCEPHALUS is remarkable for the great length of the lower spine which proceeds from the præoperculum, and which reaches almost to the edge of the elongated operculum. It is also very wide and strong, being indeed about four times as large as the spine immediately above it. There are three little spines in front of the eye. Its colour is brown, mottled and spotted on the fins with deeper and lighter shades of the same colour.

WE now come to the typical genus of this family, which is represented by several well-known British species.

The SAPPHIRINE GURNARD, so called from the fine deep blue which tints the inner surface of the pectoral fins, is one of the British fishes, and is of tolerably common occurrence upon our coasts.

SAPPHIRINE GURNARD.—*Trigla hirundo*   LYRIE.—*Agonus cataphractus.*

This seems to be the most valuable of the nine species that inhabit the British seas, being, like all the others, excellent for the table, and exceeding them considerably in size. The name of Hirundo, or swallow, has been given to this fish on account of the great size of the pectoral fins, which are almost as proportionately large to the dimensions of the fish as the wings of the swallow to the bird. Putting aside the great development of these members, and their rich blue colour, the Sapphirine Gurnard may be distinguished from the other species by the extreme smoothness of the lateral line, which may be rubbed with the finger in either direction without exhibiting the spiny roughness which is found in other Gurnards. In consequence of this structure, the fishermen sometimes call the fish the Smoothside Gurnard.

All the scales of this species are very small. The head is armed with spines, some springing from just before the eye, and others from the operculum and the shoulder. Its head is very large and flattened, and carried in a rather peculiar style, best understood from seeing a specimen or examining the illustration. The general colour of the body and head is reddish brown, the pectoral fins are very long and wide, and their colour is deep blue on the inside and brownish red on the exterior. The rays are white. When full grown it sometimes attains the length of two feet.

Several other species of Gurnard inhabit the English seas, among which may be mentioned the CUCKOO GURNARD (*Trigla pini*), sometimes called the RED GURNARD from the colour of its body. This is a very common species, and when young may be found in the rock pools at low water, measuring only a few inches in length, but perfectly exhibiting the characters of its genus. The specific title of "pini," or belonging to the pine-tree, is given to the Cuckoo Gurnard on account of the peculiar aspect of the lateral line, which is crossed with numerous short, straight, narrow, and elevated lines, which have been compared by some writers to the needle-shaped leaves of the pine. The name of Cuckoo Gurnard is given to it, because when it is first taken out of the water it emits a sound which bears a distant resemblance to the cuckoo's cry. The curious soft rays which project from the base of the pectoral fin in this and other Gurnards are evidently organs of touch, being plentifully supplied with nerves and movable at the will of the owner.

The colour of this fish is bright rosy red above, and silvery white on the sides and abdomen. These colours soon fade after the fish has been removed from the water.

The GREY GURNARD (*Trigla Gurnardus*) is also tolerably common, and is readily to be known by its short pectoral fins and the greenish brown body, spotted with white above the lateral line. On account of the peculiar sound which it utters, it is popularly known in Scotland by the name of CROONER, and in Ireland is called the NOWD.

ANOTHER curious species, the SHINING or LONG-FINNED GURNARD (*Trigla obscúra*), is sometimes found on the British shores. This remarkable fish is at once known by the great length of the second spine of the dorsal fin, which is nearly double the length of the other spines, and projects boldly with a slight curve towards the tail. It is a handsomely coloured fish, the head and upper part of the body being vermilion-red, and the abdomen white, tinged with red. The flanks are shining silvery white, and have given cause for the name of Shining Gurnard. The fins are all bright red, with the exception of the pectorals, which are deep blue.

The LYRIE, or ARMED BULL-HEAD, is found on our coasts, and is known by a great variety of names, such as ARMED BULL-HEAD, POGGE, SEA POACHER, and NOBLE.

It is a curious-looking fish, with its bony armour-plates and shielded head. It is most commonly taken near the mouths of rivers, though it is sometimes captured far out at sea. Its flesh is firm and good, but its small size and bony shields render it scarcely serviceable for the table. It feeds mostly on aquatic animals.

The body of the Lyrie is covered by eight rows of bony plates, strongly reminding the observer of the sturgeon, and the head, gill-cover, and shoulders are strongly armed with spines. The general colour of the Lyrie is brown above, crossed with several broad bands of dark brown, and the abdomen is white, with a trifling tinge of brown.

IN the remarkable genus which now comes before our notice, the body is covered with bony plates, like ancient armour, and the front part of the head is formed into a deeply cleft fork on account of the development of certain bones of the skull.

The ORIENTAL GURNARD is found in the Japanese seas, and is a good example of the genus to which it belongs, the bony plates being very large, and the forked processes of the head well developed. Between the ventral fins, each bony plate is just three times as long as it is broad. The preoperculum is furnished with a strong spine, crossed by a projecting ridge from its angle.

A VERY curious species belonging to this genus is sometimes, though very rarely, found in the British seas, and is known by the name of MAILED GURNARD (*Peristethus cataphractum*).

In this fish, the bony plates between the ventral fins are twice as long as they are broad. It mostly prefers rather deep water over rocky ground, but approaches the shallows for the purpose of spawning. Its food consists of the softer crustaceans, medusæ, and similar creatures. It is a swift swimmer, but seems to be rather reckless, as it not unfrequently strikes its forked snout against the stones, and breaks off one or both points. The flesh of the Mailed Gurnard is tolerably good, but requires some care in cooking, besides costing some little trouble in freeing it from the hard bony plates in which the body is so securely enveloped. In order to clear away these defences, the fish must be soaked in warm water, and the scales stripped off from the tail upward. In some places, such as the coasts of Spain, it is held in considerable estimation, and is especially sought by fishermen.

ORIENTAL GURNARD.—*Peristethus orientale.*

Its colour is red like that of the red Gurnard. Nearly all the rays of the first dorsal fin are extremely elongated, and, together with the mailed body, the armed head, and the double snout, give to the fish a most singular aspect. The total length of the Mailed Gurnard is about two feet.

THE two extraordinary and beautiful fishes which are represented in the accompanying illustration are remarkable, not only for the very great development of the pectoral fins, their muscles and attachments, but for the unexpected use to which those members are occasionally subservient.

These fishes, together with one or two other species, hereafter to be described, possess the power of darting from the water into the air, and by the mingled force of the impetus with which they spring from the surface, and the widely spread wing-like fins, to sustain themselves for a short space in the thinner element, and usurp for a time the privileges of the winged beings whose trackless path is through the air.

The object of exercising these strange powers seems to be, not the pleasure of the fish, but the hope of escaping from the jaws of some voracious monster of the deep, whose sub-aquatic speed is greater than that of the intended victim, but whose limited powers are incapable of raising it into the air. Foremost among these persecutors is the coryphene, often called the dolphin by sailors, and which is the so-called "dolphin" whose colours glow with such changeful beauty during its death pangs.

Little, however, do the powers of flight avail the unfortunate fish, for winged foes, known by the name of albatros, frigate-bird, and similar titles, are hovering above in waiting for their prey, and no sooner does the Gurnard launch itself fairly into the air, and so escape the open jaws of the pursuer coryphene, than the albatros swoops down with extended wings, snatches up the fish in its beak, and without altering the bold and graceful curve in which it has made the stoop, sweeps up again into its airy height, where it wheels on steady wing awaiting another victim.

Between the hungry coryphene below, and the voracious albatros above, the poor Flying Gurnard leads no very happy life, and its intermediate existence, persecuted on either side, has been often employed as a type of those unfortunate persons who are ashamed of the

3.

FLYING GURNARD.—*Dactylopterus volitans.*   INDIAN FLYING GURNARD.—*Dactylopterus orientalis.*

more lowly society in which they were born, and aspire to ascend to an elevated condition for which they are not fitted by nature.

While passing through the air, the Flying Gurnard is able slightly to change its direction, but cannot prolong its flight, by flapping its finny wings. In fact, its elevation into the air may be readily imitated by throwing an oyster-shell in a horizontal direction, taking care to throw it in such a manner that the concavity is downwards and the convexity upwards. The flight is closely analogous to that of the flying squirrels, rats, and mice among mammalia, and of the flying dragon among reptiles.

The COMMON FLYING GURNARD is brown above, passing into a beautiful rosé-colour below. The fins are black, variegated with blue spots, and on the tail fin the spots run together so as to resemble continuous bands. Its length varies from ten to fifteen inches. It is a native of the Mediterranean and warmer parts of the Atlantic, and in many parts of those seas is very common.

THE second species, the INDIAN FLYING GURNARD, is found throughout the Indian Ocean and Archipelago, and on account of its habits, its singular and striking form, and

MEDITERRANEAN URANOSCOPUS.—*Uranóscopus scaber.*

its lovely colouring, has always attracted the attention of voyagers, even though they have possessed no skill in natural history.

This beautiful fish is notable for the two long detached filaments that are planted between the head and the dorsal fin, the first being extremely elongated and the second much shorter. The first spine of the dorsal fin is solitary, and at first sight looks like another isolated filament. In all the members of this genus, the præoperculum is armed with long, sharp, and powerful spines, the scales of the body are strongly keeled, and there is no appearance of a lateral line. Four species of Flying Gurnards are known, the two which have been selected affording excellent types of their general form. In the Indian Flying Gurnard, the pectorals are covered with brown spots, and dotted rather profusely with bluish white.

WE now arrive at a moderately large family of fishes, called, from the typical genus, Trachinidæ. In these creatures the body is long and rather flattened, the gill-covers are wide, and the teeth are arranged in bands.

OUR first example of these fishes is the very remarkable MEDITERRANEAN URANOSCOPUS, a word which requires some little explanation before examining the form and habits of the species. The generic title is derived from two Greek words, literally signifying sky-gazer, and is given to the fish on account of the peculiar position of the eyes, which are set so singularly on the upper part of the head, that they look upwards, instead of sideways, as is the usual custom among the finny inhabitants of the waters.

This species lives mostly at the bottom of deep seas, and is said to angle for the smaller fish, on which it feeds, by agitating a slender filamentary appendage of its mouth in such a manner as to resemble a worm, and to pounce on the deluded victims when they hurry to the spot in hopes of a meal. Though a fish of rather repulsive aspect, its flesh is tolerably good, and is eaten in many parts of Europe and along the shores of the Mediterranean.

Its head is very large and broad, and is partially covered with bony plates, and the opening of the mouth is nearly vertical. The slender filament which has already been

s 2

mentioned is set before and below the tongue, and the shoulder and gill-covers are armed with an array of strong sharp spines.

THE typical genus of this family is represented in England by several species, of which the GREAT WEAVER is one of the most familiar.

This species is the dread of fishermen, the wounds occasioned by the sharp spine of the gill-cover, and those of the first dorsal fin, being extremely painful, and said to resemble the sting of a hornet, the evil effects extending from the hand up the arm, and even reaching the shoulder. On the first infliction of the injury, it gives little more pain than the prick of a pin or needle, but in a short time, a dull hot pain creeps up the arm, and increases in intensity for several hours. Fishermen, taught by experience, are very cautious in handling this dangerous fish, and before they place it in their basket, they cut off the whole of the first dorsal fin and the hinder part of the gill-cover. In France, this precaution is rendered compulsory by law.

GREAT WEAVER-FISH.—*Trachinus draco.*

The same remedies as those employed for the bite of the viper, such as persevering friction with hot oil, are said to be the best means of relieving the pain caused by this small but formidable fish. One of these Weavers has been known to strike three men successively, and to injure them all in the same manner.

It is very tenacious of life, and its popular title of Weaver has no connexion with loom, but is an Anglicized corruption of the French name La Vive, given to it in reference to the ease with which it endures absence from water. By our fishermen it is also known by the expressive appellations of STING-BULL and SEA-CAT ; the former name being given to it because it is supposed to be able to sting even the tough-hided monarch of the herd, and the latter from the deep pricks and scratches inflicted by its talon-like spines.

The head and body of the Great Weaver are much compressed, and its colour is deep reddish grey on the back, with a wash of brown, fading to light grey on the abdomen, and marked with oblique ochreous lines. The gill-cover is streaked with yellow. The operculum is armed with a very long and sharp spine, pointing backward. Its length is generally about one foot; but specimens have occasionally been captured that measured sixteen or seventeen inches in length.

ANOTHER smaller species of this genus is common on our shores—rather too common in fact—and is popularly known by the appropriate names of STING-FISH, OTTER-PIKE, and LESSER WEAVER, the last name being given to it on account of its comparatively diminutive form, its length seldom exceeding five inches.

Though a smaller species than the Great Weaver, it is even more obnoxious to the public, as it has an unpleasant habit of frequenting the shore, especially where the sand is mixed with mud, and there burying itself, leaving only the head exposed. In this position it awaits its prey, which consists of the smaller crustaceæ and similar marine animals, and if touched, it aims a blow at the intruder with marvellous precision and force. These peculiarities render the fish an object of detestation to timid bathers, who are afraid to venture out of their depth, as well as to those who are fond of wading in the sea in the hope of discovering some new form of animal life, or watching the native habits of known species.

Whether any venomous substance is secreted by the fish and introduced by means of the spines is a point which has not yet fairly been settled; but on a superficial view of the subject, the violently inflammatory symptoms which mostly supervene upon the wound, afford reasons for surmising that an injurious influence of some kind is exerted, though its source and action are as yet obscure.

STING-FISH.—*Trachinus vipera.*

In this, as in the preceding species, the vertical position of the mouth, awkward as it may appear when the fish is lying upon a rock or a table, is yet extremely useful and permits the fish to seize its prey while it remains half buried in the mud or sand.

The colour of the Sting-fish is reddish grey on the back, and silvery white on the abdomen and lower part of the flanks. The membrane of the first dorsal fin is black, and the tail fin is tipped with the same colour; the other fins are brown.

THE curious fish which is represented in the engraving affords a good example of a moderately large genus which is spread over many seas, being found on various shores from the Red Sea to the coast of Australia.

The INDIAN SILLAGO is easily recognised by the extraordinary length of the second dorsal spine, which in a good specimen is developed to such an extent that it equals the length of the body. The use of this structure is very obscure. As its name imports, this fish inhabits the Indian seas, and is found in the Bay of Bengal and near the mouth of the Ganges. It is held in some estimation for the table, as its flesh is light, digestible, and well-flavoured. The colour of the Indian Sillago is brown.

THE members of the genus Percophis are remarkable for the great proportionate length and cylindrical form of the body; a modification of structure which has gained for them the generic title, which is of Greek origin, and signifies Perch-Snake. The scales are of the kind denominated ctenoid.

INDIAN SILLAGO.—*Sillago dómina*.

As the different kind of scales will necessarily be occasionally mentioned while treating of the fishes, it will be useful to give a brief description of the different forms of these appendages.

THE scales are deposited by the action of the skin to which they adhere, forming part of the structure called the external skeleton, and, according to the best comparative anatomists, being closely analogous to the projecting spines that arm the body of many fishes, to the fin rays, and even to the teeth. The beautiful series of transitions through which anatomists have learned to ascertain the identity of the external skeleton with the fin rays, the spines, and the teeth, extends over a vast number of species, and is altogether too abstruse a subject to be admitted into these pages, in spite of its exceeding interest.

The reader, however, who wishes to see a specimen of transition from scales to teeth, may take one of the skate tribe—say the eagle-ray—and carefully examine the external surface of the body and interior of the mouth. On some parts of the body the scales are small, hard, and wart-like, while on others they are armed with projecting points. Following the scales over the head as far as the lips, they will be seen assuming a different shape, being there small, flattened, and aggregated together, forming a kind of bony tesselation, which, however, belongs to the skin, and can be stripped from it. When they have fairly passed the lips, they rapidly increase in size and hardness, and are then developed into the tremendous crushing-mill which has already been described.

The scales of fishes are divided by some writers into four classes. The first are termed Placoid, or flattened scales, of which the common dog-fish affords a good example. The next class is called Ganoid, or polished scales, such as those of the sturgeon.

BRAZILIAN PERCOPHIS.—*Percophis Braziliauus.*

The third series of scales are named Ctenoid, or comb-like scales, on account of the toothed projections in their posterior edge. The beautiful scales of the Perch, so largely used in the manufacture of feminine ornaments, are familiar examples of ctenoid scales. The last class are the Cycloid, or rounded scales without teeth, such as those of the carp.

On careful examination, the scales are found to be composed of two layers, which can be separated by careful manipulation, and their structure is permeated with channels, either partially closed so as to form grooves, or wholly so, and then become tubes. The comparatively straight lines which radiate towards the circumference, and the wavy lines which are drawn concentrically and cover almost the entire scale, can be referred to a common spot, called technically the "focus." This spot cannot rightly be termed the centre, because it is of variable dimensions, and, moreover, is seldom if ever placed exactly in the middle of the scale.

Many scales throw out strong tooth-like projections, which in point of fact are the teeth of the external skin, and in some species, such as several of the dog-fish, are of crystalline hardness. In some fish, such as the stickleback, a few of these processes of the skin are developed into long sword-like weapons; in others, such as the trunk-fish, they are fused together, and form a continuous bony envelope to the body, pierced for the protrusion of the tail and fins; in some, of which the diodons are good examples, they cover the body with an array of projecting spikes, like the quills of the hedgehog or porcupine, and in others, such as the sturgeons, are arranged in rows of bony plates along the body.

We return to the species now under consideration.

The BRAZILIAN PERCOPHIS is found upon the coasts of Brazil, and is apparently the sole representative of the genus in which it has been placed. The first dorsal fin is very small in proportion to the second, and the space between them is about equal to the length of the first dorsal. The ventral fins are set very far forward, being placed under the throat. The lower jaw projects considerably beyond the upper, and the cleft of the

BELTED HORSEMAN.—*Eques lanceolátus*

mouth is horizontal. The canine teeth are very large in proportion to the dimensions of the fish.

ANOTHER family, the Sciænidæ, now come before us, and will be illustrated by several examples. The members of this family are clothed with ctenoid or toothed scales ; the mouth is set in front of the snout, the teeth are arranged in bands, and the gill-covers are either unarmed or furnished with feeble spines.

The first example of this family is the BELTED HORSEMAN, a striking and boldly marked species.

This fish is found upon the Atlantic coasts of tropical America, and is perhaps the most striking of the limited genus to which it belongs. The body is oblong, and the nape of the neck is very high, its elevated line being continued by the first dorsal fin, which is short, high, and pointed, its height being just equal to the depth of the body. The second dorsal fin is long, rather low, and is covered with very thin scales. The tail fin is covered in like manner. The scales of the body are of moderate size.

The general colour of the Belted Horseman is greyish yellow, diversified with three broad belts of blackish brown, each belt being edged with whitish grey. The first band passes vertically over the head, beginning just at the base of the first dorsal fin, traversing the head behind the eye, and descending to the angle of the mouth, and the others are drawn along the body, one reaching to the base of the ventral fin, and the other to the tail. There is, however, a slight variation in different specimens.

ANOTHER species of the same genus, the SPOTTED HORSEMAN (*Eques punctatus*), is nearly, though not quite as remarkable a fish, and is notable for the bluish white spots which decorate the dorsal, ventral, and anal fins. The general colour of this fish is brown, with two vertical bands running over the side of the head, and some curved bands passing along the body from the back to the tail. This fish is found in the Caribbean seas.

CLOSELY allied to these creatures is a rather remarkable fish, called scientifically *Pogónias chromis*, and more popularly known by the name of BEARDED DRUM-FISH.

MAIGRE.—*Sciæna aquila.*

This title is given to the fish on account of the peculiar sounds produced by the fish, which are thought to bear some resemblance to the beating of a drum. The sound is apparently produced both while the fish is immersed and after its removal from the water, and probably on account of the sound-conducting powers of the water, the hearer finds great difficulty in referring the strange noises to any particular spot. These fish do not seem to thrive well in fresh water, as the drumming was invariably found to cease as soon as the boat in which the observers were sitting had left the sea-coast and entered a river.

This fish has an oblong body and a rather convex muzzle, the upper jaw being larger than the lower. The first dorsal fin is furnished with ten strong spines, and two similar spines appear on the anal fin, one being much stronger than the other. It has derived its name of Bearded Drum-fish from the large array of barbels, about twenty in number, which hang from its jaws. The generic name Pogonias signifies bearded, and is given to the fish in allusion to this peculiarity. The tail fin is rather abruptly cut short. The general colour of this fish is brownish grey, with a bold black spot on the shoulder. It is a native of the North American coasts, and is known to extend as far south as Florida.

ANOTHER noisy fish is well known under the title of MAIGRE, the strange sounds produced by this species having been heard from a depth of one hundred and twenty feet.

In one instance, perhaps in many others, the novel accomplishment has led to the destruction of its possessors, the fishermen having been directed by the sounds to the whereabouts of the utterers, and inclosed them in their nets. The flesh of the Maigre is thought to be peculiarly excellent, the head and shoulders being held in the greatest estimation.

It is a rather large fish, seldom measuring less than a yard in length, and often attaining nearly double those dimensions, and is in consequence extremely valuable to the fisherman. Although at one time it might be captured with tolerable frequency on the coasts of France, and now and then on the British shores, it is now very scarce, having shifted its localities, and being found most plentifully on the southern shores of the Mediterranean. There it seems to be hatched and to remain until it attains nearly adult age, when it crosses to the northern side of that sea, and is there found to be of considerable dimensions.

On account of its size and active habits, the Maigre struggles most powerfully when entangled in the nets, and a fine lively specimen, when lifted into a boat, will flounce about with such activity, and wield its tail with such rapid force, that it will level a human being with a blow. Warned by previous experience, the fishermen take care to quiet their energetic prey by a stroke on the head as soon as it is fairly lifted over the side of the boat.

The ear-bones of fishes have already been mentioned. In the Maigre they are of very great proportionate dimensions, and in former days were in great repute for their efficacy in charming away the colic, the only proviso being that they must be received as a gift, and not purchased, or removed by the sufferer from the head of the fish. The noise made by the Maigre is a kind of purring or grunting, which, when many of these fish are grunting in concert, can be heard at some distance.

The colour of the Maigre is delicate silver-grey, taking a slight brownish tinge on the back, and being very white on the abdomen. The fins are reddish brown, and the ruddy hue is particularly conspicuous in the first dorsal, the pectoral, and anal fins. The lateral line runs parallel with the back.

The BLACK CORVINA of the Mediterranean (*Corvina nigra*), is allied to the maigre, and is scarcely less celebrated than that fish for the excellence of its flesh.

This fish is not exclusively a marine species, but frequents salt lakes and ponds, and, though it hovers about the mouths of large rivers, probably for the purpose of feeding on the many animal and vegetable substances which are borne by their currents into the sea, does not appear to ascend their streams. In general appearance it is not unlike the maigre, and is often sold in the markets under that name.

BEFORE leaving the present family, a short notice is due to the SQUETEAGUE, or WEAK-FISH of North America (*Otolithus regális*).

This is another of the noisy fishes, producing dull sounds like those of a drum. It is plentiful about New York, and is captured in large quantities for the table. The name of Weak-fish is attributable to two causes, the one that when hooked it makes but a feeble resistance, and the other that its flesh is popularly supposed to be weakening to those who habitually live upon it. It is a useful species, for it not only affords delicate food, but its swimming-bladder can be made into isinglass which is said to be in no way inferior to that of the sturgeon. On account of its spotted skin, the French of New Orleans call it by the name of Trout.

AN example of the next family, the Polynemidæ, may be found in the MANGO-FISH (*Polynémus paradíseus*), so called on account of its fine golden yellow colour, which resembles that of the ripe mango.

The generic name of Polynemus, or Many-threaded, is given to this and others of the same group on account of the free thread-like appendages below the pectoral fin. In the present species there are seven of these appendages, the upper being of singular length, passing far beyond the tail. In another species, the MANY-THREADED POLYNEMUS (*Polynémus multifilis*), these appendages are fourteen in number.

Isinglass can also be made from the swimming-bladders of these fish, and as when quite dry one bladder will weigh from half to three-quarters of a pound, the fish might be rendered valuable to commerce. The flesh of the Mango-fish is not particularly excellent when eaten fresh, but when preserved and salted after some peculiar fashion, is in some request at the breakfast table, where it ranks with caviare and other strongly flavoured delicacies. When thus prepared, it is known by the name of "burtah."

In the Mango-fish, the first filament is twice the length of the body.

ONE example of the Sphyrænidæ, the family next in order, is the BECUNA, a rather large and tolerably ferocious fish, inhabiting the Mediterranean and many parts of the Atlantic Ocean.

This long-bodied, deep-mouthed, and sharp-toothed fish bears some resemblance to the pike both in general appearance and in habits, and is hardly less voracious than the veritable pike of our own country. It is said that from the scales of the Becuna are washed those minute crystalline spiculæ, which are so useful in the preparation of .artificial pearls, and which, when mixed and prepared for commerce, are termed *essence d'orient*. Some parts of the air-bladder are also used in the manufacture of this substance. The flesh of the Becuna is well flavoured and is often brought to table ; being capable of being dressed in a fresh state and after salting.

BECUNA.—*Sphyræna vulgáris.*

On the back, the colour of this fish is leaden blue with a wash of green, and on the abdomen it is white. The sides are in many specimens marked with dark cross-bars of the same green colour as the back. When young it is spotted with brown.

ANOTHER species of the same genus, the BARRACOUDA PIKE, is nearly as useful as an article of food, though its utility is partly impaired by the fact that at some seasons of the year its flesh is of a very unwholesome character, producing sickness, pain in the joints, sometimes accompanied with loss of the nails and falling off of the hair.

It is thought that the poisonous property is occasioned by certain food of which the fish partakes at such seasons, the fruit of the manchineel being the chief substance to which the evil effect is attributed. For many reasons, however, among which the carnivorous character of the fish is most weighty, this suggestion appears to be groundless, and naturalists have agreed that the true cause is yet to be discovered.

Whatever may be the cause, the result is too well known, and it is said that an experienced person can detect a Barracouda Pike when in this unsound condition, by making an incision in the flesh, from which a whitish fluid exudes if the flesh be poisonous. The liver also affords another test, contracting a peculiar bitter taste under such circumstances.

This fish attains a much larger size than the preceding species, and when full grown and hungry, is thought to be little less dangerous than a shark of the same dimensions. Its colour is olive-green above with a beautiful under tinting of rich green, its sides are mostly decorated with brown blotches, and the abdomen is silvery white. It is found in tropical portions of the Atlantic Ocean. About fifteen species are known.

WE now arrive at a small family of fishes, termed Trichiuridæ, or Hair-tailed fishes, in consequence of the delicate filamentary finlets which decorate the tail in some species

In all these fishes, the body is long and compressed, almost like a riband, and indeed is not at all unlike those flat "snakes" that are sold in the toy shops, and which dart in all directions when held by the tail.

The first example of these curious creatures is the SCABBARD-FISH, so called because in shape it bears some resemblance to the sheath of a sword.

On account of its shape and bright silvery whiteness, it is a most striking inhabitant of the ocean, and when writhing its way through the translucent water, in elegant undulations, it looks like a broad riband of burnished silver winding through the waves. This shining brilliancy is caused by a thin epidermis, which covers the body in place of scales, and which can be easily rubbed off by the fingers, to which it adheres, transmitting to them a portion of the metallic whiteness which it imparted to its proper owner.

In spite of the exquisite beauty of this fish, it is captured for the sake of its flesh, which is highly esteemed, and is generally sought in the months of April and May when it approaches the coasts. The drag-net is the usual instrument of capture. It seems to be a solitary fish, and lives at a considerable depth. The rapid undulations of the body are capable of propelling the creature through the water with great velocity, but, from all

SCABBARD-FISH.—*Lepidopus caudatus.*

appearances, it is not able to make much way against a rapid tide, or to overcome the dashing waves raised by a tempestuous wind.

Several specimens of this beautiful fish are in the British Museum, the largest of which measures more than five feet in length. Along the back runs a single dorsal fin, and the ventral fins are only represented by a pair of scales, a structure which has gained for the fish the generic name of Lepidopus, or Scale-fin.

ANOTHER fish that much resembles the preceding species is sometimes, but very rarely seen upon the British shores. This is the SILVERY HAIR-TAIL (*Trichiúrus leptúrus*), a species that may easily be distinguished from the scabbard-fish by the shape of the tail, which has no fin at its extremity, but tapers into a long and gradually diminishing point. This species is common in many parts of the Atlantic Ocean, and by the Spanish inhabitants of Cuba it is termed the Sabre-fish.

Another species, the SAVALA (*Trichiúrus Savala*), is found in the East Indian seas, and is sufficiently plentiful to form a recognised article of diet, and to be sold in the markets. The Savala bears salting well, and is much used for consumption when the inclemency of the weather will not permit fishing-boats to put to sea. When fresh, however, it does not suit the taste of Europeans, though in Malabar the salted fish is esteemed both by the native inhabitants and the European colonists.

BEFORE leaving this small but curious family, the ATUN (*Thyrsites atun*) deserves a passing notice.

This elegant and useful fish is found on the coasts of Southern Africa and part of Australia, and is much valued for the flakey whiteness and pleasant flavour of its flesh, which bears some resemblance to that of the cod, but is even superior in delicacy. It feeds mostly upon the cuttle-fish, the calamary being its favourite prey. So voracious is this creature that it is readily caught by making a sham calamary out of lead and leather, dressing it with projecting hooks, and flinging it into the sea. The fishermen throw this bait to some distance, and then draw it rapidly through the water, when the Atun takes it for the real calamary darting along after its usual fashion, dashes at it and is immediately hooked. In default of this bait, a strip of red cloth stuck on a hook is often a sufficient lure for this voracious fish.

MACKAREL.—*Scomber scomber.*

THE important, though not very large family of the Scomberidæ, contains many species that are almost invaluable as food, and others that are beautiful in form and interesting in habits.

Our first example of these fishes is the MACKAREL, so well known for the exceeding beauty of its colours and the peculiar flavour of its flesh. This is one of the species that are forced by the irresistible impulse of instinct to migrate in vast shoals at certain times of the year, directing their course towards the shores, and as a general rule frequenting the same or neighbouring localities from year to year. The time of their advent is rather variable, and in consequence the price of this fish varies with the scarcity or abundance. At the beginning of the season, before the shoals have made their appearance, a fresh Mackarel will fetch half-a-crown or three shillings, while in a week or two afterwards a hundred may be purchased for the same sum. As a general rule, when they are in full season, they are sold in the streets of London for three or four a shilling.

The flesh of the Mackarel is very excellent, and it possesses a rather powerful and unique flavour that has caused fennel to be looked upon as a necessary corrective in the sauce with which the fish is served. Unfortunately, it must be eaten while quite fresh, as it becomes unfit for consumption in a very short time after being taken out of the water; and in consequence of this property, the London costermongers are permitted to hawk it about the streets on Sundays, much to the discomfort of peaceable householders who long for repose and do not want Mackarel.

This fish is taken both by nets and lines, the nets being of two kinds, one called the drift-net, and the other the seine. The drift-net is, as its name implies, allowed to be drifted out by the tide, and is suspended along a cord called the drift-rope. The whole length of one of these nets when shot is sometimes a mile and a half, these enormous dimensions being attained by attaching a number of nets together at the ends. Each of these nets is one hundred and twenty feet long and twenty feet deep, and along the upper edge are fastened a series of cork floats. When the net is to be shot, a large buoy is attached to the end of the drift-rope, the buoy is thrown overboard and the sails set. As the boat dashes away from the spot, the nets, which have already been attached to the drift-rope, are thrown successively overboard, until all the nets are paid out and hang in the water like a net wall. The strain of the buoy at one end of the drift-rope and the boat at the other keeps the rope straight and the net upright.

As the Mackarel come swimming along, they are arrested by the net, which they cannot see, on account of the thin twine of which it is made, and the large meshes, which are about two and a half inches in diameter. The head slips through the meshes, but the middle of the body is too large and cannot pass. When the fish attempts to recede, the open gill-covers become hitched in the meshes, and so retain it in that uncomfortable position until the net is hauled in.

This is a delicate and difficult operation, especially when the take of fish is heavy. Mr. Yarrel mentions that in June, 1808, the nets were so heavily loaded that the fishermen could not haul them in or even keep them afloat, so that they were forced to cut the drift-ropes, and let the nets sink and be lost. The nets on this occasion were worth nearly sixty pounds, not including the value of the fish.

In the seine-net, the fish are taken by surrounding a shoal with the net, which is made with very small meshes, and either gently hauled to the surface, so that the inclosed fish can be dipped out, or even drawn ashore and then emptied.

Fishing for Mackarel with a line is also a profitable mode of taking these fish, although they cannot be caught in such multitudes as with the net. The Mackarel is a very voracious fish, and will bite at almost any glittering substance drawn quickly through the water, a strip of scarlet cloth being a very favourite bait A tapering strip of flesh cut from the side of a Mackarel is found to be the most successful of any bait, and the method of angling is simply to pass the hook through the thicker end of the strip—technically called a "lask"—and to throw it overboard a boat in full sail, so that it is towed along without trouble. The hook is kept below the surface of the water by means of a leaden plummet fixed to the line a short distance above the hook, and the Mackarel on seizing the flying bait is immediately caught. On a favourable day, when the sky is not too bright, and the wind is tolerably brisk, two or three men can take the fish as fast as they can bait and throw overboard.

The colour of the Mackarel is rich green upon the back, variegated with deep blue and traversed with cross bands of black, straight in the males; but undulating in the females. The abdomen and sides are silvery white, with golden reflections. These colours are most brilliant during the life of the fish, and as they fade soon after it has left the water, their brilliancy affords a good test of its freshness.

THE celebrated TUNNY belongs to this family, and is closely allied to the mackarel.

This magnificent and most important fish does not visit our coasts in sufficient numbers to be of any commercial importance; but on the shores of the Mediterranean, where it is found in very great abundance, it forms one of the chief sources of wealth of the sea-side population.

In May and June, the Tunnies move in vast shoals along the shores, seeking for suitable spots wherein to deposit their spawn. As soon as they are seen on the move, notice is given by a sentinel, who is constantly watching from some lofty eminence, and

TUNNY—*Thynnus Thynnus.*

the whole population is at once astir, preparing nets for the capture, and salt and tubs for the curing of the expected fish. There are two modes of catching the Tunny, one by the seine-net and the other by the "madrague." The mode of using the seine is identical with that which has already been described when treating of the mackarel, but the madrague is much more complicated in its structure and management.

The principle of the madrague is precisely the same as that of the "corral," by which wild elephants are entrapped in Ceylon.

A vast inclosure of united nets, nearly a mile in length, and divided into several chambers, is so arranged that as the Tunnies pass along the coast, they are intercepted by a barrier, and, on endeavouring to retreat, are forced to enter one of the chambers. When a number of Tunnies have fairly entered the net they are driven from one chamber to another, until they are forced into the last and smallest, called significantly the chamber of death. This chamber is furnished with a floor of net, to which are attached a series of ropes, so that by hauling on the ropes, the floor of the net is drawn up, and the fish brought to the surface. The large and powerful fish struggle fiercely for liberty, but are speedily stunned by blows from long poles, and lifted into the boats.

The flesh of the Tunny is eaten both fresh and salted. It is most extensively used, being pickled in various ways, boiled down into excellent soup, and is also made into pies, which are thought to be very excellent, and possess the valuable property of remaining good for nearly two months. The different parts of the fish are called by appropriate names, and are said to resemble beef, veal, and pork.

The food of the Tunny consists mostly of smaller fish, such as herrings and pilchards, and the cuttle-fish also form some portion of its diet.

In general shape the Tunny is not very unlike the mackarel, but in size it is vastly superior, generally averaging four feet in length and sometimes attaining the dimensions of six or seven feet. The colour of the upper part of the body is very dark blue, and the abdomen is white, decorated with spots of a silvery lustre. The sides of the head are white.

OF an allied species, the PACIFIC ALBACORE (*Thynnus Pacificus*), Mr. F. D. Bennett writes as follows, in his well-known "Whaling Voyage." "Ships when cruising slowly in the Pacific Ocean are usually attended by myriads of this fish, for many successive months.

A few days' rapid sailing is nevertheless sufficient to get rid of them, however numerous they may be; for they seldom pay more than very transient visits to vessels making a quick passage. When the ship is sailing with a fresh breeze, they swim pertinaciously by her side and take the hook greedily; but should she be lying motionless or becalmed, they go off to some distance in search of prey, and cannot be prevailed upon to take the most tempting bait that the sailor can devise.

It is probably as a protection from their chief enemy, the sword-fish, that they seek the society of a ship. I am not aware that the shark is also their enemy, but they seemed to have an instinctive dread of this large fish, and, when it approached the ship, would follow it in shoals, and annoy it in the same manner as the smaller birds may be seen to annoy those of a larger and predaceous kind, as the hawk or owl.

They are very voracious and miscellaneous feeders. Flying fish, calamars, and small shoal-fish are their more natural food, though they do not refuse the animal offal from a ship. Among other food contained in their maw, we have found small *ostracians* (*i.e.* trunk-fish), file-fish, sucking-fish, janthina shells, and pelagic crabs—in one instance a small bonita, and in a second a dolphin, eight inches long, and a paper nautilus shell, containing its sepia tenant.

It was often amusing to watch an Albacore pursuing a flying fish, and to mark the precision with which it swam beneath the feeble aeronaut, keeping him steadily in view, and preparing to seize him at the moment of his descent. But this the flying fish would often elude by instantaneously renewing his leap, and not unfrequently escaped by extreme agility."

The Bonito (*Pelamys sarda*) is a very pretty and common species that is found in the Mediterranean and many parts of the Atlantic.

This is a smaller species than the albacore, not exceeding two feet and a half in length. The flesh of this fish is eaten both fresh and when pickled, but in a fresh state is not held in very high estimation. At some seasons, it appears to contract an unwholesome quality, which is injurious to certain constitutions, causing rather a painful rash to break out on the face and body, though others can eat it with impunity. The flesh is very red in colour, and looks very like butcher's meat.

Like the albacore, the Bonito is a determined foe of that much persecuted creature the flying fish, and is often taken by means of a hook dressed with feathers so as to resemble its natural prey. It is a truly beautiful species, deserving fully the popular name of Bonito, which may be freely translated as Little Beauty. The back is deep indigo blue, mottled with a lighter shade of the same hue, and when young a number of dark streaks are drawn across the back. The abdomen is silvery white, and the cheeks and gill-covers are of the same brilliant hue.

ANOTHER species, the STRIPED BONITO (*Auxis Rochei*), inhabits the same localities, and is nearly as plentiful as the preceding fish. It may readily be known from the plain Bonito by the four dark lines which extend along each side of the abdomen and end at the tail.

THE prettily marked PILOT-FISH is frequently seen off the British coasts, but seems to be rather shy, and is not very often captured.

This little fish has long been supposed to act the part of the shark's provider, and to perform in the ocean the same actions that were once attributed to the jackal on land. Many modern writers, however, deny the truth of the statement, by saying that the Pilot-fish only follows the shark for the sake of the scraps that the larger fish is likely to leave, and that it would probably be snapped up by the shark but for its watchfulness and agility.

As is usual in such a disputation, the evidence is very conflicting, and many accounts have been published tending to throw discredit on the one side or the other, according to the particular circumstances under which the observations were made. One well-known naturalist, for example, mentions an instance where a shark was directed towards a baited hook by two Pilot-fish that accompanied him; but, on the other hand, another accomplished observer narrates an interesting anecdote of a shark being continually warned of a baited hook by his little friends, who struck their noses against his snout whenever he turned towards the bait. At last, however, he dashed at the tempting morsel and was captured, to the sorrow of the Pilot-fishes, who swam about for some time in search of their friend, and then darted down into the depths of the sea.

Mr. Bennett, in the work to which allusion has lately been made, has some very curious and interesting observations on this fish, which must be given in his own words :—

PILOT-FISH.—*Naucrates ductor.*

" Pilot-fish are almost invariably found in attendance upon the shark, though the nature of their connexion with that ferocious fish is somewhat mysterious. They will accompany ships for a considerable distance after their patron shark has been destroyed ; but I am not aware that they have ever been seen, like the remora, attending upon other large fish, whales, or miscellaneous floating bodies. The structure of their mouth and the contents of their stomach, which are usually small fish, denote that they are accustomed to seek their food in a very independent manner. We captured many of them also by hook and line, baited with flesh ; nor did they refuse the bait even when they were in company with a shark.

The reputation this fish has obtained of being the shark's pilot or provider (and which has sanctioned its trivial name) would appear to be groundless, were we guided only by the want of similar precedents in the animal kingdom. A fact, however, which came under my notice during a voyage from India, in the year 1832, led me to believe that there is some just foundation for this popular opinion.

While we were becalmed in the Atlantic Ocean, a shark was seen close to the ship, and attended by two Pilot-fish, which generally swam one above and the other below him, and occasionally went off to some distance, as if to explore the surrounding sea, although it was seldom long before they returned and resumed their former positions ; the shark, in the meantime, by its unwieldy form, slow movements, and lethargic aspect, offered a strong contrast to the sprightliness and activity of his scouts.

A baited hook was lowered from the bow of the ship ; but the shark, when alone, passed it several times without notice, and apparently without seeing it. One of the foraging Pilot-fish then approached the bait, and immediately swam off to where the shark was headed in a contrary direction, when the monster instantly turned, and followed his informant, which now swam ahead of him, in a direct line towards the

3.

T

suspended bait. He did not then hesitate a moment, but seized it and was captured. While the shark was being hauled on board, the Pilot-fish expressed the greatest concern, almost leaping out of the water in their endeavours to follow him, and swimming near the surface with every demonstration of anxiety.

These faithful little fish were observed to attach themselves to the ship, but attracted little attention until some weeks afterwards, when we spoke the *Thomas Grenville*, East Indiaman, and lowered a boat to communicate with her. One of the fish was then seen to accompany the boat to and from the stranger ship; and so devotedly did it attend upon what it might have believed to be its lost shark, as to lead the officers of the *Thomas Grenville* to remark that we had a Pilot-fish *painted* on the rudder of the boat.

Their attendance upon sharks is somewhat capricious: we have seen more than five associated with one shark, while many others of the latter tribe, and assembled in the water at the same time, have not been accompanied by one of these fishes. They have evidently nothing to dread from the voracious companion they select, but swim around, and often a few inches ahead of him, as either their convenience or caprice may dictate."

The colour of the Pilot-fish is greyish blue with a mark of silver, dark on the back, and becoming paler towards the abdomen. Five bands of dark blue pass completely round the body, and there are two faint blue bands, one on the head and the other on the tail. The pectoral fins are clouded with blue and white. The ventral fins are nearly black. The usual length of the Pilot-fish is about one foot.

EVERY one has heard of the SUCKING-FISH, and there are few who are not acquainted with the wild and fabulous tales narrated of its powers.

This little fish was reported to adhere to the bottom of ships, and to arrest their progress as suddenly and firmly as if they had struck upon a rock. The winds might blow, the sails might fill, and the masts creak, but the unseen fish below could hold the vessel by its single force, and confine her to the same spot as if at anchor. It is wonderful how fully this fable was received, and how many years were needed to root the belief out of prejudiced minds. Both scientific names refer to this so-called property, *echenéis* signifying "shipholder," and *rémora* meaning delay.

That the Sucking-fish is able to adhere strongly to smooth surfaces is a well-known fact, the process being accomplished by means of the curious shield or disc upon the upper surface of the head and shoulders, the general shape of which can be understood by reference to the engraving. This disc is composed of a number of flat bony laminæ, arranged parallel to each other in a manner resembling the common wooden window-blind, and capable of being raised or depressed at will. It is found by anatomical investigation that these laminæ are formed by modifications of the spinous dorsal fin, the number of laminæ corresponding to that of the spines. They are moved by a series of muscles set obliquely, and when the fish presses the soft edge of the disc against any smooth object and then depresses the laminæ, a vacuum is formed, causing the fish to adhere tightly to the spot upon which the disc is placed.

When the creature has once fixed itself it cannot be detached without much difficulty, and the only method of removing it without tearing the body or disc, is to slide it forwards in a direction corresponding with the set of the laminæ. In the opposite direction it cannot be moved, and the fish, therefore, when adhering to a moving body, takes care to fix itself in such a manner that it cannot be washed off by the water through which it is drawn. Even after death, or when the disc is separated from the body, this curious organ can be applied to any smooth object, and will hold with tolerable firmness. In order to accommodate the disc, the upper part of the skull is flattened and rather widened.

The Sucking-fish will attach itself to many moving objects, and has been found adhering to the plankings of ships and boats, to turtles, to whales, and to fishes of various kinds. Even the albacore, which eats the Sucking-fish whenever it can catch it, is occasionally honoured by its adhesion, and in the British seas a specimen has been captured while sticking to a cod-fish. The shark, however, is its favourite companion, and it often happens that one of these voracious creatures is attended by quite a little train of

Sucking-fishes. What object is fulfilled by this capability of adhesion, is a problem as yet unsolved. The Remora is perfectly organized and capable of procuring food for itself, and though not a swift swimmer, is able to proceed through the water with tolerable rapidity. Its mouth is moderately large, and that the creature has no difficulty in seeking a subsistence is proved by the fact that its stomach usually contains remnants of small crustacea and molluscs.

It is rather a voracious fish, and takes the hook eagerly if baited with a piece of raw flesh. When hooked, however, it is by no means secured, for as soon as it feels the prick of the sharp point and the pull of the line, it darts to the side of the vessel, dives deeply, and affixes itself so strongly to the bottom that the hook may be torn out of the mouth before the fish will relax its hold. It is, therefore, necessary to draw the Sucking-fish smartly out of the water as soon as it is fairly hooked, and in this manner great numbers can be

SUCKING-FISH.—*Echeneis remora.*

caught. The flesh is thought to be very good, and is said to resemble that of the eel, but without its richness.

The colour of this species is dusky brown, darker on the back than on the abdomen. The fins are darker than the body, and are of a dense leathery consistence. The length of this fish seldom exceeds eight inches.

THERE are about ten species of Sucking-fishes known, of which the SHIELDED SUCKING-FISH (*Echeneis scutáta*) is perhaps the most remarkable. This species may be at once recognised by the very great size of the disc, and its length being nearly one-half that of the body. At the hinder portion of the disc the laminæ are wanting, and its surface is smooth. This species attains to considerable dimensions, a fine specimen in the British Museum being nearly two feet in length.

THE well-known JOHN DORY, so dear to epicures, is found in the British seas, and is frequently seen in the fishmongers' shops, where its peculiar shape seldom fails of attracting attention even from those who are not likely to purchase it or even to have seen it on the table.

The name of John Dory is thought to be a corruption of the French name *jaune dorée,* a title given to the fish on account of the gilded yellow which decorates its body. It was called Zeus by the ancients because they considered it to be the king of eatable fish; and the name of Faber, or blacksmith, has probably been earned by the smoky tints which cloud its back. The dark and conspicuous spots on the side are thought in many places to be imprinted on the fish as a memorial of the honour conferred upon its ancestor in times past, when St. Peter took the tribute-money from the mouth of the Dory, and left the print of his finger and thumb as a perpetual remembrance of the event. Some persons, however, contend that the marks are due, not to St. Peter, but to St. Christopher; and the Greeks, who hold to the latter tradition, call the fish Christophoron.

It is a very voracious creature, feeding upon various marine animals and fishes of inferior size to itself. It will even catch and devour a cuttle-fish of great size in spite of the strength and agility of the prey, and is fond of following the shoals of pilchards for the purpose of feeding upon the young and weakly. In consequence of this habit, it is frequently captured in the same nets which are employed to take the pilchards. The flesh of the Dory is remarkably excellent, and as it is rather improved by the lapse of twenty-four hours after the fish has been taken from the sea, it is peculiarly valuable to those who live far inland, and cannot hope for the more delicate fishes, which must be eaten almost as soon as caught. Although a common fish, it always commands a high price, and as, when cooked, the head occupies so large a space, it never affords an economical dish.

JOHN DORY.—*Zeus faber.*

The shape of the Dory is very peculiar. The body is very deep, and greatly compressed. The head is oddly shaped, and the mouth can be protruded to a surprising extent. The spines of the first dorsal fin are much prolonged, and behind each ray is given off a very long waving filament, three times as long as the ray in front of it. Along the base of the dorsal and anal fins are arranged two rows of spiny scales, their points being directed backward, and one row being set at each side of the fin.

The colour of this fish is very beautiful, especially if seen immediately after its removal from the water, when golden brown, olive, white, and azure are reflected from its surface in changing tints, that rapidly vanish after death.

WE now come to a most beautiful and interesting fish, the CORYPHENE, so often erroneously spoken of as the dolphin.

This splendid fish is found in many of the warmer seas, inhabiting the Mediterranean Sea, and the Indian, Pacific, and Atlantic Oceans. The reader has, in all probability heard the old story respecting the lovely and changeful colours of the dying dolphin, and is quite

CORYPHENE.—*Coryphæna hippurus.*

aware that in the shining black and grey skin of the true dolphin no such changes take place. There is, however, more truth than usual in this tale, for the dolphin in question is really the Coryphene, whose colours are always most brilliant, and glow with changeful beauty, during the death struggle. A similar phenomenon occurs in several other fishes, of which the common red mullet is a familiar example.

The Coryphene is a most voracious creature, feeding chiefly upon the finny inhabitants of the water, and being especially celebrated for its attacks upon the flying-fish. Captain Basil Hall has given so graphic and so pleasing a description of one of their chases, that the narrative must be given in his own words. The reader will understand that whenever the word dolphin is employed by the author, the Coryphene is the species to which he alludes.

" A large Dolphin, which had been keeping company with us abreast of the weather-gangway at the length of two or three fathoms, and as usual glistening most beautifully in the sun, no sooner detected our poor little friends take wing, than he turned his head towards them, and, darting to the surface, leaped from the water with a velocity little short, as it seemed, of a cannon ball. But although the impetus with which he shot himself into the air gave him an initial velocity greatly exceeding that of the flying-fish, the start which his ill-fated prey had got, enabled them to keep ahead of him for a considerable time.

The length of the Dolphin's first spring could not be less than ten yards; and after he fell, we could see him gliding like lightning through the water for a moment, when he again rose and shot forwards with considerably greater velocity than at first, and of course to a still greater distance. In this manner the merciless pursuer seemed to stride along the sea with fearful rapidity, while his brilliant coat sparkled and flashed in the sun quite splendidly. As he fell headlong on the water at the end of each huge leap, a series of circles were sent far over the still surface, which lay as smooth as a mirror.

The group of wretched flying-fish, thus hotly pursued, at length dropped into the sea, but we were rejoiced to observe that they merely touched the top of the swell and scarcely sunk in it; at least they instantly set off again in a fresh and even more vigorous flight. It was particularly interesting to observe that the direction they now took was quite different from the one in which they had set out, implying but too obviously that they had detected their fierce enemy, who was following them with giant steps across the waves and now gaining rapidly upon them. His terrific pace indeed, was two or three times as swift as theirs, poor little things.

The greedy Dolphin, however, was fully as quick-sighted as the flying-fish which were trying to elude him, for whenever they varied their flight in the smallest degree, he lost not the tenth part of a second in shaping a new course so as to cut off the chase, while they, in a manner really not unlike that of the hare, doubled more than once upon their pursuer. But it was soon too plainly to be seen that the strength and confidence of the flying-fish were fast ebbing. Their flights became shorter and shorter, and their course more fluttering and uncertain, while the enormous leaps of the Dolphin appeared to grow more vigorous at each bound.

Eventually we could see, or fancied that we could see, that this skilful sea-sportsman arranged all his springs with such an assurance of success, that he contrived to fall at the end of each, just under the very spot on which the exausted flying-fish were about to drop. Sometimes this catastrophe took place at too great a distance for us to see from the deck exactly what happened; but on our mounting high into the rigging, we may be said to have been in at the death, for then we could discover that the unfortunate little creatures, one after another, either popped right into the Dolphin's jaws as they lighted on the water, or were snapped up instantly afterwards."

The Coryphene, however, often pays the penalty of its voracity, for, independently of the innate quality of destructiveness, which alone would induce the spectators to catch and kill the fish, the sailors are urged by the excellence of its flesh, which will afford them a welcome repast on fresh meat. They therefore bait a number of hooks in a very simple manner, by cutting some strips of glittering tin into the rude semblance of the flying-fish, attaching them to the hooks and dangling them in the air, where the indiscriminating Coryphene seizes them and pays the penalty of its voracity. Sometimes the Coryphene is captured by a strong-armed and sure-eyed sailor, who gets on the bowsprit, so as to overhang the water, and hurling at the fish a kind of barbed trident, technically called the "grains," the barbed points strike deeply into the flesh, and the fish

EYED PTERACLIS.—*Ptéraclis ocellátus.*

is triumphantly hauled on board by means of the rope affixed to the butt-end of the shaft.

Such a chase as has just been described, seems to be of comparatively rare occurrence, and any one who has witnessed it may consider himself peculiarly fortunate.

Words can hardly express the extreme beauty of the Coryphenes, as they play easily around the ship, their sides glittering in the sunbeams as if made of burnished gold and silver, and every change of attitude producing some new combination of colour. They have a curious habit of attaching themselves for a time to a passing ship, and are able, from their exceeding swiftness, to gambol around the vessel as if she were at anchor, no matter how swift her progress may be. Steam, however, is fatal to such observations, as the screw or paddles are so noisy that few fish venture within their sweep, and the water is so churned into whirling circles of froth and foam that even the glistening body of the Coryphene would be invisible below the disturbed surface.

THE very remarkable fish which is shown in the accompanying illustration, is allied rather closely to the preceding species, in spite of the great difference in form, and by some writers was placed in the same genus as that fish.

The EYED PTERACLIS is a good example of the curious genus to which it belongs, and which can always be recognised by the extreme depth of the dorsal and anal fins and their delicate tenuity of structure. The dorsal fin is moreover remarkable for the bold sweep of its extent, passing in an unbroken curve from the forehead to the tail. Owing to the development of the anal fin, the two ventrals are placed very far forward, and are seen under the throat. The members of this genus are spread over the Indian Ocean, the Sea of Marmora, and some of the American coasts.

The Eyed Pteraclis is found on the Mozambique coast. It is a very beautiful fish, the general colour being shining white as if made of polished silver, with a wash of gold upon the pectoral and tail fins, and a deepish tint of blue-grey upon the others. On the dorsal fin there is a round spot of dark blue. It seems to be a small species. About four members of this genus are known to naturalists.

BEFORE quitting this family, we must briefly notice the handsome OPAH, or KING-FISH (*Lampris luna*), which is sometimes, but rarely, found in the British seas.

This beautiful species seems to be the sole representative of its genus, it having been separated from the genus Zeus, in which it had formerly been placed, in consequence of its single dorsal fin. It sometimes attains to a considerable size, a specimen having been taken on the northern coast of England which measured five feet in total length, and weighed about one hundred and fifty pounds. The flesh of this fish is red, very good, and is said to resemble that of the salmon.

CORDONNIER, OR COBBLER-FISH.—*Caranx ciliaris.*

The colour of the Opah is bright green on the upper part of the back and sides, with reflections of purple and gold in certain lights. The fins and eyes are scarlet, and a number of round spots of pale gold are scattered upon the sides.

WE now arrive at a rather large family of fishes, which has been separated from the mackarels on account of certain anatomical variations, which will be mentioned at the end of the volume.

The CORDONNIER, or COBBLER-FISH has derived its popular name from the long sharp spines of the dorsal and anal fins, which are thought to resemble the awl and bristles employed by cobblers in their trade. This fish is a good example of the large genus to which it belongs, and in which no less than seventy species have been classed. It is found in various localities, from the Red Sea throughout all the Indian seas, and is tolerably common. The form of this fish is sufficiently curious to render it a conspicuous species, and it may be easily distinguished from its many congeners by the oblong spot on the operculum, and the six black bands that are drawn across the body and reach nearly to the abdomen.

ANOTHER species of this genus is the RUDDER-FISH (*Caranx carangus*), so called because it is fond of hovering about the rudders of vessels, apparently for the sake of picking up the refuse food that is thrown overboard. It is rather a pretty fish, the general colour being silvery white and blue. The lateral line is covered, near the tail, with a row of spinous plates. It is rather remarkable that this fish, when hooked, emits a rather loud chattering kind of noise, thought to proceed from the passage of air through the gills. The flesh of the Rudder-fish is rather coarse, but is digestible and nourishing. Another fish (*Pammelas perciformis*), found in the seas of Northern America, is sometimes called by the name of Rudder-fish.

CLOSELY allied to this fish is the well-known HORSE MACKEREL (*Trachúrus trachurus*) of our own coasts, sometimes known by the popular name of SCAD.

This species is common in the British seas, and occasionally appears in enormous shoals, almost rivalling in numbers those of the common mackerel, and crowding so closely against each other that they cannot escape if threatened by danger, and may be taken out of the sea by hand or dipped out in buckets. The flesh of the Horse Mackerel is rather coarse, and when fresh is held in very slight esteem. However, it readily takes salt, and is then much eaten, especially during the winter months.

The colour of the Horse Mackerel is dusky olive on the upper part of the back, changing in certain lights to resplendent green, which descends down the sides, and is variegated by wavy bands of blue. The sides of the head and the abdomen are silvery white. The lateral line is furnished with a row of strong and deeply keeled bony plates, which give to the hinder part of the body a somewhat squared outline.

THE well-known SWORD-FISH derives its popular name from the curious development of the snout, which projects forward, and is greatly prolonged, into a shape somewhat resembling a sword-blade. The "sword" is formed by the extension of certain bones belonging to the upper part of the head.

This fine fish is found in the Mediterranean Sea, and also in the Atlantic Ocean, and in the former locality is often very plentiful. The Sicilian fishermen are accustomed to pursue the Sword-fish in boats, and mostly employ the harpoon in its capture. The weapon is not very heavy, and by a strong and practised hand can be hurled to some distance.

The fishermen are accustomed to chant a kind of song, set to words which no one can understand, but which are supposed to be the more efficacious for their incomprehensibility. This song is thought by some writers to be a corruption of some old Greek verses, and the fishermen believe that the Sword-fish is so fond of this song that it follows the boat in which it is sung. They will not venture to speak one word of Italian, thinking that the Sword-fish would understand what they were saying, learn that they contemplated its death, and then dive and make its escape. No bait of any kind is employed, the unintelligible chant being thought to be far more efficacious than any material aid.

The flesh of the Sword-fish is always eatable and nourishing, and in small specimens is white and well-flavoured.

The use of the "sword" is not clearly ascertained. In all probability, the fish employs this curious weapon in gaining its subsistence, but the precise mode of so doing is not known. It is an ascertained fact that the Sword-fish will sometimes attack whales, and stab them deeply with its sharp beak ; and it is also known that this fish has several times driven its beak so deeply into a ship that the weapon has been broken off by the shock. In such cases, the blow is so severe, that the sailors have fancied that their vessel has struck upon a rock. Several museums possess examples of pierced planks and beams, but

SWORD-FISH. — *Xiphias gladius.*

it is possible that the fish may have struck them by accident, and not in a deliberate charge. The Sword-fish generally go in pairs.

The food of this creature is rather varied, consisting of cuttle-fish, especially the squid, and of small fishes, neither of which animals would in any way fall victims to the sword. It certainly has been said that the weapon is used for transfixing the flat fish as they lie on the bed of the sea, but this assertion does not appear to be worthy of credit.

The young and adult specimens are very different from each other. In the young, the body is covered with projecting tubercles, which gradually disappear as it increases in size, and when it has attained the length of three feet, they are seldom to be seen. Those on the abdomen remain longer than the others. The dorsal fin extends in the young specimens from the back of the head to the root of the tail, but the membranes and spines of its centre are so extremely delicate, that they are soon rubbed away, and the adult specimen then appears to have two dorsal fins.

The colour of the Sword-fish is bluish black above, and silvery white below. The whole body is rough, and the lateral line is almost invisible. The usual length of the Sword-fish is from ten to twelve feet, but specimens have been seen which much exceed those dimensions. A few examples of the Sword-fish have been captured in British waters; one, that measured seven feet in length, was taken off Margate.

THE very curious fish shown in the engraving is a representative of a genus of Sword-fishes that have been separated from the previous genus on account of the very great height of the dorsal fin.

The SAILOR SWORD-FISH is sometimes called the FAN-FISH, or SAIL-FISH, and is said to possess the power of raising or lowering the enormous dorsal fin just as a lady opens or closes her fan. Sir J. Emerson Tennent mentions this fish in the following terms:— "In the seas around Ceylon, Sword-fishes sometimes attain to the length of twenty feet, and are distinguished by the unusual height of the dorsal fin. Those both of the Atlantic and Mediterranean possess this fin in its full proportions only during the earlier stages of their growth. Its dimensions even then are much smaller than in the Indian species;

PURPLE-FINNED SAILOR-FISH.—*Histióphorus immaculátus.*

and it is a curious fact, that it gradually decreases as the fish approaches to maturity; whereas in the seas around Ceylon, it retains its full size throughout the entire period of life.   They raise it above the water whilst dashing along the surface in their rapid course, and there is no reason to doubt that it occasionally acts as a sail."

In this genus the ventral fins are reduced to one, two, or three spines, which in the present species are two in number.   The tail is very deeply forked, and the enormous dorsal fin is a uniform deep blue.

WE now arrive at the large family of the Gobies, which include many curious fish, and of which the British coasts present many representatives.

The BLACK GOBY, sometimes known as the ROCK-FISH, is a moderately common example of the enormous genus to which it belongs, and which contains more than a hundred and fifty authenticated species.   The members of this genus may easily be recognised by the peculiar form of the ventral fins, which are united together so as to form a hollow disc, by which they can attach themselves to rocks or stones at pleasure.   In fact, this disc, although differing in shape, acts on exactly the same principle as that of the sucking-fish.

The Black Goby prefers the rocky to the sandy coasts, and may be found in the pools left by the retreating tide.   Some naturalists deny that the disc is used for adhesion, but I have caught and kept many Gobies, and have frequently seen them sticking to the sides of the vessel in which they were confined.   The adhesion was achieved with astonishing rapidity, and the little fish contrived to hold itself with wonderful tenacity.   The surface of the Black Goby is very slippery, owing to the abundant mucous secretion which is poured from the appropriate glands, but after it has been in spirits for some time, the edges of the scales begin to project through the mucus, and are exceedingly rough to the touch.

Several species of Goby inhabit the British shores, such as the POLEWIG, or SPOTTED GOBY (*Gobius minútus*), a rather pretty little fish, transparent golden grey, with a multitude of tiny black dots upon the back, and generally marked with some darkish blotches upon

BLACK GOBY.—*Gobius niger.*

the sides, and a black spot on the dorsal fin. The TWO-SPOT GOBY (*Gobius Ruthen sparii*) is another British species, and may be distinguished by the two deep brown spots on either side, one just above the root of the pectoral fin, and another on the side of the tail.

In some places along the sea-coast, the Gobies are known by the popular appellation of Bull-routs, and are rather feared on account of the sharp bite which their strong jaws and pointed teeth can inflict upon the bare hand.

The general colour of this fish is blackish brown above, changing to white along the abdomen and under the chin. The length of this species seldom exceeds five or six inches.

THE pretty GEMMEOUS DRAGONET, FOX-FISH, SCULPIN, or GOWDIE, is another of the British fishes, and on account of its very remarkable shape, can easily be distinguished from any other species.

It is not a very uncommon fish, and is captured either with the hook or in a net, the latter being the ordinary method of securing it. It is rather a voracious fish, and feeds chiefly on molluscs and marine worms. The flesh of this species is firm, white, and well flavoured, and in spite its small size the Dragonet repays the trouble taken in its capture. It generally remains near the bottom of the sea, and does not often enter shallow water except when young, when it approaches the shore, and sometimes is taken in the net of the shrimper.

It is a lovely fish, well deserving its name of Gemmeous Dragonet, as its scales glitter as if set with gems, and of Gowdie, or golden, on account of the gilded lustre of its exterior. The name of Dragonet is given to it on account of the dragon-like aspect of the body and fins.

The colour of this beautiful fish is golden yellow of different shades, variegated with spots and streaks of sapphire upon the head and sides. The under surface is white. The first dorsal fin consists of four rays, the first being enormously lengthened, and reaching, if depressed, to the base of the tail. The succeeding rays rapidly diminish in length, the fourth being extremely short, barely an inch in length. The pectorals are rounded and triangular, the central ray being the largest. The length of the Gemmeous Dragonet is about ten or eleven inches.

More than twenty species of Dragonets are known, spread over a very large portion of the globe, and inhabiting the temperate seas of the Old World, and the Indian Ocean from

GEMMEOUS DRAGONET.—*Callionymus lyra.*

Mozambique to the Western Pacific islands. They are marine fishes, and inhabit the bottom of the sea at no great distance from the shore.

WE now come to a very small, but curious family, termed Discoboli, or Quoit-fishes, because the spines of the ventral fins are modified into a flattened disc, something like the quoit of the ancients. This disc has a soft, leathery margin, and enables them to attach themselves to rocks or stones after the manner of the gobies.

A very good British example of these curious fishes may be found in the LUMP-SUCKER, otherwise called the LUMP-FISH, SEA-OWL, and COCK-PAIDLE, the latter name being given to it on account of the elevated ridge along the back, which is covered with a notched and tuberculated skin not unlike the comb of the cock.

The sucker or disc of this fish is capable of very powerful adhesion, retaining its hold with such tenacity, that on one occasion, when a Lump-fish was placed in a pail containing several gallons of water, it immediately affixed itself to the bottom, and held so firmly, that when grasped by the tail and lifted, it raised the vessel in which it was placed, notwithstanding the combined weight of the water and pail.

The Lump-fish is said to make a kind of home, and to hover about the spot where the eggs are placed, for the purpose of guarding them from foes. When thus engaged it is a brave and combative fish, permitting no other finny inhabitant of the water to pass within a certain distance of its charge, and, in cases of necessity, biting fiercely with its short but sharp teeth. It is said that after the young have attained some little size, they attach themselves to their careful parent, who conveys the young family into deep water.

It is tolerably plentiful on the northern coasts of this country, and is frequently seen in the Scotch markets, where it holds a place only second to the turbot. The male is thought superior to the female, but is not so large. In the breeding season, the abdomen of the male fish assumes a bright red hue. It is a voracious creature, feeding mostly upon small fishes, molluscs, and crustaceans.

When it is freshly taken from the sea, the colours of the fish are truly magnificent, and even when suspended in the shops of the London fishmonger, its brilliant hues never fail to excite the wonder and admiration of the spectators. Blue is the prevailing tint of the upper parts of the body, but it is varied with a thousand bold and withal

LUMP-FISH.—*Cyclopterus lumpus.*

delicate shades of indigo, sapphire, and amethystine purple. The under parts of the body and the fins are rich orange-yellow. This splendid colouring is seen to greatest perfection during the breeding season.

The whole of the body is studded with little bony tubercles, which, when closely examined, are seen to be more or less star-shaped. Besides these little tubercles, there are four rows of larger and sharply pointed tubercles, one running along part of the back just behind the comb, two more along the sides, and another upon the abdomen. The dimensions of this fish are variable, but the average length is about sixteen inches.

THERE are only two genera in this small family, and both find examples in the British seas.

Of the second genus, the UNCTUOUS SUCKER or SEA-SNAIL (*Liparis vulgaris*) is a good illustration.

This species is found on most of the English coasts, but appears to be less common in the south than in the north. It derives its names of Unctuous Sucker and Sea-Snail from the soft and slime-covered surface of its body. It seems to prefer the rocky coasts, and may be found in the water-pools at low tide. The colour of this fish is pale brown streaked irregularly with a darker tint. Both the dorsal and anal fins are low, long, and reach to the commencement of the tail fin. It is a little fish, seldom exceeding four or five inches in length.

MONTAGUE'S SUCKER (*Liparis Montagui*) is remarkable for its habit of adhering to a stone or rock by the disc, and then curving its body to such an extent that the tail and the head almost meet. Even when merely lying at rest, and not employing the sucker, it assumes this remarkable attitude. It is smaller than the last species, rarely exceeding three inches in length. Its colour is rather dull orange above with bluish reflections, and white below. The fins are of a rather deep orange hue.

TOAD-FISH.—*Bátrachus grúnniens.*   FISHING-FROG.—*Lóphius piscatórius.*

ANOTHER small family now comes before us, called the Batrachidæ, or Frog-fishes, from the froggish aspect of the body and especially of the head.

The TOAD-FISH is a very curious-looking creature, with its flattened and wide head, gaping mouth, and spacious gill-cover. All the members of this genus are carnivorous fishes, and are spread through the coasts of the tropical regions, where they are mostly found on the bottom and partially buried in the sand or mud in hope of surprising the active prey on which they feed. Some species, however, are found even in the temperate seas.

The Toad-fish inhabits the East Indian seas, and has been taken at the mouth of the Ganges. Its colour is brown marked with a much darker tint, and the fins are streaked and blotched with similar colours. The body is without scales.

The FISHING-FROG, ANGLER-FISH, or WIDE-GAB, is not unfrequent on the British coasts, and has long been famous for the habit from which it has derived its popular name.

The first dorsal fin is almost wholly wanting, its place being occupied merely by three spines, movable by means of certain muscles. The manner in which these spines are connected with the body is truly marvellous. The first, which is furnished at its tip with a loose shining slip of membrane, is developed at its base into a ring, through which passes a staple of bone that proceeds from the head. The reader may obtain a very perfect idea of this beautiful piece of mechanism by taking a common iron skewer, slipping a staple through its ring and driving the staple into a board. It will be then seen that the skewer is capable of free motion in every direction.

The second spine is arranged after a somewhat similar fashion, but is only capable of being moved backwards and forwards. Fishing-Frogs are sometimes found in the shops, and the inquiring reader will find himself amply repaid if he purchases one of these fishes and dissects its head merely for the purpose of seeing the beautiful structure which has been briefly described.

The use of these spines is no less remarkable than their form.

The Fishing-Frog is not a rapid swimmer, and would have but little success if it were to chase the swift and active fishes on which it feeds. It, therefore, buries itself in the muddy sand, and continually waves the long filaments with their glittering tips. The neighbouring fish, following the instincts of their inquisitive nature, come to examine the curious object, and are suddenly snapped up in the wide jaws of their hidden foe. Many fishes can be attracted by any glittering object moved gently in the water, and it is well known by anglers how deadly a bait is formed of a spoon-shaped piece of polished metal, furnished with hooks, and drawn quickly through the water.

It is a most voracious creature, and has on several occasions been known to seize a fish that had been hooked and was being drawn to the surface. In one such case, the Angler seized on a cod-fish, and held so tightly that it would not loosen its grip until struck on the head with a boat-hook. On another occasion the fish fell a victim to its over-voracity, for having dashed at a conger-eel, just hooked, and taken it into its mouth, the eel contrived to escape through one of the gill apertures, and thus was the unconscious means of involving its captor in its own fate. Even the cork-floats on lines and nets have been swallowed by the greedy fish, and when taken in a net, it devours its fellow-prisoners with perfect unconcern.

It is impossible to mistake this fish for any other inhabitant of the ocean, its huge head—wide, flattened, and toad-like—its enormous and gaping mouth, with the rows of sharply-pointed teeth, its eyes set on the top of the head, and the three long spines, being signs which cannot be misunderstood. The general colour of this fish is brown above and white below; the ventral and pectoral fins are nearly white, and that of the tail almost black. The throat, just within the jaws, is composed of loose skin, which forms a kind of bag. The average length of the adult Fishing-Frog is about a yard.

The family in which this fish is placed may be distinguished by the peculiar structure of the pectoral fins, which are mounted on a sort of arm produced by an elongation of the carpal bones. From this peculiarity, the family is termed Pediculati, or foot-bearing fishes, as the prolonged fins enable them to walk along wet ground almost like quadrupeds.

THE very odd-looking creature, called the WALKING-FISH, which is shown in the accompanying illustration, is one of the strange and wild forms that sometimes occur in nature, and which are so entirely opposed to all preconceived ideas, that they appear rather to be the composition of human ingenuity than beings actually existing. The traveller who first discovered this remarkable fish would certainly have been disbelieved if he had contented himself with making a drawing of it, and had not satisfied the rigid scrutiny of scientific men by bringing home a preserved specimen.

In the fishes of this genus, the carpal bones, i.e. those bones which represent the wrist in man, are very greatly lengthened, more so than in the preceding genus, and at their extremity are placed the pectoral fins, which are short, stiff, and powerful, the pointed rays resembling claws rather than fins. In all the fishes of this genus the body is much compressed and decidedly elevated; but in the present species, these peculiarities are carried to an almost exaggerated extent. The first dorsal spine, with its membranous appendages, is placed as usual just above the snout, and the second ray is set immediately behind it. The third, however, is placed at a very great distance from the second, and forms part of the soft dorsal fin.

Dr. Günther remarks upon the fishes of this genus, that they are so extremely variable in form, colour, and the greater or less development of the dorsal spines, that hardly two specimens are found sufficiently alike to enable the systematic naturalist to decide upon their precise situation in the zoological scale. Moreover, their geographical range is

WALKING-FISH.—*Antennarius hispidus.*

exceedingly wide, some species ranging over the Atlantic and Indian Oceans; and the learned ichthyologist above mentioned is of opinion that many specimens which he has at present been compelled to admit into the list of separate species, will be ultimately found to be mere casual varieties.

The colour of this species is yellow, diversified with many spots and streaks of brown, some of the streaks radiating from the eye, and others extending over the dorsal fin. It is a native of the Indian seas.

THE important family of the Blennies comes next in order. They are all carnivorous fishes, many being extremely voracious, and are spread over the shores of every sea on the globe. They mostly reside on or near the bottom.

The SEA WOLF, SEA CAT, or SWINE-FISH, is one of the fiercest and most formidable of the finny tribes that are found on our coast, and has well earned the popular names by which it is known.

The Sea Wolf possesses a terrible armature of teeth, not only in the jaws, but arranged in a double band on the palate, and by means of these powerful weapons can crush with ease the hard-shelled molluscs and crustaceans on which it feeds. As may be seen from the engraving, the aspect of the Sea Wolf is far from prepossessing, its fierce head with the armed jaws, strong and cruel as those of the tiger or hyæna, and the smooth, slime-covered skin, giving it a most repulsive aspect.

Still, it is eaten in many places, especially when it is small, not more than two feet in length, and the flesh is said by competent judges to be decidedly excellent. In order to avoid disgusting the purchaser with its ugly looks, the head is mostly removed, and the skin stripped off before it is exposed for sale. The skin, though not handsome, has yet its uses, for it is strong, flexible, and durable, and is made into bags and pockets that require peculiar strength of fabric.

The Sea Wolf is sometimes taken with the hook, but is mostly found entangled in the nets together with other fish, and in either case it struggles violently as soon as it perceives the loss of its liberty. It will tear the nets to pieces with its teeth, and when hauled out of the water, it still flounces about with such vigour, and bites at every object

3. U

with such ferocity, that the boatmen usually stun it by a blow on the head before lifting it into the vessel, a very heavy stroke being required for the purpose.

The general colour of the Sea Wolf is brownish grey, with a series of brown vertical stripes and spots over the upper parts; the under parts are white. On our shores it attains a length of six or seven feet, but in the northern seas, where it thrives best, it greatly exceeds those dimensions. There is an American variety where the vertical streaks are modified into round spots of blackish brown.

THE typical genus of this family is represented by several British specimens, of which the EYED BLENNY is one of the most conspicuous.

This pretty fish is not very common, but has been taken on the southern coasts of England. From the elevated dorsal fin, and the bold dark brown spot that decorates it, this Blenny has sometimes been called the Butterfly-fish. In the Mediterranean it is tolerably common, and lives mostly among the seaweed, where it finds abundance of the smaller crustacea and molluscs.

The dorsal fin of this fish is very large, being greatly elevated and extending from the back of the head almost to the tail. The dark spot is placed between the sixth and eighth rays. The colour of the Eyed Blenny is pale brown, patched here and there with a darker tint. The dark spot on the fin is mostly edged with white or very pale yellow. The length of this fish is seldom more than three inches.

SEVERAL other British species of blenny are acknowledged, and some are very well known upon our coasts.

The SHANNY, or SHAN (*Blénnius pholis*), is a tolerably common species, and its habits have been thus recorded by Mr. Couch : " Destitute of a swimming-bladder, this fish is confined to the bottom, where it takes up its residence on a rock or stone, from which it rarely wanders far, and beneath which it seeks shelter from ravenous fishes and birds ; for cormorants, with their long and sharp beaks, drag multitudes of them from their retreats, and devour them. When the tide is receding, many of these fishes hide beneath the stones or in pools, but the larger individuals quit the water, and by the use of the pectoral fins creep into convenient holes, rarely more than one in each, and there, with the head outward, they wait for a few hours, until the return of the water restores them to liberty. If discovered or alarmed in their chambers, they retire by a backward motion to the bottom of the cavity. These circumstances show that the Shanny is retentive of life, in confirmation of which I have known it continue lively after a confinement of thirty hours in a dry box ; notwithstanding which, it soon expires in fresh water."

This species is extremely variable in colour, some specimens being mottled with different shades of brown, and others of a uniform dusky tint. It may be recognised by the deep notch or slit in the middle of the dorsal fin, and the absence of appendages on the head. The length of this Shanny is about five inches.

PASSING by the remaining blennies, all of which are very similar in habits and general appearance, we must pause for a short space to examine a very curious species belonging to the same family, called the JUMPER-FISH (*Salárias tridáctylus*).

·This odd little fish offers no remarkable beauties of colour or form, being of a simple dark brown, and without any salient points of external structure ; but it is possessed of a wonderful power of suddenly leaping out of the water, darting over the wet stones and rocks, and snapping up flies and other insects with the nimble agility of the lizard. It can scramble up a nearly perpendicular face of rock, and is so wary and agile, that on the least attempt to seize it, the little creature darts towards the sea and is nearly certain to make its escape. While engaged in this pursuit, the Jumper-fish adheres so tightly to the rock that it is not detached even by the shock of repeated waves. It is quite a little fish, not more than four inches in length. Its residence is in the seas of the East Indian Archipelago. At least fifty species of the Salarias are known to zoologists.

EYED BLENNY.—*Blennius ocellaris.*
VIVIPAROUS BLENNY.—*Zoarces viviparus.*
SEA WOLF.—*Anárrhichas lupus.*

The BUTTER-FISH, SWORDICK, or SPOTTED GUNNEL (*Centronótus gunellus*), belongs to this family, and is evidently one of the transitional species between the true blennies and those which are placed at the end of the family.

This fish is frequently captured upon our coasts, especially on the rocky shores, and is mostly found hidden under stones and seaweeds in the rock-pools left by the receding tide. The name of Butter-fish is very appropriate, and is given to it on account of the plentiful mucous secretion which is poured over its body, and which renders it so slippery that it can with difficulty be retained in the hand. It is quick and agile in its movements, and even if confined within the limits of the rocky pool, is not easily captured.

The body of this fish is much elongated and somewhat eel-shaped, the head is small, the muzzle blunt, and the dorsal fin is low and long, extending the whole length of the back. The ventral fins are very small. The colour of the Swordick is brown, in some specimens with a purple and in others with a golden wash. Along the base of the dorsal fin, and in some individuals upon the fin itself, are a number of bold black spots each with a white streak on either side. A dark brown stripe is also drawn from the eye to the lower jaw. The length of the Butter-fish is about six inches.

OUR last example of this family is the well-known VIVIPAROUS BLENNY, called also by the popular names of EEL-POUT, LUMPER, GUFFER, and GREENBONE, the last-mentioned title being given to it because, when boiled, the bones have a green hue.

As its name imports, the Viviparous Blenny lays no spawn, but produces its young alive, and able to shift for themselves.   In one case, where a female fish of about fifteen inches in length was taken, the young were about four inches long.   It is a very curious fact, that the size of the new-born young seems to depend upon that of their parent, the offspring of a Blenny of seven inches in length measuring only one inch and a half.

The flesh of this fish is tolerably good, but is not in very great repute, so that it is but seldom to be seen in the markets.   According to the most careful observations, it appears to be less common in the south than in the north of England, and while it is plentiful on the Yorkshire coasts, is scarcely to be found upon those of Devonshire.   In such cases, however, the apparent discrepancy is often attributable to the differing abilities of the observers, and not to the absolute abundance or scarcity of the species.   It generally hides itself under stones or seaweed, preferring the large heavy algæ called tang.

The body of this fish tapers gradually from the shoulders to the tail, in thickness as well as in depth, and when examined with a pocket magnifier, the surface appears to be studded with circular depressions.   Its general colour is pale brown, and its length varies between six and sixteen inches.

PASSING by several small families, we come to a very curious fish, denominated the RIBAND-SHAPED VAAGMÄR, sometimes called the DEAL-FISH (*Trachypterus árcticus*).

This singular fish is remarkable for the extreme compression of the body, a specimen three feet in length not being thicker than an ivory paper-knife.   The dorsal fin of this fish extends completely along the back, there is no anal fin, and the tail fin stands boldly erect, like the closed tail-feathers of a fan-tail pigeon.   The general colour of the Vaagmär is silvery white, and the body is covered with very small scales.   The dorsal fin is bright orange, sometimes being of a blood red, and the tail fin is of the same hue.   On each side are two oval spots of blackish grey, set obliquely on the body.   The length of this fish often reaches six feet.

It is one of the northern fishes, and is very seldom seen on our coasts.

A SPECIES even still more remarkable is, on very rare occasions, obtained on our coasts, but owing to the extreme fragility of its structure it is mostly deficient in some of its parts.

The OARED GYMNETRUS, or RIBBON-FISH (*Regalecus Banksii*), is also greatly compressed throughout its length, and is equally delicate with the last-mentioned species.   It is chiefly notable for the very odd structure of the ventral fins, which are reduced to long slender filaments, much resembling in shape the long tail-feathers of the racket-tail humming-bird.   This fish sometimes attains very great dimensions, a specimen in the British Museum measuring twelve feet in length.   Its colour is silvery grey, mottled with dusky spots of varying depth, which are most conspicuous towards the head.   The whole surface of the skin is plentifully studded with bony tubercles, and on the line of the abdomen each tubercle is furnished with a hooked point directed backwards.   Along the lateral line runs a row of elongated flat scales.

IN the next family, the tail is mostly armed with one or more bony spines or plates, small in the young, but increasing in size with the dimensions of the fish.

The SEA SURGEON is a good type of these fishes, and derives its popular name from the sharply pointed and keen-edged spine on the side of the tail, which cuts and wounds like a surgeon's lancet.   The generic name signifying Thorn-tail is given to it in consequence of this structure.   This species is found on the Atlantic coasts of Tropical America and Africa, and is tolerably plentiful in the Caribbean seas.   The scales of this fish are very small, and the single spine on each side of the tail is movable and set in a longitudinal groove.   Its food is of a vegetable nature.

In colour it is rather variable, but the ground tint is usually of a brownish hue, and the operculum has a black edge.   In some specimens the end of the tail is marked with a white band, which encroaches on part of the tail fin, and there is also a narrow white edge to that fin. There are in certain individuals a few darkish streaks drawn across the body, some black longitudinal stripes on the dorsal and anal fins, and in the young the sides are

SEA SURGEON.—*Acanthurus chirurgus.*

marked with darkish waving lines. This fish sometimes attains a rather large size, a specimen in the British Museum being nineteen inches in length. The genus is rather comprehensive, containing between forty and fifty known species.

NEARLY allied to the surgeon-fish is a very curious species, called the UNICORN THORN-TAIL (*Naseus unicornis*), on account of the singular structure of the forehead, which is modified in front into a long and horn-like protuberance, rather conical in shape, and projecting forwards in a line with the body. This horn is not to be seen in the young fish, and only attains its full dimensions when its owner has reached adult age. Sometimes the horn is longer than the snout, but in most specimens it is slightly shorter. Each side of the tail is furnished with two lancet-bearing plates, which are not movable.

This species is found from the Red Sea to Japan and Polynesia. Its colour is brownish grey, and the dorsal and anal fins are marked with longitudinal blue stripes. The largest specimen in the British Museum measures twenty-two inches in length, and its horn is three inches long.

THE extraordinary fish, called, from its habits, the CLIMBING PERCH, is a native of Asia, and is remarkable for its apparent disregard of certain natural laws.

This singular creature has long been celebrated for its powers of voluntarily leaving the failing streams, ascending the banks, and proceeding over dry land towards some spot where its unerring instinct warns it that water is yet to be found. There are several fish which are known to have this power; the common eel of England, for example, which has frequently been observed crossing the fields in its passage from one stream to another. I have even seen the eels creeping over rocks, and contriving, in some mysterious manner, to crawl along the flat horizontal surface of an overhanging rock as easily as a fly walks on the ceiling. But I believe that the eel only passes over moist ground, whereas the Anabas seems quite indifferent to such considerations, and takes its journey over hard, dry, and dusty roads, heated with the burning beams of the noonday sun, without appearing to feel much inconvenience from the strange nature of the transit.

Several species, of which the Anabas Scandens has been chosen as the best example,

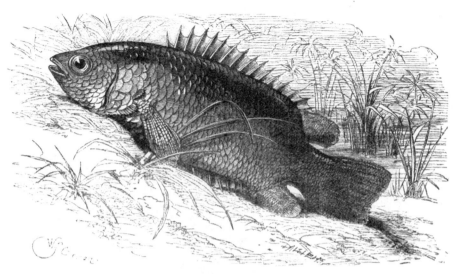

CLIMBING PERCH.—*Anabas scandens*.

possess this singular property of walking over dry ground, so that the old proverb of a fish out of water is, in these cases, quite inapplicable. Several instances of this remarkable propensity have been collected by Sir J. Emerson Tennent, and have been inserted in his valuable work on the Natural History of Ceylon. The following account is written by Mr. Morris, the Government agent in Trincomalee :—

"I was lately on duty inspecting the bund of a large tank at Nade-cadua, which being out of repair, the remaining water was confined in a small hollow in the otherwise dry bed. Whilst there, heavy rains came on, and as we stood on the high ground, we observed a pelican on the margin of the shallow pool gorging himself: our people went towards him, and raised a cry of Fish ! fish ! We hurried down, and found numbers of fish struggling upward through the grass, in the rills formed by the trickling of the rain. There was scarcely water to cover them, but nevertheless they made rapid progress up the bank, on which our followers collected about two baskets of them at a distance of about forty yards from the tank. They were forcing their way up the knoll, and had they not been interrupted first by the pelican and afterwards by ourselves, they would in a few minutes have gained the highest point, and descended on the other side into a pool which formed another portion of the tank.

. . . As the tanks dry up, the fish congregate in the little pools, till at last you find them in thousands in the moistest parts of the beds, rolling in the blue mud, which is at that time about the consistence of thick gruel.

As the moisture further evaporates, the surface fish are left uncovered, and they crawl away in search of fresh pools. In one place I saw hundreds diverging in every direction from the tank they had just abandoned, to a distance of fifty or sixty yards, and still travelling onwards. In going this distance, however, they must have used muscular exertion enough to have taken them half a mile on level ground, for at these places all the cattle and wild animals of the neighbourhood had latterly come to drink, so that the surface was everywhere indented with footmarks in addition to the cracks in the surrounding baked mud, into which the fish tumbled in their progress. In those holes which were deep, and the sides perpendicular, they remained to die, and were carried off by kites and crows.

My impression is, that this migration must take place at night or before sunrise, for it was only early in the morning that I have seen them progressing, and I found that those I brought away with me in the chatties appeared quiet by day, but a large proportion managed to get out of the chatties by night—some escaped altogether, others were trodden on and killed.

One peculiarity is the large size of the vertebral column, quite disproportioned to the bulk of the fish. I particularly noticed that all in the act of migrating had their gills expanded."

It is known of the Climbing Perch that the fishermen of the Ganges, who subsist largely on these fishes, are accustomed to put them into an earthen pan or chatty as soon as caught; and although no water is supplied to them, they exist very well without it, and live this strange life for five or six days.

On opening the head of this fish, the curious structure which enables it to perform such marvellous feats is clearly seen. Just within the sides of the head, the "pharyngeal" bones, *i.e.* the bones that support the orifice between the mouth and gullet, are much enlarged, and modified into a series of labyrinthine cells and duplications, so that they retain a large amount of water in the interstices, and prevent the gill-membranes from becoming dry. Some writers say that this fish is capable of climbing up the rough stems of palm-trees, in search of the water that lodges between the bases of the dead leaves and the stem, but this account is now held unworthy of belief. In the Tamoule language it is called Paneiri, or Tree-climber.

THE small genus Atherinidæ has a British representative in the SAND SMELT (*Atherina presbyter*), a pretty little fish, and one that is of great use to fishermen, both for sale and for bait.

It is extremely plentiful upon the southern coast, and in many places is sold as the true smelt, which it somewhat resembles in flavour and the peculiar odour as of cucumber. Owing to the small size of this fish the net is the usual mode of capture, the fashion of which varies according to the locality. On some coasts the net is about ninety feet in length and eighteen in depth, and is drawn along the sands by the united aid of one party in a boat and the other on the shore. In other places, however, it is circular and supported on an iron hoop. It is then baited with broken crustacea and lowered into the water. At intervals it is raised smartly to the surface, and the entrapped Sand Smelts removed.

The colour of the Sand Smelt is the palest pink, diversified with a broad belt of shining silvery white, which is drawn along the side. The cheeks, gill-covers, and the base of each pectoral fin are of the same white hue. Upon the upper part of the back and head are a great number of little black spots. The length of the fish is from six to seven inches.

WE now come to the important family of the Mugilidæ, of which the common GREY MULLET is a good example. In all these fish there are two dorsal fins, the first having four stiff spines. They are spread over all sea-coasts and fresh waters of the temperate and tropical regions. The mode of feeding is rather curious. These fish live chiefly on the soft organic substances that are found mixed with weed and sand, and in swallowing the food a considerable amount of sand is taken into the mouth. The fish, however, is furnished with a kind of self-filtering apparatus, by means of which the heterogeneous mass is raked and sifted, as it were, and the indigestible portions rejected.

The Grey Mullet deserves notice as being one of the most daring and ingenious of the finny race, and is, in fact, a very fox for artfulness. The idea of constraint is most obnoxious to it, and its instincts of freedom are so strongly developed that it endeavours to recover its liberty in the most extraordinary ways.

If, for example, it has been inclosed in a net, it will at once dart to the side and try to leap over the head-rope into the open sea. Moreover, if one fish succeeds in the attempt, the remainder immediately follow their leader, like a flock of sheep jumping over a hurdle. If the net is raised so high that the leap is impracticable, the fish tries to

GREY MULLET.—*Mugil cápito.*

creep under it; and if that mode of escape be cut off, it examines every mesh, in hopes of finding some defective spot through which it may insinuate itself. Mr. Couch mentions that he has seen a Grey Mullet, after trying all other modes of escape, deliberately retire to the greatest possible distance from the wall of net, and then dash furiously at the meshes, as if to break through them.

It likes a frequent change from salt to fresh water, and often proceeds up rivers to some little distance, returning, however, with the tide. It has even been taken from the sea and placed in a fresh-water pond, where it has grown well and obtained a great weight in proportion to its length. While feeding, it may be seen rooting in the mud in a very swine-like fashion; and can mostly be captured by using a bait composed of boiled meat or vegetables. It is, however, not easily hooked, as its peculiar mode of feeding causes it to detect the hard hook and to keep it so slightly within the mouth that the lip only is caught, and will mostly yield to the struggles of the fish. The fly, however, is often used successfully; and it is found that the Grey Mullet will bite even at the great gaudy salmon-flies, so large and so unlike any fly of any climate that they would hardly be thought capable of deceiving a mole.

The colour of the Grey Mullet is bluish grey above, and the sides and abdomen are white with a silvery lustre, and traversed with several longitudinal lines of darkish grey. There is a dark spot towards the base of the pectoral fin.

The genus Mugil is very large, containing between sixty and seventy species, several of which are found in British waters.

THE fishes belonging to the family of the Ophiocephalidæ, or snake-headed fishes, are able to leave the water for a time and to crawl upon land, deriving their power from a curious structure of the breathing organs. It has already been stated that a fish can breathe as long as the delicate membranes of the gills are wet; and that in those fishes which are able to live out of water for any length of time, a peculiar modification of the breathing organs is requisite in order to supply the needful moisture. In the family to which the climbing perch belongs, a series of thin laminated plates are arranged in a cavity above the gills, thus retaining a sufficient supply of water between the laminæ. In the present genus, however, there are none of these laminæ, but the water is retained in a simple cavity which communicates with the gills.

Of this family, the CORA-MOTA, or GACHUA (*Ophiocéphalus gachua*), is a good example.

This fish is a native of the fresh waters of Eastern India and its archipelago, and in its general shape and movements is so very snake-like that Europeans will seldom eat it. The Cora-mota is common in the ponds and dykes of Bengal; and is one of the fish popularly supposed to be rained from the clouds, as it is generally to be found on the grass after a heavy shower. However this may be in other instances, it is tolerably clear that the Cora-mota has been in concealment during the drought, and ventures into the fresh wet grass as a welcome change from the muddy ditches in which it has been forced to reside. It can also find a plentiful supply of food on the moist herbage; and as on account of its peculiar formation it is able to move on land with considerable ease, its migrations will often extend to considerable distances.

The Cora-mota is remarkably tenacious of life, and can survive the severest wounds for a wonderfully long period. The natives of India take advantage of this peculiarity, and with the disregard of inflicting torture that seems to be inherent in the Oriental mind, are in the habit of selling the fish piecemeal, and cutting it up for sale while still living. Indeed, the *habitués* of the market will not pay the best price if the fish does not flinch from the knife.

The colour of this species is brown crossed with several dark bars. Its length seldom exceeds a foot.

ANOTHER species of this genus, the BARCA (*Ophiocephalus barca*), is a much handsomer fish, attains a considerable size, and is considered to be useful for the table. This fish is one of the mud-lovers, living for the most part in holes excavated in the banks of Indian rivers, and only putting out its head in search of prey.

The colour of this species is violet spotted profusely with black, and the fins are marked with sundry bold bars and dots. In length it often attains three feet.

THE remarkable BAND-FISH, or SNAKE-FISH (*Cépola rubescens*), is a British example of a curious family, consisting of one genus only, and about seven species.

The Band-fish is not uncommon in the Mediterranean, though it is seldom taken off the English coasts. Its body is long and much compressed, like that of the vaagmär, already described; and when winding its way through the translucent water, its carmine body with the glittering scaly mail have earned for it the popular names of FIRE-FLAME and RED RIBAND.

Little is known of its habits, except that it is a shore-loving fish, delighting to bask under the heavy masses of sheltering seaweed, and that it feeds mostly on molluscs and crustacea. Several specimens of this fish have been found on the beach after a storm; and Mr. Yarrell remarks, with some acumen, that all the fish formed after this pattern, with their compressed bodies affording little resistance to the water, and their length preventing the concentration of muscular force upon a single centre of motion, are ill fitted for combating tempestuous waters, and are flung about at the mercy of the waves.

The head of the Band-fish is small, and the eye is full and very large, its diameter being nearly half the depth of the head. The body is greatly compressed, slender, and very smooth; the scales being minute and glittering in the sunbeams. The dorsal fin extends from the top of the head to the end of the tail, and the anal fin is nearly as long. Its colour is rather variable, shades of purple and orange exhibiting themselves in certain specimens. In all examples, however, red is the predominant hue. The length of the adult Band-fish is usually about fifteen or twenty inches.

IN the curious species which belong to the genus Centriscidæ, or spike-bearing fishes, the body is much compressed, and one of the spines of the first dorsal fin is long, sharp, and powerful. The bones which form the front of the head are greatly prolonged, and are modified into a kind of long tube, at the end of which is placed the narrow mouth. It is thought that the fish obtains its food by sucking it along the tube, the needful vacuum being formed by the dilatation of the throat.

The BELLOWS-FISH, sometimes called the TRUMPET-FISH and the SEA SNIPE, is most common in the Mediterranean, though a few casual specimens have been taken off the British islands. It prefers to reside in moderately deep water, and is mostly found where the bottom of the sea is muddy. Its food is not precisely known, but is thought to consist of minute marine animals. The first spine of the dorsal fin is enormously large, strong, sharply pointed, and armed on its under surface with a row of saw-like teeth, that

BELLOWS-FISH. — *Centriscus scolopax.*

must render it a very efficient weapon of offence. This spine is also movable. The flesh of this fish is eatable; but as the head occupies so large a portion, the amount of flesh is rather small when compared with the size of the fish.

When adult, the colour of the Bellows-fish is bright red on the back, becoming lighter on the sides, and changing to silvery white and golden yellow on the abdomen. The sides of the head are of the same colour as the abdomen. When young, the red of the back and sides is not visible, and the whole body glitters with a silvery lustre. It is not a large fish.

THE family which now comes before our notice is in many ways remarkable, and deserves some little attention before proceeding to the remaining fishes.

In the Fistularidæ the snout is greatly prolonged, as in the preceding family, and bears the mouth at the end of a bony tube. The body, however, is extremely long and snake-like, and there is no long spine to the dorsal fin. There are only two genera in this family, the one being covered with scales, and the other destitute of those appendages.

The TOBACCO-PIPE FISH is found in several parts of the tropical Atlantic, and is notable for its very peculiar form. The body is without scales, and the tail fin is deeply forked, the two central rays being sometimes united and prolonged into a lengthened filament, and at others separate, but still elongated. The outer edge of the tube is either smooth, or very slightly notched. The colour of this fish is greenish olive, and the upper parts of the body are marked with blue streaks and spots. In some specimens the back takes a reddish brown hue.

THERE is a curious family of fishes, termed the Mastacembelidæ, in which the body is long and eel-like, covered with little scales, and remarkable for the odd-looking snout and its appendage.

In these strange-looking fishes, of which the SPOTTED MASTACEMBELUS (*Mastacémbelus maculátus*) is a good example, the dorsal fin is very long, its front portion consisting of a number of short free spines. The anal fin is also furnished with similar spines, and the ventral fins are altogether wanting. The gill-openings are reduced to a narrow slit, and the movable appendage of the upper jaw is smooth on its under side. The jaws are furnished with minute teeth, and the lower jaw is but slightly movable. In all the

TOBACCO-PIPE FISH.—*Fistuláriu tabaccaria.*

species of this genus, with the exception of the Spotted Mastacembelus, the præoperculum is armed at its angle with small teeth.

This species is found in the fresh waters of Java and Sumatra. The dorsal fin joins that of the tail, which is again joined by the long anal fin. The colour of the fish is brown, diversified with darker blotches, and the fins are edged with yellow.

THE Flat Fishes, as they are popularly called, or the Pleuronectidæ, as they are named scientifically, are among the most remarkable of the finny tribe. The latter name is of Greek origin, and signifies side-swimmer, in allusion to the mode of progression usually adopted by these fishes.

The popular but erroneous idea of these creatures is, that their bodies are flattened so that the abdomen rests on the ground and the back remains uppermost; but a brief examination suffices to show that the form of these fishes is really compressed, so that when a turbot or a sole is placed on the ground, it lies upon one side or the other. Though varying in colour, it is found that the upper side is always of a dark tint, the under side being quite if not wholly white. This arrangement is most useful in guarding them against the attacks of enemies, their flat dark upper surface bearing so great a resemblance to the sand on which they love to creep, that they can scarcely be distinguished, even when the eye is directed towards them.

While at their ease, they slide themselves over the bed of the sea in a kind of creeping fashion, and have an odd custom of lying with the head raised in a manner that irresistibly reminds the observer of the cobra. If alarmed, they start at once into the vertical position usually assumed by fishes, and dash off with astonishing speed. As they swim, the flat fishes undulate through the water in a most graceful manner, and it is very

SOLE — *Solea vulgaris*

interesting to watch one of the common flat fishes, such as the plaice or the sole, swim with serpentine ease and elegance, and then suffer itself to sink slowly to the bottom, where it sits, with raised head and watchful eyes.

It is evident that if the eyes of the flat fishes were placed in the manner customary among the finny race, one eye would be rendered useless as long as the fish was lying upon its side. This difficulty is therefore met by a most extraordinary modification of the bones of the head, by which means both the eyes are brought to that side which remains uppermost, and are thereby enabled to command a wide view around. There have been one or two instances where the eyes have been placed one on each side, but these may be considered as simple variations from the ordinary rule.

The COMMON SOLE is one of the most familiar of our British flat fishes, and is found on all our coasts, those of the southern shores being the most plentiful, and attaining the largest dimensions.

The Sole can be taken by the line, but the fishermen always use the trawl-net, a kind of huge dredge, with a mouth that often exceeds thirty feet in width. As these nets are drawn along the bed of the sea, the great beam which edges the mouth scrapes the mud and sand, and alarms the fish to such an extent that they dash wildly about, and mostly dart into the net, whence they never escape. Vast numbers of Soles are taken by this method of fishing, and as the trawls bring to the surface enormous quantities of crustaceans, molluscs, zoophytes, and other marine inhabitants, the energetic naturalist cannot employ his time better than in taking a sail in one of these boats, and enduring a few hours' inconvenience for the sake of the rich harvest which he is sure to reap. Some of the rarest and most valuable British animals have been taken in the trawl-nets.

The Sole is in condition throughout the greater part of the year, the only time when it is not worth eating being from the end of February to the last week in March, when the fishes are full of roe, and the flesh is rather soft and watery. It is a hardy fish, and can soon be acclimatized to live in fresh water; and it is said that under such circumstances the fish can be readily fattened, and becomes nearly twice as thick as when bred in the sea. Sometimes the Soles venture into the mouths of rivers, passing about four or five miles into the fresh water, and depositing their multitudinous eggs in such localities.

The colour of the Sole is almost always brown on the right side and white on the left, but examples of reversed Soles are not uncommon, where the left side is brown and the other is white. The scales are small, and give a rough, rasp-like sensation to the hand. The dimensions of this fish are very variable, an average specimen weighing about a pound or eighteen ounces. Much larger examples, however, occur occasionally, and Mr. Yarrell mentions one instance where a Sole measured twenty-six inches in length, eleven and a half inches in width, and weighed nine pounds.

ZEBRA SOLE.—*Solea zebrina.*

The ZEBRA SOLE is a native of Japanese waters, and is remarkable for the waving dark streaks with which its body is covered, and which bear a great resemblance to the stripes upon the zebra's hide. In habits it appears to resemble the common species.

SEVERAL species of Sole are found upon the British coast. The LEMON SOLE, or FRENCH SOLE (*Solea pegusa*), derives the former of these titles from the lemon-yellow colour of its upper surface, and the latter from the localities in which it is most commonly found. It is found generally about sixteen miles off our coasts. The colour of this fish is orange, mixed with light brown, and mottled with little round spots of wood-brown. It is wider in proportion to its length than the common Sole. Another species, the VARIEGATED SOLE (*Solea variegata*), is sometimes, though rarely, taken off our shores. It may be known by the reddish brown colour, clouded with dark brown. The body is rather thick in proportion to its length.

The last British species is the SOLENETTE, or LITTLE SOLE (*Monochirus linguátulus*), a small species, seldom more than five inches long, and of a reddish brown colour, without cloudings.

PERHAPS the most remarkable of these fishes is the TRANSPARENT SOLE (*Achírus pellúcidus*).

This rare and interesting fish is a native of the Pacific Ocean, and is notable for the extreme pellucidity of its body, which is so marvellously transparent, that when swimming in a vase of water, or lying on the bottom, the algæ or stones can be distinctly seen through its structures. It is quite colourless, except a very slender and very delicate pink streak on the edge of the back, and several similar lines upon the sides; the perfect but glass-like skeleton is hardly to be detected, and even the viscera are almost invisible. It is a very little fish, appearing not to exceed two inches in length; but its width is proportionately great, so that the fish assumes a nearly circular form. The eyes are silvery white, and the pectoral fins are wholly absent.

THE well-known TURBOT, so widely and so worthily celebrated for the firm delicacy of its flesh, inhabits many of the European coasts, and is found in tolerable abundance off our own shores. Like all flat fishes, it mostly haunts the sandy bed of the sea, but will sometimes swim boldly to the surface of the water. It is a restless and wandering fish, traversing considerable distances as it feeds, and generally moving in small companies.

Two modes of catching the Turbot are employed by fishermen, namely, the trawl-net and the long line. As long as the fish remain on the banks, or tolerably near the shore, the net is used, and in its capacious mouth is taken a strange medley of fishes, among which the Turbot is generally plentiful.

Should, however, the Turbot retire into deep water, or should the weather be too rough for the management of the net, the fishermen employ the line for its capture. The Turbot is gifted by nature with a fine and discriminating appetite, and voracious as it is, it refuses to touch any bait that is not quite fresh, and is said to reject it if any other fish has even bitten it. Certain small fishes are in great repute, especially those which glitter with a silvery lustre. Formerly the lampern of the rivers was extensively used as bait, as its skin is smooth and shiny, and it can be kept alive for a considerable time. The atherine, sea scorpion, and father-lasher are now, however, the principal favourites with the fishermen. The Turbot feeds upon molluscs and crustaceans besides fish.

The Turbot is known in Scotland by the title of BANNOCK FLEUK, or SPAWN FLEUK, the former name being given to it on account of its flat shape, which resembles a bannock or oatcake, and the latter because it is thought to be at the best while in roe. After spawning, *i.e.* about August, its flesh loses its peculiar firmness, but in a very short time the fish regains its condition.

The colour of the Turbot is brown of different shades on one side, usually the left, and the whole of that side is spotted with little round bony tubercles, which may be found in the skin after boiling. The size of this fish is extremely variable. The average weight is six or seven pounds, but Turbots are often taken of far greater dimensions. The largest specimen of which an authentic notice is preserved, was taken near Plymouth in the year 1730, and weighed seventy pounds.

ANOTHER flat fish, the BRILL (*Pleuronectes rhombus*), called in Scotland the BONNET FLEUK, and in Devonshire and Cornwall known by the names of KITE and BRETT, is held in much estimation for the sake of its flesh, which is but little inferior to that of the turbot, and is, indeed, sometimes fraudulently substituted for that fish. The Brill resembles the turbot in food and habits as well as in appearance, but does not attain the same dimensions, seldom exceeding seven or eight pounds in weight. The skin of the dark side is devoid of the bony tubercles which are found in the turbot. Its colour is reddish brown, mottled with a darker tint of the same colour, and variegated with numerous round white spots of a pearly lustre. On account of these spots the Brill is sometimes called the PEARL. When young, the pale reddish brown is covered with spots of black or very dark brown.

PASSING by the two species of Topknots, we come to the PLAICE, so well known by the bright red spots which are scattered over its dark side.

This is one of the commonest of the British flat fishes, and, happily for the poor, is taken in such quantities that it supplies nutritious aliment at a very low rate of purchase. It is taken chiefly with the trawl-net, but can be captured with the line, as it bites freely at a bait, generally the common lugworm, and is one of the fish that is most usually caught by amateur sea-fishers. Even the shrimpers take large quantities of small Plaice in their

TURBOT.—*Pleuronectes máximus.*   PLAICE.—*Pleuronectes platessa.*   FLOUNDER.—*Pleuronectes flesus.*

nets; and along the coast this fish is so numerous, that at low water it may be seen in great numbers darting over the sandy flats, the white surface glittering in the light as the little creatures dash wildly along in their terror of the approaching enemy.

I have caught numbers of Plaice, some measuring six or seven inches in length, by merely wading into the muddy sand, holding them down with the feet, and picking them out with the hands. Their terrified wriggle is easily felt by the bare feet, as the fishes find themselves pressed into the sand whither they had fled for refuge, and by a little dexterous management they may be captured by inserting the fingers under the foot, and seizing them firmly across the body.

The colour of the Plaice is light brown, variegated with a number of bright red spots upon the body and the dorsal and anal fins. When young, the Plaice has often a dark spot in the centre of each red mark.

The FLOUNDER, MAYOCK FLEUK, or BUTT, is quite as common as the plaice, and is found in salt, brackish, or fresh water, sometimes living in the sea, sometimes inhabiting the mouths of rivers, and sometimes passing up the stream for many miles.

In former days the Flounder has been known to ascend the Thames as high as Hampton Court, and has there been observed actively chasing the minnows and driving them into shallow water. As this fish is capable of living in fresh water, it has often been transferred to ponds, and will there fatten rapidly.

The colour of the Flounder is usually brown, taking a darker or lighter shade according to the nature of the ground on which the fish rests, those that inhabit the muddy shores being nearly black, and those which prefer the sand taking a yellower hue. Generally the eyes and the colour are on the right side, but reversed specimens are very common, and in some instances the fish has been entirely white or wholly brown. The average weight of the Flounder is three or four pounds.

ONE or two other examples of the British flat fishes deserve a passing notice.

The COMMON DAB (*Platessa limanda*) is plentiful upon all the sandy coasts of Great Britain, and may at once be recognised by the roughness of its surface, or structure, which has gained for it the specific title of limanda, or file-backed—the Latin word *lima* signifying a file. Its flesh is very good, and is thought to be in best condition from the end of January to April. Its colour is pale brown, and its length seldom exceeds eight inches Three or four other species of this genus are known on the British coasts.

A VERY large species of flat fish, the HOLIBUT (*Hippoglossus vulgáris*), is also captured off our shores, but the specimens which are exhibited in the London markets are usually brought from the northern fisheries. The flesh is tolerably good, but is rather dry and without much flavour. It is rather longer in proportion to its width than is generally the case among flat fishes. Its colour is brown of different shades, and the surface smooth, the small oval-shaped scales which cover it being soft and without projections. This fish attains a large size, specimens of five feet in length not being uncommon. The largest example on record measured above seven feet in length, and weighed more than three hundred pounds.

THE well-known COD-FISH is a native of many seas, and in some localities is found in countless legions.

This most useful fish is captured in vast numbers at certain seasons of the year, and is always taken with the hook and line. The lines are of two descriptions, namely the long lines to which a great number of short lines are attached, and the simple hand-lines which are held by the fishermen. The long lines sometimes run to an extraordinary length, and shorter lines, technically called snoods, are affixed to the long line at definite distances. Whatever may be the length of the snoods, they are fastened at intervals of double their length, so as to guard against the entanglement of the hooks. For example, if the snoods are six feet long, they are placed twelve feet apart on the line ; if four feet long, eight feet apart, and so on.

To the end of each snood is attached a baited hook, and as the sharp teeth of the fish might sever a single line, the portion of the snood which is near the hook is composed of a number of separate threads fastened loosely together, so as to permit the teeth to pass between the strands. At each end of the long line is fastened a float or buoy, and when the hooks have been baited with sand launce, limpets, whelks, and similar substances, the line is ready for action.

The boat, in which the line is ready coiled, makes for the fishing-place, lowers a grapnel or small anchor, to which is attached the buoy at one end of the line, and the vessel then sails off, paying out the line as it proceeds, and always "shooting" the line across the tide, so as to prevent the hooks from being washed against each other, or twisted round the line, which is usually shot in the interval between the ebb and flow of the tide, and hauled in at the end of about six hours.

As soon as the long line has been fairly shot, and both ends firmly affixed to the grapnels, the fishermen improve the next six hours by angling with short lines, one of which is held in each hand. They thus capture not only Cod-fish, but haddock, whiting, hake, pollack, and various kinds of flat fishes. On favourable occasions the quantity of fish captured by a single boat is very great, one man having taken more than four hundred Cod alone in ten hours.

The Cod is a most uncertain fish in its habits, sometimes haunting the same locality for a number of successive years, and then suddenly leaving it and repairing to some spot where not a fish might be found on the preceding year. New fishing-grounds are frequently discovered, and it sometimes happens that the fishermen are fortunate enough to alight on

COD.—*Gadus morrhua.*

a spot hitherto untouched, where, to use the graphic description of a sailor, the Cod are "as big as donkeys, and as common as blackberries."

Rockall is one of the latest discoveries of this nature. It is a sandbank in the North Atlantic, about 136 miles from St. Kilda, and only distinguishable by a small rock like a rude haystack. The Cod are there so plentiful and so large that each fishing-boat sold her five days' catch for £140; and after due preparation, the fish were disposed of at nearly double that price.

A great part of the estimation in which this fish is held depends upon the perfect manner in which it takes salt and the length of time during which it can be preserved in an eatable state. Salted Cod is to many persons a great dainty, but to others, among whom I must be reckoned, is insufferably offensive, and even with all the additions of sauce and condiment is barely eatable.

The Cod is sometimes sent away in a fresh state, but is often split and salted on the spot, packed in flats on board, and afterwards washed and dried on the rocks. In this state it is called Klip-fish or Rock-fish. The liver produces a most valuable oil, which is now in great favour for the purpose of affording strength to persons afflicted with delicate lungs or who show symptoms of decline. The best oil is that which drains naturally from the livers as they are thrown into a vessel which is placed in a pan filled with boiling water. The oil is then carefully strained through flannel, and is ready for sale.

The roe of the Cod is useful for bait, the sardine in particular being very partial to that substance. Much of the roe is stupidly wasted by the fisherman, who carelessly flings into the sea a commodity of which he can sell any amount, and for which he can obtain ten or eleven shillings per hundredweight. In Norway the dried heads of the Cod are used as fodder for cows, and, strange to say, the graminivorous quadrupeds are very fond of this aliment.

Like several other marine fish, the Cod can be kept in a pond, provided the water be salt; and if the pond should communicate with the sea, these fishes can be readily fattened for the table. Several such ponds are in existence, and it is the custom to transfer to them the liveliest specimens that have been caught during the day's fishery, the dead

3.   X

or dying being either sold or cut up as food for their imprisoned relatives. These fishes are extremely voracious, and will eat not only the flesh of their kinsmen, but that of whelks and other molluscs, which are abundantly thrown to them. It is found that under this treatment the Cod is firmer, thicker, and heavier in proportion to its length than if it had been suffered to roam at large in the sea.

The colour of the Cod is ashen green rather mottled with deeper tints, and the abdomen is white. The head is very large, there is a long fleshy barbule on the chin, and the pupil of the eye is blue. Varieties in colour and even in form are not uncommon, and in some cases are thought to be produced by difference of diet and locality. The average length of an adult Cod-fish is about three feet, and its weight twelve pounds.

SEVERAL other species of this genus are found in British waters, such as the DORSE (*Morrhua callarias*), the HADDOCK (*Morrhua æglefinus*), a well-known and very valuable fish, mostly found along the north-eastern coast, and the WHITING POUT, SMELTIE, or KLEG (*Morrhua lusca*), so often manufactured into whitings by the simple process of slicing off certain parts of the fish, skinning it, and pushing its tail through the head. In this state it is sold and consumed as whiting; and as one fish is just as good as the other, the consumer suffers no injury, and the enterprising vendor is recompensed for his trouble. The Pout is graphically termed by the fishermen the Stinkalive, because it becomes putrid so soon after death. While living, various iridescent colours play over the surface of the fish, but as soon as it is dead the colours and the dark bands disappear, and the whole upper surface becomes of a dull yellow-brown, the abdomen being whitish with a tinge of blue-grey.

THE common WHITING (*Merlangus vulgaris*) is closely allied to the fishes of the preceding genus, and is too well known to need description. The COAL-FISH (*Merlangus carbonarius*), and the POLLACK (*Merlangus pollachius*), belong to the same genus as the whiting; and the HAKE (*Merlucius vulgaris*) is closely allied to them.

IN the large and important group of fishes to which our attention is now drawn, the ventral fins are wholly wanting, the body is long, snake-like, smooth and slimy on the exterior, and in many cases covered with very little scales hidden in the thick soft skin.

SAND LAUNCE.—*Ammodytes lancea.*

OUR first example is the SAND LAUNCE, a very common fish on many of the British coasts, and usually found wherever the shore is of a sandy character. The generic name Ammodytes signifies sand-diver, and is given to this fish in consequence of its habit of burying itself in the wet sand, where it remains hidden and secure from marine foes.

The rapidity with which it achieves this feat is really remarkable. As the waves of the ebbing tide recede, the fish pushes its projecting under-jaw well into the sand, scoops backward and forward like a pig grubbing in soft soil, gives a wriggle of the glistening body and a twirl of the slender tail, and vanishes as if by magic. Caring not for the absent waters, and finding a sufficiency of moisture in the wet sand, the fish remains uninjured in its retreat, where it lies safe from the many aquatic foes who chase it in the waters, and from whom the shore-sand affords the only refuge.

The Sand Launce is extremely valuable for bait, especially for such fastidious fish as the turbot, and is abundantly taken by the fishermen, who persecute the glittering little creatures, and by means of a many-toothed rake drag them from their sandy refuge. In some places the Sand Launce is taken in small-meshed nets that are dragged through the sand just at the water's edge, and in many localities the children of the fishermen are sent regularly to the shore for the purpose of hooking the Sand Launce out of their retreats by means of certain instruments of iron, curved sicklewise.

The colour of the Sand Launce is glittering silvery white, and its length, when adult, is six or seven inches. On account of its active movements it is in some places popularly known by the name of the WRIGGLE.

ANOTHER species of this genus, the SAND EEL or HORNELS (*Ammodytes Tobianus*), is common on the English shores, and is sometimes mistaken for the preceding species, from which, however, it may be distinguished by its greater size, its larger head, the farther setting back of the dorsal fin, the browner colour, and more opaque body. When full grown, the Sand Eel will reach the length of a foot or thirteen inches.

THE two Eels represented in the engraving are examples of some very common and useful British fish.

The SHARP-NOSED EEL derives its name from the shape of its head, and by that structure may be distinguished from the second species. In their habits the Eels are so similar that the present species will be taken as an example of the whole genus.

Eels are found in almost all warm and temperate countries, and grow to a very large size in tropical regions. They are, however, impatient of cold, and in the extreme northern or southern parts of the world are not to be found. In many of the Pacific islands these fish are held in great estimation, being preserved in ponds and fed by hand, and in New Zealand they afford one of the staple articles of consumption. In some parts of the world, however, and even in many portions of Great Britain, a strong prejudice exists against Eels, probably on account of their resemblance to snakes, and even a hungry man will not eat one of these wholesome and nutritious fish.

The Eel is one of the most mysterious of our river fishes, and although much is now known that formerly was involved in obscurity, there is still much to learn respecting its habits, and more especially its mode of reproduction. It is probable that difference of locality may influence the Eel and cause difference of habit, but it is certain that if a number of practical observers in different parts of England set themselves to watch the Eel and its customs, their accounts would vary in the most perplexing manner, and to build a theory upon so unsafe a basis is quite impossible.

Of the general habits of the Eel, the Hon. Mr. Grantley F. Berkeley has given the following short and interesting account :—

" During hot, still, sunny weather, day and night, in the month of June, the Eels are chiefly on the top of the water. Wherever masses of weeds lie, and what is called the cow-weed grows the longest, there Eels do congregate, to bask in the sun by day, to enjoy by night the warmth left in the weeds by the sun, and there, while thus luxuriating, to snap at and catch the myriads of gnats, moths, flies, and other insects that seek the weeds for food or rest, and by damping their wings become an easy prey to their ambushed assailants. In waiting for the otter, or watching the river, I have often sat in my boat embayed in weeds, and seen and heard the Eels thus occupied , and near and within these weeds, in the particular weather alluded to, the wire-traps, nets, and snig-pots take best.

SHARP-NOSED EEL.—*Anguilla acutirostris.*   BROAD NOSED-EEL.—*Anguilla latirostris.*

The haunts of Eels are quite as variable as the weather. In warm, still weather, seek them on the rapids and near weeds either waving on the surface of the water or in floating masses of detached weeds that the eddies of the stream have wound and kept in one place. In blowing, cooler, or rainy weather, then look for them in the still, deep ditches. If a flush of water comes, and a little shallow stream running from or into the main river becomes fuller than usual, then let all the capturing gear be set to take them on, to them, this delicious change of ground, for against this stream they will work as long as it is freshened. In one night, in a little stream of this sort, I took thirty pounds weight of Eels."

Like several fishes which have already been mentioned, Eels are very tenacious of life, and are able to live for a long time when taken out of water, owing to a simple but beautiful modification of structure, which retains a sufficient amount of moisture to keep the gills in a damp state and able to perform their natural functions. These fishes have been seen crawling over considerable distances, evidently either in search of water, their own dwelling-place being nearly dried, or in quest of some running stream in whose waters they might descend to the sea after the manner of their race.

At the Dargle, near Dublin, I have seen multitudes of little Eels crawling up the banks, and have been much amused by watching them wriggle themselves, without any apparent purpose, over the smooth surface of an overhanging rock, to which they clung and upon which they moved as freely as a fly on a ceiling. These little Eels were about eight or ten inches in length, and were so active as to escape the grasp unless the hand was moved with extreme rapidity.

Vast multitudes of these little Eels are in the habit of proceeding up the rivers in the springtime, and in some places are known as Elvers. They are caught in great quantities, scalded and pressed into masses termed Eel, or Elver cake. When dressed, these little Eels afford a luxurious repast; and before dressing, the effect of the myriad tiny black eyes that speckle the macaroni-like mass of white bodies is most peculiar.

Towards the latter end of summer, the Eels migrate towards the sea, and it is found that these fishes can live either in fresh or salt water with equal ease, the mouths of rivers being favoured localities. It sometimes happens that even in our seaport towns and marine watering-places, the common river Eel is caught by those who are angling in the sea for marine fish. This quality is peculiarly valuable in the Eel, as it enables the Dutch fishermen, who annually supply our markets with vast numbers of these fish, to bring them across the sea in vessels that are fitted with "wells" pierced for the transmission of the sea-water through which the vessel is sailing.

Numbers of these Dutch Eel-boats may be seen about Gravesend, not daring to ascend the river on account of the polluted state of its water. In the wells of these boats the Eels remain for ten days, and require no food.

Eels are captured in various modes. "Bobbing," or "clodding" as it is sometimes termed, is a very common and successful mode, consisting in making a bunch of earthworms strung on worsted, and lowering it near the place where the Eels are known to be feeding. The voracious fish seize eagerly on the bait, and bite so fiercely, that they are pulled out of the water before they can disengage their teeth from the worsted. Another plan is by night-lines, which are laid in the evening and taken up in the morning.

One of the most successful methods, however, is by spearing, and is extensively adopted by bargemen, many of whom always have an Eel-spear on board. The spear is not unlike the conventional trident of Neptune, except that the prongs are four in number, flattened, slightly barbed on each edge, and spread considerably from their junction with the shaft. This is pushed at random into the muddy banks where the Eels love to lie, and whenever it encounters one of these fish, the long snake-like body is caught between the jagged prongs and lifted into the boat before it can extricate itself.

The food of the Eel is extremely various, for the creature is most voracious, and eats every living being that it can master, whether aquatic or terrestrial. Even mice and rats fall a prey to this hungry fish, and on one occasion an Eel was found floating dead on the water, having been choked by a rat which it had attempted to swallow, but which was too large to pass down its throat. It has even been caught with a fly while the fisherman was angling for trout.

The tenacity of life possessed by this fish is really remarkable; and it is worthy of notice that the best mode of killing Eels is to grasp them by the neck and slap their tails smartly against a stone or post. The muscular irritability of the body is wonderfully enduring, and after the creature has been cut up into lengths each separate piece moves about as if alive, while at the touch of a pin's point it will curve itself as if it felt the injury. When all such irritability has ceased, the portions will flounce about vigorously if placed in boiling water; and even after they have remained quiet under its influence, the addition of salt will make them jump about as vigorously as ever. Of course there can be no real sensation, the spinal cord having been severed.

The reproduction of the Eel has long been a subject of discussion, some persons thinking that the young are produced in a living state, and others holding that they are hatched from eggs. This question has, however, been set at rest by that universal revealer, the achromatic microscope, which has shown that the masses of oily-looking substance generally called fat are really the aggregated clusters of eggs, and that these objects, minute though they may be, not so large as the dot over the letter i, are quite perfect and under the microscope are seen to be genuine eggs.

The BROAD-NOSED EEL is at once to be distinguished by the greater breadth of its head, bluntness of its nose, and soft unctuousness of its body. It does not seem to attain so great a size as its sharp-nosed relative. Besides these species, a third British Eel, the SNIG, is found in some parts of England, and is known by its olive-green back and the golden yellow of the under parts. The Grig is a term applied by fishermen to any Eel of a small size, and even the name of Snig is employed in a very vague fashion.

THE well-known CONGER EEL is a marine species, very common in our seas, and being most usually found on the rocky portion of the coast.

CONGER.—*Conger vulgaris.*

This useful fish has, of late years, come into more general use than formerly, and its good qualities are more appreciated. The flesh, though not very palatable if dressed unskilfully, is now held in some estimation, and for the manufacture of soup is thought to be almost unrivalled. The fishermen can now always obtain a ready sale for the Congers; and those which are not purchased for the table are mostly bought up and made into isinglass.

Congers are chiefly caught with the line, and it is found that, voracious as they are, they are fully as fastidious as the turbot, and will not touch a bait that is in the least tainted. Small fish, such as young dabs and plaice, are among the favourite baits, but the sand launce affords the most irresistible of lures, its bright glittering surface tempting the voracious fish to its own destruction. The arms of cuttle-fish cut into lengths is another bait used for taking the Conger.

As the teeth of this fish are sharp, and the muscular power very ·great, the lines are made of proportionate strength, the portion to which the hooks are attached being twisted in such a manner that the Conger's teeth cannot cut through them. The hooks, too, are very strong, and are made of soft iron, so that if the fish has gorged the bait, the hook can be drawn out by jerking until it straightens, when a few minutes' labour with a hammer suffices to restore its form. The line is generally a long one furnished with "snoods" at regular intervals, as has already been related of the cod-fish. For Conger fisheries the snoods are about nine feet in length, and the principal line rather more than four hundred feet.

When the Congers are being hauled into the boat, they plunge about with the most desperate efforts to escape; and should their sharp teeth seize the fishermen, the result is far from agreeable. The men, therefore, always kill the large Congers by a blow on the underside of the body, where they are far more vulnerable than on the head. The sailors will sometimes kill the Conger by squirting the juice of their "quids" into its mouth. In very cold weather the Congers are apt to be seized with a curious malady, which causes them to rise to the surface of the water, and there to float, unable to sink. On dissection, the swimming bladder is found to be enormously swollen

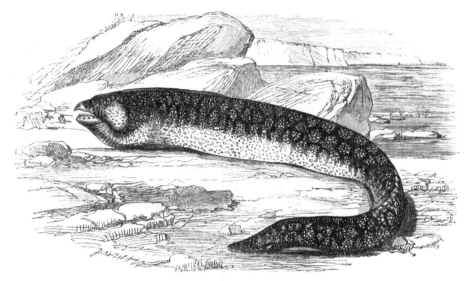

MURÆNA.—*Muræna Helena.*

but the reason of this distension seems rather obscure. In 1827, thousands of Congers were seen lying dead on the shore, at Eastbourne, during a severe frost; and in January, 1855, a similar circumstance occurred on several of our coasts.

The colour of the Conger is pale brown above and greyish white below. It often attains to a very great size, measuring ten feet in length and weighing more than a hundred pounds.

THE beautifully mottled MURÆNA is tolerably common in the Mediterranean, but is so scarce towards the British coast that it can hardly be considered as one of the true English fishes.

In former days the Muræna was held in great distinction by epicures; and the wealthy were accustomed to preserve them in ponds built for that special purpose. In these ponds the Murænæ were fattened, and several of the aristocrats laboured under the imputation of feeding them with an occasional slave, whenever an ill-fated domestic had the misfortune to offend them. The flesh is very white in colour, and of a peculiar and very delicate flavour. This fish can live either in salt or fresh water, but appears to prefer the sea.

The colour is golden yellow in front and purple towards the tail; and the whole body is covered with bands, irregular rings, and spots of deep and pale gold, purple, and brown. The dorsal fin begins a little behind the head and runs to the tail, where it is united with the anal fin. Both these fins are, however, low and fleshy, and not at all conspicuous. The length of this fish is extremely variable; one specimen captured off the British shores measured four feet four inches in length.

The ELECTRIC EEL is even more remarkable for its capability of delivering powerful electric shocks than the torpedo, but as it is never found in the British seas it is not so well known as that fish.

The Electric Eel is a native of Southern America, and inhabits the rivers of that warm and verdant country. The organs which enable it to produce such wonderful effects are double, and lie along the body, the one upon the other.

ELECTRIC EEL.—*Gymnótus eléctricus.*

The reader will remember that in the torpedo the electric effect was produced by a number of little columns; in the Electric Eel, the corresponding organ consists of a great number of divisions, technically called "septa," which are again subdivided by lesser transverse membranes. One organ is always larger than the other; and it was found that in a fish measuring about two feet four inches in length, there were thirty-four septa in the larger organ and fourteen in the smaller. On an average two hundred and forty transverse membranes are packed in each inch, thereby giving a vast extent of electricity-producing surface. It was calculated by Lacepède, that the expanse of this organ in an Electric Eel of four feet in length is equivalent to one hundred and twenty-three square feet, while that of a large torpedo only equals fifty-eight feet.

In the native country of these fishes they are captured by an ingenious but somewhat cruel process. A herd of wild horses are driven to the spot and urged into the water. The alarmed Gymnoti, finding their domains thus invaded, call forth all the terrors of their invisible artillery to repel the intruders, and discharge their pent-up lightnings with fearful rapidity and force. Gliding under the bellies of the frightened horses, they press themselves against their bodies, as if to economize all the electrical fluid, and by shock after shock generally succeed in drowning several of the poor quadrupeds.

Horses, however, are but of slight value in that country, hardly, indeed, so much valued as pigeons in England, and as fast as they emerge from the water in frantic terror, are driven back among their dread enemies. Presently the shocks become less powerful, for the Gymnotus soon exhausts its store of electricity, and when the fishes are thoroughly fatigued they are captured with impunity by the native hunters. A most interesting account of this process is given by Humboldt, but is too long to be inserted in these pages.

Several of these wonderful fish have been brought to England in a living state; and many of my readers may remember the fine Gymnotus that lived in the Polytechnic Institution. Numbers of experimenters were accustomed daily to test its powers; and the fatal, or at all events the numbing, power of the stroke was evident when the creature was supplied with the fish on which it fed. Though blind, it was accustomed to turn its head towards the spot designated by the splashing of the attendant's finger, and as soon

HAIR-TAILED GLASS EEL.—*Tilurus trichiurus.*
PIG-NOSED GLASS EEL.—*Hyoprórus Messinensis.*
ROUND-HEADED GLASS EEL.—*Leptocéphalus tǽnia.*

as a fish was allowed to fall into the water the Gymnotus would curve itself slightly, seemed to stiffen its muscles, and the victim turned over on its back, struck as if dead by the violence of the shock.

When full grown, the Electric Eel will attain a length of five or six feet, and is then a truly formidable creature. The body is rounded, and the scales small and barely visible. According to Marcgrave, the native name for this fish is Carapo.

WE have already seen some examples of fishes where the body is extremely transparent, and now come to an entire family where this peculiarity is the chief and most obvious characteristic.

The skeleton of the Leptocephalidæ, or Glass Eels as they are termed, from their Eel-like shape and singular translucency, is very imperfect, merely consisting of cartilage, and so slight, that even in the head, where the greatest strength is required, the brain can be seen through the translucent skull in which it lies. Their bodies are always extremely compressed and mostly leaf-like, so transparent that when lying in a vessel containing water they would hardly be noticed, and the lateral line is formed by the intersection of the muscles, as may be seen by reference to the illustration.

The PIG-NOSED GLASS EEL may be known by the lengthened form of its head and snout, which are far longer in proportion to the dimensions of the fish than in any other member of the family. The generic term Hyoprorus literally signifies swine-beaked, and in former days was applied to a certain kind of galley which had a long and slightly turned-up beak. The sudden height of the body just behind the head is very remarkable, and on close examination, a row of mucous pores will be found along the jaws and on the head. The eyes are not very large, and the general length of the species is between four and five inches. As its specific name imports, it has been taken at Messina.

The HAIR-TAILED GLASS EEL is much longer in proportion than the last-mentioned species, and its body is so extremely compressed that it is hardly thicker than the paper on which this account is printed. This species is also found at Messina. The jaws are short and round, the eye rather small, and the tail tapers away to a hair-like point. The length of this fish is rather more than a foot, and a row of minute points runs along each edge of the body.

The typical genus Leptocephalus is a rather large one, containing more species than the four preceding genera together.

The ROUND-HEADED GLASS EEL derives its specific name of Tænia, or tapeworm, on account of its resemblance to that unpleasant internal parasite. Its head is, as its name denotes, short and much rounded, and the eyes are globular, projecting, and extremely large. The jaws are tolerably well furnished with small teeth. In shape it is long and rather rounded, and the absence of fins renders its resemblance to a tapeworm extremely striking. It seems to be an Asiatic species, having been captured in India and the neighbouring islands.

AN example of this genus, the ANGLESEY MORRIS (*Leptocephalus Morrisi*), has been taken on our own coasts. In this species the head is blunt, the eye moderate, the body much compressed, and deepest at the latter third of its length. When living, its polished surface reflects gleams of iridescent light as it winds its graceful way through the sea-weeds among which it loves to sojourn, like a ribbon of animated nacre. But when dead and placed in spirits, all the delicate opalescence of its body fades, and soon deteriorates into an opaque dull whiteness like wet parchment.

BLIND-FISH.—*Amblyopsis spelæus.*

THE reader will remember that on several occasions it has been deemed expedient to give examples of remarkable deviations from the ordinary system, and to call attention to the wonderful economy of nature, which is most averse to wastefulness, and declines to expend its powers on organs that if existing would be in abeyance. A recent example of such modification has been given in the proteus, on page 192, that curious reptile, or semi-reptile, which inhabits caves wherein penetrates no ray of light, and which, having no need of external eyes, is altogether devoid of such useless organs.

The BLIND-FISH of America affords another instance of similar economy in structure. Living, like the proteus, in a subterranean and perfectly dark grotto, it needs no eyes, and

ANCHOVY.—*Engraulis encrasicholus.*
SHAD.—*Alósa vulgáris.*

in consequence possesses none, their place being merely indicated by two minute black dots on the sides of the head. The head is naked, but the body is covered with scales, and the jaws are furnished with some small but sharp teeth. Its colour is whitish grey, as is, indeed, mostly the case with animals that have been long deprived of the colour-giving sunlight. The grotto which contains this very remarkable little fish is in Kentucky.

WE now come to that most valuable family of fishes, the Herring tribe, called technically Clupeidæ, from the Latin word *clupea*, a herring.

THE well-known ANCHOVY is properly a native of the Mediterranean Sea, though it often occurs on our coasts, and has once or twice been captured in our rivers. Indeed, one practical writer on British fishes thinks that the capture of the Anchovy off our shores is a task that would be highly remunerative if properly undertaken, and that, with proper pains, the British markets might be fully supplied with Anchovies from our own seas.

This little fish has long been famous for the powerful and unique flavour of its flesh, and is in consequence captured in vast quantities for the purpose of being made into Anchovy sauce, Anchovy paste, and other articles of diet in which the heart of an epicure delights. Unfortunately, however, the little fish is so valuable, that in the preparations made from its flesh the dishonest dealers too often adulterate their goods largely, and palm off sprats and other comparatively worthless fish for the real Anchovy.

As the head is always removed before the process of potting is commenced, the deception is not easily detected—the long head with its projecting upper jaw and deeply cleft gape affording so clear an evidence of the identity of the fish, that no one would venture to pass off one fish for the other if the heads were permitted to remain in their natural places. The flavour of the veritable Anchovy is rudely imitated by various admixtures and its full rich colour is simulated by bole armoniac and other abominations.

The very long generic title of this fish was given to it in ancient times, and is still retained, as being at once appropriate and sanctioned by the verdict of antiquity. Its literal signification is "gall-tinctured," and the name has been given to it on account of the peculiar bitter taste of the head, in which part the ancients supposed the gall to be placed. The colour of the Anchovy is bluish green on the back and upper part of the head, and the remainder of the body silvery white; the fins have a tinge of green, and are beautifully transparent. The scales are large and fall off almost at a touch. The length of the Anchovy varies from five to seven inches.

The COMMON or ALLICE SHAD is extremely plentiful on some of our coasts, but appears to be a rather local fish, and while it abounds in some places to be wholly absent from others.

The Shad is fond of ascending rivers, especially if the water be clear; and while the Thames was still unstirred by the paddles of multitudinous steamboats, and unpolluted by the contents of countless sewers, this fish would ascend the river for a considerable distance, and has been taken in good condition near Hampton Court. Some persons think that the flavour of the fish improves in proportion to its proximity to the river source. Except in size, the Shad bears a very close resemblance to a herring, and in some places is called the King of the Herrings.

The colour of the Shad is dark blue on the upper part of the head and back, variegated with glosses or reflections of brown and green, either colour predominating according to the angle at which the light falls upon the surface. The remainder of the body is white. There is another British species of this genus, the TWAITE SHAD (*Alosa finta*), which is about half the size of the Allice Shad, weighing on an average about two pounds. Both these fish may be at once distinguished by a deep cleft or notch in the centre of the upper jaw.

The HERRING is undoubtedly the most valuable of our British fishes, and the one which could least be spared. It is at once the luxury of the rich and the nourishment of the poor, capable of preservation throughout a long period, easily packed, quickly and simply dressed, and equally good whether eaten fresh or salted, smoked or potted.

During the greater part of the year, the Herring lives in deep water, where its habits are entirely unknown. About July or August, the Herring is urged, by the irresistible force of instinct, to approach the shores for the purpose of depositing its spawn in the shallow waters, where the warm rays of the sun may pour their vivifying influence upon the tiny eggs that will hereafter produce creatures of so disproportionate a size, and where the ever-moving tides may fill the water with free-oxygen as the waves dash on the shores and fall back in whitened spray, thus giving to the water that sparkling freshness so needful for the development of the future fish.

The Herrings, when they once begin to move, arise in vast shoals, and direct their course towards some part of the shore. In their choice of locality they are most capricious fish, sometimes frequenting one spot for many successive years, then deserting it for a length of time, and again returning to it without any apparent reason for either course of proceeding. They are essentially gregarious while on the move; and each shoal is so closely compacted, and its limits so well defined, that while one net will be filled almost to bursting with Herrings, another net, only a yard or two distant, will be left as empty as when it was shot.

The months of spawning are October and November, and until the Herring has performed the office for which it came to the shore, its flesh is in its best condition. As soon, however, as the spawning is completed, the fish, then technically called "shotten"

HERRING.—*Clupea harengus.*   SPRAT.—*Clupea sprattus*

Herrings, retire to the deep whence they issued, and there remain until the succeeding year summons them to a repetition of the same duties.

The usual method of catching Herrings is by drift-nets, which are spread, or "shot," in innumerable lines of complicated cordage, forming a veritable labyrinth of ropes and meshes, the back lines adding to the general complexity of the structure.

The night view of a Herring fishery is singularly beautiful, owing to the phosphorescent properties of the fish, a phenomenon common to many of the finny tribes, and which is well described by Mr. W. H. Maxwell, in his "Wild Sports of the West." "The darkness of the night increased the scaly brilliancy which the phosphoric properties of these beautiful fish produce. The bottom of the boat, now covered with some thousand Herrings, glowed with a living light which the imagination could not create and the pencil never imitate. The shades of gold and silvery gems were rich beyond description; and much as I had heard of phosphoric splendour before, every idea I had formed fell infinitely short of its reality."

The Herring is one of the fish that cannot endure absence from water, and dies almost immediately after it is taken out of the sea, thus giving rise to the familiar saying, as "dead as a Herring."

The food of the Herring is extremely varied, even in the comparatively shallow waters, and its subsistence during the time it is submerged in the deep is necessarily unknown. In the stomach of the Herring have been found crustacea of various kinds, molluscs, the spawn and fry of other fish, and even the young of its own kind. It can be taken with a hook, and has been known to seize a limpet that was used as bait.

Few fish have so many foes as the Herring, its marvellously gregarious habits rendering it an easy prey to finned and feathered foes ; and its shoals are so perseveringly preyed upon by the birds of the air and the fishes of the sea, that even the havoc annually made by man falls probably short of that caused by the ceaseless attacks of the myriad enemies that surround them.

Several of the cetaceans are in the habit of following the Herrings to the shore, and gorging themselves during the whole of their stay.   Various members of the shark tribe rove ceaselessly among the shoals, and by their peculiar habit of snapping the fish asunder before swallowing it, cause a large quantity of oil to escape, and to produce a peculiar effect upon the surface.   The chimæra and many other of the large-sized fish take the opportunity of enjoying a boundless feast, and prowl around the shoals like roaming banditti.   The sea-birds above are quite as voracious as the fish below, and as actively engaged in the pursuit of their prey ; so that when man, birds, and fishes have had their share, it seems quite wonderful that the whole Herring race is not exterminated.

In the Norwegian fisheries, the boatmen do not like to shoot their nets unless they are sure that a shoal of Herrings is passing.   During the daytime, therefore, they watch for their prey with an instrument called a water-glass, which is nothing more than a trumpet-shaped tube of wood, with a simple plate of glass let into the broad end.   This is pushed under water, and when the eye of the observer is applied to the upper extremity, enables him to see to a considerable depth, the vision not being disturbed by the shifting lines of the surface.   At night the men lower a line, to which a weight is attached, and as soon as they feel the line jarred by the passing fish, they shoot their nets, in the certainty of effecting a capture.

The colour of the Herring is blue above with greenish reflections, and the rest of the body is silvery white.   After the fish has been dead for some hours, the cheeks and gill-covers become red, as if from injected blood.

ANOTHER species of this fish, called LEACH'S HERRING (*Clupea Leachii*), is taken off our coasts during the winter months ; the roe being well developed at the end of January, and the spawn deposited in February.   It is a small species, between seven and eight inches in length.

THE common SPRAT is another very useful fish, though not so extensively valued as the herring.

Like that fish, it swims in vast shoals during the spawning season, which immediately succeeds that of the herring, so that from July to February and March the public can command a continual supply of fresh sea-fish, which can be purchased at so cheap a rate as to be within the reach of all classes, and are, nevertheless, of such excellent flavour, that if they were as scarce as they are plentiful, they would be held in high estimation at the tables of the wealthy.   To the taste of many persons, however, the Sprat is too rich and too strongly flavoured to be in much request.

This fish is captured in nets of various kinds, the nature of the net mostly depending on that of the locality ; and as it swims in shoals quite equal in numbers to those of the herring, it is taken in countless multitudes when the boats happen to be fortunate in their selection of a fishing-ground.   Now and then the "take" is so enormous that even the London markets, which usually absorb every eatable article which can be brought for sale, and often anticipate the future crops or supplies, are at times so overstocked with Sprats that the fishermen can find no ordinary sale for their perishable goods, and are perforce obliged to dispose of them to the farmers, who spread them over their lands for manure, most unfragrant but exceedingly fertilizing.

At one time the Sprat was thought to be the young of the herring, pilchard, or shad, a mistake occurring in all probability from the vague manner in which the word Sprat is employed in many seaside villages, any little whitish fish being called by that name. It can, however, be distinguished even in the dark from the young of either of these fishes, by means of the sharply notched edge of the abdomen.   In colour it is very like the herring.

FLYING-FISH. — *Exocœtus volitans*

ONE or two more members of this genus demand a brief notice.

The PILCHARD, or GIPSY HERRING (*Clupea pilchardus*), is another of the gregarious fish, and is taken about the month of August by a wonderfully intricate system of boats and nets that seem capable of sweeping every fish out of the sea. Though very like the herring, it may easily be distinguished by the position of the dorsal fin, which is set so far forward that if the fish be held by the first ray of that fin its body slopes upward, whereas in the herring it is nearly balanced and slightly inclines downward.

ANOTHER species of this genus is peculiarly dear to the metropolitan epicure under the name of WHITEBAIT. The scientific name of this fish is *Clupea alba*. This most delicate little fish, which looks so exquisitely beautiful in the opalescent translucency of its living state that the mind almost recoils from defiling it with the appliances of cookery, is taken in the mouths of several British rivers, and is sometimes known to ascend for many miles up the stream. It was long thought to be the fry of some other fish, but this opinion has long been disproved. The colour of this pretty little fish is all silvery white, with the exception of the back, which is tinged with ashen green.

THE far-famed FLYING-FISH exists in many of the warmer seas, and derives its popular name from its wonderful powers of sustaining itself in the air.

The passage of this fish through the atmosphere can lay no just claim to the title of flight, for the creature does not flap the wing-like pectoral fins on which it is upborne, and is not believed even to possess the power of changing its course. As much of the history of the Flying-fish has been given while treating of the coryphene, the reader is referred to the description of that fish on page 277, where may also be seen an illustration of the attitudes assumed by the Flying-fish as it speeds its course through the air while attempting to avoid its deadly foe beneath.

In allusion to the habits of this remarkable fish, Mr. F. D. Bennett, in his " Narrative of a Whaling Voyage," has the following valuable remarks :—

"The principal external agents employed in this mode of locomotion are the large lobe of the tail fin and the broad transparent pectoral fins, which, on this occasion, serve at least as a parachute, and which, being situated close to the back, place the centre of suspension higher than the centre of gravity. It is also curious to notice how well the specific gravity of the fish can be regulated, in correspondence with the element through which it may move. The swim-bladder, when perfectly distended, occupies nearly the entire cavity of the abdomen and contains a large quantity of air; and in addition to this, there is a membrane in the mouth which can be inflated through the gills; these two reservoirs of air affording good substitutes for the air-cells so freely distributed within the bones of birds, and having the additional advantage of being voluntary in their function.

The pectoral fins, though so large when expanded, can be folded into an exceedingly slender, neat, and compact form; but whether they are employed in swimming in the closed or expanded state, I have been unable to determine.

Under some circumstances of excitement, these fish will leap over the bulwarks of a ship and be killed by the violence with which they strike against the deck or spars. This usually occurs at night, or early in the morning, and a light displayed from the chains of a vessel on a dark night will bring many of them on board in the same manner. Their flesh is the *bonne bouche* of voyagers; it bears some resemblance to that of the herring. Although the Flying-fish excites so much commiseration for its persecuted state, it is itself predaceous, feeding chiefly on smaller fishes."

The ancients were well acquainted with the Flying-fish, and in their narrative seven improved upon its powers, as was customary with the voyagers of those days, and asserted that, as soon as night came on, this fish left the ocean, flew ashore, and slept until morning safe from the attacks of its marine enemies. The generic name of exocætus, literally a "sleeper-out," refers to this supposed habit.

THE reader will doubtlessly remember that the power of sustentation in the atmosphere for a more or less prolonged period is exhibited in the three preceding orders of vertebrated animals, and that, in every case, this object is attained by the modification of parts already existing, and not by the addition of special members.

In the bats, for example, the lengthened bones of the fore-limbs, together with the extension of the skin, form a flying apparatus of wonderful perfection, and in the creatures that are popularly, though erroneously, called "flying" squirrels and "flying" rats, the capability of passing through considerable distances, upborne by the air, is achieved by a development of a similar nature but of less extent, the skin of the sides being much widened, though the limbs retain their usual comparative dimensions. These, as well as other creatures who move through the air on the same principle, ought rather to be termed "skimmers."

In the birds the power of flight is physically owing to the development of the fore-limbs and the modification of the structures which clothe the skin. In the two succeeding orders of vertebrates, no truly flying species are at present known to exist. The only reptile that in this period of the world's history is enabled to sustain itself in the air, even for a limited space, is the little flying dragon, a creature which has its ribs nearly straight instead of curved, and by means of this formation is enabled to sweep from one tree to another just like the flying squirrel. But in the older ages of the world, flying reptiles were abundant, with "wings" that measured some thirty feet from tip to tip, and evidently both agile and strong of flight like those of the bat, to which they bear a great resemblance. Lastly, we have several examples among the fishes, where the pectoral fins, answering to the fore-limbs of the higher vertebrates, are so greatly expanded and enlarged, that they can be spread horizontally, and bear their owner on a short course through the air.

In any case, the power of flight is wonderful, but in the fish it seems almost to partake of the miraculous, inasmuch as these creatures inhabit a different element, and do not even breathe the atmosphere into which they are capable of launching themselves. Yet, when more closely examined, the flying power of the fish is not one whit more remarkable than the diving powers of the otter and penguin, both of which creatures are able to leave their own element for that of the finny race, to dive through the water with such marvellous address that they can chase and capture even the swift and active fishes, and can remain submerged for a much longer period than the Flying-fish can remain in the air. Flying and swimming are indeed convertible terms, as are wings and fins, wings being the fins of the air, and fins the wings of the waters.

It is well known that the flight of this fish is short and intermittent, the creature being obliged to dip into the sea after its sweep through the air, and the cause of this necessity is said to be the drying of the fin-membranes, which prevents the fish from sustaining itself in the air, inasmuch as it leaps with fresh vigour into the atmosphere after being refreshed by a dip in the water. I cannot but think, however, that one reason of the intermittent flight is simply that the original impetus is exhausted, and that the fish requires to seek the water in order to obtain a fresh start. There may, perhaps, be another reason.

It has already been mentioned that the climbing perch, and other fish of similar habits, possess certain reservoirs of water, which constantly bathe the gills, and thus oxygenize the blood in the same method that is adopted when the body is entirely submerged in the waters. Now, the Flying-fish possesses no such reservoir, and the question therefore arises, whether the passage through the air may not act powerfully on the blood by supplying it with an excess of oxygen, and, through the blood, upon the nervous system, producing a kind of temporary delirium or intoxication. It would be an interesting experiment to catch a Flying-fish after its flight, and test the temperature of the blood with a trustworthy instrument, and also to ascertain the length of time that a Flying-fish can survive when removed from the water.

Perhaps some of my readers may have the opportunity of making these investigations, and of solving some of the countless divine enigmas which surround us. It may be that if the full history of the Flying-fish were discovered, we should find therein not only the operations of nature as exemplified in the individual species, but discover enshrined within the limits of that delicate framework the deepest mysteries of vitality, health, and a material locomotion of which we have as yet hardly dreamed.

About thirty species of Flying-fish are known, mostly belonging to the Mediterranean Sea, but others occur in the North Sea, the Atlantic and Pacific Oceans.

BEFORE proceeding to our next figured example of the finny tribes, we must briefly notice a curious fish which seems to be a kind of balance to the sword-fish already mentioned, the "sword" in this instance belonging to the lower instead of the upper jaw, and being formed by a prolongation of its bones. It is known by the scientific name of *Hemiramphus argenteus*, and is found near the surface of the water in the Pacific Ocean. Its colour is uniform silvery white, and its average length is only four inches.

THE odd-looking GAR-FISH is known by a vast variety of names, such as SEA PIKE, MACKAREL GUIDE, SEA-NEEDLE, LONG-NOSE, GORE-BILL, HORNFISH, and GREENBONE, the last-mentioned title being given to it because, when it is boiled, its bones are of a bright green hue. The name of Mackarel Guide is owing to the fact that its spawning season exactly precedes that of the mackarel, and the other names explain themselves.

This is one of the marine fish, and is sometimes taken and sent to market, generally causing some little excitement as its long pointed head and brightly coloured body lie shining on the dealer's table. It is not, however, extensively captured, on account of a senseless prejudice which exists in many parts against the fish, the green hue of the spine being its probable cause. Despite of prejudice, the fish is an excellent one, and when properly dressed is not unlike eel, but is not so rich.

It is a voracious and bold-biting fish, taking almost any animal substance used as bait, and seizing it so strongly that it hooks itself without any trouble to the angler. To those who fish for their living, and not merely for sport, the Gar-fish behaves in a very agreeable manner; for instead of plunging about when it feels the hook, and by its struggles frightening all other fishes away, it gives one strong pull when it finds itself checked, and then resigns the contest, hanging quietly until released by the fisherman. As a number

3.   Y

of hooks are fastened by snoods to each line, the advantages of a quiet capture can hardly be overrated.

The colour of the Gar-fish is dark bluish green on the back and upper part of the head, fading gradually on the sides into the silvery white of the abdomen. Its usual length is about two feet.

GAR-FISH.—*Belóne vulgaris*.

THE fierce and voracious PIKE has well earned its titles of Fresh-water Shark and River Pirate, for though perhaps not one whit more destructive to animal life than the roach, gudgeon, and other harmless fish, the prey which it devours are of larger size, and its means of destruction are so conspicuous and powerful, that its name has long been a by-word for pitiless rapacity.

The Pike is found in almost every English river, and although supposed to have been artificially introduced into our country, has multiplied as rapidly as any indigenous fish. The Pike is the master of the waters in which it resides, destroying without mercy every other fish that happens to come near its residence, none seeming able to escape except the perch, whose array of sharp spines daunts even the voracious Pike from attempting its capture. As if to show that the Pike really desires to eat the perch, and is only withheld from doing so by a wholesome dread of its weapons, there is no better bait for a Pike than a young perch from which the dorsal fin has been removed. It will even feed upon its own kind, and a young Pike, or Jack as it is then called, of three or four inches in length, has little chance of life if it should come across one of its larger kindred.

At the beginning of spring, the Pike leaves the larger rivers, and ascends the creeks and narrow ditches in order to deposit its spawn. Many fine fish are captured at that time of year by penning them in with a couple of nets, which are gradually approached towards each other until the fish is inclosed between them.

After hatching, the growth of the young Jack is extremely rapid, and according to Bloch, it will attain a length of ten inches in the first year of its life. If well fed, the growth of this fish continues at a tolerably uniform rate of about four pounds per year, and this increase will be maintained for six or seven successive years.

The voracity of the Pike is too well known to need much comment. A tiny Jack of five inches in length has been known to capture and try to eat a gudgeon of its own size, and to swim about quite unconcernedly, with the tail of its victim protruding from its mouth. Had it been suffered to live, it would probably have finished the gudgeon in course of time, as the head was found to have been partially digested. Three water-rats have been found in the stomach of one Pike, accompanied by the remains of a bird too far decomposed to be recognisable, but supposed to be the remnants of a duck. An opinion was once prevalent, and still exists in some places, that the Pike would not eat the

PIKE.—*Esox lúcius.*

gold-fish, being scared by the burnished glitter of their scales. This idea, however, is incorrect, as the Pike has been seen to devour gold-fish in spite of their brilliant armour. So universal is the appetite of this fish, that it has even been known to seize the paste bait which had been used for other and less voracious inhabitants of the waters.

When the Pike attains a tolerable size, it takes possession of some particular spot in the bank, usually a kind of hole or cave which is sheltered by overhanging soil or roots, and affords a lair where it can lurk in readiness to pounce upon its passing prey. It is rather remarkable that these fish seem to be well acquainted with the most "eligible residences," and that if a large Pike is taken from a hole, another is sure to take possession in a very short space of time.

The Pike seems to have no limit to its size, for it is a very long-lived fish, and seems always to increase in dimensions provided it be well supplied with food. A fish of ten or twelve pounds' weight is considered to be a fine specimen, though there have been examples where the Pike has attained more than five times the latter weight. These huge fishes of sixty or seventy pounds are, however, of little value for the table.

The colour of the Pike is olive-brown on the back, taking a lighter hue on the sides, and being variegated with green and yellow. The abdomen is silvery white.

The SALMON is undoubtedly the king of British river-fish ; not so much for its dimensions, which are exceeded by one or two giant members of the finny tribe, but for the silvery sheen of its glittering scales, its wonderful dash and activity, affording magnificent sport to the angler, the interesting nature of its life from the egg to full maturity, and last, but not least, for the exquisite flavour and nutritive character of its flesh.

In former days, before civilization had substituted man and his dwellings for the broad meadows and their furred and feathered inmates, the Salmon was found in many an English river. Now, however, there are but few streams where this splendid fish can be seen, for, in the greater number of British rivers, the water has been so defiled by human agency that the fastidious Salmon will not suffer itself to be poisoned by such hateful mixture of evil odours and polluted waters ; and in the few streams where the water is still sufficiently pure for the Salmon to venture into them, the array of nets, weirs. and all kinds of Salmon traps is so tremendous, that not one tithe of the normal number are now found in them.

The ingenuity which has been exhibited in the invention of these "infernal machines," as the fixed nets have been justly termed, and the amount of labour which has been expended in their manufacture, are worthy of a better cause; for in their arrangement the habits of the fish have been carefully studied, and, in their manufacture, its capabilities have been foreseen. The evil has, of late years, arisen to so great a height, that the Salmon would soon have been extirpated from our rivers, had not the nation wisely interfered to prevent the loss of so much national wealth, and given the fish a fair chance of re-establishing itself in its former plenty.

The shortsighted persons who plant all these obstructions forget that by this wholesale destruction of the Salmon they are acting against their own interests, and that if they destroy the ill-conditioned and young fish, as well as the adult and healthy Salmon, they condemn themselves to the probability of eating bad fish for the present, and the certainty of total deprivation for the future. The fact, however, seems to be, that each petty proprietor of a fishery is jealous of the neighbours above and below him, and indiscriminately slaughters all fish that he can capture in his own waters, simply that they may not pass into those of his neighbour.

The preservation of this noble fish is truly a subject of national importance, and it is to be hoped that, by judicious legislation and active administration of the law, the Salmon may no longer be the rich man's luxury, but again hold its legitimate place as the poor man's cheap subsistence. That it should ever re-enter the Thames, from which it has been banished for more than forty years, is a dream that perhaps may never be realized. But as the increasing facilities of transport become more developed, a Salmon stream in the far north is virtually brought within a few miles of any railway station in the kingdom, and every portion of our island may perchance procure this delicious fish even before the well-known "curd" has vanished.

While speaking of this curd, which is to the Salmon what the fin is to the turbot, and the green fat to the turtle, it may be mentioned that the practice of "scoring" is destructive of this delicacy, and indeed is one of the most ingenious methods of spoiling the fish that can be invented.

The life history of the Salmon is very interesting, and in many parts not a little mysterious. In the short space which is allowable for the subject, I will endeavour to trace the life of a Salmon from its earliest entrance into the world to its exit therefrom; putting forward no particular theories, but merely enumerating the accredited observations that have been made on this curious subject.

We will begin with the cradle that is prepared for the expected brood. This is a groove in the gravelly bed of a river, and is scooped out by one or both of the parents. Even here a discrepancy exists between practical observers, some of whom aver that the groove is made by both parents by means of rooting with their noses in the ground; others that the male Salmon scoops out the gravel with a hook-like appendage that is developed on his chin during the breeding season; while others declare that the male never troubles himself about the labour of scooping the groove, his duty being to watch over his mate and to fight any other fish of his own sex and species who may intrude upon their home, and that the whole task devolves upon the female, who executes it by twirling her tail and not by grubbing with her snout.

The whole process of depositing the numerous eggs occupies on the average about ten days, and, after it is accomplished, the parent fish leave the eggs to be hatched by surrounding influences, while they themselves quit the spot and remain in the river for a short period while they recover from the exhaustion caused by the process. During this period they are unusually ravenous, and vast quantities of the young of their own kind, which are about that time abundant in the river, fall victims to their insatiable appetite. After a time, and about the months of March and April, they drop down from pool to pool, in any flood which may seem favourable to them, until they reach the sea, where they are supposed to remain from six weeks to three or four months, when they again seek the river, vastly increased in weight and improved in condition.

SALMON.—*Salmo Salar.*

While they are occupied in this migration, the abandoned eggs are gradually approaching maturity under the influence of warmth and the rushing waters, and after a period, varying according to the temperature of the water, the young Salmon bursts through its prison. It is then a tiny and almost transparent creature, hardly to be recognised as a fish; and being too feeble to employ the mouth in obtaining subsistence, bears a portion of the egg still adhering to the abdomen like a transparent amber-coloured sac flecked with tiny blood-vessels; and by gradually absorbing this material into the system, preserves its life until its increased dimensions permit it to seize prey with the little mouth, afterwards to be so formidably arrayed with teeth.

In this stage of their existence, the little Salmon are called by a great variety of names and are marked with eight or ten dark patches upon the sides. It was for a long while a moot question as to how long these little fish remained in the river previous to their first migration to the sea. Some authorities announced that, from repeated experiments, they had found that these fish did not stay more than one year in the river; others as positively averred that on similar grounds they were enabled to show that they remained two years. Subsequent and more satisfactory experiments, conducted at Stormontfield, on the Tay, have shown, however, that both parties are right; for, singular to say, some of the fish go down after one year and some only after two years' confinement in the fresh water. And stranger still is the fact that hitherto all experiment, attention, and inquiry have failed to afford a satisfactory explanation of this apparent incongruity or irregularity of nature, and it remains a problem unsolved to this day. When the fish are about to depart for the sea, their mottled coat is exchanged for a covering of bright silvery scales, under which, however, the dark patches still exist, and can be seen by holding the fish in certain lights, or by rubbing off the lightly clinging scales. At this period the fish is called a Smolt.

It now prepares itself for an excursion to the sea, and, urged by an irresistible instinct, finds its way down the stream, until it at last emerges into the ocean. What may be the course of its marine life is not known, the fish being lost in the wide expanse of ocean; but, in the course of the autumn, it returns to the river whence it came, and forces its way up the stream. The technical name for the fish is now Grilse or Salmon Peal, and after its second visit to the sea it is called a Salmon. After spawning, the fish is in very bad condition, and is known by the name of Kelt. The flesh is then white, evil-smelling, and loathsome to a degree; nevertheless the French consume vast quantities of it disguised by the arts of cookery, and hundreds of tons are annually exported to Paris during the "close" season, to the great injury and destruction of our fisheries.

An animated controversy has raged at intervals respecting the identity of so changeable a fish at the several epochs of its life, and few persons seem to be able to decide positively whether certain fishes are young Salmon, or a separate species called by the name of Parr.

It has been thought by many practical observers that, as in the so-called Parr—which is marked just like the young Salmon—the milt, or soft roe as it is more popularly called, is fully developed, the creature must have reached adult age. But the female Parr has never yet been found with spawn in the same perfect condition, and experiments have been proved that the very young male Salmon, when only weighing an ounce and a half, and being about the size of a man's finger, has the milt fully developed, and capable of vivifying the eggs of the adult female Salmon. It may be casually observed that the young of the Salmonidæ bear a great resemblance to each other, and that the word Parr is used in a very vague and loose manner.

The perseverance shown, and the bodily efforts made by the Salmon in passing up the stream are really wonderful. No rapid seems too powerful to be overcome, and even falls of a considerable height are surmounted with marvellous force and address, the fish shooting from the bed of the river and concentrating all its forces into a simultaneous effort which drives it high into the air. It often happens that the fish leaps short of its mark, or alights in a wrong place, and is swept down again by the falling waters. Nothing daunted, however, it recommences its efforts, and is sure at last to succeed. An ingenious arrangement of alternate stages is now often affixed by proprietors to a fall where it is too high for the Salmon to pass it by a single spring. The fish soon find out this water staircase, and flock to the spot in preference to attempting the passage in any other locality.

Traps and nets of various kinds are often fixed at the falls, so as to catch the fish that have not made good their leap, or who happen to shoot out of the right direction.

During their whole lives the Salmon continue to migrate to the sea, invariably if possible returning to the identical river in which they were born. This migration is useful in the extreme, as the fish is liable to be infested while in the river with various parasitic animals, which cannot endure salt water, and fall off on its entrance into the sea, while the marine parasites are in a similar manner killed by fresh water, and die when the fish re-enters the rivers.

The Salmon is a most voracious fish, and its mouth is supplied with a tremendous array of teeth. In the upper jaw, the edge of the jaw-bone is supplied with a row of teeth; inside which runs a second row based on the bones of the palate. Along the roof of the mouth is placed a single straight row of teeth, set on a bone technically called the "vomer." The lower jaw-bone is furnished with its row of teeth to match those of the upper jaw, and each edge of the tongue is also toothed. When the jaws are closed, these complicated teeth lock into each other in a most admirable fashion, the single row of the under jaw fitting between the two ranks of the upper jaw, and the straight row of teeth on the vomer exactly coming between the two sets of tongue teeth. The vomerine teeth are, however, only to be found in perfection in the young fish, as they are gradually lost during the growth of their owner, and in an old fish are mostly reduced to two or three in number, while in some very old specimens one solitary tooth remains, set on the very front of the vomer, the sole survivor of the former array.

The food of the Salmon is extremely varied, as must necessarily be the case with a fish that passes its life alternately in fresh and salt water, and is certainly of an animal nature, but the precise creatures that form its sustenance are not accurately known; it is, however, known to feed largely on some varieties of the echinus, or sea-urchin. Fastidious as the Salmon may be in some cases, it is sadly indiscriminate in others, as is shown by the artificial fly which is made expressly for its capture. This extraordinary composition of gaudy feathers, gold and silver thread, and various heterogeneous materials, that are fastened on a large hook, and termed a "Salmon fly," has not the least resemblance to any insect that ever existed, and it is hardly possible to believe that the fish can be deluded into the idea that such an object belongs to the insect race. Perhaps the Salmon may be attracted by the very novelty of the object, and be induced to snap at it under the idea of securing a new dainty.

The Salmon is one of those fish that must be eaten fresh, in order to preserve the full delicacy of its flavour. If it be cooked within an hour or two after being taken from the water, a fatty substance, termed the "curd," is found between the flakes of flesh. If, however, more than twelve hours have elapsed from the death of the fish, the curd is not to be seen, and the Salmon is much deteriorated in the judgment of epicures.

It has already been mentioned that the colour of the Salmon changes greatly during the course of its life. In the adult fish, the back and upper part of the head are dark blackish blue, the abdomen is glittering white, and the sides are of an intermediate tint. On the body, and especially above the lateral line, a few dark spots are scattered. During the breeding season, the male Salmon assumes its most vivid hues, an orange golden tint spreading over the body, and the cheeks being marked with bright orange streaks. The size of this fish is extremely variable, some specimens having been caught that weighed sixty pounds, and Mr. Yarrell mentions one case where a female Salmon was captured about the year 1821, and was remarkable for weighing eighty-three pounds. This great weight was owing more to the depth and thickness of the fish than the length.

NEXT to the salmon, the bright-scaled carmine-speckled active TROUT is perhaps the greatest favourite of anglers, and fully deserves the eulogies of all lovers of the rod; its peculiarly delicate flesh, its fastidious voracity, and the mixture of strength, agility, and spirited courage with which it endeavours to free itself from the hook, forming a combination of excellences rarely met with in any individual fish.

The Trout is found in rapid and clear-running streams, but cares not for the open and shallow parts of the river, preferring the shelter of some stone or hole in the bank, whence it may watch for prey. Like the pike, it haunts some especial hiding-place, and, in a similar manner, is sure to take possession of a favourable haunt that has been rendered vacant by the demise of its predecessor or its promotion to superior quarters. Various baits are used in fishing for Trout, such as the worm, the minnow, and the fly, both natural and artificial, the latter being certainly the neatest and most artistic method. The arcana of angling are not within the province of this work ; and for information on that subject, the reader is referred to the many valuable works which have been written by accomplished masters of the art.

There is a curious method of catching Trout, much in vogue among the juvenile fishers. This process is called " tickling," and is managed as follows : The tickler gets quietly into the stream, and walks slowly along the banks, feeling carefully for any depression or cavity. One hand is then introduced very gently, while the other is placed over the entrance of the hole, the fingers being spread so as to prevent the exit of any fish that may happen to be resident in that locality. Several such cavities may be tried without success, but at last the smooth side of a fish is felt by the finger-tips.

TROUT.—*Salmo fario.*

The startled fish gives a great flounce on being touched, and tries to dash out of the hole, but, being checked by the spread hand, retires to the recesses of its cavern. The finger-tips are then gently brought against the abdomen of the fish, which soon endures the contact, and permits the hand gradually to inclose it. As soon as that is the case, the fish is suddenly grasped, snatched out of the hole, and flung ashore before it can find time to struggle from the captor's hold. Some accomplished ticklers aver themselves to be capable of thrusting the fore-finger into the gill and out at the mouth, and hooking out the fish in this singular manner.

The colour of the Trout is yellowish brown above, speckled with dark reddish brown, and a number of carmine spots are scattered along each side of the lateral line. The abdomen is silvery white, and the lower part of the sides rich golden yellow. There is, however, considerable variation in the colour of the Trout, the locality having considerable influence upon the tints.

One or two other species of this genus are found in British waters, and require a passing notice.

The BULL, or GREY TROUT (*Salmo eriox*) is found in several of our rivers, and in the Tweed is tolerably common. It often attains a very large size, but a specimen weighing more than fifteen pounds is not very common.

The SALMON TROUT (*Salmo trutta*), is another British species, and in general habits is very like the salmon, migrating to the sea, and returning to the rivers in a similar fashion.

The CHARR (*Salmo salvelinus*), the well-known and delicately flavoured SMELT (*Osmerus eperlanus*), called also the SPIRLING or SPARLING, the GRAYLING (*Thymallus vulgaris*), the VENDACE (*Coregonus Willoughbii*), and the ARGENTINE (*Scopelus Humboldtii*), so useful for bait, all belong to the same family as the salmon and the trout.

The PIRAYA, or PIRAI, has been removed from the salmonidæ and placed in another family on account of certain structural differences.

This fish is very plentiful in the rivers of Guiana and Brazil, where it swims in large troops, and is, according to many accounts, a very unpleasant neighbour. It is a most voracious being, with teeth nearly as sharply edged as those of the shark, and a boldness little short of that fish's well-known audacity. It is said, according to Spix, that if

PIRAYA.—*Serrasalmus Piraya.*

even so large an animal as an ox happens to get into one of their shoals, it is immediately assailed, and bitten so severely that it may succumb under its injuries before it can cross a stream thirty or forty feet in width. According to some authors, one of the South American tribes are in the habit of placing their dead in the streams, leaving them to the attacks of the Piraya, which in a single night will clear away the whole of the soft parts, and leave a clean skeleton ready for their peculiar mode of sepulture. Even living human beings seem to enjoy no immunity from this hungry fish, but to be liable to severe bites while bathing,

Be these stories literally true, or only exaggerations of reality, the jaws and teeth of the Piraya are perfectly capable of inflicting such injuries as have been briefly described. The teeth are nearly flat, triangular, and with edges sharp as those of lancets, and are employed by the Macoushi Indians to sharpen the points of those fearful wourali-poisoned arrows so well known to fame since they were brought by Mr. Waterton from Guiana. A part of the jaw containing five or six teeth is carefully cleansed, a hole is bored through the jaw-bone, and a string is passed through the hole and fastened to the edge of the quiver. The arrows are readily sharpened by placing the points between any two teeth and drawing them rapidly through the edges. There are now before me several of these arrows, kindly given me by Mr. Waterton, and which have been sharpened by this process.

IN a neighbouring family is placed a very remarkable fish, called the LUMINOUS SCOPELUS (*Scopelus stellátus*).

Mr. Bennett, who seems to have discovered this curious creature, writes as follows respecting its appearance :—

"Though the discovery of these (luminous) medusæ was a satisfactory explanation of the phosphorescent appearance of the water, I had yet to learn that the latter effect was partly produced by living, bony, and perfectly organized fish. Such fish were numerous in the sea this night, and a tow-net captured ten of them in the space of a few hours. They were a species of Scopelus, three inches in length, covered with scales of a steel-grey colour, and the fins spotted with grey. Each side of the margin of the abdomen was occupied by a single row of small and circular depressions of the same metallic grey hue as the scales, a few similar depressions being also scattered on the sides, but with less regularity.

The examples we obtained were alive when taken from the net, and swam actively upon being placed in a vessel of sea-water. When handled, or swimming, they emitted a vivid phosphorescent light from the scales or plates covering the body and head, as well as from the circular depressions on the abdomen and sides, and which presented the appearance of so many small stars spangling the surface of the skin. The luminous gleam (which had sometimes an intermittent or twinkling character, and at others shone steadily for several minutes together) *entirely disappeared after the death of the fish.* In two specimens we examined, the contents of the stomach were small shrimps."

The head of this fish is large and blunt, the eyes are remarkably large, and of a silvery whiteness. The scales are very loose, and fall off with the least touch.

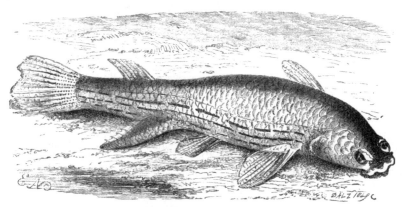

STARGAZER.—*Anableps tetrophthalmus.*

THE fish which is represented in the accompanying illustration may fairly take rank as one of the oddities of the finny race.

Flat headed, round bodied, and strong scaled, with projecting eyes of most remarkable formation, the STARGAZER has long attracted the attention of naturalists, and given the anatomical investigator much trouble in unravelling the intricate mechanism of its eyes. At a first glance, the fish appears to possess four distinct eyes, each of these organs being divided across the middle, and apparently separated into two distinct portions. In fact, an opaque band runs transversely across the cornea of the eye, and the iris, or coloured portion, sends out two processes which meet each other under the transverse band of the cornea, so that the fish appears to possess even a double pupil. Still, on closer investigation, the connexion between the divisions of the pupil are apparent, and can readily be seen in the young fish. The lens is shaped something like a jargonelle pear, and is so arranged that its broad extremity is placed under the large segment of the cornea.

The Stargazer is a native of Surinam, and is one of the viviparous fish. Three species of this genus are now known to naturalists.

THOUGH not so brightly spotted as the trout, nor so desperately active when hooked, and very inferior in flesh, the CARP is yet in much favour with anglers, on account of its extreme cunning, which has earned for the fish the name of Fox of the waters. As the number of British fish is so great, and our space so small, it will be needful to compress the descriptions as much as possible, and to omit everything that does not bear directly on the subject.

The Carp is found both in rivers and lakes, and in some places, among which the royal palaces of France may be mentioned, will often grow to an enormous size, and become absurdly tame, crowding to the bank on the least encouragement, and poking their great snouts out of the water in anxious expectation of the desired food. It is most curious to watch these great creatures swimming lazily along, and to see how completely they have lost the inherent dread of man by the exercise of their reasoning powers, which tell them that the once feared biped on the bank will do them no harm, but, in all probability, will be the means of indulging their appetite with favourite food.

The Carp is one of the fish that retains its life for a lengthened period even when

CARP.—*Cyprinus cárpio.*   BARBEL.—*Cyprinus barbus.*

removed from the water; and if carefully packed in wet moss so as to allow a free circulation of air, will survive even for weeks. Anglers never seem sure of the Carp—taking plenty on one day and none at all for a week afterwards, the fish having been aroused to a sense of their danger, and declining to meddle with anything that looks as if it might hide a hook. Even the net, that is so effectual with most fish, is often useless against the ready wiles of the Carp, which will sometimes bury itself in the mud as the ground line approaches, so as to allow the net to pass over it; or, if the ground be too hard for such a manœuvre, will shoot boldly from the bottom of the water, leap over the upper edge of the net, and so escape into the water beyond.

A fine Carp, say of six or seven pounds, is a truly handsome fish, its large shining scales lying on its body in the most beautiful regularity, and gleaming with olive-brown washed with gold. The abdomen is white, with a tinge of yellow.

THE beautiful GOLD-FISH (*Cyprinus aurátus*), so familiar as a pet and so elegant as it moves round the glass globe in which it is usually kept, is another member of this large and important genus. It seems to have been brought to this country from China, and has almost acclimatized itself to the cold seasons of England. Its habits and splendid clothing are too well known to need description.

ANOTHER well-known member of the same genus is the BARBEL, a fine but not brilliant fish, which is common in many of the English rivers.

This fish may easily be known, from the four fleshy appendages, called beards or barbules, which hang from the head, two being placed on the nose and the other two at each angle of the mouth. It is one of the mud-loving fish, grubbing with its nose in the soft banks for the purpose of unearthing the aquatic larvæ of various insects which make their home in such places, and being, in all probability, aided by its barbules in its search after food.

The following remarks on the habits of the Barbel are by Mr. F. T. Buckland :—

"After fishing, without any sport, a place in the river near Old Windsor, which had been previously well baited, I gave up the idea in consequence of the information of a man who was cutting willows. He said he 'could show me where the Barbel were.' Accordingly, we dropped quietly down in the punt a few yards from where we had been fishing, and pointing down under a bush he said, 'Look there, sir!' Underneath a thick hedge was a deep bank, hollowed out into ledges. The roots of a willow-tree grew out of this bank some six yards into the stream, and formed an archway large enough to wheel a wheelbarrow through. Peering down into the water, I saw at the lower entrance of the archway a dark moving mass, which I soon made out to consist entirely of Barbel.

There they were, great fellows, the patriarchs of the river, swaggering and pushing about their neighbours, of four or five pounds weight ; and little fellows not larger than roach, quietly reposing on the bottom, or rolling their fat sides round and round as their quarters became too crowded. After watching them some time, and ascertaining for a fact that the Barbel when at rest does more or less 'walk upon the bottom by means of his fins,' and trying them with every possible kind of bait (which, of course, they would not touch), I lowered the rod among them.

In an instant they all rushed up into their vegetable archway. Into this, which was too small to hold them all, they pushed, and crowded, and hustled, exactly like a flock of sheep driven out of a pen between two hurdles by the butcher's dog. They were evidently fighting for good places under the arch, as the tails of the hindermost moved in a most vigorous and energetic way, and every now and then one of them would lose his place, get pushed out, and would come sailing swiftly down away from the crowd ; then, turning head to stream, would charge right into the crowd again.

I have visited this Barbel's home several times since, and have always witnessed the same performance. What they get to eat there I cannot tell, unless they go out, like the cats, at night, and forage in the deeps."

The Barbel is sometimes so sluggish in its movements, and so deeply occupied in rooting about the bank, that an accomplished swimmer will sometimes dive to the bed of the river, feel for the Barbel along the banks, and bring them to the surface in his bare hand. From this habit of grubbing in the mud, the Barbel has earned the name of Fresh-water Pig.

The colour of the Barbel is brown above with a green wash, and yellowish green on the sides. All the scales have a metallic lustre, and the cheeks and gill-covers have also a polished look as if covered with very thin bronze. The abdomen is white. The Barbel is somewhat long in proportion to its weight, which is extremely variable, seldom, however, exceeding eleven or twelve pounds.

THE three fish that are represented in the engraving are well-known inhabitants of the British rivers, and are all in some favour with anglers.

The TENCH is hardly so common as the other two species, preferring the slowest and muddiest rivers, and thriving well in ponds and lakes or even clay pits. No water

ROACH.—*Leuciscus rútila*
BREAM.—'*Abramis branu.*
TENCH.—*Tinca vulgáris.*

indeed, seems to be too thick, muddy, or even fetid, for the Tench to inhabit, and it is rather curious that in such cases, even where the fishermen could scarcely endure the stench of the mud adhering to their nets, the fish were larger sized and of remarkably sweet flavour.

In the winter months the Tench is said to bury itself in the mud, and there to remain, in a semi-torpid condition, until the succeeding spring calls it again to life and action. The colour of the Tench is greenish olive, darker above than below, and with a fine golden wash.

The BREAM is mostly found in rather large lakes or in slowly running rivers, the lakes of Cumberland being favourite resorts of this fish. Although the flesh of the Bream is not held in any great estimation, being poorly flavoured and very full of bones, so that, in spite of the great depth of its body, there is scarcely sufficient flesh to repay the trouble of cooking, still, the fish was formerly in much repute as a delicacy; so that either the fish seems to have deteriorated, or the present generation to have become more fastidious. Spring and autumn furnish the best Bream, and the flesh can be dried something like that of the cod-fish.

The colour of the Bream is yellowish white, except the cheeks and gill-covers, which have a silvery lustre without any tinge of yellow. Sometimes the Bream attains a considerable size, reaching a weight of twelve or fourteen pounds.

SLY SILURUS.—*Silúrus glanis.*

THE last of the three is the ROACH, a fish especially dear to scientific anglers, on account of its capricious habits, and the delicate skill required to form a successful Roach-fisher.

An angler accomplished in this art will catch Roach where no one without special experience would have a chance of a bite, and will succeed in his beloved sport through almost every season of the year, the winter months being the favourites. So capricious are these fish, and so sensitive to the least change of weather, that a single hour will suffice to put them off their feed, and the angler may be suddenly checked in the midst of his sport by an adverse breeze or change in the temperature.

The Roach is a gregarious fish, swimming in shoals, and keeping tolerably close to each other. It is not a large species, all over a pound being considered as fine specimens, and any that weigh more than two pounds are thought rare. It is a pretty fish, the upper parts of the head and body being greyish green glossed with blue, the abdomen silvery white, and the sides passing gradually into white from the darker colours of the back. The pectoral, ventral, and anal fins are bright red, the former having a tinge of yellow, and the dorsal and tail fins are brownish red.

CLOSELY allied to the roach is the DACE (*Leuciscus vulgaris*), a common and small species that inhabits most of our streams. The well-known CHUB (*Leuciscus céphalus*) also belongs to this genus, as does the BLEAK (*Leuciscus alburnus*), in many counties called the TAILOR BLAY by the ignorant, from the idea that whenever any other fish, especially the pike, wounds its skin, it immediately seeks the aid of the Bleak, which, by rubbing its body against the wound, causes the torn skin to close. The beautifully white crystalline deposit beneath the scales was much used in the manufacture of artificial pearls, hollow glass beads being washed in the interior with a thin layer of this substance, and then filled with white wax. The scales of the whitebait were also used for the same purpose. The MINNOW (*Leúciscus phoxínus*) is another member of this large genus, and is too well known to need description.

WE now come to another family, selecting as an example a tolerably well-known species. The SLY SILURUS, sometimes called the SHEAT-FISH, is found in many rivers in different parts of the world.

As may be seen by the engraving, it is a curious-looking fish, and easily recognisable by the six tentacular appendages of its mouth, the two that are situated on the upper lip being of very great length. The precise object of these tentacles is not quite clear, though some persons believe them to be used as decoys, like the fin rays of the fishing frog, and to be employed in enticing unwary fish within reach of the mouth. Dr. Günther has kindly informed me, that he has often seen these fishes at liberty in their native streams, and that they are capable of directing the points of the tentacles towards any object that they seem anxious to examine. It is, therefore, probable, that these curious appendages are employed as organs of touch. It is one of the mud-loving fishes, and has a custom of hiding itself in holes, or nearly burying itself in the soft alluvium of the river's bed.

The flesh of the Silurus is not held in very high estimation, although its flavour is good, for it is so fat and gelatinous that it is difficult of digestion, and not to be eaten by persons of small assimilative powers. A kind of coarse isinglass, or very fine glue, is made from the swimming-bladder of this fish. The eggs of the Silurus are not very numerous in proportion to the size of the adult fish, and are of a greenish colour. They are much eaten by various fish.

The rather peculiar shape can be seen by the illustration, so that no verbal description is necessary. The colour is dark green above the lateral line, and of a paler tint below it, and a number of spots are scattered over the body without any apparent arrangement. The abdomen is of a yellowish colour, and the fins are tinted with blue and yellow. The Silurus sometimes reaches a considerable size, specimens of seven feet in length and weighing from seventy to eighty pounds having been captured.

HORNED TRUNK-FISH.—*Ostrácion cornútus*

A VERY curious order of fishes now comes before our notice. These creatures are called Pectognathi, because their jaws are coalescent.

THE remarkable family of the Trunk-fishes, or Sclerodermi, are known by the curious structure of the external surface, which is composed of a series of hard scales forming a continuous bony armour.

In the genus Ostracion, of which the HORNED TRUNK-FISH, or COFFER-FISH, is a good example, the body is either three or four-sided, and covered with a solid coat of mail formed of six-sided plates or scales, and pierced with holes, through which protrude the mouth, the tail, and the fins. The whole of the interior structure is modified in accordance with this external and inflexible cuirass; and on comparing the general

TRIGGER-FISH.—*Balistes conspicillum.*

form of this creature with that of certain reptiles, the analogy between the Coffer-fish and the tortoise is too close to escape observation. None of these fishes are in request as articles of food, their flesh being small in quantity, and in some species even thought to have a poisonous effect; but the liver is very large, and yields a tolerable supply of oil. All the Coffer-fishes are natives of the tropical seas, and but few species are known.

THE very curious TRIGGER-FISH is an example of a moderately large genus, inhabiting the warmer seas of many parts of the world, and which, on account of their rough and fierce exterior, are sometimes called FILE-FISHES, or LEATHER-JACKETS.

The name of Trigger-fish is derived from the peculiar structure of the dorsal fin. When the fin is erected, the first ray, which is very thick and strong, holds its elevated position so firmly that it cannot be pressed down by any degree of force; but if the second ray be depressed, the first immediately falls down like the hammer of a gun-lock when the trigger is pulled. The mechanical structure of these curious rays is extremely interesting, but the description would occupy too much space to be inserted in this work, beside requiring careful illustrations.

A strong feeling against the flesh of this fish exists among seafaring men, but, like many other nautical prejudices, is quite without foundation, the flesh being sweet and nutritious, though rather coarse. On the abdomen there is a bony keel, and on each side of the tail there are several rows of horny spines. This species is found in Japanese waters. In colour it is one of the most striking of its genus, being tolerably large, and black in colour, diversified by some large pale yellow or white spots upon the sides of the abdomen. Two other species are also given, in order to exhibit the curious variety of form and colouring found in these remarkable fish. The BRISTLY TRIGGER-FISH is notable from the quantity of bristle-like appendages to the tail, and the general appearance of the UNARMED TRIGGER-FISH can best be understood by reference to the illustration. All the fishes of this genus (which has been divided by some authors into several other genera) are inhabitants of the tropical seas, where they haunt the rocky coasts, and make the ocean

HAIRY URCHIN-FISH.—*Diodon pilósus.*   URCHIN-FISH.—*Diodon hystrix.*

radiant with their vivid tintings.   To all appearance they are vegetable feeders, as nothing but crushed seaweed has been found in their stomachs.

In the members of the family Gymnodontes, or Naked-toothed fishes, the jaws project from the mouth, and are covered with a kind of ivory or bony substance, composed of very little teeth fused together.

The URCHIN-FISH, or SEA HEDGEHOG, is a good example of the genus Diodon, or Two-toothed fishes ; so called because their jaws are not divided, and only exhibit one piece of bony substance above and another below, looking as if the creature only possessed two large teeth.

This curious fish is remarkable for the tremendous array of spiny points which it bears on its skin, and for the power of inflating its body into a globular form, and thus causing the spines to project in every direction, like the quills of an irritated porcupine or a hedge-hog that has coiled itself into a ball.   From this custom of inflating its prickly body it is sometimes termed the Prickly Globe-fish.

Mr. Bennett gives a curious account of the Diodon, from which the following remarks are extracted : "This fish lived for several hours after it had been removed from the water, and, as often as it was handled, it inflated its body, erected its spines, gnashed its teeth, and produced sounds by an emission of air through its mouth.   It inflated its body by pumping air into the gullet by frequent gulps or gasps.   The contents of the stomach were fragments of the shells of *Hylæa tridentata*, and a sucking-fish, the head of which was detached from the body, and much mutilated.   The sailor who brought me the specimen, reported that it had also thrown up a sucking-fish at the time it was captured.

3.   z

The resemblance which the spines bear to the similar erectile weapons of the porcupine and hedgehog among quadrupeds, is much increased by the existence of a thick layer of muscle (or *panniculus carnosus*) lining the entire skin of the fish, and firmly attached to the roots of the spines. From the extent of surface this muscle occupies, its power is evidently great, and its use is discovered by handling the inflated fish. The spines are then necessarily protruded, and press forcibly against any object with which they are in contact. But their action does not cease here, for the hand holding the inflated animal feels a constant boring motion of the spines, which is extremely painful, and very different from that produced by the passive contact of a thorny body.

Immediately beneath the muscular lining of the skin, there is a thick opaque membrane, covering the belly of the fish, and this, when distended with air, produces the inflated appearance of the body. It can be distended or emptied at the option of the animal, and is evidently supplied with air through the mouth and throat. I noticed, that by blowing into the mouth of a dead fish, I could inflate this membrane, and by passing a finger into the throat could as readily empty it.

To produce the inflation of the body, which must be regarded as a mode of defence, it would appear that the fish should have access to atmospheric air. Two nostrils, provided with large fleshy margins at the extremity of the snout, may supply this want when the fish is on the surface of the sea, and it is questionable whether water may not be employed as a substitute for air in the ocean's depths."

When full-grown, a fine specimen of this fish will measure more than a foot in diameter.

THE smaller figure on the same illustration represents the HAIRY URCHIN-FISH, a creature that is easily recognised by the bristle-like fineness of the spines.

STRIPED SPINEBELLY, OR BALLOON-FISH.—*Tétrodon lineatus.*

CLOSELY allied to the diodons are the Tetrodons, or Four-toothed fishes, so called because both jaws are divided in the middle, giving them the appearance of possessing four teeth, two above and two below. The spines of these fishes are comparatively small. The Tetrodons are popularly known by the title of BALLOON-FISH, as, like the diodons, they have the power of distending themselves with air, and causing all the spines to erect themselves. When inflated, they necessarily turn on their backs. The STRIPED SPINE-BELLY is a good example of this genus.

BESIDES the tetrodons, this group includes another genus called Triodons, or Three-toothed fishes, the upper jaw being divided into two parts, and the lower remaining entire. The spines are short, and a moderately large sac is seen beneath the body. The POUCHED TRIODON (*Triodon bursárius*) may be selected to represent the genus.

OUR last example of this curious order is the well-known SUN-FISH, which looks just as if the head and shoulders of some very large fish had been abruptly cut off, and a fin supplied to the severed extremity.

SUN-FISH.—*Orthragoriscus mola.*

Several specimens of this odd-looking fish have been captured in British waters, and, in almost every case, the creature was swimming, or rather floating, in so lazy a fashion, that it permitted itself to be taken without attempting to escape. In the seas where this fish is generally found, the harpoon is usually employed for its capture, not so much on account of its strength, though a large specimen will sometimes struggle with amazing force and fury, but on account of its great weight, which renders its conveyance into a boat a matter of some little difficulty.

The flesh of the Sun-fish is white and well-flavoured, and is in much request among sailors, who always luxuriate in fresh meat after the monotony of salted provisions. In flavour and aspect it somewhat resembles that of the skate. Its liver is rather large, and yields a considerable amount of oil, which is prized by the sailors as an infallible remedy against sprains, bruises, and rheumatic affections.

One of the most curious peculiarities of this fish is the structure of the eye, which is bedded in a mass of very soft and flexible folds belonging to the outer membranous coat, while it rests behind on a sac filled with a gelatinous fluid. When the creature is alarmed,

PEGASUS.—*Pegasus natans.*   SEA DRAGON.—*Pegasus draco.*

it draws the eye back against the sac of fluid, which is thus forced into the folds of skin, and distends them so largely as nearly to conceal the entire organ behind them.

While swimming quietly along, and suffered to be undisturbed, it generally remains so near the surface that its elevated dorsal fin projects above the water. Only in warm, calm weather is it seen in this attitude, and during a stormy season it remains near the bed of the sea, and contents itself with feeding on the seaweeds which grow so luxuriantly at the bottom of the shallower ocean-waters.

The colour of the Sun-fish is greyish brown, darker upon the back than on the sides of the abdomen, and the skin is hard and rough. It often attains a very great size, one that was harpooned on the equator measuring six feet in length. Several species of Sun-fish are known.

In the strange-looking fishes which are represented in the accompanying illustration, we have farther instances of the inexhaustible variations of form and structure with which this world teems, and which seem to be more plentiful, more bizarre, and more incomprehensible in the ocean than on the earth, in the air, or even in the rivers and other fresh waters of the globe.

The order to which these creatures belong is known under the name of Lophobranchii, or Crest-gilled fishes ; so called on account of the form of the gills, which are composed of little round tufts, and nearly hidden by the gill-cover. There is but little flesh upon the bodies of these remarkable fishes, which are protected by a hard bony armour, which, when examined, is found to be most beautifully constructed, so as to protect the animal and to allow of annual increase of dimensions.

The habits of the Lophobranchiate fish are not less remarkable than their forms, more especially those instinctive customs which relate to the continuation of their species. In almost every case, the eggs are given into the care of the male fish, who is provided with some apparatus suitable for the onerous task which he undertakes. In some species, the eggs are placed in certain folds or pouches of skin on the abdomen, breast, or tail ; in others they are carried between the ventral fins ; and in others again they are deposited in rows on the abdomen or breast, and there retained until the young escape from their prison.

SEA HORSE.—*Hippocampus brevirostris.*　　　　GREAT PIPE OR BILL-FISH.—*Syngnathus acus.*

In the family to which belongs the SEA DRAGON, the breast is developed in a wonderful manner, being always broader than deep, and in some cases the breadth very much exceeding the depth. The mouth is set under the projecting snout in a manner like that of the sturgeon, and the pectoral fins are extremely large and strong, reminding the observer of the same members in the flying gurnards. This seems to be a rather variable species both in form and colour.

The PEGASUS does not possess pectoral fins of such great size as the preceding species, but is yet a very remarkable fish. It is rather long-bodied, and the tail is composed of twelve rings. The much elongated snout is flat and thin, and is furnished on its upper edge with short spines directed backwards. The general aspect of its form can be understood by reference to the engraving; its colour is yellowish brown. The Pegasus is a Javanese creature.

The family of the Syngnathidæ is represented by several British species.

The SEA HORSE is common in many European seas, and is sometimes captured on the British coasts. In all these fishes there is only one dorsal fin, set far back, and capable of being moved in a marvellous fashion, that reminds the observer of a screw-propeller, and evidently answers a similar purpose. The tail of the Sea Horse, stiff as it appears to be in dried specimens, is, during the life of the creature, almost as flexible as an elephant's proboscis, and is employed as a prehensile organ, whereby its owner may be attached to any fixed object. The two specimens represented in the engraving are shown in the attitude which the creatures are fond of assuming. The head of the Sea Horse is wonderfully like that of the quadruped from which it takes its name, and the resemblance is increased by two apparent ears that project pertly from the sides of the neck. These organs are, however, fins, and when the fish is in an active mood, are moved with considerable rapidity. It is rather a remarkable fact, that the Sea Horse, like the chameleon, possesses the power of moving either eye at will, quite independently of the other, and therefore must be gifted with some curious modification in the sense of sight, which enables it to direct its gaze to different objects without confusing its vision.

The colour of this interesting little fish is light ashen brown, relieved with slight dashes of blue on different parts of the body, and in certain lights gleaming with beautiful iridescent hues that play over its body with a changeful lustre. About twenty species of Sea Horses are known.

In the foreground of the illustration is shown the GREAT PIPE-FISH, also called the BILL-FISH and NEEDLE-FISH, one of the commonest species of its genus.

This creature is found along our shores, and can mostly be captured at low water among the seaweed that has been left in the rock pools. To watch these remarkable fishes is an interesting occupation, for they assume such odd attitudes and perform such curious movements, that they never fail to arrest the attention, and never tire the observer. Sometimes they may be seen swimming about with tolerable speed like other fishes, their curious dorsal fins working like an Archimedean screw, and their long snouts being poked into every crevice. Sometimes, assuming a perpendicular attitude, they put their noses to the ground, and hold their tails aloft, while with their beak-like snouts they stir the sand, or, by ejecting water from their mouths, blow little hollows in it, probably for the purpose of disturbing the minute crustaceans and other marine creatures that find refuge in such localities.

The colour of the Great Pipe-fish is pale brown, diversified with transverse bars of a dark tint. The average length of a fine specimen is about eighteen inches, but it is said that the fish sometimes attains a length of two or even three feet.

Several species of this genus inhabit British waters, a rather curious example being the SNAKE PIPE-FISH (*Syngnathus anguineus*). This little fish is remarkably slender, and altogether snake-like in form, its length being about fourteen inches, and its thickness scarcely exceeding that of a common goose-quill. The dorsal fin is set very far forward. The tail fin is very tiny, and might easily escape observation altogether. A curious account of the Pipe fishes, and their mode of protecting and maturing their eggs, may be seen in the *Zoologist*, p. 7052.

THE RATHER quaint-looking species which is represented in the accompanying illustration, is a good example of a remarkable order of fishes, where the body is covered with hard bony scales that do not overlap each other, but are arranged side by side, like the tiles of a pavement, or the cubes of mosaic-work. This bony armour is very hard and smooth externally, being covered with a thin layer of a kind of enamel.

Although popularly called the BONY PIKE, from the mailed exterior and the lengthened wide-jawed form, which has some resemblance to that of a pike, this fish belongs to a totally different order, and in most points of its construction is formed after a different fashion. The general structure, indeed, of the Bony Pike is very remarkable, and affords another instance of the difficulty with which the fish are classed. The body is elongated, and the jaws are also lengthened and well furnished with teeth, looking very like an exaggerated pike's mouth, or the head of the common gavial of the Ganges. In each jaw there is a single row of sharp and conical teeth, and between them, and on the palate, are numerous other teeth, much smaller in size.

The scales of the Bony Pike are rhombic in form, very like the flat porcelain tiles with which certain ancient chimney-pieces were wont to be decorated, and hardly inferior to those tiles in the polished hardness of their exterior. They are very regularly arranged, being set so as to form a series of oblique rows, extending from the back to the abdomen. As in the sturgeons and sharks, the vertebral column runs along the upper edge of the tail fin. This fish is found in the lakes of America, and sometimes attains a considerable

BONY PIKE.—*Lepidosteus osseus.*

size, being often captured measuring three or four feet in length, and is said sometimes to attain a length of seven feet. Several species are said to inhabit the same waters; but when the remarkable diversity of form and colour which often reigns among the fishes is considered, it is highly probable that the supposed species may be nothing more than well-marked varieties. The flesh of the Bony Pike is said to be good.

THE well-known LAMPREY and its kin are remarkable for the wonderful resemblance which their mouths bear to that of a leech.

They are all long-bodied snake-like fish, and possess a singular apparatus of adhesion, which acts on the same principle as the disc of the sucking-fish, or the ventral fins of the goby, though it is set on a different part of the body. If all had their rights, indeed, the title of sucking-fish ought more correctly to be applied to the Lamprey than to the creature which is at present dignified by that appellation; as the one really applies its mouth to any object to which it desires to adhere, and forms a vacuum by suction, whereas the sucking-fish attains the same object by pressing the edges of the disc against the moving object to which it wishes to attach itself, and forms the needful vacuum by the movement of the bony laminæ.

Several fishes are popularly known by the name of Lamprey, but the only one to which the title ought properly to be given is the larger of the two species in the engraving.

The Lamprey is a sea-going fish, passing most of its time in the ocean, but ascending the rivers for the purpose of spawning. April and May are the months in which this fish is usually seen to enter the rivers of England, but in more northern countries the time is postponed according to the climate. In Scotland, for example, the usual month for spawning is June, and, as a general rule, the latter end of spring and the spawning of the Lamprey are synchronous.

The flesh of the Lamprey is peculiarly excellent, though practically unknown to the great bulk of our population, and the juvenile student in history is always familiar with the fatal predilection of British royalty for this fish. Though it spends so much of its time in the sea, it is seldom captured except during its visit to the rivers, and even in that case is only in good condition during part of its sojourn. Practically, therefore, the Lamprey is less persecuted than most of the finny tribe who are unfortunate enough to possess well-flavoured flesh, and whose excellencies are publicly known.

It has already been mentioned, that in many places there is a strong prejudice against the eel, and it is remarkable that a fish like the Lamprey, which is even better flavoured than the eel itself, should share a similar opprobrium. Many a hardy and hard-living fisherman flings out of his net, in disgust, a Lamprey which would hold a high place at a nobleman's table, and for which many a *bon vivant* would liberally open his purse. But it is always found that tastes vary with the locality, and that even nationalities have much to do with the great question of food.

A mere strip of water, over which a man can pass in less time than he occupies in strolling for his morning's walk or ride, marks out a radical difference in the food of two great nations; and even the veteran cosmopolite, well versed as he is in the customs of various lands, can seldom divest himself so fully of his nationality that he can bring himself to eat the objects which his childhood was taught to detest. Logic, for example, proves that snails ought to be equal or superior to oysters, as they are cleaner feeders, and live ever among green leaves; and philosophy reminds us that we ought to take advantage of any addition to the food of man. Forcible, however, as the one may be, and irrefragable the argument of the other, their combined influence will be insufficient to eradicate the early ideas which connect snails and loathsomeness by a link, wholly illogical, but not the less binding.

When the Lamprey deposits its spawn, it is obliged to form a hollow in the bed of the river, in which it can leave the eggs in tolerable safety, and performs this operation with great speed and no small skill. The fish is not gifted with any great power of fin, and cannot make much head against a sharp current, needing to rest at intervals, and for that purpose fastening on to some large stone over which the stream has no control.

But when it sets to work upon its nursery, it takes advantage of the current to help it in its labours, and, by the mingled force of the stream and its own muscular action, soon contrives to carry away the pebbles that would interfere with the well-being of its future young.

The process is simple enough. When the Lamprey has fixed upon the convenient spot to which it is urged by its unfailing instinct, it surveys the locality for a short time and then sets vigorously to work. Fastening itself to one of the obnoxious pebbles, and disposing its body so as to gain the strongest hold upon the rushing stream, it " backs water " with wonderful energy, and fish and stone are soon seen tumbling together down the current. In this way, the Lamprey will remove stones of such a magnitude that a fish of three times its dimensions would appear unable even to stir them. As soon as the stone has been moved a yard or two away, the Lamprey wriggles its way back again, and takes possession of another stone. By a repetition of this process, the hollow is soon made, and the industrious fish is able to deposit its eggs therein.

The respiration of the Lamprey is effected after a very curious manner, the chief internal organs being bathed with water by means of seven round holes at each side of the neck. The movements of this fish are very graceful, and although it is not powerful of fin, it can ascend very rapid currents by making little dashes upward, fastening itself to some fixed object, resting for a while, and then repeating the same manœuvre.

The colour of the Lamprey is olive-brown, spotted and mottled with dark brown and deep greenish olive. Its ordinary length is from sixteen to twenty inches.

THE lesser figure represents the LAMPERN, called in some counties the LAMPREEN, with that curious faculty of transposition which induces the rustic to speak of thursting instead of thrusting, and to call birds' nests, brids' neesuns.

The Lampern is plentiful in many of the English rivers, and if the generality of residents near the water were only aware of its excellence for the table, would soon be thinned in numbers. The prejudice that exists against the eel and the lamprey is absolutely mild when compared with the horror with which the Lampern is contemplated in many parts of England. Not only do the ignorant people refuse to eat it, but they

LAMPREY.—*Petromyzon marinus.*　　　LAMPERN.—*Lampetra fluviátilis.*

believe it to be actually poisonous, and would sooner handle an angry viper than a poor harmless Lampern.

It is fortunate for the fish that its evil reputation is so widely and firmly established, for, under shelter of its name, it passes scathless through many a stream, from which it would be nearly extirpated if its right character and good qualities were better known.

Granted the bad reputation, the creature certainly behaves in a manner well calculated to strengthen any unfavourable reports. For, as soon as grasped, it writhes about in a viperine, not to say venomous fashion, and is sure to fix its sucker of a mouth on the imprisoning hand. Few uninitiated captors can endure to any farther extent, and when they feel the cold lips pressed to the skin, and the quick suck by which the fish attaches itself, they generally utter a scream of terror, and fling the Lampern away as far as their arm can jerk it. Yet the creature has no idea of using its mouth as a weapon of offence, and when it affixes itself to the hand, is only seeking for a point of support as a fulcrum for its struggles.

Certainly, it has teeth, and under proper circumstances can use them in the task for which teeth were made, but it seems either to be unable or unwilling to employ them as weapons. I have caught thousands of these fish with the bare fingers, and had six or seven affixed on my hand at the same time; but they never did the least harm, and though I am afflicted with a peculiarly delicate skin, they did not even leave the least mark of their presence.

It is a pretty sight to see these fish come working their way up a rapid stream, especially when the water is high and the current strong. They are very cunning in seeking the spots where "dead water" is produced, and steer for them as for so many havens of refuge. In the shallower parts of the river they may be seen in their fitful progress from stone to stone, sometimes appearing singly, sometimes in bands of three or four, and sometimes in actual shoals. In the deeper water the Lampern is very seldom seen, as it is a ground-loving fish, and keeps itself mostly to the bed of the river. Now and then, however, a solitary specimen may be seen near the surface, wriggling its way painfully along; and in muddy spots, where it finds no hold, is sometimes swept back repeatedly by the current before it succeeds in passing the inhospitable spot.

Like the sea lamprey, it scoops hollows in the pebbly bed of some stream for the purpose of depositing its eggs, and removes the stones in like manner. Sometimes a pair of Lamperns settle upon one spot, and by dint of tugging and hauling, make a cradle for their special benefit. But it often happens that a great number of these fish, fifty or sixty for example, will settle themselves in the same locality, and make a hollow as large as the rim of an ordinary pail.

They present a very quaint appearance while thus engaged. An inexperienced observer would hardly take the black wriggling mass to be composed of fish, but rather to be a collection of decaying weeds, agitated by the ripple of the stream. Soon, however, these masses can be resolved into their constituent elements, and are gradually analysed into the long, slender bodies of the Lampern, all holding by their mouths to the pebbles, and writhing their tails at a wonderful rate, reminding the observer of a lamb's tail while the little creature is taking nourishment.

Every now and then one of their number becomes detached, and is swept down the stream, carrying with it a stone which it has found to be objectionable; and after it has dropped the stone some few yards down the river, it may be seen to work its way up anew, and regain its former position. Certain parts of rivers are particularly favoured by the Lampern, and year after year these fish may be found in precisely the same spots.

The flesh of the Lampern is remarkably excellent, and in many places, remote from its habitation, is in great repute, and is indeed admired by many who have not the least idea of the fish they are eating. A large part of the eel pies so famous in the metropolis is composed of Lampern flesh, and in the opinion of competent judges the substitute is better than the reality. It can be dressed in a variety of ways, stewing and potting being the favourites. Yet, as a general rule, the poorer portion of the community refuse to eat the fish, and suffer the pangs of cruel hunger rather than avail themselves of the rich banquet at their very doors.

It is pitiful to see how much nutritious and palatable food is annually wasted through prejudice and ignorance; and none could confer greater benefits on the country than he who would teach the poor, by precept and example, how to avail themselves of the food that lies wasting at their feet. It would be a work of many years, and the rooted prejudice of the old would never be overcome; but the coming generation might receive the lesson, and if the community can be taught to appreciate the value of edibles already existing, they receive just the same benefit as if they were taught how to produce an entirely new article of consumption.

The Lampern was once in great request as bait for cod and turbot, but seems of late years to have gone out of fashion.

Along each side of the neck are seven round holes, leading to the respiratory apparatus, and a single aperture on the top of the head is also connected with that organ. The eye is round, and set rather far back. The mouth is nearly circular, and is well supplied with teeth. The general shape of the body is cylindrical for the greater part of its length, but flattened towards the tail, and the colour is slaty blue above and white beneath. Its length is usually about one foot or fifteen inches.

A BEAUTIFUL adaptation of structure to circumstances is seen in the POUCHED LAMPREY, an inhabitant of the fresh waters of Southern Australia.

This remarkable fish possesses many points of interest, among which the enormous throat-pouch is the most conspicuous. In the common sea lamprey of Europe, the throat is dilatable below, but in the present species the skin is distended so greatly as to form a large pouch, such as is represented in the engraving. Taking into consideration the frequent droughts that take place in the country where this creature lives, it is almost certain that the pouch is intended to hold a supply of water, which will enable respiration to be carried on during the days of drought.

The mouth of this species is very large and filled with formidable teeth, the whole interior of the disc being studded with them, and the tongue armed with two long and sharp fangs that seem calculated to do good service to their owner. A fine specimen of this fish is in the British Museum.

POUCHED LAMPREY.—*Geotria Australis.*

A VERY remarkable species of lampern, termed the SAND PRIDE, or MUD LAMPREY (*Ammocœtes branchiális*), is found in many rivers, and has sometimes been mistaken for the young of the sea lamprey.

It may, however, be distinguished by the form of the mouth, which is of a horse-shoe shape, and incapable of adhesion like that of the fishes belonging to the preceding genus. The Sand Pride, although tolerably common, is not very often seen, owing to its habit of burying itself in the muddy or sandy beds of rivers. It is taken plentifully near Oxford, and is common in the upper parts of the Thames. Its colour is yellowish brown, the latter tint preponderating on the back and the former beneath. It is a very small species, seldom exceeding six or seven inches in length.

The MYXINE, or GLUTINOUS HAG-FISH, is so remarkably worm-like in its form and general appearance that it was classed with the annelids by several authors, and was only placed in its proper position among the fishes after careful dissection.

The Myxine is seldom taken when at large in the sea, but is captured while engaged in devouring the bodies of other fish, to which it is a fearful enemy in spite of its innocuous appearance. It has a custom of getting inside the cod and similar fishes, and entirely consuming the interior, leaving only the skin and the skeleton remaining. The fishermen have good reason to detest the Myxine, for it takes advantage of the helpless state in which the cod-fish hangs on the hook, makes its way into the interior, and if the fish should happen to be caught at the beginning of a tide, will leave but little flesh on the bones. The cod thus hollowed are technically called "robbed" fish. Six Myxines have been found within the body of a single haddock.

The name of Glutinous Hag-fish is derived from the enormous amount of mucous secretion which the Myxine has the power of pouring, from a double row of apertures, set along the whole of the under surface, from the head to the tail. It is said that the fish is accustomed to envelop itself in a cloud of glutinous matter whenever it is alarmed, and under cover of this substance to escape the view of its enemies; thus presenting a curious parallel to the well-known habit of the sepia and the aplysia, or sea-hare, which, when startled, stain the water with their protective secretion and shoot off under shelter of the sudden darkness.

Around the lips of the Myxine are eight delicate barbules, which are evidently intended as organs of touch; the mouth is furnished with a single hooked tooth upon the palate, serving apparently as an organ of prehension, and the tongue is supplied with a double row of smaller but powerful teeth on each side, acting on the principle of a rasp. The Myxine can scarcely be said to possess any bones, the only indication of a skeleton being the vertebral column, which is nothing more than a cartilaginous tube, through which a probe can be passed in either direction. The structure of the breathing-organs is very remarkable, but can be hardly understood without the employ-

HAG-FISH, OR MYXINE.—*Myxine glutinosa.*

ment of figures. Suffice it to say that a double row of branchial cells take the place of gills or lungs, and are supplied with water through a spiracle in the upper part of the head, and two little apertures on the under surface.

The colour of the Hag-fish is dark brown above, taking a paler tint on the sides, and greyish yellow below. Its length is generally about a foot or fifteen inches.

THE last of the fishes is a creature so unfishlike that its real position in the scale of nature was long undecided, and the strange little being has been bandied about between the vertebrate and invertebrate classes. Between these two great armies the LANCELET evidently occupies the neutral ground, its structure partaking with such apparent equality of the characteristics of each class, that it could not be finally referred to its proper rank until it had been submitted to the most careful dissections. In fact, it holds just such a position between the vertebrates and invertebrates as does the lepidosiren between the reptiles and the fishes.

It has no definite brain, at all events it is scarcely better defined than in many of the insect tribe, and only marked by a rather increased and blunted end of the spinal cord. It has no true heart, the place of that organ being taken by pulsating vessels, and the blood being quite pale. It has no bones, the muscles being merely attached to soft cartilage, and even the spinal cord is not protected by a bony or even horny covering. The body is very transparent, and is covered by a soft delicate skin without any scales. There are no eyes, and no apparent ears, and the mouth is a mere longitudinal fissure under that part of the body which we are compelled, for want of a better term, to call the head, and its orifice is crossed by numerous cirrhi, averaging from twelve to fifteen on each side. Altogether, it really seems to be a less perfect and less developed animal than many of the higher molluscs.

The habits of this remarkable fish are very curious; and it will be better to give the original accounts in the words of the narrators, than to condense or paraphrase them.

The first history of the Lancelet is given by Mr. Couch, who was the first captor of the fish on the British coasts. He saw its transparent tail projecting from beneath a stone on the shore at low tide, and swept it into his hand together with some water "When alive," writes Mr. Couch, " this fish had a very evident though diaphanous fin, extending from near the snout round the extremity of the tail, which it encircled in the manner of the same organ in the eel, and terminating at the vent. This specimen was not found in a pool, but lay buried in a small quantity of sand, at about fifty feet from the receding tide; and on turning over a small flat stone that was on the sand the tail of the fish appeared exposed.

When moved, it exhibited signs of great activity, so that the head could not readily be distinguished from the tail; and as there can be no doubt that the fish had sought the shelter of the sand in which it was found, there is little question that mud is its usual habitation; a circumstance still more probable by its want of eyes.

It was discovered on the 21st of December, 1831, after a heavy storm that had torn it from its native situation, which, from its rarity, we may suppose to be in deep water. In February, 1838, I obtained two other specimens, which had been thrown up by a tempest. The largest measured two inches and three-tenths in

LANCELET.—*Amphioxus lanceolatus.*

length, which enabled me to discern still more of the internal structure of this fish."

Several other specimens have been obtained, mostly scraped up in the dredge. On the coasts of the Mediterranean this fish is not uncommon; and the following interesting account of some of its habits is given by Mr. Wilde, in his narrative of a voyage to the Madeiras, Teneriffe, and along the shores of the Mediterranean. After describing the general appearance of the fish, he proceeds as follows :—

"These little animals had a power of attaching themselves to each other in a remarkable manner, sometimes clustering together, and at others forming a string six or eight inches long; the whole mass seemed to swim in unison and with great rapidity, going round the vessel in a snake-like form and motion. They adhered to each other by their flat sides, when in line, the head of one coming up about one-third on the body of the one before it; no doubt those sides are of use in forming this attachment.

The mouth was a circular disc surrounded by cilia that continued in constant motion. When put into a tumbler of water it moved round the glass, and although no eyes were perceptible, it carefully avoided the finger or any substance put in its way, stopping suddenly, or turning aside from it. Both these animals, when taken out of the water, kept up a strong pulsatory motion for some time. The small one by this means pumped out of its interior a quantity of air and water; and they could be seen coming to the surface to inhale, and a globule of air was observed floating through the internal cavity.

In the larger species the internal tube was perfectly distinct and of a blue colour. When put into spirits and water it died almost immediately, and turned opaque; a number of circular bands also appeared on it."

The general aspect of the Lancelet is not unlike that of the leptocephalus already described, the delicate transparent body and the diagonal arrangement of the muscles causing a considerable resemblance between the two. But the leptocephalus is at once distinguished by its head, which, although very small in proportion to the body, is yet perfect, possessing well-developed eyes, gill-covers, jaws, and teeth, whereas the Lancelet has no particular head, and neither eyes, gill-covers, jaws, nor teeth.

CHEQUERED SWINE-FISH.—*Halichœres hortulanus*

THE following species have been inadvertently omitted in the preceding descriptions, on account of recent alterations in scientific systems, and are now inserted as a brief appendix.

The great and important family of the Labridæ, or Lip-fishes, so called in allusion to the thick and fleshy lip with which their mouths are furnished, are spread over the greater portion of the globe, few coasts being without several representatives of the group.

These fishes are not only remarkable for the full fleshiness of their evidently sensitive lips, but for the endless variety of rich and vivid tints with which their bodies are decorated, hues pure as the bright patterns of cathedral windows, and often arranged with a symmetrical regularity of outline and a daring harmony of contrasting colours that do not wholly lose their pictorial effect when translated into the simple black and white of a wood engraving, and when seen on the healthy and living fish appear as if flung on its scales direct from the kaleidoscope. Of the three primary colours, red seems to retain its purity of tone more perfectly than either the blue or the yellow, the former usually being mingled with yellow, and forming greens of varying qualities, while the latter frequently takes a slight tinge of red, and becomes warm orange. These tints are extremely variable, ranging through every tone of the secondary colours, and even in different individuals of the same species the colour is so uncertain that no dependence can be placed upon it as a means of determining the particular species.

Even on our own coasts the Labridæ are most lovely creatures, but it is in the tropical and warmer seas of the world that they are to be seen in their full brilliancy. No artist can transfer to paper the radiant hues that glow on these favoured members of the finny race, and no pen can do justice to their wondrous splendour, as they dart through the crystalline water like living meteors, or leisurely traverse the forests of moving algæ, balancing themselves among the submarine foliage like humming-birds of the sea.

The CHEQUERED SWINE-FISH is one of the singular species which are so frequent in the hotter seas, and which exhibit a surface at once coloured with the brightest hues, and decorated with a pattern as geometrically regular as if drawn by the aid of rule and compass. This fish is found on the coasts of Ceylon. Its colour is bright green on the back, gradually changing to golden yellow towards the abdomen. The chequered marks on the body are purple and grey, and the stripes and spots on the head are rich glowing orange. In allusion to the wicker-like markings and the brilliant colours of the fish, the natives call it by the name of Hembili Girawah; the former word signifying the little basket in which betel is carried, and the latter a parrot.

Though edible, this fish is not regularly captured for the markets. It generally frequents the rocky portions of the coast. It never attains any great dimensions, none seeming to measure more than fifteen inches in length.

SLY EPIBULUS.—*Epibulus insidiator.*

THE curious fish which is represented in the accompanying illustration, has derived its popular and scientific titles from the crafty manner in which it obtains its food.

In this fish, the mouth and jaws exhibit a very remarkable modification of structure, which enables the creature to protrude its mouth with great rapidity, and to such a degree, that when pushed forward to its greatest extent, it is even longer than the head. Aided by this apparatus, the SLY EPIBULUS captures its prey as follows. It feeds mostly upon small fishes, and instead of alarming them by charging among their ranks, and so giving itself a tedious and possibly an unsuccessful chase, it quietly withdraws itself to some sheltered spot, and waits, still and motionless as the watchful kingfisher, and no less dangerous to the smaller members of the finny tribe. No sooner does an unwary fish pass near the cunning enemy, deceived by its apparent harmlessness, than the movable mouth is suddenly projected with the rapidity of a serpent's stroke; and the victim is seized and swallowed in a moment.

The entomological reader will doubtlessly perceive the analogy between the Epibulus and the larva of the dragon-fly. Each is aquatic, each is furnished with a peculiar apparatus capable of being projected and withdrawn with great speed, and each depends more on this structure for food than on its swiftness, activity, or strength.

The movement of the jaw is not merely accomplished by the elongation of the lips and soft parts of the mouth, but by a most curious development of certain bones, technically called the "intermaxillaries." These bones are considerably lengthened, and lie in corresponding grooves of the skull, so that they glide backwards or forwards whenever the fish desires to protrude or withdraw its mouth. The Sly Epibulus is a native of Java, Sumatra, and the Moluccas.

BALLAN WRASSE.—*Labrus bergylta.*

FROM the many species of the genus Labrus that are found on our coasts, the BALLAN WRASSE has been selected. This fish, otherwise known by the popular names of ANCIENT WRASSE, or OLD WIFE, is not uncommon on the shores of the British Islands, and is one of those species which is mostly found haunting the rocky portions of the shore.

It is not in any great estimation as an article of food, the flesh being too soft, and not possessing any particularly good flavour. It generally frequents the deep rocky gullies, where the water is tolerably tranquil, except when the waves are beaten into foam by a storm. Here it may be seen swimming about in the clear element, concealing itself among masses of seaweed, and ever and anon darting forth to secure some tempting morsel, such as a passing crab or prawn.

Mr. Couch remarks concerning this species, that " it takes a bait freely, and fishermen remark that when they first fish in a place, they take but few, and those of a large size ; but on trying the same spot a few days after, they catch a greater number, and those smaller ; from whence they conclude that the large fish assume the dominion of a district, and keep the younger at a distance."

The Wrasse deposits its spawn in spring or towards the beginning of summer, and, as is usual with most fishes, assumes its brightest apparel previous to performing that duty. There is much uncertainty about the colouring of the Ballan Wrasse, but in general the colour is red above and on the sides, variegated with oval spots of rich bluish green ; the fins are green, sparely spotted with red, and the abdomen is pale orange. It does not attain to any great size, seldom having been known to exceed sixteen or eighteen inches in length, and two or three pounds in weight.

TESSELATED PARROT-FISH.—*Scarus harid.*

SEVERAL other British species of the same genus are known, such as the GREEN STREAKED-WRASSE, or GREEN-FISH (*Labrus Donovani*), a rather rare but very beautifully coloured fish, almost wholly green and slightly streaked. Some naturalists think that this is only the young of the preceding species. The most curiously decorated British species is, however, the THREE-SPOTTED WRASSE (*Labrus trimaculatus*). This fish is decorated with a rich ruddy orange over the greater part of its body, becoming slightly paler on the sides, and changing to golden yellow on the abdomen, with the slightest possible dash of red. On the upper part of the back, and occupying portions of the dorsal fin, are three large spots of deep rich purple, between which are placed four similarly shaped spots of pale rose. These spots, however, are rather variable in number.

IN the course of the preceding pages our notice has been drawn to many remarkable forms of fishes, some terrible in their fearful armatures of spines and teeth, some repulsive from their slimy exterior and coldly malignant aspect, and others almost bordering on the grotesque from the odd and eccentric manner in which various parts of their structure are modified. The TESSELATED PARROT-FISH of the Ceylonese seas, though not strikingly unique in its external appearance, as many of the species already described, is, when closely examined, one of the most wonderfully coloured fish in the world. As may be seen by the engraving, the whole body is covered with a beautifully drawn pattern of elongated hexagons, as perfect and regular as those of a honeycomb.

The colours of this remarkable fish are as follow: The general hue of the Tesselated Parrot-fish is azure-blue, covered with a hexagonal network of golden yellow. The oddly shaped head is bright yellow, streaked and spotted with blue. The dorsal and anal fins are brown edged with green, and the pectorals and ventrals are brown with the front rays green. The tail fin is wholly green. The natives call this fish by the name of Laboo Girawah, the former term being the name of a certain gourd or pumpkin which is marked in a somewhat similar fashion.

The species belonging to this genus are very numerous, and have received their rather appropriate title of Parrot-fishes from the rich beauty of their colours and the peculiar

3.

UNARMED TRIGGER-FISH.—*Balistes vetula.*   BRISTLY TRIGGER-FISH.—*Balistes tomentósus.*

form of their jaws, which are very strong, covered with great numbers of mosaic-like teeth, and curved in a manner that greatly resembles the beak of a parrot. As the fish wears out the teeth rapidly while crushing the corallines and other hard substances on which it feeds, a provision is made for insuring a continual supply of new teeth to replace those which are worn away and rendered useless. The young teeth are perpetually being developed towards the base of the jaws, and by a beautiful yet simple adaptation of existing parts, which cannot be made intelligible without the use of diagrams, advance in orderly succession towards the front, and take their places with unfailing certainty in the densely compacted mosaic-work which arms the jaws.

THE two curious fish represented in the engraving belong to the genus Balistes, described on page 336, and the illustration was omitted in consequence of a mistake of the artist in separating them from the Trigger-fish (*Balistes conspicillum*).

The name of Balistes is derived from the ancient weapon of war, termed the Balista, which projected a spear or heavy stone with exceeding violence. The curious spine with which the back is armed can be suddenly erected and depressed, as if shot with a spring, and has been compared to the weapon above mentioned. It has been thought that the flesh of these fishes is poisonous, but the truth of this opinion is very dubious. They are all decorated with bold and sometimes beautiful markings, black, ashen grey, blue, and yellow being their usual colours.

INVERTEBRATE ANIMALS.

MOLLUSCS.

WE now come to the second great division into which all animated beings have been distinguished. All the creatures which we have hitherto examined, however different in form they may be, the ape and the eel being good examples of this external dissimilarity, yet agree in one point, namely, that they possess a spinal cord, protected by vertebræ, and are therefore termed Vertebrated animals.

But with the fishes ends the division of vertebrates, and we now enter upon another vast division in which there is no true brain and no vertebræ. These creatures are classed together under the name of Invertebrated animals; a somewhat insufficient title, as it is based upon a negative and not on a positive principle. Whatever may be its defects, it has been too long received, and is too generally accepted, to be disturbed by a new phraseology, and though it be founded on the absence and not the presence of certain structures, it is concise and intelligible.

Numerous as are the species of the vertebrated animals, those of the invertebrates outnumber them as an army outnumbers a company. Although many species of mammals, birds, reptiles, and fishes, are at present known to science, and the yet unrecognised species are necessarily extremely numerous, there is some hope of obtaining an approximate calculation of their respective numbers. But with the invertebrates, any approach to a census even of known forms is well-nigh impracticable; and as it is evident that the ocean alone contains within its fathomless depths myriads of beings as yet hidden from mortal eyes, the reader may conceive the utter impossibility of offering the slightest conjecture respecting their numbers.

It is evident, therefore, that in a work of this limited extent, no complete and detailed history of the invertebrates can be attempted, and that the utmost that can be achieved is, to give a general outline of each class and order, to describe and figure the leading or typical forms, and to give occasional detailed accounts of certain species, selected for the beauty or singularity of their form, for the interesting nature of their habits, or for their direct usefulness to man.

IN entering upon a wholly new class of beings, we are forced to abandon the characteristics on which we lately depended for the classification, and to seek for new structures whereon to build our system. The peculiar formations which are now accepted for this purpose will be mentioned as we proceed with the work.

THE first order of Invertebrated animals is called MOLLUSCA, a name given to these creatures on account of the soft envelope which surrounds their bodies. . Most of the Molluscs are farther protected by a larger or smaller shell, and in former days, when mere externals were rated beyond their value, with the great mass of naturalists, or rather of virtuosi. the shell was the only portion of the animal that was considered, and the study of these beings was therefore termed conchology, or "shell-lore." Beautiful as the shells are, and capable as are the present school of naturalists of appreciating the gorgeous richness or tender delicacy of their colours, and the strange eccentricity or exquisite grace of their forms, those who study these beings have now learned to comprehend that the shell is, despite of all its beauty, but a secondary consideration, and that the true naturalist cares not so much for the lifeless shell as for the creature which inhabited it, and from whose tissues it was formed.

In the course of the few pages which can be given to the Molluscs, the reader will observe that, wherever practicable, the form of the inhabitant has been given together with that of the shell. There are, moreover, many most curious and interesting forms among the Molluscs, where not a vestige of shell exists, and which, by the former plan, would be entirely excluded from the attention of the "conchologist."

Before proceeding to describe the different species of Molluscs, I will briefly recount the principal characteristics which designate these. beings, and separate them equally from the apparently less organized forms above them and the apparently more organized forms below.

In the first place, it is worthy of notice, that—except in the cephalopods, popularly known as the cuttle-fishes, which are thought by many anatomists to form a distinct class by themselves—the symmetry of arrangement which has been found throughout the previous order is now seen no more; the corresponding pairs of limbs and other structures now vanish, and the only indication of its existence is found by close examination of the nervous system.

There is no true brain in these animals, the whole nervous system being composed of certain masses, or knots of nervous matter; from which radiate the needful filamentous conductors of the mysterious life-power. The visual organs of the Molluscs are, with a few exceptions, but slightly developed, and are far inferior to those of the crustaceans, insects, and other creatures which take a comparatively low rank in the animal kingdom. Many Molluscs are clearly destitute of these organs when adult, though on their primary entrance into life they are gifted with a tolerably well developed, though evanescent organ of sight. Whenever the true eyes are present, they are only two in number, and are sometimes placed in very singular situations, the well-known black-dotted tentacular eyes of the common snail affording a familiar example. The little coloured points which exist in some species are not here taken into consideration.

The power of progression in the Molluscs is greatly varied, being in some species almost absent, while in others it is developed to a wonderful degree. Many of these creatures, such as the mussel, the limpet, and the oyster, scarcely stir from the spot where they have once fixed their habitation; the snail and those of a similar form glide slowly along by means of the curiously developed mass of muscular fibres, technically called, from its use, the foot; the scallop drives itself through the water in short jerks or flights, caused by slowly opening and then rapidly shutting its valves; several species are known to jump by a sudden stroke with the foot; the nautilus urges its shell through the waves by the violent expulsion of water from its interior, and is driven along on just the same principle by which a sky-rocket soars into the air; and the flying squid, one of the cuttle-fishes, is able to rival even the flying-fish in its aërial journeys, shooting through the air to considerable heights, and even leaping fairly over both bulwarks of a ship and alighting in the water on the opposite side of the vessel.

The old fable of the nautilus and its sails has long been rejected, but the fabricators of this legend need not have visited the ocean for an example of a molluscan boat. Any one who is in the habit of watching the streamlets that irrigate while they drain our meadow lands, must have seen the common water-snails come floating down the current, lying on their backs, their shells submerged, and the edges of their fleshy foot turned up on all sides so as to convert that organ into a miniature flat boat.

That the Molluscs, or at all events some of the species, possess the sense of hearing, is tolerably evident from an examination of the structures. Near the nervous knobs, or ganglia as they are scientifically termed, of the head, are placed some little vesicles, each filled with a transparent fluid, and containing a tiny knob or spikelet of chalky matter, very similar to the well-known ear-bones of fishes, and probably serving a similar purpose. These "bones" appear to be perpetually in motion within their crystalline cell.

The circulation of the Molluscs is tolerably defined, especially in the higher and best-developed species, where the blood is urged on its course by a definite heart, and ramifies through the body by means of well-developed vessels. In the lower forms, however, these vessels can no longer be distinguished, and the blood circulates through a system of little cavities distributed in the body. So completely is this the case, that many Molluscs can be successfully injected, by introducing a fine-nosed syringe at random into the body, and pressing the heated substance very gently into the system.

The body is surrounded by a soft membrane termed the mantle, which is flexible extensile, and contractile to a marvellous degree. The movements of the Molluscs are mostly performed by means of the mantle, and through this structure the shell is secreted and moulded into form.

Of the secondary services rendered to man by the Molluscs we know but little, owing to the localities in which the greater number of species live, their nocturnal or darkness-loving habits, and their extreme dislike to intrusion. Several species, such as the pholas or burrowing shell, and the teredo, so notorious as the shipworm, are well known to be actively injurious to man, by destroying the foundations on which his edifices are built, or the vessels in which he trusts his life and property to the waves. Yet even these insidious enemies may have their uses to man, and by destroying the wrecks on which many a noble vessel might be driven and share the same fate, may be the salvation of costly property and invaluable lives.

As to those which are known to be directly useful to mankind, it will be sufficient to give a brief enumeration at present and to mention particulars when we come to the individual species.

Usefulness to man is, when reduced to its essence, the capability of affording him food, and therefore the edible species must take first rank as regards use. The oyster is familiarly known to rich and poor, the latter being often more practically cognisant of its value than the former. It affords at once a refined luxury and a health-giving nutriment ; it can be eaten uncooked, or opens a wide field for culinary art ; and it has the farther advantage of being very plentiful, very cheap, very accessible, and very easily preserved in a living state until needed. The mussel is another largely consumed Mollusc, especially among the lower classes who cannot afford to buy oysters ; and in some parts of the world is cultivated and bred in millions, the ever-increasing numbers, together with the peculiar accommodation which they require, threatening to obliterate many a natural harbour, and causing a well-grounded apprehension among ship-owners that their vessels may be deprived of their accustomed refuge by means of this simple Mollusc. The scallop again, with its classically famous shell and coral-red foot, is another of the edible species, as is the cockle, another well-known bivalve of our coasts. Some of the foreign bivalve Molluscs are considered as very great delicacies, among which the clam takes a very high rank. There is also the huge giant clam, formerly rare, but now perfectly familiar ; which has to be cut away from the rock by hatchets, and whose contents are equivalent to a large round of beef, very well flavoured, but rather tough and stringy.

The single-valved species furnish many edible examples, such as the whelk and the periwinkle, so largely consumed by the poor, and even the cuttles are capable of affording a tolerably good repast when properly dressed. As a general rule, however, the bivalves are most esteemed, as they are not so fibrous in texture, and therefore not so tough as the univalves.

The shells of the Molluscs are also of much service to mankind. Putting aside the well-known money cowry, perhaps the most infinitesimally divided currency in the world, many species are of exceeding value for the materials furnished by their shelly coverings. Some species, where the shell is of that lovely nacreous nature which we popularly term mother-of-pearl, are extensively employed in the manufacture of " pearl " buttons, handles to pocket-knives, ornamental utensils, and in the inlaying of costly furniture ; and even pearls themselves, the most precious offspring of the ocean, are composed of the same substance as the nacreous coating of the shell ; other shells are largely used in the manufacture of cameos, their alternate coats of creamy white and rich red or warm brown giving beautiful artistic effects when skilfully handled.

In former days, one of the univalve shells, known now as the purpura, and little heeded except by the owners of marine aquaria, was employed for the purpose of producing the celebrated purple of the ancients, which none but the imperial family were permitted to wear.

Having now taken a superficial glance at the Molluscs and their uses, we will proceed to the description of individuals, and examine closer into details.

THE highest of the molluscs are those beings which are classed together under the title of CEPHALOPODA. This is a term derived from two Greek words, the former signifying a head, and the latter a foot, and it is applied to these creatures because the feet, or arms as they might also be called, are arranged in a circular manner around the mouth.

In these animals, which are, as has already been mentioned, thought by many naturalists to be above the molluscs, the organization is highly developed. The nervous system is more like that of the vertebrates than is the case with any other kind of mollusc, the knot of ganglia in the head bearing no small resemblance to a real brain. The Cephalopods breathe by means of a pair of gills or branchiæ, one set on each side of the body, and the circulating system by which the blood is driven through those organs and thence to the remainder of the structures is very complete.

They are all animals of prey, and are furnished with a tremendous apparatus for seizure and destruction. Their long arms are furnished with round, hollow discs, set in rows, each disc being a powerful sucker, and, when applied to any object, retaining its hold with wonderful tenacity. The mode by which the needful vacuum is made is simple in the extreme. The centre of the disc is filled with a soft, fleshy protuberance, which can be withdrawn at the pleasure of the owner. When, therefore, the edges of the disc are applied to an object, and the piston-like centre withdrawn, a partial vacuum is formed. and the disc adheres like a cupping-glass or a boy's leather sucker.

These discs are all under the command of the owner, who can seize any object with an instantaneous grasp, and relax its hold with equal celerity. The arms are almost as movable and as useful to the cuttle-fish as the proboscis to the elephant, for beside answering the purposes which have been mentioned, they are also used as legs and enable the creature to crawl on the ground, the shell being then uppermost.

We will now proceed to a few selected species of Cephalopods, and in the course of describing the several individuals, will examine the curious points of structure which are common to all.

Our first example is the celebrated ARGONAUT, or PAPER NAUTILUS, the latter title being given on account of the extreme thinness and fragility of the shell, which crumbles under a heedless grasp like the shell of an egg, and the former in allusion to the pretty fable which was formerly narrated of its sailing powers. It is rather remarkable, by the way, that the shell of the Argonaut is, during the life of its owner, elastic and yielding, almost as if it were made of thin horn.

Two of the arms of the Argonaut are greatly dilated at their extremities ; and it was formerly asserted, and generally believed, that the creature was accustomed to employ these arms as sails, raising them high above the shell, and allowing itself to be driven over the surface by the breeze, while it directed its course by the remaining arms, which were suffered to hang over the edge of the shell into the water and acted like so many oars. In consequence of this belief, the creature was named the Argonaut, in allusion to the old classical fable of the ship *Argo* and her golden freight.

Certainly, the Argo herself could not have carried a more splendid cargo than is borne by the shell of the Argonaut when its inhabitant is living and in its full enjoyment of life and health. The animal, or "poulp" as it is technically called, is indeed a most lovely creature, despite of its unattractive form. "It appeared," writes Mr. Rang, when describing one of these creatures which had been captured alive, "little more than a shapeless mass, but it was a mass of silver with a cloud of spots of the most beautiful rose-colour, and a fine dotting of the same, which heightened its beauty. A long, semi-circular band of ultramarine-blue, which melted away insensibly, was very decidedly marked at one of its extremities, that is of the keel. A large membrane covered all, and this membrane was the expanded velation of the arms, which so peculiarly characterises the poulp of the Argonaut.

The animal was so entirely shut up in its abode, that the head and base of the arms only were a very little raised above the edges of the opening of the shell. On each side of the head a small space was left free, allowing the eyes of the mollusc some scope of vision around, and their sharp and fixed gaze appeared to announce that the animal was watching attentively all that passed around it. The slender arms were folded back from their base, and inserted very deeply round the body of the poulp, in such a manner as to fill in part the empty spaces which the head must naturally leave in the much larger opening of the shell."

ARGONAUT, OR PAPER NAUTILUS.—*Argonauta Argo.*

Mr. Rang then proceeds to show the real use of the expanded arms, which is to cover the shell on its exterior, and, as has since been definitely proved, to build up its delicate texture and to repair damages, the substance of the shell being secreted by these arms, and by their broad expansions moulded into shape. The larger figure on the right hand represents the Argonaut while thus within its shell. The expanded extremities of these arms are seen covering nearly the whole shell, and their bases, set with suckers, are bent bridge-like over the rest of the animal. The large eye is seen just protruding out of the shell, the bases of the arms are curved over and behind it, and some clusters of eggs are seen sheltered under the arch of the expanded arms.

The modes of progression employed by the Argonaut are to the full as wondrous as its fabled habits of sailing. Its progression by crawling has already been casually mentioned. While thus engaged, the creature turns itself so as to rest on its head, withdraws its body as far as possible into its shell, and, using its arms like legs, creeps slowly but securely along the ground, sometimes affixing its discs to stones or projecting points of rocks for the purpose of hauling itself along.

When, however, it wishes to attain greater speed, and to pass through the wide waters, it makes use of a totally different principle.

As has already been mentioned, the respiration is achieved by the passage of water over the double gills or branchiæ; the water, after it has completed its purpose, being ejected through a moderately long tube, technically called the siphon. The orifice of the siphon is directed towards the head of the animal, and it is by means of this simple apparatus that the act of progression is effected. When the creature desires to dart rapidly through the water, it gathers its six arms into a straight line, so as to afford the slightest possible resistance to the water through which it passes, keeps its velated arms stretched tightly over the shell, and then, by violently ejecting water from the siphon, drives itself, by the reaction, in the opposite direction.

The uppermost figure in the illustration shows the Argonaut in the act of swimming. The extremity of the siphon is seen projecting immediately below the eye, and the reader will at once understand that if the action forcibly ejects water from the tube, the effect will be to drive the animal rapidly in the contrary direction, *i.e.* from right to left, as seen in the engraving. The third figure represents an empty shell, and is given for the purpose of showing the partially spiral and deeply grooved keel, as well as the extreme tenuity of the shell itself.

The animal, or poulp, is very slightly connected with the shell, and, when captured in a net, will sometimes voluntarily leave its home. Many persons have therefore thought that the poulp was not the fabricator of the shell, but only an intruder on the premises of the rightful owner, having taken possession of the shell as a defence for its soft body, just as the well-known soldier or hermit crab arms its soft and unprotected tail with the shell of a whelk, periwinkle, or a trochus. This opinion, ingenious and deserving of examination as it was, has, however, been proved erroneous by a number of experiments, which have shown that the Argonaut is not only the occupier, but the architect of its graceful dwelling, and that the expanded arms are at once the furnishers of the material and the executors of the work.

The precise food of the Argonaut is not ascertained, but Mr. Bennett presumes that, as he always found the globular and translucent, but empty shells of the hyalea, one of the wing-footed marine species, adhering in numbers to the discs on the Argonaut's arms, these creatures must have been captured and devoured by the more powerful mollusc.

As the various cephalopods are so numerous as to preclude all possibility of figuring and describing each species, we must therefore content ourselves with a typical form of each family, and a general account of its members.

The species belonging to the family of the Octopodidæ, or Eight-armed Cuttles, possess no external shell like that of the nautilus, its place being taken by two short styles or "pens" in the substance of the mantle. There are eight arms, unequal in length, and furnished with double or single rows of the suckers which have already been described.

They are solitary beings, voracious to a degree, and so active that they find little difficulty in capturing their prey, or in escaping from the attacks of their enemies. Even when pursued into the narrow precincts of a rock pool, the creature is not easily caught. When threatened, or if apprehensive of danger, the Polypus, as the animal was formerly called, darts with arrowy swiftness from one side of the pool to the other, and at the same time so discolours the water with the contents of its ink-bag, that its course is not perceptible, nor, until the water has become clear again, can the precise locality of the creature be discovered. Even if detected, it is not easily captured, as it has a knack of forcing its unprotected body into some crevice, so that no hold can be taken of it, and then affixing itself by its suckers to the rock with such wonderful tenacity that it can hardly be detached as long as life remains.

One example of this family is the WEBBED SEPIA, a very curious animal, found on the coast of Greenland. Its colour is violet, and the arms are united by a web almost to their tips. The suckers are set in single rows. Only one species of this genus is known.

In the Octopods the suckers are set directly upon the arms, and the eyes are fixed in their orbits; but in the Decapods, another section of these creatures, the suckers are placed on footstalks, and armed with a bony ring on each. The eyes are movable, and the shell is internal, lying loosely in the mantle. This so-called shell has, however, no real title to the name, being either a spear-shaped body of a horny substance, such as is popularly known by the name of sea-pen, or a curious aggregation of chalky particles, familiar under the title of "cuttle-bone." This "bone" is not attached to the animal by any muscles, but lies loosely in a kind of sac in the mantle, and will drop out if the sac be opened.

THE family of the Teuthidæ, popularly known as Calamaries, or Squids, are distinguished by their elongated bodies, their short and broad fins, and the shell or pen which is found in their interior. All the Squids are very active, and some species, called FLYING SQUIDS by sailors, and *Ommastrephes* by systematic naturalists, are able to dash out of the sea and dart to considerable distances. Mr. Beale mentions that he has seen tens of thousands of these animals dart simultaneously out of the water when pursued by dolphins or albacores, and propel themselves through the air for a distance of eighty or a hundred yards. While thus engaged, they have a habit of moving their long tentacles with a rapid spiral motion, which may possibly help them in their flight, as it undoubtedly does in their propulsion through the water. This peculiar action has been compared by the writer to that of an eight-pronged corkscrew. An interesting account of the Flying Squid may be seen in Bennett's "Whaling Voyage," where it is casually mentioned that these creatures frequently leaped on the deck of the vessel in their daring flight, and sometimes struck themselves violently against the bows, and fell back injured into the

WEBBED SEPIA.—*Cirrhoteuthis Mulleri.*   LITTLE SQUID, OR SEPIOLA.—*Sepiola Atlántica*
COMMON SEPIA.—*Sepia officinális.*

sea. This Squid has even been known to fling itself fairly over the ship, and to fall in the water on the other side.

One species of Squid is tolerably common on our coasts, and is often used for bait by the Cornish fishermen.

OUR present example of this family is the LITTLE SQUID, or SEPIOLA, of which genus six species are known, inhabiting most parts of the world, and living on our own shores.

ONE species, the ROCK SQUID, which sometimes attains a large size, may be considered as a formidable antagonist if irritated, as may be seen by the following account by Mr. Beale :—

"While upon the Bonin Islands, searching for shells upon the rocks which had just been left by the receding sea-tide, I was much astonished at seeing at my feet a most extraordinary-looking animal crawling towards the surf, which had only just left it. I had never seen one like it under such circumstances before; it therefore appeared the more remarkable. It was creeping on its eight legs, which, from their soft and flexible nature, bent considerably under the weight of its body, so that it was lifted by the efforts of its tentacles only a small distance from the rocks.

It appeared much alarmed at seeing me, and made every effort to escape, while I was not much in the humour to endeavour to capture so ugly a creature, whose appearance excited a feeling of disgust not unmingled with fear. I, however, endeavoured to prevent its career by pressing on one of its legs with my foot, but although I made use of considerable force for that purpose, its strength was so great that it several times quickly liberated that member, in spite of all the efforts I could employ in this way on wet slippery rocks.

I now laid hold of one of the tentacles with my hand, and held it firmly, so that the limb appeared as if it would be torn asunder by our united strength. I soon gave it a powerful jerk, wishing to disengage it from the rock to which it clung so forcibly by its suckers, which it effectually resisted. But the moment after, the apparently enraged animal lifted its head, with its large eyes projecting from the middle of its body, and, letting go its hold of the rocks, suddenly sprung upon my arm, which I had previously bared to my shoulder for the purpose of thrusting it into holes of the rocks to discover shells, and clung with all its suckers to it with great power, endeavouring to get its beak, which I could now see between the roots of its arms, in a position to bite.

A sensation of horror pervaded my whole frame when I found this monstrous animal had affixed itself so firmly upon my arm. Its cold slimy grasp was extremely sickening, and I immediately called aloud to the captain, who was also searching for shells at some distance, to come and release me from my disgusting assailant. He quickly arrived, and taking me down to the boat, during which time I was employed in keeping the beak away from my hand, quickly released me by destroying my tormentor with the boat knife, when I disengaged it by portions at a time. This animal must have measured across its expanded arms about four feet, while its body was not larger than a large clenched hand. It was that species of Sepia which is called by the whalers Rock Squid." This formidable creature was probably one of the ARMED CALAMARIES (*Enoploteuthis*).

These molluscs possess the property of changing their colours at will, and rival even the chameleon in the rapidity with which the varied tints flit over their surface like cloud shadows over the fields. The structure by which this process is achieved is very interesting. Over the surface of the animal is spread a thin integument, containing a vast number of circular or oval vesicles filled with a coloured fluid. This fluid is of various tints, such as black, brown, yellow, indigo, azure, and red, and in many species three or four of these colours are found. The vesicles are gifted with a most mysterious power of contraction and expansion, though they have no apparent connexion with the nervous system. The least touch will set them in motion, and exposure to light has the same effect, even after the death of the animal, or if a portion of the skin be removed from the body. It is evident that the tint of the surface is caused by the particular vesicles which are expanded, and its depth by the degree of expansion.

The celebrated " ink " of these creatures, from which the valuable colour called " sepia " was formerly obtained, deserves a brief notice.

This substance is liquid, and is secreted in a sac popularly termed, from its office, the " ink-bag." The sac is filled with a spongy kind of matter, in which the ink lies, and from which it can be forcibly expelled at the will of the animal. The ink-bag is not always in the same position, but some species have it in the liver, others near the siphon, and others among the viscera. There is a communication between the ink-bag and the siphon, so that when the ink is ejected, it is forcibly thrown out together with the water, so that the very effort for escape serves the double purpose of urging the creature away from danger and discolouring the water in which it swims. The animal can eject the ink with such force that it has been known to dedecorate a naval officer's white duck trousers with its liquid missile, the aggrieved individual always asserting that it took deliberate aim for that purpose.

Generally, the animal throws out its ink on the least alarm, a circumstance of some importance in geology. It was discovered by Dr. Buckland that in many specimens of fossil cephalopods, called scientifically Geoteuthis, *i.e.* Earth Squid, the ink-bag remained in the animal untouched by its long sojourn within the earth, and even retaining its quality of rapid mixture with water. A drawing was actually made by Sir F. Chantrey, with a portion of "sepia" taken from a fossil species, and the substance proved to be such excellent quality that an artist, to whom the sketch was shown, was desirous of learning the name of the colourman who prepared the tint.

THE Belemnites, those curious cucumber-like fossils, popularly called Thunderbolts, which are found in various strata, are now known to be the remains of ancient Calamaries, of which the entire animal, with its mantle, fins, ink-bag, siphon, eyes, and tentacles, has been discovered.

THE third, and largest figure in the engraving, represents the common SEPIA, whose wonderful chalky "bone" is so frequently thrown on our shores after the death of the animal in which it was developed.

This so-called bone was formerly in great repute for various purposes, but is now merely employed in the manufacture of pounce and dentifrice, for which latter purpose, however, prepared chalk is quite as effectual, being indeed the same substance, though in the form of powder. It is composed of a vast number of nearly horizontal layers, supported by innumerable little pillars or fibres of the same substance. If one of these shells be snapped across, the structure will be well shown even to the naked eye, while with the help of a common pocket-lens, even the minutest details can be examined. The upper coat will mostly scale off so as to show its smooth surface, while the successive ranges of glittering pillars look like a copy of the Giant's Causeway in miniature, as the irregular fracture breaks up their ranked columns into deep caverns and bold projecting rocks. A diagonal cut with a knife will farther expose the hard horizontal strata with their myriad pillars; but the method by which the structure exhibits itself in its greatest beauty is to make a very thin transverse section, mount it in Canada balsam on a glass slide for the microscope, and employ polarized light in its examination.

In consequence of its peculiar formation, the cuttle-bone is extremely light when dry, and admits so much air into the interstices that it swims easily in water. One of these shells is seen in the illustration, lying on the ground near the Sepia.

In the right-hand corner, and lying on the ground, is seen a group of dark oval bodies, looking something like a bunch of purple grapes, and from this resemblance termed Sea Grapes by the fishermen. These are the eggs of the Sepia, which may often be found on the seashore, flung there by the retiring tide, and left to perish unless rescued by some friendly hand. If these bunches of eggs be placed in a vessel of sea-water, and guarded from danger, they may be seen daily changing in appearance, until at last they burst asunder and let loose the inmates on the world. Nothing can exceed the nonchalant demeanour of the tiny creature not two minutes old. It deliberately makes the tour of its glassy prison, examines every detail with minute attention, and having quite satisfied its curiosity, poises itself for a moment just above the ground, blows out a circular hollow in the sand with a sharp expulsion of water from the siphon, and settles quietly into the bed thus prepared for it.

The Sepia, as well as others of its family, is a terribly voracious creature, and is consequently hated by the fishermen, who sometimes see the Cuttles busily at work on the fishes that are hanging on their hooks, holding the bodies firmly in their arms, and tearing away the flesh with astonishing quickness, as if they knew the shortness of their tenure. Their vigilance is, however, not to be dulled by their appetite; for although they will hold on to the last moment, they are far too wary to allow themselves to be caught, and as soon as the fatal net approaches, they cast off their animated grapnels, and dart away swiftly from the threatened danger.

BEFORE proceeding to another large group of cephalopods, it is needful to mention the curious animals called, from the shape of their shell, Spiralidæ. These singular creatures form a distinct though very small family, containing only three species.

CHAMBERED NAUTILUS.—*Nautilus Pompilius*

In them, the shell is very delicate, and is rolled into a spiral form, something like the proboscis of an elephant when curled up. These shells are very common on the shores of New Zealand, where they are scattered in thousands, and are sometimes thrown on the southern shores of England by the waves of the Gulf Stream. Yet the animal which formed the shell is extremely rare, and is very seldom found except in a very fragmentary and battered condition.

ANOTHER order of cephalopods is called by the name of Tetrabranchiata, or Four-gilled animals, because the organs of respiration are composed of four branchiæ. These creatures possess a very strong external shell, which is divided into a series of gradually increasing compartments connected together by a central tube called the siphuncle. As the animal grows, it continues to enlarge its home, so that its age can be inferred from the number of chambers comprising its shell.

In former times these creatures were very abundant, but in our day the only known living representative is the CHAMBERED or PEARLY NAUTILUS, a section of which is shown in the illustration in order to exhibit the beautiful spiral home in which the creature resides, and the structure of the chambers, together with their connexion by means of the siphuncle.

While the animal still lives, the short tubes that pass through the walls of the chambers are connected by membranous pipes, and even in a specimen that has been long dead, these connecting links hold their places, provided that the shell has not been subjected to severe shocks. In one of these shells now before me, which I have very cautiously opened, the whole series of membranous tubes can be seen in their places, black and shrivelled externally, but perfect tubes nevertheless.

At the left-hand of the illustration the animal may be seen as it appears when lying in its shell, the eye just peering from the edge, the points of the retracted tentacles visible, and the hood covering all. Although the shell of the Chambered Nautilus is common enough, the animal is very rarely seen, often, in all probability, escaping notice by its utter dissimilarity to the popular ideas of a shell. While living, it incloses the shell in its mantle, and on several occasions has been mistaken for a dead cat or a lump of blubber, until a sudden alarm induced the creature to throw aside the mask, and to address its best energies to escape. Moreover, it is one of the deep-water species, and does not seem to ascend to the surface unless driven by the force of a storm.

Before quitting the subject of this animal, I cannot resist the opportunity of inserting the exquisite little poem by Dr. Holmes, on a broken shell of the Chambered Nautilus :—

> " This is the ship of pearl, which poets feign
>   Sails the unshadowed main—
>   The venturous bark that flings
> On the sweet summer wind its purpled wings,
> In gulfs enchanted, where the siren sings
>   And coral reefs lie bare,
> Where the cold sea-maids rise to sun their streaming hair.
>
> Its webs of living gauze no more unfurl;
>   Wrecked is the ship of pearl !
>   And every chambered cell,
> Where its dim dreaming life was wont to dwell,
> As the frail tenant shaped his growing shell,
>   Before thee lies revealed—
> Its irised ceiling rent, its sunless crypt unsealed !
>
> Year after year beheld the silent toil
>   That spread his lustrous coil;
>   Still, as the spiral grew,
> He left the past year's dwelling for the new,
> Stole with soft step its shining archway through,
>   Built up its idle door,
> Stretched in his last-found home, and knew the old no more.
>
> Thanks for the heavenly message brought by thee,
>   Child of the wandering sea,
>   Cast from her lap, forlorn !
> From thy dead lips a clearer note is born
> Than ever Triton blew from wreathèd horn !
>   While on mine ear it rings,
> Through the deep caves of thought I hear a voice that sings :—
>
> Build thee more stately mansions, O my soul,
>   As the swift seasons roll !
>   Leave thy low vaulted past ;
> Let each new temple, nobler than the last,
> Shut thee from heaven with a dome more vast,
>   Till thou at length art free,
> Leaving thine outgrown shell by life's unresting sea."

In many of the cephalopods already mentioned, the only specimens ever captured belonged to the female sex, and seemed almost to pass, like the Amazons of old, an existence which may be termed a misandrous life. At last the male was discovered, but in a form so unlike that of the female, and apparently so utterly unworthy of her, that although its existence had long been known to naturalists, it was set down as a parasitic worm, and called by the title of Hectocotyle. That of the argonaut is hardly more than half an inch in length ; has two rows of little suckers, arranged alternately, forty-five being placed on each side ; there is a thread-like appendage in front, nearly as long as the whole animal, and the skin contains changeable spots of red and violet. It is a very curious and suggestive fact, that the newly hatched female argonaut is very similar in appearance to the fully developed male. Similar instances of discrepancy between the sexes are found in the lac insect, and many crustaceans, the male being nearly one thousand times less than his mate. This subject is, however, extremely mysterious.

There are many fossil species of this order, but as the limits of this work are far too circumscribed to admit even a tithe of the important existing species, no space can be found for the fossils. It must, therefore, suffice in the present instance to remark that the well-known ammonites, rams'-horns, and other similar fossil shells, belonged to the same order as the Chambered Nautilus.

The colour of the shell is very beautiful. The ground is white, over which are drawn, as with single dashes of a painter's brush, sundry bold streaks of reddish chestnut, mostly coalescing above, and reaching nearly to the centre of the spiral. This porcelain-like material is, however, only an outer coat laid on the real pearly substance of the shell, which is seen on looking into the hollow or into any of the chambers. The Chinese avail themselves of this double coating, and, with the untiring perseverance of their laborious nature, take the greatest trouble to spoil the finest shells, by covering them with their grotesquely unperspective carvings of figures and landscapes, cut so as to relieve the deep colour of the raised figures by the white pearly background. Unlike the shell of the argonaut, which is almost as fragile as if made of sugar, that of the Nautilus is firm and strong, and will bear a considerable amount of rough handling before betraying any signs of injury.

WE now take leave of these highly developed molluscs, and pass to another order where the organization is not nearly so perfect, and where the habits are either so common-place as to be devoid of general interest, or the animals so shy that they never can be seen performing any act which is likely to attract the attention of an unprofessed naturalist.

The last order of molluscs are named cephalopods, because their feet surround their head. The order which now comes before our notice is composed of animals which crawl upon a broad muscular organ, termed, from its use, the foot. It is an enormously large order, containing all the snails, whether terrestrial, aquatic, or marine, the whelks, limpets, and similar animals not so familiarly known. Many species are much used as food, while others are of great service in the arts, furnishing employment to many hundreds of workmen. As the shell of these creatures consists of one piece or valve only, they are sometimes termed univalves, in contradistinction to the oysters, muscles, scallops, and similar shells, which are termed bivalves, in allusion to their double shell.

Passing by, for the present, the mere anatomical peculiarities of these creatures, and leaving them to be mentioned in the Table of Generic Differences at the end of the volume, we will proceed at once to our first example of the gasteropods, the BEAKED SPINDLE-SHELL, so called from the rather distant resemblance which its long and pointed form bears to a spindle, and the elongated beak-like process which is seen pointing downwards to the ground as the animal walks along. The right-hand figure represents the Beaked Spindle. In the family to which this mollusc belongs, the lip of the shell is always extended and deeply notched.

The animal possesses large and well-developed eyes, set on stout and moderately long footstalks, from each of which a small tentacle proceeds. The foot is small, and not well calculated for crawling, but more adapted for the abrupt leaping style in which these animals proceed, their weighty shell swaying from side to side in rather an awkward manner.

There is a structure belonging to these animals which must be described before proceeding further, inasmuch as its shape and comparative dimensions often afford valuable indications by which a species, or even a genus, may be distinguished. · This structure is called the "operculum," and its use, when fully developed, is to close the aperture of the shell, when the animal has withdrawn itself into the recesses of its home.

The operculum can be well seen in the water-snails, where it attains its full size, and exactly fits the opening which it is intended to protect. The material of which the operculum is essentially composed is a horny substance, but in some species the horn is strengthened by layers of the same nacreous matter which lines the shell, and becomes so thick and heavy that, when found separate from its owner, it is often mistaken for some species of shell. The operculum is very variable both in its form and comparative dimensions, and even in its presence or absence. Sometimes it is circular, like a flat plate, and composed of concentric circles, while in some species it assumes a regularly spiral form, like a flattened watch-spring.

Whatever be its shape, it is always developed from a well-defined spot, termed the nucleus, which is in many cases, but not in all, the centre of the structure. The forms of this most variable plate are too numerous to be separately described, but in the course of the following pages the reader will often find the operculum represented in its place on the mollusc, and will be able to form a better idea of it than would be gained by mere description. In many instances the operculum is only rudimentary, and in others is so very small in proportion to the dimensions of the shell that it is quite incapable of closing the opening. This is mostly the case with shells which have very wide-spreading mouths. It is a curious fact that the operculum is developed at the same time with the shell, and even within the egg this plate may be detected upon the inclosed mollusc.

On reference to the engraving, the reader will see the animal represented as in the act of walking, a feat which, on account of the small size of the foot, is not achieved with much ease. In front of the foot may be seen the siphon pointing downwards to the ground, and on each side of the siphon are set the footstalks, which carry the eyes at their extremities and a tentacle about the centre. The colour of the Beaked Spindle-shell is a pale brownish chestnut, with a slight wash of yellow. The beak, a long spiny process of the shell, is edged with black, and the inside is pure white. The surface of the shell is smooth, and its length is about five inches in a good specimen.

THREE-HORNED STROMB.—*Strombus tricornis.*   **BEAKED SPINDLE.**—*Rostellária curvirostris.*

On the left-hand of the engraving is another shell belonging to the same family, the THREE-HORNED STROMB, represented empty, and lying on the ground in order to show the curious form of its lip.

The Strombs form a large genus, containing about sixty species, and are found in almost every warm sea. They do not appear to be deep-water lovers, being mostly found on the reefs at low water, and seldom extending their range beyond ten fathoms. The operculum of the Strombs is rather curious in its structure, the nucleus being set at one extremity, and the operculum being made up of a succession of horny plates or scales overlapping each other like the tiles of a house, or the successive steel layers of a carriage-spring.

Some species of Strombs attain a considerable size, and are much used in the arts, as, for example, the GIANT STROMB, or FOUNTAIN SHELL (*Strombus gigas*), one of the West Indian species, which sometimes attains the weight of four or five pounds, and is imported into Europe in several thousands for the use of engravers, who cut the well-known cameos from its beautifully tinted substance. Three hundred thousand of these shells were brought to Liverpool alone in a single year. As the animal increases in age, it gradually fills up the hollow apex and spines with solid matter, and thus materially adds to the weight of the shell without improving its value to the engraver. In some parts of the world, such as Barbadoes, the Giant Stromb is eaten, and sold regularly for that purpose. Pearls of a delicate pink colour have sometimes been found in this shell, but their occurrence is not frequent, probably on account of the careless and unobservant habits of the negroes who clean the shells. Pearls are also found in other species belonging to this genus.

The teeth of the Strombs are extremely beautiful and most complicated, but cannot well be described without the aid of illustrations.

The colour of the Three-horned Stromb is brown and yellow of different shades, richly mottled with pale saffron. Its average length is about four inches.

COMMON SPIDER SHELL.—*Pteróceras lambis.*   ORANGE-MOUTHED SPIDER-SHELL.—*Pteróceras aurantius.*

THE two species represented in the illustration derive their popular names from the generally spider-like contour of their form.

When adult, the outer lip is furnished with several horny appendages, always curved, and not precisely of the same shape in the same species, although the general character of their form is sufficiently well marked to distinguish the species. One of these horns is always close to the spine, and is rolled in such a manner as to form a posterior canal. About ten species of these animals are known, and seem to be confined to the Chinese and Indian seas.

The COMMON SPIDER-SHELL is seen on the left of the illustration, the animal being represented in the act of crawling, the foot being applied to the ground, the operculum trailing behind, and the two eyes in their long footstalks projecting in front. The colour of this shell is very handsome, consisting of boldly mottled chestnut, like the hue of old rosewood, variegated with white, and traversed by lines of orange. The interior of the shell is pale brown, with a dash of yellow. The average length is three or four inches.

BRITISH WOODCOCK-SHELL.—*Murex erináceus.*   THORNY WOODCOCK.—*Murex tenuispinis.*

On the right-hand may be seen the empty shell of the ·ORANGE-MOUTHED SPIDER-SHELL, so disposed as to exhibit the curious projecting horns, with their sharp points and bold curves.   It is worthy of notice that in all the Spider-shells these projections are not developed until the creature has attained adult age, the young Spider-shell resembling that of the stromb.   From the peculiar shape of some of the species, these creatures are sometimes called SCORPION-SHELLS.

The colour of the Orange-mouthed Spider-shell is creamy white on the exterior. and rich orange within.   The curved spines are white and shining, and bear no small resemblance to the poison-teeth of serpents.

THE shells that are included in the family of the Muricidæ may readily be distinguished by the straight beak or canal in front, and the absence of any such canal behind.   The eyes of these animals are not placed on long footstalks, as in the preceding family, but are set directly on the tentacles, without any supporting stalk or projection.   All the animals belonging to this family are not only carnivorous, but rapacious, preying on other molluscs, and destroying them with the terrible armature called the tooth-ribbon, and which, when examined with the microscope, proves to be a set of adamantine teeth, sharp-edged and pointed as those of the shark, and cutting their way through the hard shells of their victims as the well-known cordon saw passes through thick blocks of hard wood.

About one hundred and eighty species are known to belong to the typical genus, and there is hardly a portion of the world where a Murex of some kind may not be found.

THE larger of the two specimens represents the shell which is popularly known under the name of THORNY WOODCOCK, the latter title being given to it, in common with several

3.   B B !

of its congeners, on account of the long beak, which is thought to bear some resemblance to that of the woodcock, and the former in allusion to the vast number of lengthened spines or thorns which are arranged regularly over its surface. It has also received the equally appropriate and more poetical name of VENUS' COMB.

This shell is found in the Indian Ocean, and varies greatly in dimensions, four or five inches being about the average length. It is evident that as nothing is ever made in vain, or to be wasted, the wonderful array of external spines must play some important part in nature, if not in the economy of the particular species. But what that part may be, and what may be the object of these beautiful structures, is a problem which seems almost insoluble, at all events, with our present means of discovery.

The colour of the shell is very pale brown, each ridge being slightly tuberculated and edged with white. The spines are uniform drab, or very pale brown, with an almost horny translucence.

ANOTHER species is given in the same illustration, in order to show the animal and the position of the eyes, to which reference has already been made. This is the BRITISH WOODCOCK or HEDGEHOG SHELL, a creature which, as its popular name imports, is a native of our seas. It is very much smaller than the thorny woodcock, and affords a good example of the contrast that can often be effected by different animals which yet belong to the same genus. Its length is hardly more than an inch and a half, and its colour is a pale yellowish brown.

IF the so-called tongue-ribbon be carefully removed from the owner, and mounted on a glass slide for microscopic purposes, it can be examined with the attention it deserves, and by its great beauty will be sure to repay the labours of the observer. This beautiful structure contains a vast series of the teeth already mentioned ; and although the general formation of these organs and their arrangement is alike in all the Muricidæ, the teeth themselves are varied in an almost incredible manner, and are so differently shaped, that the particular species can be recognised by one who is conversant with these structures, merely from a glance at the tooth-ribbon. By means of this curious and complicated apparatus, the carnivorous animal is able to attack other molluscs with unfailing certainty.

In vain do they retire into the recesses of their shelly homes, for the rasp-like teeth of their assailants soon bore a hole through the opposing walls, and enable the marauder to kill and devour the victim at its leisure. Perhaps it would be more correct to transpose the two verbs, and say, devour and kill, for in the molluscs the life retains its hold of the body with such tenacity, that the greater part of the substance may be actually eaten by the slayer, while the remainder is still quivering with life, or even endeavouring, with vain instinct, to crawl away from its foe.

The reader has, doubtlessly, while wandering on the seashore, picked up many shells in which a round hole had been made, as perfectly circular as if cut with a drill. This hole is an unfailing indication that the quondam owner of the shell has fallen a victim to some mollusc such as we have just been describing, and whose task seems to be the continual restoration of the balance of nature, which would be disturbed by the unlimited increase of the vegetable-eating shells.

ONE or two other species belonging to this genus require a passing notice. The WOODCOCK'S HEAD (*Murex haustellum*), remarkable for its long peak and rounded shell, inhabits the same localities as its more beautiful neighbour, being found in the Indian and Chinese seas. It has but few of the spines which decorate the thorny woodcock in such profusion, and even those which are seen upon the surface are comparatively short. The rounded body of the shell, however, together with its long beaked process, does really bear some resemblance to the head and bill of the bird from which it takes its popular name.

The ROYAL MUREX (*Murex regius*) is a very fine example of this genus, and is valued, not only for its rarity, but for the extreme beauty of its form and colouring, which render it an ornament to any cabinet. In colour it resembles the thorny woodcock.

TWISTED TRITON.—*Triton distortus*   SEA TRUMPET.—*Triton variegatus.*   WRINKLED TRITON.—*Triton anus.*

THE large empty shell lying in the centre of the engraving represents the SEA TRUMPET, or CONCH-SHELL, so familiar from the use to which it has been put for ages, and which has rendered it a classical appendage to the marine deity whose name it bears.

The Sea Trumpet sometimes attains to a large size, a foot or more in length; and, when it has attained its full dimensions, is employed among the South Sea Islanders and Australians as a trumpet. In order to fit the shell for this purpose, a round hole is bored in the side, at about one-fourth the length from the tip, and the required sound is elicited by laying the shell to the lips, and blowing across the hole as a performer blows the flute, The note—if the noise produced can be called by that name—is hollow and disagreeable; but as it is loud and unlike any other sound, it answers the purpose of those who employ it. While blowing the conch, the performer introduces his right hand into the cavity, much in the manner of a player upon the French horn.

About a hundred species of Triton are known, and are found spread over the greater part of the world, except the colder seas. They do not seem to frequent the deep water, from ten to twenty fathoms appearing to be their favourite depth.

The ground colour of this handsome shell is white, mottled, in rather irregular spiral rows, with ruddy brown and yellow, and warming to ruddy chestnut at the point. Within, it is white; and the bold zebra-like streaks which may be seen upon the top of the lip are composed of smooth white ridges on a black ground. It must be understood, however, that many shells are so variable in their colouring, that a collector may possess fifty examples of a single species, and not find any two specimens with precisely the same colouring. The colours that are here given are those of the specimens examined in the

SPINED FROG-SHELL.—*Ranella spinosa.*   FROG-SHELL.—*Ranella Rana.*   BULL-FROG SHELL.—*Ranella bufonia*

course of writing the descriptions; and I must here return many thanks to Mr. G. B. Sowerby for his liberality in giving me access to his collection, in order to describe the individual specimens from which the illustrations were taken.

BELOW the Sea Trumpet lies another shell, which would hardly be taken for a Triton until turned over, so as to show the whole of the contour. This is the WRINKLED or OLD WOMAN TRITON, so called because the corrugated and rudely oval mouth, with its white crumpled folds, is thought to bear some distant resemblance to the face of an old woman surrounded with a close cap. The Wrinkled Triton is comparatively a small species, as may be seen from the proportions preserved in the figure.

BEHIND the larger figure is seen the TWISTED TRITON, represented in the act of crawling, and given, not so much to exhibit any peculiarity of its shell, which is hidden behind that of the larger species, as to show the form of the animal, its large foot, and eyes placed at the bases of the tentacles. The operculum of this animal is small and leaf-shaped, the nucleus being at one end.

As the spider-shells, which have been recently described, have received their popular names from the distant resemblance which they bear to the arachnidian race, so the shells upon the accompanying illustration derive their titles from a still more distant resemblance to the batrachians, and go popularly by the name of Frog or Toad Shells. Even in the short space that has been given to the molluscs, the reader must have been struck with the singular perceptive powers of conchologists, who discover analogies and detect resemblances in creatures so dissimilar in shape and colour, that few are capable of appreciating them, even after they have been pointed out, or of precisely comprehending the grounds on which the distinguishing names are given.

The FROG-SHELL is the central and uppermost of the three species represented in the illustration, and seems to have been gifted with its popular name on the same principle that caused a well-known dramatic character to detect in a cloud an equal resemblance to a whale and a camel. All the members of this genus possess two rows of ridges, technically called "varices," upon the shell, one row being placed on each side. There are about fifty species of Ranella, spread over all the warm seas. Like the preceding shells, they prefer the shallow to the deep waters, and may be found at almost all depths—from the bare rocks left waterless by the receding tide to a depth of eighteen or twenty fathoms.

The colour of the Frog-shell is not very striking at a distance, but elegant and delicate when closely examined. The surface of the shell is variously mottled with brown of differing tones and intensities, and is traversed by multitudinous rows of tiny raised dots or tubercles of pure white, like porcelain.

The animal of this shell is represented as crawling, for the object of displaying the long tentacles, the position of the eyes, the broad foot, and the small oval operculum, with its layers of bony substance, and its nucleus placed at one side.

On the right hand is shown the BULL-FROG SHELL, exposed so as to show the roughly tuberculated surface, with its deep hollows and bold ridges of thick shelly substance, together with the projecting horns on either side. The colour of this shell is extremely variable. In the handsomest specimens the ground colour is creamy white, largely mottled with bold tints of deepest brown and purest white. But in many instances the entire shell is of a very pale tone, yellow predominating, and the brown entirely subservient, and presenting the same contrast to the full-coloured shell as the albino to the negro.

The third figure represents the SPINED FROG-SHELL, a name which, in this instance, is partially appropriate on account of the sharp and rather long spines or projections with which the shell is furnished. None of these shells are of very great size, their average length being about two inches.

As the family of the Muricidæ is a very large one, and comprehends a vast variety of curious forms, which, though differing in many unimportant details, are yet identical in those characteristics which determine the genus, it is needful to illustrate it by several examples of these apparently discrepant forms.

LITTLE FIG-SHELL.—*Pyrula ficus.*

In the accompanying illustration is given a very pretty shell, termed indifferently the LITTLE FIG or LITTLE PEAR SHELL, its general outline being thought sufficiently pear or fig like to warrant the application of the name. Both scientific names refer to this far-fetched resemblance, *pyrula* signifying a little pear, and *ficus* meaning a fig.

The left-hand figure shows the creature as it appears when crawling, and is given for the purpose of showing one of the strange peculiarities of the genus. The long siphon is seen protruding from the projecting canal, whose structure is shown in the right-hand figure; the slender body, with the round black eyes set at the base of the tentacles, is placed under the siphon; the broad leaf-like foot sustains the shell and serves it as an organ of progression, and the mantle forms lobes on the sides which nearly meet over the back of the shell. The foot is abruptly cut off, or truncated in front, and modified so as to form a short horn or partial crescent at each side.

Nearly forty species of Pear-shell are known to conchologists, and are spread over the warmer seas of the world, living in moderately deep water, varying from sixteen to thirty-five fathoms of depth.

This is a thin and delicate shell, the large expanded lip being especially so, and, in consequence, is very light when the inmate has been removed. The colour is very pale yellow, with brown and white arranged in wavy mottlings. Its average length is about four inches.

The delicate thinness of the shell is not, however, a character common to the entire genus, for another species, the BAT-LIKE PEAR-SHELL (*Pyrula carnária*), is quite as remarkable in the opposite direction, its shell being peculiarly large and ponderously constructed. This shell is found in the Indian Ocean, and its general colour is dark bay. In all these shells, however, the long canal which projects from the front of the shell is always open, not being filled up with solid matter as the animal increases in age; and the columella, or pillar of shelly substance, which runs up the centre of the whorls, like the solid centre of a screw, is always smooth.

WITH the two shells represented in the accompanying illustration, our examples of the Muricidæ end.

THE large and boldly mottled shell is popularly known by the really appropriate name of TULIP WHELK, its rich and variegated colouring bearing some analogy to that of the flower from which it derives its name; while the general shape is sufficiently like that of the whelk to warrant its use, even though the two shells belong to different families. The generic name of this shell is derived from a Latin word signifying a band, and is given to it on account of the boldly banded stripes in which the colours are disposed. As in the last-mentioned genus, the canal, though not so elongated, is always kept open.

Comparatively few living species of Tulip-shells are known to conchologists, sixteen or seventeen being their utmost limit. These shells inhabit the warmer seas, and some of them attain a great size, such as the GREAT TULIP-SHELL, which sometimes reaches a length of nearly two feet.

Although extremely variable, the colouring of the Tulip-shell is always rich and striking. In the present species the ground colour is mostly cream-white, on which are dashed bold mottlings of deepest mahogany-brown alternating with purest white. The interior of the shell is orange-brown, and the point is black. In some specimens, however, the creamy white ground is replaced by warm orange, and the lip is streaked within. All the Fasciolariæ are fond of muddy grounds, and are mostly taken within a depth of seven fathoms. The siphon of this animal is very long, the eyes are set at the base of the tentacles, and the small and practically useless operculum is placed at the end of a short and stout footstalk.

THE lower figure represents the empty shell of the SPINDLE or DISTAFF SHELL, so called in allusion to its form. Its scientific names are both given in consequence of its general resemblance to these objects, the former signifying a spindle, and the latter a distaff.

At least a hundred species of Spindle-shells are known, and their range extends over the greater part of the globe. One large species (*Fusus antiquus*), called, from its colour, the RED WHELK, is common on our shores, and off some of the coasts of Scotland is extensively captured for sale. When the empty shell is held to the ear, the reverberations

TULIP WHELK. — *Fasciolária Tulipa.*   SPINDLE-SHELL.—*Fusus colus.*

of sounds are gathered in its wide lip, and, being returned to the ear in a broken and confused manner, give forth a monotonous sound, rising and falling at intervals, and are thought by the uneducated to be the imprisoned murmurs of the waves. For this reason, the shell is popularly known as ROARING BUCKIE. In some places the empty shell is used as a lamp, the cavity containing the oil and the wick being drawn through the canal, thus producing a charmingly elegant lamp, which even exceeds in beauty the classical forms of the ancients, and quite equals them in efficacy.

Another species, the GIANT SPINDLE (*Fusus colósseus*), is remarkable as being one of the largest living examples of the gasteropods.

In the illustration, the animal of the Spindle-shell is seen just in the act of protruding itself from its habitation, in order to show the position of the eyes and the spotted mantle. The foot of the animal is moderately broad, and the operculum is small, and shaped not unlike a sea mussel-shell. The colour of the Spindle-shell is nearly white, and almost uniformly tinted, but darkening slightly towards the point.

WE now arrive at another and rather larger family, of which the common WHELK is a familiar example.

This is one of the most carnivorous of our molluscs, and among the creatures of its own class is as destructive as the lion among the herds of antelopes. Its long tongue, armed with row upon row of curved and sharp-edged teeth, harder than the notches of a file, and keen as the edge of a lancet, is a most irresistible instrument when rightly applied, drilling a circular hole through the thickest shells as easily as a carpenter's centre-bit works its way through a deal board.

The front of the tongue often has its teeth sadly broken, or even wanting altogether, but their place is soon supplied by others, which make their way gradually forward, and are brought successively into use as wanted. As a general rule, there are about a hundred rows of teeth in the Whelk's tongue, each row contains three teeth ; and each tooth is deeply cleft into several notches, which practically gives the creature so many additional teeth.

WHELK.—*Buccinum undatum.*   CONCHOLEPAS.—*Concholepas Peruviana.*

Vast quantities of Whelks are taken annually for the markets, and are consumed almost wholly by the poorer classes, who consider them in the light of a delicacy. They are, however, decidedly tough and stringy in texture, and, like the periwinkle, which is also largely eaten, are not particularly digestible. The mode of taking these molluscs is very simple: Large wicker-baskets are baited with the refuse portions of fish, and lowered to the bottom of the sea by ropes. The ever-hungry Whelks instinctively discover the feast, crowd into the basket by thousands, and are taken by merely raising the laden basket to the surface, and emptying it into a tub. Sometimes the Whelk is captured by the dredge, but the baited basket is the quickest and surest method. Besides its use as an article of human consumption, it is sometimes employed by the fishermen as bait for their hooks.

The reader will doubtlessly have observed on the seashore considerable masses of little yellowish capsules, mostly empty, and so light as to be drifted on the surface of the sea like so many masses of corks These are the empty egg-cases of the Whelk, and in the illustration a small group of them is represented adhering to the oyster-shell on the right-hand of the cut. At the proper season of the year, when the egg-clusters are flung on the shores by the gales, the little Whelks can be discovered within the capsules, several shells being found in each case. When hatched, the young escape into the sea through a round hole in the capsule.

The empty shells of this mollusc are often seized by the well-known hermit or soldier-crab, who is obliged to content himself while in his youth with the shell of the tops, advances with increasing size to the larger habitation of the purpura, and when he

LITTLE HARP-SHELL.—*Harpa minor.*   IMPERIAL HARP-SHELL.—*Harpa imperialis.*

has attained his full dimensions, searches for an empty shell of the Whelk and therein ensconces himself, safe from almost every enemy except man. The colour and general appearance of the Whelk are too well known to need description.

BELOW the whelk are seen two shells, one empty, and the other still retaining its inhabitant, both belonging to the same species.

This is a most remarkable shell, with hardly a vestige of whorl, and so wide-mouthed and so shallow that it almost looks as if it were the half of some bivalve shell.

This animal is called the CONCHOLEPAS, and appears to be the only species belonging to that genus. It is a native of Peru, as its name imports, is found along the shores in great plenty, and sometimes attains considerable dimensions. As may be seen by reference to the illustration, the foot of the animal is very wide and thick, and its eyes are set half-way up the tentacles. The spot where the animal is attached to the interior of the shell is shown in the right-hand figure. The shell is very thick and solid, deeply grooved and ridged on its upper surface, and extremely heavy in proportion to its dimensions. Its colour is very dark sooty brown on the exterior, and dull white within, something like the peculiar white of inferior porcelain, or the tint of skimmed milk.

THE sweeping curves, broad swelling lip, and regular ridges, of the next genus of shells, have earned for them the popular title by which they are known.

About nine or ten species belong to this pretty genus, some of which are rare and costly. The IMPERIAL HARP-SHELL, which is represented on the left-hand of the engraving, is still a valuable shell; but in former days, when the facilities of commerce were far less than at present, it could only be purchased at a most extravagant rate. A small specimen is now valued at about fifteen shillings or a pound, and a fine one will cost about three pounds; but, in former days, as much as fifty pounds have been paid for a specimen which can now be obtained for one-twentieth of the sum. A similar diminution has taken place in the cost of nearly all shells.

The Harp-shells are only found in the hottest seas, and are taken mostly on the shores of the Mauritius, Ceylon, and the Philippine Islands. They frequent the softer and more muddy parts of the coast, and prefer deep to shallow water. None of the Harp-shells possess the operculum.

The colour of the Imperial Harp is pale chestnut and white, with a dash of yellow arranged in tolerably regular and slightly spiral bands.

THE smaller figure on the right-hand, having the animal attached, is the LITTLE HARF-SHELL, a darker species, and one that seldom attains a greater length than an inch and a half. This figure is given for the purpose of showing the peculiar foot, which is very large, broad, and leaf-shaped, and, as will be seen by reference to the illustration, has a deep fissure just behind the tentacles, nearly cutting the organ asunder. It is said that, when the animal is irritated, the fissure becomes widely expanded. Some writers say that, if the animal is very much terrified, it withdraws itself into its home with such rapidity that the expanded front of the foot is unable to contract sufficiently, so that the fissure is caught against the sharp front edge of the shell, and thus undergoes involuntary amputation.

The general colours are tolerably similar throughout the Harps, but each species always preserves its peculiar individuality. One species, for example, has the spaces between the ridges pencilled in elaborate wavy markings of chocolate on white, and the ribs themselves barred at regular intervals by lines of deep brown; while another, known by the name of VENTRICOSE HARP (*Harpa ventricósa*), has the spaces filled with a succession of arches, one within the other, and of a rich brown colour.

A VERY common shell may often be found on the seashore, looking like a small whelk with a smooth whitish shell, boldly banded with reddish brown. This is the COMMON PURPLE, or PURPURA (*Púrpura lapillus*), another member of this genus, and worthy of notice as being one of the shells which furnish the celebrated Tyrian purple of the ancients. This colour, which, by the way, contains so little blue as to be unlike the tint which we now call by the name of purple, is evidently the analogue of the ink found in the sepia, and is secreted in a little sac by the throat, containing only one small drop.

For the very best dye this material was extracted carefully from the individual shells, but for an inferior kind it was obtained by pounding a quantity of the Purpuræ in a mortar, and straining off the juice, which was thus mixed with the blood and general moisture of the animals, and consequently of less value than the pure dye. So expensive was the dye obtained by this latter process, that a pound of wool stained with it could not be purchased under a sum equalling thirty pounds of present money. Any one can try the experiment of dyeing a little strip of linen with the matter obtained from a single shell. After breaking the shell carefully so as not to crush the inhabitant, the cell containing the colouring matter will be seen lying across the head or neck of the animal, and can be removed by opening the sac and taking up the yellowish white contents with a small camel's-hair brush, or the point of a new quill-pen. When the linen is imbued with this liquid and placed in the rays of the sun, it immediately begins to change its colour, and passes through a series of tints with such rapidity that the eye can hardly follow them, unless the slanting rays of the rising or setting sun are chosen for the purpose.

Light green is the first tint, which soon deepens until the yellow is driven out by the mysterious power of light, and the substance assumes a blue colour. After a while another primary colour makes its appearance, and a gradually deepening red being diffused through it, the blue changes to rich purple. Even the blue, however, has a comparatively feeble hold, for if the empurpled linen be well washed with soap and water, and again exposed to the sun's rays, the blue gradually fades before the increasing red, and a deepish crimson will be the ultimate result. Thus, we have the curious phenomenon of a semi-liquid, almost destitute of colour while shielded from light, assuming in regular succession the three primary colours, first mixed so as to form the secondary tints, and gradually separating and retiring before the superior power of their successors.

The eggs of the Purpura may often be found adhering to stones, shells, or even to each other, by means of little footstalks, which give the object a singular resemblance to an egg-cup with the egg placed in it.

Many other molluscs secrete a colouring fluid, such as *Murex trunculus*, which was another of the shells employed to furnish the imperial purple; and *Murex brandaris* seems to have been used for a similar purpose. One of the staircase shells, *Scalaria diadema*, also secretes a purple-producing substance; and *Cerithium telescopium* has been known to stain a piece of paper a permanent green.

MAGILUS.— *Magilus antiquus*

ONE of the strangest, though not the most beautiful, of shells is the MAGILUS, a native of the Red Sea and the Mauritius.

On reference to the illustration, the reader will see two figures, one representing a group of madrepores, in which a small and delicate shell is lying, and the other a long, crumpled, and partly spiral tube, with a shell at one end and an opening at the other. Strange as the assertion may seem, these two figures represent the same animal in two stages of its development.

For the purpose, apparently, of carrying out some mysterious object, the Magilus resides wholly in the masses of madrepore, and in its early youth is a thin and delicate shell without anything remarkable about it. As it advances in age, it enlarges in size, as is the case with most creatures; but its growth is confined to one direction, and, instead of enlarging in diameter, it merely increases in length. The cause of the continual addition made to its length is probably to be found in the growth of the madrepore in which it is sheltered, and which would soon inclose the Magilus within its stony walls did not the mollusc provide against such a fate by lengthening its shell and taking up its residence in the mouth.

The most curious point, however, in the economy of the Magilus is, that, as fast as it adds a new shell in front, it fills up the cavity behind with a solid concretion of shelly matter, very hard, and of an almost crystalline structure, so as to leave about the same amount of space as in the original shell. The animal is always to be found in the very front of the shelly tube, and closes the aperture with a strong operculum that effectually shields it against all foes.

The colour of the Magilus is whitish. Only one species is known.

In the peculiar formation of the shell there is an evident analogy with the successive chambers formed by the pearly nautilus. In both cases the animal is of small dimensions

APPLE TUN-SHELL.—*Dolium pomum.* SPOTTED NEEDLE-SHELL.—*Terebra maculata.*
SPOTTED IVORY-SHELL.—*Eburna areolata.*

when compared with the magnitude of its dwelling, and in both cases the creature continually advances forward, taking up its residence in a chamber formed in the front of the shell, and closing the passage behind in proportion to its advance. The chief difference, however, between the two is, that the Magilus, being a fixed shell and inhabiting a stony tunnel, needs not the delicately structured shell required by the active nautilus, and therefore merely fills up the useless portions of the shell with solid matter, requiring no hollow chambers and no tube of communication.

The three shells represented in the illustration belong to the same comprehensive family.

The SPOTTED NEEDLE-SHELL, or SPOTTED AUGER, derives its name from the long and sharply pointed form of the shell. More than one hundred species of this genus are known, all inhabitants of the warmer seas, and the greater part resident within the tropics. In all these shells, the aperture is very small and the canal short. As may be seen by reference to the engraving, the operculum is small and pointed, having the nucleus at the smaller extremity. In many species the animal is entirely blind; and even in those cases where eyes are present, they are very small, and set at the end of the minute tentacles. The Spotted Needle-shell occupies the extreme right of the illustration.

The colours of this shell are very pretty. A broken band, composed of very dark and large spots of rich brown, winds round the edge of each spire, and the rest of the shell is covered with bands of pale brown and white, and longitudinal wavy streaks of cream-white.

JUST below the needle-shell is seen the beautiful SPOTTED IVORY-SHELL, also a native of the hotter latitudes.

Few species, not more than eight or nine in number, are known to exist at the present day. They are all very smooth and polished on the exterior, and their substance is so thick and solid that they seem almost to be made of earthenware. They reside at a moderate depth, being generally found in twelve or fourteen fathoms of water. It is worthy of notice that the rich spotted markings of the shell are repeated upon the body of the animal. The members of this genus possess tolerably large eyes, set at the base of the long tentacles. As in the preceding genus, the operculum has its nucleus at the pointed end.

The colour of the Spotted Ivory-shell is pure porcelain-white, richly spotted with deep brownish red, something like the tint known to artists as burnt sienna. It is not a very large shell, being about two inches in length.

THE last figure, which occupies the extreme left of the illustration, represents the APPLE TUN-SHELL, one of a moderately strong genus, deriving their popular name from the rounded and barrel-shaped outlines of the shell.

The animal is shown as it appears when crawling, for the purpose of exhibiting the curious striped edges of the feet, and the manner in which the siphon is carried bent over the front of the shell like the uplifted proboscis of an elephant. In these shells the spire is comparatively small and short, and the aperture very large, thus producing a great contrast to the needle-shell on the same engraving. About fourteen species of Tun-shells are known, all inhabiting the warmer seas.

The colour of the Apple Tun-shell is nearly white, diversified here and there with a few spots of the palest saffron. The interior, however, is much more deeply coloured than the outside, being of a rich yellow with a slight dash of orange.

THE beautiful HELMET-SHELLS also belong to this useful family. Several of these shells are employed by the engravers in the manufacture of cameos, the differently coloured layers producing most exquisite effects when cut by a judicious operator. The colours vary greatly in the different species, and sometimes there is a slight variation even in different individuals belonging to the same species. Cameos, for example, that are cut from the HORNED HELMET-SHELL (*Cassis cornúta*) are white, upon a ground of rich orange; those that are made from the WARTY HELMET-SHELL (*Cassis tuberósa*) are white, on deep dark red; the cameos formed from the shell of the RUDDY HELMET (*Cassis rufa*) are saffron-yellow on warm orange. Another beautiful species, called the QUEEN CONCH (*Cassis Madagascariensis*), furnishes a white cameo on a claret-coloured ground.

These shells are tolerably thick and solid, and their external surface is covered with bold ridges, marking the periodical growth. These ridges are technically called "varices." All the Helmet-shells are natives of the tropical seas, and appear to prefer the shallow waters near the coast.

ON the extreme left of the next illustration is a dark smooth shell, represented as crawling on the ground, and partially enveloped in the spotted textures of the living creature.

This is the BLACK OLIVE, so called on account of the jetty blackness of its exterior, and the oval, rounded form, which is not unlike that of the fruit whose name it bears. The genus Oliva is a very large one, comprising more than one hundred species, and found in all the warm and tropical seas. As may be seen by the figure, the mantle is furnished with two large lobes, that nearly meet over the back while the animal is moving, and which throw out certain filamentary projections, that look very like tentacles in the wrong place. The foot is very large—so large, indeed, that the shell is partly buried in its soft material—and the eyes are, as may be seen in the figure, placed before the middle of the tentacles.

Owing, probably, to the great development of the foot and mantle, the Olives are active creatures, gliding about with tolerable speed, burying themselves in the sand when the

BLACK OLIVE.—*Oliva mauritana.*   LIGHTNING DOVE-SHELL.—*Columbella fulgurans.*
BISHOP'S MITRE.—*Mitra episcopalis.*

tide leaves the shores on which they are creeping; and if laid upon their backs, they can easily resume their original position by the use of the spreading foot. In spite of their elegant and harmless aspect, the Olives are predaceous and hungry creatures, and can readily be captured by the simple process of tying a piece of meat to a line, lowering it towards the spot where the Olives are creeping, and hauling it up at intervals, carrying with it the various molluscs that have attached themselves to the bait, and do not think of loosening their hold until too late.

The shell of the Black Olive is beautifully polished and of a deep rich black, through which a slight tint of brown can be observed in certain lights. The inside is porcelain-white, and the average length is not quite two inches.

TOWARDS the right-hand of the illustration may be seen a little shell represented as crawling upon the frond of a seaweed, and being covered with zigzag white streaks. This is the LIGHTNING-COLOURED SHELL or DOVE SHELL, its popular name being derived from the peculiar appearance of its markings.

The numbers of the Dove-shells are very great, at least two hundred living species being known. They are all inhabitants of the warmer seas, and inhabit the flat sandy shores where the water is shallow, or congregate about the stones scattered on rocky coasts. All the Dove-shells are small, and prettily marked, the present species being chosen on account of the quaint beauty of the stripes with which its shell is decorated. The ground colour of the Lightning Dove-shell is jetty black, and the bold zigzag streaks or flashes upon its surface are shining yellowish white.

BEFORE describing the long and pointed shell in the centre of the engraving, we must pass temporarily to the next illustration, representing two of the Cone-shells, or *Conidæ*, a family so called on account of their form. All the Cones have a similar external outline: the aperture is long and narrow, the head of the living animal is more or less lengthened, the foot is splay and abruptly cut off in front, the tentacles are rather widely separate, and the eyes are placed upon these organs.

The right-hand figure represents the TEXTILE CONE-SHELL, brought from the Mauritius. This handsome species is about four or five inches in length, and its markings are curiously disposed, so that it is impossible to say which is the ground colour. The dark, narrow, angular lines are dark brown, accompanied by white, and variegated by dashes of yellow umber. The bold triangular spots are pure white, and the inside of the shell is of the same colour.

ADMIRAL CONE —*Conus ammirális.*   TEXTILE CONE.—*Conus téctilis*

THE empty shell lying on the ground, at the left of the illustration, is the ADMIRAL CONE, and is placed so as to exhibit the peculiarities of the long and narrow aperture. This species, in common with the other members of the genus, haunts the fissures and holes in rocks, and the warmer pools in coral reefs. They all take a moderate range of depth, varying from one to forty fathoms. Though slow in their movements, they are extremely voracious, their formidable teeth being well adapted for their predatory habits, and sometimes, it is said, being used effectively upon the hand of their captor. *Conus aulicus* has a bad reputation for such conduct, rather unexpected on the part of a shell-bearing and apparently helpless mollusc.

The Admiral Cone is not so large as the preceding species, but its comparatively small dimensions are outweighed by the elegance of its colouring. The light-coloured bands that surround the shell are very pale yellow, and the remainder is brown of different shades, boldly mottled with white. The interior of the shell is white. This is one of the very large genera, comprising about two hundred and seventy species, all inhabiting the hotter seas, and being most common in the tropics. Two species only are known in the Mediterranean.

RETURNING to the illustration on page 382, we find a long pointed shell with regular spiral markings, occupying the centre of the engraving. This is the BISHOP'S MITRE, a shell belonging to another family, termed the Volutidæ. In these shells the aperture is rather deeply notched in front; the animal has its siphon recurved, and the foot is very large, in some species partly hiding the shell. The eyes are either placed upon the tentacles or near their base.

The shell of the Bishop's Mitre is spindle-shaped, long-spired, and stout in substance. As may be seen by the illustration, the proboscis is very long. This mollusc possesses, in common with many others of its class, the capability of protecting itself when alarmed

MUSICAL VOLUTE.—*Voluta musica.*   NEPTUNE'S BOAT.—*Cymba Neptuni.*   BAT VOLUTE.—*Voluta vespertilio.*

by the sudden emission of a purplish liquid, having to human nostrils a peculiarly nauseous odour.

The Mitres, &c. are a very numerous genus, about three hundred and fifty living species being known and named.   All the large species inhabit the tropics ; and although there are some which are found in cooler regions, they are of very small dimensions, and mostly frequent the moderately shallow waters, though a few species are found at a depth of eighty fathoms.

The colour of the Bishop's Mitre is very pleasing, being pure shining white on the back-ground, and the spots being of a rich warm bay, the red predominating.

AT the bottom of the accompanying illustration may be seen two curiously marked shells, one empty and the other containing the animal.   These belong to the typical genus of the Volutidæ, which contains about seventy species, and is spread over most of the warm seas.

The Bat Volute is represented on the right-hand of the engraving, and exhibits not only the bold markings of the shell, but the curious form of the animal which made it. At each side of the large siphon may be seen a lobe projecting from its base, and the eyes are set on lobes projecting from the base of the tentacles.   When the tooth-ribbon of the Volute is examined under the microscope, its armature is seen to consist a series of three-pointed teeth, forming a very powerful engine of destruction.

The colour of the Bat Volute is extremely variable, some individuals being very dark in proportion to others ; and even the shape of the variegations is most uncertain.   In the specimen from which the figure was taken, the markings are boldly contrasting tints of brown and white; but in other examples the large brown dashes are replaced by delicate wavy pencillings, something like the back of pencilled pheasants. It is rather remarkable that, as the Bat Volute increases in age and dimensions, it fills up the hollow of the spine with solid shell, thus adding considerably to the weight of the specimen.

On the left hand is shown an empty shell, called, from its peculiar markings, the MUSICAL VOLUTE, the series of lines being supposed to represent the clefs, and the spots doing duty for the notes.

As in the preceding instance, this shell is most variable in the shape and colour of its markings, and even the number of lines differs considerably. In this specimen is found the normal number of five lines and four spaces; but in some examples there are only four lines, while in others their number is increased to seven. The colour of the specimen which is figured in the illustration is a mixture of grey neutral tint and pale brown, the lines being nearly black, and the interior of the shell a very pale drab.

THE large, uniformly coloured shell occupying the upper part of the engraving represents the species called NEPTUNE'S BOAT, a rather pretty, though simple looking, shell.

But few species of the genus Cymba are known, nine or ten being their apparent number; and these creatures appear to be found mostly in Western Africa, though Europe can boast of an example. The peculiar form of the animal can be seen from the illustration, its oddly-shaped proboscis and recurved siphon giving it a very curious aspect. The foot is, as may be seen, of very great size, and deposits a thin enamel on the under side of the shell. When first born, the young animal is of very great size when compared with its shell. The nucleus is large and globular, and in the youth of the animal is sufficiently conspicuous, but as the inhabitant increases in age, and the home increases in size, the nucleus becomes partly concealed by the growth of the shell, the whorls of which form a flattish ledge around it.

Although not a very large shell, nor remarkable for the variety of its colouring, the Neptune's Boat has yet a pleasing effect to the eye, and, when examined, is really an elegant and delicate shell. Its walls are very thin in proportion to its dimensions, and the bold, sweeping curves of the surface always call forth admiration. Its colour is uniform palish drab on the exterior, and the inside is beautiful pinky white, like that of a blush rose.

On the right hand of the following illustration, and occupying the central portion, the reader will perceive a curious-looking shell represented as crawling upwards, the animal having a very broad and flat foot, and its shell almost covered with the striped mantle. This is the MARGINELLA, our last example of the Volutidæ.

About ninety species of Marginella are known to zoologists, all belonging to the tropical or warm seas. As may be seen by the engraving, the animal is very large in proportion to the size of its home, and the mantle is so formed that the two lobes almost meet over the back of the shell, nearly concealing it from view. The tentacles are long, and the eyes are placed upon them near their base. The shell is smooth and polished, and when adult the outer lip has its edge considerably thickened, thus gaining the generic title of Marginella.

The colour of the shell is grey, streaked with black lines, and the animal itself is of a pinkish hue, diversified by red rays.

WE now come to the family of the Cowries or Cypræidæ, three representatives of which family are seen in the same engraving. As in the last genus, the mantle is expanded into two lobes, which nearly meet over the back of the shell; but in many species these lobes are covered with filaments, like so many tentacles. The eyes are either near the base or middle of the tentacles, and the tooth-ribbon is powerfully armed.

THE most familiar example of these shells is the common ENGLISH COWRY, which may be seen on the upper left-hand of the engraving, crawling diagonally upwards

3.   C C

and remarkable for the great length and breadth of the foot, and development of the mantle and tentacles.

All the Cowries are lovers of the shallow waters near shore, and are carnivorous in their habits, feeding mostly upon the numerous zoophytes that inhabit the same coasts. These shells change their forms in a truly remarkable manner. When young, the shell is very like that of a volute, having a prominent spire and a rather wide-spreading lip ; but in process of time the lobes of the mantle expand over it on either side, and by degrees deposit so thick a layer of smooth, shining substance, that the spire is entirely hidden. The pale streak which generally exists along the back of the Cowries indicates the line where the edges of the mantle nearly meet.

This little Cowry is so well known as to need no description.

The celebrated MONEY COWRY (*Cyprœa monéta*) belongs to this genus. These little white shells are well known as being the medium of barter in many parts of Western Africa ; and vast multitudes of them are gathered from their home in the Pacific and Eastern seas, and imported into this country for the purpose of immediate exportation to the African coast. Sixty tons' weight of Money Cowries have been freighted at a single British port in one year.

IN the left-hand bottom corner of the engraving may be seen the beautiful PANTHER COWRY, represented as it appears while living, its mantle covered with the curious appendages which look very like the tentacles of the sea anemones. This species derives its name from the rich mottling of the surface. A larger species is called the TIGER COWRY. One of these shells is largely used by the natives of the Sandwich Islands as sinkers for their nets, and a singularly ingenious bait is made from the same shell for the capture of the cuttle-fish.

A number of Cowries are cut into fragments and so fitted together as to form an oval ball of considerable size, with a smooth and mottled surface. Something by way of a tail or balance is fastened to one end of the ball, and the fishing-line tied to the other. The bait is now complete, and is quietly lowered near the spot where the cuttle is known to live, and drawn slowly along the ground. The ever-watchful cuttle is immediately attracted by this novel object, and thinking it to be some hitherto unknown delicacy, darts at it, and arrests its progress by attaching one of its arms to the smooth surface. The fisherman then gives a slight jerk to his line, and the deluded cuttle, fancying that 'ts prey is trying to escape, makes fast another arm. By repeated jerks the cuttle is induced to cling with all its force to the bait, when the fisherman rapidly hauls up the line, and flings the sprawling mollusc on the shore before it is aware of its danger.

Several of these large Cowries can be successfully employed in the manufacture of cameos, especially when human heads form the subject, as the dark mottlings of the shell can be used with singular effect in expressing the deep warm shadows of wavy tresses. The various articles of ornament that are made from these shells are too multitudinous even to be enumerated, much less described. About one hundred and fifty species of this genus are known.

THE grooved or wrinkled edges of the lips are well known to every one who has handled a Cowry, and these ridges assume a remarkable development in the DEEP-TOOTHED COWRY, a figure of which may be seen in the right-hand bottom corner of the engraving, the empty shell being laid so as to exhibit the opening and the lips. The colour of this shell is extremely variable, but is mostly a mottled wood brown, sometimes diversified with bands, and dark inside. It is not a very large species.

THREE examples of the curious Egg-shells are to be seen in the same engraving. The upper central figure represents the POACHED EGG, a popular and appropriate name, as the peculiar shape and colour of the shell bears a singular resemblance to the contour and tints of a well-poached egg as it trembles on the toast. Thirty-six species of the Eggs

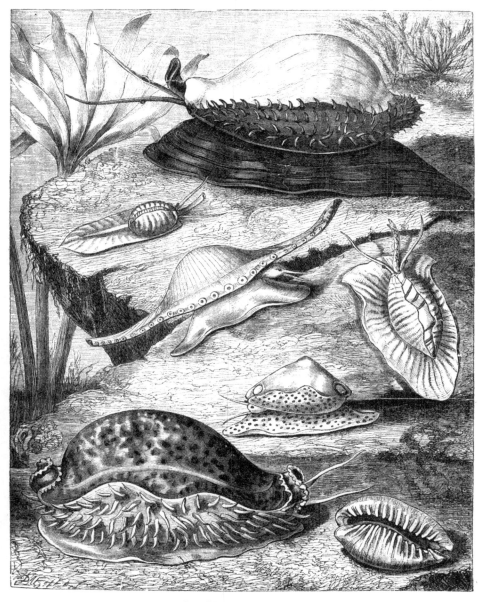

POACHED EGG.—*'Ovulum ovum.*

ENGLISH COWRY.—*Cypræa Europæa.*   WEAVER'S SHUTTLE.—*'Ovulum volva.*   MARGINELLA.—*Marginella diádocha.*

PANTHER COWRY.—*Cypræa pantherina.*   WARTY EGG.—*'Ovulum verrucósum.*   DEEP-TOOTHED COWRY.—*Cypræa caurica*

are known, spread sparingly over the greater part of the world. The under surface and opening of these shells are not unlike those of the cowries, except that in the Eggs the inner lip is without the ridges.

A very curious elongated shell occupies the centre of the engraving. This is the shell probably known by the name of Weaver's Shuttle on account of its peculiar shape. It is, in fact, one of the Eggs, but has the aperture lengthened into a long canal at either end. The foot of this species is narrower than in the other members of the same genus, but is especially adapted for crawling over the stems of the gorgonia, one of the zoophytes on which the mollusc feeds.

Our third and last example of these shells is the Warty Egg, remarkable, not so much for the tuberculated exterior of the shell, as for the richly spotted foot and mantle.

We now arrive at a vast army of shells called the Sea Snails, and distinguished by having the edges of the aperture without notches, the shell spiral or limpet-shaped, and the operculum either horny or covered with hard, smooth, shelly matter.

Our first example of this family is the Natica, the large shell in the centre of the engraving.

As will be seen, on reference to the illustration, the mantle of this species is very large, and the front of the foot is developed into a fold, which turns backward over the head and serves as a kind of protection. As the animal is without eyes, this curious structure causes no inconvenience. All the Naticæ, of which about ninety species are known, are found upon the sandy beds of the sea, and sometimes are taken at a depth of nearly six hundred feet from the surface. They are very predaceous in their habits, feeding principally on little bivalves, which they can assault with their short but strongly armed tongue-ribbon. The eggs of these creatures are very remarkable. They are compacted into a kind of spiral roll, broad and rather short, which is suffered to be flung about at the mercy of the waves, and is sometimes found resting on the sands when the tide has retreated.

The colours of the Naticæ are marvellously permanent, and even in the fossil state they are preserved and retain some degree of their original brilliancy. The species which is here represented is yellowish, and marked with grey bands.

An empty shell may be seen lying in the right-hand corner of the engraving. This is the Naticella-shell, closely allied to the preceding, and given in order to exhibit the shape of the aperture.

Just above the naticella may be seen a shell, together with the animal, represented as crawling.

This is an example of another family, the Neritidæ, and is popularly known by the name of the Smooth Nerita. As may be seen on reference to the engraving, the foot is moderate, the tentacles are exceedingly large, and the eyes are set on footstalks just behind the base of these organs. The Neritas are all inhabitants of the warmer seas, and are found plentifully within the tropics.

The colour of the present species is slightly variable, but in the individual specimen the shell is marked with bold zigzag streaks of white and pale buff, and the interior is pure white at the lip, changing to beautiful canary yellow in the interior. The operculum is thick, hard, solid, and highly burnished, as if overlaid with glass; its edge is regularly and finely grooved.

Several allied shells are inhabitants of the fresh instead of the salt waters, and are known as Neritines.

One of the most curious of these shells is the Spined Neritina, which may be seen on the extreme left of the illustration. The animal of the Neritina is not unlike that of the preceding genus, but there are one or two minute differences which cannot readily be explained without figures. The operculum is shelly, with a flexible border, and has some small teeth on its straight edge. All the Neritinæ are globular in their general shape, darkly spotted or banded with black and purple, and covered with a polished

SPINED NERITINA.—*Neritina spinósa.*
CROWN NERITINA.—*Neritina corona.*
NATICA.—*Nática Chemnitzii.*
SMOOTH NERITA.—*Nerita polita.*
NATICELLA.—*Naticella porcelána.*

bone-like epidermis. The colour of the Spined Neritina is deep green-black on the exterior and blackish white within. The shell is thick and solid at the aperture, but becomes thinner towards the interior.

THE topmost figure represents another species of this genus, the CROWN NERITINA, given for the purpose of showing the animal as it appears while crawling. The colour of this shell is grey, diversified with dark streaks. One species of this genus, the RIVER NERITINA (*Neritina fluviátilis*), is found in the rivers of England.

THE two curious shells represented in the accompanying illustration are examples of another family, that of the Clubs, or Cerithiadæ. The shell of the Cerithites is spiral, more or less elongated, and the operculum is horny and spiral. The tentacles are placed rather far apart, and the eyes are set on very short footstalks. These creatures inhabit either marine, brackish, or fresh water.

THE left-hand figure represents the PELICAN'S FOOT, sometimes called the SPOUT-SHELL, on account of the manner in which the aperture is lengthened into a kind of spout in front. This shell has a rather elongated spire, and is considerably tuberculated on the exterior. As the animal approaches maturity, it adds fresh substance to the lip, until it bears some resemblance to the webbed foot of an aquatic bird. The animal has a short and rather abrupt muzzle, and moderately long, cylindrical tentacles, having the eyes set on protuberances near their base. Only three species of this genus seem to be at present known, but they have a wide range of locality, being spread over the greater part of the world, and found at various depths, sometimes being taken in a hundred fathoms of water.

The colour of the Pelican's Foot is white, with a tinge of pink, and white inside. The shell is thickly and strongly made, and heavy in proportion to its weight. It is not a large species, seldom measuring more than two inches in length.

On the right hand of the illustration is seen an elongated shell, shown together with the animal.

This is the GREAT CLUB-SHELL, a species belonging to the typical genus of the family. It is rather a large genus, containing at least one hundred known species, and ranging over the whole world.   The largest species are, as is usually the case, to be found within

PELICAN'S FOOT.—*Aporrhäis pes-pelicáni.*   LARGE CLUB.—*Cerithium palustre.*

the tropics.   The shell is considerably elongated, and with many whorls, and the " varices " or marks of growth are partially visible on the exterior.   The aperture is decidedly small when compared with the dimensions of the shell, and has a somewhat twisted canal in front.   The outer lip is rather wide, and the inner is much thickened.

One of these shells, the MARSH CERITHIUM (*Cerithium palustre*), is supposed by some persons to produce the strange sub-aquatic musical sounds that exist in several Eastern lakes.   A detailed account of these sounds, together with the reason for this conjecture, may be found in Sir J. E. Tennent's "Natural History of Ceylon."

The colour of the Great Club-shell is deep chocolate-brown on the exterior, slightly mottled with varying tints, and the interior is brown, but without the chocolate hue.

In the family of the Turritellidæ, the shell is either tubular or spiral ; the aperture is not waved, notched, or formed into canals ; the foot is very small, the muzzle is short, and the eyes sunk rather deeply into the base of the tentacles.

THE figure on the left-hand of the accompanying illustration represents a species belonging to the typical genus of this family, the COMMON TURRITELLA.

In all the Turritellas the shell is long, pointed, and with many whorls ; the aperture is rounded and its edge thin ; the operculum is horny and with many whorls, and with a slightly fringed edge.   About fifty species of these shells are known, spread over the whole world, and inhabiting the moderately deep waters of the shores, ranging from a depth of one to fifty fathoms.   They are supposed to be carnivorous.   The colour of the Common Turritella is whitish.

THE curious WORM-SHELL, which derives its name from its long and twisted form, is a very remarkable shell, and, if carefully examined, affords much instruction as to the mode in which the molluscs build up their wonderful homes. It looks, indeed, much as if it were in the preliminary stage of shell-making, and had completed its arrangements with the exception of pressing the whorls together. When young, the spiral form is tolerably regular, but as it grows in years its regularity decreases, and the shell exhibits the form represented in the figure which occupies the centre of the engraving.

The aperture of the Worm-shell is round, and the operculum is consequently circular, and fits the opening with tolerable closeness. Its external face is concave. When not open, the tube is found to be supplied with many partitions of the same material as its walls. The colour of the Worm-shell is greyish yellow.

TURRITELLA.—*Turritella communis.*   SNAKE-SHELL.—*Siliquaria anguina.*
WORM-SHELL.—*Vermetus lumbricalis.*

ON the extreme right of the engraving may be seen a shell of somewhat similar construction, but readily distinguishable by the longitudinal slit which extends throughout its entire length. About seven species of the Siliquaria are known, all of which are carnivorous in their habits, and are found within sponges. As in the last species, the SNAKE-SHELL is regularly spiral at its commencement, where it was constructed by the animal in its youth, but loses its regularity in exact proportion to its age. Its colour is whitish. The small head is shown just protruding, and exhibiting the stopper-shaped operculum.

THE lower figure in the engraving on page 392 represents the empty shell of the STAIRCASE or PRECIOUS WENTLETRAP, in former days one of the scarcest and most costly of the specimens of which a conchologist's cabinet could boast. There was hardly any sum which a wealthy connoisseur or virtuoso, as the fashion was then to call those who were fond of natural history, would not give for an especially large and perfect example of this really pretty shell. Now, however, its glory has departed, for a tolerably good specimen may be procured for a few shillings, and a Wentletrap which would a few years ago have been sold for fifty pounds can now be purchased for fifteen shillings.

Putting aside, however, the question of rarity or cost, this shell is a very interesting one, both for its beauty and the mode of its construction. It is purely white and partly transparent, the elevated ridges being of a more snowy white than the body of the shell, on account of their superior thickness, which does not permit the light to pass through them as in the case of the thinner body. The whorls of this shell are separate from each other, and apparently bound together only by the projecting ridges, so that the general appearance is as if the whorls of a worm-shell had been pressed nearly together, and then kept in their place by a succession of shelly elevations. This beautiful shell is found in the Indian and Chinese seas.

STAIRCASE WENTLETRAP.—*Scalária pretiósa.*
COMMON WENTLETRAP.—*Scalaria commúnis.*

THE smaller figure, showing the shell attached to the animal, represents the COMMON or FALSE WENTLETRAP, a species tolerably common upon our coasts.

In this shell the whorls are united together and furnished with a number of circular elevations, which, however, are not nearly so bold as those of the preceding species, but thick in proportion to their height, set obliquely on the shell, and smooth. The animal has a proboscis-like mouth, which can be retracted at the will of the owner; the tentacles are tolerably long, placed near together; and the eyes are set near the base of the tentacles. The foot is triangular, with the front rather obtuse, and supplied with a fold. When disturbed or alarmed, the creature is capable of exuding a dark purple fluid. Nearly one hundred species of Wentletrap are known, all the largest examples being found in tropical regions. They live at a considerable depth, sometimes being captured in eighty fathoms of water, and little seems to be known of their habits.

The colour of the Common Wentletrap is rather varied. Sometimes it is dull white, sometimes it is very pale brown, and in a few specimens the shell is reddish violet, with the ribs purple.

WE now arrive at another family, termed the Litorinidæ, or Shore Molluscs, because the greater number of them frequent the coasts, and feed upon the various algæ. The shell is always spiral and never pearly, by which latter characteristic it may be distinguished from certain shells belonging to another family, but somewhat similar in external appearance. The aperture is rounded. The animal has its eyes set at the outer bases of the tentacles, and the foot is remarkable for a longitudinal groove along the sole, so that in the act of walking each side advances in its turn. The tongue is rather long, and is armed with a formidable series of sharp teeth that serve admirably for the purpose of scraping away the vegetable matter on which the animal feeds. The operculum is horny, and rather spiral.

THE common PERIWINKLE (*Litorína litórea*) is the most familiar example of this family, and is too well known to need any detailed description. The Periwinkle is found upon our rocks in great profusion, occupying the zone between high and low water, and always being found near the edge of the tide. There is, however, another species (*Litorína rudis*) which occupies a rather higher zone than the previous species, and which, though very plentiful, is not eaten, in consequence of its young obtaining their shells before eggs are laid, and having a gritty and unpleasant effect upon the teeth. Sea birds, however, are

INDIAN PHORUS.—*Phorus 'Indicus.*   STAIRCASE OR PERSPECTIVE TROCHUS.—*Solárium perspectivum.*

not very particular about this drawback, neither is the thrush, which, in winter, when the snails are hidden away in their dark recesses, finds a meal easier to be obtained on the sea-shore than in hunting for its usual prey.

ONE of the prettiest members of this family is the WINDING STAIRCASE-SHELL, or PERSPECTIVE TROCHUS, so named on account of the peculiar formation of its whorls.

If the shell be held with its top downwards, it looks exactly as if it had been wound around a conical centre which had afterwards been withdrawn, and the projecting edges of the whorls have a wonderful resemblance to the perspective view of a winding staircase seen from below.

Perhaps the most remarkable point about this genus is the singular operculum of some of the species, which differs from that of any other mollusc. Instead of being a nearly flat plate of horny or shelly substance, it is a conical structure of shelly matter with a riband of membranous substance wound round it, and projecting like the mechanical form so well known as Archimedes' screw. The object of this singular variation is quite unknown.

The colour of the shell is rather variable, but consists of mottlings with brown, ochre, and white.

A VERY curious member of this family is the LOOPING SNAIL (*Truncatella truncátula*), a little species that is remarkable for the habit which has earned for it its popular name. All these creatures inhabit the space between tide marks, and can live for many weeks without water. Their mode of progression is very peculiar, and closely resembles that of the leeches or looping geometric caterpillars with which we are so familiar. When they

SHELL-COLLECTING PHORUS.—*Phorus conchylóphórus.*

walk they fix the head firmly, then draw up the body in an arch, fix the foot, and then push the head forward. The foot is short and rounded at each end.

The shell is very small, about the size of a split sweet pea, and would escape the eye of ordinary observers. The animal is furnished with short and diverging tentacles, the head is divided into two lobes, and the eyes are placed in the centre behind the tentacles.

THE second figure in the engraving is the INDIAN PHORUS, or MINERALOGIST, a name given to the creature in allusion to its extraordinary habit of agglutinating bits of stones and other substances to its shell. Sometimes it prefers other shells, either in fragments or entire, and is then termed the CONCHOLOGIST. In one example shown to me by Mr. Sowerby, the creature had selected a number of shells of a tiny bivalve, and had stuck them round the edges of its own shell in such a manner that they form a spiral line, marking the growth of the shell. One or two little bits of stone accompany them, and they all lie with the hollow upwards. The animal has a rather long proboscis, and long tentacles, with the eyes set at their outer bases. The foot is long and narrow behind.

The outer lip is very curious in its structure, being extremely thin, projecting above and receding below. The operculum is horny, and formed by overlapping scales. The colour of the Indian Phorus is yellowish brown above, and pearly white within. The edges of the lip are ragged and crumpled like those of a withered leaf.

IN the accompanying illustration is represented a magnificent specimen of the SHELL-COLLECTING PHORUS, where the animal has contrived to group itself around with the different shells that are figured. The long-pointed shells are clubs, or cerithinæ, a Venus-shell is seen at the mouth, and a lucina at the base. The name Phorus is of Greek origin, and signifies a carrier. The movements of the Phorus are said to be very clumsy, the animal staggering and tumbling about like the stromb-shells already described.

WE now arrive at the Top-shells, or Turbinidæ, a rather large and important family. In all these creatures the shell is spiral, and beautifully pearly in the interior, the nacre appearing when the outer coating is removed. The animal has a short head, rather long tentacles, with eyes mounted on footstalks at their base, and the head and sides are decorated with fringed lobes. They are all inhabitants of the sea and are vegetarians in their diet, their array of sharp teeth being very useful in rasping away the substances on which they feed.

IN former days, the PHEASANT-SHELLS were articles of great price and rarity, some specimens almost rivalling the precious wentletrap in the enormous sums asked and obtained for them. Now, however, that their habitations have been discovered, and more frequent voyages are made, they have become comparatively plentiful, although, from the fragility of their structure, a perfect specimen is not at all common, and will still bring a good price in the conchological market.

The Pheasant-shells are now found in great numbers on the sandy beaches of several shores, being especially plentiful on the coast of Port Western, in Bass's Straits. The high tide sweeps them towards the shore, where they are left by the receding waters, and seek for shelter beneath the masses of sea-weed that are always flung on the beach by the tide. On lifting these sheltering weeds, the Pheasant-shells may be found crowded together under their wet fronds. They can move with some speed, the duplicate nature of the foot aiding them greatly in progression.

COMMON TOP.—*Trochus sizyphinus.*
AUSTRALIAN PHEASANT-SHELL.—*Phasianella Austrális.*

Although vegetable feeders to a great extent, the Pheasant-shells will eat animal substances, and can be captured in great numbers by the simple process of lowering a net to the bottom of the sea after baiting it with bits of meat tied to the meshes. They are active creatures, always on the move, and therefore free from the various parasitic plants and animals of the sea which invariably take possession of all stationary or sluggish objects. This very activity, however, is injurious to the perfection of the fragile shell, its delicate lip being mostly notched and broken by coming in contact with hard substances.

The colour of the Pheasant-shells is extremely variable, and after death the shell often assumes quite a different tint to that which it bore during life. Several shells, which presented tints of brown and green during the life of the inhabitant, changed to a reddish hue after death and when the rays of the sun had poured their strange influence on them.

As may be seen by the engraving, the animal is of a very curious shape. The tentacles are extremely long, pointed, and slender, and the head is supplied with certain fringed lobes. The foot is large, rounded in front and pointed behind, and is divided by a deep groove along its centre so as to form two lobes, each of which is alternately slid forward in the act of walking. The animals are mostly possessed of rather bright colours, some specimens being wholly grass-green, others brown dotted with green, and others with the foot spotted with white or reddish violet. The operculum is very hard and shelly, slightly convex, and, when viewed from the inside face, exhibits very perfectly its spiral structure.

The colour of the Australian Pheasant-shell is very beautiful, consisting of delicate wavy pencillings and mottlings of pink, carmine, grey, and brown. These marks are more decided and the carmine purer towards the apex of the shell, the colours becoming bleared in proportion as they approach the mouth. These are the colours of the specimen from which the figure was taken, but this is an exceedingly variable species, the tints differing greatly in tone and distribution, and some specimens being almost wholly ochrous and grey, with hardly a trace of the beautiful carmine and rose-pink of other individuals.

THE little pointed shell on the left of the illustration is the well-known TOP of our own shores.

This little shell, which is here represented of the natural size, is one of the most plentiful species of the British coasts, and may be found by hundreds either crawling among the sea-weeds at low water, or flung upon the sands by the tide. The shell of this creature is beautifully pearly, and when the outer coating is removed the iridescent nacre below has a very lovely appearance. Jewellers and lapidaries employ these shells largely in their art, polishing them carefully and then stringing them together so as to form bracelets and necklaces, or affixing them as ornaments to various head-dresses. Another little shell, called TURBO VERSICOLOR, which is brought from Southern America, is also used for similar purposes. The specimens of Top-shells which are found in the sands are seldom quite perfect, the apex of the spine being usually worn down and rubbed so as to display the sub-lying nacre.

About one hundred and fifty species of Trochus are known, some of them attaining considerable dimensions, and all possessing shells of exceeding beauty. The peculiar form of the animal is shown in the illustration. The tentacles are rather long, and the eyes are seen at the extremity of the little footstalks, at their base. The neck-lappets are rather large, and the sides are furnished with lobes and tentacular projections. The operculum is horny, flat, and spiral. Trochi are found all over the world, and have a considerable water range, being captured at all depths, from the shallow waters of the shore to a depth of a hundred fathoms.

ANOTHER beautiful species of Trochus is the NILOTIC TOP, a shell which is remarkable for the rich contrast of scarlet flashes on a white ground. One of the rarest species of this genus is the IMPERIAL TOP (*Trochus imperialis*), a shell which has hitherto been found only in New Zealand, and may probably be confined to that strange land. It is a handsome as well as a rare species, and is notable for the bold rounded projections which radiate from the whorls. Its colour is violet-brown above and white below. Some authors, however, separate this shell from the Trochi, and place it in a separate genus, on account of the toothed whorls.

The DOLPHIN-SHELL affords another instance of the entire discrepancy between the shell and the popular name that is given to it, this species bearing no more resemblance to a dolphin than to a roach, a cow, or a peacock.

Twenty species of this genus are known, all being inhabitants of the hotter seas, and found chiefly near the tropics. There are but few whorls in this shell, and they are formed in so singular a manner, that they look as if they had been crushed down by a heavy weight and thus prevented from rising into a pointed spire. In this species the tentacles are surrounded with blunt spines, but in others they are edged with sharp teeth or with wrinkled projections. The aperture is round and pearly, and the operculum is horny, flat, and spiral. When the shell is inspected from below the centre, or "umbilicus"

NILOTIC TOP.—*Trochus Nilóticus.*   DOLPHIN-SHELL.—*Delphinula laciniata.*

as it is technically called, is seen to be hollow, and looks somewhat like that of the solarium.

The animal has no head lobes, but the sides are furnished with lobes and cirrhi, *i.e.* the tentacular projections already mentioned. The Dolphin-shells prefer the shallow water, and at low tide may be picked by hand off the reefs. The colour of the Dolphin-shell is purplish tending to black, and the aperture is pearly.

In the next family of molluscs, the shell assumes a very remarkable shape, for although it is constructed on the spiral principle, the spire is so very small and the aperture so very large that many persons would take it for one of the bivalves. Most of these shells are shaped after the form of the human ear, which has earned for them the title of Haliotidæ, *i.e.* Sea-ears. The outer lip is either notched at the edge, or perforated with a series of oval holes, and the general substance of the shell is pearly, and coloured with tints. vivid as those of the rainbow, and refulgent as those of the humming-bird's throat. The outer angle of the shell is perforated by a series of oblong holes, which are generally filled up as they approach the spine and at last are wholly obliterated. Some-times, as in a very fine specimen now before me, several of these perforations coalesce, so as to form a long rounded slit, nearly two inches in length. The animal has slender and conical tentacles, with the eyes set on footstalks at their outer bases; and as the operculum which would be needed to close so large a mouth would be of unwieldy dimensions, that organ is almost rudimentary, and appears to serve no practical purpose.

The Ass's Ear is one of the larger species of the genus Haliotis, and is one of the most beautiful among the shells. Even when rough and unpolished, just as it appears after the removal of the animal, the rich iridescence of its interior is almost dazzling in the intense brilliancy of its colouring; and when, by the use of acids, the rough outer coat is removed and the nacreous substance of the shell exposed, there is hardly any marine production that approaches it and none that surpasses it in beauty.

This is a very useful shell to the manufacturer, its thick solid substance, with its lovely iridescence, rendering it well adaptable for the construction of buttons and similar articles, and also for inlaying in the darker woods. Very beautiful sleeve-links are cut out of the muscular impression, its heavy material giving the requisite strength, while its peculiarly corrugated structure produces a very beautiful effect, either when ground and polished or suffered to retain its ordinary contour.

As may be seen by the illustration, the foot is large and rounded, and when the animal is crawling upon the rocks, can be formed into a sucker that enables the creature to adhere to the stones almost as firmly as the limpet.

THE smaller perforated empty shell is the GUERNSEY EAR-SHELL, popularly known throughout the Channel Islands by the name of ORMER.

This shell does not attain to so great a size as the preceding, but is, if possible, even more beautiful when polished and the opaque outer coat removed by means of acids and hard labour. The growth of each successive year is marked by a bold ridge, sweeping in a curve from the spine to the edge, and rapidly enlarging towards the margin. These ridges are caused by a regular series of furrows, in reality very shallow, but, on account of the peculiar manner in which they reflect the light, appearing to possess considerable depth. The effect presented by these ridges is really marvellous, the rich iridescence of delicate pink, green, and blue, with the slightest imaginable lines of golden light marking them, being quite beyond the powers of description or even of artificial colours. Each ridge is perforated by a single hole near its extremity, and their course is marked even on the interior of the shell.

The animal of the Guernsey Ear-shell is largely eaten, but requires careful management in the cookery, as it is liable to be tough and stringy if badly handled. Before being subjected to the culinary art, it is well beaten, like a beef-steak, and is then cooked in various ways.

A SIMILARLY shaped shell, but without any perforations on the edge, may be seen creeping along the ground. This is the STOMATIA, or FURROWED EAR-SHELL, so called because the place of the holes is supplied by a single groove or furrow. This shell is a native of the hotter seas. Its colour is pale reddish grey on the exterior, and pearly within.

AT the top of the same illustration, a very curious snail-like shell may be seen, floating on the water, and attached to a kind of raft. This is the VIOLET SNAIL, so called from the beautiful violet-blue of the shell.

The Violet Snail inhabits several seas, and is most common in the Atlantic Ocean, though it is also found in the Mediterranean ; and a few specimens are occasionally driven upon our shores. Though in the look of the shell there is nothing sufficiently remarkable to attract notice, the habits and structure of the animal are most curious and interesting. The Janthina is essentially a surface species, always floating about, incapable of directing its course, and not even able to sink when threatened with danger. Being quite at the mercy of the winds and waves, it is often seen floating in great numbers, thus denoting the existence of some aërial or marine current, and may in such cases be swept up by thousands.

The most remarkable peculiarity in this creature is the so-called "raft," a projection from the foot, which is probably a modification of the operculum, and is capable of discharging some of the duties of that organ. This raft, as may be seen from the engraving, is of great size when compared with the shell, and is composed of a vast number of small air-sacs, which render the whole structure extremely buoyant. Over these sacs the creature has no control, its utmost power over the raft being limited to its partial withdrawal into the aperture of the shell so as to close its mouth after the manner of the operculum.

Beneath the raft may be seen a great number of little pendent vesicles. These are the egg-capsules ; and it is evident that those at the extremity of the raft are the first formed, as the capsules at the farther end are often found to have dismissed their inhabitants, while those in the middle contained young nearly ready for entrance into the

VIOLET SNAIL.—*Janthina communis.*

EAR-SHELL.—*Stomátia rubra.*   ASS'S EAR.—*Haliótis asininus.*   GUERNSEY SEA-EAR.—*Haliótis tuberculátus.*

world, and in those at the base were found only the undeveloped eggs. This raft retains the air for a considerable time after the death of the animal, and even if removed and dried will not lose its expanded form for several weeks. In a month or so, however, it generally collapses. A similar result takes place if the creature be placed in spirits, so that to prepare a specimen permanently is a task of great difficulty.

The attachment of the raft to the shell is exceedingly slight, and in rough weather is frequently broken by the tossing waves, so that the raft is seen floating as usual, while the animal to which it belonged has probably sunk deeply beneath the surface. Whether the creature possesses the power of reproducing the raft is a mooted question. When floating at ease, the raft is the only portion that is visible, the shell itself being under water.

The food of the Janthina is said to consist mostly of the small blue velellæ, but as the animal is without eyes, and is incapable of directing its course, it cannot be very rapacious. Some minute brown shells have been found in the stomach of several specimens.

The Janthina secretes a rather richly coloured fluid, respecting which many conflicting opinions have been given. Mr. F. D. Bennett, who has made some valuable observations on this curious mollusc, has the following remarks upon the fluid :—

"The body of this mollusc contains a very blue liquid, which, when the animal is punctured, exudes to the amount of three or four large drops. It is readily diffused through water or colourless spirit—to the former it communicates a faint tinge of its own peculiar hue, and to the latter a pink colour, with a purple shade. It communicates its colour to paper, and may be conveniently used as a blue ink; several memoranda and pages of my journal, written with this fluid, have, after a lapse of more than five years, retained their original appearance both in colour and intenseness. For this use, however, it must be employed from the recent animal, as it will not keep in any quantity, but becomes thin and discoloured.

It is believed that this fluid is analogous in use to the black secretion which the cuttle-fish pours forth to obscure the water and elude the pursuit of its enemies; but this opinion must be received with some qualification. The living examples of Janthina which I have irritated when they have been confined in a vessel containing sea-water have not emitted any of the coloured fluid; when taken in hand, they would sometimes allow a little to exude; but the entire quantity obtained from one animal by artificial means was never sufficient to cloud or obscure, although it would stain about half a pint of pure water."

When attached to the living animal, the shell is covered with a delicate and rather slippery membrane, and the violet hue possesses a peculiar liveliness, which soon vanishes after the death of the inhabitant. The raft is delicate white. Its length is about one inch, and the shell is very similar in shape and dimensions to that of our common garden-snail.

THE well-known univalves, so familiar under the name of Limpets, are divided into several families, on account of certain variations in the structure of the shell. The first family is termed Fissurellidæ, on account of the fissure which appears either at the apex cr in the front edge of the shell.

All the Limpets are strongly adhesive to rocks, as is well known by every one who has tried to remove one of these molluscs from the stony surface to which they cling. The means by which the animal is able to attach itself with such firmness is analogous to the mode in which the suckers of the cuttle-fish adhere to the objects which they seize, the formation of a vacuum. and the consequent pressure of the atmosphere, being the means employed. The foot cf the Limpet is rounded, broad, thick, and powerful; and when the animal wishes to cling tightly to any substance, it presses the foot firmly upon the surface, and retracts its centre, while its edges remain affixed to the rock. A partial vacuum is therefore formed, and the creature becomes as strongly attached to the rock as a boy's leathern sucker to the stone on which he has pressed it.

The Limpets, although inactive and comparatively stationary, are able to glide slowly by means of the foot, and leave very curious tracks in their progress. In the course of long adhesion to a single spot, the animal works a slight hollow, into which the edges of the shell sink when the creature is alarmed and presses itself to the rock, and thus protects itself very effectually from being dislodged by the insertion of an animate or inanimate lever under the shell. Several sea birds, and one or two of the land birds, are in the habit of feeding on Limpets, which they dislodge by the simple process of driving their sharp beak between the shell and the rock before the mollusc takes the alarm and presses itself against the stone to be moved. When the Limpet is quite at ease, it has a custom of relaxing the foot and raising the shell from the rock, and at such times it may be detached even by a quick stroke with the finger.

Most, if not all, of the Limpets are edible, and can be eaten either cooked or in a fresh state as they come off the rocks. Shipwrecked mariners have frequently saved themselves from perishing with hunger by gathering the Limpets from the rocks on which the vessel was cast away; and I have made many a luncheon on these creatures when engaged in searching the shores, and when time has been too precious to be wasted in leaving the spot for refreshment. An empty shell of the mussel forms an admirable spoon for the extemporized banquet.

The KEYHOLE LIMPET is so called on account of the aperture at the top of the shell, which serves as a passage through which is expelled the water that has passed over the gills. This aperture is found in all the species of the genus Fissurrella, but varies greatly in form and comparative dimensions, being, in some cases, a mere rounded hole in the

KEYHOLE LIMPET.—*Fissurella maxima.*

CUP-AND-SAUCER LIMPET.—*Calyptræa rudis.* (Empty shell below.)   HUNGARIAN BONNET LIMPET.—*Pileopsis Ungaricus.*
LADY'S BONNET.—*Calyptræa equestris.*   DUCK-BILL LIMPET.—*Parmóphorus austrális.*   Empty shell on the left.

shell, while in others it is a long and curiously shaped aperture, very like the keyhole of a lock. The aperture increases with the shell, being hardly perceptible when the animal is young, but encroaching rapidly until it removes the whole of the sharp apex. These animals are mostly found at the same depth with the great tang seaweeds, but are sometimes to be taken in fifty fathoms of water. The genus Fissurella is a large one, comprising about one hundred and twenty species.

THE curious DUCK-BILL LIMPET inhabits the hotter seas, and is found on the shores of New Zealand, the Red Sea, and the Cape. It belongs to a small genus, containing about ten species.

This shell derives its name from its peculiar shape, which certainly does bear some distant resemblance to the beak of a duck. The animal is of very great comparative dimensions, and while living covers the shell with its mantle. Its colour is black, and its sides are edged with short fringes. The eyes are set on the outer bases of the tentacles The colour of the shell is very pale yellow.

A NUMBER of nearly allied shells, belonging to the same family as the preceding species are called CUP-AND-SAUCER LIMPETS, from the peculiar cup-shaped process on the interior, the shell itself taking the place of the saucer. This process forms the base, to which are attached the muscles which draw the animal to the rock. None of these Limpets appear to be active, seldom quitting the spot on which they have settled themselves in their infancy. The form of the shell is extremely variable, depending greatly on the substances to which it adheres, and the colour seems to be quite as mutable as the form. A specimen in my possession has an exceedingly thick shell, with very deep ridges, and a boldly waved edge. Its colour is brown, of various shades, diversified with a little ochreous yellow. The "cup" is very much lighter than the interior of the shell, and is of a greyish white with a slight yellow tinge, and marked with wavy streaks that give it a singular

3.   D D

resemblance to chalcedony. The substance of the cup is very delicate, hardly thicker than the paper on which this account is printed.

ANOTHER species of the same genus, the LADY'S BONNET, is seen on the same engraving. The generic name is derived from the word *calyptra*, which signifies a lady's cap. The food of these molluscs seems to be rather varied, as they are known to eat the minute algæ, and one specimen has been observed in the act of devouring a little sea-slug which was placed in the same vessel.

It may be here remarked, that all the Limpets seem to alter the shape of their shell to suit the surface to which they adhere; and even the common Limpet of our own shores will be frequently found to have modified the edge of its shell into very curious outlines, for the purpose of accommodating it to certain hard projections or depressions of the rock to which it clings. As these molluscs possess and exercise the power of loco-motion, it is remarkable that they should mould their habitations into eccentric forms that can only suit one particular spot. For example, if a Limpet be attached to a concave surface, the shell becomes convex in a precisely corresponding degree; should it prefer a convex body, the shell is concave; and in many cases where the shell is affixed to a cylindrical object, such as a submerged stick, or perchance the stem of a large seaweed, it adapts itself so thoroughly to its support, that it clasps the cylinder in a partial embrace. It is possible that the power of adaptability may be rapid in its operation, and that the animal can supply new matter to the shell when needed, so as to render it concave or convex at pleasure, or to fill up the grooves and notches which have existed when it clung to an irregular surface.

Two specimens of the HUNGARIAN BONNET LIMPET are given in the same illustration. This shell is almost invariably found adhering to oysters in a moderate depth of water, varying from five to fifteen fathoms, though it sometimes prefers a greater depth. The finest specimens are, however, taken in the shallower waters. The popular name is sufficiently appropriate in this instance, as the shell is exceedingly like the celebrated Phrygian bonnet of the ancients, or the republican cap of a later period.

The COMMON LIMPET is so familiar that it need not be figured nor described. One species of its genus attains to an enormous size, measuring a foot in diameter, and having a shell of very great thickness.

THE next family, called appropriately Dentalidæ, or the Tooth-shells, have long puzzled zoologists to assign their right position in the scale of nature, and even baffled the wide experience and penetrative acuteness of Cuvier himself. The general opinion of the systematic naturalists of his time referred the Dentalidæ to the annelids or worms; but Cuvier always expressed his doubts as to the accuracy of their views, and remarked that the solution of the problem would be found in the nervous and respiratory systems.

This conjecture has been followed up by later anatomists, and the Dentalidæ have not only been recognised as molluscs, but have actually been referred to their proper position in that order. Let one of the perforated shells, such as the fissurella, be enormously lengthened in proportion to its width—drawn out, in fact, like glass under the hand of a workman—and it will form a very good idea of the Dentalium. The animal must necessarily undergo certain modifications to enable it to accommodate itself to the elongated habitation, and in consequence is so long and worm-like as to have induced many experienced naturalists to refer it to a class very far below that which it rightly occupies.

In all these creatures the shell is formed very much like an elephant's tusk, except that a small perforation exists at the small end. The animal is very curiously formed, being remarkably worm-shaped, and having a conical foot with a rather sharp point, and holes at the side. The animal is attached to the shell close to the little orifice at the point, and has no definite head, no tentacles, and no eyes. The creature seems to be almost wholly carnivorous.

PRICKLY CHITON.—*Chiton aculeátus.*
SHORT-SPINED CHITON.—*Chiton brevispinósus.*

MARBLED CHITON.—*Chiton marmóreus.*
BANDED CHITON.—*Chitonellus fasciátus.*

ELEPHANT'S TUSK.—*Dentálium arcuátum.*

WE now come to the curious family of molluscs called appropriately Chitonidæ, or Mail-shells, because their shells are jointed like the pieces of plate armour. When separated from each other, the plates bear a strong resemblance to the joint of a steel gauntlet, and overlap each other in a similar fashion, a thick and strong mantle taking the place of the leather. There are eight of these plates, and all of them have a somewhat saddle-like shape. A similar arrangement may be observed in the lower abdominal plates of many beetles. Each of these plates is fixed to the mantle by certain rounded processes from their front edge, and when the plates are examined separately the processes will be plainly seen, white and pearly as the interior of the shell.

The last plate is thought by Mr. Gray to be analogous to the shell of the limpet, and the seven other plates to be "portions of its anterior slope successively detached." At all events, the hinder plate is fixed more firmly to the mantle than any of the others; and while its companions fall off with handling, or give way at a slight pull, the hinder plate holds firmly in its place.

The genus Chiton is an extremely large one, containing more than two hundred species, which are spread over the whole of the globe, wherever the sea is not absolutely frozen. The greater number of species are found on the rocks about low water, and on our own coasts the naturalist may fill his bottle with specimens, provided he examines the proper spots. Some of them are found at a depth of ten or fifteen fathoms, and are obtained by dredging, while a few of the smaller species are found in eighty or a hundred fathoms of water.

The Chitons are able to roll themselves up in a partial kind of manner, and present a curious resemblance to the well-known armadillo or pill woodlouse. In all these creatures the shell is extremely beautiful, on account of the minutely graceful markings with which it is decorated, and which, in most cases, require the aid of a magnifying glass to exhibit their true beauties.

SEVERAL species are represented in the engraving. At the top is shown the PRICKLY CHITON, remarkable for the array of rather long spines with which the moveable plates are armed, and which, when the creature contracts itself, give it a curious resemblance to the hedgehog. Its colour is reddish brown on the exterior, and pinky white within. Although this shell attains a very great size, a large specimen measuring about five or six inches in length, it is not as valuable as in its youth, the curious spines being gradually lost as it approaches old age, just as human beings lose their hair, and the shell being by degrees rubbed tolerably smooth in some places and encrusted in others with corallines, calcareous matter, and the shelly coatings of various marine zoophytes. Sometimes the seaweeds find a lodgment on the shell, as is often the case with other comparatively stationary molluscs, such as the common limpet; and in that case the algæ not only find a home, but conceal their protector by their waving fronds.

THE smaller and smooth figure represents the MARBLED CHITON, a rather prettily coloured shell, its exterior being rusty red mixed with brown and yellow, and edged with brown. The SHORT-SPINED CHITON may be seen with the shell covered with short spines, and partly turned up so as to show the head of the animal. Its colour is sooty black, but this dull uniformity of a sombre hue is more than redeemed by the beautiful and minute pencilling with which its surface is engraved. The last figure represents the BANDED CHITON, or CHITONELLA, a creature that has been removed by modern naturalists into a separate genus on account of the formation of the armour. The plates do not cover the entire surface as in the preceding genus, as only a portion is seen above the mantle. The defence is, however, nearly as perfect as in the previous genus, as the projections approach each other beneath the surface of the mantle, and would act as effectually in shielding the internal organs as if the plates had met on the surface. These creatures are generally found in the clefts of canal rocks.

The animal is more active than that of the limpet, but does not appear to be very locomotive in its habits. Its broad creeping disc adheres very strongly to the rocks, and holds the animal so firmly that, if it should happen to have taken up its abode within a crevice, to extract it without tools would be an impracticable task. Like the dentalium, this creature possesses neither eyes nor tentacles.

PASSING from the sea to the land, we come to those gasteropods which breathe atmospheric air, and are furnished with respiratory organs suited to the lower element in which they live. These creatures fall naturally into two large sections, the one being destitute of an operculum and the other possessing that remarkable appendage. They are respectively called inoperculate and operculate gasteropods, and it is with the former that we have now to deal. The inoperculate are generally furnished with large shells; but in some, such as the slugs, the shell is either very small or wholly absent. The shell of these animals, when present, is not nearly so hard and porcelain-like as that of the sea-snails, and contains a much larger proportionate amount of animal matter. It is worthy of notice, that in order to prevent the waste of moisture in those species which live on land, and the entrance of water in those which inhabit the ponds and rivers, the respiratory passage is small, and closed with a kind of valve.

THE first family is that of the Snails, or Helicidæ, containing a vast number of species. Most of the Snails have a shell large enough to permit the animal to withdraw itself wholly into the protecting domicile. During the time when they are active these creatures require no closure of their shells, and accordingly have no vestige of an operculum, as may be seen by looking at a common Snail. In the winter, however, when they retire from active life, and need that the aperture of their domicile shall be closed, the place of the operculum is supplied by a layer of hardened mucus, sometimes strengthened with the same substance of which the shell is composed, and always being perforated with a little hole to permit the inhabitant to respire. Any one may see this structure, called technically the epiphagus, by examining a Snail drawn from the crevice in which it ensconces itself during the winter months.

The animal has a rather short head, furnished with four tentacles, the upper pair being the largest, and bearing at their tips the little black specks which are supposed to

CHRYSALIS-SHELL.—*Pupa uva.*   LEMON BULIMUS.—*Bulimus citrinus.*
TESTACELLA.—*Testacella haliotoïdes.*   GREAT GREY SLUG.—*Limax antiquorum*
(Empty shell on left.) EDIBLE SNAIL.—*Helix pomatia.* (Empty shell beneath.)

act as eyes. These tentacles are retractile; and it is very interesting to watch them drawn back or pushed out like the finger of a glove, and to see the curious manner in which the eye speck is shot, as it were, through the tentacle attached to the slender black thread which runs up its centre.

THE genus Helix, which is universally accepted as the type of this family, is of enormous extent, both in numbers and range of locality, containing more than fourteen hundred species, and spread nearly over the whole earth. The common garden Snail is a too familiar instance of this genus, but is so well known, that it is not figured in the engraving, nor will it be described. I may, however, mention, that its depredations can, in a great measure, be checked by searching for it in the winter months, and taking it from the crevices in which it hides itself, or even by destroying the eggs which it lays just under the surface of the soil, and which look like pellucid peas. The much-maligned thrush, too, is a mighty hunter of Snails, and, in spite of its autumnal raids on the fruit, does such good service in Snail-killing before the world is astir, that it ought to be encouraged by the gardener, and the fruit which it eats considered as the wages paid for killing the Snails.

OUR present example is the great EDIBLE SNAIL, which is even now largely consumed in many parts of the world, and is regularly fed and fattened for that purpose. In England it is not very common, but in certain localities may generally be found. It is a remarkable fact, that in many spots where the Romans—great connoisseurs in Snails—had fixed their establishments, the Edible Snail is still to be found. I have seen many of these great Snails caught near Oxford. Regular houses were built for the purpose of fattening the Snails, which were bred to an enormous size by constant feeding with a mixture of meal and new wine. There are even now on the Continent several snaileries, where the inmates are abundantly supplied with food, though they are not fattened with the elaborate precaution of the Roman times.

Even the common Snail is thought a delicacy by those who are sufficiently strong-minded to eat it; and it is quite common to see, even in Paris, the poorer orders dressing their dinner of Snails on an iron plate, heated over burning charcoal. I once knew an old woman, one of the few surviving wearers of scarlet cloaks, who used daily to search the hedges for Snails, for the purpose of converting her milk into cream. This cheap luxury was obtained by crushing the Snails in a piece of linen, and squeezing their juice into the milk. She showed me the whole process, which I afterwards imitated as far as the mixture with the milk, but could not bring myself to test the result by taste.

Towards the end of autumn, the Edible Snail ceases to feed, withdraws itself to some sheltered spot, mostly under grass, moss, or dried leaves, and then sets about making its winter habitation. This process is very curious, and is thus described by Mr. Bell:—

"A large quantity of very viscid mucus is secreted on the under surface of the foot, to which a layer of earth or dead leaves adheres; this is turned on one side, and, a fresh secretion being thrown out, the layer of earth mixed with mucus is left. The animal then takes another layer of earth on the bottom of the foot, turns it also to the part where he intends to form the wall of his habitation, and leaves it in the same manner, repeating the process until the cavity is sufficiently large, and thus making the sides smooth, even, and compact. In forming the dome or arch of the chamber, a similar method is used, the foot collecting on its under surface a quantity of earth, and the animal, turning it upwards, leaves it by throwing out fresh mucus; and this is repeated until a perfect roof is formed.

As I have very often watched this curious process, I am certain of the facts. On removing very carefully a portion of the roof soon after its completion, I was enabled to see the formation of the operculum. In about an hour, or even less, the whole surface of the collar of the mantle instantaneously pours out the calcareous secretion in considerable quantity. This is at first a fluid or thick cream, but very soon acquires exactly the consistence of bird-lime, being excessively adhesive and tenacious; and in about an hour after it is poured out, it is perfectly solid."

On the same engraving may be seen a shell, represented as climbing up a daffodil. This is the Lemon Bulimus, an example of another large genus, containing more than six hundred species. These shells can be distinguished from those of the snails, to which they are closely allied, by the greater comparative length, the oval shape of the aperture, and the thickness of the outer lip. The last whorl is always very large. Some of the exotic species, such as *Bulimus ovatus*, attain a large size, and lay eggs even larger than those of the chaffinch, the young animal having a shell measuring an inch in length when hatched. Several species are very beautifully coloured. Many species of Bulimi are excellent food, and are sold in the markets.

The largest of all the land snails are the great Agate-shells (*Achatina*) of Africa, which attain a length of eight inches, and lay eggs larger than those of the bulimus, and with hard calcareous shells.

We now come to the Chrysalis-shell, shown as creeping up a sapling. This shell belongs to a large genus, containing about one hundred and sixty species, and has received its popular name from its shape, which bears some resemblance to that of a chrysalis. This animal has always a short foot, pointed behind, and very short lower tentacles.

We now arrive at the great family of Limacidæ, or Slugs, a race of beings which many a gardener doubtlessly wishes extinct.

In these creatures the foot and body are indistinguishable from each other; the head is retractile; and the whole creature can be gathered into a short rounded mass, looking so like a pebble that it would escape a casual glance. At the first view, the Slugs appear to be destitute of shell, but, on a closer examination, the shell is found upon the fore part of the body, and either entirely or partially buried beneath the integuments. When removed, it is not unlike the operculum of many molluscs, being small, flattish, and with an evident nucleus. They have four tentacles, like those of the snails, the eye-dots appearing, as in those molluscs, on the tips of the upper and longer pair. The respiratory orifice is placed on the right side of the body.

The GREAT GREY SLUG is the largest of the British species, and when furnished with abundant food, on which it can fatten itself during the night, and a secure hiding-place, whither it can retreat during the day, often attains an enormous size. The careless gardener, who has suffered heaps of old rubbish to collect in his dominions, is often horrified, when he at last removes the stones or sticks, to find under them a number of huge Grey Slugs, that have been silently consuming his flowers and vegetables, and lie slimy and obese at his mercy, bewildered with the unaccustomed light, and unable to escape their impending and deserved fate. It is true that Slugs, snails, and all similar creatures, must have been created for some useful purpose, and in their proper place discharge the duty for which their forms were designed and their instincts implanted; but it is clear that a garden is not the proper place for Slugs, and that if they make their appearance within its precincts, they must be extirpated; just as rats, which are useful in a sewer, are noxious in a house, and must pay with their lives the penalty of their intrusion.

In the fields, this Slug is very common, though not often seen, as it hides itself by day, and only comes out to feed at night. While engaged in "sugaring" trees for moths in the summer evenings, I have often been surprised at the number of Great Grey Slugs which came crawling down the trunks of the trees, attracted apparently by the scent of the sweet mixture, and induced by its odour to leave the hiding-place in which they had been cunningly ensconced during the daytime. Hollow trees, especially those where the entrance is very small, are great haunts of this Slug, and the crevices found at the junction of branches are also in favour with these destructive molluscs. Pollard willows are generally very populous with these Slugs.

Just below the figure of this animal may be seen its little shell, represented as it appears when removed from the body.

THE well-known BLACK SLUG (*Limax ater*) belongs to the same genus as the preceding species, and is very common during the summer, coming out of its hiding-place during the evening, and making its appearance along the sides of roads, in hedgerows, and similar situations. It is nearly, but not quite so large as the grey species.

THE common RED SLUG, or LAND SOLE (*Arion rufus*), is another member of this family. It may be known by the deep red-brown of its body, which sometimes approaches to black. It is very plentiful in gardens, and as, on account of its colour, it is not readily seen in the dark, it escapes observation, and does much damage without being discovered. Those who desire to rid their gardens of these pests will find that a very effectual plan is to search the grounds after dark, by the aid of a "bull's-eye" lantern.

The greater number of the Slugs can thus be discovered, and the remainder can mostly be destroyed by watering the ground with a very weak solution of ammonia. This substance is fatal to slugs as well as to worms; causes them to pour out a white frothy mucus, so as to betray their presence; and, if employed in sufficient quantity, will kill the mollusc without injuring the vegetables. The Slugs and worms that are thus captured should be thrown into a vessel containing a strong solution of the same fatal substance, when they will perish almost immediately, and after they have been left in the vessel for twelve hours, so as to make their death absolutely certain, may be used as a valuable manure.

Another useful plan is to make an attractive shelter from boards or tiles, and place cabbage-leaves within. The Slugs will crowd to the bait during the night, and, finding themselves sheltered, will remain there during the daytime, when they may be captured and destroyed.

ON the left of the illustration on page 405 may be seen a little Slug with a semi-spiral shell, represented as crawling up a short stump.

This is the TESTACELLA, one of the very few carnivorous land molluscs. The Testacella, although plentiful, is seldom seen, on account of its peculiar habits. It feeds almost wholly on earth-worms, which it pursues through all the windings of their retreats, its long lithesome body enabling it to insinuate itself wherever the worm can burrow, and its hard little shell securing it from danger by stopping up the tunnel behind its progress. This curious slug can be obtained in gardens by digging up the loose soil, but, on account of its services to the gardener, should be released, and permitted to resume its destructive avocations.

The tooth-ribbon of this creature is most formidably armed, having about two thousand teeth arranged in fifty rows. The teeth are needle-shaped, barbed, sharply pointed, slightly curved, and converge towards the centre of the ribbon, thus forming a weapon which no worm is capable of resisting. Only three species of Testacella are known ; our British species is supposed to have been introduced from Southern Europe.

IN the following illustration is shown a group of Water-snails, several of which can be found in every large pond or stream.

The two central figures represent two species of APPLE-SNAILS, belonging to a genus remarkable for several peculiarities of formation. Although placed with the pond-snail and planorbis for the convenience of bringing the various water-snails into a single illustration, the Apple-snails belong more properly to the gill-bearing molluscs, and in the systematic arrangement follow the phorus, described on page 394.

The Apple-snails are found throughout the warmer parts of the world, inhabiting the lakes and rivers, and, in case of drought, burrowing deeply into the mud and remaining buried for a lengthened period, sometimes for a term of years, until a fresh supply of water arouses them from their strange torpor and urges them again to seek the upper regions.

In his "Natural History of Ceylon," Sir J. Emerson Tennent mentions this curious habit. "The *Ampullaria glauca* is found in still water in all parts of the island, not alone in tanks, but in rice-fields and the water-courses by which they are irrigated. When, during the dry season, the water is about to evaporate, it burrows and conceals itself till the returning rains restore it to activity and reproduce its accustomed food. There, at a considerable depth in the soft mud, it deposits a bundle of eggs with a white calcareous shell, to the number of one hundred or more in each group.

The *Melania paludina*, in the same way, retires during the droughts into the muddy soil of the rice-lands, and it can only be by such an instinct that this and other molluscs are preserved when the tanks evaporate, to reappear in full growth and vigour immediately on the return of the rains.

A knowledge of this fact was turned to prompt account by Mr. Edgar S. Layard, when holding a judicial office at Point Pedro in 1849.

A native who had been defrauded of his land complained before him of his neighbour, who, during his absence, had removed their common landmark, diverting the original watercourse and obliterating its traces by filling it up to a level with the rest of the field. Mr Layard directed a trench to be sunk at the contested spot, and discovering numbers of the Ampullaria, the remains of the eggs, and the living animal which had been buried for months, the evidence was so resistless as to confound the wrong-doer and terminate the suit." After a few hours of rain, the Apple-snails may be observed emerging from their muddy retreat as if to welcome the newly found moisture.

As may be seen from the illustration, the animal of the Apple-snail is very curiously formed. The long siphon, formed by a development of the neck-lappet, is seen on the left. Projecting just without the shell are seen the eyes, set at the extremities of short

RAM'S-HORN APPLE-SNAIL. *Ampullária cornu-ariëtis.*
WATER-SNAIL.—*Limnæa stagnalis.*

POUCH-SHELL.—*Physa hypnórum.*
COMMON PLANORBIS. —*Planorbis córneu:*
STAIRCASE APPLE-SHELL.—*Ampullária scaláris.*   RIVER LIMPET.—*Ancylus lacustris.*

and stout footstalks, and the enormously long tentacles are placed just in front of the eyes. At the first glance the creature appears to have four tentacles, but; on a closer examination, the front pair are seen to be merely developments of the muzzle. In one respect, the Ampullaria seems to be a connecting link between the gill-bearing and lung-bearing molluscs, being said by high authorities to possess a pulmonic or lung sac, in addition to its gills.

On the left hand of the illustration the common POND-SNAIL, or LIMNÆA, is shown in the act of climbing up the stem of a water-plant. In all the members of this family the shell is thin, and sufficiently capacious to contain the entire animal when it desires to withdraw itself into its home. The aperture is simply rounded, without notches or ridges, and the lip is sharp.

The Pond-snail may be found in almost any fresh water, and, if carefully watched, proves to be quite an interesting creature. It can creep with tolerable rapidity after the usual manner of snails, and has besides a curious method of progression without making any exertion of its own. In streams, when the animal has a mind to change its locality without needing to exert itself, it achieves the task of converting the journey into a voyage, and its foot into a boat. This transformation is soon effected, the animal first crawling up some plant that projects out of the water, reversing its position, so that the shell lies undermost, and then hollowing the foot so as to form it into a shallow boat-like shape. It then looses its hold of the supporting plant, and boldly launches itself on the surface.

Hundreds of these creatures may often be seen floating in successive squadrons along the surface of a stream, knocking against all kinds of obstacles without fear, and travelling farther in one hour than they could otherwise have achieved in a week. If, however, the spectator throw a stone into the water all is changed in a moment, the expanded foot is suddenly contracted as the circling ripples reach the voyagers, and the mimic fleet sinks to the bottom as if by magic.

The tentacles of this animal are rather short, broad, compressed, and triangular. The eggs of the Pond-snail are laid in ribbons of transparent gelatinous substance, and the process of their gradual development can easily be watched by means of the microscope. Most of the Pond-snails feed on vegetable substances, more especially decaying leaves, but the present species is carnivorous in its habits.

On the right-hand of the engraving, and exactly opposite the pond-snail, is an equally common British shell, called from its flattened whorls the Planorbis. In this animal, the foot is short and round; the tentacles are long, slender, and leave the edges at their inner bases. Both this and the preceding species are in the habit of burying themselves in the mud during a drought, and there passing a semi-torpid existence.

Below the planorbis is a very remarkable species, called, from its peculiar shape, the Fresh-water or River Limpet. This appears not to be a British shell, but is found in various parts of America, Madeira, and some portions of Europe. It inhabits swiftly running streams, and is mostly seen attached to stony and aquatic plants. Although the shell is so limpet-like, the animal does not partake of the resemblance, being very like that of the pond-snail, and having triangular tentacles with the eyes at their bases. The generic term, *Ancylus*, is of Greek origin, and signifies a small round shield or target. An empty shell is represented lying on the ground near the animal.

The little elongated shell shown crawling over the body at the right-hand top corner of the engraving is the Pouch-shell, a British species of a rather small genus, extending over the greater part of the globe. In the greater number of species, the mantle is fringed with long filamentary appendages, but in the present example the edges are quite plain. It is, however, always flat and much expanded. The tentacles are long and slender, and the eyes placed at their bases. The shell is thin, spiral, polished, and the aperture is rounded in front.

In the accompanying illustration are represented some of the strange and almost grotesque forms which are assumed by many of the molluscs. These belong to a fresh order, in which the shell is sometimes altogether wanting, and even when present is of very small dimensions, and is almost, if not wholly, concealed by the soft parts. In fact, they may be considered as the marine analogues of the common land slugs. The gills of these animals are rather curiously formed, not being placed in a definite cavity as is the case with the previous species, but projecting boldly from the surface of the animal, and set towards the rear of the body. On account of this position of the gills, the animals are termed Opistho-branchiæ, or Rearward-gilled molluscs. The whole internal structure of these creatures is fully as curious as their external form, and well repays dissection, the organs of digestion especially being rather complicated, and possessing many points of interest.

In the centre of the engraving is seen an empty and rounded shell, with the exterior spotted and the interior of the lip shown. This is the Bubble-shell, one of a moderately large genus of molluscs, that are found in almost all tropical and temperate seas, and may generally be captured where the bed of the ocean is of a sandy nature. Another species is drawn on the upper left-hand corner, for the purpose of showing the manner in which the shell is enveloped in the tissues of the animal. In all the family to which the Bubble-shells belong, the shell is very thin, globular, and cylindrical, and the aperture is long and rounded. The large side lobes are said to be often used as fins for swimming.

BUBBLE-SHELL.—*Aplustrum.*

WOODY BUBBLE-SHELL.—*Bulla lignaria.*   LOBIGER.—*Lobiger Philippii*
   BUBBLE-SHELL.—*Bulla ampulla.*   BOATMAN-SHELL.—*Philine aperta.* (Empty shell below.)
                        CYLICHNA.—*Cylichna cylindrácea.*
                        HATCHET-SHELL.—*Dolabella Rumphii.*

In both these species, and in all that belong to the same genus, the foot is squared, the head is furnished with a large disc, blunted in front, and divided into two lobes behind. The eyes are never of great size, and in many species are altogether wanting. The organ called the gizzard is furnished with three plates of hard chalky matter, in form somewhat resembling the plates from the back of a chiton, and thought to answer the purpose of teeth.

On the top of the engraving is shown a closely allied species, formerly placed in the same genus, but now separate on account of several structural differences. The APLUSTRUM has a shell rounded and slightly coloured, but small in proportion to the size of the animal. The foot of the Aplustrum is exceedingly large, capable of concealing the shell entirely in its folds. Behind the tentacles are four large and flattened lobes, and the eyes are very small and set at the inner bases of the tentacles.

A very curious creature is seen nearly in the centre of the engraving, lying horizontally below the empty shell of the Bulla Ampulla.

This is an example of a genus termed CYLICHNA: it is remarkable for one or two peculiarities. The animal is smaller in proportion than is generally the case with this family, being short and broad; and not able to wrap its lobes over the shell. The head is flat, blunt in front, and the eyes are deeply sunk in the tissues. The shell is cylindrical and mostly smooth, and the aperture is narrow.

Just below the head of the aplustrum a curious animal is seen, represented as it appears when crawling over the rocks. This is the Boatman's Shell, a specimen of a very small genus, containing only five species, but yet represented in England. It is a sand-loving creature, mostly remaining in a rather considerable depth, and being found in fifty fathoms of water. This animal has no eyes ; and although the side lobes are turned up, and are very large, they do not envelop the shell. In order to show the formation of the shell, an empty specimen is given just below the living animal.

We now come to another family, in which the shell is never visible, being either very small and covered by the mantle, or wholly wanting, and in some instances being translucent and flexible as horn. The common Sea Hare (*aplysia depilans*) is a well-known example of this curious genus, which has given its name to the entire family. The Sea Hares possess the power of throwing out at will a rather large amount of a peculiar coloured fluid, mostly of a deep violet, which is secreted by part of the mantle. This habit formerly caused the Sea Hare to be held in great dread, the popular opinion attributing to the violet fluid the most virulent properties, such as staining the skin indelibly with the purple dye, injuring its texture like strong caustic, and causing the hair to fall off. In allusion to the last-mentioned idea, the animal has received its specific title of *depilans*.

We will now briefly examine a few of the more remarkable species.

At the bottom of the engraving is seen an oddly-shaped creature, looking as if it were made of some rough membrane covered with little projections, and tacked loosely over an egg-shaped centre. This is the Dolabella, or Hatchet-shell, so called on account of the form assumed by the shell, which is, however, so entirely hidden under the softer parts that it is not visible until the lobes have been put aside.

In order to show its peculiar shape, one of the shells is represented as it appears when separated from the animal. As may be seen, it is small in proportion to the size of the creature which formed it, and as it is attached only to the hinder part of the body, it is of little use in protecting the soft parts from injury. In colour the shell is pure shining white and pearly ; its substance is very thick, and it is covered with a tough membrane, technically called the epidermis. The colour of the animal is dull olive-green.

Our last example of this family is the animal which is shown on the upper right-hand corner of the illustration, and represented in the act of swimming, for the purpose of showing the manner in which the side lobes can be used as fins, and employed in urging the animal through the water. This creature is called Lobiger, or Lobe-bearer, in consequence of the rounded and flattened lobes that project from each side of the body, much like the four wings of a butterfly. The tentacles are also flattened and rather oval, and the eyes are very small and set on the sides of the head without any footstalks. The foot is small, and the hinder part of the body is lengthened and pointed so as to resemble a tail. The shell is small, oval, transparent, flexible, and set on the body so as to act as a shield to the plume-like gills. This species is found on the coasts of Sicily.

There is a curious animal belonging to the next family of molluscs, called, from the mode in which the shell is carried, the Indian Umbrella. In this creature the body is large, roundish, and covered with tubercles, and in shape resembling something a great limpet ; while the flat, white, pearly shell is perched horizontally on the very middle of the back, just like an Eastern umbrella held over the palanquin of some great potentate. The colour of the animal is dull ochreous yellow.

We now arrive at a very remarkable series of molluscs which have been separated by systematic naturalists into a distinct section, appropriately called Nudibranchidæ, or Naked-gilled Molluscs, because their gills are always external and placed on the back or

GLAUCUS.—*Glaucus Pacificus.*

DORIS.—*Doris Johnstoni.*

DOTO.—*Doto coronata.*

DENDRONOTUS.—*Dendronotus arborescens.*

EOLIS.—*'Eolis coronata*

sides of the animals. Many of these strange creatures are to be found on the British coasts; and if the reader should wish to gain a further insight into their habits, and to examine the marvellous forms which the different genera assume, as well as their exquisitely delicate and varied colouring, he is referred to the magnificent work of Messrs. Alder and Hancock, published by the Ray Society. The entire structure of the Nudibranchs is most curious and well worthy of examination, but is too purely anatomical for admission into these pages. A few, however, of the more notable structures will be mentioned in the course of our description.

THE slug-like animal which is represented crawling on the frond of a laminaria, at the left-hand of the engraving, is the common DORIS of our own shores. All the members of the family to which this creature belongs may be known by the plume-like gills set in a circle on the middle of the back, like the feathery coronet with which the Blackfoot Indian adorns the head of his horse, and the two tentacles placed more towards the front. In the skin are imbedded a vast number of little spiculæ.

In the genus Doris both the tentacles and the gills can be drawn within the body at the will of the animal; and the mantle, which is pierced to permit the passage of the gills and tentacles, is very large and covers the foot, as well as the whole upper surface. All the nudibranchs are carnivorous, and feed chiefly on zoophytes. The sea-anemones are favourite prey of these voracious creatures, and many an enthusiastic naturalist has gained a knowledge of their habits at the expense of his special favourites; for whenever a nudibranch is placed in the same aquarium with specimens of actiniæ it is sure to attack them, eating great holes in their substance, and, if not removed, will be as destructive as a hungry fox in a poultry-yard.

THE next family is represented by two species, each of which will be briefly described. In this family, called Tritonidæ, the gills are arranged in lines along the sides of the back, and the tentacles can be withdrawn into their sheaths.

The DENDRONOTUS derives its very appropriate name from two Greek words, the former signifying a tree, and the latter the back. The beautiful branched gills are set in a very shrub-like fashion upon the back, and even the tentacles and appendages of the head are branched so as to correspond with the gills. In this and many other species the digestive organs are covered with prolongations, which are carried into the little tuber-culated appendages which stud the surface, and have given cause to M. de Quatrefages to establish his theory of "phlebenterism," which he defines as "an anatomical arrange-ment which is generally characterised by ramified prolongations of the digestive tube, in virtue of which the digestive apparatus, to a certain extent, supplies the place of the circulatory apparatus, and aids in the process of respiration."

In his "Rambles of a Naturalist," Vol. I. note xv., the same author makes the following valuable remarks : "At the same time I found in the Eolidinæ a heart and a tolerably well developed arterial system, but I could not distinguish any veins ; nevertheless, I was so imbued with the ideas of Cuvier, that it was long before I was willing to admit the non-existence of these vessels. Nor was I convinced that there were really no veins in the Eolidinæ, until I had followed the same blood-corpuscle throughout its entire circuit—until I had seen it issue from the heart and pass through the arteries by whose walls its movements were regulated, and until it was conveyed through every part of the general cavity and finally carried back to the heart." The whole subject is too complicated and detailed to be here investigated, but it is of too great importance to have been left without a brief notice.

THE central figure at the bottom of the illustration represents the beautiful DOTO, here shown in the act of making prey of a sea-anemone, which in vain withdraws its tentacles and contracts itself into a shapeless mass at the touch of an enemy. It may be here remarked that the word Doto is the name of one of the sea nymphs of mythology, and that in consequence of their exquisite colouring and beautiful forms, the names of nearly all the nymphs have been given to different species of nudibranchs. The tentacles of this animal are slender and can be retracted into certain trumpet-like sheaths. which are seen projecting from the body. In this creature the processes of the digestive system pass into the large appendages on the back ; and it is a curious fact that although they fall off when the animal is handled, they are soon reproduced, and the creature seems to suffer little inconvenience from their loss. Examples of this creature can be found on the British coasts.

THE remaining figures are representatives of another family, called the Eolidæ. In these creatures the theory of phlebenterism finds its best proofs, as the processes of the digestive organs extend throughout the beautiful projections on the back, even though, as in one genus, they are placed on footstalks.

The beautiful EOLIS is common on our own coast, and may be seen moving over the plants and stones with tolerable activity, and always keeping their tentacles and papilla in motion, sometimes contracting and sometimes extending them, while the movement of the water causes them to wave in a very graceful manner. These papilla possess the property of discharging a milky kind of fluid when the animal is irritated. The fluid, however, is quite harmless, at all events to the human skin. As in the previous case, the papillæ are liable to fall off at a touch. While using the dredge, the naturalist is sure to bring plenty of nudibranchs to the surface ; but owing to their habit of contracting themselves into a shapeless mass, an uninitiated observer will probably fail to notice them and fling them overboard again, together with the seaweeds, stones, and other refuse substances. The Eolis is a voracious being in spite of its delicate beauty, and if several of them are kept in a vessel and not supplied with the sertularia and other zoophytes on which they feed, they will attack and devour each other.

OUR last and most remarkable example of the nudibranchs is the GLAUCUS, or SEA LIZARD, the strange creature in the upper left-hand corner. In this animal the gills are slender, cylindrical, and supported on three pairs of lobes or footstalks.

The Sea Lizard is very common in many parts of the Atlantic, where it is found in vast numbers during a calm and when the sea is smooth. Mr. F. D. Bennett writes as follows about this strange and eccentrically formed being. "These creatures obtain in greatest number where currents most prevail; they are active and very predatory in their habits, and would appear, from the observations of my brother, which I have already confirmed, to subsist chiefly upon the soft parts of the defenceless genera Velella and Porpita. The specimens we captured and kept in sea-water contracted their bodies into many convulsive attitudes, but seldom employed their branchial fins, and floated buoyantly while passive. When immersed in fresh water they contracted themselves into a very small compass, assumed a globular form, cast the tentacles from off their branchial fins, lost their colour, and expired in a very few moments.

When handled, the white colouring matter of the body is easily detached, and leaves the corresponding portions of the integument transparent; when a considerable accumulation of air, in the form of small bubbles, is visible among the viscera, and fully accounts for the hydrostatic power the animal possesses. The colour of the entire upper surface of the body is dark blue, with the exception of a broad silvery streak extending from the head to the extremity of the tail; the inferior surface of the body and fins is pearl-white. The head also is white, but has a vertical blue line on its centre. The entire animal is inclosed in a tough, but smooth and delicate integument."

WE now arrive at a new order of molluscs, if possible stranger than that which has just been briefly described. The animals of this order are inhabitants of the sea, but differ from their kind in living almost wholly on the surface of the waters instead of crawling upon the stones or plants of the ocean bed. They are termed Nucleobranchiata, because the organs of respiration and digestion are gathered into a mass or nucleus upon the hinder portion of the back.

The first family of these creatures is represented by the CARINARIA, which may be seen swimming at the upper part of the engraving. In this genus, the gills are protected by a small and very delicate shell of glassy translucence, bearing but little proportion to the size of the animal. The creature itself averages two inches in length, and is very transparent, permitting the vital functions to be watched by the help of a microscope. When swimming, the Carinaria reverses its attitude, and, as seen in the engraving, keeps the tiny shell downwards. Exactly opposite the shell is placed the curiously modified foot which is now no longer useful for creeping, but is formed into a fin wherewith the creature can propel itself through the water, or a rudder by which it can guide its course. The curious mouth is seen in front, and the extremities of the gills are shown just peeping from the edge of the shell.

It is a tolerably quick swimmer, using both tail and fin in its progress, and sinking or rising with equal facility. Sometimes the animal is caught without its shell, and the empty shell is occasionally seen floating on the surface.

ANOTHER species of the same family, the FIROLA, is extremely like the Carinaria, but is without the glassy shell. It attains a considerable size, being sometimes five inches in length. It seems not to be so active as the last-mentioned species. In the second family of this order, namely, the Atlantidæ, the animals are furnished with a well-developed shell, sufficiently large to contain them when they desire to contract themselves within its precincts.

A SMALL, but important, group of molluscs now comes before us. These are the Pteropoda, or Wing-footed Molluscs, so called from the fin-like lobes that project from the sides, and are evidently analogous to the similar organs in some of the sea-snails. These appendages are used almost like wings, the creature flapping its way vigorously through the water, just as a butterfly urges its devious course through the air. They are found in the hotter seas, swimming boldly in vast multitudes amid the wide waters, and one species (*Clio boréalis*) has long been celebrated as furnishing the huge Greenland whale with the greater part of its subsistence. We will now examine in detail our representative species.

CARINARIA.—*Carinaria Mediterranea.*
HYALEA.—*Hyalea tridentata.* (Empty shell below.)   CYMBULIA —*Cymbulia Peronii.*
CLEODORA.— *Cleodora pyramidata.*   SPIKE-SHELL.—*Creseis aciculata.*

The curious figure on the left-hand of the illustration is the HYALEA, remarkable not only for the two wide fins which are found in all the family to which it belongs, but for the long appendages which pass through certain apertures in the shell, and trail behind as the creature proceeds on its course. It will be also seen that the wings are united by a nearly semicircular lobe. The empty shell is placed below in order to show its curious structure.

JUST on the right of the Hyalea is a smaller creature, with an odd-looking three-pointed shell, hanging as it were from the wings. This is the CLEODORA, a very beautiful and interesting animal, of which Mr. F. D. Bennett writes as follows: "On that part of the body which is lodged in the apex of the shell, there is a small, globular, pellucid body, resembling a vesicle, and which at night emits a luminous gleam, sufficiently vivid to be visible even when it is opposed to the strong light of a lamp. It is the only example of a luminous shell-fish I have ever met with; nor would the luminosity of this species be of any avail, did not the shell possess a structure so vitreous and transparent. Examples were chiefly captured at night or in the evening."

AN allied genus, notable for the straight-pointed shell, is represented by the SPIKE-SHELL, the little creature on the right of the Cleodora. The fins of this animal are rather narrow, and the apex of the shell soon loses its sharpness, being by degrees divided into compartments and gradually broken off. The Spike-shell is mostly found near floating seaweed

YELLOW LAMP-SHELL.—*Terebrátula flavescens.*
(Open shell on right.)

PARROT-BILL LAMP-SHELL.—*Rhynchonella psittácea.*

SNAKE-HEAD LAMP-SHELL.—*Terebrátula caput serpentis*

GOOSEBILL.—*Lingula anatina.*

OUR last example of the Pteropoda is the CYMBULIA, the large-winged animal placed above the spike-shell. Though greatly resembling the carinaria in general appearance, it is divided from that creature by many important structural differences. Its shell is flexible, and in shape and translucency somewhat like the glass-slipper of fairy mythology, the point, or toe, being set forward. Only three species of this genus are known.

As group after group of molluscs passes before our notice, each seems to be more extraordinary than its predecessor, and to present us with stranger and more unexpected forms. The molluscs of the next group are the first of the bivalves, but stand alone in many particulars, and evidently form a transition between the gasteropoda and the ordinary bivalves. They are all inhabitants of the sea, and, when adult, are found attached to rocks, coral branches, and even other shells; but in their earlier stages are apparently able to swim freely through the water, as is the case with many other molluscs.

In the ordinary bivalves, the two shells correspond with the right and left side of the animal; but in the Brachiopoda, as these creatures are called, the one covers the upper and the other the lower portion, and are called accordingly the dorsal and ventral valves. Of these, the former is smaller than its companion, to which it is jointed by means of certain interior sockets, which receive corresponding hooks in the ventral valve, and lock them together so tightly, that they cannot be separated without something being broken. The ventral valve is large, and is marked by a decided beak, not unlike the bill of a parrot. In most instances the beak is perforated with a round hole, through which passes the peculiar organ by which the animal attaches itself to the substance on which it rests; and when this is not the case, the hooked beak itself answers that purpose.

In the interior is a rather complicated internal skeleton, a good example of which is shown in the empty shell of Terebratula. The food is obtained in a singular manner. The animal is furnished with a pair of rather long arms, covered with vibrating fibres, or

3.

cilia, and by means of the constant action of the cilia a current is caused, which drives a continual stream over the mouth, and enables the animal to seize the minute animals that dwell in the sea and are distributed throughout the waters.

WE will now proceed to the examination of our selected examples of these curious molluscs.

With the exception of the elongated shells or footstalks, and the dark shell in the centre, all the specimens shown in the illustration belong to the genus Terebratula, this name being derived from a Latin word signifying a wimble, and given to the animal in allusion to the round hole which perforates the beak. The popular name of LAMP-SHELL also refers to the same aperture, because it looks like the round hole through which the wick of an ancient lamp is drawn. The structure of the shell itself is very curious, being made up of innumerable flattened prisms laid side by side and arranged in a slightly oblique position, so that their ends project over each other, something like the slates in a house-roof. The substance of the shell is also perforated by multitudes of very minute circular apertures.

BETWEEN the Yellow Lamp-shell and the empty valve may be seen the PARROT-BILL LAMP-SHELL, so called from the shape of the beak, which is long and hooked in a manner which much resembles the beak of the bird whose name it bears. The colour of this species is black.

OUR last example of these remarkable molluscs is the GOOSE-BILL LAMP-SHELL, seen on the extreme right of the engraving. All the members of the family to which this animal belongs are known by the long and comparatively narrow valves, and the foot-stalk which attaches them to the rocks, and which passes from between the valves. The substance of the shell is rather soft and perforated. The valves are slightly open at each end, and blunted in front. Very little is known of its habits in the living state, but it is worthy of notice that the Goose-bill Lamp-shell is the oldest known form of organic life.

THE next great group of molluscs is that which is known by the technical term of Conchifera, and includes all the ordinary bivalves, i. e. where each valve corresponds with the right or left side, and not with the upper and under surface, as in the brachiopods just described.

Though not possessing so many species as the gasteropoda, this group surpasses it in point of numbers, the bivalves being produced in countless myriads, and, perhaps, less exposed to the attacks of foes than most of the race. They are extremely useful in both salt and fresh water, feeding on the particles that would otherwise pollute the element in which they live. Their mode of feeding is somewhat similar to that of the last-mentioned group, the water being driven over the mouth by the continual action of certain appendages, and there cleared of all its solid portions. So completely does a bivalve effect this purpose, that it even intercepts the microscopic plants and animals which are invisible to the naked eye, and conveys them to the stomach with marvellous certainty.

If the stomach of a scallop, for example, be opened, and its contents touched with nitric acid and heat, myriads of flinty skeletons will be found in it, once supporting the minute structures of diatoms and other organisms well known to any microscopist. Even if some water be coloured with indigo, and a healthy bivalve placed in the same vessel, the water will gradually lose its dark blue tint, and finally become clear and colourless, the particles of indigo having been removed by the mollusc as the water is driven over the gills and mouth.

The whole structure of a bivalve is very curious, and can be readily studied by any one who chooses to purchase a few oysters, or mussels, or cockles. Those parts of the structure which are of most importance will be briefly described while treating of individual species.

Systematic naturalists are rather perplexed with regard to the particular characteristics which must be employed in the arrangement of the molluscs; and the necessary consequence of this uncertainty is, that almost each investigator has invented a fresh system,

CHINESE WINDOW.—*Placuna placenta.*　　　SADDLE-SHELL.—*Anómia ephippium.*
LONG-HINGED OYSTER.—*Ostrea Canadensis.*

based on those characteristics which in his own judgment are most valuable. In the arrangement, however, which has been followed in these pages, the bivalves are divided into two sections, founded on the presence or absence of the siphons through which the water passes for the double purpose of creating the blood, and supplying the digestive organs with the necessary food. In the first section, termed Asiphonidæ, are included those bivalves in which the siphons are wanting.

In the first family of which the common OYSTER is a very familiar instance, the two valves are unequal in size, and the animal inhabits the sea. The Oyster is too well known to need description; but it may be mentioned, that practical naturalists have for some years been carefully studying its habits, for the purpose of breeding the valuable mollusc artificially, and so of securing a constant supply throughout the four months of the year during which the creature is out of condition. In this country the system is being gradually carried out, but in France it is developed to a very large extent, and with great success.

The details of the process are too elaborate to be here described, but the general idea may be given in a few words. The very young spawn, or "spat," as it is technically named, is removed from the natural beds, and is dispersed in shallow "banks," so that each tiny Oyster has plenty of room, and can affix itself to the bed of the bank without being injured by the pressure of its fellows. Fascines, made of slender branches, and sunk into banks paved with stone, birch, and broken earthenware, are found to be most useful for this purpose. In the banks near Dieppe, the Oysters are seen lying in regular rows like the tiles of houses, and are at all times ready to be taken from the bed and sent to market.

This process possesses a double value, inasmuch as an oyster-bed, if left to itself, would increase to such an extent as to endanger navigation; and these inland banks are always accessible, whatever may be the weather. In some cases, when confervoid growths are rife, the Oysters attain a decided green hue, and are thought very valuable by connoisseurs in such matters. In all improvements, however, there is always some drawback. The Oysters produced by artificial culture are acknowledged to be fatter and finer than those which are suffered to grow in the open seas; but their artificial size is said to be a poor compensation for their comparative want of flavour, the artificially bred Oyster being to the marine mollusc what the capon is to the pheasant.

In the sea, thousands of Oysters perish by the attacks of a strange enemy. The reader has doubtlessly remarked that the shells of many Oysters are partially perforated by little round holes. These are the marks left by a kind of sponge, called Cliona, which burrows into and gradually destroys the shells of this mollusc, causing them to fall to pieces by its ravages.

In the centre of the engraving on the previous page is seen a species of the genus Ostrea, which is remarkable for its very long hinge.

THE uppermost figure represents the CHINESE WINDOW-SHELL, a curious and rather valuable shell, which is found in the country from which it takes its name.

This shell is extremely flat, and of a beautiful translucence, and in many parts of China is employed for windows, just as is ground-glass among ourselves, the nacreous substance permitting the light to pass through, but effectually preventing an inquisitive eye from distinguishing objects within the apartment. Very small pearls are found in this shell, too minute and too opaque to be employed by jewellers. They are, however, collected and exported to India, where they are calcined and formed into lime for the use of wealthy betel chewers. They are also burned in the mouths of the dead.

The shell is of great use in commerce, affording the substance from which is cut those large flat "pearl" buttons that were formerly so fashionable, but seem now to have descended to the denizens of the stable. The button "moulds" are cut from the shell by an instrument that somewhat resembles the trephine, by which portions of the skull are removed in case of severe injuries, and in their rough state look like gun-wads. They then pass through a series of processes in which they are polished and pierced, and made ready for sale.

THE lowest figure represents the SADDLE-SHELL, remarkable for the way in which the shell is attached to other substances. Two specimens may be seen in the illustration placed on an oyster. The contrivance by which it is attached is most remarkable. The animal deposits a plug or peg of shelly matter on the oyster, and in the right valve there is a hole or notch into which the peg fits, much after the fashion of a button. When the left valve is in its place, this contrivance is hidden. The shell of this creature is beautifully thin—hardly thicker, indeed, than the paper on which this account is printed —and elegantly waved. It inhabits the British seas.

THE three objects in the accompanying illustration represent different views of the same species.

The LIMA, or FILE-SHELL, is worthy of notice on account of the curious refuge which it constructs by binding together a large mass of shells, corals, sand, and other materials, by means of the silken threads or "byssus" which it is capable of secreting. On the left hand is seen a specimen almost entirely buried in the mass of nullipores which it has gathered around its shell. The upper figure on the right hand exhibits the same shell as it appears when free and in the act of swimming. The long tentacular appendages are kept in constant movement, possibly without the will of the animal, keeping up their writhing contortions just as our hearts continue to beat without our knowledge. Even after the death of the animal, and when they have been separated, the filaments continue to move, twining and twisting like so many worms.

The File-shell can pass through the water with some rapidity, urging itself along by the sharp closing of its valves. Its colour is crimson. with the exception of the mantle,

LIMA.—*Lima hians.*

which is orange. The shell is pure white, so that a living and healthy specimen is a most beautiful creature.

THORNY OYSTER.—*Spondylus Americanus.*

A VERY curious example of this family is found in the THORNY OYSTER, a species that is remarkable for the singularly long projections from the shell. The object of these

HINNITES.—*Hinnites pusi* PINNA.—*Pinna pectinata.* VARIEGATED SCALLOP.—*Pecten varius.*
HAMMER-SHELL.—*Malleus vulgaris.*

spines is rather obscure, but is said to answer a double purpose, the one being to act as a *chevaux-de-frise*, whereby the attacks of marauding fish or other foes may be repelled, and the other to aid in fixing the animal to the spot on which it has established itself. Any fish, however, that would be strong-jawed enough to crush the shell, even without the spikes, would care no more for them than does a donkey for the prickles of a thistle; and the smaller and more insidious enemies would receive no check from the hedgehog-like array of bristling points. The animal of the Thorny Oyster is eatable, and in many places is looked upon as a delicacy.

The last example of this family is the HINNITES, a shell remarkable for its exceeding variability of form. When young, it wanders freely through the ocean; but when it finally settles down in life, it acts like weak-minded men, and moulds itself to the locality in which it happens to reside. If its gets among scoriæ, as is not unfrequently the case, the shell follows all the irregularities of its resting-place; and in one instance, where one of these shells had settled upon a group of serpulæ, it had accumulated itself to them in the most curious manner, actually overlapping the shell so as to form its edge into the half of a hollow cylinder. The colours are red, brown, and white, but their relative amount and the manner of their disposal are as variable as the form.

THE next family are termed Wing-shells, or Avicularidæ, because the apices, or "umbones" as they are called, are flattened and spread on either side something like the wing of a bird. The interior of the valves is pearly, and the exterior layer is composed of a kind of mosaic work of five or six sided particles. This structure is easily to be seen by means of a moderately powerful simple lens, merely by holding up a scallop or other shell before the window, so as to allow the light to pass through it.

THE lowest figure in the illustration represents the curious Hammer-shell; a dark, whitish species, which, but for the oddity of its form, which somewhat resembles that of the hammer-headed shark, would attract but little attention. As it lies on the ground, it would hardly be taken for a shell by one who was not acquainted with it, the enormously expanded ears and strangely crumpled valves giving it a most unshell-like aspect. This strange form is, however, only to be seen in the adult specimens, or when young; the shape of the Hammer Oyster is very like that of the pearl Oyster, presently to be described.

THE floating figure at the right hand of the engraving represents the VARIEGATED SCALLOP, shown as it appears when passing through the water.

In common with many other molluscs, this creature is able to move with considerable swiftness by means of repeated strokes of its valves, a single stroke carrying it for several yards.

The animal is very beautiful, its colour being orange or fine scarlet, and the mantle marbled with brown of different hues. A series of round black dots, called ocelli, and thought to answer the purpose of eyes, are ranged around its edge, and surrounded by long tentacular filaments. Like the Oyster and mussel, the Scallop is considered as a delicacy, and eaten dressed in various ways. The shell is of little value, its chief use in these days being as a vessel in which Oysters are "scalloped;" but in the ancient times it was in great request, as the sign of one who had made a pilgrimage to the shrine of St. James. When at rest, the Scallop lies on the right valve.

The well-known PEARL OYSTER is one of the most valuable of the shell-bearing molluscs, furnishing the greater part of the pearls which are set by jewellers and worn by ladies. The specimens represented in the engraving were brought from Ceylon; but these creatures are also found in other parts of the world, such as Madagascar, Swan River, Panama, &c. Not only the pearls themselves are valuable, but the shells are of great importance in the commercial world, furnishing the best "mother-of-pearl," as the nacreous lining of the valves is called.

The pearls are secreted by the animal in precisely the same manner as the nacre of the shell, and are, indeed, the same substance, formed into a globular shape, and disposed in concentric layers so as to give that peculiar translucency which is quite indescribable, but is known among jewellers by the name of "water." As to the precise method and object of their formation, opinion differs, the general impression being that they are morbid secretions, often stimulated originally by a grain of sand or some such substance finding admission into the shell. These objects may be obtained by introducing into the shell certain extraneous bodies, around which the nacre is secreted so as to form very good imitations of the pearls formed after the usual manner. Examples of such artificial pearls will be mentioned in the course of the following pages.

The Pearl Oyster does not produce its costly harvest under six or seven years of age, and it is, therefore, a matter of importance that the bed should be so managed that the young Oysters may be suffered to remain in peace until they have attained an age which renders them capable of repaying the expense of procuring them, and that no part of the bed should be harried where the Oysters are too small to produce pearls. It is hoped that the increasing knowledge of the mollusc and its habits will enable proprietors to sow the sea with pearls just as they sow a field with grain, and that the harvest may be equally certain in either case.

The Oysters are now obtained by means of men who are trained to the business, and who can remain under water for a considerable time without being drowned. Each diver takes with him a net bag for the purpose of holding the Oysters, puts his foot into a stirrup, to which hangs a stone weighing about thirty pounds, and after taking a long breath is swiftly carried to the bottom. He then flings himself on his face, fills his bag

PEARL OYSTER.—*Meleagrina margaritifera.*

as fast as he can, and when his breath begins to fail, shakes his rope as a signal, and is drawn up together with the bag.

Very exaggerated accounts have been given of the time passed under water by the divers, from two to seven minutes being mentioned as the usual periods. The real fact, however, is, that one minute is the ordinary average; a few men being able to endure an immersion of a minute and a half. This is a long period, as any one will confess who has attempted to repeat the feat. Yet, with a little practice, it can be achieved, even by those who can lay no claim to extraordinarily capacious lungs; and I have more than once performed it with tolerable ease. If the lungs be thoroughly filled four or five times in succession, and emptied to the last gasp, so as to expel all foul air that may be lingering in the tiny vesicles, the blood becomes so well oxygenized, that a further supply of breath will not be needed for some time, and a deep inspiration will serve to keep the blood in a healthy state for a marvellously long time. All swimmers who are fond of diving will find that they can remain under water nearly twice their usual time by taking this simple precaution.

The best plan for procuring the Pearl Oyster is evidently the employment of the diving bell, so that the best shells might be leisurely selected, the spot left undisturbed, and the sharks outwitted. In the illustration are given specimens of the shell in various stages, as well as the interior, showing the pearls as they appear when the animal is removed.

REVERTING to the illustration on page 422, we see a large, flattish, wedge-shaped shell lying on the rock, and moored to the stone by a number of short threads. This is the PINNA, so called from the Latin word, signifying a wing.

The aggregate mass of these threads is termed the byssus, and is, indeed, a very curious object. The threads are spun by the foot, and are attached to the centre of each valve, thus forming a powerful cable by which the shell is moored to the rock. The threads are wonderfully strong, silken in their texture, and, had the mollusc been sufficiently plentiful, might have been employed in various manufactures. In the British

NOAH'S ARK.—*Arca Noœ.*   MAGELLANIC MUSSEL.—*Mýtilus Magellánicus.*
FORK-TAILED DATE-SHELL.—*Lithódomus caudígera.*
EDIBLE MUSSEL.—*Mýtilus edúlis.*   FINGER DATE-SHELL.—*Lithódomus dáctylus.*

Museum may now be seen a pair of gloves that have been woven from the byssus of the giant Pinna, a species which sometimes attains the length of two feet, and has a most singular appearance when old, owing to the mass of parasitic creatures, such as serpulæ, balani, and sundry zoophytes, that always congregate on such substances.

It is remarkable that a little crab, called, from its habits, Pinnotheres, is often found within the shell of this mollusc, and was formerly thought to have entered into a tacit agreement with its host to act as sentinel and to bring in food as a return for the hospitality afforded to it. This, however, is not a solitary instance of such strange alliance, several other molluscs being known to shelter their particular crustacean guest. When at rest, the Pinna is mostly buried in the sand, with the exception of the upper edges of the shell, which are permitted to protrude just above the substances in which the rest of the creature is immersed.

WE now come to the large, useful, and even beautiful family of the Mussels, although, in most cases, their beauty is not perceptible until the shell has been polished and the rich tints thereby brought out. Rough and polished mahogany are not more unlike each other than the Mussel-shell before and after the polishing process. Some species are marine, while others inhabit the fresh water, and all may be known by the peculiar shape of the shells.

The EDIBLE MUSSEL, so common in the fishmonger's shop and the costermonger's barrow, is found in vast profusion on our coasts, where it may be seen moored to rocks, stones, and fibres, alternately covered with water or left dry according to the flowing and ebbing of the tide. The heedless bather is sometimes apt to come unexpectedly upon a collection of these molluscs, and if he once meets with that misfortune, his lacerated limbs, cut in all directions by the knife-like edges of the shells, will serve as effectual warnings not to repeat the same imprudence.

At some periods of the year the Mussel is extremely injurious as an article of food, though the effects seem, like those produced by eating the bonita, to depend greatly on the constitution of the partaker, some being able to eat it with impunity, while others who have shared the same meal are visited with asthma, violent rash, nausea, and many other symptoms which, though not absolutely dangerous, are peculiarly annoying. The Mussel is largely used for bait as well as for human consumption, more than thirty millions being collected annually in one locality for that purpose. Little ill-shapen and badly coloured pearls are often found in this mollusc, but are quite useless for the market. Attempts have been successfully made to propagate the breed of Mussels; and the vast plantations, as they may be called, of these creatures have increased to such an extent, that they threaten to obliterate several useful bays for all maritime purposes.

The specimen which is represented in the engraving was found by Mr. Sowerby, clinging to a wooden pile, at Harwich, some eighteen inches above the water, and having six or seven pieces of stone, bits of shell, and similar substances, attached to the byssus, among which it lay ensconced as is shown in the illustration.

An allied species, the Dreissena, inhabiting the fresh waters, has of late years rapidly overrun our island, having been originally imported into the Surrey Docks, whence it has spread with astonishing fertility, passing from one river to another, getting into all the little rivulets that trickle between meadows, and even obtaining entrance into artificial basins by means of the water that feeds them through pipes. About Oxford its spread was most notable, the many waters with which that city is surrounded literally teeming with the mollusc very soon after its advent was first discovered. The shell is like that of the edible Mussel, but shorter, and without the beautiful nacreous lining.

The reader may observe in the upper portion of the engraving a number of little shells buried deeply in some substance. These are specimens of the Fork-tailed Date-shell, a little ochre-coloured shell, without any peculiar beauty of form or colour, but yet as remarkable a creature as any that has been or will be mentioned.

This little being has the power of burrowing deeply into the hardest stone; and in the present instance, the substance in which the Lithodomi are imbedded is a shell of the gigantic limpet from Madagascar, measuring about six inches in diameter and half an inch or so in thickness. The specimen, which I have carefully examined, is a really wonderful one, the thick, hard, and solid substance of the shell being literally riddled with the holes of the Lithodomus, whose forked processes just project from the circular aperture much like the eggs of the common scatophagus from the substance in which they are sunk. The form of the holes may be seen by reference to the illustration, and the reader will doubtlessly notice that their direction is by no means uniform; the generality being at right angles with the surface, but many arranged in an oblique direction, and some driven almost horizontally.

The method by which this little mollusc contrives to excavate its chamber is a complete mystery. It is known that in its earlier stages it spins a byssus, and attaches itself to substances like the common Mussel, but that in process of time it begins to bore its way into the object to which it is moored. As the shell increases in size, the chamber is enlarged in dimensions; but the original aperture remains of the same diameter as when first bored, and therefore effectually prevents the animal from making its exit.

Some persons have suggested that the animal employs an acid for the purpose of dissolving the rocks; but if such were the mode of operation, the shell would suffer equally with the stone. A continual current of water forms the basis of another theory; and provided that the animal were sufficiently long-lived, there is no doubt but that the constant action of water would in process of time wear away the stone, however hard it might be. But as yet no theory has sufficiently accounted for the fact that the creature excavates these chambers with wonderful rapidity, and that, in all cases, the chamber corresponds with the shape and size of the shell. It is evident, also, that the shell itself is not the means by which the chamber is bored, as the peculiar shape of the hole prevents the shell from rotating.

The Lithodomus seems to drive its curious tunnels through everything that comes in its way, for, in one case, a specimen has bored through the upper part of the limpet-shell, broken into a chamber already excavated by another individual, and forced its way fairly through the inhabitant as well as the habitation.

The animal is slightly luminous, as is the case with most of the burrowing molluscs. The colour of the shell is uniformly pale brown.

The FINGER DATE-SHELL, which is seen lying on the ground just under the Mussel, is a rock burrower, and so beautifully decorated that it seems a sad pity to bury so lovely a shell in so dark a recess. At a little distance it is quite ordinary in appearance, being apparently a plain, mahogany-coloured shell; but when examined closely, it is found to be elegantly formed, coloured with a peculiarly rich ruddy brown hue, and sculptured with myriads of minute waved ridges and channels drawn crosswise over the shell, which give wonderful effects of light and shade, and heighten the tints materially. The animal is edible, and is eaten like that of the common Mussel.

ON the upper part of the illustration, and towards the left hand, may be seen a rather curious-looking shell, which, from its rude resemblance to the familiar toy of childhood, is called the NOAH'S ARK.

The Ark-shells are found all over the world, hidden under stones, in the crevices of rocks, or even within the forsaken burrows of the pholas or the date-shell. Owing to their retiring habits, and the nature of the localities in which they live, they are mostly distorted or damaged. They can move themselves very fairly by means of a curious conical byssus, composed of a series of thin plates, which can be cast off or re-formed at the will of the animal.

IN the two species represented in the following illustration, we have examples of pearl-bearing molluscs inhabiting the fresh waters.

The BRITISH PEARL-MUSSEL was once a valuable inhabitant of our rivers, on account of its contents. It is now, however, seldom sought except for bait, and in the latter capacity is more useful than in the former, as it is estimated that not more than one per cent. contain any pearls, and not more than one per cent. of the pearls is of any commercial value. The older and more irregular the shell, the better chance is there of finding a pearl; and a diligent collector may soon obtain a tolerable series of these objects for his cabinet. Now and then, however, a really fine pearl is found; and one, that was obtained from the Conway, now holds a place in the crown of England.

This Mussel is tolerably active, and, if laid on the sandy or muddy floor of an aquarium, will soon assume its usual attitude. It begins by gently opening the shell and pushing out its foot, and, after a series of apparently aimless movements, succeeds in setting itself on its edge. It then forces itself forward, leaving a deep groove behind it, or wriggles itself downwards until only an inch or so of its shell protrudes. The edges of the shell are sharp as knives; and many a timid bather, who is afraid to swim, and contents himself with paddling about in shallow water, has suffered severe cuts on the foot from these treacherous inhabitants of the water.

The interior of this Mussel is of a lovely azure, and the exterior exhibits beautiful colourings when cleaned with a very weak solution of muriatic acid and then polished. It is a very prolific species, the outer gills of the female containing a vast number of tiny Mussels not yet able to obtain their own living, and waiting until they have gained sufficient strength to be launched on the world. One specimen will produce nearly three hundred thousand young in the course of a year.

THE genus to which the CHINESE PEARL-MUSSEL belongs is distinguished by the thin elastic wings into which the valves are produced.

From this species the Chinese, those incorrigible tricksters, are in the habit of producing imitation-pearls by a very simple process. A string of small shot is introduced between the valves, and the animal restored to its native element. The irritation caused

BRITISH PEARL-MUSSEL.—*Unio margaritiferus.*   CHINESE PEARL-MUSSEL.—*Dipsas plicatus*

by the presence of the foreign body forces the mollusc to deposit the nacreous secretion upon the intruding substances, and after a while the shot are covered with layer upon layer of pearly substance, the thickness of the coating depending upon the length of time occupied in the construction.

The same ingenious people are also accustomed to make curious little pearl-covered josses, by stamping them out in thin bell-metal, slipping them into the shell, and leaving them between the valves until they are sufficiently coated with pearl. The example from which the illustration was taken is in Mr. Cuming's collection together with several others.

In the Thorny Clam, a curious member of another family, the shell is covered with long and branching projections, something like the horns of a young roebuck. All the Clams are natives of the warmer and tropical seas, especially among coral reefs, and their colour and shape are extremely variable. Mr. Broderip writes of them as follows: "The shells are attached by their external surface to submarine bodies, such as coral rocks, and shells have been observed at depths varying from points near the surface to seventeen fathoms. These shells appear to be subject to every change of shape, and often of colour; that the accidents of their position may bring upon them. Their shape is usually determined by the body to which they are fixed; and the development of the foliated laminæ which form their general characteristic is effected by their situation; and their colour most probably by their food, and their greater or less exposure to light. The Chama that has lived in deep and placid waters will generally be found with its foliations in the highest state of luxuriancy, while those of an individual that has borne the buffeting of a comparatively shallow and turbulent sea will be poor and stunted." The Clams are generally attached by the upper valve. The animal is edible, and is considered a great delicacy. About fifty species of Clams are known.

THE Tridacnidæ are easily known by their deeply waved shells, with the indented edges fitting into each other, and the overlapping foliations of the surface. Although separated from the true Clams, they are popularly called by the same name. The YELLOW CLAM is represented as it appears when buried in a mass of white madrepores, and is given in this attitude for the purpose of exhibiting the interlocking edges and the bold foliations. A well-known species, called from its enormous dimensions the GIANT CLAM (*Tridacna gigas*), was formerly rare, but is now tolerably plentiful. It attains to a gigantic size, sometimes weighing more than five

THORNY CLAM.—*Cham Lázarus.*

hundred pounds, and containing an animal which weighs twenty pounds, and can furnish a good dinner to nearly as many persons. The natives of the coasts on which it is found—

SPOTTED BEAR'S-PAW CLAM.—*Hippopus maculáta.*　　　YELLOW CLAM.—*Tridacna crocea.*

namely, those of the Indians seas—are extremely fond of this creature, and eat it without any cooking, just as we eat oysters. The substance of the shell is extremely thick and solid, and enables it to be used for many ornamental purposes.

HEART-COCKLE.—*Cardium cardissa.*

SPIRAL HEART-COCKLE.—*Isocardia cor.*   TUBERCULATED COCKLE.—*Cardium tuberculatum.*

In former days, when this species was very rare, a magnificent specimen was presented to the church of St. Sulpice, in Paris, where it may now be seen, the valves being set up as *bénitiers* for containing the holy water. This shell dates from the time of Francis I. It is evident that the byssus by which so enormous a shell is moored to the rocks must be of great size and strength, and, indeed, is so strong as to require an axe for its severance. The muscles, too, by which the animal contracts its shell are enormously powerful; and it has been remarked by Mr. Darwin, that if a man were to put his hand into one of these shells, he would not be able to withdraw it as long as the animal lived.

On the left hand of the illustration on the previous page is seen the SPOTTED BEAR'S-PAW CLAM, which has been placed in a separate genus on account of a difference in the number of projections on the hinge, technically called hinge-teeth. The animal has been represented in this peculiar manner for the purpose of displaying the mouth, which is marked by the coronet-like circlet around it, and the foot, which is seen below just projecting from its groove. This animal also spins a byssus, which is, however, weak and slight compared with that of the gigantic species just described.

The family of the Cockles, or Cardiadæ, so called from their heart-like shape, is well represented by the common COCKLE (*Cárdium edúle*) of our British shores. Generally, the Cockle is a marine animal; but it sometimes prefers brackish water to the salt waves of the ocean; and a small variety is found in the Thames nearly as high as Greenwich, when the water is sensibly flavoured with salt at each high tide. Another species, the PRICKLY COCKLE (*Cardium aculeátum*), is found on the southern coast, and regularly brought to market.

The Cockle is gathered in great numbers for the purpose of being eaten, although, as tne greater number are consumed in the open air, they can hardly be said to be procured for the table. According to Mr. Maxwell, " a crowd of the more youthful description of the peasantry are collected every spring tide to gather Cockles on the sands by daylight when the tide overruns. The quantities of these shell-fish thus procured would almost exceed belief; and I have frequently seen more than would load a donkey collected in one tide by the children of a single cabin. They form a valuable and wholesome addition to the limited variety that the Irish peasant boasts at his humble board; and afford children, too young for other tasks, a safe and useful employment."

This mollusc frequents sandy bays, and remains about low-water mark, burying itself in the sand by means of the powerful foot, which also enables it to leap to a surprising height.

IN the front of the opposite engraving is seen the common HEART-COCKLE, and on the left is the remarkable SPIRAL HEART-COCKLE, notable for the boldly spiral umbones. In the illustration, the partly opened valves permit a portion of the animal to be seen, including the two orifices of the siphon, through which water is injected and expelled, and the foot, which is seen protruding in a tongue-like fashion shortly below the siphon.

This animal is in the habit of burrowing in the sand, leaving only the openings of the siphon above the surface. These organs are also represented in the figure of the TUBERCULATED COCKLE, where the double ends of the siphon are shown projecting towards the left hand, and are at once recognisable by their fringed edges; and the large foot is seen below, carrying the superstructure along. Even when taken out of the water, the Cockles are very lively; and if placed in a pan or basin they tumble about with great energy, knocking their shells against each other and the sides of the vessel with remarkable activity.

BRITISH CYTHEREA.—*Cytheréa chione.*   BANDED VENUS-SHELL.—*Venus fasciáta.*

WE now come to a group of these shells where the siphons are extremely long. The first family is represented by the BANDED VENUS-SHELL, so called on account of its beautiful colours and elegant form, and the bands which traverse its surface. All the Venus-shells are handsome, and have well deserved their name. The shells are extremely hard in texture, thick, and smooth, and are mostly found in the warmer seas.

It is from one of these shells that the American Indians make the well-known wampum. They examine the sea-coast, pick up the fragments of Venus-shells that have been broken and battered by the waves, shape them rudely into form, bore holes through them, and thread them on slender thongs of raw hide. Now-a-days, however, the metallic coinage of the European has superseded the wampum of the red man, just as the

Birmingham gun has taken the place of the Indian bow, the Sheffield hatchet the place of the tomahawk, and the Witney blanket that of the original bear skin.

About one hundred and seventy species of Venus-shells are known, spread throughout all parts of the world, and ranging from low-water mark to a depth of one hundred and forty fathoms. In the figure the animal is shown as it appears when crawling, the slightly diverging ends of the double siphon appearing in front, and unequal in size, one of the peculiarities of the genus.

THE beautiful BRITISH CYTHEREA is seen in the foreground of the illustration, its elongated siphons projecting towards the right like the tubes of a double-barrelled gun, and their orifices surrounded with a fringe of tentacular appendages, and the thick sturdy foot appearing behind. This genus is closely allied to the preceding, and is therefore appropriately named Cytherea, that being one of the classical epithets applied to Venus in consequence of her predilection for the island of Cytherea in the Ægean Sea. In this animal the two portions of the siphon do not diverge.

IN the family of the Mactridæ, or Trough-shells, the valves are of equal dimensions, and rather triangular in shape. The animal has the two channels of the siphon united as far as the extremity, and the foot is ample and strong.

The common Trough shell is a British species, and is found on many of our coasts, always preferring those of a sandy nature, where it can hide itself by sinking just below the surface. The foot is capable of considerable motion, and can be extended to some length ; and when the movements are rapidly performed, it enables the creature to jump about nearly as actively as the cockle. The Trough-shells are found in all parts of the world, and in some coasts of our own islands are so plentiful that they are gathered for the purpose of feeding pigs. The species which is usually employed for this purpose is *Mactra subtruncáta*, and, like the cockle, it is taken at low water. Although so usually inhabiting the zones just below and above low-water mark, these shells are sometimes found as low as thirty fathoms beneath the surface.

THE very remarkable shell from which protrude two enormously long siphon tubes is the SCROBICULARIA, an example of the family Tellinidæ, all the members of which are notable for the length and divergence of these tubes.

These creatures are spread over the greater part of the globe, seldom inhabiting deep water, and being fond of localities where the bed of the sea is of a sandy or muddy nature, as they are able to bury themselves deeply in those substances, and to carry on respiration, and obtain nutriment by means of the long siphon tubes. The larger tube is that which is employed for the ingress of water, the shorter for its exit ; and when the animal is at rest, it remains in an upright position in the sand or mud. The orifices of the siphons are without fringes, and the tubes can be extended to five or six times the length of the shell. The specific title of piperita, or peppery, is given to the animal on account of its peculiarly pungent flavour, which, however, does not prevent it from being rather extensively consumed as an article of food in some parts of the world.

WE now come to the well-known Solenidæ, or Razor-shells, so called on account of their shape.

These curious molluscs always live buried in the sand in an upright position, leaving only an opening shaped like a keyhole, which corresponds with the two siphon tubes. Those who are fond of examining the sand and rocks at low water will doubtlessly have been startled and amused by little jets of water which spirt some few inches in height, but never reappear.

These are caused by the RAZOR-SHELL ; and if the locality whence the jet started be watched, the little keyhole-like orifice will be seen. To catch the mollusc that emitted the water is no easy task, but may be managed in two ways. The simplest but roughest method is to take an iron rod hooked at the end, plunge it into the sand like a harpoon, and pull it out smartly in an oblique direction, bringing with it the shell. This method,

SCROBICULARIA.—*Scrobicularia piperita.*
SWORD-BLADE RAZOR-SHELL.—*Solen ensis.*

COMMON TROUGH-SHELL.—*Mactra stultorum.*
COMMON RAZOR-SHELL.—*Solen vagina.*

however it may answer for those who only want the creature for the purpose of eating the animal, or using it as bait, is by no means suited to those who wish to capture the inhabitant uninjured and to experimentalize upon it. These, therefore, must employ a different plan.

It is sufficiently remarkable, that although the Razor-shell lives in saline water, it has the greatest objection to salt in an undiluted state, and is as much annoyed by that substance as a worm by ammonia or a slug by lime. To catch this creature unhurt is easy enough, and merely requires that a spoonful of salt should be dropped into the burrow, and the animal adroitly seized as it protrudes itself in anger at the liberty which has been taken with its home. If, however, the intended captor should bungle the attempt, the Razor-shell immediately slips to the very bottom of the burrow, and, even though a pound of salt be expended, will not be induced to show itself a second time. These creatures are generally found at a depth of one or two feet, and when they make their burrows, as they are often in the habit of doing, among the rocks, not even the hooked iron can draw them from their retreat.

The three figures of the Razor-shells require some explanation.

At the right hand, and at the bottom of the engraving, the upper part of the common Razor-shell is shown as it appears when at rest and in full feed, the water passing into the system through one of the fringed apertures, and out at the other. The upper figure, lying horizontally upon the rock, represents the same species, for the purpose of showing the foot, by means of which the rapid ascent and descent in its burrow is accomplished.

3.   F F

The third figure represents another species, the SWORD-BLADE RAZOR-SHELL, with the ends of the siphon-tubes just protruding from the extremity.

IN the next family, called Gaper Shells, because the valves when closed do not unite completely, but leave a moderately wide aperture at the hinder part, the shell is strong, thick, and opaque ; the foot is comparatively small, and the siphons are united and retractile.

GAPER SHELL.—*Mya arenaria*

The GAPER SHELL inhabits sandy and muddy shores, and is especially fond of frequenting the brackish waters of river-mouths, where the streams are sure to bring with them a soft deposit of mud and sand. The species which is represented in the engraving burrows nearly a foot in depth into the sand, and is able to breathe and gain subsistence by the long siphons, which just protrude above the surface. In looking at this animal, and observing its habits, the entomologist is forcibly reminded of the manner in which the rat-tailed maggot, *i.e.* the larva of *Eristalis tenax*, the great bee-like fly, with enormous eyes, is in the habit of hovering for a moment over a flower or leaf, settling for a moment, and then darting off again with lightning speed. Like the Gaper Shell, this larva spends its life deeply buried in the mud, carrying on the business of respiration by means of a long tube which, like the siphon of the mollusc, can be retracted or extended at will.

The Gaper Shell is much sought after in many places as an article of food, not only by man, but by birds and beasts, such as the walrus and the blue fox.

THE nearly cylindrical shell seen lying on the ground in the front of the illustration on the opposite page is the WATERING-POT SHELL, a curious creature found in some of the hotter seas.

This species is a good example of a family termed the Gastrochænidæ, in which the valves are thin, gaping, and when adult, often connected with a rather long calcareous tube, as in the present instance.

The Watering-pot Shell derives its name from the curious perforated disc which closes its lower extremity, and bears no small resemblance to the rose of a watering-pot. In allusion to the same peculiarity, the French writers call the animal by the name of Arrosoir. All the species are burrowers, some into coral, some into stone, some into shells, and others into sand, as is the case with the creature which we are now examining. From the other end of the tube the siphons can be protruded to some extent, and withdrawn when the animal is alarmed.

One species belonging to this family, the *Gastrochæna modiolína*, has been known to drive its burrow fairly through some oyster-shells into the ground below, and then to make a permanent home by cementing all kinds of materials into a flask-like case and fixing its neck into the perforated oyster-shell.

PIDDOCK.—*Pholas dáctylus.*   PAPER PHOLAS.—*Pholas papyracea.*
WATERING-POT SHELL.—*Aspergillum pulchellum.*

THE two upper figures are examples of some very curious and common shells, popularly called PIDDOCKS, and found in profusion along the sea-coast.

The common Piddock may be found in vast numbers in every sea-covered chalk rock, into which it has the gift of penetrating so as to protect itself from almost every foe. The two specimens on the upper left of the engraving belong to this species, and are represented so as to give two views of the same shell, the one to show the peculiar curve of the shell and the protruding foot, and the other to exhibit the umbonal valves.

Every one is familiar with the beautiful white shell of the Piddock, crossed by a series of elegantly curved projections, something like the teeth of a file. According to some writers, it is by means of these projections that the creature is able to burrow into the rock; and the possibility of such a feat has been proved by the simplest possible means, namely, by taking a Piddock into the hand and boring a similar hole with it. Mr. Robertson, who kept these creatures alive in their chalky burrows, devoted much time to watching them, and finds that during the process of burrowing they make a half turn to the right and then back to the left, never turning completely round, and, in fact, employing much the same kind of movement as is used by a carpenter when boring a hole with a bradawl.

Mr. Woodward remarks very justly, that "the condition of the Pholades is always related to the nature of the material in which they are found burrowing; in soft sea-beds they attain the largest size and greatest perfection, whilst in hard and especially gritty rock, they are dwarfed in size, and all prominent points and ridges appear worn by friction. No notice is taken of the hypothesis which ascribes the perforation of rocks, &c.

to ciliary action, because, in fact, there is no current between the shell, or siphon, and the wall of the tube." As soon as the animal has completely buried itself it ceases to burrow, and only projects the ends of the siphon from the aperture of the tunnel.

Some species of Piddock are eaten, *Pholas costáta*, one of the South American species, being a good example. In England it is seldom used except for bait, its fine white foot, which looks, when fresh, as if cut out of ice, answering that purpose admirably, its glittering whiteness serving to attract the attention of the fish, and its toughness causing it to adhere strongly to the hook.

Several other genera are worthy of notice, among which the Martesia is, perhaps, the most curious, shells belonging to this genus having been found in cakes of wax floating on the waves off the Cuban coast, and others in masses of resin on the shores of Australia. The PAPER PHOLAS, represented on the right hand of the illustration, is shown as it appears when boring into the red earth of Brighton.

IN the accompanying illustration we have an example of a creature, which, though useful enough in many ways, and doing good service in transmuting dead and decaying substances into living forms, is yet the dread of mariners and the terror of pier builders.

The SHIP-WORM, as this mollusc is appropriately called, from its depredations on the bottoms of ships and all submerged wooden structures, is found in most seas, and on our own coasts works fearful damage by eating into piles, planks, or even loose wood that lies tossing about in the ocean. I have now before me a portion of the pier at Yarmouth, which is so honeycombed by this terrible creature that it can be crushed between the hands as if it were paper, and in many places the wood is not thicker than ordinary foolscap. This piece was broken off by a steamer which accidentally ran against it; and so completely is it tunnelled, that although it measures seven inches in length and about eleven in circumference, its weight is under four ounces, a considerable portion of even that weight being due to the shelly tubes of the destroyers.

I have also a block of oak, picked up off Broadstairs, where the Ship-worm has been nearly, though not quite so destructive as in the former instance. This specimen is notable, as giving an example of a principle on which many piers, &c. have been protected from this mollusc. A large iron bolt passes through the midst of the block, and the rust of the projecting head has spread itself for some distance over the wood. Multitudes of holes, large and small, surround the bolt, but not one has pierced that portion over which the rust extends. Knowing the objection entertained by the Ship-worm to rust, engineers have been in the habit of driving a number of short iron nails, with very wide heads, into the timber, arranging them in regular rows, with their heads at no great distance from each other. The action of the salt water soon causes the rust to spread over the spaces between the heads, and upon these spots the Ship-worm refuses to settle.

Another plan, and a very effective, though rather expensive one, consists in forcing a solution of corrosive sublimate into the pores of the wood. This salt of mercury is very destructive to animal life, and M. Quatrefages asserts that one twenty-millionth part of corrosive sublimate is enough to destroy all the young Ship-worms in two hours, and that a ten-millionth part would have the same effect in forty minutes. He therefore proposes that ships should be cleared of this terrible pest by being taken into a closed dock, into which a few handsful of corrosive sublimate should be thrown and well mixed with the water. The salts of copper and lead have a similar effect, but are not so rapid in their operation. The wooden piles on which jetties and piers are supported can be preserved in the same manner. Iron, however, is now rapidly superseding wood for such structures, and is quite impervious to the attacks of any mollusc, no matter how sharp its teeth.

When removed from the tube, the Ship-worm is seen to be a long greyish white animal, about one foot in length and half an inch in thickness. At one end there is a rounded head, and at the other a forked tail. The curious three-lobed valves are shown in the foreground as they appear when separated from the animal. The burrow which the creature forms is either wholly or partially lined with shell, and it is worthy of notice that the Shipworm and its mode of burrowing was the object that gave Sir I. Brunel the idea of the Thames Tunnel.

SHIP-WORM.—*Teredo navális.*

The Teredo did not always lead this fixed and darkling life, but at one time of its existence it swam freely through the ocean, having organs of sight and hearing for the purpose of guarding itself against the dangers of the deep. The following description of the successive stages of this mollusc is from the pen of the able naturalist whom we have already mentioned, and who has devoted much time to the study of the Ship-worm and its habits.

" The larva, which is at first almost spherical and entirely covered with vibratile cilia, may be compared to a very minute hedgehog, in which every spine acts as a natatory organ. It swims in all directions with extreme agility, and this first state continues about a day and a half. Towards the end of this time the external skin bursts, and after being encrusted with calcareous salts, becomes a shell, which is at first oval, then triangular, and at last very nearly spherical. While the shell is being formed the vibratile cilia disappear, but the little animal is not on that account condemned to inactivity. In proportion as the external cilia diminish we observe that another equally ciliated organ becomes developed, which widens and extends in such a manner as to form a large collar or ruff margined with fringes. This new organ of locomotion may be entirely concealed within the shell, or may be extended from it, and acts in the manner of the paddle-wheel of a steamboat.

By means of this apparatus the young larva continues to swim with as much facility as in its earlier age, but it now moreover requires another organ, which enables it to advance upon a resisting plane, as, for instance, along the side of a glass vessel. This is a sort of fleshy foot, very similar to a long and very mobile tongue, which can be extended and contracted at will. The larva of the Teredo possesses, moreover, organs of hearing similar to those of several molluscs, and eyes analogous to those of certain annelids.

During this period of existence, our mollusc therefore enjoys to a high degree the characteristic faculties of a living being. It moves, and is brought in relation with the external world by organs of the special senses. Now, however, comes a last metamorphosis, and this same Teredo loses its organs of motion and sensation, becoming a kind of inert mass in which vegetative life replaces almost entirely the active spontaneity of the animal."

While we cannot but admire the painstaking research which has brought to light these curious processes in the life of the Teredo, we must dissent from the opinion expressed in the last paragraph. All animals are created to enjoy the life which the divinely implanted instinct forces them to lead, and there is no reason whatever for supposing that the Teredo when sunk in its burrow, without any apparent organs of sight or hearing, is not enjoying its life as thoroughly as when it roamed the ocean in full possession of both faculties. It seems to me that in speaking of the perfect Teredo as an inert being, more like a vegetable than an animal, M. de Quatrefages loses sight of true philosophy, and falls into the same error as the earlier naturalists, who wrote with scornful compassion of the miserable lives led by the sloth, the mole, and the woodpecker. They fell into the mistake of judging all other beings by their own standard, and knowing that to be for ever buried underground like the mole, crawling among branches like the sloth, or perpetually hammering at tree-trunks with their mouths like the woodpecker, would be a most wretched and painful life to themselves, hastily concluded that the creatures above mentioned must be equally miserable under similar circumstances.

It is, moreover, hardly in accordance with common sense that the animal should be less happy in its perfect than in any preliminary stage of existence, inasmuch as then only can it fully carry out the great law of nature. No one fancies that the butterfly is less happy than the caterpillar, or the bee than the grub, simply because in these particular cases the elements of happiness are more palpable to human senses, or more analogous to our own ideas upon the subject. We have now begun to learn the wondrous powers of adaptation, whereby every living being is rendered happiest in its proper place; and it surely is but philosophical to infer that whatever might be the case with ourselves, who were not made to burrow in wood, the Teredo, which was created for that express purpose, will enjoy its life like every other being. It is true that man would be very wretched if deprived of eyes and ears, and compelled to spend his existence in a hole just large enough to contain his body; but he would not be more miserable than the Teredo would be, if removed from its congenial tunnel and forced to live in a warm drawing-room. "What is one man's food is another's poison;" and the conditions which would inflict unendurable torture upon one being, afford the only means by which another can be made happy. We may as well pity the polar bear for the bitter cold of its native land, or the tropical animals for the burning heat of their midday sun, as waste our compassion upon the Teredo because it lives in a burrow beneath the waters.

An enormous species of this genus, called from its dimensions the Giant Teredo (*Teredo gigantea*), has been found at Sumatra. This huge mollusc sometimes attains the length of six feet, and a diameter of about three inches, but fortunately for timber, does not make its habitation in that substance, contenting itself with boring into the hardened mud of the sea-bed. The colour of the shelly tube is pure white externally and yellow within. On account of its mud or sand burrowing habits, the specific title of *arenária* has been applied to this species.

The strange-looking objects that are represented in the accompanying illustration have long perplexed systematic naturalists, and even now, although they have been the subject of careful examination by accomplished zoologists, many parts of their economy are enigmatical in the extreme. The order to which they belong is called by the name of Tunicata, because the animals possess no shell, but are covered with an elastic tunic. Some of them are transparent and really beautiful, while others are apparently little more than shapeless masses of gelatinous substance, studded with minute stones, fragments of shells, and coarse sand, overgrown with seaweeds, and perforated by certain bivalve molluscs.

PELONÆA.—*Pelonæa corrugata.*
SEA-SQUIRT.—*Ascidium hyalinum.*
BOLTENIA.—*Boltenia reniformis.*
SYNTETHYS.—*Syntethys Hebridicus.*
CYNTHIA.—*Cynthia morus.*
BOTRYLLUS.—*Botryllus polycyclus.*
CLAVELLINA.—*Clavellina lepidiformis*

So strange are many of the structures, and so unexpected are many of the habits of these creatures, that zoologists are often in great doubt as to their true position in the natural kingdom. For, while many anatomical details are closely analogous to those of the molluscs, others are equally analogous to those of the higher zoophytes, and it is not always easy to hold the balances with an even hand.

In these curious beings we find the first indications of propagation by buds springing from the body of the parent just as a bud springs from the branch of a tree, and the perplexing alternation system is sometimes found in the Tunicata. Both these systems will be explained when we come to describe the zoophytes.

Without enumerating the various details of anatomical structure in these creatures, we may mention that the animal is composed of two sacs, one inside the other, and that the functions of nutrition and respiration are performed, as in the bivalves, by means of two orifices, one of which admits and the other expels the water, which laves the respiratory apparatus and brings food to the organs of nutrition.

In our present arrangement, which is that followed by Mr. Woodward, the Tunicata are divided into three families, easily distinguishable by the eye. The members of the first group are called Solitary Ascidians, those of the second Social, and those of the third, Compound Ascidians, their characters being sufficiently explained by their names. The name Ascidian is of Greek origin, and is the diminutive of a word signifying a water skin, such as is used in the East.

THE simple or solitary tunicates are classed together under the name of **Ascidiadæ**. The common SEA-SQUIRT, figured on the left hand of the illustration, is a good example of the typical genus.

This animal, in common with all its kin, feeds mostly, if not wholly, upon the minute vegetable organisms, such as the desmids, diatoms, &c. which abound throughout the water, and the manner in which these substances are brought to the digestive organs is equally simple and beautiful. "The mouth," writes Mr. Rymer Jones, "is quite destitute of lips or other extensile parts, and situated, not at the exterior of the body, but at the very bottom of a capacious bag inclosed in the interior of the creature.

It is obvious, then, that whatever materials are used as aliment, must be brought into the body with the water required for respiration; but even when thus introduced, the process by which they are conveyed to the mouth still requires explanation.

A truly miraculous apparatus is provided for this purpose. The whole surface of the respiratory chamber is covered over with multitudes of vibratile and closely set cilia, arranged in millions, which by their united action cause currents in the water, all of which flow in continuous streams directly towards the mouth. It is sometimes possible, in very young and transparent specimens, by the aid of a good microscope, to witness the magnificent scene afforded by these cilia when in vigorous action.

—— ' salientia viscera possis
Et perlucentes numerare in pectore fibras.'

The effect upon the eye is that of delicately toothed oval wheels revolving continually from left to right, but the cilia themselves are very much closer than the apparent teeth, the illusion being caused by a fanning motion transmitted along the ciliary lines, producing the appearance of waves, each wave representing a tooth of the supposed wheel.

Whatever little substance, alive or inanimate, the entering water brings into the branchial sac, if not rejected as unsuitable, lodges somewhere on the respiratory surface, along which each particle travels horizontally, with a steady, slow course, to the front of the cavity, where it reaches a downward stream of similar materials, which hurries onward, receiving accessories from both sides, until at last the whole is brought into the gaping mouth."

Some of the Ascidia are beautifully coloured with orange, crimson, and white, though others are plain and almost repulsive when dredged from the sea-bed. Several species attain to a considerable size, being five or six inches in length. They do not seem to like deep waters, being mostly found near low-water mark, and seldom at a greater depth than nineteen or twenty fathoms.

AT the right hand of the illustration is seen another British tunicate. This is the CYNTHIA, one of a rather numerous genus, not uncommon on our coasts. The AGGREGATED CYNTHIA (*Cynthia aggregáta*) is to be found on almost any substance that has remained for any length of time below low-water mark, and stones, rocks, wooden piles, or even the larger seaweeds, are frequently covered with these curious creatures, sometimes set in solitary state, and sometimes gathered together in groups by means of the interlacing of the fibres by which they attach themselves. Some species are eaten, *Cynthia microcosmus* being the most in favour, and regularly brought to market for sale. This animal derives its specific title from the multitude of animal and vegetable parasites that grow upon it, and so transform it into a little world.

THE animal which is represented at the upper part of the engraving is the PELONÆA, so called from two Greek words, the former signifying mud, and the latter to inhabit. This animal, as its name imports, is in the habit of burying itself in the mud, where it remains fixed and nearly motionless, respiring and obtaining nutrition by means of the two open tubes seen at the smaller end. Only two, or perhaps three, species of this genus are known, and the animal is found in northern Scotland and Norway.

THE curious BOLTENIA, so called after Dr. Bolten, a naturalist, of Hamburg, is found in rather deep waters, being sometimes drawn up by fishermen's lines from a depth of seventy fathoms. The animals of this genus are attached to long footstalks, at the end of which the creature sways like a fritillary on its slender stalk. The two orifices by which water is admitted into and ejected from the system are seen, and their remarkable four-cleft openings are well displayed. When very young, the Boltenia is often found affixed to the stem of its parent.

WE now arrive at the Social Ascidians, two examples of which are given in the illustration. The CLAVELLINA may be seen lying upon the leaf of the laminaria, the different individuals being united by long filaments. The blood circulates through these channels of communication, passing to and fro through separate tubes. It is a small creature, and extremely transparent, the latter characteristic making it a valuable species to the physiologist, who is enabled to watch its structure, and the methods in which the different organs perform their duties, without needing to dissect it. The Clavellina may be found on the English shores at low water, adherent to rocks, stones, or seaweed, to which it attaches itself by means of the tiny root-like projections which are developed from the outer tunic, something like the little rootlets by which ivy clings to a wall.

OUR second example is the SYNTETHYS, another British species, which is seen affixed to the stone on the right hand of the illustration, spreading flower-like from a small base. When full grown, a group of these creatures forms a largish mass, nearly six inches in diameter, and as many in height, each member of the group being about two inches long. They are rather transparent and of a greenish colour, and, when touched, they will contract themselves violently, and vanish into the common mass on which they are seated. These animals are propagated both by eggs and buds, the buds being produced on offshoots of the creeping tube. Sometimes the young one severs its connexion with the parent, and fixes upon some fresh locality, there to form the basis of a new colony, but it frequently remains on the same spot, and only serves to increase the general mass.

THE Botryllidæ, or Compound Ascidians, are represented in our list by the common BOTRYLLUS, the star-shaped animal which is shown under and a little to the right of the syntethys. The "tests," or equivalents of the shell of these animals, are fused into a common mass in which these individuals are imbedded. In the present genus the animals are arranged in a star-like form, each group consisting of a number of individuals, not less than six, and not more than twenty, in number. Many of these groups, or systems as they are technically called, are found upon the common test. The branchial orifices are simple, and the other orifice is common to all the members of the group, and forms, as it were, the centre of the radiating star. Six British species are known, which may be found on stones and seaweed at low-water mark.

A VERY beautiful and curious mollusc, called from its luminous appearance the PYROSOMA, i.e. Fire-body, is an example of the next family. This is one of the compound tunicates, and looks like a gelatinous cylinder, open at one end, and closed at the other, and having its body covered with numerous zoïds grouped in whorls. A large Italian-iron tube, studded with daisies, will give a good idea of its general shape.

The ejecting orifices of the aggregated animals all open into the hollow interior of the cylinder, and the consequence of this structure is, that by the constant flow of the rejected water, the whole mass is driven slowly and regularly through the waves. When seen at night they look just as if they were made of glowing white-hot iron, and they are at times so numerous as to choke up the nets of the fishermen, and diffuse so strong a light around them that even the fishes are rendered visible when they happen to swim within the sphere of its radiance. There is generally a greenish hue about the light.

Of the appearance presented by these animals when existing in great numbers, Mr. F. D. Bennett gives the following vivid and valuable account : " When assembled in the sea, and, as is usually the case, near the surface, these creatures present a gorgeous spectacle ; their vivid phosphoric light being sufficient to illuminate, not only the extent of ocean they occupy, but also the air above, rendering all surrounding objects visible during the darkest night, and permitting a book to be read on the deck, or near the stern cabin-windows of a ship. They are occasionally collected together in incredible numbers. On two occasions, at midnight (in lat. 20° and 40° N. Atlantic Ocean), the ship sailed over many miles of water which they had illuminated, and in which they were so densely crowded as to be taken in any amount by buckets or nets.

When captured, they exhibited no signs of animation, and emitted a peculiar half-fishy odour. When left in a vessel of sea-water, and allowed to be tranquil, their light was withheld, or only sparingly displayed; but when they were handled, or the water in which they were contained was agitated, their body instantly became one blaze of phosphoric light, which, upon close examination, could be observed to proceed from myriads of luminous dots, occupying the situations of the small brown specks, noticeable in the fleshy structure of the mollusc. Upon the irritating cause being removed, the phosphoric light gradually expired, and the Pyrosoma remained in darkness until again disturbed, when it once more illuminated objects with its vivid gleam; and this was repeated until after the death of the animal, when no luminous effect could be produced.

When living specimens were immersed in fresh water, they not only existed for some hours, but emitted a constant light. Even after they had been so much enfeebled as to cease to give light in sea-water, or after they had been seriously mutilated, their phosphorescence invariably reappeared when they were put into fresh water, which appears to act as a peculiar stimulus in reproducing the phosphoric light of these, as well as of most other marine luminous animals.

The Pyrosoma does not communicate its luminosity to water, nor to any object in contact with it (like many luminous Medusæ), its body being enveloped in a membrane that has no luminous secretion. But when the mollusc is cut open in water, some of the brown specks before mentioned will escape, and, diffusing themselves through the fluid, shine independent of the animal: in this respect, as well as in their structure and colour, bearing some resemblance to the luminous scale on the abdomen of the small fire-fly of Bengal."

OUR last example of these remarkable molluscs is the SALPA, which is mentioned on account of the curious phenomenon called "alternate generation," which is exhibited by this creature.

The Salpa takes two distinct forms, so entirely unlike each other that no one who was unacquainted with the circumstance would imagine that they could possibly belong to the same species. Sometimes the Salpæ are seen united in long chains, and swimming through the ocean with a beautifully graceful movement that greatly resembles the undulations of a swimming serpent. Sailors often call these chains of Salpæ by the name of Sea Snakes.

The remarkable characteristic in this creature is, however, that the solitary Salpa produces a chain of united individuals, and that each of the united Salpæ becomes the parent of a solitary one. So that, as Mr. Rymer Jones happily remarks, "a Salpa mother is not like its daughter or its own mother, but resembles its sister, its granddaughter, and its grandmother." When swimming at ease through the water, the Salpa, like many other inhabitants of the ocean, is hardly perceptible, on account of the extreme transparency of its structure, the only indication of its presence being a kind of iridescence as the light plays upon the delicate membranes. The motive power is obtained by regular contractions of the body, by which the refuse water is rejected with some force; and thus drives the creature along by direct action, just as a rocket is propelled through the air. It is a remarkable fact, that in the chain of united Salpæ, each individual expands and contracts in exact unison, so that the force is applied to the water in the strongest possible manner. Sometimes the chains become broken up, but the fragmentary portions do not seem to be at all inconvenienced by the change in their condition, swimming about as actively as before. The creature is very slightly luminous, giving forth its phosphorescent light when touched, and especially when pressed.

POLYZOA.

The very remarkable beings which now come before our notice are appropriately termed Polyzoa, from two Greek words, signifying "many animals," because a large number of individuals are massed together in groups of various forms and textures. The continental naturalists mostly designate them by the term of Bryozoa, or "moss animals," on account of their frequent resemblance to the various mosses; but as this term has been employed in far too wide a sense, grouping under one common designation a number of beings belonging to different classes, the more recent observers have decided on the more appropriate title of Polyzoa.

For very many years—indeed, from the earliest days of natural history until comparatively modern times—the Polyzoa were ranked among the vegetables; and a learned Italian observer who ventured to express his opinion that they partially, at least, partook

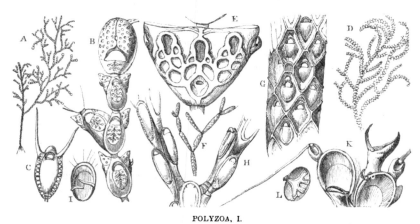

POLYZOA, I.

A. *Catenicella lorica*. B. *Catenicella hastáta*. × C. *Catenicella cornúta*. × D. *Calpidium ornatum*.
E. *Calpidium ornátum*. × F. *Salicornária farciminoides*. G. *Salicornaria farciminoides*. × H *Cellulária Peachii*. ×
I. *Menipea Fuguensis*. × (Mouth of a cell.) K. *Scrupocelláría ferox*. × L. *Canda reptans*. × (Mouth of cell.)
The sign × signifies that the object is magnified.

of the nature of animals, was persecuted by the professors of the day with the usual acrimony excited by a discoverer who is in advance of his time. Even the acute and experienced Linnæus could not receive the new doctrine, which was for a while "exploded" by the researches of another naturalist, who announced that he had seen corals in flower, thus setting the question at rest in the minds of those who desired to be so convinced.

Truth, however, stood its ground, and though for a time suppressed by those who had a personal interest in maintaining the theories which they had so long promulgated, in the due course of events became triumphant.

The true animal nature of these and many other beings, which had been formerly classed among the vegetables, was at length fairly proved by the researches of two eminent men, Trembley and Ellis, the latter of whom may lay claim to the honour of having produced the best and most comprehensive work of his time; a work, indeed, which is valuable even at the present day, owing to the invariable clearness and occasional brilliancy of the descriptions, and the number and accuracy of the engravings.

Ellis called all these creatures by the name of Corallines, a title now given to one of the true vegetables, but discovered many anatomical and physiological details, and set their animal nature beyond a doubt. All his researches were conducted with the aid of instruments which in our day would be thought almost useless, the microscope employed being only a simple lens mounted on a stand, and devoid of the complicated apparatus for magnifying and illuminating that now afford such aid to the observer.

After the animal nature of the Polyzoa had been fairly established, they were confounded with many other marine and aquatic inhabitants, such as the corals and the various zoophytes, in consequence of the superficial resemblance between their external forms. Lately, however, their true place in the animal kingdom has been discovered, and their affinity with the lower molluscs clearly proved, the tunicates forming the connecting link between the molluscs proper and the molluscoids, as these animals are sometimes called.

Having glanced at the general history of these curious and really beautiful animals, we will proceed to examine the form and characteristics of the individual species.

Should the reader obtain from the sea or fresh water a being which is evidently either a zoophyte or one of the Polyzoa, he may set his doubts at rest by examining the tentacles, and if he finds that they are furnished with cilia, or minute filaments, he may assure himself that they belong to the group of animals on which we are now engaged.

The forms assumed by the general mass of the various species of Polyzoa are extremely different, some resembling twigs or mosses; others looking like lumps of spongy substance adhering to sticks, stones, or leaves, or even lying freely in the water; others being flat and ramified, like broad-leaved seaweeds; others spreading film-like over leaves, stones, shells, or similar objects; while a few are able to crawl at liberty, the entire organism being animated by some wonderful instinct, which urges all the myriad individuals of which it is composed to employ their force in the same direction.

The number of these creatures is so vast that it is impossible to give more than a brief description of them; but in the following pages it will be found that a careful selection has been made of the typical forms, and that sufficient details of their structure will be given to enable the reader to form a general idea of the subject, and in most cases to refer any specimens which he may find to their genera or families. Those who desire further information on the subject will find it in Busk's elaborate catalogue of the Marine Polyzoa, in the British Museum, and the large work by Allman on the Fresh-water Polyzoa, published by the Ray Society.

Putting aside the classification of the polyzoa until the termination of the work, we will proceed at once to the description of the many species on our illustration, delineated by the accurate pencil of Mr. Tuffen West.

The first family of the polyzoa (according to the arrangement employed in the British Museum) is known by the manner in which the cells are arranged around an imaginary axis, and connected with each other by flexible stalks. The general shape of the whole group or "polyzoary," as it is termed by some authors, is very shrub-like, standing boldly erect, and giving out branches by two and two, after the fashion called by botanists "dichotomous."

Fig. A, Plate I., is an example of one of these beings, the LITTLE CHAIN, or BREAST-PLATE, one of those beings that are so plentiful in the sea, and are properly called zoophytes. This figure is of the natural size; but in order to show the peculiarities of structure, two examples of species belonging to the same genus are given as they appear when considerably magnified. Fig. B is the *Catenicella hastata*, wherein is seen the shape of the cells, the form of their mouths, the method in which they give out their branches, and the peculiar organs called technically "avicularia" and "vibracula;" the former being processes that in many species bear an almost absurdly close resemblance to the heads of birds; and the latter, curious hair-like projections, which move regularly backward and forward as if impelled by machinery. These remarkable organs will be presently described more fully.

The members of the present genus are found most commonly in the Australian seas, seldom in the southern hemisphere, while in the northern hemisphere they are almost entirely unknown. Many specimens have been taken from Bass's Straits, at a depth of forty-five fathoms. As a general rule, however, the polyzoa prefer the shallower waters, and are most commonly found a little below low-water mark.

Fig. C. is another species belonging to the same genus, and is remarkable for the long pointed spines that project from the margin, like a pair of cow's horns. In allusion to this peculiarity it is called *Catenicella cornúta*.

At fig. D is shown another curious polyzoon, termed *Calpidium ornatum*, also found in Bass's Straits, at the same depth as the preceding species. A magnified figure is given at E, in order to show the singular method of its construction.

As the reader may observe, each cell is extremely wide in proportion to its depth, and instead of possessing but one mouth, is pierced with three apertures shaped something like keyholes. It is conjectured that each cell is inhabited by three separate individuals, a supposition which is strengthened by the great comparative dimensions of the cell and the thickness of its walls. Still, no sign of internal partitions have been discovered, although some remains, apparently of the inhabitants, have been seen at the bottom of the cell. In some cases there are only two apertures to each shell.

An example of the typical genus of this family is given at fig. F, where the *Salicornaria farciminoides* is represented of the natural size. The strange specific name of this creature is given to it on account of its external resemblance to the Farciminaria, another genus of polyzoa which will be presently described. In this genus the cells assume a kind of honeycombed aspect, being almost hexagonal in their shape and pressed closely together.

In this place it may be as well to mention that in all the species belonging to the first subdivision of the polyzoa the mouth is not quite at the extremity of the cell, is of a somewhat crescentic form, and furnished with a movable lip or door, which closes the aperture when the animal retreats. In many cases this lip is membranous. All the marine polyzoa are termed Infundibulata, or Funnel-shaped animals, on account of their form ; their fresh-water relations being called Hippocrepia or Horseshoe animals, because the tentacles are arranged in a shape resembling that of a horseshoe. It will be, perhaps, hardly necessary to apologize for the introduction of so many technical terms, the fact being that the minute dimensions of the objects have caused them to escape popular observation, and to depend for their nomenclature upon the learned and scientific. Still, the technical phraseology is never employed where its use can be avoided, and when circumstances render its introduction inevitable, its meaning and the reasons for its employment are always given.

We now arrive at another family, the Cellularidæ, where the general shape resembles that of the preceding family, but the cells, instead of being arranged round an imaginary axis, and so forming cylindrical branches, are arranged on the same plane. Fig. H is a magnified example of this family, the *Cellularii Peachii*, so called in honour of the eminent naturalist, Mr. Peach.

At fig. I is shown the mouth of a single cell, belonging to the genus Menipea, found in Tierra del Fuego, and termed from its habitat, *Menipea Fuguensis*. The object of giving this example is to show the curious " operculum " which closes or rather guards the mouth of the cell, and in this genus is in the form of a simple spike. This species is found at low water.

At fig. K we have good examples of the avicularia and vibracula, already mentioned, and this species has been chosen on account of the enormous dimensions of the former organ, which is nearly as large as the cell to which it is attached. Many instances of the avicularia will be seen in the succeeding plates.

The avicularium is an object which is set somewhere about the middle of a cell, and always upon its outside, and assumes various shapes in the different species of polyzoa. What may be the precise nature of the avicularia is at present rather a mystery, and no one can definitely pronounce them to be actual portions of the cell, or merely parasites that remain affixed to the same spot. In all cases there is a decided resemblance to the head of a bird, though in some species the similitude is closer than in others. Only one avicularium is to be found on a single cell, though many cells do not possess these strange appendages.

By close examination, it will be seen that the avicularium can be roughly distinguished into three portions; namely, a base by means of which it is attached to the cell, a rather large head, and a movable spine like the lower mandible of a bird's beak. In those examples where the avicularium is seated directly upon the cell, the only movement is that of the lower mandible, which opens and shuts with a continual motion, as if it were a veritable head of a hungry bird snapping at its food. In those cases, however, where the base is lengthened into a neck, the entire head is endowed with motion, nodding up and down in the most lively manner, very like those wooden birds sold in the toyshops, whose head and tail are alternately raised and depressed by means of strings and a weight. But, whether the head moves, or is still, the jaws continually open and shut, and will often inclose between their parts any small worm that may happen to come across their path, and have even been known to seize each other in their grasp.

In Plate II. of the polyzoa, fig. N, may be seen an instance where the beak has seized a victim. When the mandibles have thus closed upon any object, they retain their grasp with astonishing tenacity, and when, as sometimes happens, two avicularia have seized the same worm, the unfortunate victim is rendered entirely helpless by the grasp of its foes.

The purpose of these objects seems to be rather dubious, but two conjectures have been offered, which at all events are worthy of notice.

According to the opinions of some observers, the avicularia answer the purpose of police, and force intruders to leave the spot where their presence might do harm to the creature on which they are placed. This duty seems, however, to be performed by the vibracula, and we must search for another theory for the true object of the avicularia. Mr. Gosse has put forward a conjecture which is not only highly ingenious, but bears with it the elements of probability.

"More than one observer," he remarks, "has noticed the seizure of small roving animals by these pincer-like beaks, and hence the conclusion is pretty general, that they are in some way connected with the procuring of food. But it seems to have been forgotten, not only that these organs have no power of passing the prey thus seized to the mouth, but also that this latter is situated at the bottom of a funnel of ciliated tentacles, and is calculated to receive only such minute prey as is drawn within the ciliary vortex. I have ventured to suggest a new explanation.

The seizure of a passing animal, and the holding it in a tenacious grasp until it dies, may be a means of attracting the proper prey to the vicinity of the mouth. The presence of decomposing animal substance in water invariably attracts crowds of infusory animalcules, which then breed with amazing rapidity, so as to form a cloud of living atoms around the decaying body, quite invisible in the aggregate to the unassisted eye; and these remain in the vicinity, playing round and round until the organic matter is quite consumed. Now, a tiny annelid or other animal caught by the bird's head of a polyzoon and tightly held, would presently die; and though in its own substance it would not yield any nutriment to the capturer, yet by becoming the centre of a crowd of busy infusoria, multitudes of which would constantly be drawn into the tentaculean vortex and swallowed, it would be ancillary to its support, and the organ in question would thus play no unimportant part in the economy of the animal."

In the engraving a solitary avicularium may be seen perched upon the cell, having its beak widely expanded. Avicularia of various shapes and sizes may be seen upon the sides of several other magnified specimens on this and succeeding plates.

We now proceed to the vibraculum, one of which organs is seen at fig. K, attached to the left-hand cell, and shown as it appears when bent downwards upon its base.

The vibraculum is hollow, the interior being filled, during the life of the animal, by a fibrous contractile substance, which enables the organ to perform its curious movements. These movements are very irregular as regards time, but very regular in their directions, each vibraculum sweeping slowly over the whole surface within its reach, first moving in one direction and then in the other, and it is sufficiently notable that these movements will continue for several days after the death of the polype to which it is attached.

The last figure on the engraving, marked L, represents the mouth of a cell belonging to another polyzoon, and is given to show the curious operculum, with its branched form, like the horn of a fallow deer, and may be contrasted with the simple spiny operculum shown at fig. I.

Passing to the second plate, we shall find several curious, and one or two bizarre forms of Marine Polyzoa. Fig. A represents the Bull's-horn Coralline of Ellis—the Ladies' Slipper, as it is more elegantly and equally appropriately named at the present day. The cells of this species bear a considerable resemblance to a series of delicate, slender-toed slippers, adherent to each other, while from the opening protrudes the beautiful bell-shaped circle of tentacles. Sometimes a rudimentary cell may be found, but always below the aperture.

At fig. B is shown a creature called the Snake-head Coralline, from the extraordinary similitude between it and the reptile from which it takes its name. It is a common species.

At fig. C is given a single cell of *Beánia mirábilis*, for the purpose of showing the series of thorns or spines which surround the mouth. This species is found mostly on

POLYZOA, II.

A. *Scrupária cheláta.* × (Ladies' Slipper.)   B. *Ætéa Anguina.* × (Snake-head Coralline.)   C. *Beánia mirábilis.* ×
D. *Farciminaria aculeáta.* ×   E. *Gemellária loricáta.* ×   F. *Gemellária loricáta.*   G. *Dimetópia spicáta.* ×
H. *Notámia bursária.* × (Shepherd's-purse Coralline.)   I. *Notámia avicularium.* ×   K. *Caberéa Patagónica.* ×
L. *Bicellária ciliáta.* ×   M. *Bicellária tuba.* ×   N. *Búgula aviculária.* ×   O. *Búgula aviculária.*
P. *Búgula* (an avicularium.) ×
The sign × signifies that the object is magnified.

shells. Each cell is united to its predecessor and successor by a slender tube, a part of which may be seen in the figure.

Fig. D is the curious Farciminaria, remarkable for the array of short and stout spines with which its surface is thickly studded. It is a New Zealand species, and appears to be the sole representative of its family. It grows in slender branches, which are dichotomous.

In the family to which the *Gemellária loricáta* belongs, the cells are arranged in pairs and opposite each other, the orifices of the pairs looking in the same direction. This species, represented on fig. E as it appears when magnified, is the Coat-of-Mail Coralline of Ellis, deriving its name from the shape of the cells, which bear no slight resemblance to steel corslets. Fig. F represents the same species of the natural size.

The succeeding family, of which the *Dimetópia spicáta*, fig. G, is an example, may be known by the arrangement of the cells, which are in pairs, but with their mouths placed at right angles to each other. When growing, it is a very pretty species, being white, nearly transparent, and attaining a height of about three inches. It grows in thick tufts, and is found in Bass's Straits.

Fig. H represents the Shepherd's-purse Coralline of Ellis, *Notámia bursária*, a common British species, and at I is shown its peculiar avicularium, with its tobacco-pipe-like head. Several smaller birds' heads are to be seen on the main figure.

Fig. K is given in order to show a most curious vibraculum, which is toothed like a saw, and belongs to a creature called *Caberéa Patagónica*, living in the country from which it takes its specific name.

A very remarkable species, *Bicellária ciliáta*, is given at fig. L, in order to exhibit its general aspect, and the manner in which the cells are surrounded by long processes. At fig. M is shown a greatly magnified figure of the avicularium belonging to another species of the same genus, conspicuous for the enormously long stalk of the head, and the three finger-like appendages at the base.

The three remaining figures represent a tolerably common British species, popularly called the Bird's-head Coralline (*Búgula avicularia*), on account of the number, shape, and activity of the avicularia. Fig. O shows the natural size of the polyzoa, which may be found lining the sides of rock-pools, or affixed to shells, and even to living crustaceans, the spider-crab being often enveloped in its soft plumy branches to such an extent that it marches beneath their shade, like Macduff's army under its leafy disguise. At N is seen a magnified portion of the same species, with the avicularia in various attitudes, one being shown in the act of intercepting a minute worm that has happened to cross its sphere of action; and at P is seen one of the birds' heads greatly magnified, so as to show its wonderful resemblance to the object whence it derives its name. I know scarcely a more wonderful sight than is presented by a living specimen of the Bugula, with its wonderful appendages in full action. As if moved by machinery, they nod up and down like automata, sometimes throwing themselves back like the head of a fantail pigeon; the mouth opens slowly, with a wearied kind of air, that almost forces the observer to yawn in sympathy with the deliberate movement, while ever and anon the jaw suddenly closes with a snap so sharp that the ear instinctively watches for the sound.

EVERY one who has walked along the seashore must have observed the pretty leaf-like Sea Mats strewn on the beach, and admired the wonderful regularity of their structure, perceptible to the naked eye; but when magnified even by a pocket lens, their beauty increases in proportion to the power employed, and the marvellous arrangement of the cells, and the orderly system in which they are placed, are almost beyond belief. Beautiful, however, as they are in this state, they are but the dead and lifeless habitations of the creatures who built the wondrous cells, and the only method of showing the Sea Mat in its full glory, is to take a living specimen from the stone or shell to which it is affixed, and watch it under the microscope while the creatures are still in full activity. At Plate III. fig. A, is shown a portion of the common Sea Mat, sometimes called the Hornwrack, of its natural size, and fig. B represents a few cells of the same species rather highly magnified.

On the upper right-hand cell may be seen the funnel-shaped group of tentacles belonging to one of the animals, and in the centre is a curiously shaped cell, which is analogous to the birds' heads which we have so lately examined, the place of the lower jaw being supplied by a kind of lid.

The peculiar manner in which the polypes of the Sea Mat protrude themselves is quaintly and accurately described by Mr. Gosse in his "Evenings at the Microscope." After pointing out the cradle-like shape of the cells, he proceeds as follows: "Suppose that a coverlid of transparent skin were stretched over each cradle from a little within the margin all round, leaving a transverse opening just in the right place, viz. over the pillow, and you would have exactly what exists here. There is a crescent-form slit in the membrane of the upper part of the cell, from which the semicircular edge and lip can recede if pushed from within.

Suppose, yet again, that in every cradle there lies a baby with its little knees bent up to its chin, in that zigzag posture that children, little and big, often like to be in. But stay, here is a child moving! Softly and slowly pushes open the semicircular slit in the coverlid, and we see him gradually protruding his head and shoulders in an erect position, strengthening his knees at the same time. He is raised half out of bed, when lo! his head falls open, and becomes a bell of tentacles. The baby is the tenant polype."

Fig. C is the Toothed Sea Mat, given in order to show the curious toothlike appendages from which it derives its name.

POLYZOA, III.

A. *Flustra foliacea.* (Sea Mat.)	B. *Flustra foliacea.* ×	C. *Flustra denticulata.* × (Toothed Sea Mat.)
D. *Carbasea episcopalis.* × (Bishop's Mitre.)	E. *Carbasea episcopalis.*	F. *Diachoris crotali.* ×
G. *Diachoris magellanicus.* ×	H. *Membranipora pilosa.*	I. *Membranipora pilosa.* ×	K. *Leprália Landsborovii.*
L. *Leprália spinifera.*	M. *Leprália spinifera.* (Ovicell. ×)	N. *Leprália trispinosa.*

The sign × signifies that the object is magnified.

At E is seen a curious polyzoon, found in Bass's Straits at a depth of forty-five fathoms, and bearing the name of *Carbasea episcopalis.* This species is found in two forms, either parasitic on sertularia and various polyzoa, and then of small size, or leading an independent existence, and reaching considerable dimensions. It is chiefly remarkable for the singular form of the ovicells, which bear a wonderfully close resemblance to bishops' mitres, and have earned for the species the title of *episcopalis.* A group of these ovicells may be seen at fig. D.

Two specimens of another genus may be seen at figs. F and G, the former representing a single cell of *Diachoris crotali,* for the purpose of showing the remarkable appendages which guard the mouth; and the latter, *Diachoris Magellanicus,* exhibiting the method in which each cell, except at the margins of the fronds, is connected with six others, something like the stellate cells in pith. In fact, the Diachoris is a flustrum dissected, the cells being drawn away from each other and connected by stalks. In the illustration, the connecting stalks are well shown, by which the central cell is united to the six that surround it, and several of the avicularia are also given.

Any one who picks up a piece of a dark seaweed, will find that many parts of its structure are covered with a peculiar growth, that looks as if a portion of Sea Mat had been cemented upon it. This substance is indeed closely allied to the Sea Mat, and is chiefly to be distinguished by the membranous nature of the polyzoary, which will not permit it to stand boldly erect after the manner of the true Sea Mat. At fig. H is shown a small mass of the *Membranipora pilosa*, as this species is called, surrounding the stem of a seaweed, and at I, a small portion of the same creature is given in order to show its aspect when magnified.

As may be seen from the illustration, the feathery plume of tentacles is extremely graceful, and, when the creature is living, has a remarkably elegant effect. In a specimen now before me, viewed by a power of only thirty diameters attached to the binocular microscope, the polypes of the Membranipora are beautifully exhibited, some shut up closely in their homes, some just putting forth their heads from the cells, others half protruded, and a few with the plumy tufts displayed in all their beauty. It is as well to view this and other polyzoa with different kinds of illumination, both as opaque and transparent bodies; artificial light is, however, to be avoided.

Fig. K represents a magnified portion of an allied species, *Lepralia Landsborovii*. In the Lepralia, as well as the Membranipora, the process of development is very interesting, especially as it can be readily watched under the microscope.

Towards the end of May, specimens of Lepralia, Flustra, and Membranipora should be procured and placed in shallow glass vessels containing sea-water. After a little time, minute beings, much resembling the ordinary infusoria, are seen swimming about. Presently, the "gemmules," as these creatures are technically named, become stationary, affix themselves to some definite spot, and develop a feather-covered polype, being now similar in shape to a single cell of the species from which it was produced. Buds, or projections, are soon formed at the sides, which are rapidly developed into new cells, and in their turn are the means of putting forth new cells. Thus it will be seen that each polyzoary spreads from a centre; and that, although a free gemmule is capable of producing stationary cells, the greater number of cells have never passed through the state of their original progenitor. When a polyzoary has attained a considerable size, it is not unfrequent to find the margins of the group filled with vigorous and lively polypes, being those last produced, while the centre is composed of empty cells, the original inhabitants having died out from old age.

At figs. L and M we observe two portions of *Lepralia spinifera*, given for the purpose of showing the short sharp thorns with which the edges of the cell are guarded, and the curious ovicell, notable for its beautiful sculpturing of ridges radiating around a centre. Fig. N is the *Lepralia trispinosa*, distinguished by the three long spines from which it derives its name.

On Plate IV. in figs. A to J inclusive, are represented some of the forms assumed by this large and interesting genus, which comprises between forty and fifty known species. Among these especial attention must be drawn to *Lepralia monoceros*, or the Unicorn Lepralia (fig. C), so called on account of the single horn, or club, with which it is furnished; the *Lepralia alata*, or Winged Lepralia (fig. B), remarkable for the classically elegant sculpturing and the projecting "wings;" and the *Lepralia variolosa* (fig. G), so called on account of the peculiar mottlings which are thought to bear a resemblance to the face of a person seamed with the small pox.

At fig. K is shown the spoor-like avicularium of the *Cellepora fusca*, a polyzoon notable for its urn-shaped and chalky, stiff cells, arranged either irregularly or in the form of a quincunx, *i.e.* like the cinque spots on a die, ∴ It is a native of Bass's Straits. A tolerably common British species, *Eschara foliacea*, so called from its superficial resemblance to the scar left by a deep wound, is given at fig. L, and a very curious polyzoon of the same genus, *Eschara flabellaris*, may be seen at fig. M, and is remarkable for the hood or helmet-like ovicells. About eleven species of this genus are known.

At fig. N is shown one of the most curious of the polyzoa, named, from the external resemblance which its apertures present to the meshes of a net, the *Retepora*, or Net-pored animal. The polyzoary of this species is hard, chalky, is only pierced on one surface, and has so much the appearance of the true stony corals, that it might easily be mistaken for one of these objects. It is a British species, and the specimen is represented of the natural size.

The next few figures represent examples of an interesting family of polyzoa called Selenariadæ, because they are round as the full moon, or Norval's famous shield. In the circular form of the margin they all agree, but differ considerably in their curvatures, some being with one side plane and the other convex, while others are convex-concave, like a watch-glass, or, to speak more accurately, like the lens technically called a meniscus. They are all remarkable for their very large vibracula.

At fig. O is one of these creatures, *Cupularia Lowei*, so called on account of its resemblance to a dome. It is represented of the natural size, but at fig. P is a magnified portion of the same species, exhibiting the form of the cells and the enormous vibracula with which their mouths are swept. This species is found at Madeira. Fig. Q is the *Lunulites capulus* of the natural size, and fig. R represents the same species highly magnified, for the purpose of showing the regular arrangement of the cells and the curious vibracula, which in this genus are usually bifid or even trifid at their extremities. Another species of the same genus, *Lunulites cancellatus*, is given at fig. S.

At fig. T is shown a very shield-like genus of this family, its name being *Selenaria maculata*, and a section of the same specimen is given below, at fig. U. The reader will not fail to remark its exact resemblance to the target-shield used by many nations, ancient and modern. Fig. V is a portion of the same species, as it appears when highly magnified, and is given for the purpose of showing the curious vibracula. It may be here mentioned that in this last-mentioned family the vibracula are thought by some naturalists to act as locomotive organs.

Our space is so rapidly drawing to a close, that it will be hardly possible to give much more than a rapid mention of the names of the next few engravings, together with a brief account of one or two of the more interesting examples.

POLYZOA, V.

A. *Crisia eburna.*   B. *Crisia eburna.* ×
C. *Idmónea atlántica.*
D. *Idmónea atlántica.* ×   (Front view.)
E. *Idmónea atlántica.* ×   (Side view.)
F. *Pustulópora delicátula.*
G. *Pustulópora delicátula.* ×   H. *Hórnera.*
The sign × signifies that the object is magnified.

In the group of polyzoa given in the accompanying illustration, the reader will not fail to notice the great external similarity between their forms and those of the true stony corals so familiar to us. It will, moreover, be found, that in many details of their structure, there is a decided analogy between them and the true zoophytes which will be described in the latter part of the work.

Fig. A, Plate V. represents the *Crisia eburna* of the natural size, and just below, at fig. B, is given a portion of the same species as it appears when magnified. The arrangement of the cells is simple and elegant, and the various branches are connected with each other by means of certain horny joints. The use of these joints is, in all probability, to enable the polyzoary to resist the action of the waves, and so to avoid the fractures which would probably result if the joints were as stiff and inflexible as the cells. A similar provision will be seen in the *Gorgonia*, a zoophyte which will be described on a future page.

A remarkably constructed species, the *Idmonea atlantica*, is shown at fig. C, where it may be seen looking something like a many-legged spider, with its branches protruding from a rounded centre which represents the body of that animal. Two magnified representations of this curious species are to be seen at figs. D and E, the former giving the front view and the latter the lateral aspect, in order to show the peculiar construction of the polyzoary, and the firm manner in which the cells are set upon their common basis. For the figures in this and several of those of the succeeding plates, I am indebted to the kindness of Mr. Busk, who allowed them to be taken from his forthcoming part of the "Catalogue of Marine Polyzoa."

At fig. F is seen a small portion of a polyzoon termed *Pustulopora delicatula*, the generic name being on account of the minute dots with which the surface of the cells and polyzoary is studded, and the specific title in reference to the delicate structure and soft brown hue with which it is coloured. The same specimen is represented at fig. G, much magnified, and is remarkable for the profusion and great comparative length of the cells. The last figure in the engraving, marked H, is a stoutly-built polyzoon named *Hornera*, and is notable for its resemblance to several corals.

In the species given in Plate VI. the resemblance to the corals, the madrepores, and even the stony habitations of certain marine worms, is very close and striking.

Fig A represents a polyzoon fancifully entitled *Alecto*, in honour of one of the Furies of that name. It is seen of its natural size as it appears while spreading itself over the inner surface of a shell. At fig. B, a portion of the same species is shown as it appears when magnified, and is given to exhibit not only the method in which the cells are sunk into the polyzoary, but the mode in which the branches are developed from each other.

At fig. C is delineated a portion of the appropriately named *Tubulipora serpens*, a being which has the cells even more elongated than in the *Pustulopora* which has been lately described. The singular resemblance between the lengthened cells of this species and the hard shelly tubes of the well-known *Serpula*, so familiar on account of its scarlet and white plumes and marvellously engraved stopper, must be evident to every one who has seen the little creature, or even noticed its empty habitation.

Fig. D represents a curious species, called from its shape *Discopora patina*, the former word being of Greek origin and given in allusion to its disc-like form, and the latter being a Latin word signifying a flat dish like our present champagne glasses. The numerous pores or orifices through which the animal protrudes, are seen upon the surface. A magnified example of the same species is seen at fig. E, having been broken asunder in

POLYZOA, VI.

A. *Alecto dichótoma.* B. *Alecto dichótoma.* × C. *Tubulipora serpens.* D. *Discópora pátina.*
E. *Discópora pátina.* × F. *Discópora pátina.* × (Contorted.)
The sign × signifies that the object is magnified.

order to show the manner in which the cells are massed together, so as to produce a honeycomb-like aspect at their mouths. If the reader will compare this with the last-mentioned species, the evident connexion between the two will be readily perceptible.

A very remarkable modification of the same species may be seen, rather magnified, at fig. F. The original gemmule from which the whole mass sprang had made a mistake in its settlement, having fixed itself upon a slender stem where it could find no space for its expansion into the normal circular form. Being fixed, it could not move, but philosophically made the best of the situation, and finding itself unable to spread into a single disc-like body, and equally unable to extend beyond the supporting substance to which it was affixed, adopted a compromise, and coiled itself into the singular form here represented.

OUR next group of polyzoa exhibits some very remarkable forms. At fig. A is seen a specimen of the *Serialaria lendigeri*, a species which without the aid of a lens presents no particular points of interest, but, when magnified, is seen to be a really curious being. Two branches of this creature are represented at fig. B, as seen when moderately magnified, and the further details of its structure are given at fig. C, where the polypes are shown protruding from their bases, and the peculiar dottings of the cells are seen. It will be noticed that the cells are gathered into groups, connected with each other by the stalk-like processes of the polyzoary.

Fig. D represents a sprig of the *Bowerbankia imbricata* of the natural size, and the same species is shown much magnified at E.   In this species the cells are also placed upon the footstalk formed by the polyzoary, but they are not grouped together as in the last-mentioned species.   The polypes are long and slender, and the walls of the cells are delicately transparent, thus allowing the observer to examine the structure of the polype through the walls.

A part of the digestive organs of the *Bowerbankia* is deserving of a passing notice. Like all the other species, it possesses a feathery crown of tentacles sprouting around the mouth, and directing the minute objects which serve it for food from the mouth into the œsophagus, popularly called the gullet.   In this genus, however, a further provision is made, for immediately below the œsophagus comes a kind of contractile gizzard, lined with a marvellous pavement of teeth arranged in a tesselated formation, and capable of

POLYZOA, VII.

A. *Serialária lendigeri.*   B. *Serialária lendigeri.* ×   C. *Serialária lendigeri.* × ×
D. *Bowerbánkia imbricáta.*   E. *Bowerbankia imbricáta.* ×   F. *Buskia nitens.*

The sign × signifies that the object is magnified.

bruising and crushing the food before it is passed into the stomach and thence to the intestine.   One species of this genus, *Bowerbankia densa*, is common on our coasts, being found parasitic on *Flustra foliacea* in patches of about an inch or so in diameter, and may be readily obtained by those who are desirous of studying its habits and structure.

At fig. F is given a small portion of a polyzoon named *Buskia nitens*, the former title being in honour of the eminent naturalist to whom we are indebted for most of our accurate knowledge of these creatures, and the latter alluding to the shining appearance of the species.

WITH the accompanying plate our examples of the Marine Polyzoa are terminated.

Fig. A is a small specimen of the natural size of the *Alcyonidium gelatinosum*, popularly called the Sea Ragged Staff, Mermaid's Glove, or more commonly, Dead Man's Fingers, in allusion to the cold dampness of its surface.   Great numbers of this polyzoon can be found on the sea-shore, especially after a storm, when it may be seen lying among the masses of seaweed and other débris that are flung on the beach by the angry waves.

In its natural state it is affixed by its base to stones, shells, and other supports, and is always extremely irregular and variable in its form, no two specimens being alike. When picked up, its aspect is anything but attractive, but when placed in sea-water and suffered to remain at rest for a while, it becomes a most beautiful object. From each of the tiny pits with which its surface is thickly studded, projects a polype, with a beautiful crown of waving tentacles, and so numerous are these polypes, that they densely cover the surface and render microscopic observation rather difficult.

As in other species, fresh colonies of the Alcyonidium are formed by gemmules, which are given forth from the general mass, swim about freely for a time, by means of the cilia with which their surface is thickly studded, and when they have attained a proper age, settle down and at once begin to develop fresh cells on all sides. The little vesicles wherein the gemmules are originally formed, may be seen in the spring scattered through the transparent substance of the polyzoary, and looking like little white points. Each vesicle contains about five or six gemmules, and as it can be easily isolated, its rupture and the consequent escape of the gemmules can be easily seen in a moderately powerful microscope. A head of a single polype with its crown of tentacles is shown at fig. B.

Referring to fig. D on the same plate, the reader will see a group of little objects on footstalks, looking wonderfully like the common moss that grows so plentifully on walls. This is the *Pedicellina echinata*, of which a magnified group is seen at C, and the extremity of a tentacle still more highly magnified, at E, in order to show the cilia by which the necessary currents are formed in the water for the purpose of obtaining food.

POLYZOA, VIII.

A. *Alcyonidium gelatinósum.*
B. *Alcyonidium gelatinósum.* × (Tentacles of a single polype.) C. *Pedicéllina echinata.* ×
D. *Pedicéllina echinata.* (Natural size.)
E. *Pedicéllina echinata.* × × (Part of tentacle.)
The sign × signifies that the object is magnified.

PASSING to Plate IX. fig. A, we come upon one of the most remarkable polyzoa that at present are known to exist. As may be seen by reference to the engraving, which represents a specimen slightly magnified, the entire polyzoary is not only free and unattached to any object, but even possesses the power of locomotion. In the present instance it is shown in an attitude which it frequently assumes, namely, crawling over the stem of some aquatic plant. In order to qualify it for this process, the lower surface of the polyzoary is modified into a flattened disc, which thus becomes analogous to the foot of the gasteropodous molluscs already described. The substance of the disc is contractile.

To an ordinary eye, that any creature should crawl, would not appear a very surprising fact, but to the mind of a naturalist, the whole phenomenon is full of wonder. It is easy enough for a single being to advance in a given direction, and even though it has a very army of legs, like a centipede or a julus, the limbs are all directed by the same mind. But in the present case, there is no common centre to which the wills of the myriad polypes that compose the group can be referred; and the locomotive capacities of the Cristatella remain one of the many unsolved mysteries with which nature abounds.

In all respects, this is a remarkable species. Instead of hiding in darkness and coveting the shade, as is the case with nearly all the polyzoa, the Cristatella exults in light, and loves to crawl in shallow waters where it is exposed to the full blaze of the meridian sun. The ordinary length of the Cristatella is from one to two inches, and its general aspect reminds the observer of a yellowish brown hairy caterpillar, softer than the well-known woolly bear, or larva of the tiger-moth, and indeed looking much as if it were made from the soft silken substance denominated " chenille."

It is one of the fresh-water species, and, as may be seen by reference to the illustration, the plume of tentacles is not funnel-shaped, but formed as if set in a horseshoe.

Immediately below the Cristatella at fig. B, may be seen a rather formidable object surrounded with grapnels, and termed a "statoblast." This again is a most marvellous production, and requires some explanation. The statoblasts are developed within the

POLYZOA, IX.

A. *Cristatella mucédo.* ×   B. *Cristatella mucédo.* (Statoblast. ×)   C. *Lóphopus crystallinus.* ×
D. & E. *Lóphopus crystallinus.* (Statoblast. ×)   F. *Lóphopus crystallinus.* (On rootlets of Duckweed.

The sign × signifies that the object is magnified.

cavity of the parent, where they may be seen of different sizes, and in most cases arranged like beads on a string. They consist of two nearly hemispherical or oval discs, which are united at their edges, and the line of junction strengthened by a more or less deep ring, so that the general aspect of the statoblast is not unlike that of the planet Saturn, as may be seen by reference to figs. D and E, and which give two views of the statoblast of the animal beneath which they are placed.

In the Cristatella and another genus, the Pectinatella, the statoblast is armed with a double series of hooks, starting from the edges of the discs, those of the opposite discs alternating with each other and extending well beyond the ring. The spines are gradually developed, and force their way through an enveloping substance which surrounds the statoblast. After a while it often happens that the soft, gelatinous envelope is washed away and the statoblast assumes the appearance exhibited in the illustration.

These wonderful objects contain the future young; and the process of development is briefly as follows: Within the walls of the parent they attain their full size, and when the parent dies at the end of the season, they are liberated and pass from its body. They then attach themselves to subaquatic substances, such as vegetables, sticks, stones, &c. and at the beginning of the next season the two discs separate, and out comes the young Cristatella, ready to take upon itself the tasks for which it was created. It often happens that the two discs of the statoblast cling to the young for some time after it has given up its contents, and the little creature carries about the separate halves in a manner that reminds the observer of a bean newly sprouted from the ground and bearing the two halves of the seed which was planted. The Cristatella also produces buds, and in fact, the statoblast is a kind of bud of rather peculiar construction. The disc of the statoblast is brown.

At fig. F is an example of an interesting polyzoon found in our ponds and streamlets, adhering to the rootlets of duckweed, and deserving of peculiar interest as being the first species of polyzoa that was detected. The honour of its discovery rests with Trembley, who named it appropriately "Polype à panache," the plume-like group of tentacles being sufficiently large to be seen with the naked eye. In this creature, the place of the external wall or ectocyst, is taken by a soft gelatinous envelope, as may be seen by reference to fig. C, where the creature is represented on an enlarged scale.

Mr. Allman remarks that in the interior of the Lophopus are often to be seen a vast number of little glittering particles of a pear-like shape, which move about through a series of tubes connected with each other like the capillary vessels of the vertebrates. After much investigation of the subject, he came to the conclusion that they were merely parasitical.

WITH Plate X. is concluded the history of the polyzoa. At fig. A is shown a portion of a fresh-water polyzoon called scientifically the *Alcyonella fungosa*, and popularly the Fresh-water Sponge, because when dry it has a very sponge-like aspect. It is found in masses of considerable size, sometimes weighing as much as a pound, adherent to various substances which are constantly beneath the waters in which it lives. It frequently develops itself round the pendent twigs of the weeping-willow and other trees which dip the extremities of their branches into the water. I have seen the timbers of locks quite encrusted with the Alcyonella in many places, and I have in my possession a very fine specimen which I took out of the reservoir at Swindon.

When carefully removed and placed in fresh water, it gradually develops a kind of white downy appearance over its entire surface, which disappears with the rapidity of magic if a hand is moved quickly over the vessel in which it is lying. This downy appearance is caused by the tentacles which protrude themselves in vast numbers, and instantly retract when the creatures are alarmed.

At fig. B is given a magnified view of an Alcyonella which has been cut in order to show the shape and arrangement of the tubes, and it will be noticed that two of the tubes have run at right angles with the others. A few statoblasts are also seen occupying the lower part of the tubes. At fig. C is represented one of the polypes, showing its beautiful plumy tentacles arranged in horseshoe fashion.

A polyzoon already mentioned, and named *Plumatella repens*, is seen at fig. D, shown of the natural size. Mr. Allman has made some curious observations on this species and the manner in which the food is passed through the digestive organs. The process is too complicated to be described at length, but is very remarkable, the food being tossed backwards and forwards in a singular manner, then a portion pinched off by the end of the stomach, which assumes a kind of hour-glass shape, and in fine pushed here and there in so rapid a manner that it soon loses all the appearance which the various particles presented when taken into the stomach, and becomes of a brown colour, evidently caused by the secretions of the stomach.

Even in swallowing its food, the Plumatella displays considerable powers of discrimination, accepting some particles as they pass over the mouth, and rejecting others as unworthy of reception. This is a tolerably common British species, and is plentiful in Ireland. Its usual places of abode are under stones, submerged branches, floating leaves, and similar substances. Sometimes it attains a considerable size, spreading over a square foot of surface and having some branches more than three inches in length. It is in best condition towards the end of summer.

Fig. E represents a portion of the Plumatella considerably magnified, showing the polypes in various stages of protrusion, some being seen withdrawn wholly into their cells. Fig. F is a statoblast of the same species, and fig. G represents a side view of the same object. In fig. E the position of the statoblasts is seen as they lie in the tubes of the polyzoary, connected with the vital organs of the inclosed polypes. It will be noticed that this statoblast is without spines.

POLYZOA, X.

A. *Alcyonella fungosa.*   B. *Alcyonella fungósa.* × C.   *Alcyonella fungósa.* (Tentacles of a polype.)   D. *Plumatella repens*
E. *Plumatella repens.* ×   F. *Plumatella repens.* (Statoblast. ×)   G. *Plumatella repens.* (Statoblast, side view. ×)
H. *Fredericella Sultána.* ×   I. *Fredericella Sultána.* (Statoblast. ×)   K. *Paludicella.*   L. *Paludicella.* ×
M. *Paludicella.* (Tentacles. ×)   N. *Paludicella.* (Tentacles. ×)   O. *Paludicella.* (Tentacle. × ×)
The sign × signifies that the object is magnified.

At fig. H is shown the tentacular crown of a lovely polyzoa, called *Fredericella sultána ;* the former title being given to it in honour of M. Fr. Cuvier, the celebrated naturalist, and the latter being earned by its graceful and queenly beauty. This is a common British species, and is found plentifully in the Regent's Canal, where it may be detected growing in tufts on submerged stones, plants, sticks, and similar objects. It also inhabits rivulets, but seems to prefer tolerably still waters. As it is tolerably hardy, it is useful to microscopists, who can keep it alive in a common vial of water and place it under the microscope whenever they choose. At fig. I the statoblast of this species is shown.

Our last example of the Polyzoa is the *Paludicella Ehrenbergii,* a specimen of which is shown of the natural size at fig. K. This curious species is shown slightly magnified at fig. L, and exhibiting the manner in which the polypes protrude from their cells ; and at fig. M the head and tentacular plume of a single polype is delineated, so as to show its peculiar form has lost the horseshoe outline and assumed the funnel-like shape of the marine polyzoa. This arrangement of the tentacles seems to be unique among the fresh-water species, for although the tentacles of Fredericella appear at first to assume the circular form, a more careful examination will show that this is not really the case.

The mechanism by which the floating particles contained in the water are inevitably driven towards the mouth is seen in fig. O, which represents the tip of a single tentacle greatly magnified, and is given in order to show the arrangement of the cilia, all those of one side pointing upwards and those of the opposite side downwards. Fig. N represents the tentacular plume viewed from the front, and is given in order to show that the arrangement of these organs is really circular. Only the bases of the tentacles are delineated.

INSECTS.

THE INSECTS, to which we must devote a few pages, afford the first examples of the Articulata, i.e. the jointed animals without vertebræ. Their bodies are composed of a series of rings, and they are separated into at least two and mostly three portions; the head being distinct from the body. They pass through a series of changes before attaining the perfect form; and when they have reached adult age they always possess six jointed legs, neither more nor less, and two antennæ, popularly called horns or feelers.

In most instances their preliminary forms, technically called the larva and pupa, are extremely unlike the perfect Insect; but there are some in which, at all events externally, they retain the same shape throughout their entire life. The whole of the growth takes place in the preliminary stages, so that the perfect Insect never grows, and the popular idea that a little Insect is necessarily a young one is quite incorrect. It is true that smaller and larger specimens occur in every species, but this difference in size is due to some external influences that have acted on the individual; and we find large and small examples of an Insect, say a wasp, or a beetle, just as we find giants and dwarfs among mankind.

Insects breathe in a very curious manner. They have no lungs nor gills, but their whole body is permeated with a network of tubes through which the air is conveyed, and by means of which the blood is brought in contact with the vivifying influence of the atmosphere. These breathing tubes, technically called tracheæ, ramify to every portion of the creature and even penetrate to the extremities of the limbs, the antennæ, and even the wings, when those organs exist. Their external orifices are called spiracles, and are set along the sides.

They have very little internal skeleton, the hard materials which protect the soft vital organs being placed on the exterior, and forming a beautiful coat of mail, so constructed as to defend the tender portions within, and yet to permit perfectly free motion on the part of the owner. Certain projections of this substance are often found in the interior, especially in the thorax, a central portion of the creature, and are used for the attachment of muscles where considerable power is needed.

This external skeleton is quite unique in its chemical composition, being made almost entirely of a substance called chitine, to which are added several other materials, such as animal matter, albumen, and the oil which gives the bright colours so prevalent in most of the species.

There are many other interesting points in the structure of the Insects, such as the eyes, the wings, the tracheæ, &c., which will be described in the course of the following pages.

The systems on which the Insects have been arranged are as perplexing as numerous, differing according to the characteristics chosen by their authors. In this work the system employed is that of Mr. Westwood, which seems to combine many advantages to be found in the different arrangements of various authors, and is sufficiently intelligible to be understood without any painful exercise of the memory.

The first order, according to this author, is called the Coleoptera, a word of Greek origin, signifying sheathed-winged animals, and includes all those insects which are more popularly known under the title of Beetles. In these insects the front pair of wings are modified into stout horny or leathery cases, under which the second pair of wings are folded when not in use. The hinder pair of wings are transparent and membranous in their structure, and when not employed are arranged under the upper pair, technically called the elytra, by folds, in two directions, one being longitudinal and the other transverse. On examining these wings carefully it will be seen that their supporting nervures are furnished with hinge-like joints, which permit them to be folded in the right direction and no other. One of the best examples of a folded wing among the beetles is to be found in the common Cocktail Beetle (*Staphylinus*), where the large and beautiful wings are packed away under two little square elytra, just as a folded map is packed into its covers. In other instances where the elytra are very long, as in the common Musk Beetle (*Cerambyx*), the wings are first folded longitudinally and then a little piece doubled over at the tip, so as to fit within the cover.

The mouth is furnished with jaws, often of considerable power, which move horizontally.

The last character that must be considered in the beetles is the mode of the meta-morphosis or change of form which is undergone by them before they attain their perfect state. After being hatched from the egg they take the form that is popularly known under the title of grub, and is quite unlike the shape of the perfect insect. In this state they remain for various periods according to the species and the climate, and then pass into the second, or pupal state, when they look much like the perfect insect, but are unable to move about. This characteristic seems to separate them from the earwigs, cockroaches, and grasshoppers, which would otherwise have been included in the same order with the beetles, but are now placed in separate orders on account of the character of their preliminary stages, where the shape of larva, pupa, and perfect insect are very similar, and the pupa is active.

The reader must not imagine that in every insect which may be classed among the beetles he will find all the organs which have been mentioned, for in many cases they seem to be entirely wanting. The female glow-worm, for example, has no wings at all; and in many genera the second pair of wings are wanting and the elytra fused together at the edges. Still, in most instances the organs, or, at all events, the indications of their existence, can be found; while the character of the metamorphosis will always settle the true rank of an insect whose position seems doubtful.

Passing over, for the present, the details of classification, we come to the first family of insects, scientifically called the Cicindélidæ, and popularly known by the name of Tiger Beetles, or Sparklers, both names being very appropriate; the former on account of their exceeding voracity, their ferocious habits, and the wonderful activity of their movements; and the latter in allusion to the brilliancy of their colours as they flash along in the sunshine. These Beetles are represented by several British species, among which the common TIGER BEETLE (*Cicindéla campestris*) is the most common, and perhaps the most beautiful. Well does this little creature deserve its popular name, for what the dragon-fly is to the air, what the shark is to the sea, the Tiger Beetle is to the earth; running with such rapidity that the eye can hardly follow its course; armed with jaws like two reapers' sickles crossing each other at the points; furnished with eyes that project from the sides of the head and permit the creature to see in every direction without turning itself; and, lastly, gifted with agile wings that enable it to rise in the air as readily as a fly or a wasp. Moreover, it is covered with a suit of mail, gold embossed, gem studded, and burnished with more than steely brightness, light yet strong, and though freely yielding to every movement, yet so marvellously jointed as to leave no vulnerable points even when in full action, and, in fine, such a suit of armour as no monarch ever possessed and no artist ever conceived. True, to the naked or un-

Tricóndyla Wallacei.
Manticora latipennis.
Harpalus cáspicus.

Lébia crux-minor.

Cicindéla octonotata.
'Anthia sulcáta.

observant eye it seems to be but a dully green beetle with a blue abdomen, but if placed under the microscope, and a powerful light directed upon it, it blazes out with such gorgeous brilliancy that the eye can scarcely endure the glory of its raiment. The groundwork of its upper surface seems to be burnished and encrusted thickly with emeralds, sapphires, diamonds, and rubies of unspeakable brightness; while the whole of the under surface is as if made of watch-spring steel, so hard and so shining is its texture, and so rich a purple blue is its colour.

Even in its larval state the Tiger Beetle is a terror to other insects, snapping them up as they pass by its burrows and dragging them into the dark recesses of the earth to be devoured. Several American species inhabit trees, and are quite as destructive among the branches as their congeners upon the earth. The typical species which is represented in the illustration is the EIGHT-SPOT TIGER BEETLE of India.

The British Tiger Beetle is remarkable for exuding a powerful scent, much resembling the odour produced by a crushed verbena leaf.

An example of a very large genus belonging to this family is given in the engraving under the title of *Manticora latipennis*, the generic title being given to it because its great dimensions and ferocious habits are thought to bear some analogy with those of the fabled Manticora, a beast which the older naturalists were accustomed to describe with great zest, and in an engraving now before me had figured with the face of a human being, with hair carefully parted, six rows of shark's teeth, and a tail armed with a very arsenal of projectile spikes. Two species of this genus are known, and their very great comparative dimensions may be readily seen by comparing the Manticora with the Cicindela, both being drawn of the natural size, and the Cicindela being rather a large species. At the left-hand upper corner is shown a species belonging to the same family, which is remarkable for its very great comparative length and the globular centre of its thorax.

A VERY large and important family of Beetles, the Carabidæ, now comes before us, which is represented in England by very many species, the common Ground Beetles being familiar examples. The celebrated BOMBARDIER BEETLE (*Brachinus crépitans*) belongs to this family. This little beetle is plentifully found in many places, though it seems to be rather local, the banks of the Thames being one of its favourite resorts. When this beetle is handled, a sharpish explosion is heard, and on looking at the creature, a tiny wreath of bluish vapour is seen to issue from the body. This vapour has a very pungent odour, and when discharged against the skin, leaves a yellow mark like that produced by nitric acid. Originally, it is a liquid, secreted by certain glands, but as soon as it come in contact with the atmosphere it becomes suddenly volatilized, thereby producing the explosion and causing the smoke-like vapour to arise. The insect can fire off its miniature artillery seven or eight times in succession. Even after the death of the insect, the explosion can be produced by pressure.

One species of this genus, *Lebia crux-minor*, is given in the illustration. It is one of the British beetles, and notable for the cross-like mark from which it derives its name. The largest species belonging to this family are to be found in the exotic genus Anthia, an example of which is given in the illustration. The males are remarkable for the enormous size of the mandibles, and the thorax, or chest, seems to be divided into two parts. Most of the species are found in Southern Africa, *Anthia sulcata* is a native of Senegal.

It may be here remarked that the very largest of the Carabidæ is a Javanese beetle, named MORMOLYCE, which is mostly found under the branches of trees. Mr. Westwood mentions that one of these insects in his possession has attained the extraordinary length of three inches and a half. It is a very odd-looking insect, hardly recognisable as a beetle, and more resembling the mantis than the beetles, having a very long neck, a very flat body, elytra wide and flattened like those of the leaf-insect, and a thorax also flat and deeply toothed at the edges.

Passing by the group of Carabidæ known as the Scaritidæ, a sub-family of beetles which are mostly found under stones or in holes near the sea-shore, we come to the Harpalides, of which the *Harpalus caspicus* is our present example. They are mostly rather small, and seldom bright coloured, with the exception of a few species, such as the well known SUN BEETLES, which are so familiar to us as they run actively over gravel walks or roads as if enjoying the blazing sunbeams. It is worthy of notice that the dark coloured species seldom voluntarily show themselves by day. The species of this sub-family are extremely numerous, and it is no easy matter to distinguish with certainty between them. One very remarkable insect, belonging to the genus Blemus, inhabits the sea-shore, and resides under stones just above low-water mark, and in situations which are only left bare at very low tides. How the creature contrives to exist under such apparently adverse conditions is something of a mystery.

WE now come to the large group of WATER BEETLES, which are divided into several families. The fresh waters of our rivers, ponds, and lakes are very populous with the Water Beetles, which may be seen by thousands on a summer day, swimming, diving, rising to the surface, and evidently enjoying life to the utmost.

In order to enable them to perform the various movements which are necessary for their aquatic existence, their hind legs are developed into oars with flattened blades and stiff hairy fringes, and the mode of respiration is slightly altered in order to accommodate itself to the surrounding conditions. It has been already mentioned that in all insects the respiration is conducted through a series of apertures set along the sides, and technically called spiracles. In the Water Beetles, the spiracles are set rather high, so as to be covered by the hollowed elytra, and to be capable of breathing the air under those organs. When, therefore, the beetle dives, it is in noways distressed for want of

Vespa Mandarinia.   Athalia spinarum.   Nómada signáta.
                                                         Libéllula trimaculáta.

                                  Chrysantheda dentáta.

Xylotrúpes dichótomus   Sagra Boisduvdlii.

                              Bombus funerárius.

air, as it carries a tolerable supply beneath the elytra When, however, that supply is exhausted, the beetle rises to the surface, just pushes the ends of the elytra out of the water, takes in a fresh supply of air and again seeks its subaquatic haunts. Any one may see in almost any ditch the Great Water Beetle (*Dyticus marginalis*), ever and anon rising to the surface, poking its tail out of the water, and then diving to the bed of the stream.

Towards evening, this, in common with many other Water Beetles, is accustomed to leave the streams, to spread its wide wings and to soar into the air. In the early morning it again seeks its watery home, and is accustomed to save time and exertion by closing its wings and dropping like a stone as soon as it perceives the water below. Sometimes, however, it is direfully disappointed, mistaking the glitter of glass, such as the sashes of cucumber frames, the tops of green-houses, and similar objects, for the water into which it is accustomed to plunge, and in consequence receiving no small shock when it comes in contact with the hard glass. Even when kept in an aquarium, the Water Beetles are apt to act in this manner, and to leave their residence after dark. The larva of the Dyticus is a terribly ferocious creature, both in aspect and character. It inhabits the waters, and is a very hyæna in the terrible grasp and power of its jaws. The perfect insect is quite as voracious, and when a number are kept in a single vessel, they are sure to attack and kill each other. No one who cares for the animated inhabitants of his aquarium should permit a Dyticus to be placed among them, as a fox makes no more havoc in a chicken roost than a Dyticus in an aquarium.

WATER BEETLES.

Ilybius ater.　　　　　*Dyticus dimidiatus*

The courage and ferocity of the creature are so great that it will attack insects much larger than itself, and not even the fish can escape the jaws of this insatiable devourer. Before the character of the Dyticus was as generally known as is now the case, a gentleman had put a pair of these Beetles into a vessel of water, thinking that they would be sure to agree and permit him to watch their habits. Next morning, however, when he came to inspect his protégés, he was shocked and startled to find that the male was dead and already half eaten by his disconsolate widow.

The male of the Great Water Beetle, in common with other species, is specially notable for the singular development of the fore-legs, the tarsi of which are developed into a most extraordinary apparatus caused by the dilation of the three first joints, which are flattened so as to form a nearly circular disc, covered on its under surface with a multitude of wonderfully constructed suckers, one being very large, another about half its size, and the others very small, and set on pear-shaped footstalks. This structure may be seen on reference to the illustration. This forms a beautiful object for the microscope, but requires some care in its preparation, as the air contained in the suckers refuses to permit the Canada balsam to enter, and unless it be wholly expelled, a series of unsightly black blots is the result.

The larger specimen in the engraving is the *Dyticus dimidiatus*, and the smaller is the *Ilybius ater*, both British species.

To this group belong the WHIRLWIG BEETLES or GYRINIDÆ. so plentiful on the surface of our rivers and ponds, but always choosing a still spot, where they are overshadowed by the bank or an overhanging tree, for the locality wherein they perform their mazy dance. These insects are very hardy, and even on a winter's day the Whirlwigs may be seen taking advantage of the least gleam of sunshine and wheeling around their complicated maze as merrily as if the warm winds of summer were breathing on them.

PASSING by several large and interesting families, we come to the curious creatures which will at once be recognised by reference to the illustration. These beetles are popularly known by the name of ROVE BEETLES, or COCKTAILS, the latter name being given to them on account of their habit of curling up the abdomen when they are alarmed or irritated. The common BLACK COCKTAIL has, when it assumes this attitude, standing its ground defiantly with open jaws and elevated tail, so diabolical an aspect that the rustics generally call it the devil's coach-horse. It has, moreover, the power of throwing out a most disgusting odour, which is penetrating and persistent to a degree, refusing to be driven off even with many washings.

These beetles are termed Staphylinidæ, or Brachelytra, the latter term signifying short elytra, and being a very apposite name, as the elytra are short, square, and not more than one-fourth the length of the abdomen. These beetles are very ready at flight, the smaller species being especially obnoxious on account of their habit of getting into the eyes and nostrils on the summer evenings. Popularly they are then thought to be flies, but if captured, their

ROVE BEETLES,
'Ocypus olens. Creóphilus maxillósus.

true nature is easily discoverable. If we watch one of these Beetles settling after its flight, we shall see the object of its flexible tail. The wings are so large, and the elytra so small, that the process of folding the delicate membranes could not be completed without some external aid. When the insect alights, it suddenly furls its wings into loose folds that irresistibly remind the observer of a half-furled topsail, and then by means of its tail, it pushes the wings under the elytra, which are then shut down. This process, although rather elaborate, is effected in a very rapid manner. The forceps of the earwig are employed for the same purpose, the elytra being as proportionately small, and the wings as proportionately large, as those of the Cocktail.

The Cocktails are carrion feeders, and the bite of the larger species is often rather poisonous, on account of the putrid matter which is apt to adhere to them. I have known a man to have his hand swollen and rendered useless for a week in consequence of the bite of one of these insects.

The two species which are represented in the illustration are common in England. The lower figure represents the Creophilus maxillosus, which is plentiful in and about drains or dead animal matter, and may be known by the grey hairy look of the elytra. There is a smaller species (Staphylinus erythrópterus) which has the elytra of a dusky red, and is not so common as the preceding insect. I have often remarked that the red-backed shrike is very fond of this insect, and used to find the nests of the shrike by means of the beetles that the bird had stuck upon the thorns near its home.

3. H H

The Staphylinidæ include a vast number of species that may be found in almost every imaginable locality, and live on almost every imaginable kind of food. Their dimensions are as variable as their habits, some species exceeding an inch in length, while others are so minute as almost .to escape the naked eye, and best to be caught by waving a sheet of wet paper in the air.

The upper figure represents the Devil's Coach-horse, shown of the natural size.

NEXT to the Staphylinidæ are placed some insects that have become quite famous for their curious and valuable habits. These are the Necrophagæ, popularly and appropriately termed Burying Beetles.

It is owing to the exertions of these little scavengers that the carcases of birds, small mammals, and reptiles are never seen to cumber the ground, being buried at a depth of several inches, where they serve to increase the fertility of the earth instead of tainting the purity of the atmosphere. These beetles may easily be captured by laying a dead mouse, mole, bird, frog, or even a piece of meat on the ground, and marking the spot so as to be able to find the place where it had been laid. It will hardly have remained there for a couple of hours before some Burying Beetle will find it out, and straightway set to work at its interment. The plan adopted is by burrowing underneath the corpse and scratching away the earth so as to form a hollow, into which the body sinks. When the beetles have worked for some time they are quite hidden, and the dead animal seems to subside into the ground as if by magic.

BURYING BEETLES.

Hister cadaverinus.　　　*Necróphorus vestigátor.*　　　*Silpha opáca.*

The strength and perseverance of these beetles are so great that a very short time suffices to bury the creature completely below the ground, and the earth being scraped over it, the process is complete. The object of burying dead animals is to gain a proper spot wherein to deposit their eggs, as the larvæ when hatched feed wholly on decaying animal substance. If the bait be dug up the day after it is laid, a pair of Burying Beetles will in all probability be found, and if the animal be of large dimensions a dozen or more will have congregated round it.

I have known a dead sheep that had been flayed and the carcase thrown into a dense spinney, absolutely swarm with these beetles and their larvæ. Owing to the great size of the animal they had not been able to inter it in spite of all their efforts, but had crowded in such numbers upon so vast a feast that I captured almost every species of British Burying Beetle upon or near that sheep. Many of the larvæ are wonderfully active, black, flattish, and running with great speed.

In the accompanying illustration two figures are given of one of the common Burying Beetles, showing the creature as it appears while walking and during flight. The right-hand figure represents a flattish black species very common on any decaying animal substance, and producing wonderfully active grubs. I have found them by hundreds in old bones. On the left hand is shown an odd little beetle, black and shining as if carved out of jet, which belongs to a large genus found near and upon similar substances.

WE now come to the Lamellicorn beetles, so called from the beautiful plates, or lamellæ, which decorate the antennæ. This family includes a vast number of species, many of which, as, for example, the Common Cockchaffer, are extremely hurtful to vegetation both in the larval and adult form. In this family are found the most gigantic specimens of the Coleoptera, some of which look more like crabs than beetles, so huge are they and so bizarre are their shapes. In all these creatures the lamellæ are larger and more beautiful in the female than in the male insect.

The COMMON COCKCHAFFER is too familiar to need any description of its personal appearance, but the history of its life is not so widely known as its aspect. The mother beetle commences operations by depositing the eggs in the ground, where in good time the young are hatched. The grubs are unsightly looking objects, having the end of the body so curved that the creatures cannot crawl in the ordinary fashion, but are obliged to lie on their sides. They are furnished with two terribly trenchant jaws like curved shears, and immediately set to work at their destructive labours.

They feed mostly upon the roots of grasses and other plants, and when in great numbers have been known to ruin an entire harvest. To turf they are especially destructive, shearing away the roots with their scissor-like jaws and killing the vegetation so effectually that the turf withers, turns yellow, then brown, and can be taken up

COCKCHAFFER.—*Melolontha vulgáris.*   STAG BEETLE.—*Lucánus cervus.*

and rolled by hand. Several birds, among which the rooks, crows, and blackbirds are our best allies, are very fond of these grubs, which are little more than soft lumps of fat enclosed in a thin but tough skin, and by the aid of their delicate senses are enabled to detect the larva as it carries on its insidious task below the ground. For three years it continues in its larval state; and after a brief sojourn in the pupal condition changes its skin for the last time, and emerges from the ground a perfect Cockchaffer.

Even in its perfect state it is a terribly destructive insect, working sad havoc among the foliage of trees. I have seen whole rows of trees entirely stripped of leaves by the Cockchaffers—"hannetonnées," as the French call the results of the Cockchaffer's depredations. In the evening, however, when the Cockchaffers begin to fly abroad, the nightjar issues from its home, and spends the night in flying round the trees, snapping up with unerring aim every Cockchaffer that ventures within its swoop. The larvæ, or grubs, of the Cockchaffer are popularly known among agriculturists by the names of White Worm and Connaught Worm. In some countries, however, the fat oily bodies of the grubs are turned to account, being boiled for the purpose of extracting the oil, which is used for greasing wheels and other similar purposes.

The STAG BEETLE is the largest of our British Coleoptera, and when it has attained its full dimensions is an extremely powerful and rather formidable insect, its enormous mandibles being able to inflict a very painful bite, not only on account of the powerful muscles by which they are moved, but in consequence of the antler-like projections with which their tips are armed. These horn-like jaws only belong to the male, those of the female being simply sharp and curved mandibles, in no way conspicuous.

The larvæ of the Stag Beetle reside in trees into which it burrows with marvellous facility, and as after they have emerged from their holes they appear to cling to the familiar neighbourhood, they may be found upon or near the trees in which they have been bred.

From the formidable shape of the mandibles it might be supposed that the Stag Beetle was one of the predacious species. This, however, is not the case, the food of this fine insect consisting mostly, if not wholly, of the juices of vegetables, which it wounds with the jaws so as to cause the sap to flow. It is true that specimens have been detected in the act of assaulting other insects, but they never seem to have been observed in the act of feeding upon their victim. Whether the food be of animal or vegetable nature, it is always liquid, and is lapped, or swept up, by a kind of brush which forms part of the mouth, and looks like a double pencil of shining orange-coloured hairs.

It seems that during the winter the Stag Beetle hibernates, as there is in the Ashmolean Museum, at Oxford, an earthen cell, or cocoon, in which was found a Stag Beetle very neatly packed, with its horns bent over its thorax. A popular name for this beetle is Hornbug.

IN the accompanying illustration is represented a beetle that has been rendered for ever famous by the honours which the ancient Egyptians paid to it, and the frequency with which it is represented upon their hieroglyphs, and even sculptured on a gigantic scale in the hard granite which that wondrous race could work so easily. This is the Sacred Scarabæus of the Egyptians, an insect which deserves a passing notice on account of its curious habits.

The reader will remember that the burying or sexton beetles are in the habit of interring the dead bodies of various small animals in order to form a convenient nidus in which to deposit their eggs, and insure for their young a bountiful supply of food as soon as they enter the world. The Scarabæus is urged by a similar instinct, but exercises it upon different materials. Every one who has walked in the field must have noticed the singular rapidity with which patches of cowdung disappear, and many may have observed that this phenomenon is caused by the efforts of sundry beetles, which burrow beneath the mass and convey the substance deep into the ground. The common watchman beetle, so well known from its habit of flying on droning wings in the evening, is one of the best known of these valuable beetles; and it is worthy of notice that, despite of the nature of the substance in which they work, not a speck adheres to their bright and polished armour.

The Egyptian beetle employs similar substances for the cradle of its future young, but not in the same manner, kneading it into irregular balls in which it deposits its eggs, and then rolling it away by means of its odd-looking hind legs. After it has made the ball, which is often larger than itself, the beetle sets to work to roll it to a convenient spot where the earth is soft, and performs this curious operation by a retrograde motion, the hind legs directing the ball, while the four other legs are employed in locomotion. During this operation the beetle seems to be standing on its head, the hind legs being necessarily much elevated in order to guide the ball, which by dint of much rolling becomes nearly spherical. A tolerably deep hole is then excavated in a suitable spot, the ball rolled into it, and the earth filled in.

Many beetles perform this useful operation; and even in several European countries where the beauty of the climate is only equalled by the uncleanliness of the inhabitants, these beetles are of inestimable service, and are, perhaps, the only means whereby the towns and villages are rendered endurable, at all events to unaccustomed eyes and nostrils. Fortunately these insects fly by day as well as by night, and being gifted with

SACRED EGYPTIAN SCARABÆUS.—*Scarabæus sacer.*   ATLAS BEETLE.—*Chalcosóma atlas.*

extraordinary powers of scent, are sure to be on the spot as soon as their labours are required.

There are few parts of the globe where beetles possessing similar habits are not to be found, and although they do not display equal skill in the construction of egg-containing balls, they are equally efficacious in the results. It may be here mentioned that the watchman beetle of England (*Geotrúpes stercorárius*) is the "shard-borne" beetle mentioned by the poets, the title being due to the shelly elytra which are held aloft during its flight. They are marvellously tenacious of life; and as an example of this property I may mention that I once caught a Geotrupes in the air which had been mulcted of one elytron, lost several of its limbs, and the whole of its abdomen, the contents having been evidently scooped out by some bird. Yet it was quite strong on the wing, and seemed little the worse for its injuries.

THE right-hand figure in the illustration represents a very fine Lamellicorn, the ATLAS BEETLE, a native of the Philippines and part of India. The form of this insect can be seen from the drawing, and its colours are as follows :—The male is of a brilliant metallic olive green, brightly polished and shining; but the female is of a much duller hue, having the thorax and the base of the elytra rough, and the green of a blackish cast. The length of the male is about three inches.

The very odd looking beetle on the left hand of the illustration belongs to the family Rutelidæ, the members of which belong entirely to the hot countries of the globe, and are most plentiful in the tropics. They do not seem to attain the gigantic dimensions which are found among the allied families, such as the Dynastidæ, but are all very beautiful insects on account of the extreme brilliancy of their colouring. The CHRYSOPHORA has been chosen as the example of this family, on account of its curious form and glowing colours. The specimen is a male, and is given to show the extraordinary development of the hind legs, which seem disproportionately long and stout when compared with the moderately sized body. Another point of interest in this beetle is the structure of the "tibia" of the hind leg, *i.e.* the joint immediately preceding the jointed foot. On reference to the engraving, it will be seen that the lower part of this joint is prolonged

CHRYSOPHORA.—*Chrysóphora chrysochlóra.*   POLYPHEMUS BEETLE.—*Chelorhina Polyphémus.*

into a stout and sharp spur, not unlike that on the leg of most gallinaceous birds. The object of this curious modification is not known.

The POLYPHEMUS BEETLE is an example of the family termed Dynastidæ, or powerful beetles, on account of their enormous size and strength. They are the giants among insects ; for although many others exceed them in length or width, these creatures are so stoutly made, that any other insect becomes dwarfed when placed by their side.

In this family, the males are remarkable for the strange and often grotesque horny processes which are developed from the head and thorax, the females being destitute of these ornaments. Most of the Dynastidæ inhabit tropical regions, only a very few species being found in Europe. They are generally night fliers, ascending to considerable elevations, and during the day they hide themselves in holes in the earth, in hollow trees, or similar situations. Their food seems to be nearly, if not wholly, of a vegetable nature.

PASSING by one or two families of more or less importance, we arrive at the Buprestidæ, a family of beetles remarkable for the extraordinary gorgeousness of their tints, almost every imaginable hue being found upon these brilliant insects.

They are found in many portions of the globe, but, as is generally the case with insects, their colours take the greatest intensity within the tropics. They fly well, and seem to exult in the hottest sunshine, where the bright beams cause their burnished raiment to flash forth its most dazzling hues. They are, however, slow of foot, and, when alarmed, have a habit of falling to the ground with folded limbs, as if they were dead.

The species that is given in the illustration is one of the finest of this splendid family. The sides of the thorax are covered with little round pits, something like the depressions on the head of a thimble, and are of a fiery copper hue. The head and middle of the thorax are light burnished blue, like that of a well-tempered watchspring, and the elytra are warm cream-coloured, diversified with a patch of deep purple blue at each side, and another at the tip. The CHRYSOCHROA is a native of India.

The larvæ of this family are wood borers, and there is a curious account by Mr. Marsham of an insect of this family (*Buprestis splendens*), which suddenly made its appearance out of a desk which had remained in one of the Guildhall offices for more than twenty years. The wood was afterwards planed carefully away, and the track of the larva was laid open. In what state the creature passed so long a period, whether as egg, larva, pupa, or perfect insect was not ascertained. The word Buprestis literally signifies Bull-burster, and was given to these beetles by the ancients, who thought that when they settled upon cattle their presence was so injurious as to cause the bodies of their victims to burst asunder.

CHRYSOCHROA.—*Chrysochróa Bugnetii.*
GLOWWORM.—*Lampýris noctilúca.* (Male.)   CUCUJO.—*Pyróphorus luminósus.*

THE right-hand figure on the illustration represents the celebrated CUCUJO, or FIRE-FLY OF BRAZIL. At each side of the base of the thorax may be observed two light patches, which in the living insect are of a pale yellow, and at night burn with a lustre far surpassing that of the glowworm of England. When the insect expands its wings for flight two more fire-spots are seen beneath the elytra ; and when the creature approaches near the observer, the whole interior of its body seems to be incandescent. These insects are nocturnal in their habits, and at night in the forests, when the air is filled with myriads of blazing stars, crossing and recrossing in every direction, making the deepest glades luminous with their flaming lamps, and appearing and vanishing as if suddenly brought into existence and as suddenly annihilated, they present a sight almost too magnificent for description. So splendid are these beetles, that the ladies are often in the habit of catching them and trimming their dresses with these living diamonds, taking care to fasten them in such a way as not to injure them.

When in full glow, the light is so intense that a letter or a book may be read by its aid, provided that the insect be slightly squeezed so as to excite it to throw out the luminous element. There are very many species of Fire-flies, but this is the best known, and one of the most luminous of its kind. Mr. Westwood mentions that one of these insects was brought in a living state to London, and was kept alive by continually moistening the woodwork of its cage.

The Elateridæ, or Spring Beetles, so well known from their habit of jumping with a slight clicking sound when laid on their backs, are allied to the Buprestis beetles.

THE celebrated GLOW-WORM belongs to the typical genus of its family.

Contrary to the usual rule among insects, where the male absorbs the whole of the beauty, and the female is comparatively dull and sombre in colour and form, the female carries off the palm for beauty, at all events after dusk, the male regaining the natural ascendancy by the light of day. Either through books, or by actual observation, almost every one is familiar with the Glow-worm, and would recognize its pale blue light on a summer's evening. Many, however, if they came across the insect by day, would fail to detect the brilliant star of the night in the dull, brown, grub-like insect crawling slowly among the leaves, and still fewer would be able to distinguish the male, so unlike are the two sexes.

It has often been said that the female alone is luminous. This, however, is an error, as I have caught numbers of these beetles of both sexes, and always found that the males were gifted with the power of producing the peculiar phosphorescent light, though in a much smaller degree than their mates, the light looking like two small pins' heads of phosphorus upon the end of the tail.

Seen by day, the male is a much handsomer looking insect than the female, being soft brown in colour, long bodied, and wide winged, altogether beetle-like ; while the female is more like a grub than a perfect insect, has no wings at all, and only the slightest indications of elytra.

The larva of the Glow-worm feeds upon molluscs, especially upon the smaller snails, which it is able to devour even when retracted within the walls of the shell. In order to cleanse its body from the slime which exudes from the molluscs on which it feeds, and which is poured out with double abundance when they are injured or irritated, the larva of the Glow-worm is furnished with a kind of brushy appendage near the tail, which is thrust from its concealment when wanted, and vigorously applied to the body by means of the flexible joints of the abdomen.

NEXT to this family is another, called the Telephoridæ, which is represented in England by the well known beetles, popularly called, from their red or bluish colours, SOLDIERS and SAILORS. They are found in great quantities in the spring, and upon the umbelliferous flowers they assemble plentifully. They are carnivorous, voracious and combative to a degree, and in my school-days the fashionable spring amusement consisted in setting Soldiers and Sailors to fight with each other. They fly readily, but slowly, and only to short distances, and may be known while in the air by their peculiar attitude, the long body hanging nearly vertically from the wings.

A VERY destructive family, termed Ptinidæ, must now be briefly noticed. To this family belong the insects which are so well known by their labours, though themselves are mostly hidden from sight. Among the Ptinidæ are placed the little beetles that eat holes in our furniture, books, &c., and do such irremediable damage in so short a time. Mr. Westwood mentions one instance where a new bedpost was wholly destroyed by one species of these beetles (*Ptilinus pectinicornis*) in a space of three years. I have seen books fearfully damaged by these insects, which have a habit of boring in a straight line as long as they can find material, and have been known to begin at one end of a shelf full of books, and drill a hole so completely through them, that a string could be threaded through the hole and all the books lifted by it. They do not, however, push their way completely through the substance, in which they have been bred, until they have attained their perfect form, but always leave a slight film of uneaten substance by which they may be concealed.

The celebrated Death Watch belongs to this family—this name being popularly given to several species, such as *Anobium striatum* and *tesselatum*, on account of the ticking sound which is made by knocking their heads against the woodwork, and which is used as a signal to their mates.

Other species feed upon wool, flax, leather, &c., and are terrible pests to the collector, who often finds his museum half-ruined before he has the least suspicion that anything is wrong with it. There is an odd little beetle belonging to this family, called *Mezium sulcátum*, a spider-like looking insect, with a globular body, covered with soft plush-like hair of light brown. These insects may be found abundantly in old cupboards, and I once found a forgotten china jar filled nearly two inches deep with the bodies of these little insects.

TOWARDS the middle of spring and for the next month, may be found certain very handsome looking beetles of a deep rich red colour, and remarkable for the beautifully toothed antennæ. This insect is to be seen mostly upon flowers, and is popularly known by the name of CARDINAL BEETLE. The scientific title is *Pyrochróa rubens*. This is the only British genus of the family to which it belongs, and which is called Pyrochroidæ, in allusion to the typical genus. The word Pyrochroa, or Flame-coloured, is given to this beetle on account of its bright red exterior.

A succeeding family, the Mordellidæ, is chiefly remarkable for the curious fact that the larvæ of several of its genera, those of *Ripiphorus* for example, inhabit the nest of the common wasp, undeterred by the poisoned stings of their involuntary hosts from taking possession of their home. It seems that each specimen of this beetle monopolizes a single cell, and entomologists are of opinion that the nurse wasps feed the intruders, together with the rightful owners of the cells, not being able to distinguish between them.

THE two insects represented in the accompanying illustration are found in England, and are here given as examples of the family Cantharidæ, of which the BLISTER FLY, sometimes called the SPANISH FLY, is the typical species.

In the whole of this family, certain noxious elements are strongly developed, which, like all other noxious things, can be transmuted and modified into benefits by those who know how to use them. There is a certain substance secreted within these creatures technically called Cantharidin, and looking, when separated from extraneous matter, like minute crystalline flakes of snowy whiteness. It can be dissolved in spirit, but not in water.

The Blister Fly is by no means a common species in England, though it has occasionally appeared in considerable numbers.

OIL BEETLE.—*Meloë violáceus.*

BLISTER, OR **SPANISH FLY.**—*Cántharis vesicatória.*

In such cases, however, it is extremely local, and does not appear to be disseminated through the country. Spain is famous for the multitudes of Blister Flies which are found within its limits, and the whole of South-western Europe is prolific in this remarkable beetle. Whenever it may be present, its vicinity is known by the powerful odour which it exhales, just as the musk and tiger beetles may be detected by the nostril, though unperceived by the eye. On account of its peculiar properties, it is not easily prepared, the dust which flies from the dried and drying insects being light, searching, pungent, and inflammatory to the last degree.

The larva or grub of this beetle is said to reside under ground, and to feed upon the roots of vegetables.

The Spanish Fly is a handsome insect, nearly an inch in length, and of a rich silken green, with a gold gloss in certain lights. It is a very remarkable fact that fish will eat the Cantharis without injury, and anglers have found, rather to their surprise, that if they could fix a Cantharis on their hook, it proved to be a very effectual bait for fish, the chub seeming particularly fond of this very stimulating food. The common hedgehog has been known to eat these insects with impunity.

THE second figure represents an insect belonging to the same family, and very common in England. It popularly goes by the appropriate name of OIL BEETLE, because, when handled, it has the property of pouring a yellowish oily fluid from the joints of its legs.

As may be seen by reference to the illustration, the abdomen is extremely large in proportion to the rest of the body, and the short diverging elytra descend but a very little way below the thorax. Insects of this genus—especially the males, where the elytra are longer than in the other sex—are used by unprincipled druggists for the purpose of mixing with the true blister fly, which they resemble sufficiently to deceive an inexperienced eye. In some parts of the world, however, they are always employed in connexion with the blister beetle, or even used instead of that insect. The oily matter that is poured from the joints is considered in some countries to be a specific for rheumatism, and is expressed from the insect for medicinal purposes.

The Oil Beetle is represented of the natural size, and its colour is dull indigo blue.

A FEW other insects of this family are rather remarkable in their habits. One of these is the SITARIS, the larva of which is found in the nests of several of the mason bees (*Anthophora* and *Osmia*) and the general opinion of naturalists is that they feed upon the larvæ of those insects. Some, however, think that their only object in this intrusion, is to eat the provision of pollen that has been laid up for the young bee.

IT is of course impossible in a work of so limited an extent to give more than an outline of each class of animals. Before proceeding to the insects which are figured in the next illustration, we will casually notice a few of the more interesting species belonging to the intermediate families.

The MEAL-WORM, so well known to bird-fanciers as a wholesome diet for nightingales and other birds; to millers, for its ravages among the grain; and to sailors, for its depredations among the biscuit, is the larva of a beetle named *Tenebrio molitor*, the former word being given to it in allusion to its love of darkness, and the latter to the damage which it occasions to the miller. This is one of the maggots which have caused sailors to knock the edge of a biscuit upon the table before eating it, an action which in many old voyagers has become so deeply rooted a habit, that they are actually unable to resist the movement. These larvæ are terribly sharp toothed, eating their way through the sides of casks while in search of food. Some species of the same genus have the power of ejecting an acrid fluid to the distance of more than a foot; the one most remarkable in this respect being a Brazilian insect, *Tenebrio grandis*.

WE now arrive at a vast group of beetles, embracing several thousand species, which are popularly classed under the name of Weevils, and may all be known by the peculiar shape and the very elongated snouts. Many of these creatures have their elytra covered with minute but most brilliant scales, arranged in rows, and presenting, when placed under the microscope, a spectacle almost unapproached in splendor. They are mostly slow in their movements, not quick of foot, and many being wholly wingless.

Many of these creatures are extremely injurious to vegetables, both while growing and when stored up in barns or granaries. Most persons are too familiar with the little maggots that infest peas, and frequently ruin whole pods at a time, each pea containing a single white grub. These are the larvæ of the PEA WEEVIL (*Bruchus pisi*), which feed upon the soft substance of the pea, and make their escape just about the time when the vegetable is sufficiently ripe for gathering. One of the CORN WEEVILS (*Bruchus granarius*), so destructive to grain, also commits great ravages among the peas. One species of this genus inhabits the cocoa-nut, and the creatures are infinitely more abundant in tropical than in temperate climates. It is thought, indeed, that several species of these destructive insects have been imported into England in cargoes of grain, and finding the country suitable to their habits, have thriven here.

Another species of Weevil, the GRASS WEEVIL or Lisette (*Rhynchites Bacchus*) commits terrible devastations among the growing vines, sometimes stripping the bushes of their leaves, which it rolls up and lines with silk.

The most brilliant of the Weevils are to be found in the typical family Curculionidæ, to which belong the well-known Diamond Beetles, in such request as objects for the microscope. Magnificent, however, as are these insects, some of our common little field Weevils, which may be found abundantly on peas, nettles, and other vegetables, yield to them not a single jot, when properly magnified and illuminated, the successive rows of glittering scales with their numerous facets being quite as splendid as the scale-lined pits which cover the elytra of the Diamond Beetle.

The maggots that are so frequently found in nuts, and which leave so black and bitter a deposit behind them that the person who has unfortunately tasted a maggot-eaten nut

Xenócerus semiluctuósus.
'Eupholus Linnéi.
Balaninus proboscideus.

is forcibly reminded of the Dead Sea apple, with its inviting exterior and bitter dusty contents, also belong to the Weevils, and are the larvæ of the NUT WEEVIL (*Balaninus nucum*). All the members of this genus are remarkable for the extraordinary length of the snout, at the extremity of which are placed the small but powerful jaws. An example of a foreign species, a native of Cayenne, is shown in the illustration, and is given in order to show the inordinate length of the snout.

While the nut is yet soft and undeveloped, the female Weevil bores a hole at the base of the fruit, deposits an egg therein, and makes the best of her way to another nut, which she treats in a similar manner. As the nut increases, the young grub feeds on the interior of the nut, which is at first soft and milky, so as to suit its infant needs, and by degrees hardens into a fruity substance more fit for it after its jaws and digestive organs have acquired strength. After it has attained its full growth, it gnaws a round hole through the shell of the nut, allows itself to drop to the ground, buries itself below the surface, and in the ensuing autumn emerges in the perfect form.

The common CORN WEEVIL (*Calandra granaria*) is perhaps the most destructive of its tribe, its depredations far exceeding those of the insects that destroy nuts, acorns, apples, cherries, flowers, and other vegetables. This pest of corn-dealers is of very small size, not larger than the capital letter at the beginning of this sentence, and is therefore able to make its way through very small crevices. Like the preceding species, it passes its larval existence within the grain on which it feeds. devours the whole of the interior.

and then, gnawing its way through the shell, becomes transformed in process of time into its perfect shape, which is that of a little long-beaked Weevil of dull red colour, which, however, under the microscope, is singularly beautiful.

Many species belonging to this destructive genus are equally plentiful all over the world, and equally injurious. There seems, indeed, to be no vegetable substance that is not eaten by the Weevils, which appear to have a peculiar liking for those that are used for human food. Almost every article has its peculiar Weevil. There is the RICE WEEVIL, for example (*Calandra oryzæ*), known from the previous species by the four red spots on the elytra, which is nearly as destructive towards rice and Indian corn as the Corn Weevil towards wheat.

One of the largest species is a native of the West Indies, and is known by the name of the PALM WEEVIL (*Calandra palmarum*). This huge Weevil sometimes attains the length of two inches, and its colour is a dull, velvet-like black. The larva of this large beetle is a great fat white grub, called gru-gru by the negroes, and considered by them to be a great dainty. The more educated inhabitants know this grub by the name of *Ver palmiste*. This grub is especially fond of the newly planted canes, and is sometimes so terribly destructive among them that a fresh planting becomes necessary. When this creature is about to attain its pupal condition, it weaves for itself a kind of cocoon formed from the fibres of the plant in which it lives.

BEFORE noticing the long-horned insects, we must briefly mention a terribly destructive family of beetles, that are certainly allied to the Weevils, but whose precise degree of relationship does not seem to be very accurately understood.

To this family belongs the far-famed *Scolytus destructor*, a little dull coloured insect, insignificant in appearance, but able to lay low the loftiest elm that ever reared its leafy head. Hundreds of our finest trees have fallen victims to the devouring teeth of this tiny beetle, a creature hardly the sixth of an inch in length. These insects not only burrow into the trees for the purpose of obtaining food, but therein they deposit their eggs, and therein are the young larvæ hatched.

The mother beetle deposits the eggs in a row, and the young, immediately upon entering the world, begin to eat their way through the wood, all diverging at right angles from the burrow in which they were laid, and all increasing the diameter of the burrow in exact proportion to their own growth. Hundreds of these quondam dwellings may be seen on roadside fences and railings, and so numerous are they on many trees that the bark falls off in flakes, the course of the sap becomes arrested, and at last the tree dies from the injuries to which it has been subjected by these minute but terrible foes, who work in darkness, unseen and secure. The grubs or larvæ may often be found in these tunnels. They are thick, round, and fat, without feet, and of a whitish colour, except the horny head with its powerful jaws.

WE now come to the Longicorn beetles, so called on account of the extraordinary length of the antennæ in many of the species. These insects are well represented in England by many species, though none have the antennæ of such wonderful length as are shown in the illustration on page 475.

As in the preceding family, the Longicorn beetles pass their larval state in wood, sometimes boring to a considerable depth, and sometimes restricting themselves to the space between the bark and the wood. The grubs practically possess no limbs, the minute scaly legs being entirely useless for locomotion, and the movements of the grub being performed by alternate contraction and extension of its ringed body. In order to aid in locomotion the segments are furnished with projecting tubercles, which are pressed against the sides of the burrow. Those of the common wasp beetle (*Clytus arietis*) may be found at the beginning of summer in fir trees, or in palings and posts of that wood.

The XENOCERUS, so remarkable for the inordinate length of the antennæ, is one of the finest examples of the Anthribidæ, not only for the long and thread-like antennæ, but for the beauty of its colouring. It is a native of Amboina, where it was found by Mr. Wallace. The male is jetty black diversified with small white stripes on the head and thorax. The elytra are boldly decorated with the same contrasting hues. The female is also white and black, but the former colour greatly predominates, the black being reduced to marks on the sides of the head and thorax, the tips of the elytra, and four black spots, two on the middle of the elytra and the other two on the thorax.

The well-known MUSK BEETLE (*Cerambyx moschâtus*) belongs to this group. The scent, which more resembles attar of roses than musk, is extremely powerful, and is often the means of betraying the presence of the insect as it lies hidden among the leaves. The larva is a wood-borer, and I have taken numbers out of old willow trees, which I split with wedges for the express purpose.

Two more examples of the Longicorn beetles are given in the accompanying engraving, in order to show the variety of form exhibited by these beautiful insects.

On the left hand is seen a beetle with a large tuft of hair on each of the antennæ. This is the *Disaulax Cayennensis*, a native of the country whence it takes its name. These curious tufts present a very striking appearance, being jetty black with white tips.

Disaulax Cayennensis.
Chrysoméla cereâlis.   *Aspidomorpha amplissima.*   *Plectodèra scalâtor*

The stout bases of the elytra are yellowish orange. The whole of the body is boldly marked with deep black and snowy white of a silvery lustre. The *Plectodera scalator*, a much larger species, belongs also to the Longicorns, and like the preceding species, is marked with black and white, though the arrangement of the tints is different.

THE broad and flat insect in the centre of the engraving is the *Aspidomorpha amplissima*, a beetle that is found in the Philippines, and is the largest of the Tortoise Beetles, or Cassididæ. These insects derive their popular name from the tortoise-like shape of the body, which is so expanded that the whole of the limbs are concealed under its shelter. Many of these beetles are a light green, or greenish brown, and when they are stationary upon a leaf they can with difficulty be distinguished. The larva is remarkable for possessing a large forked appendage upon the end of the tail, which turns over the back and is loaded with excrementitious substances, so that the creature can hardly be seen under the load which it bears.

In the present species the body is chestnut-brown, and the elytra are furnished with wide, thin, and semi-transparent margins. Their centre is spotted with black.

Passing by several families, we come to our last example of the Coleoptera, the *Chrysomela cerealis*, a British example of a very large family.

All the Chrysomelidæ are round-bodied, and in most cases are very brilliantly coloured with shining green, purple, blue and gold, of a peculiar but indescribable lustre. They are slow walkers, but grasp the leaves with a wonderfully firm hold. The British species of Chrysomela are very numerous. One of the genera belonging to this family contains the largest British specimen of these beetles, commonly known by the name of the BLOODY-NOSE BEETLE (*Timarcha tenebricosa*) on account of the bright red fluid which it ejects from its mouth and the joints of its legs when it is alarmed. This fluid is held by many persons to be a specific in case of toothache. It is applied by means of permitting the insect to emit the fluid on the finger and then rubbing it on the gum, and the effects are said to endure for several days. The larva of this beetle is a fat-bodied, shining, dark green grub which may be found clinging to grass, moss, or hedgerows in the early summer. They are so like the perfect insect that their identity cannot be doubted.

THE family of the Coccinellidæ, or Ladybirds, is allied to the Chrysomelidæ, and is well known on account of the pretty little spotted insects with which we have been familiar from our childhood. Though the LADYBIRD is too well known to need description, it may be mentioned that it is an extremely useful insect, feeding while in the larval state on the aphides that swarm on so many of our favourite plants and shrubs. The mother Ladybird always takes care to deposit the eggs in spots where the aphides most swarm, and so secure an abundant supply of food for the future offspring.

EARWIGS.

TAKING leave of the beetles, we now proceed to a fresh order, distinguished by several simple characteristics, among which may be mentioned the soft and leathery elytra or fore-wings, the wide and membranous hind-wings, and the forceps with which the tail is armed. The insects belonging to this order are popularly known by the name of Earwigs, and are represented in this country by several species of different dimensions.

Before proceeding to the description of individual species, it is necessary to remark that the word Earwig is slightly incorrect, and owing to a popular notion that the insects crawl into the human ear, thence into the brain, and complete their work by causing madness in the minds of those who are afflicted by their presence.

The fact is, as all must know who have the least smattering of anatomy, that the insect never could gain admission to the brain by means of the ear. In the first place the cerumen which is secreted in the ear and serves to preserve the natural moisture of the tissues, is so inexpressibly bitter, and so entirely opposed to the habits of the Earwig, that if one of these insects should by chance happen to crawl into the ear, its first impulse would be to retreat. In the second place, the drum of the ear would present an impassable obstacle, and in the third place, supposing the drum to have been ruptured, and the Earwig to have passed the spot where it existed, the complicated bony passages through which the auditory nerve passes would be too small to admit of its passage, even if the nerves which fill the channels were removed.

In point of fact, the correct name of this insect is the Earwing, so called because its spread wings have an outline somewhat resembling that of the human ear.

The membranous wings of the Earwig are truly beautiful. They are thin and delicate to a degree, very large and rounded, and during the day-time packed in the most admirable manner under the little square elytra. The process of packing is very beautiful, being greatly assisted by the forceps on the tail, which are directed by the creature with wonderful precision, and used as deftly as as if they were fingers and directed by eyes. The Earwigs seldom fly except by night, and it is not very easy to see them pack up their wings. Some of the smaller species, however, are day-fliers, and in spite of their tiny dimensions, may be watched without much difficulty.

Earwigs feed on vegetable matter, especially preferring the corollas of flowers. Pinks, carnations, and dahlias are often damaged greatly by these insects, which sometimes occur in vast quantities, and ruin the appearance of a well-tended flower-bed. Fortunately for the florist, the habits of these insects are constant, and they can be destroyed in great numbers by those who desire to kill them. Being intolerant of light, they avoid the sunshine by every means in their power, and creep into every crevice that may hide them from the unwelcome light.

In consequence they are fond of crawling among the thick and shadowy petals of the dahlia, and are frequently found in the slender spur of the nasturtium, so that any one

FIELD CRICKET.—*Gryllus campestris.*
MOLE CRICKET.—*Gryllotalpa vulgáris.*　　　　GIANT EARWIG.—*Forficula* (or *Labidura*) *gigantea.*

who is about to eat one of these flowers will do well to examine the spur before he makes the attempt. Knowing these habits, horticulturists catch them by hundreds by the simple plan of placing lobsters' claws, bits of hollow reed, and similar objects, on the tops of sticks, knowing that the Earwigs will crawl into them at the dawn of day, and may be shaken out and killed when the gardener goes his rounds.

The Earwig is remarkable for a parental affection quite exceptional in the insect race, the mother watching over her eggs until they are hatched, and after the young have entered the world, taking as much care of them as a hen takes of her chicken.

There are about seven or eight British species, some of them being of very small size. I have often seen them flying about at midday, when they might easily be mistaken for beetles. They have several times alighted on the sleeve of my coat, and afforded good opportunities of watching the curious manner in which the wings are tucked under their cases. The largest British species is that which is given in the illustration. It is of very rare occurrence, and seldom seen, as it only inhabits the seashore, and never shows itself until dusk. I have a fine specimen that was caught on the sands near Folkestone, in the month of July.

ORTHOPTERA.

A LARGE and important order succeeds the Earwigs, containing some of the finest and, at the same time, the most grotesquely formed members of the insect tribe. In this order we include the grasshoppers, locusts, crickets, cockroaches, and leaf and stick insects, and its members are known by the thick parchment-like upper wings, with their stout veinings and their overlapping tips. As in all the orders, there are exceptional species, wherein one or more of these attributes are wanting. But the characters are in themselves constant, and in most cases the indications of the missing member can be found. For example, many species never obtain wings at all, in many others the males only are furnished with these organs, and in others they are so small as to escape a casual notice.

THE first family of Orthoptera is the Blattidæ, a group of insects familiar under the title of Cockroaches.

In these insects the body is flattened, the antennæ are long and thread-like, and the perfect wings are only to be found in the adult male. The common COCKROACH, so plentiful in our kitchens, and so well known under the erroneous name of black-beetle— its colour being dirty red, and its rank not that of a beetle—is supposed to have been brought originally from India, and to have found itself in such good quarters that it has overspread the land in all directions.

The Cockroaches are particularly fond of heat, and are found in greatest abundance in kitchens, bakehouses, and other places where the temperature is always high. They are nocturnal in their habits, very seldom making their appearance by daylight, but leaving their hiding places in swarms as soon as darkness brings their day. On board ship they become an almost intolerable nuisance, pouring out of the many hiding places afforded to them by a ship's timbers as soon as the lights are put out, and drive sleep far away by their pestilent odour and their continual crawling over the face and limbs of those who are vainly endeavouring to seek repose.

Together with the rats and mice, these insects sometimes increase to such an unbearable extent, that when the vessel comes to a port, the crew are sent on shore, pots of lighted sulphur are placed in the hold, and the hatches battened down for four-and-twenty hours. This severe treatment kills all the rats and mice, and all the existing generation of Cockroaches, and is so far a temporary relief. But the eggs, which are laid in great profusion, retain the elements of life in spite of the sulphureous fumes, and in a few months the ship will be nearly as much overrun as before with these pests.

There are several means of destroying the Cockroaches in houses, and if they are perseveringly carried out, a dwelling may be kept comparatively free from them. The common red wafers, if scattered over the floor, are rapid and effectual poison to these insects, and meal mixed with plaster of Paris has the same effect. Traps, too, can be readily made by twisting a funnel of paper, putting it into the neck of a jar with a little sugar and water at the bottom, and laying slips of wood or pasteboard as ladders by which the Cockroaches can reach the treacherous banquet. Those that enter will never escape with life, and the quickest way of killing them is to pour boiling water into the jar.

A hedgehog is also a good remedy against Cockroaches, and if allowed the run of the kitchen during the night, will be wonderfully efficacious in keeping down their numbers.

The eggs of the Cockroach are not laid separately, but enclosed in a hard membranous case, exactly resembling an apple puff, and containing about sixteen eggs. Plenty of these cases may be found under planks or behind the skirting boards where these insects love to conceal themselves. Along one of the edges of the capsule there is a slit which corresponds with the opening of the puff, and which is strengthened like that part of the pastry by a thickened margin. The edges of the slit are toothed, and it is said that each tooth corresponds with an egg. When the young are hatched, they pour out a fluid which has the effect of dissolving the cement which holds the edges together, the newly-hatched Cockroaches push themselves through the aperture, which opens like a valve, and closes again after their exit, so that the empty capsule appears to be perfectly entire.

The shape of the young much resembles that of the perfect insect, except that in neither sex are the wings in existence. In the pupal stage the resemblance is preserved, the creature is active, and exhibits the rudimentary wings. The reader may often have seen white, brown, and mottled Cockroaches. These are the insects that have lately changed their skins; and if one of these creatures be taken, it will be found that in a day or two it will attain the same reddish brown colour as its companions.

The Cockroach is a very active insect, running both backwards and forwards with astonishing speed, and is furnished at the extremity of the abdomen with two short projections resembling miniature antennæ, and popularly regarded as such.

Turning to the engraving on page 479, the reader will observe in the centre of the illustration a figure of a short, stumpy insect with large hind legs. This is the FIELD CRICKET, a noisy creature, inhabiting the sides of hedges and old walls, and making country lanes vocal with its curious cry, if such a word can be applied to a sound produced by friction. The Field Cricket lives in burrows, made at the foot of hedges or walls, and sits at their mouth to sing. It is, however, a very timid creature, and on hearing, or perchance feeling, an approaching footstep, it immediately retreats to the deepest recesses of the burrow, where it waits until it imagines the danger to have gone by.

Despite of its timidity, however, it seems to be combative in no slight degree, and if a blade of grass or straw be pushed into its hole, it will seize the intruding substance so firmly that it can be drawn out of the burrow before it will loosen its hold. The males are especially warlike, and if two specimens be confined in the same box, they will fight until one is killed. The vanquished foe is then eaten by the victor. In White's "Natural History of Selborne" there is a careful and interesting description of the Field Cricket and its habits.

The well-known HOUSE CRICKET ('Acheta doméstica) is a near relation of the above-mentioned species, and is so familiar as to need no description.

ON the same illustration, and occupying the left hand, is one of the oddest looking of the British insects, the MOLE CRICKET, so called on account of its burrowing habits and altogether mole-like aspect. This insect is represented of the natural size, and, as may be seen, attains considerable dimensions. Those who like to give the needful time and trouble will find the internal anatomy of the Mole Cricket to be highly developed, remarkably interesting, and easily dissected.

Like those of the mole, the fore-limbs of the Mole Cricket are of enormous comparative size, and turned outwards at just the same angle from the body. All the legs are strong, but the middle and hinder pair appear quite weak and insignificant when compared with the gigantic developments of the front pair. This insect is rather local, but is found in many parts of England, where it is known by sundry popular titles, Croaker being the name most in vogue near Oxford, where it is found in tolerable plenty.

The wings of the Mole Cricket are large and handsome; and when folded, their hardened outer edges project along the back like two curved spines. Some persons have thought that this insect is the cause of the well-known phenomenon called the Will of the Wisp, or Jack o' Lantern, because in a locality where one of these deceptive lights was fluttering after its uncertain wont, a Mole Cricket was captured on the wing.

The food of the Mole Cricket is chiefly of a vegetable nature; but the insect will eat animal food when offered, having been known to feed upon raw beef with great zest. Like the field cricket it is very combative, and when it has vanquished its foe is sure to eat him. As may be imagined from the tasks which it performs in driving burrows

3   I I

through the earth, the muscular strength of the Mole Cricket is exceedingly great; and when the insect is held in the hand, its struggles for escape are apt to inflict rather sharp scratches on the skin of the captor.

The colour of the Mole Cricket is brown of different tints, darker upon the thorax than on the wing-covers, both of which organs are covered with a very fine and short down.

As might be surmised from the extraordinary muscular power of the fore-legs, the Mole Cricket can burrow with great rapidity. The excavation is of a rather complicated form, consisting of a moderately large chamber with neatly smoothed walls, and many winding passages communicating with this central apartment. In the chamber are placed from one to four hundred eggs of a dusky yellow colour; and the roof of the apartment is so near the surface of the ground that the warmth of the sunbeams penetrates through the shallow layer of earth, and causes the eggs to be hatched.

The food of the Mole Cricket is mostly of a vegetable nature, but it has been known to feed upon raw meat, upon other insects, and even to exhibit a strong cannibalistic propensity when shut up in company and deprived of the normal food.

THERE is a singular species, called *Schizodáctylus monstrósus*, now common in the insect cases sent from India, which is notable for the manner in which the enormously long wings and their covers are rolled at their tips into spiral coils. This belongs to the same family as the mole cricket, and like that insect is a burrower, making holes nearly a yard in depth.

The MIGRATORY LOCUST is a well-known instance of a very large family of insects represented in our own land by many examples. All the Locusts and Grasshoppers are vegetable feeders; and in many cases their voracity is so insatiable, their jaws so powerful, and their numbers so countless, that they destroy every vestige of vegetation wherever they may pass, and devastate the country as if a fire had swept over it.

Such is the case with the Migratory Locust, so called from its habit of congregating in vast armies, which fly like winged clouds over the earth, and, wherever they alight, strip every living plant of its verdure. So assiduously do they ply their busy jaws, that the peculiar sound produced by the champing of the leaves, twigs, and grass-blades can be heard at a considerable distance. When they take to flight, the rushing of their wings is like the roaring of the sea; and as their armies pass through the air, the sky is darkened as if by black thunder clouds.

The warm sunbeams appear to be absolutely necessary for the flight of the Locust, for no sooner does the sun set than the Locusts alight and furl their wings. Woe to the ill-fated spot where they settle, for they consume everything that their jaws can sever, and are not content with eating the green herbage, but devour even linen, blankets, or tobacco. At the approach of the aerial hosts every one is in fear except the Bushman, who welcomes the Locust with all his heart; for he has no crops to lose, no clothing to be destroyed, and only sees in the swarming insects his greatest luxury, namely, an abundant supply of food without any trouble in obtaining it. In the path of the Locust he kindles large fires, and the insects, being stifled with the smoke, and having their wings scorched by the flames, fall in thousands, and are gathered into heaps, roasted, and eaten. Those that remain, after the Bushman has eaten his fill, are then ground between two stones into a kind of meal, which is dried in the sun, and can be kept for a considerable period without becoming putrid. This substance does not seem to be very palatable to Europeans, but its distastefulness is probably owing to the careless way in which the insects are scorched over the fire, as Dr. Livingstone speaks quite highly of the Locust as an article of food, thinking it superior to shrimps. Honey is always eaten together with the Locusts, whenever that sweet condiment can be obtained, as it serves to render the insects more digestible.

Some persons dry the Locusts for winter food, and catch them in the early dawn before the sunbeams have heated the air sufficiently to impart animation to these chilly insects. They are swept into sacks as they hang in great clusters from the branches, and then carried away on the backs of oxen to the place where the process of drying takes place. During the transit the living cargo makes a great buzzing inside the sacks. Herodotus, in Book IV. c. 172, mentions the fact that Locusts are used for food : " When they have caught Locusts, they dry them in the sun, reduce them to powder, and, sprinkling them in milk, drink them." The flavour of the Locust is rather variable, and is regulated by the food on which the insect has lately lived.

WALKING-STICK INSECT.— *Bacteria trophinus* COCKROACH.—*Blatta orientalis*.
MIGRATORY LOCUST.—*Locusta migratória*.

In Southern Africa, the herds of larval Locusts are little less to be dreaded than their mature and winged hosts. They pass over the ground in a broad torrent, several inches in depth, and are popularly known by the title of " voet-gangers," *i.e.* foot-goers. At night they crawl up trees and hang like clusters of grapes from the boughs. Nothing seems to impede their march. Wide and deep trenches are overpassed as if the ground were quite level ; should they come to a stream or canal, the vast hosts which form the van march boldly into it and with their dead bodies form a bridge over which their companions can march ; and should fires be lighted across their course, the flames are quickly stifled by the dense masses of insects that fling themselves unhesitatingly into the burning heap.

In such vast numbers do the Locusts assemble, that after a severe storm, which is always fatal to these insects, a continuous bank of dead Locusts has been formed, three or four feet in height, nearly fifty miles in length, and exhaling a most pestilential odour from the decomposing bodies.

There are so many and so familiar accounts of these Locust armies that little need be said upon the subject. The following passage, however, taken from Cumming's well-known work on Southern Africa, will give the reader a just and vivid idea of the multitudinous hosts that pass through the air :—

" On the following day I had the pleasure of beholding the first flight of Locusts that I had seen since my arrival in the colony. We were standing in the middle of a plain of unlimited length, and about five miles across, when I observed them advancing. On they

EYED PTEROCHROZA.—*Pterochroza ocellata.*

came like a snow-storm, flying slow and steady, about a hundred yards from the ground. I stood looking at them until the air was darkened with their masses, while the plain on which we stood became densely covered with them. Far as my eye could reach, east, west, north, and south, they stretched in one unbroken cloud; and more than an hour elapsed before their devastating legions had swept by. . . .

Locusts afford fattening and wholesome food to man, birds, and all sorts of beasts; cows and horses, lions, jackals, hyænas, antelopes, elephants, &c. devour them. We met a party of Batlapis carrying heavy burdens of them on their backs. Our hungry dogs made a fine feast on them. The cold frosty night had rendered them unable to take wing until the sun should restore their powers. As it was difficult to obtain sufficient food for my dogs, I and Isaac took a large blanket which we spread under a bush, whose branches were bent to the ground with the mass of Locusts which covered it, and having shaken the branches, in an instant I had more Locusts than I could carry on my back; these we roasted for ourselves and our dogs."

Our common English grasshoppers belong to the true Locusts.

Now and then is found in our fields a very large locust-like insect, of a beautiful grass-green hue, and having at the end of its tail a long, flat-bladed instrument called an ovipositor, and used for the purpose of boring holes in the earth and placing its eggs below the surface. This is the GREAT GREEN GRASSHOPPER (*Phasgonúra*, or *Acrida viridíssima*), which unfortunately loses its soft light green colour soon after death, and as it dries becomes a dirty yellowish brown. It is a very fine insect, often measuring two inches in length and three inches and a half over the expanded wings. It seems to be rather capricious in its appearances, in some years being quite plentiful, and in others hardly to be seen. The jaws of this insect are wonderfully powerful, and its captor will act wisely to keep his finger out of their reach. The internal structure of this grasshopper is extremely interesting, and on account of its large dimensions are easily studied. The gizzard is especially worthy of notice.

THE singular insect which is represented in the accompanying illustration is one of those beings in which are found a strong resemblance to other parts of creation. In this insect, we have an example of a member of the animal kingdom reproducing with startling fidelity the forms, colours, and even the accidental variations of leaves and flowers, thus exhibiting another phase of that wonderful adaptive power which gives to many flowers, such as the orchids, a striking resemblance to bees, butterflies, and other insects. In this instance, the resemblance to leaves is not only due to the peculiar outline and the leaf-like nervures, but to the presence of certain spots which look exactly like the tracks of leaf-mining or leaf-devouring caterpillars. These creatures belong to the same family as the locusts, and their habitation is Brazil.

In the upper left-hand corner of the illustration on page 483 may be seen a strange-looking insect, with an attenuated body and long slender limbs. This is a WALKING-STICK INSECT, one of a most remarkable family of Orthoptera, none of which are found excepting in the hottest parts of the earth. This insect belongs to the family of Phasmidæ, an appropriate title, derived from a Greek word signifying a spectre, many of these creatures being as it were the mere unsubstantial visions or shadowy outlines of insects.

The chief point of interest in these creatures is their marvellous external resemblance to certain portions of the vegetable kingdom, some assuming the forms of a broken branch and twigs with such extraordinary fidelity that the most practised eye is often deceived, and others taking not only the flat outline and half curl of fallen leaves, but even reproducing their peculiar nervures and soft vegetable green with such marvellous exactness, that those who see them for the first time can hardly be made to believe that they are not the objects which they so faithfully represent. As if to add to the singularity of these creatures and to keep up the illusion, the eggs of several species are ribbed and coloured precisely like the seeds of certain plants.

In the illustration on page 486 is represented a LEAF INSECT, one of the singular species which have such a wonderful resemblance to fallen leaves. The peculiar leaf-like elytra may be seen on reference to the engraving, as also the singular manner in which the limbs are furnished with wide flattened appendages, in order to carry out the leafy aspect. Only the females possess the wide, veined wing-covers, those of the male being comparatively short. The wings, however, are entirely absent in the female, while in the opposite sex they are very wide and reach to the extremity of the body. One of these insects has been brought to England, and lived for a considerable time in a greenhouse.

THE Mantidæ, or Praying Insects, also belong to the Orthoptera. These creatures derive their name from their habit of sitting with their long and flattened fore-legs held up and joined as if in the attitude of prayer. So remarkable an insect could not fail to be the subject of many wild fables, some of which may take rank as popular superstitions. For example, it was long thought that if any one lost his way in a forest and met with a MANTIS, he had only to ask the insect to direct him on his road, when the obliging creature would stretch out one of its arms and point out the proper direction. According to old legends, one of these insects, being met by St. Francis Xavier and commanded to chant a prayer as well as to act it, responded to the request of the saint by singing a canticle—we presume in the Latin language.

Unfortunately for the character of the Mantis, the real reason for holding up its feet is, to be in readiness for seizing its prey or to defend itself from an enemy, the creature being voracious as a wolf and combative as a gamecock. It feeds chiefly upon other insects, stealing upon them quietly and catching them in its claws by a rapid movement, just as the loris takes its winged prey; and should it meet with another of the same sex and species, the two begin to fight with dauntless courage, cutting at each other with their fore-legs with the skill of practised swordsmen, and making their strokes so truly and with such force, that they have been known to sever the body of their antagonist with a single blow. The winner, that is to say the survivor, generally consummates his victory by devouring the body of his slaughtered foe.

The Chinese are fond of keeping these insects in cages and matching them against each other like gamecocks or bulldogs. These creatures are said by some authors to be cowardly, because, if ants are put into their cages, they endeavour to escape in all directions. True as the fact may be, the inference is quite unwarrantable, the Mantis

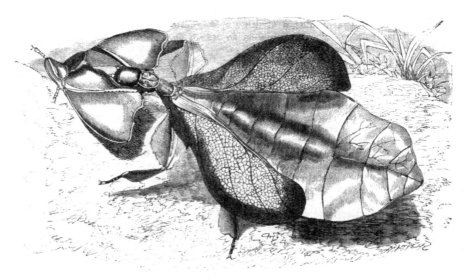

LEAF INSECT.—*Phyllium scythé.* (Female.)

being entirely justified in trying to escape from such direful foes as the ants of its own country. During the late war in India, a picket of soldiers contrived to disturb a large wasps' nest, and were forced to scatter in all directions in order to avoid the attacks of their small but formidable antagonists, for whose assaults they, being Highlanders, were very ill prepared. Yet no one would impugn the courage of the soldiers (the officer in command, an old pupil of my own, having won the Victoria Cross); and the ants are even more terrible foes to the Mantis than the wasps to human beings, their dimensions being quite disproportionate, and their usual prey being insects whom they overpower by numbers and united action, so that the size and courage of the Mantis are impotent when opposed to such foes.

THYSANOPTERA.

The next order, according to Mr. Westwood's arrangement, is that called the Thysanoptera, or Fringe-winged Insects, on account of the manner in which the wings are edged with long and delicate cilia. They are all little insects, seldom exceeding the tenth or twelfth of an inch in length, but, although small, are capable of doing considerable damage. They are mostly to be found on plants and flowers, especially those blossoms where the petals are wide and deep and afford a good shelter. The convolvulus is always a great favourite with them. Greenhouses are sadly liable to their inroads, and owing to their numbers they are very injurious to melons, cucumbers, and similar plants, covering their leaves with a profusion of decayed patches, that look as if some powerful acid had been sprinkled over them. Only one family of these insects is acknowledged by entomologists.

NEUROPTERA.

WE now come to an order of insects containing some of the most beautiful and a few of the most interesting members of the class. They are known by the possession of four equal-sized membranous wings divided into a great number of little cells technically called areolets. The mouth is furnished with transversely movable jaws, and the females do not possess a sting or valved ovipositor. In this order are comprised the ant-lions, the dragon-flies, the termites, the lace-wings, and the May-flies.

THE first family in Mr. Westwood's arrangement is that of the Termites, popularly but erroneously known by the name of WHITE ANTS, because they live in vast colonies, and in many of their habits display a resemblance to the insect from which they take their name. All the Termites are miners, and many of them erect edifices of vast dimensions when compared with the size of their architects. For example, the buildings erected by the common White Ant (*Termes bellicósus*) will often reach the astonishing height of sixteen or seventeen feet, which in proportion to the size of the insect would be equivalent to an edifice a mile in height if built by man. The dwelling is made of clay, worked in some marvellous manner by the jaws of the insect-architects; and is of such astonishing hardness, that, although hollow, and pierced by numerous galleries and chambers, they will sustain the weight of cattle, which are in the habit of ascending these wonderful monuments of insect labour for the purpose of keeping a watch on the surrounding country. A full-sized habitation of the warlike Termite resembles a large irregular cone, having a diameter about equal to its height, and covered with turrets and smaller cones. Nor is this all, for the subterranean excavations are every whit as marvellous as the building, consisting of galleries, chambers, and wells some fourteen inches in width, and penetrating about five feet into the earth. These excavations serve for homes, for nurseries, and for roads of communication between the several portions of the vast establishment.

To give a complete history of the Termites would be a task demanding so much time and space, that it cannot be attempted in these pages; and we must, therefore, content ourselves with a slight sketch of their general history, premising that many parts of their economy, and especially those which relate to their development, are still buried in mystery.

The most recent investigations give the following results :—

Each Termite colony is founded by a fruitful pair, called the king and queen, who are placed in a chamber devoted to their sole use, and from which they never stir when once enclosed. These insects produce a vast quantity of eggs, from which are hatched the remaining members of the colony, consisting of neuters of both sexes, the females being termed workers and the males soldiers, the latter being distinguished by their enormous heads and powerful jaws; of larvæ of two forms, some of which will be fully developed, and others pass all their lives in the worker or soldier condition; of pupæ of two forms; and, lastly, of male and female perfect insects, which are destined to found fresh colonies. The neuters of either sex are without wings.

In founding a colony, the order is as following:—The parent pair are taken possession of by the workers, who enclose them in a chamber which is intended as the nucleus of the infant establishment. The walls of this chamber are pierced by holes which will suffer the workers to pass, but are far too small to afford exit for the king or queen. Shortly after they have been fairly installed, the royal pair lose their wings, and a wondrous change soon takes place in the female. Though her head, thorax, and legs retain their normal dimensions, her abdomen begins to swell in the most preposterous manner, until it is as long as a man's finger and about twice its thickness, thus precluding its owner from advancing a single step.

The queen, thus developed and for ever fixed in her home, is truly the mother of her subjects, producing nearly eighty thousand eggs in twenty-four hours. The eggs are carried off by the workers as soon as laid, and conveyed to suitable places in the nest, where they are guarded until they are hatched, and are then fed and watched until they have passed through their preliminary stages of existence.

The great bulk of a Termite establishment is composed of workers, who outnumber the soldiers in the proportion of a hundred to one. By the mysterious instinct which is implanted in these insects, the soldiers and workers confine themselves to their respective occupation, the former doing nothing but fight and the latter nothing but labour.

The untiring energy and wonderful skill of the labourers are evident by the structures which they raise; and the courage of the soldiers is quite as remarkable in its way. These born gladiators seem to exult in war, and no matter what may be the size of the assailant, they hurry to the attack with reckless fury, biting fiercely with their sharp jaws; and when once they have taken hold of anything that they fancy to be inimical to them, retain their grasp with the proverbial tenacity of the bulldog. They may be pulled to pieces without loosening their hold, and the body may be dragged away, leaving the head still affixed by the jaws. The bare-limbed negro is soon forced to quit the field, for although the Termite cannot sting, it can draw, at least, its own weight of blood; and even the fully clad European will bear gory proofs of its prowess.

There are many species of Termite, and all are fearfully destructive, being, indeed, the greatest pest of the country wherein they reside. Nothing, unless cased in metal, can resist their jaws; and they have been known to destroy the whole woodwork of a house in a single season. They always work in darkness, and at all expenditure of labour keep themselves under cover, so that their destructive labours are often completed before the least intimation has been given. For example, the Termites will bore through the boards of a floor, drive their tunnels up the legs of the tables or chairs, and consume everything but a mere shell no thicker than paper, and yet leave everything apparently in a perfect condition. Many a person has only learned the real state of his furniture by finding a chair crumble into dust as he sat upon it, or a whole staircase fall to pieces as soon as a foot was set upon it. In some cases the Termite lines its galleries with clay, which soon becomes as hard as stone, and thereby produces very remarkable architectural changes. For example, it has been found that a row of wooden columns in front of a house have been converted into stone pillars by these insects.

The greater number of species belong to the tropical regions, where they are useful in destroying the fallen trees that are so plentiful in those latitudes, and which, unless speedily removed, might be injurious to the young saplings by which they are replaced. Two species, however, are known in Europe, namely, *Termes lucifugus* and *Termes ruficollis*, and have fully carried out their destructive character; the former species devouring oaks and firs, and the latter preferring olives and similar trees. At La Rochelle, these insects have multiplied so greatly as to demand the public attention.

M. de Quatrefages, who visited one of the spots in which these destructive insects had settled themselves, gives the following account of their devastating energy:—"The Prefecture and a few neighbouring houses are the principal scene of the destructive ravages of the Termites, but here they have taken complete possession of the premises. In the garden not a stake can be put into the ground, and not a plank can be left on the beds, without being attacked within twenty-four or forty-eight hours. The fences put round the young trees are gnawed from the bottom, while the trees themselves are gutted to the very branches.

Within the building itself, the apartments and offices are all alike invaded. I saw, upon the roof of a bedroom that had been lately repaired, galleries made by the Termites which looked like stalactites, and which had begun to show themselves the very day after the workmen left the place. In the cellars I found similar galleries, which were either half-way between the ceiling and the floor, or running along the walls and extending, no doubt, up to the very garrets; for on the principal staircase other galleries were observed, between the ground floor and the second floor, passing under the plaster wherever it was sufficiently thick for the purpose, and only coming to view at different points where the stones were on the surface; for, like other species, the Termites of La Rochelle always work under cover wherever it is possible for them to do so. It is generally only by incessant vigilance that we can trace the course of their devastations and prevent their ravages.

At the time of M. Audoin's visit, a curious proof was accidentally obtained of the mischief which this insect silently accomplishes. One day it was discovered that the archives of the department were almost totally destroyed, and that without the slightest external trace of any damage. The Termites had reached the boxes in which these documents were preserved by mining the wainscoting, and they had then leisurely set to work to devour these administrative records, carefully respecting the upper sheets and the margin of each leaf, so that a box which was only filled by a mass of rubbish seemed to enclose a file of papers in perfect order.

The hardest woods are attacked in the same manner. I saw on one of the staircases an oak post, in which one of the clerks had buried his hand up to the wrist in grasping at it for support, as his foot accidentally slipped. The interior of the post was entirely formed of empty cells, the substance of which could be scraped away like dust, while the layer that had been left untouched by the Termites was not thicker than a sheet of paper."

It is most probable that these insects had been imported from some vessel, as they have attacked two opposite ends of the same town, the centre being at present untouched. M. de Quatrefages tried many experiments on these insects with the view of discovering some method of destroying them, and came to the conclusion that if chlorine could be injected in sufficient quantities, it would in time have the desired result.

One good quality is, however, attributable to the Termite. The insect is eatable, and even by Europeans is pronounced to be peculiarly delicate and well flavoured, something like sweetened cream. The Termites are prepared for the table by various methods, some persons pounding them so as to form them into a kind of soft paste, while others roast them like coffee beans or chestnuts.

For further information respecting these wonderful insects, the reader is referred to Smeathman's elaborate investigations recorded in the *Philosophical Transactions* for 1781, and to M. de Quatrefages' "Rambles of a Naturalist," Vol. II.

Passing by, for the present, several families of the Neuroptera, we come to the Libellulidæ, or Dragon-flies. These insects are very familiar to us by means of the numerous Dragon-flies which haunt our river sides, and which are known to the rustics by the very inappropriate name of Horse-stingers, they possessing no sting and never meddling with horses or any other vertebrate animal. The name of DRAGON-FLY, on the contrary, is perfectly appropriate, as these insects are, indeed, the dragons of the air, far more voracious and active than even the fabled dragons of antiquity.

Even in their preliminary stages the Dragon-flies preserve their predatory habits, and for that purpose are armed in a most remarkable manner. During the larval and pupal states the Dragon-fly is an inhabitant of the water, and may be found in most of our streams, usually haunting the muddy banks, and propelling itself along by an apparatus as efficacious as it is simple, and exactly analogous to the mode by which the nautilus forces itself through the water. The respiration is carried on by means of the oxygen which is extracted from the water; and the needful supply of liquid is allowed to pass into and out of the body through a large aperture at the end of the tail. On taking one of these creatures from the water the extremity of the tail seems to be pyramidal, but on examination will be seen to consist of several pointed flakes which can be separated and then disclose the aperture above mentioned.

By means of this apparatus, water is admitted into the body, and, after giving up its oxygen, is violently expelled, thereby forcing the insect forward with a velocity proportioned to the power of the stroke. If one of these creatures be put into a glass vessel, it appears at first to move by simple volition; but if a little sand be allowed to settle at the bottom, the disturbance caused among the grains by the ejected water will show the mode of progression. If the larva be allowed to take in the water, and then suddenly moved into the air, the force with which it expels the contained water will drive it to a distance of three or four inches.

Such are its means of locomotion ; those of attack are not less remarkable or less efficacious.

The lower lip, instead of being a simple cover to the mouth, is developed into a strange-jointed organ, which can be shot out to the distance of nearly an inch ; or when at rest, can be folded flat over the face, much as a carpenter's rule can be shut up so as to fit into his pocket, and can be rapidly protruded or withdrawn very like the instrument called a " lazy-tongs." Like that instrument, it is furnished at its extremity with a pair of forceps, and is able to grasp at passing objects with the swiftness and certainty of a serpent's stroke.

The creature remains for some ten or eleven months in the preliminary stages of existence, and when the insect is about to make its final change, the undeveloped wings become visible on the back. When its time has come, the pupa leaves the water, and crawls up the stem of some aquatic plant until it has reached a suitable elevation ; it clings firmly with its claws, and remains apparently quiet. On approaching it, however, a violent internal agitation is perceptible, and presently the skin of the back splits along the middle, and the Dragon-fly protrudes its head and part of the thorax. By degrees, it withdraws itself from the empty skin, and sits for a few hours drying itself, and shaking out the innumerable folds into which the wide gauzy wings have been gathered. After a series of deep respirations of the unwonted air, and much waving of the wings, the glittering membranes gain strength and elasticity, and the enfranchised insect launches forth into the air in search of prey and a mate.

VARIEGATED DRAGON-FLY.—*Libéllula variegáta.*

There are very many species of Dragon-flies, all very similar in their habits, being fiercely predaceous, strong of wing, and gifted with glittering colours. Unfortunately, the rich azure, deep green, soft carnation, or fiery scarlet, of these insects fade with their life, and in a few hours after death the most brilliant Dragon-fly will have faded to a blackish brown. The only mode of preserving the colours is to remove all the interior of the body, and to introduce paint of the proper colours. This, however, is but an empirical and unsatisfactory sort of proceeding ; and no matter how skilfully it may be achieved, will never be worth the time bestowed upon it. In many species, the sexes are of different colours, as, for example, in the beautiful DEMOISELLE DRAGON-FLIES, where the male is deep purple, with dark spots on the wings, and the female a rich green, with the wings uncoloured. Another form of Dragon-fly may be seen on the illustration at page 463. and is given as an example of the restricted genus Libellula.

THE singular group of insects termed SCORPION-FLIES or PANORPIDÆ, also belong to this family. These insects derive their popular name from the curious appendage with which the abdomen of some of the species is armed. The male of the common Scorpion-fly has the sixth and seventh rings of the abdomen rather slender, and capable of

DEMOISELLE.---*Calépteryx splendens.*   GREAT DRAGON-FLY.---*Ictinus pugnax.*

movement in every direction; while the last ring is modified into a stout, thick, rounded form, furnished with a pair of forceps not unlike those of the earwig. While at rest, the creature sits with the abdomen curled quietly over the back like a pug-dog's tail; but when irritated or alarmed, it brandishes the tail about in a very alarming manner, snapping at the same time with the forcipated extremity, and, if it seizes the finger, can inflict a very perceptible nip. Few persons, indeed, who are not accustomed to the insect can summon up sufficient moral courage to hold it while its scorpion-looking tail is being flourished in so menacing a manner.

THE beautiful LACE-WING FLIES, or Hemerobiïdæ, are also members of this order. Several species of the Lace-wings are also called by the appropriate name of Golden Eyes, on account of the extreme brilliancy of the large and projecting eyes, which glow as if with internal fires, and give forth flashes of gold and ruby light. Unfortunately, there is a sad drawback to their beauty, for, when handled, they exhale a most powerful and indescribably odious stench, unlike any imaginable combination of evil savours, but quite unique, and never to be forgotten after a single experience. The Lace-wings may be taken in the evening as they fly from tree to tree, and in the day time may be found clinging to the under side of leaves.

The eggs of these creatures are even more remarkable than their parents. The mother-insect begins operations by drawing a fine, but viscous, thread from some twig or leaf, the thread becoming hard, elastic, and transparent as it comes in contact with the air. At the end of this thread she then places a little white egg, and, having fixed it firmly, proceeds to repeat the process until she has set some twelve or fifteen of these curious objects. The footstalks on which the eggs are placed are so firm that they will hold their burden in a nearly upright attitude. The real character of these eggs has only been ascertained in comparatively late times, they having formerly been placed among the mosses, to which indeed they bear no slight resemblance.

THE far-famed ANT-LION is one of the insects that are more celebrated in their preliminary than in the perfect stage of existence. As may be seen by reference to the illustration, their perfect form is very light and elegant, and closely resembling that of the dragon-flies, save that the wings are lighter, softer, and broader. In their larval condition, however, they are by no means attractive-looking creatures, somewhat resembling flattened maggots with rather long legs and very large jaws, the legs being apparently useless as organs of progression, all movements being made by means of the abdomen. Slow of movement as is this creature, and yet predaceous, feeding wholly on living insects, the mode of obtaining its food seems to be rather a problem. The solution, however, is simple enough, the creature digging a pitfall, and lying ensconced therein while the expected prey approaches.

Mr. Westwood, who kept several of these insects in a living state, has given the following description of them :—" It is in very fine sand that the larva makes its pitfall. When placed upon the surface, it bends down the extremity of the body, and then pushing or rather dragging itself backwards by the assistance of its hind legs, but more particularly of the deflexed extremity of its body, it gradually insinuates itself into and beneath the sand, constantly throwing off the particles which fall upon, or which it shovels with its jaws or legs upon its head, by suddenly jerking them backwards.

'Ossaque post tergum magnæ jactata parentis.

Proceeding in this manner, in a spiral direction, it gradually diminishes the diameter of its path, and by degrees throws so much of the sand away as to form a conical pit, at the bottom of which it then conceals itself, its mandibles widely extended, being the only parts that appear above the surface ; with these, any luckless insect that may happen to fall down the hole is immediately seized and killed.

When the fluids of the victim are exhausted, the Ant-Lion, by a sudden jerk, throws the dry carcase out of the hole. Should, however, the insect by chance escape the murderous jaws of its enemy, the latter immediately commences throwing up the sand, whereby not only is the hole made deeper and its sides steeper, but the escaping insect is probably hit, and again brought down to the bottom of the pit. It is chiefly upon ants and other soft-bodied insects that these larvæ feed. They are, however, capable of undergoing long fasts, for one of my larvæ remained from October till March without food.

Previous to assuming the pupa state, the larva forms a globular cocoon of less than half an inch in diameter, of fine sand, glued with silken threads spun from a slender, telescopic-like spinneret, placed at the extremity of the body, and is lined with fine silk. The pupa is small, not being half an inch long, inactive, and with all the limbs laid upon the breast. When ready to assume the perfect state, it uses its mandibles, which are quite unlike those of the larva and imago (i. e. the perfect form), to gnaw a hole through the cocoon, and pushes itself partly through the aperture, in which it leaves its pupa skin. Immediately on assuming the perfect state, the abdomen is almost immediately extended to nearly three times its previous length."

In order to enable the Ant-Lion to extract the juices of the insects on which it feeds, the inner curve on each mandible is deeply grooved, and another portion of the jaws, technically called the maxilla, plays within the groove. The larva, half-sunk in its pitfall, is shown in the left-hand lower corner of the illustration.

THE beautiful NEMOPTERA COA belongs to an allied family of this order, and is here given in order to show the curious development of the hinder pair of wings, a peculiarity which is repeated, though not on so extensive a scale, in many of the butterflies.

IN the centre of the same engraving may be seen a flying insect, which will at once be recognised as the MAY-FLY, or EPHEMERA, the best and most familiar type of the family to which it belongs, and which is scientifically called the Ephemeridæ.

This insect has long been celebrated for its short space of life, a single day sometimes witnessing its entrance into the perfect state and its final departure from the world. The popular idea concerning these insects is, that the whole of their life is restricted to a single day. This, however, is an error, as they have already passed at least two years in their preliminary stages of existence. In the larval and pupal states, they are inhabitants of the water, and are fond of hiding themselves under stones, or burrowing into the muddy banks. Under the latter circumstance they make a very curious tunnel, something like a double-barrelled gun. It is said that the larva feeds upon mud, and, as a proof of this assertion, it may be mentioned that Swammerdam always found mud within those specimens which he dissected. I can personally vouch for the accuracy of his remarks, but would not like to assert that, although mud was always found in the stomach and intestines of those larvæ which I have dissected, it might not have been swallowed with the food rather than composed it.

Nemoptera Coa.

ANT-LION.—Myrméleon translátus.

MAY-FLY.—Ephémera vulgáta.

The May-fly is peculiarly notable for a stage of development which seems to be quite unique among insects. When it has passed through its larval and pupal state, it leaves the water, creeps out of its pupa case, and takes to its wings. After a period, varying from one to twenty hours, it flies to some object, such as the trunk of a tree or the stems of water-plants, and casts off a thin membranous pellicle, which has enveloped the body and wings, the dry pellicle remaining in the same spot, and looking at first like a dead insect. After this operation, the wings become brighter, and the three filaments of the tail increase to twice their length. Some authors call the state between the leaving the water and the casting the pellicle the "pseudimago" state.

Some of these insects are well known to fishermen under the names of green and grey drake, the former being the pseudimago, and latter the perfect form of the insect, which is represented in the illustration. Sometimes these insects occur in countless myriads, looking like a heavy fall of snow as they are blown by the breeze, and having on some occasions been so plentiful, that they have been gathered into heaps and carted off to the fields for manure.

The Perlidæ, known to anglers by the name of STONE-FLIES, belong to the Neuroptera. Several species of the same family are popularly called Yellow Sally and Willow-fly.

They may be known by the large folded front pair of wings, and the two bristle-like appendages at the tail.

QUITTING the Neuroptera, we must give a few lines to another order of insects, the TRICHOPTERA, popularly known by the name of CADDIS-FLIES.

These insects, of which there are many species, are chiefly remarkable in their larval state, on account of the curious portable habitations which they construct. All anglers are familiar with the Caddis, and the singular variety of form and material employed in the construction of its home. Being a soft, white grub, totally unarmed, and presenting a most delicate morsel to every river-fish, the Caddis is forced to conceal itself in some way from its innumerable foes. For this purpose, it builds around itself a nearly cylindrical tube, open at each end, and composed of substances varying according to the locality and the species. Sometimes these tubes are made wholly of short pieces of stick, laid sometimes side by side, and sometimes in a partly spiral form, something like the wires of the submarine telegraph. Sometimes the tubes are made of sand or little stones, while the deserted shells of the planorbis, and other fresh-water shells, are very common materials.

More than once I have known the Caddis to affix living shells to its case, and in consequence to be sadly bewildered when it wanted to move in one direction and the fettered shell tried to move in another. Fragments of reeds, patches of dead leaves, straw, seeds, and all kinds of similar substances, are employed for this purpose by the Caddis. In my own collection, I have several specimens where the creature has made use of the opercula once belonging to dead molluscs, and one instance where the Caddis has pressed into its service the chrysalis of some moth that had fallen into the water. In some species, the shape of the tube is precisely like that of the elephant tusk-shell, described on page 403, and is built up of fine sand. I found great numbers of them in a deserted stone quarry in Wiltshire.

When the larva is about to enter the pupal state, it prepares for the expected change by fixing the tube to some firm object, and spinning a sieve-like net across each extremity, so as to permit water to pass, but to exclude all enemies. There is a regular pattern in this net, and each species seems to have a pattern peculiar to itself. Cases thus prepared may be procured near the beds of streams and rivers, where they may be seen anchored to the submerged plants and stones. The reader will not fail to notice the analogy between the moveable tube of these insects and that of the too common clothes-moth. The word Trichoptera signifies hairy-winged, and is given to these insects in allusion to the soft hairs with which the front pair of wings are usually coated.

WE now come to a vast order of insects, technically called the HYMENOPTERA. In these insects the wings are four in number, transparent, membranous, the veins comparatively few, and the hinder pair smaller than the others. Their mouth is furnished with powerful horny jaws, and with a tongue guarded by the modified maxillæ. The females are armed with a many-valved sting or ovipositor. In this enormous order are included all the bees, wasps, and their kin, the great family of saw-flies, the ichneumons, the gall-flies, and the ants, each single family being so large, and presenting so many points of interest, that an entire volume could be devoted to them with great profit. Our space, however, prohibits us from attempting more than a slight sketch of each family, together with descriptions of a few typical species. Without, therefore, enumerating the various arrangements of this large order, or the characteristics on which they are founded, we will proceed at once to the family of the Tenthredinidæ, or Saw-flies, the first in Mr. Westwood's system.

In this and the next family, the females are furnished with a peculiar ovipositor, composed of several pieces, and which, though connected with a gland secreting an irritant fluid, are not envenomed as in the bees, wasps, and their kin. All these insects are comprised under the general term of Terebrantia, or borers, and fall easily into two large groups, in one of which the abdomen proceeds directly from the thorax, and in the other is connected with the thorax by means of a footstalk. Each of these groups is further subdivided, as will be seen in the course of the following pages.

The true Saw-flies are known by the curious piece of animal mechanism from which they derive their name. The females of this family are supplied with a pair of horny saws, placed side by side on the lower extremity of the abdomen.

These saws are of various forms, according to the particular species to which they belong, and may be seen even in the dried specimens, the top of their sheath slightly projecting, and their shapes plainly visible after the removal of a portion of the abdomen. When taken from the insect and placed under the microscope, they present a very pretty appearance, owing to the gently-curved ribs with which their sides are strengthened and decorated. The saws act alternately, one being pushed forward as the other is being retracted. Their object is to form a groove in some plant, in which the eggs of the mother insect can be deposited, and wherein they shall find a supply of nourishment in order to enable them to complete their development; for it is a most remarkable fact that, after the egg is deposited in the groove, it rapidly increases in size, obtaining twice its former dimensions.

In the genus Cimbex, of which an example is given in the illustration, the larvæ possess twenty-two feet, and have the power of discharging a translucent greenish fluid

Cimbex femorata.
Rhyssa persuasória.

Urócerus gigas.
Ichneumon grossárius.

from certain pores placed on the sides of the body just above the spiracles. This feat they can repeat six or seven times in succession. When they have eaten their way to the next stage of existence, they spin a cocoon of a brownish colour and of a stringy, tough consistency, and either suspend it to the branches of the tree on which they have been feeding, or hide it under fallen leaves. In this cocoon they remain for a comparatively short time, and then emerge as perfect insects.

The terrible TURNIP-FLY (*Athalia centifóliæ*) belongs to this family. The larva of this species is popularly called the Nigger, on account of its black colour. A species of Athalia may be seen in the engraving on page 463, and is the smallest of the three flying insects at the upper part of the illustration. The larva of this species feeds upon the various cabbages, eating away the whole of the soft green parts of the leaves, and only rejecting the thick nervures. It makes no cocoon, but retires into the ground, excavates a kind of oval cell, which it lines with a slimy substance, and there awaits its final change.

The well-known black GOOSEBERRY-FLY (*Nématus grossuláriœ*) is another of the Saw-flies. Its larva, so destructive to the fruit, is blackish grey. These tiresome creatures are often seen in great numbers, more than a thousand having been taken on a single gooseberry-bush, and there are two broods in the course of a year. Without going into further details, it is sufficient to say that there is hardly a plant without its especial Saw-fly, and that any one who can discover a really effectual mode of checking their ravages, will confer no slight benefit on mankind.

THE fine insect on the same illustration, which is known by the name of the GIANT ICHNEUMON, is an example of the next family, in which the ovipositor is converted into a gimlet instead of a double saw. With this powerful instrument, the female is enabled to drill holes into living timber for the purpose of depositing the eggs. When they are hatched, the young grubs immediately begin to gnaw their way through the wood, boring it in every direction, and making burrows of no mean size. Those of the present species prefer fir and pine, and I have had specimens of the wood sent to me which have been riddled by the grubs until they looked as if they had harboured a colony of the ship-worm. The perfect insects often make their appearance in houses, the larvæ having been concealed in the timbers and rafters; and I know of one case where a gentleman who had built a wooden garden-house, was sadly annoyed by the multitudes of the Sirex which emerged from the timber. In such cases the insects do not seem to attain their full dimensions, but appear dwarfed and stunted. All wood-boring insects are, however, extremely variable in size.

The next group of the Terebrantia is called Entomophaga, or Insect-eaters, because the greater number of them are parasitic upon other insects, just as the Saw-flies are parasitic upon vegetables. In these insects the ovipositor is furnished with two delicate spiculæ, and the last segments of the abdomen are not formed into a telescope-like tube.

The first family is that of the Cynipidæ, or Gall insects, the creatures by whose means are produced the well-known galls upon various trees, the so-called oak-apple being perhaps the best known, and the Ink-gall (also found on the oak) the most valuable. These Galls are formed by the deposition of an egg in the leaf, branch, stem, twig, or even root of the plant, and its consequent growth. The well-known Bedeguar of the rose, with its soft mossy envelope and delicate green colour, relieved by bright pink, is caused by one of these insects (*Cynips rosœ*); and the celebrated Dead Sea-apples are nothing but galls formed by the *Cynips insána*. The spherical oak-galls, which contain a single insect, and are about the size of a large marble, are closely allied to the true Ink-galls; and if one of these objects be cut with a knife, the action of the astringent juice upon the iron of the blade will produce a kind of ink. The best galls are those which are gathered before the insect makes its escape, as the astringent quality is then more powerful.

The true Ichneumons, of which a specimen is given in the illustration, form a vast group of insects, the British Ichneumonidæ alone numbering many more than a thousand described and acknowledged species. In them the ovipositor is straight, and is employed in inserting the eggs into the bodies of other insects, mostly in their larval state. In some cases, this slender and apparently feeble instrument is able to pierce through solid wood, and is insinuated by a movement exactly like that which is employed by a carpenter when using a bradawl. When not engaged in this work, the ovipositor is protected by two slender sheaths that enclose it on either side.

Were it not for the Ichneumons, our fields and gardens would be hopelessly ravaged by caterpillars and grubs of all kinds, for practical entomologists always find that when they attempt to rear insects from the egg or the larval state, they must count upon losing a very large percentage by the Ichneumons.

Take, for example, three or four caterpillars of the common white cabbage butterfly place them under water, and open the body from end to end. It will be found that, in almost every case, the caterpillar bears the seeds of death within its body in the shape of tiny white grubs, like very minute grains of rice. These creatures are the young of an Ichneumon-fly (*Microgaster glomerátus*), and retain their place within the caterpillar until the time for it to change into the perfect form. They then simultaneously eat their way out of the skin, spin a number of bright yellow silken cocoons, and in process of time change into tiny flies and set out on their destructive mission. The caterpillar never survives their attacks, and is seldom able to move away from the spot whereon it happened to be when the Ichneumons make their escape, the body being enveloped in their yellow cocoons.

All the Ichneumon-flies may be distinguished by their fussy restless movements, as they run up and down any object on which they may settle, and the continual quivering of their antennæ. The two lower figures in the illustration belong to this family, that on the left showing an example of the long ovipositor with which several species are furnished, and the other being given in order to show the wasp-like abdomen and the curled antennæ.

The *Rhyssa persuasoria* is the largest British Ichneumon, and is parasitic on *Urocerus juvencus*, another species of the same genus as that to which the giant Ichneumon belongs. The larva on which it preys bores deep holes in fir-trees, and, in consequence, the Rhyssa may be seen running up and down the trunks in search of some spot where the ovipositor may be introduced so as to lodge in the hidden larva. So deeply does the insect contrive to force its weapon into the wood, that it sometimes is unable to withdraw it, and may be seen hanging dead and dry to the tree in which it has buried the ovipositor too firmly.

PASSING by several families belonging to this group, we must briefly mention the beautiful RUBY-TAIL FLIES, or CUCKOO FLIES, or Chrysididæ, so plentiful in summer about old walls and similar localities. These are distinguished by the fact that, in the females, the last segments of the abdomen are formed into a telescopic tube, which can be projected or retracted at pleasure, and is furnished with a minute sting. These are, perhaps, the most brilliant in colour of any British insect, and are veritable humming birds of the insect tribes, their bodies literally flashing with ruby, sapphire, and emerald, as they flit restlessly in the sunbeams. They are parasitic insects, and haunt the walls for the purpose of depositing their eggs in the larvæ of sundry solitary bees and wasps.

Crabro cribrárius. *Philanthus triángulum.*

IN the next great division of Hymenopterous insects, the ovipositor of the female is changed into a sharply pointed weapon, popularly called a sting, and connected with a gland in which is secreted a poison closely analogous to that which envenoms a serpent's tooth. These are again divided into the Insectívora, or those which have fore-wings not folded, and the larvæ solitary and feeding on other insects ; the Sodáles, where the fore-wings are not folded, and the larvæ are social ; and the Diplóptera, where the fore-wings are folded, and the larvæ (in the social species) inclosed in separate cells.

The first of these sections comprises all those curious and interesting insects known popularly by the names of Sand Wasps and Wood Wasps. These creatures are in the

3. K K

habit of making burrows into the ground or in posts, and placing therein their eggs, together with the bodies of other insects which are destined to serve as food for the future progeny. Spiders are sometimes captured and immured for this purpose. In many instances the captured insects are stung to death before they are placed in the burrow, but it is often found that they only receive a wound sufficient to paralyse them, so that they lead a semi-torpid life until they are killed and eaten by the young grub. Two of these Sand Wasps are given in the illustration. That on the left is one of the wood-borers, drilling its burrow into posts, palings, and similar substances, and feeds its young with the larvæ of one of the leaf-rolling caterpillars that lives in the oak, and is scientifically known by the name of *Tortrix chlorána*. It also employs for this purpose several two-winged insects. One species of these burrowing wasps prefers the well-known cuckoo-spit insect for this purpose (*Aphróphora spumária*), pulling it out of its frothy bed by means of its long legs.

The right-hand figure represents a species that is in the habit of provisioning its burrow with the hive-bee, which it contrives to master in spite of the formidable weapon possessed by its victim, and then murders or paralyses by means of its sting. M. Latreille mentioned that he saw from fifty to sixty of these insects busily engaged in burrowing into a sandbank not more than forty yards long; and as each female lays five or six eggs, and deposits a bee with each egg, the havoc made among the hives is by no means inconsiderable.

Monédula signata.   *Pompílus nóbilis.*   *Scólia pratórum.*

In the accompanying illustration is shown a Brazilian species, belonging to a genus which is represented in England by more than twenty species. In these insects the legs are very long and spider-like, enabling their owners to run about among grass with great vivacity, their wings quivering all the while with violent agitation. Some of the species are in the habit of catching spiders, and provisioning the burrows with them. It is worthy of notice, that the largest specimens of Hymenoptera are to be found in exotic insects belonging to this family, the genus Pepsis being most remarkable for the great dimensions of its members.

The right-hand figure represents an insect which, though common in Southern Europe, has not yet been satisfactorily proved to be an inhabitant of England. Judging by the habits of those species which have been studied, the whole of the family to which it belongs are sand-burrowers, and seem to be cruelly predacious, mastering insects of considerable size, and dragging them into their burrows. One of these insects (*Scólia bicincta*) has been known to capture and inter a large locust, the tunnel being some eighteen inches in depth and very wide at the mouth.

The fine insect called *Chlorion lobatum,* which is shown in the centre of the illustration on page 501, is a formidable but useful creature, waging fierce war against cockroaches, those pests of Oriental houses. Its services are fully appreciated by the natives, none of whom would kill one of these insects on any account, or permit any one to injure it. With the slaughtered cockroaches it stocks its nest as a provision for the young when they escape from the egg. These insects are tolerably numerous, and are all remarkable for the bright and yet deep purple and green of their bodies, and sometimes of their wings.

AT the right-hand of the accompanying illustration may be seen a curious wingless insect, with head disproportionately large, when the size of its body is taken into consideration. This is an example of a family where the females, although armed with a powerful sting, are quite destitute of wings. Most of the Mutillidæ are exotic, requiring a large amount of heat to preserve them in health, only a very few being natives of our own country. In some of the larger species the sting is fearfully poisonous, a single insect having been known to make a man so seriously ill that he lost his senses a few minutes after being stung, and his life was despaired of for some time. A child has been known to die from the effects of the sting inflicted by the Scarlet Mutilla of North America, an insect whose weapon is as long as the abdomen. All these insects appear to be sand-borers.

GIANT ANT.—*Pónera grandis.*   *Myrmécia forficáta.*   LARGE-HEADED MUTILLA.-- *Mutílla cephalótes*
RED ANT.—*Formica sanguinea.*

THE last-mentioned insect evidently affords a transitional link between the previous families of Hymenoptera and the true Ants, or Formicidæ. These insects, as is well known, associate in great numbers, and as is peculiarly the case with the bees, the great bulk of their numbers is composed of workers, or neuters, which are destined to perform the constant labours needful to regulate so large a community. The perfect insects of either sex take no part in the daily tasks, their sole object being to keep up the numbers of the establishment. In the Ants, moreover, the neuters are without wings, and even the perfect insects only retain these organs for a brief period of their existence.

Every one has heard of the objects called ants' eggs, which are so strongly recommended as food for the nightingale and other birds, and many persons though they have seen them, have believed them really to be the objects which their popular name would infer. In truth, however, they are the cocoons of the stingless ants, in which the insects are passing their pupal state before emerging in their winged condition. It has been already mentioned, that only the perfect males and females possess wings.

K K 2

As soon as they gain sufficient strength, they fly upward into the air, where they seek their mates and soon descend to earth. The males, having now nothing to do, speedily die, as they ought, but the females begin to make provision for their future households. Their first proceeding is a rather startling one, being the rejection of the wings which had so lately borne them through the air. This object is achieved by pressing the ends of the wings against the ground, and then forcing them suddenly downwards. The wing then snaps off at the joint, and the creature thus reduced to the wingless state of a worker, is seized upon and conveyed to a suitable spot, where she begins to supply a vast quantity of eggs. These are carefully conveyed away and nurtured until they burst forth into the three states of male, female, and neuter, the precise method by which the development is arrested so as to produce the neuter condition not being very accurately known.

The remaining three figures on the illustration represent different species of Ants, the two larger species being natives of Brazil. In the tropics, the Ants are alternately curses and blessings to the inhabitants. They are terribly destructive, they eat everything softer than stone or metal, they swarm in houses, on the plains, and in woods, and occasionally they march in vast armies, taking a line as direct as the old Roman roads, and not to be stopped by any less obstacle than a river. They pass through houses, and at their approach all the human inhabitants vacate the premises, none daring to oppose so redoubtable a foe. In this case, however, the visits of the Ants are greatly beneficial, for in a very short time the column will have passed fairly through the house, and left no living creature within its walls ; beetles, spiders, scorpions, centipedes, reptiles, and even the rats and mice, being torn to pieces by their powerful jaws.

In our own country they do little harm, except in houses, where they sometimes swarm to an unpleasant extent. In gardens, too, they are often unpleasantly numerous, but can be easily destroyed by pouring boiling water or naphtha into their tunnels. The RED ANT is remarkable for being an English example of the slave-making insects. These creatures invade the nests of the Brown Ant (*Formica fuscus*), carry off the pupæ, and hatch them in their own nests, where they labour with perfect cheerfulness, unacquainted, indeed, with the fact that they are in captivity. The well-known WOOD ANT (*Formica rufa*) is a very interesting insect, its large nest, composed externally of bits of hay, twigs, &c. being fully as wonderful as the combs of the bees or wasps. If one of these nests be broken into, the powerful acid smell of the formic acid secreted by the insects is strongly apparent; and if the hand be held within an inch or two of the insects, they will cover it with this acid, the first feeling being something like the contact of a nettle, but the slight prickling sensation going off in a few minutes. All their habits are very interesting, and well worth examination. Through lack of space, however, we must now leave the Ants and proceed to the next family.

WE now come to the Wasps, in which the wings are folded throughout their entire length when at rest. The left-hand figure in the illustration represents an Australian example of the Solitary Wasps, many of which are found in England. The curious nest of this insect is shown immediately above, suspended to a branch. The creature makes a separate nest for each egg, the material being clay well worked and the shape as is represented in the engraving. The nest is stocked with the larvæ of moths or butterflies.

To this family belongs that wonderful Burrowing Wasp, which is a builder as well as an excavator, and which erects a tubular entrance, often more than an inch in height, with the fragments of sand which it has dug from the tunnel. It is thought, and probably with correctness, that the object of the insect in making this edifice is to deter its parasitic foes from entering so long and dark a channel. The tube is always curved. When the burrow is completed, the Wasp lays its egg in the tunnel, and packs in it a series of little green caterpillars, which serve as food for the larva. When the arrangements are completed, the Wasp takes down her tube, and employs the materials in closing the mouth of the tunnel. The technical name of this insect is *Odynerus muraria*. Another species is also known to possess this curious faculty.

The true Wasps, or Vespidæ, come next in order. These insects are gregarious in their habits, building nests in which a large, but uncertain number of young are reared.

Eúmenes arcuátus.

Chlórion lobátum.

Vespa aucális.

The common Wasp makes its nest within the ground, sometimes taking advantage of the deserted hole of a rat or mouse, and sometimes working for itself. The substance of which the nest is made is a paper-like material, obtained by nibbling woody fibres from decayed trees or bark, and kneading it to a paste between the jaws. The general shape of the nest is globular, and the walls are of considerable thickness, in order to guard the cells from falling earth, a circular aperture being left, through which the inhabitants can enter or leave their home.

The cells are hexagonal and laid tier above tier, each story being supported by little pillars, made of the same substance as the cells, and all the open ends being downwards, instead of laid horizontally, as is the case with the bees. It will thus be seen that, on account of this arrangement, the nurse-wasps are enabled to get at the grubs as they lie, or rather hang, in their cells, with their heads downwards.

The grubs are fat, white, black-headed creatures, very well known to fishermen, who find them excellent bait after they have been baked, so as to render them sufficiently hard to remain on the hook. When they are about to enter the pupal state, they close the mouths of their cells with a silken cover, through which the black eyes are plainly visible, and there wait until they emerge in the perfect state. The grubs are fed with other insects, fruit, sugar, meat, or honey, the mingled mass being disgorged from the stomachs of the nurses and thus given to their charge.

There are separate cells for males, females, and neuters, the two former classes only being produced towards the end of autumn, so as to keep up a supply for the succeeding year.

A very fine species of Wasp is shown in the engraving. This is a native of China, and another magnificent species may be seen at the upper part of the engraving on page 463.

Many species of Wasp inhabit England, the HORNET (*Vespa Crabro*) being the largest, and, indeed, being nearly equal in dimensions to any tropical species. This formidable insect makes a nest very similar to that of the wasp, but the cells are necessarily much arger. The nest is generally placed in hollow trees, but I have known a colony of these insects to establish themselves in an outhouse, and to cause great annoyance before they could be expelled.

A very pretty nest is also found in this country, the work of the *Vespa Britannica.* It is suspended to branches, is nearly globular in shape, and extremely variable in size, some specimens being nearly a foot in diameter, while others are comparatively small. A very pretty specimen in my possession is about the size of a tennis ball. Some exotic species make nests, the covering or outer case whereof is thick and tough as pasteboard, and nearly white in colour. One of these nests, which is found in the Brazils, is popularly called the Dutchman's pipe, its shape somewhat resembling an exaggerated pipe-bowl, the aperture for ingress and egress doing duty for the mouth, and the branch on which it is suspended taking the place of the stem. I believe that the insect which forms this curious structure belongs to the genus Chartergus. The central orifice penetrates through all the layers of combs.

The left-hand figure on the engraving at page 498 represents a fine insect, a native of Brazil, belonging to the Bembecidæ. Of this family we have no British examples. This species is in the habit of catching grasshoppers of considerable size, carrying them off, and stocking with these insects the habitation made for its young. A very fine species of Chrysis is parasitic upon it.

THERE are, perhaps, few insects so important to mankind as those which procure the sweet substance so well known by the name of honey. Nearly all the honey-making Hymenoptera are furnished with stings, and in many species the poison is fearfully intense. Some of these insects, such as the HIVE BEE, make waxen cells of mathematical accuracy, the larvæ being placed in separate cells, and fed by the neuters. In some cases, such as the common HUMBLE BEE, the cells are egg-shaped, each cell being either occupied by a larva, or filled with honey; while in some species the eggs are placed parasitically in the nests of other bees, so that the larvæ feed either upon the stores of food gathered for the involuntary host, or upon the body of the deluded insect itself.

The Hive Bee is the typical example of the honey-gatherers, but its general economy is too well known to need much description. Suffice it to say, that, as in the ants, the community consists of males, females, and neuters, but that in the Bees, all the members of the establishment are winged, and the wings are permanent. In each hive there is one fully-developed female, called the queen, several others in process of development, and intended to be the heads of future establishments, a limited number of males, and a vast band of neuters, *i. e.* undeveloped females. The males have no sting, but both the females and neuters are armed with this tiny, but formidable weapon. Since in civilized countries the Hive Bees are kept in habitations of limited size, their numbers soon outgrow their home, and a large number accordingly quit the hive under the government of the old queen, the rule of the hive being taken up with one of the young queens which has burst from its cell in the meanwhile. A fresh colony is founded as soon as the Bees can meet with shelter, and their new residence is speedily filled with honey and young. The cells of the Bee-comb are set back to back, and each comb hangs like a thick curtain from the top and sides of the hive, so that the cells lie nearly horizontally.

In gathering honey, the Bees lick the sweet juices from flowers, swallow them, and store them for the time in a membranous cup, popularly called the honey-bag. When this cup is filled, the Bee returns to the hive, and discharges the honey into cells, closing its mouth with wax when it is filled. The structure of the Bee-cell, its marvellous adaptation to the several purposes for which it is intended, its mathematic accuracy of construction, whereby the best amount of material is found to afford the greatest amount of space and strength, are subjects too complicated to be here described, but may be found in many works which have been written upon the Hive Bee.

THE members of the genus NOMADA, a specimen of which may be seen in the right-hand upper corner of the engraving on page 463, are very wasp-like in their general aspect, are not hairy, and are, indeed, often taken for small wasps by inexperienced observers. They are, however, true bees. Their habits are rather obscure, but they are thought to be parasitic insects.

Bombus oriéntalis.

Apis fasciátus.

Xylócopa Caffra.

The right-hand insect in the engraving is the CARPENTER BEE of Southern Africa, one of those curious insects which construct a series of cells in wood. After completing their burrow, which is open at each end, they close the bottom with a flooring of agglutinated sawdust, formed of the morsels bitten off during the operation of burrowing, lay an egg upon this floor, insert a quantity of "bee-bread," made of the pollen of flowers and their juices, and then cover the whole with a layer of the same substance that was used for the floor. Upon this is laid another egg, another supply of bee-bread is inserted, and a fresh layer of sawdust superimposed. Each layer is therefore the floor of one cell and the ceiling of another, and the insect makes on the average about ten or twelve of these cells.

AT the left hand of the engraving is shown one of the numerous HUMBLE BEES, a group of insects readily recognised by their thick hairy bodies and general shape. Their nests are placed underground, often in banks, and contain a variable number of cells, sometimes not more than twenty in number, and sometimes exceeding two, or even three, hundred. The cells are loosely connected together, and are of an oval shape, their texture being tougher and more paper-like than those of the hive bee. In these, as in the ordinary bees, there are the three kinds of inhabitants; but with the Humble Bees, both the females and neuters take part in the labours of the establishment, while the number of perfect females is comparatively large.

The honey made by these insects is peculiarly sweet and fragrant, but to many persons, myself included, is rather injurious, always causing a severe headache. Some of the Humble Bees (*Bombus muscórum*) employ moss in the construction of their nests, and pass it, fibre by fibre, through their legs, in a manner that reminds the observer of carding cotton. They are on that account popularly called CARDER BEES. Others, again (*Bombus lapidarius*), prefer to make their nests in heaps of stones, or similar localities, and these are the fiercest of their kind. Generally, the Humble Bees are quiet and inoffensive, even permitting their nest to be laid open and the cells extracted without offering to molest the invader. The ORANGE-TAILED HUMBLE BEE, however, is large and fierce; and possessing a powerful sting, with a very large poison-gland, becomes no despicable foe to those who offend it, or whom it chooses to consider as foes.

The nests of the Humble Bees are not permanent like those of the hive bee, but perish during the winter, the only survivors being a few females, who are destined to found fresh colonies in the succeeding year.

The central figure of the plate represents the BANDED BEE, the species which is so greatly in use in Egypt, and which is fed by being placed on board of barges, and transported down the Nile, so as to insure a bountiful supply of honey. The owners of the hives pay a small sum to the owners of the boats, and, in return, their bees are carried along the fertile stream during the honey season, and afterwards returned with full combs. Payment is mostly in kind, thus insuring the proper fulfilment of the compact.

For want of space, we are compelled to pass by many interesting Hymenoptera, such as the Leaf-cutter Bees, the Wood-borers, and the Mason Bees, each of which creatures would demand more space than can be given to the whole of the insects.

Another species of Humble Bee is given at the lower part of the engraving on page 463. The remaining figures represent two beetles, one belonging to the Dynastidæ, and given in order to show the remarkable doubly-forked projection of the head, and the other, belonging to the Crioceridæ, and represented on account of the astonishing development of the hinder legs. It is nearly related to the beautiful little Donacia, so common on aquatic plants, and is equally brilliant in its golden green armour.

STREPSIPTERA.

A VERY small, but very remarkable order now comes before our notice, the STREPSIPTERA, comprising insects of very minute proportions, all of which are parasitic upon the bodies of different bees and wasps, five, and even six, having been discovered within a single wasp. Their presence may generally be discovered by the peculiar swollen aspect of the abdomen, and, in many cases, the heads of the parasites may be seen protruding from between the segments.

The name Strepsiptera signifies, literally, twisted wings, and is given to these creatures because the front pair of wings are transformed into short and twisted appendages, quite useless for flight or for defending the second pair of wings. These are almost disproportionately large, membranous, and with a kind of milky look as the insect flies through the air. The eye is composed of a very few lenses, in some species only fifteen on each side, two or three thousand being the ordinary average among insects. The antennæ are of a remarkable form, branched and forked like the horns of a stag. The thorax is enormously large, and the abdomen of very small size ; but as the creature does not appear to take food during its life in the perfect state, this is of little moment. Curiously enough, the larvæ of these insects are themselves subject to internal parasites ; and it is very possible that they, in their turn, may be infested by other creatures less than itself, and equally disagreeable.

LEPIDOPTERA.

We now come to an order in which are included the most beautiful of all insects, namely, the Butterflies and Moths. On account of the feather-like scales with which their wings are covered, and to which the exquisite colouring is due, they are technically called Lepidoptera, or scale-winged insects.

The wings are four in number, and it is occasionally found that the two pairs are connected together by a strong bristle in one, and a hook-like appendage in the other, so that the two wings of each side practically become one member, in a manner similar to the formation of many hymenopterous insects. Those species which take any nourishment

Mechanitis Lysimnia.	Gynœcia Dirœ.	Catagramma Marchdlii.
Thecla.		
Helicópis cupido.	Papilio Thoas.	
Mesosémia misipsa.	Epicália ansœu	Papilio Protesilaus.

subsist entirely upon liquid food, which is drawn into the system by suction, and not by means of a brush, as is the case with the liquid-feeding beetles and bees. The wings are strengthened by nervures, which are of great use in determining the position of the insects.

The scales with which the membranous wings are at once protected and adorned are of various shapes, sometimes broad, flat, and overlapping each other like the tiles of a house-roof, sometimes long and hair-like, sometimes drawn out like a set of park palings with notched tops ; while others assume the most fantastic forms, and perhaps resemble negro hands with spread fingers, as in the well-known death's-head moth, or battledores, as in the little blue butterfly of the meadows. Their surfaces are always sculptured in some way, and the markings on these minute objects, most of which are singly invisible to the naked eye, are so bold and determinate, that, in most cases, an entomologist can name the genus, and in some the species, from which a scale has been taken.

The series of changes undergone by the Lepidoptera are, perhaps, better known than those of any other order, on account of the large dimensions and conspicuous habits of the insects. The larva is popularly known under the title of caterpillar, and is always a vegetable feeder, sometimes devouring the bark, now and then the solid wood, but mostly the leaves of the plants on or in which it resides. During its life in the caterpillar state, it eats almost incessantly, laying up in the interior a store of fat on which it may sustain existence during its pupal stage, and increases in size with wonderful rapidity. The larva, for example, of the death's-head moth is larger and longer than the middle finger of an ordinary man, and yet when first hatched from the egg is no longer than the capital letters at the head of this page. The skin cannot keep pace with the growth, and accordingly splits as soon as its expanding properties are exhausted, permitting the caterpillar to crawl from its old envelope, when it resumes eating with great vigour.

After the skin has been cast several times, the creature exchanges the caterpillar dress for that of the pupa, or chrysalis, and remains in that state for a variable time without taking food, and almost without motion. The form of the chrysalis is mostly spindle-shaped, but in many cases, especially the butterflies, it is angular, and altogether oddly formed. The locality chosen by the creature during this strange period of its life is even more variable than its form. Some chrysalids remain within the trunks of trees, some burrow deeply into the earth, some hang themselves up by their tails, some sling them-selves horizontally in hammocks, while many spin cocoons, or build strong wooden edifices in which they may await their last change. At the proper time, the pupa skin also splits, and permits the perfect insect to emerge from its prison, and take to the wide world. If the creature retains memory, the change of habit must be indeed wonderful. From being a slow, crawling, grub-like creature, gnawing coarse vegetable substances with ceaseless voracity, it becomes a denizen of air, fluttering gaily over space, and daintily sipping its little modicum of sweet juices from the odorous flowers.

Having given this general glance at the order, we will now proceed to our examples.

In the system which is adopted in this work, the Lepidoptera are divided into two sections, the Butterflies and Moths, technically called Rhopalocera and Heterocera, which may generally be distinguished from each other by the form of the antennæ, those of the Butterflies having knobs at their tips, while those of the Moths are pointed. The first family is that of the Papilionidæ, in which are included the largest and most magnificent specimens of this order. The fanciful names with which so many of these insects have been honoured are chiefly due to Linnæus, who was so struck with the splendid dimen-sions and gorgeous colouring of these insects that he deemed them worthy of Homeric titles, called them Equites, or knights, and, separating them into two divisions, gave to all those which have red spots on the sides of the breast the names of the Trojan heroes, and to those which were without the red spots, and had an eye-shaped mark on the lower wings, the names of the Greek warriors who fought against Troy. Unfortunately, for this division, it happens that the two sexes of many species are very diverse, and cause great confusion, so that Polycaon, one of the Greeks, and Laodocus, a Trojan hero, have been found to be the two sexes of the same insect, the latter being the female.

AMPHRISIUS. *Ornithóptera Amphrisius*

The splendid insect which is shown in the illustration is one of a genus which, by common consent, takes the first rank among the Lepidoptera, in consequence of their great size, the elegant boldness of their shapes, and the richness of their colouring. In them, the two fore-legs are not stunted, as is the case with so many butterflies, but are large, strong, and can be used in walking; and the tips of the antennæ have a very slight bend upwards. The caterpillar of this insect is furnished with two retractile tentacles, placed on the neck in a fork-like shape, and is able to protrude these organs at will, at the same time emitting a very unpleasant odour. The chrysalis is hung up by means of silken threads from the tail, and the body is prevented from swinging about too rudely by a pair of very stout silken cords, one of which is affixed to each side, and moors the pupa firmly to the substance against which it hangs. The shape of the pupa is rather peculiar, being angular in its outline, and having also an irregular curve.

The perfect AMPHRISIUS is a boldly marked insect, though without much variety of colouring. The upper wings are rich blackish brown, and the lower are fine king's yellow, edged with jetty black, and having a fringe of long hair-like scales upon their inner edges. The under side is nearly of the same colours, except that a few dashes of chalky white are seen upon the upper wings, as if dashed in with a quick sweep of a dry brush, and a similar dash of yellow is seen upon the lower wings. The abdomen is bright yellow below and dark brown above, and round the neck is a narrow collar of fiery crimson, rich and silken as " chenille."

Another species of this genus, the POSEIDON, is a great favourite with the natives of the Darnley Islands, who are accustomed to catch several of these fine insects, to tie one end of a fine thread to the butterfly, and the other to their hair, so as to permit the insects to flutter about their heads. This style of head-dress is much admired, and, indeed, is equally poetical and artistic. Many of these insects differ greatly, according to sex, the upper wings of the male PRIAM, for example, being velvety black, striped with silky green, and the hinder wings entirely silky green, spotted with black and orange; while the female is dark brown, spotted with white.

IN the genus Papilio we find the insects to be of nearly as magnificent proportions as in the former genus, though none of them reach the enormous size of the Priam, which will sometimes measure nearly eight inches across the spread wings. The colours are, however, more varied, and quite as brilliant, while a curious feature is often added by the prolongation of the hinder wings into two long tail-like appendages. The larva is of varied form, sometimes smooth, sometimes covered with fleshy protuberances, sometimes long, and able to throw out or to withdraw at pleasure the two first segments of the body, sometimes short, thick. and grub-like, and in one or two instances marvellously resembling snails in the general form. The genus is a very comprehensive one, including between two and three hundred known species, among which may be found almost every imaginable tint in every gradation, and exhibiting bold contrasts of colour which scarcely any human artist would dare to place together, and which yet produce a result equally striking and harmonious.

Our first example is the SARPEDON, one of the most common of the genus, being found plentifully throughout many parts of Asia, Australia, and the Sandwich Islands. Its flight is rather swift, and easily recognisable, and, in common with many allied species, it has regular beats, traversing the same ground time after time with almost mechanical regularity. Entomologists take advantage of this habit, and if they see one of these butterflies pass over a certain spot, they just go and sit down where they saw the insect, and catch it as it comes round on its next circuit.

The colouring of the Sarpedon is simple, and yet brilliant. On the upper side the ground colour of the wings is brown, with a very slight tinge of chestnut; and the bold band which traverses the wings is rich iridescent green, having much of the azure reflection that is seen in green nacre. The little crescentic marks on the lower wings are of the same hue. Below, the broad band is very pale green, and on the lower wings are several streaks and dashes of crimson.

The HECTOR forms a fine contrast to the preceding insect, its colours being almost wholly black and flaming crimson. On its upper surface, the front pair of wings are sooty black, with a broad dash of grey-brown over the centre, and a little pencilling of the same colour near the tips; and the lower wings are deep velvety black, diversified with spots of intense crimson. The wings have a very narrow edging of white. The chest and part of the abdomen are black, and the head and rest of the abdomen of the same rich crimson as the spots on the wings. The under surface is coloured much in the same way, except that the crimson spots are larger.

On the large engraving at page 505, and in the right-hand lower corner, may be seen a butterfly, with two very long straight tails to its under wings. This insect belongs to the genus Papilio, and is known by the name of Protesilaus. The colours of this species are comparatively dull, but when examined, their soft contrasts are so pleasing to the eye, that any change would only be for the worse. Above, the upper wings are partly transparent, a large patch of white scales being set near their base. The under wings are mostly white, with the exception of a little dash of scarlet on their inner edge, a few half moons of dull yellow near their tips, and streaks of blackish brown on their edges and along the centre of the tails. On the under side is a little more variety, the shining

SARPEDON.—*Papilio Sarpedon.*   HECTOR.—*Papilio Hector.*

transparent membrane of the upper wings being crossed with dark bars, and the central stripe of the lower wings being edged with scarlet. It is a native of Demerara.

Our last example of this genus is the THOAS, a very striking insect, whose colours are almost wholly black and yellow. This insect is to be seen in the same illustration as the last, and its colours can be well imagined from the fact that, excepting a very small spot of orange-red on the inner edge of the lower wings, all the dark parts are black, and all the light are rich yellow. Below, it is almost wholly yellow, but of an ochreous and duller cast. In England we have but one acknowledged example of the genus Papilio. This is the beautiful SWALLOW-TAILED BUTTERFLY (*Papilio Machaon*), a rare and brilliant creature. The flight of this insect is rather high, swift, and straight. I have once seen it crossing a field near Oxford, on the banks of the Cherwell.

TWO examples yet remain of the Papilionidæ. The first is the very remarkable insect which is known by the name LEPTOCIRCUS, and which, until comparatively late years, was as rare as it is singular. This insect is not of great size, the expanded wings seldom exceeding an inch and a half, and being usually rather less in their measurement. The

LEPTOCIRCUS. — *Leptocircus Cùrius*   THAIS. — *Thais Hypsipyle*.

general colour of this butterfly is brown, with the exception of a moderately broad greenish band along the centre of the wings. In the female the band is nearly colourless, and the light patch on the upper wings is transparent. The under parts are nearly of the same colours, except that the outer edges of the tails are fringed with a narrow line of glittering white, like burnished silver. The insect is a native of Siam and Java.

Lastly, we come to the prettily-marked Thais, one of a genus of Papilionidæ, which can always be known by peculiar markings of their wings. The colours are, in all the species, yellow, black, and red, and the wings are edged with a series of bold festooned marks. The inner edges of the hinder wings are deeply scooped, as if to permit free motion of the abdomen. The larva of this insect feeds on the Aristolochia; its form is short and cylindrical, and its surface is covered with short spines and hairs. It has the forked filamentous appendage of the neck. The colour of this species is rather complicated, but may be described as follows:—On the upper surface, the first pair of wings are yellow, marked with black, and the lower pair are also yellow, but have a row of crimson spots just within the black festoons. On the under surface, the upper wings are paler, and are marked with four red spots along their edge; the under wings have also red spots, but the festoons are deeper and more angular than on the upper surface.

WE now come to another family, called the Pieridæ, which may be known at once by the manner in which the inner edges of the hinder wings are folded, so as to form a kind of gutter in which the abdomen rests. In all these insects, the colours are comparatively sober, the upper surface being generally white and black, and the under surface sparingly coloured with red and yellow. To this family belong our common white butterflies, together with the well-known Brimstone Butterfly, the harbinger of spring; all the Marbled Butterflies, the Orange-tip, and the now scarce Veined-white, which last mentioned insect belongs to the typical genus. The EPICHARIS is almost wholly white and black above, a slight tinge of rose-colour appearing on the lower edge of the hinder wings, and being due to the rich orange-red spots on the under surface. All the colour is concentrated upon the under surface of the lower wings, the groundwork of which is bright yellow traversed by black nervures, and which are adorned by six large oval spots of orange-red. Our well-known Orange-tip Butterfly is a familiar example of a similar gathering of the colour upon the under surface of the lower wings.

There is a pretty butterfly, called the SPIO, which also belongs to this large family, and may be distinguished from the succeeding group by the angulated front wings. It is a native of tropical America, Java, and India, in all of which countries it is tolerably plentiful. The colours of the upper surface are deep black, largely mottled with yellow and orange. The under surface is washed with pale yellow, purple, and brown of various depths.

ON the accompanying illustration are seen some specimens of a beautiful group of butterflies placed in the family Heliconia, because their graceful forms and elegantly disposed tints are presumed to render them worthy of the companionship of Apollo and the Muses. No British species belongs to this pretty family.

EPICHARIS.—*Pieris Epicharis*

The uppermost figure represents the Phono, a native of Jamaica, Brazil, and the neighbouring parts. The wings of this curious insect are almost wholly transparent, the opaque and coloured portions being confined to a narrow band round the edge, and a few spots and streaks upon the wings. All these markings are blackish brown, except on the under side, where the edge of the hinder pair of wings is tinged with yellow, and sometimes marked with a series of little white spots. An allied butterfly, the Transparent Heliconia (*Helicónia diáphana*), so closely resembles this species, that the two are often confounded together.

The lowermost figure at the right hand is the ERATO, a native of Surinam. In this insect there is always some variation in colour, and the sexes are so different that they might easily be supposed to belong to separate species. In the male the upper wings are rich brownish black with large spots of yellow, and the lower wings are also blackish brown, streaked in a radiating manner with blue, and edged with little oval spots of pure white. The female has the ground colour of the same hues as her mate, except that the base of the upper wings is boldly striped with rusty red, and the radiating streaks on the lower wings are of the same warm tint. In both sexes the under surface is brown, with pale yellow spots on the upper wings, and narrow streaks of pale red on the lower wings. The spread of wings is about three inches.

One species of this genus (*Helicónia Charitónia*) is very gregarious in its habits, great numbers gathering in some particular spot, and playing about like the gnat assemblies that are so common in the summer time. So plentiful are they, that when tired they can hardly find a place to rest upon, as crowds are continually settling upon the neighbouring trees, and as continually driving off the crowds which have just sat down to rest.

The last figure represents the MARSÆUS, a very elegantly shaped butterfly, a native of tropical America. The ground colour of the wings is black, diversified with many bold stripes and patches of orange, and a large golden yellow mark across the extremity of each upper wing. In the illustration, the white patch on the upper wings represents the golden yellow of the insect.

PHONO.—*Ithónia Phono.*

MARSÆUS. -*Mechanitis Marsœus.*　　SPIO.—*Léptalis Spio.*

ERATO.—*Helicónia Erato.*

In the upper left-hand corner of the engraving on page 505 is another example of this genus. The upper wings of the LYSIMNIA are chestnut at the base, and thence black to the tip, with the exception of two bold patches of nearly transparent membrane. The under wings are chestnut, edged with black, and having a jagged black streak across them, above which is a transparent stripe. The under surface is coloured in nearly the same manner, except that a row of white spots runs round the edge.

To the same elegant family belong the butterflies of the genus Euplœa, a good example of which is the MIDAMUS. This insect strongly reminds the British entomologist of the purple emperor, the sober brown of the wings changing to rich shining purple when the light falls at a particular angle. At first sight, the butterfly appears to be quite a dull and inconspicuous insect, its colours being hardly more attractive than the simple black and white of the engraving. But if it be moved so that the light falls diagonally on its wings, the dull brown suddenly changes as if by magic into imperial purple of a richness exceeding the power of man to imitate, and more than realizes the metamorphosis achieved by the fairy godmother's wand. This transformation is confined to the upper wings, the lower retaining their simple brown hue. The upper wings are sprinkled with some pale spots. The under side is greyish brown, marked with spots similar to those on the upper surface.

PERHAPS the most interesting of these butterflies is the now celebrated BUGONG (*Euplœa hamáta*), the so-called "moth" on which the aborigines of New South Wales are in the habit of feeding.

The Bugong is found chiefly upon a range of granite hills called the Bugong Mountains, and it is rather remarkable that the insects congregate upon the outcropping granite masses in preference to the wooded sides of the mountains, and are found in greatest plenty at a consider-able elevation. Mr. G. Bennett visited these mountains for the express purpose of investiga-ting the subject, and in his "Wanderings" gives the follow-ing interesting account of the Bugong :—

"The Bugong moths, as I have before observed, collect on the surfaces and also in the crevices of the masses of granite in incredible quantities. To procure them with greater facility, the natives make smothered fires underneath those rocks about which they are collected, and suffocate them with smoke, at the same time sweeping them off, frequently in baskets-full at a time. After they have collected a large quantity, they proceed to pre-pare them, which is done in the following manner.

A circular space is cleared on the ground, of a size pro-

MIDAMUS.—*Euplœa Midamus.*

portioned to the number of insects to be prepared. On it a fire is lighted, and kept burning until the ground is considered to be sufficiently heated, when, the fire being removed, and the ashes cleared away, the moths are placed upon the heated ground, and stirred about until the down and wings are removed from them; they are then placed on pieces of bark and winnowed, to separate the dust and wings mixed with the bodies. They are then eaten, or placed in a wooden vessel and pounded by a piece of wood into masses or cakes resembling lumps of fat, and may be compared in colour and consistence to dough made from smutty wheat mixed with fat. The bodies of the moths are large,

3. L L

and filled with a yellowish oil resembling in taste a sweet nut. These masses, with which the 'netbuls' or 'talabats' of the native tribes are loaded during the season of feasting upon the Bugong, will not keep above a week, and seldom even for that time; but by smoking they are able to preserve them for a much longer period. The first time this diet is used by the native tribes, violent vomiting and other debilitating effects are produced, but after a few days they become accustomed to its use, and then thrive and fatten exceedingly upon it.

These insects are held in such estimation among the aborigines, that they assemble from all parts of the country to collect them from the mountains. It is not only the native blacks that resort to the 'Bugong,' but the crows also congregate for the same purpose. The blacks (that is, the crows and the aborigines) do not agree about their respective shares, so the stronger decides the point; for when the crows, called 'Arabul' by the natives, enter the hollows of the rocks, and feed upon the insects, the natives stand at the entrance, and kill them as they fly out, and they afford them an excellent meal, being fat from feeding upon the rich Bugong. So eager are these feathered blacks, or Arabuls, after this food, that they attack it even when it is preparing by the natives; but as the aborigines never consider any increase of food a misfortune, they lie in wait for the Arabuls with waddies, or clubs, kill them in great numbers, and use them as food."

The colour of the Bugong is dark brown, with two black eye-like spots on the upper wings. The body is rather stout, filled with a yellow oily substance, and covered with down. It is not a large insect, the spread of wing averaging an inch and a half.

ARCHIPPUS.—*Dánais archippus.*

WE now arrive at another family, of which the ARCHIPPUS affords a good example of the typical genus. This fine insect measures about four inches and a quarter between the points of the outspread wings, of which the entire contour is bold and sweeping. There is but little diversity of colouring in this butterfly; rich chestnut striped and streaked with black being the ground tint, and relieved round the edges with white spots, arranged in a rather irregular double series. The under surface presents similar hues, but of a paler cast. The head, thorax, and abdomen are deep velvety black, decorated with small spots of snowy white.

THE large and important family of the Nymphalidæ contains a vast number of species, most of which are notable for their brilliant colouring, and many of which are well-known natives of England. These insects are, indeed, so numerous, that only a very slight sketch can be given of them. The large and boldly marked insect in the lower left-hand corner of the engraving is the DIDO, a native of Brazil and Guiana, and is here represented of the natural size. The ground colour of its wings is blackish brown, and all the lighter parts are soft leafy green, with a slight pearly gloss. On the under surface, the ground colour is chocolate, the green marks are much paler, and rather more opalescent than on the upper surface, and are edged with silvery white. There are, besides, several bands of the same delicate hue on various parts of the wings. The caterpillar of this insect is green, diversified with a red and white stripe on each side of the body, and covered with several rows of short spines, besides two rather long appendages to the tail.

THYODAMAS.—*Cyrestis Thyódamas.*
DIDO.—*Cethósia Dido*

THETIS.—*Marpésia Thetis.*
Agraulis monéta.

The uppermost figure in the same engraving represents the Thyodamas, an insect marked in a very unique fashion. Having a ground colour of greyish white, the whole surface is scribbled over with lines and streaks of brown, differing greatly in width, some being fine as if traced with a crowquill, and others broad and decided as if drawn with a brush. Along the edges of the wings are a few double lines of rusty brown. The under side of both wings is much paler, and the markings are finer and farther apart.

The right-hand upper figure is an example of the genus Marpesia, and is remarkable for the bold contour of wing, and the elongated tail with which it is decorated. The colour of the Thetis is by no means various, but has, nevertheless, a decided and pleasing effect. The upper surface is uniform ruddy chestnut, over which are drawn several narrow stripes that traverse nearly the entire wings, passing from the edge of the upper pair, to the extremity of the lower. From the lower margin of each under wing start two projections or tails, one being rather short, and the other very long, narrow, and slightly enlarged at the tips. The under side is pale rusty red, with a very slight gloss of blue when seen in certain lights.

The last figure in this illustration represents the *Agraulis moneta*, an insect that closely resembles the well known Adippe Fritillary of our own island, save that the colour is deeper, and the metallic spots of the under surface larger and brighter. The upper surface of this handsome insect is rich ruddy chestnut, and on the under side of the wings are a number of large spots which shine as if they had been plated with silver, and then carefully burnished. It is necessarily impossible to represent this peculiar metallic lustre in a simple engraving, but a good idea of its real beauty may be formed by imagining the ground colour of the upper wings to be pale chestnut, that of the under wings wood-brown, and all the spots to be composed of highly burnished silver leaf.

To this family belongs the brightly coloured genus Vanessa, of which the common PEACOCK BUTTERFLY is a familiar British example. This insect, which is one of the finest of our British butterflies, is very common in our own country, and may be seen very plentifully in fields, roads, or woods, when the beauty of its colouring never fails to attract admiration.

One of the most notable peculiarities in this butterfly is the uniform dark hues of the under side, which present a great contrast to the varied shades of blue and red which decorate the upper side. The object of this arrangement seems to be that the insect may be able to conceal itself from its foe at will, a purpose which is readily attained by a very simple manœuvre. When the Peacock Butterfly thinks itself in danger, it flies straightway to some shaded spot, such as a tree trunk or old palings, closes its wings over its back, and remains motionless. The effect of this proceeding is, that the wide expanse of bright colours is suddenly replaced by a flat, dark, leaf-like object, which looks more like a piece of bark torn from the tree than an insect. The apparent vanishing of the butterfly has always a rather startling effect even to those who are accustomed to it, the large brilliant creature disappearing as mysteriously as if annihilated or covered with the cap of darkness.

PEACOCK BUTTERFLY—*Vanessa Io.*

The caterpillar of the Peacock Butterfly feeds upon the stinging-nettle, in common with others of the same genus, and therefore the insect is worthy of our protection. Its general shape and appearance may be gathered from a reference to the illustration; its general colour is black, studded with tiny white points. The chrysalis is one of those which hang suspended during the time of their nonage, and is frequently found to be infested with the ichneumon-fly.

The beautiful SCARLET ADMIRAL, so well known by the broad scarlet stripes that are drawn over the wings; the LARGE and SMALL TORTOISESHELL BUTTERFLIES; the COMMA BUTTERFLY, so called from a comma-shaped white mark on the under wings, and the rare and beautiful CAMBERWELL BEAUTY, are all British members of this genus.

THE insects which are represented in this engraving are all members of the genus Catagramma, and can easily be distinguished by the remarkable manner in which the under surface of the lower wings is coloured. There is in all a somewhat circular arrangement of lines, which in many species take the form of a figure of 8 more or less distinctly outlined. The generic name Catagramma refers to this peculiarity, and is derived from a Greek word signifying a delineation. They are all inhabitants of the warmer portions of the New World.

The uppermost figure in the illustration is the Catagramma Peristera (or the Pigeon Catagramma), so called because the changing shades of the wings bear some resemblance to the opaline hues of a pigeon's neck. The ground colour of the upper surface is black, with two large patches of scarlet in the centre of each wing, the scarlet changing to violet when the light falls obliquely on the wings. The under surface of the upper wings is of paler tints, but coloured in a similar manner, except a slight streak of blue on the edge, and a stripe of buff across the tip. The under wings are yellowish buff, variegated with two black patches in the centre, each of which is garnished with a pair of azure spots. Just above these marks are two black streaks, and a curved blue stripe edged with black runs round the lower margin.

The butterfly represented in the central figure is very differently coloured, the upper surface being black, and barred with a curved band of shining azure shot with green. The under surface of the first pair of wings is white at the base, then comes a broad band of rose-pink, then a very broad band of black, and the tip is white. The lower wings are greyish white like frosted silver, and inscribed with a pattern something like that of the preceding species, except that the lines and spots are very fine, and are of a deep black-brown hue.

The lowest figure in this illustration is the Mionina, which is coloured in a totally different fashion. The upper surface of both pairs of wings is black, diversified in the first pair by a rather long patch of orange-yellow, and in the second pair by a large mark of rich blue shot with green. On the under surface the groundwork of the wings is also black, but on the first pair the orange patch is very large, and there is a curved streak of the same hue towards the tip, followed by a tiny curved stripe of blue. The bold curved marks on the second pair of wings are ochreous yellow, and the oval spots are glittering azure with white centres. The narrow line which runs round the margin is blue, edged with yellow.

If the reader will turn to the engraving on page 505 he will find a figure in the upper right-hand corner, that represents the *Catagramma marchalii*, an insect that is marked after the same fashion, but more boldly than the last-mentioned species. The upper surface is black, with a short azure band on the upper wings, and a very narrow grey-blue streak round the lower edge of the second pair. The under surface of the first pair of wings is scarlet from the base nearly to the edge, where a broad band of black streaked with white completes the wing. The markings of the under wings are blackish brown or very pale wood-brown, except one tiny patch of scarlet on the upper edge.

Catagramma peristera.
Catagramma Metiscus.
Catagramma mionina

THE large and boldly coloured insect called the POLLUX is a native of Ashantee and Guinea. As is evident by the enormous dimensions of the thorax, which contain the muscles that work the wings, so wide and strongly made, the butterfly is swift and enduring of flight. The upper surface of both wings is deep rich black-brown, and the body is of a similar, but rather paler hue. The somewhat indistinct markings on the upper wings are ochreous yellow, and those at the base of the lower wings are likewise yellow, which fades into white towards the base. The slight edging of the lower wings is blue, except the little streak at the angle, which is yellow. The under surface is very richly mottled, though without any brilliant colours. The basal half of the wings is jetty black with streaks and rings of white; then follows a broad white belt changing gradually·into buff, and on the upper wings the remainder is brown, marked indistinctly with shades of grey. In the lower wings the white belt is followed by a broad stripe of chocolate, then by festoons of grey upon brown, then of a row of deep blue spots, then by a waved band of yellow, and lastly by a border of black. The legs are black and white like the base of the wings, from which they can hardly be distinguished when folded.

POLLUX.—*Nymphalis Pollux.*

ACONTHEA.—*Adolias aconthea*

THE butterfly which is known by the appropriate name of ACONTHEA (which word is of Greek origin, signifying thorny), is a native of Java and India, and tolerably plentiful.

Although not remarkable for any brightness of hue, its tint being peculiarly sober, the regular shape of the larva and pupa render it worthy of observation. The caterpillar is mostly found on some species of Bryonia, and may be at once recognised by the wonderfully long projections from its body, which are evidently analogous to, though far surpassing in size, those upon the caterpillar of the peacock-butterfly, which is represented on page 516. When it has cast its skin for the last time, and is about to change into the pupa state, it prepares for the coming event by spinning a large web of stout and shining silken threads, which often nearly cover the under surface of the leaf to which it is afterwards suspended. It then bursts through the caterpillar-skin, hitches itself to the silken web, and hangs there until its final change into the perfect form. As may be seen from the illustration, the shape of the pupa is very remarkable reminding the observer of an ancient jousting-helmet with the visor down.

THE family of the Erycinidæ comes next in order, and, as may be seen from the specimens upon the illustration, embraces insects of very differing forms and colours. The strange-looking insect on the left hand derives its name from Mr. Bates, who discovered it, and who has given the following interesting account of its habits :—

Zeonia Bátesii.   Erycina Julia.   Calydna calamita.
                         Eurygone opalina.

" The beautiful Zeonia, of which I sent you a fine series last July, I met with in a forest near Ega, which I had traversed and examined before many times in all seasons. The first specimen I found was a straggler in a different part of the forest. On July 21st, after a month of excessively dry and hot weather, in ascending a slope in the forest by a broad pathway, mounting from a moist hollow choked up with monstrous arums and other marsh plants, I was delighted to see another of what has always been so exceedingly rare a group of butterflies. It crossed the path in a series of rapid jerks, and settled on a leaf close before me. Before I had secured it I saw another, and then shortly after a third.

I mounted to the summit of the slope, followed a branch pathway which led along the brow of the ridge, without seeing any more, but returned again to examine well the exact spot where I had captured the three ; for it very often happens that a species is confined to a few square yards of space in the vast forest, which, to our perceptions, offer no difference through its millions of acres to account for the preference. I entered the thicket from the pathway, and a few yards therein found a small sunny opening, where many of the Zeonia were flitting about from one leaf to another, meeting one another, gambolling and fighting, their blue transparent tinge, brilliant crimson patch, and long tails, all very visible in the momentary intervals between the jerks in their flight.

I was very busy, you may imagine, at first, in securing a supply of specimens; I caught perhaps 150, two-thirds of which fell to pieces in the bottom of the net, so fragile is their texture. I then paused to look around the locality, and endeavour to find the larvæ and pupæ. I walked through the thicket in all directions, and found the space peopled by the species was not more than from twenty to thirty square yards in extent, but within this space they were innumerable. Up the trees, as far as the eye could reach, the leaves were peopled with them. It is possible the brood belonged to some one tree. The only two pupæ I could find were, it is true, on two distinct kinds of trees, but this is no proof that the larva may not have fed on one tree only. I was disappointed at not finding the larva, although I searched well during this and the three following days. On the second day the butterflies were still coming out; on the third day they were much fewer and nearly all worn; and on the fourth day I did not see a single perfect specimen, and not a dozen altogether."

The white portions of the wings are membranous and transparent, and the dark portions are nearly all black, except that the base of the projecting portions of the lower wings is deep blue. The light-coloured bar is rich scarlet. This specimen represents a male; the hinder wings of the female are closer together, and the tails are nearly straight.

THE central upper figure represents the *Erycina Julia*, a native of tropical America. This insect, though not so peculiar as the preceding, is nevertheless a handsome and even striking creature. The upper surface of both wings is deep sooty black, over which are drawn two bars of flaming scarlet, of that fiery tint so familiar in the flower of the scarlet verbena. The under side is even more splendid than the upper. Appearing at a first glance to be simple dark brown, with a little scarlet patch on the inner edge of the lower wings, it only needs a side-light to flash into dazzling azure, which again changes to equally dazzling green, not unlike the gorget of the humming-bird. The place occupied by the scarlet bands of the upper surface are black below, and serve to add to the richness of effect.

ON the extreme right of the illustration is the small but elegant *Calydna calamita*. The upper surface of this insect is black, diversified with numerous blue and white spots. Below, the ground colour is brown, spotted profusely with black and white, and having some short transverse lines of yellow. This insect inhabits the regions about the Amazon.

The central lower figure represents a butterfly of simple, but extremely beautiful colouring. Unless held in a favourable light, the insect seems to be of a simple orange colour, but if held with its head towards the observer's eyes, and the sun being behind his back, its wings glow with a golden effulgence that surpasses all power of description. As the insect is gently turned or held so as to communicate a quivering motion, all the tints of the rainbow play over the trembling wings, and the glory reflected from its surface is almost intolerable to the eye. As is the case with all the butterflies, this insect is represented of its natural size.

ON the illustration at page 505, two more examples of this family may be seen. The first is placed in the centre of the left-hand side, and immediately under the tip of the left wing of the great Thoas butterfly. This is the *Helicopis cupido*, an insect which if only viewed on its upper surface seems, except for the long and slender projections of the hinder wings, to be hardly worthy of much observation, the colour being pale and dull brown, changing to pale rusty red towards the base of the wings, and having a rather large whitish spot in the centre of the upper wings. But on turning it over, so as to bring the under surface into view, it proves to be a really wonderful insect. The upper wings have little remarkable about them, their colour being brown, becoming paler towards the edge, and having a sharply defined whitish yellow mark in the centre. But it is on the lower wings that the chief interest is concentrated. On a ground of ochreous yellow are a number of large spots which look exactly as if they were made of gold-leaf artificially affixed to the wings.

the resemblance being so close, that without the aid of a magnifier which shows their real structure, a person who had seen them for the first time might well imagine that they had been veritable pieces of gold-leaf, and fastened to the wing by cement. This butterfly is a native of Demerara.

Our last example of this family is the MISIPSA, which may be seen in the left-hand lower corner of the same engraving. The colour of this pretty little insect is silvery blue, over which are drawn a number of black bands, thus producing a very bold effect. The under surface is simply light brown, with some bands of a darker hue. The Misipsa inhabits the regions about the Amazon.

THE magnificent insect called the NEOPTOLEMUS belongs to the genus Morpho, in which are contained some of the most resplendent beings to be found in the world, all being beautiful, and some endowed with a gorgeousness of colouring that is almost inconceivable. In the present species the upper wings are of the richest azure, glittering like burnished metal, and iridescent as the opal, but with far greater intensity of hue. In some lights the colours are sombre enough, being only pale grey and darkish brown; but when the light falls favourably upon the wings, their colours are truly magnificent. Around the edges of the wings is a broad belt of black, very deep towards the tips, and narrowing towards the angle. The under side is soft brown, decorated with many irregular stripes of yellowish grey, and besprinkled with a number of eye-like spots arranged in a tolerably regular row, three on each of the upper wings, and of nearly equal size, and four on each of the lower wings, one being very large and separate from the rest, and the remaining three small and close together. In the centre of each eye there is a little white spot, round which is a broad ring of black, then a narrower ring of buff, then a line of black, and lastly a grey line.

NEOPTOLEMUS.—*Morpho neoptolemus*

JUST above the left-hand corner of the Thoas' wing in the illustration on page 505, may be seen a little butterfly of simple colouring. This is one of the HAIR-STREAK butterflies, belonging to another family called the Lycænidæ. In this family are contained the beautiful blue butterflies so common in the fields, and whose exquisitely spotted under surface never fails to attract admiration. All the Copper Butterflies belong to the same family.

The present species is a native of Demerara, and is very scarce, not yet having received a name in the catalogue of the British Museum. The colour of the upper wings is brown, with slight streaks of blue radiating from the bases, and that of the lower wings is blue, edged with brown. Below it is brownish grey, with a single narrow line of rusty red crossing both pairs of wings, and a dash of the same colour on the hinder edges.

BEFORE taking a final leave of the butterflies, it is necessary to mention a family of Lepidoptera, which possess so many of the characteristics belonging to the butterflies, and so many of those belonging to the moths, that entomologists find some difficulty in placing them in their proper position, some considering them as members of the Rhopalocera, and others as belonging to the Heterocera. These insects are popularly known by the name of SKIPPERS, on account of their short and irregular flight. Several of these insects inhabit Great Britain, and may be found mostly along hedge-banks towards the end of the day. They do not seem to fly very high, but pass in their peculiar jerking fashion along the banks, flitting in and out of the herbage with restless, eager movements, which can never be mistaken for the flight of any other insect. All these creatures have rather large heads, their antennæ have a slight hook at the tip, and their wings are small when compared with the dimensions of the body, thus producing the peculiar flight.

DUMOLIN'S HAWK-MOTH.—*Smerinthus Dumolinii.*

THE second great division of the Lepidoptera is that of the Moths, distinguishable by means of the pointed tips of their antennæ, which are often furnished with a row of projections on either side, like the teeth of a comb; and in the males are sometimes supplied with branching appendages. In most instances the wings are conjoined by means of the bristle and loop which have already been mentioned.

The first family of the Moths is the Sphingidæ, a group which contains a great number of swift-winged insects, popularly and appropriately called Hawk-moths, from the strength and speed of their flight. In many instances the proboscis is of great length, sometimes equalling the length of the entire body, and in such instances it is found that the insect is able to feed while on the wing, balancing itself before a flower, hovering on

LIME HAWK-MOTH.—*Smerinthus Tiliæ.*   DEATH'S-HEAD MOTH.—*Acheróntia Atropos.* (And larva.

tremulous wing, and extracting the sweets by suction. In some cases, however, such as the well known death's-head moth, the proboscis is very short, barely exceeding the length of the head. In the long-tongued Hawk-moths the chrysalis is furnished with a distinct horny case, in which the elongated proboscis can be packed during the period occupied in development. In the genus Smerinthus the wings are sharp and angulated, and the tongue is short.

ONE of the commonest species of this genus is the LIME HAWK-MOTH, so called because the larva feeds on the leaves of the lime-tree. It is a green caterpillar, thick bodied, covered with little protuberances, and upon each side are some whitish streaks edged with red or yellow. Just at the end of the tail there is a short knobby protuberance, and the fore part of the body is rather narrow. When the larva has completed its time of feeding, it descends to the ground, and buries itself about eighteen inches deep in the earth, whence the chrysalis may be extracted in the winter by the help of a pickaxe and trowel. Beside the lime, the elm and birch are favoured residences of this insect. The Lime Hawk-moth is shown in the curious attitude which it assumes while in repose, the wings being held well away from the object to which it clings.

Although very common in some places, it seems to be rather local, being scarcely if ever found in many spots where the trees which it loves are abundant. About Oxford it is extremely plentiful, and I have reared hundreds of the insects, both from the larva taken from the trees, and the pupa obtained by digging. The colour is very variable, but the general tints are leaf brown and green, with a few blackish spots and stripes, the brown being towards the base and the olive-green towards the tips of the wing.

THE splendid insect appropriately named the DEATH'S-HEAD MOTH, is tolerably common in our island, though, from its natural habits, and the instinct of concealment with which the caterpillar is endowed, it is not so frequently seen as many rarer insects. Owing to the remarkably faithful delineation of a skull and bones upon the back of the thorax, the insect is often an object of great terror to the illiterate, and has more than once thrown a whole province into consternation, the popular idea being that it was some infra-natural being that was sent upon the earth as a messenger of pestilence and woe, if not indeed the shape assumed by some witch residing in the neighbourhood.

I once saw a whole congregation checked while coming out of church, and assembled in a wide and terrified circle around a poor Death's-head Moth that was quietly making its way across the churchyard-walk. No one dared to approach the terrible being, until at last the village blacksmith took heart of grace, and with a long jump, leaped upon the moth and crushed it beneath his hobnailed feet. I keep the flattened insect in my cabinet, as an example of popular ignorance, and the destructive nature with which such ignorance is always accompanied.

Although in itself a perfectly harmless creature, it yet has one unpleasant habit, and is said to make its way into beehives for the purpose of feeding on the honey. Still, its numbers are so inconsiderable, that it could do but little harm in an apiary, and need not be dreaded by the owner.

The caterpillar of this moth is enormously large, sometimes measuring five inches in length, and being very stoutly made. It feeds on various plants, the jessamine and potato being its favourites, and may be best found by traversing potato-grounds in the night, and directing the light of a bull's-eye lantern among the leaves. It can be readily kept and bred, but requires some careful tending, and it must be remembered that it will only eat the particular food to which it has been accustomed, and if bred among the potato will refuse the jessamine leaf, and vice versâ. When the caterpillar is about to change into its chrysalis state, it should be placed in a vessel containing seven or eight inches of earth, which should be kept moderately damp by means of a moist sponge or wet piece of moss laid on the top. If this precaution be not taken, the shell of the chrysalis is apt to become so hard that the moth is unable to break its way out, and perishes in the shell. I have several specimens where the moth has thus perished. The caterpillars are also much infested by ichneumon-flies, so that the collector often finds his hopes of a fine insect destroyed by these small and fatal flies. It is worthy of remark that when this moth first emerges from the chrysalis shell, its wings, legs, and antennæ are enveloped in a fine and delicate membrane, which soon dies when exposed to the air, and falls off in pieces, permitting the limbs to unfold themselves. Mr. Westwood regards this membrane as analogous to the pellicle upon the pseudimago of the may-fly, described at page 493.

One of the most curious points in the history of the Death's-head Moth is its power of producing a sound, a faculty which is truly remarkable among the Lepidoptera. The noise is something like the grating, squeaking cry of the field-cricket, but not nearly so loud. The mode of producing the sound is rather doubtful, but modern investigations seem to confirm the opinion of Huber and Rösel, who thought that the sound was produced by friction of the abdomen against the thorax just at the junction. At all events it is certain that the moth always bends its abdomen downwards whenever this squeak or cry is heard, and a circular tuft of orange-coloured hairs below the wings is seen to expand at the same time.

The colour of the caterpillar is bright yellow, and the body is covered with many small tubercles. Along each side run seven oblique bands of a fine green. At the end of the tail is a granulated kind of horn, and upon the back are many spots of black and blue. The colour of the moth is briefly as follows :—On the upper surface the front pair of wings are blackish brown covered with waved stripes and dashes of deep black and powdered with white. There are also some stripes of rusty red on the edges. The lower wings are ochre yellow, and marked with two bands of deep bluish grey, the upper band about half the width of the lower. The thorax is blackish brown, and has on its surface a marvellously accurate semblance of a human skull and collar-bone. The plumes or lengthened scales of which this is composed are beautifully soft, with a rich

UNICORN HAWK-MOTH.—*Sphinx convólvuli.*

deep pile, and feel like velvet under the fingers. A fine specimen of the Death's-head Moth is the largest of our British insects, and, indeed, almost the largest insect found in Europe, the spread of wing sometimes reaching nearly six inches. The antennæ are remarkable for their stiff and sturdy make and the curious hook with which they are terminated.

WE now arrive at the typical genus of the family, of which the CONVOLVULUS HAWK-MOTH affords a good example. It may be mentioned that the term Sphingidæ is derived from the peculiar attitude sometimes assumed by the caterpillars, which have a custom of raising the fore part of the body so as to bear a fanciful resemblance to the well known attitude in which the Egyptians were accustomed to represent the mysterious Sphinx.

The fine insect which has been chosen for an illustration of this genus is scarce throughout England, but seems to be found sparingly in most parts of the country, especially towards the south. As is the case with many of the nocturnal moths, its eyes shine brightly at night, and on account of their great size are very conspicuous in this respect. The specific name of the moth has been given to it because the caterpillar is known to feed on the common field convolvulus or bindweed, and it is sometimes known by the title of Convolvulus or Bindweed Hawk-moth. The caterpillar is mostly green, spotted and splashed with black and brown, and having a row of oblique stripes on each side. Generally the stripes are yellow, and edged with black, but they are sometimes wholly of the bolder colour, while the entire caterpillar sometimes assumes a brownish hue. Upon the end of the tail there is a sharp curved horn, quite harmless, and whose use is at present unknown. The colour of the wings is mostly wood-brown, chequered with ash, grey, and white, and the abdomen is ringed with broad bands of rose-colour and narrow stripes of black, while down its centre runs a broad streak of grey.

We have several other fine British insects belonging to this genus, such as the PRIVET HAWK-MOTH (*Sphinx Ligustri*), and the PINE HAWK-MOTH (*Sphinx Pinastri*).

THE beautiful OLEANDER HAWK-MOTH belongs to another genus, in which the caterpillar has the power of prolonging or withdrawing the head and neck like the proboscis of an elephant, a faculty which has earned for another insect the name of elephant hawk-moth.

This species is very rare in this country, probably because the plant on which the caterpillar feeds does not belong to the British flora. It has, however, been captured on one or two occasions. The caterpillar is greyish green except the four first segments, which are pale yellow, and along each side runs a streak of white. On the sides alone are seen those oblique streaks which are found in so many of the caterpillars belonging to this family. The colour, however, changes considerably with the period of growth, so that while at one time it takes the colouring above-mentioned, at another it is almost uniform yellow, and at another becomes quite dark. The white stripe along the sides, however, always retains its place.

The general colour of the moth is olive-green, streaked and shaded with rose-colour, grey, and white. As is unfortunately the case with many brightly coloured insects, the Oleander Hawk-moth is very apt to fade if exposed to the light, and if the cabinet in which it is placed be not carefully darkened, the delicate green is sure to take an unpleasant dusky hue. The expanse of wing is about four inches.

ALTHOUGH not gifted with the brilliant hues which decorate so many of the Hawk-moths, the HUMMING-BIRD MOTH is a more interesting creature than many an insect which can boast of treble its dimensions and dazzling richness of colour. This insect may be readily known by its very long proboscis, the tufts at the end of the abdomen, and the peculiar flight, which so exactly resembles that of the humming-bird, that persons accustomed to those feathered genus have often been deluded into the idea that England actually possesses a true humming-bird.

Its flight is not unlike that of the horse-flies. Balancing itself upon wings that quiver with such rapidity that their outline is hardly distinguishable, this moth hovers before any flower that may take its fancy, and plunging its long tongue into the deepest recesses of the blossom, extracts the sweet contents without needing to alight. Indeed, the whole of its flight is a strange mixture of apparent vagueness and swift accuracy, as it darts hither and thither with such lightning speed that the eye can scarce follow its course, ever and anon poising itself so that the body remains as still as if pinned to the spot, while the whirring wings are scarcely visible. I have lately seen one of these moths surveying the whole extent of a new brick wall, but after watching it for a long time, could not make out what might be its object.

Owing to this arrowy flight and the piercing vision with which such flight is always accompanied, the capture of the moth is a matter of no small difficulty, and when it settles, the quiet sober hues of its plumage render it so similar in colour to the objects on which it rests, that the eye can hardly distinguish its outline ; and, being gifted with an instinctive appreciation of the objects best suited for its concealment, it is sure to alight on some surface which presents hues akin to those of its body and wings.

The caterpillar of this moth is of a pale green colour, with a whitish stripe along the side, and sprinkled with numerous little white spots. On its tail it has a short and sharp horn, the tip of which is of an orange yellow colour. It feeds on many plants, preferring those of the stellate order, and from that circumstance the moth derives its specific name of *stellatarum*. The Galeum seems to be its favourite plant.

The moth is seldom seen during the hot hours of the day, but comes from its place of concealment in the early morning and towards evening, and as it has a curious custom of frequenting gardens and cultivated grounds, affords many opportunities for observation. It is bold as well as cautious, as if aware of its marvellous powers of flight, and will even enter greenhouses in search of some blossom that may have taken its fickle fancy.

The colour of the upper wings is dusky brown, with a few blackish waved bars, and that of the under wings brownish yellow, with darker edges. The abdomen is nearly of the same colour as the upper wings, and the segments are decorated with tufts of black and white hairs at the sides. The spread of wings is rather under two inches.

IN the curious moths of which the HYLAS is a good example, the wings are as transparent as those of the bee tribe, and, indeed, the hymenopterous idea seems to run

OLEANDER HAWK-MOTH.—*Chærocampa Nerii.* (And larva.)

HYLAS BEE HAWK-MOTH.—*Sesia Hylas.*   HUMMING-BIRD HAWK-MOTH.—*Macroglossa stellatarum.*

through the whole of these creatures so thoroughly, that the shapes of their bodies, the mode of flight, and even the manner in which they move the abdomen, are so bee and wasp like, that an inexperienced observer would certainly mistake them for some species of the hymenoptera. Others there are which bear an equal resemblance to the gnats, and are of correspondingly small dimensions.

These insects fly in a manner somewhat resembling the movements of the humming-bird moth, and dart about with considerable speed, though they are not so craftily wary as that insect, and can be captured with comparative ease.

The wings of this creature are transparent, with the exception of the nervures and the tip of the fore-wing, which are dark brown, and the hinder margin of both, which are ochreous green. The general colour of the body is also ochreous green, and a broad, dark red band, edged with black, crosses the middle of the abdomen. The under surface of the thorax is white, and the tail-tufts are black. This insect is a native of Western Africa.

In the next family, the Anthroceridæ, we find a number of moths of no great dimensions, but possessing great brilliancy of colouring, and flying by day. A very familiar example of this group is found in the GREEN FORESTER, a pretty little insect, not exceeding an inch and a quarter in the spread of wing, but coloured with extremely pure hues. It may be found plentifully in the month of June, and is most common on the outskirts of woods. The caterpillar of this insect feeds on the common dock and several allied plants, and like the perfect insect is of a green colour, but diversified with two rows of black dots along the back, and a row of red dots on either side. The colour of the moth is very simple, the upper wings being of a soft golden-green, with a peculiar silken gloss, and the under wings brown. The body is green, but with reflections of blue.

The well-known BURNET-MOTH, so familiar on account of the rich velvety-green, spotted with scarlet, which decorates its wings, also belongs to this family. The caterpillar feeds on many plants, and is notable for making a spindle-shaped cocoon in which it passes through its pupal state. This cocoon is of a light brown colour, and is usually fastened to an upright stem of grass. The species which is given in the illustration is marked after the same manner as the Burnet Moth, except that the spots are white.

In the Ægeriidæ, the wings are as transparently clear as in the Sesiadæ, and the general aspect is equally unlike that of a moth. The species which is shown in the illustra-

LUNAR HORNET CLEAR-WING.—*Trochilium bembeciforme.*   *Syntomis phegea.*
GREEN FORESTER.—*Ino tátices.*
CURRANT CLEAR-WING.—*Ægéria tipuliformis.*

tion is very common in England, and is fond of haunting currant-bushes, where it may be captured without much difficulty, being rather dull and sluggish in taking to flight, though when once on the wing it is quick and agile in its movements. On account of its resemblance to the large gnats, it is popularly called the GNAT CLEAR-WING. The caterpillar of this insect feeds upon the pith of the currant-trees.

THE large insect in the same illustration represents the LUNAR HORNET CLEAR-WING, an insect which is of tolerably, but not very frequent occurrence. Its popular name is given to it in allusion to its singular resemblance to a hornet, the similitude being so close as to deceive a casual glance, especially when the insect is on the wing. In common with all the members of this genus, the Hornet Clear-Wing is a rather sluggish insect, being oftener seen at rest than on the wing, and being mostly found while clinging to the trunks or leaves of the trees on which they lived in the larval state. Their flight is rather slow and heavy, and as their tongues are comparatively short, they are not able to poise themselves on the wing, and sip the sweets of flowers while balancing themselves in the air.

The larva of the present species feeds upon the willow, boring into the young wood and sometimes damaging it to a serious extent. All these insects inhabit, while in the larval state, the interior of branches or roots, and make a kind of cocoon from the nibbled

Urania Sloanus.   Cástnia licus.

fragments of the wood. Just before undergoing the transformation, the larva turns round so as to get its head towards the entrance of the burrow, and after it has changed into the pupal form, is able, by means of certain projections on the segments, to push itself along until the upper half of the body protrudes through the orifice, and permits the perfect moth to make its escape into the open air.

The wings of this insect are transparent, with orange-red nervures and dusky fringes. The head and thorax are shining brown-black, with a yellow collar, and the abdomen is ringed with orange and dark brown.

THE Uraniidæ form a curious and somewhat doubtful family, some authors having considered them to belong to the butterflies rather than the moths. Many of these insects are of most gorgeous colouring; their form, including the tailed wings, is very like that of a butterfly, and they are diurnal in their habits. Still, the preliminary stages of the caterpillar and pupa are such that they prove the insects really to belong to the moth tribe. All these insects are inhabitants of the hotter parts of the earth, and are most plentiful within the tropics.

The *Urania Sloanus* is a native of Jamaica. The colour of this beautiful insect is rather complicated, but may be briefly described as follows. The ground colour is intense black, and the upper wings are striped with glittering golden green, the middle band being the broadest. There is a broad transverse band near the middle of the hinder wings, copper-coloured in its centre, fading into grey towards the ends of the wings, and broken up by a number of black patches. The under side is pale shining gold, shot with bluish green, and profusely marked with black. The wings are fringed with white.

NEW ZEALAND SWIFT.—*Hepialus virescens*

The *Castnia licus* comes from Brazil and Central America. Its colouring is bold and yet simple. The upper surface of the first pair of wings is dark blackish brown shot with green, the latter colour being best seen by looking along the wing from point to base. Near the outside edge of the hinder wings is a row of azure spots, and the narrow fringe is white and brown. A bold white band runs through the centre of both pairs of wings.

THE curious moth in the engraving is a foreign example of a genus well known in England by some curious though common insects belonging to a family called the Hepialidæ. In the typical genus the larva is entirely subterranean, feeding on the roots of plants, and, as in some of the preceding insects, the chrysalis is able to ascend its burrow when near the time of assuming the perfect form. All these moths are very quick of wing, darting in a nearly straight line with such swiftness that they look like mere light or dark streaks drawn through the air. Yet they are captured with comparative ease, as they are not so agile as swift, and can be taken by quickly striking a net athwart their course. From their great speed, they are known by the popular name of Swifts.

The NEW ZEALAND SWIFT is a truly curious insect, not so much for its form or colours, but for the strange mischance which often befals the larva, a vegetable taking the place of the ichneumon-fly, and nourishing itself on the substance of the being which gives it support. A kind of fungus affixes itself to the larva, and becomes developed on its strange bed, taking up gradually the fatty parts and tissues of the caterpillar, until at last the creature dies under the parasitic growth, and is converted almost wholly into vegetable matter.

WOOD LEOPARD.—*Zeuzera æsculi.*   GOAT-MOTH.—*Cossus ligniperda.* (And larva.)

The common Swift Moths of England are well known, more especially the GHOST-MOTH, an insect so called from a rather unique habit in which it indulges. The two sexes are very different, the upper surface of the male being of a bright silvery whiteness, while the female is dull brown. In both sexes the under surfaces are brown. The female lurks in the grass towards even-tide, and is there rendered invisible by her sombre colouring. The male hovers just above her, remaining in one spot for a wonderful length of time, its white wings glittering ghost-like in the air, and altogether presenting a decidedly spectral aspect, which is increased by the fondness of this insect for haunting church-yards, where the grass is mostly green and luxuriant. If the creature be alarmed, it disappears like magic, vanishes utterly from sight like an extinguished spark, leaving no trace of its presence, nor giving the least sign of the direction in which it has departed. After a little pause, the white unearthly meteor is seen in precisely the same position, having become visible as mysteriously as it vanished.

The method by which this alternate vanishing and reappearance is managed is simple enough, consisting merely in taking advantage of the different colours of the upper and under surface of the wings. As long as the moth is undisturbed, it hovers quietly, exhibiting the white upper surface. But when alarmed, it drops to the ground or settles on some plant, where it hangs so as to display only the brown under surface, and is consequently as invisible as its hidden mate. Many persons have been greatly terrified by this moth.

THE well-known GOAT-MOTH is, next to the death's-head moth, one of the largest of the British Lepidoptera, its body being thick, stout, and massive, and its wings wide and spreading.

The reader may perhaps have observed certain large round holes in the trunks of trees into which a finger can be readily thrust, and out of which an empty chrysalis case often projects. These are the burrows made by the caterpillar of the Goat-moth, and often are very destructive to the trees. The larva itself is but little smaller than that of the death's-head moth, and is by no means an attractive-looking creature. Its body is smooth and shining, mostly of dull mahogany-red tinged with ochreous yellow, and having a large oval patch of chestnut on the back of each segment. It is gifted with a curiously wedge-shaped head, and its muscular power is enormous, as may be proved by actual experiment during the life of the creature, or inferred from the marvellous arrangement of muscles which are made visible upon dissection.

It exudes a liquid of powerful and fetid odour, thought by some to resemble the unpleasant effluvium exhaled by the he-goat. Its influence extends to a considerable distance, and a practised entomologist will often detect the presence of a Goat-moth caterpillar simply by the aid of the nostrils. In spite, however, of the repulsive aspect and unpleasant odour, this creature is thought to be the celebrated Cossus of the ancients, a grub which was found on trees, and, when dressed after some particular fashion, was looked upon as a very great dainty. The whole structure of the caterpillar is very interesting, and has been thoroughly worked out in the elaborate treatise by Lyonnet, to which the reader is referred for further information on the subject.

The caterpillar passes three years in the larval stage of existence, and when the time approaches for its change, it ceases to burrow, and scoops out a convenient cell in the tree, lining it with a fabric of mixed wood-scrapings and silken threads. Before it emerges from this retreat, it pushes itself through its burrow like a sweep ascending a chimney, protrudes about half of the body, and then emerges from the chrysalis shell, which it leaves within the burrow. The texture of the wings is soft and downy, and the colours are of a very sober order, being greys and browns of different tones, with a very little ochreous yellow and a number of narrow black waving streaks.

THE last figure in the illustration represents the WOOD LEOPARD-MOTH, a very prettily marked insect, though without the least brilliancy of colour. The caterpillar of this insect feeds upon the interior of many trees, seeming to prefer the wood of the apple, pear, and other fruit trees. It is a naked, fleshy-looking larva, of a light yellow colour, and having a double row of black spots upon each segment. Like the goat-moth, it prepares a cocoon-like cell when it is about to take the pupal form, but the lining is of stronger materials, cemented firmly together with a glutinous substance secreted by the insect. The moth is seldom seen until July, and is tolerably plentiful in some places, appearing to be decidedly local and rather intermittent in its visits.

THE family of the Bombycidæ includes several insects of inestimable value to man-kind, the various silk-producing moths being included in its ranks. The common silk-worm is too familiar to need any notice, but as it is not generally known that upwards of forty silk producing moths exist in different parts of the world, a short history will be given of some of them, together with a figure and a brief description of one of the finest species.

All these insects secrete the silk in two large intestine-like vessels in the interior, which contain a gelatinous kind of substance, and become enormously large just before the caterpillar is about to change into a pupa. Both the silk organs unite in a common tube at the mouth, technically called the spinneret, and through this tube the semi-liquid is ejected. As soon as it comes into contact with the air it hardens into that soft, shining fibre with which we are so familiar. If a single fibre of silk be examined through a good microscope, it will be seen to consist of two smaller fibres laid parallel to each other, like the barrels of a double gun, this structure being due to the double secreting vessels. The goodness of silk chiefly consists in the manner in which these semi-fibres are placed together. Silk-worm "gut," as it is called by anglers, is made by steeping the caterpillars in strong vinegar for a time, and then pulling them suddenly until they elongate into the well-known threads to which hooks are attached.

The caterpillar employs the silk for the purpose of constructing a cocoon in which it can lie until it has assumed the perfect form ; and proceeds with wonderful regularity and dispatch in its work, its head passing from side to side, always carrying with it a thread, and the cocoon being gradually formed into the oval shape which it finally assumes. The few outermost layers are always rough and of poor quality ; these are stripped off, and the end of the thread being found, it is fastened to a wheel, and spun off into a hank of soft

AILANTHUS SILK-WORM.—'*Attacus Cynthia.*

yellow fibre. The colouring matter is very variable, sometimes being hardly visible, and at others giving the silk a bright orange tint. It fades much on exposure to light.

Among the many silk-worm moths may be mentioned the DASEE-WORM of Bengal (*Bombyx fortunátus*), an insect that makes an inferior silk, with which the bales are often adulterated unless the owner or purchaser is very careful in examining them. The silk is yellow, and there are several crops annually. A much more valuable insect is also cultivated in Bengal, by the name of BORO POLOO (*Bombyx textor*). The caterpillar is small, and the cocoon of proportionate dimensions. The silk is very good, and of a pure white. One of the commonest insects reared by the same nation is the TUSSER or TUSSEH of the Bengalese (*Anthérea Páphia*), called by different names by the various tribes which cultivate it. It is very abundant, and as it is hardy and feeds on many kinds of food, is a truly valuable insect. It supplies the natives with great part of their clothing, and is even imported into England. There are several large manufactories of this silk, the most important of which is at Bhagulpore. The habits of this insect seem to vary much according to the locality.

The AILANTHUS SILK-WORM has lately attracted great attention in this country, and appears likely to supersede the ordinary silk-worm in many respects. It is a native of China, and has been largely used for the purpose of supplying clothes for the people. As the name implies, the caterpillar feeds upon the Ailanthus tree (*Ailanthus glandulosus*), which, although imported from warmer climates than our own, grows well and fast in this country, and has been firmly acclimatised. Rearing the Ailanthus-moth is one of the easiest of processes, the caterpillars remaining quietly on the trees and spinning their cocoons amid the branches. The eggs are hatched in a similar manner to those of the common silk-worm, and after being fed through their first moult with picked leaves, are transferred to the trees, and there left. It is of course necessary to cover the trees with netting in order to prevent the birds from feeding on such delicate morsels.

The colour of the caterpillar is green, marked with black, except the head and the last segment, which are yellow. The general colour of this moth is greyish yellow above, with splashes and markings of dull violet, black, and white. The transparent crescent is worthy of notice. The silk is strong, and takes dye easily, but does not possess the peculiar gloss which has long been proverbial. It is a truly fortunate circumstance that this insect has been so opportunely brought into notice, as it is wonderfully hardy, not subject to many diseases to which the common and delicately constituted silk-worm is liable, and being apparently free from that strange fungoid parasite which occasionally commits such fearful ravages, and has been known to depopulate a whole district in a single night.

An allied species, the ERIA SILK-WORM (*Attacus ricini*), has long been in use in many parts of Asia, where it is cultivated by the peasants, and affords them raiment of a marvellously enduring character, and yet sightly. Although the cloth that is woven from the silk of this insect is loose and seemingly flimsy of texture, it is so wonderfully durable, that a garment is said to last during nearly an entire lifetime. A large amount of condensed information on this subject can be obtained from a paper by F. Moore, Esq. in the "Technologist" for July, 1862.

THE family of the Arctiidæ. so called because some of the hair-covered larvæ have a bear-like look, is represented in England by many examples, some being really handsome insects, and others remarkable for some peculiarity in themselves or the larvæ.

HOUSE-BUILDER MOTH.—*Oiketicus Saundersii.*

PERHAPS the most curious example of this family is the species which is represented in the accompanying illustration, and which derives its name from its habits.

The HOUSE-BUILDER MOTH is common in many parts of the West Indies, and is in some places so plentiful as to do considerable damage to the fruit-trees. As soon as the larva is hatched from the egg, it sets to work in building its habitation; and even before it begins to feed, this industrious insect begins to work. The house is made of bits of wood and leaves, bound together with silken threads secreted in the interior. When the creature is small, and the house of no great weight, it is carried nearly upright ; but when it attains size and consequent weight, it lies flat and is dragged along in that attitude. The entrance of this curious habitation is so made that the sides can be drawn together, and whenever the creature feels alarmed, it pulls its cords and so secures itself from foes.

In this domicile the transformations take place, and from its aperture the male insect emerges when it has assumed its perfect form, and takes to flight. But the female behaves in a very different manner. According to the ancient maxim, she stays at home and takes care of her house, from which she never emerges, nor indeed can she emerge, as she has no external vestige of wings, and looks more like a grub than a moth; the head, thorax, and abdomen being hardly distinguishable from each other. Love and courtship with this insect are carried on quite in an Oriental fashion, pushed to extremes ; for whereas the Oriental in many cases never sees the face of his veiled bride until after the nuptial ceremony is completed, the House-builder never sees his mate either before or after marriage, and so is obliged either to love blindly or not at all. Perhaps, considering the peculiar ungainliness of his spouse, he is rather fortunate than otherwise in the fate which forbids him to contemplate the charms that lie hidden behind the dense curtain that shrouds the nuptial couch, and which, but for the mystery that surrounds them, might inspire any feeling rather than that of affection.

The grub-like female is seen lying on the ground, just below the flying figure of the male insect. It will be noticed that, except for the feathered body, the creature looks more like a larva than a perfect insect. Owing to the resemblance which these remarkable insects bear to the fasces which were borne by the lictors before Roman consuls, one species has been termed the Lictor-moth. The Singhalese appropriately call them by a name that signifies billets of firewood, and believe that the insects were once human beings who stole firewood while on earth, and are forced to undergo an appropriate punishment in the insect state. About five species of House-builder Moths are known.

The LOBSTER-MOTH derives its name from the grotesque exterior of the caterpillar. As may be seen by reference to the illustration, this larva is one of the oddest imaginable forms, hardly to be taken for a caterpillar by one who was not acquainted with it. The apparently forced and strange attitude in which this caterpillar is represented is that which it assumes when at rest. The second and third pair of legs are much elongated. The moth itself displays no very notable points of structure except the raised tufts on the disc of the fore wings.

The second example of this family belongs to the typical genus of the first sub-family, and is one of the Prominent-moths, so called on account of the prominent tufts on the inner margin of the upper wings. Several species of this genus are found in England, and that which is figured has only lately been discovered in this country. Its general colour is white. In most of these insects the larva is decorated with hairs or projections, and the structure of the perfect insect is so uncertain that systematic entomologists have been and still are greatly troubled about their proper arrangement.

The well-known TIGER-MOTH (*Arctia caja*), with its red and brown colouring, is a well known example of this family, and its caterpillar is no less familiar under the name of Woolly Bear. This is a very harmless creature, feeding almost wholly on the dead nettle, but some of its allies are terrible plagues to the agriculturist, or even to the country at large, having been known to inflict serious damage to crops, and in some parts of Germany even to strip whole forests of their foliage.

One of these insects, called the VAPOURER-MOTH (*Orgyia antiqua*), is especially remarkable for the strange contrast between the sexes, the male being a wide-winged moth of the ordinary kind, and the female a fat grub-like creature with hardly a vestige of wing, and scarcely stirring from the spot on which it is placed. The well-known PUSS-MOTH (*Cerura vinula*), so called because its markings bear some resemblance to those of a tabby cat, belongs to this family. The caterpillar of this moth is a handsomely coloured creature, remarkable for the odd, sphinx-like attitude which it assumes when at rest, the pink St. Andrew's cross which is drawn over the back, and the forked appendage at the

Notodonta bicolor.   Hypercompa Hera.

PALE TUSSOCK-MOTH.—*Dasychira pudibunda.* (And larva.)   GIPSY-MOTH.—*Hypogymna dispar.* (And larva.)

LOBSTER-MOTH.—*Stauropus fugi.* (And larva.)

extremity of the body, from which a pair of long and delicate filaments can be thrust or withdrawn at pleasure. This caterpillar constructs a cocoon of wonderful strength, composed of bits of wood cemented together, and of such hardness that a penknife cannot penetrate it without risk of being snapped in the attempt.

The GIPSY-MOTH differs much in its colouring, according to the sex, the male being blackish brown and the female greyish white. The upper wings of both sexes are marked with four waved transverse bands of moderately light brown, and a dark brown mark near the middle of the front edge like the letter $\bigvee$, inside of which is a blackish spot. On the Continent this moth is very abundant, and the caterpillar is often extremely injurious to the trees.

At the upper left-hand corner of the engraving may be seen another moderately winged moth, called the PALE TUSSOCK-MOTH, the name being derived from its colour and the tufts of hair that decorate the body of the caterpillar like tussocks of grass upon a field. The caterpillar goes by the popular name of the Hop-dog. The colour of the Pale Tussock-moth is light brownish grey, the fore wings being diversified with several marks of blackish brown, the shape and dimensions of which may be seen by reference to the engraving. The hinder wings are much paler, and the band is dark brown.

The last specimen in the illustration is the HERA, a moth belonging to a genus which is known in England by the beautiful DOMINULA or SCARLET TIGER-MOTH, with its rich green and scarlet wings. In the present instance, the fore wings are cream-coloured with broad markings that look at first sight as if they were black, but when viewed in a good

STING-MOTH.—*Doratifera vúlnerans.* (With larva and cocoon.)
ISSE.—*Pericopis Isse.*   HELICONIA-MOTH.—*Hyclósia hclicónides*

light are seen to be of the deepest imaginable green with a velvety lustre. The hinder wings are rich crimson scarlet, decorated with three or four black spots. This species is found in several parts of Europe, though not apparently a native of England. It is shown in the upper right-hand corner of the illustration.

THE moth which is represented in the accompanying illustration presents while in its perfect state but few points of interest, and would not be likely to attract the attention of any but an entomologist searching for new species. But in its preliminary stage of existence, and especially during the larval life, it is a very remarkable creature, both for its odd shape and its offensive qualities. As may be seen by reference to the engraving, the larva is stout, broad, and strong, and furnished with four protuberances in front and an equal number behind. From these protuberances the creature is able to project bundles of slight filaments of a bright yellow colour, which are capable of piercing the skin like so many darts; and whether through mechanical action, which irritates the nerves, or through any poisonous substance with which they are imbued, they act on the skin like venomed stings, and inflict exceedingly painful wounds.

The reader may possibly be aware that many caterpillars, common in England, possess a highly irritating power, and are able to cause no small inconvenience to those who handle them roughly. The well-known and handsome palmer-worm caterpillar (the larva of the gold-tailed moth) possesses this property in a very high degree, as I can testify from personal experience. In the course of dissecting a series of these creatures, my face and the back of my hands were swollen into hard knots, as if moderate-sized potatoes had been inserted beneath the skin, and the sensation caused was not unlike the dull pain that ensues after the sting of a wasp, mingled with the intense irritation of nettle-rash.

There are, besides this insect, several other British species of caterpillars which have a greater or less power of injuring those who handle them incautiously. None of them, however, appear to possess any poisonous secretion, the unpleasant and even painful effects being apparently due simply to the irritating mechanical action of the sharp hairs upon the nerves. All persons do not suffer alike from these larvæ, the tender-skinned often enduring severe pain from touching a caterpillar which a person of stouter integuments can handle with perfect impunity.

The STING-MOTH is a native of New South Wales, and the caterpillar feeds on the leaf of the stringy bark-tree. About the month of February it changes into the pupal state, and resides for some time in a curious kind of habitation. Just before it throws off the last larval skin, the caterpillar weaves a small and close cocoon or case, of an egg-like shape, which it suspends to the stem of a leaf, and therein awaits its final change.

The colour of the moth is simple, but rather pretty. The fore wings are chestnut, edged with green and white, and the hinder wings are bluish grey, edged with yellow and marked with green, yellow, and brown.

The family of the Lithosiidæ is represented in England by several moths, of which the CINNABAR-MOTH (*Callimorpha Jacobæa*) is perhaps the best known, on account of its vermilion and scarlet wings of precisely the same colour on both sides. The two examples which are given in the illustration are chosen on account of their singular resemblance to the Heliconia butterflies; both of them are natives of Brazil and the neighbouring countries. The first of these species, the ISSE, has the upper wings black, beautifully diversified by some red longitudinal stripes at the base, succeeded by two broad yellow patches. Near the edge there are some white spots. The hinder wings are red, veined with black and bordered with a broad black band on which are some red spots.

The second figure in the illustration represents a very pretty moth, a native of Brazil, which derives its specific name from the resemblance already mentioned. The ground colour of the fore wings is black, upon which are placed two large oblique patches of transparent membrane which is of a dull yellowish hue, and a few white spots are set upon the outer edge. The hinder wings are almost wholly transparent, of the same yellowish tint as the spaces on the fore wings, and are edged with a broad black border. There are some white spots near the edge.

WE now come to the large family of the Noctuidæ, containing a very great number of species, many of which so closely resemble each other that to distinguish them is not a very easy matter.

The delicately coloured PEACH-BLOSSOM MOTH derives its name from the colours upon the wings, which closely resemble the soft pink upon the peach blossom. Although spread over the southern parts of England, it does not appear to be very plentiful, and does not assemble near one spot, as is the case with many rare and local moths. The caterpillar is easily known on account of a large hump that projects just behind the head, the summit of which is cleft into two bands, and also by the series of triangular elevations along the back, upon which runs a pale zigzag line. The colour of this larva is originally dark brown, but it sometimes assumes a paler hue. It may be found sparingly upon the common bramble about May. The fore wings of the Peach-Blossom Moth are soft brown, with a few waved lines running nearly parallel with the edges, and having five spots of delicate pink, the shape and position of which can be seen in the illustration. The hinder wings are simple greyish yellow, with a single waved line running across the middle.

The LARGE SWORD-GRASS MOTH derives its generic name from the handsome appearance of the caterpillar, the word calocampa being derived from two Greek words, the former signifying beautiful, and the latter a grub. The larva may be found in summer and autumn upon many plants, but especially on spinach, lettuce, and asparagus, and is not very common. Its colour is rich green; a double row of white spots runs along the back, the rows being divided by a yellow line, then a row of white spots arranged in groups, and lastly a line of scarlet. The moth itself, although of pleasing tints, is not nearly so handsome as the caterpillar. The general colour is brown, in some individuals marked with yellow and in others with chestnut. The curiously shaped marks upon the wings are brown-black. The hinder wings are grey, and the fringe is yellow. When this moth is alarmed it has a habit of falling to the ground, with the upper wings drawn closely round the body and the antennæ and legs folded. In this attitude it looks

CLIFDEN NONPAREIL.—*Catócala fráxinus.*

SWORD-GRASS MOTH.—*Calocampa exoléta.* (And larva.)   PEACH-BLOSSOM MOTH.—*Thyatira batis.*

more like a stray piece of stick than a moth, and would escape any one who was not searching carefully for it and was acquainted with its habits.

The largest insect in the illustration is the CLIFDEN NONPAREIL, a fine and rare example of the Underwing-moths, so called because the hinder pair of wings are mostly of some bright colour while the upper pair are of comparatively sober tints. All these insects have a habit of settling on trunks of trees, or objects of similar dark hues, and drawing their upper wings so closely over each other as to conceal the brilliantly coloured lower wings entirely beneath their shelter. When so seated, or rather suspended, as they always hang in a vertical attitude, it is almost impossible to discover them, even though they be marked down to the very tree on which they alight. They require some little care on the part of the pursuer; for although they depend much on their dull colouring for concealment, they are very alert on the wing, and the moment that they take alarm they speed away with wonderful alacrity.

The caterpillar of the Clifden Nonpareil is very scarce in this country, and when found is almost invariably in one of the southern counties. Upon the Continent, however, it is tolerably plentiful. It feeds on various trees, such as the poplar, oak, elm, and ash, the last mentioned seeming to be its favourite. Its general colour is ashen yellow, with a multitude of minute black spots. Nearly at the end of the body there is a bluish black protuberance with a few yellow spots, and on the next segment there is an oblique black line. Before it changes into the pupal form, it makes a loose silken cocoon among leaves. This moth, although the largest British example of the genus, is not the most brilliantly coloured. The general colour of the upper wings is grey, with festoon-like markings of brown. The hinder wings are deep blackish brown, with a curved band of light blue across the middle. Both pairs of wings are surrounded by a narrow fringe of pure snowy white.

The family of Noctuidæ comprehends at least four hundred British species, and twice the number belongs to Europe.

THE next family derives the name of Geometridæ from the mode of walking adopted by the larva. These creatures have no legs on the middle of the body, and are in consequence unable to crawl in the usual manner. Their mode of progression is popularly and appropriately termed " looping," and the caterpillars are called " loopers." When one of these larvæ desires to advance, it grasps the object firmly with its fore feet, and draws the hind feet close to them, forming the body into an arched shape, not unlike the attitude of a cat which meets a strange dog. The hinder feet then take a firm hold, the body is projected forward, until the fore feet can repeat the process. The whole action of the larva reminds the observer of the leech when crawling.

The power of grasp and general strength of muscle enjoyed by these larvæ are really surprising. Many of them can seize a branch with their hind feet, stretch out the body nearly horizontally, and remain in that position for hours. Some slight idea of the muscular force required to perform this action may be imagined by grasping an upright pole with the hands, and trying to hold the body out horizontally. Several of these caterpillars are of dull brown hues, and being furnished with sundry projections in different parts of their bodies, they resemble dried sticks so closely that they can hardly be distinguished from the branches to which they cling. In order to avoid the danger of falling, or for the sake of escaping birds and other foes to caterpillars, they always take the precaution of spinning a thread from the branch, so that if alarmed they instantly drop, and lower themselves gradually to the ground. Entomologists take advantage of this habit by shaking the branches of trees, and catching the caterpillars as they descend, in order to escape the foe which they imagine is assailing them from above.

The SWALLOW-TAILED MOTH is a well-known British species, very common in woods, and being mostly found among the underwood, whence it may be dislodged by beating the branches. The caterpillar feeds on many shrubs, but prefers the willow, the lime, and elder trees, the elder being its chief favourite. Two specimens of the caterpillar are given for the purpose of showing its attitudes while at rest and while walking after the "looper" manner. The cocoon is also shown, made of withered leaves, and hanging from an ivy twig.

The colour of the larva is reddish brown, with some dark longitudinal lines, and before it changes into the pupal state it makes a slight silken cocoon among leaves. The colour of the wings is delicate sulphur yellow, fading to shining white at the base of the wings, and diversified with narrow streaks of brown.

The PEPPERED MOTH derives its name from the colour of the wings, which are white, covered with little black dots, that look as if they had been shaken out of a pepper-castor. The stripes on the fore wings are black.

The **V**-MOTH, another of a very common species of this family, is so called on account of the dark brown mark upon the fore wings, which much resembles the letter after which it is named. The general colour of this moth is greyish white, variegated with dark brown marks. All these species are very sluggish in their habits, flying slowly even in the evening, and in the daytime hiding among the densest underwood ; and if disturbed, only flying for a few yards, and then diving into their shelter. Many species of this family have broad waved marks upon the wings, and are therefore called carpet moths.

The colours of the ARGENT AND SABLE MOTH are explained by its name, the ground colour being white with a delicate creamy tint, and the markings all jetty black.

Our last example of this family is the WINTER-MOTH, a species which is remarkable for the shape of the female, which, as may be seen by reference to the illustration, is without wings, those organs existing only in a rudimentary state. On many parts of

ARGENT AND SABLE MOTH.—*Melanippe nastáta.*

V-MOTH.—*Halia Vaudria.* SWALLOW-TAIL MOTH.—*Ourápteryx sambucária.* (With two specimens of larva.)

PEPPERED MOTH.—*Biston betulária.* WINTER-MOTH.—*Cheimatóbia brumáta.* (With larva and wingless female.)

the Continent this moth is exceedingly plentiful, and is very destructive among the fruit-trees, devouring the leaves and tender shoots as they spring forth in the early part of the year. The general colour of the male is dusky brown, mottled with darker tints; the hinder wings are also brown, but of a paler hue.

Before proceeding further, it may be as well to observe that the wings of insects are very remarkable organs, almost anomalous in their character, and having little apparent analogy with the corresponding members in the higher animals. In the birds the wings are simply the development of the first pair of limbs, which are thus sacrificed to terrestrial locomotion, and in the bats and flying-fishes the so-called wings are still modified from the limbs. But in the insects the full number of limbs is preserved, so that the wings are to be attributed to another source. The nearest analogy to these organs seems to be found in the flying dragon, the little lizard which has already been described, and in which the ribs are widely expanded, carrying with them the skin of the sides, and partly answer the purpose for which wings were intended. But even in these curious creatures the analogy fails, inasmuch as the expanded ribs are not furnished with sets of muscles by which they can be moved, and the only aërial office that they can perform is the sustentation of their owner for a very brief period, without any capability of renewing the impetus, and with but trifling power of altering the course.

On the last illustration is a group containing specimens of several families of moths, many of which contain numerous species, and some of which are very small and apparently insignificant, though their vast numbers often give them powers of destruction which are unequalled by the larger but scarcer insects.

The PEBBLE HOOK-TIP MOTH, seen towards the middle of the illustration, is a curious insect, and one that has greatly perplexed systematic entomologists to place it in its proper position. The Geometridæ, as a rule, have the antennæ perfectly simple and thread-like, but the male of this insect has those organs in a feathery form, like those of other families. The larva, again, is of rather eccentric shape, with projections along its back, with tufts of stiff hairs, and assuming an attitude very like that which is characteristic of the puss-moth larva already described.

The popular name of this moth is derived from the hook-like tips of the wings. Its colour is reddish buff, over which are drawn a number of waved dusky streaks. In the centre of the wing there is a dusky spot, and an orange-brown stripe is drawn from the inner margin to the extremity.

THE family of the Yponomeutidæ contains many very beautiful species, none of which are of great size, but which, when magnified and a strong light thrown upon them, seem to be among the loveliest of the moth tribes. The graceful LONG-HORN MOTH is one of the prettiest and most elegant of these insects. If an observer be walking in the woods, and should keep a careful watch among the leaves of the shrubs and underwood, he will often see sundry delicate filaments, like the threads of the gossamer-spider, waving in the light, but having an iridescent surface, which shows that they could never have derived their origin from the spider. On following these filaments to their source, he will find that they belong to a little reddish coloured moth, which sits on the branch with closed wings, and permits the long thread-like antennæ to wave freely in the breeze. It is then probable that other specimens will be seen flitting about in the sunbeams, or even threading their way through the thick foliage with an ease that seems truly wonderful when the great length and tenuity of the antennæ are taken into consideration.

The caterpillar of this, and others of the same family, makes for itself a kind of case of leaves, which it carries about wherever it goes, and which serves equally to conceal and to protect the body. I have bred many species of the Yponomeutidæ from their cases, which seem to be most common upon the oak leaves, though they may be found on many trees. The reader may perhaps wonder what becomes of the antennæ during the pupal state. In order to exhibit the curious provision which is made for this purpose, a figure is given of the chrysalis, showing how the antennæ are coiled up in a spiral form at the extremity of the abdomen.

The colour of this pretty moth is reddish brown of many tones, taking in some places a golden hue, and having a rich gloss that has earned for it the name of Japan-moth. When placed under the microscope, it affords a magnificent spectacle, its scales literally blazing as if made from burnished gold, copper, and silver.

The pretty LITTLE ERMINE-MOTH belongs to the same family, and derives its name from its delicate white colour and black spots. The larva of this insect is of a grey colour spotted with black, and in some places is found in almost incredible numbers. These caterpillars live in society, and spread themselves over the ill-fated tree on which they live, weaving strong webs as they go, and lowering themselves to the ground when alarmed. During the spring of the present year I saw a large blackthorn tree standing in a garden that was covered with these caterpillars and their webs. The tree swarmed with them. Thousands were seen marching in battalions towards the few twigs on which some green leaves still remained; the air below the tree was filled with caterpillars that were hanging at the end of their threads, and the walls of the garden and adjacent house were covered with them. It was not possible to enter the garden without being covered with their delicate but marvellously strong threads, for as soon as one set was broken away, another set took their place.

Fortunately they do not conceal themselves while in the pupal state, and can therefore be easily destroyed, so as to prevent a second and a more numerous generation from appearing to complete the work of rapine. As they have lived together, so together they undergo their transformation, making their cocoons in the webs which they had spun

LITTLE ERMINE MOTH.—*Yponomeuta padellus.*
(With caterpillar in web.)

LILAC-LEAF ROLLER.—*Gracillária syringellc*
(With a rolled leaf.)

PEBBLE HOOK-TIP.—*Platypteryx falcatária.*

WHITE PLUME.—*Pteróphorus pentadáctylus.*

APPLE MOTH.—*Tortrix pom4na.*

LONG-HORN.—*'Adela De Geerella.* (With pupa.)

beforehand. One ingenious entomologist contrived to force these insects to spin their webs on paper, and thus obtained a quantity of silk, which he afterwards had manufactured into different articles.

The larvæ of the Tortricidæ frequently reside within leaves, and either roll them up in the manner which has just been described, or get into the middle of a bud or cluster of leaves, and draw them together with their silken nets. The leaves of the willow are very liable to the attacks of these insects. A very pretty and well known species of these moths is the OAK-LEAF ROLLER (*Tortrix viridana*), a moth of a beautiful apple-green upon the upper wings. In some places, these moths swarm to a fearful extent, stripping whole trees of their leaves. I have known the oaks to be surrounded with whole clouds of these moths, fluttering about like gnats, and forming an exhaustless banquet to the empis-flies, which were catching them by thousands, embracing them in their long legs, and flying about with their prey, sucking their juices like so many winged vampires.

Other species live beneath the bark of trees, or even burrow into the wood, while others are hatched in the interior of fruits, and live unsuspected in their retreats until they are on the point of changing to the pupa form, when they eat their way out, and leave a round hole as a memento of their presence. The CODLING-MOTH is one of the commonest of these tiresome insects, living in the middle of the fruit from which it takes its name, and giving rise to the condition which is termed "maggoty." The larva is a round, fat, white grub, which may too often be found in the interior of an apparently sound and ripe apple, and which gives to every part which it has touched a very bitter and nauseous flavour, like that of a worm-eaten nut. None of the Tortricidæ are of very brilliant colours, the Oak-leaf Roller being one of the most conspicuous. The fore wings are dark greyish brown, striped transversely with a darker tint. On the outer part of each wing there is a dark brown space streaked with golden bars. The hind wings are simple dusky brown.

The rose suffers sadly from the ravages of several of these moths, some of which feed within the bud, and others tie the young leaves together and feed upon the interior.

THE Tineidæ form a very large family of moths, all of which are of small dimensions, and some exceedingly minute. From several points in their structure, Mr. Westwood seems doubtful whether they ought not to be united to the Yponomeutidæ; the general narrowness of their wings, and the rare occurrence of labial palpi, being the points by which they have been separated. The larvæ of most of the species form portable cases of various materials, in which they reside, some feeding upon animal, and others upon vegetable substances. The too well-known CLOTHES-MOTH (*Tinea tapetzella*) belongs to this family. There is another species of the same genus, popularly called the WOLF-MOTH (*Tinea gravella*), which haunts granaries and malt-houses, and does great damage by feeding on the grains and fastening them together with its silken web.

The pretty little LILAC-LEAF ROLLER MOTH belongs to this family. Those who possess gardens have doubtlessly noticed that many leaves of the common lilac are rolled into a cylindrical form, bound together by silken threads, and that if this little case be opened, out tumbles a small whitish caterpillar with a black head, who loses no time in letting itself to the ground by means of a silken fibre spun from its mouth. How the larva rolls the leaf is quite a mystery, and though it has been watched by many careful observers and seen to fasten its threads, the precise force which makes the leaf assume its cylindrical form is as yet undiscovered. The caterpillar lives within, and feeds upon the rolled portion of the leaf, thus surpassing the feat of Ascanius recorded in the "Æneid." The fore wings are golden brown, with pale brown transverse markings, and the hinder pair are edged with long greyish fringes. The larva feeds on the ash and privet as well as on the lilac.

Another British species belonging to this family is the HONEY-MOTH (*Galleria*), which invades the hives, and contrives to devour the combs, without caring for the poisoned weapons with which the legitimate inhabitants are armed. This creature drives silken tunnels through the combs in every direction, always working under cover, and never exposing more than its shining black head, which is so hard and polished, that it defies all the stings in the hive. These voracious creatures are injurious to the entomologist as well as the apiarian, for if a few of their eggs should happen to be in any comb which forms part of his collection, he will soon find that the entire comb will be converted into one mass of silken tunnels, through which the active little larvæ are passing and repassing with great rapidity. These creatures do not need to turn, as they can travel backwards or forwards with equal speed.

OUR last example of the Lepidoptera is the beautiful WHITE-PLUME MOTH, an insect which never fails to attract attention, on account of the singular elegance and beauty of its form.

This insect belongs to as mall family which is remarkable for the fact that, except in one genus, the wings, instead of being broad membranous structures, are cleft into narrow rays, feathered in a most soft and delicate manner. The White-plume Moth is to be seen in the evenings, flying in a curious uncertain manner, and looking not unlike a snow-flake blown casually by the wind. It seems never to fly to any great distance, settling quite openly on leaves or plants, without taking the precaution of clinging to the under side, as is the custom with so many of the smaller moths. When it rests, it folds the wings so that they only look like a single broad ray. The legs of this moth are very long and slender. The colour of this insect is pure white.

An allied insect, the TWENTY-PLUME MOTH (*Alúcita hexadactyla*), has its wings cleft into a great number of plumes, thus giving rise to its popular name. In reality there are twenty-four plumes, each of the fore wings being cleft into eight divisions, and the hinder wings into four. It is much smaller than the White Plume, and is fond of haunting houses, where it may be seen moving up and down the window panes with much agility. Autumn is the best time for finding this little moth. Its general colour is ashen grey, with two darkish bands, and a white fringe.

Hotinus maculatus.

Pœcilóptera circulátu.
Cicáda flosfólia.

HOMOPTERA.

In the next order are comprised some very grotesque insects, some of which have been thought to belong to other orders, and a few not being known to be insects at all until comparatively late years. They have rounded bodies, not more than three joints in the tarsi, and their wings are four in number, wholly membranous, the fore pair being larger than the hinder, but not overlapping in repose. The mouth forms a kind of tube, sometimes nearly as long as the body, and often sufficiently hard and stiff to pierce the skin.

In this curious order are placed the Aphides, those little green insects that swarm upon roses and other plants, and are termed "blights" by gardeners, who employ that term in a strangely wide sense; the Cicadæ, with their beautiful membranous wings, their large heads, and their loud voices; the tribe of Hoppers, of which the Cuckoo Spit insect, known in its perfect state under the name of Frog-hopper, and the beautiful Scarlet Hopper, are familiar British examples; the wonderful Lantern-flies, also leapers, which are found only in hot climates; the Wax Insects of China; and lastly, the Scale Insects, or Coccidæ, from which the "lac," so important in commerce, is obtained.

The Cicadæ have three joints to their feet, these members affording useful character-istics in settling the precise position of the various species. They are very large insects sometimes measuring more than six inches between the tips of the expanded wings. Their mouth or beak is three-jointed and very long, being tucked under the body when not required. The females are furnished with a curious apparatus, by which they are enabled to cut grooves in the branches of trees for the purpose of depositing their eggs therein, and which is clearly analogous to the instrument possessed by the saw-flies. On the under side of the body, and nearly at the extremity, are seen a pair of jointed valves, which form the scabbard to the boring instrument. At first sight, the borer appears like a spear-head deeply notched along both edges; but on a closer examination it is seen that this apparently single instrument is composed of three pieces, namely, two saw-edged blades, set back to back, and a central support in which they both slide. There seems little doubt that these instruments work alternately, like the saws of the tenthredo.

3.

The slits made by these curious saws are wonderfully deep considering the instruments with which they are cut, and look as if little splinters of wood had been partially detached by a penknife, but left adherent at one end. Each of the burrows under these elevations is about a third of an inch in length, and contains from four to ten eggs. Altogether, each female deposits between six and seven hundred eggs. As soon as the young are hatched, they emerge from the cell, and make their way to the ground At this period of their existence they are not unlike the common flea, both in size and shape. They grow rapidly, and when they are changed to the pupal form exhibit but little alteration in form, except that the rudimentary wings are visible externally. They live for some time in the preliminary stages, and guard themselves against the frosts of winter by burrowing into the ground to a depth of nearly a yard. When the perfect insect makes its escape, it leaves the empty pupal shell nearly entire, except a slit along the back through which the creature has passed.

The male Cicada has the power of producing a shrill and ear-piercing sound, so loud in many species that it can be heard at a considerable distance, and becomes a positive annoyance, like the same tune played for several hours without intermission. The organ by which the sound is produced is internal, but its position may be seen externally by looking at the under side of the body, just behind the last pair of legs, where a pair of horny plates may be seen. These plates are the protecting covers of the sound-producing apparatus, which consists of two drum-like membranes and a set of powerful muscles. The colour of the perfect insect is mostly of a yellowish cast, and the wings are firm, shining, and membranous, somewhat resembling those of the dragon-fly in texture, but having larger cells or spaces between the nervures.

One species of Cicada is a native of England (*Cicada Anglica*), and is to be found in the New Forest. Generally, however, the Cicadæ are tropical insects, or, at all events, inhabit the warm countries, those in the cooler parts of the world being comparatively small. Several species of Cicada are eaten like the locusts.

THE wonderful LANTERN-FLIES are known by the three-jointed antennæ and the two ocelli beneath the eyes.

It may here be remarked that the eyes of insects are of twofold character, namely, the compound and the simple, the former being constructed of a variable number of facets, so arranged that each, though a separate eye, with its own optic nerve, is made to coincide with the others, and to produce a single image in the sensorium. Many insects, especially those which fly or run rapidly, have a vast number of facets in the compound eye, the common peacock butterfly possessing about thirty-four thousand of these lenses, seventeen thousand on each side. The average number, however, is about six or eight thousand. The ocelli, or simple eyes, are round, lens-like objects, mostly set in front of the head; and it is imagined that the two sets of eyes perform distinct offices, the compound eyes for the purpose of observing distant objects, and the ocelli in order to examine the food or any substance within close proximity.

In many of the Fulgoridæ, the head is formed into the oddest imaginable shapes, sometimes lengthened into a curved horn, like that of the Lantern-fly, sometimes broad, with a deep keel above, and sometimes with a raised edge of knife-like sharpness. The head is said to emit a phosphorescent light, similar to that of the fire-flies.

The Wax Insects belong to this family. These creatures are plentiful in China, where the waxen secretion is manufactured into many useful articles, and is equal, if not superior, to that obtained from the bee. That this creature should produce wax is thought to be very marvellous, but there is no reason to consider the fact more wonderful than that the bee should secrete a similar substance. There is this difference, that the bee produces the wax from six little pockets arranged along the abdomen, whereas the Fulgora pours it from various parts of the body, just as the oil is emitted by the meloë-beetle already described.

THE Cercopidæ, or Hoppers, are well known in this country, mostly from the habits of the larva, and the saltatorial powers of the perfect insect. The CUCKOO-SPIT, or FROG-HOPPER, is very plentiful in this country, and is often a great annoyance to amateur gardeners, who dislike to find their hands or faces suddenly wetted with the frothy exudations in which the creature lives enshrined. The larva fixes itself upon various plants, and sucks their juices through its long beak, which it plunges into the soft substance. When the accumulation of froth is very great, which usually happens in the heat of the day, a drop of clear water begins to form at the lowest part, into which the froth drains itself, and is presently relieved by the falling of the drop. The scientific name of this insect is *Aphrophora spumaria*. Another species of Frog-hopper (*Aphrophora Goudotii*), a native of Madagascar, pours out clear water without the preliminary process of forming the liquid into froth. In its perfect state it can leap to an extraordinary distance, the spring being so smartly made as to cause a sharp tap on the object from which it leaps. As it alights it often tumbles over, and loses some little time in kicking about before it can again get on its short legs. The species which is represented in the engraving belongs to the same genus as the beautiful SCARLET HOPPER of England, so frequently found on ferns in the outskirts of woods.

MANY insects of this family are remarkable for the endless diversity of shape into which the first segment of the thorax, technically called the "prothorax," is modified. In some of these insects it is shaped something like a helmet, or rather like an exaggerated helmet plume, twice as large as the insect to which it belongs, and extending far beyond the extremity of the body. Another species has the prothorax moulded into a pair of branches, under which the creature walks as if it were labouring under some adventitious load. The same portion of another species is shaped something like a mussel shell, and exactly covers the whole of the insect, the sharp keel being uppermost.

None, however, are more strange than that which is exhibited by the BELL-BEARER, a Brazilian

ROSE APHIS.—*Aphis Rosæ.*   SCALE INSECT.—*Coccus Westwoodii.*
SPOTTED HOPPER.—*Cercopis dorsivittáta*
BELL-BEARER.—*Bocÿdium globuláre.*
FIVE-BANDED TETTIGONIA.—*Tettigónia quinquesignáta.*

species. In this insect the prothorax is furnished with a strangely-constructed appendage standing boldly on a footstalk, throwing out a single branch backwards, and a curved cross-bar in front, on which are set four little globular nodules, the two centre being placed on short footstalks. The object of these curious appendages is quite unknown. The colour of the Bell-bearer is blackish upon the body, glossed with red on the abdomen, and the wings are mottled with the same hue. The branched organ is black, and the little globes are covered with fine hairs.

The TETTIGONIÆ are not uncommon in this country, where they are much admired for the rich beauty of their colouring, and from their peculiar shape are often mistaken for beetles. The larvæ of these insects are mostly found among shrubs and on trees, and feed like the frog-hoppers, by thrusting their beaks into the leaves and young bark. The insect preserves a similarity of shape throughout its transformation ; only the larva is entirely wingless, and the pupa presents these organs in a rudimentary state.

Passing by the Psyllidæ, another family of this order, we come to the Aphides, a family comprising a great number of species. The whole history of these insects is remarkable in the extreme, presenting many points which seem almost incredible, which destroy several old-established opinions, and in all probability will serve, when fully investigated, in establishing a new basis on which to found a more perfect system. The Aphides are wonderfully prolific, crowding upon plants until they completely hide them from view, and all employed in sucking the juice by means of the peculiar beak. They haunt every part of the plant, the leaves and their stalks, the branches, and even the roots being infested by these persevering destroyers, which often do great damage, and even force the leaves and branches to twist themselves into extraordinary contortions. Some species raise certain excrescences which serve as habitations for the insects.

In many species there is a pair of tubercles towards the extremity of the insect, which exude a sweetish liquid in a manner analogous to the frothing of the cuckoo-spit. This liquid falls upon the leaves of trees and is then known by the name of honey-dew. Bees are very fond of this substance, and, wherever it is present in any quantities, may be seen licking up the sweet secretion. Ants are equally fond of honey-dew, but they go to the fountain-head at once, and lap it as it flows from the tubercles. Whole regiments of ants may be seen ascending trees in search of the Aphides; and it is very amusing to see how they will search every atom of a tree on which the Aphides live, so as not to allow a single insect to escape them.

The white, cottony substance that is found upon the trunks of apple-trees, and is popularly called the Apple Blight, is produced by one of the Aphides called popularly the American Blight, and known to naturalists under the title of *Lachnus lanigerus*. In some species the beak or proboscis is of enormous length, three times as long as the body, and projecting like a tail as it is held folded under the body.

Many species of Aphis are winged, although every individual does not require those organs, being born, developed, producing a family, and then dying without requiring wings, or even moving from the spot whereon it lived. One of the most curious circumstances connected with these insects is, that although, like all insects, the sexes are distinct, it often happens that the presence of a male is quite needless in the production of a family, at least nine generations of fertile females having been successively produced within three months. One colony of the Aphis Dianthi continued to increase for a space of four years, without a single male among their number; and it seems generally to be the case that the males are not produced until the end of the summer, the only espousals required being those of the last generation of the year. These lay eggs, which are hatched in the succeeding spring, and from them pours an uninterrupted succession of females, a veritable colony of Amazons, who have no husbands, and millions of which die the mothers of countless descendants, without ever having seen a member of the opposite sex.

THE right-hand upper figure in the illustration represents the winged male of a small but very remarkable insect, known scientifically as the Coccus, and popularly as the SCALE INSECT, or MEALY BUG, the former title being applied to the exterior of the female, and the latter given on account of the white mealy substance that is found within her body. These insects are sad pests to gardeners, infesting various fruit trees, and increasing with such rapidity that their progress can scarcely be checked. The young, too, are of such minute size that they can hardly be seen or destroyed. It appears, however, that the most effectual way of checking their depredations is to make a kind of semi-liquid paste of fine clay and water, and with a brush to wash it well into the bark of the affected trees, so as to cover the insects, deprive them of air, and debar them from removing. Three or four coats are necessary, in order to stop up the minute cracks which are sure to take place in the drying clay, and which would afford ample opportunities of egress to these tiny creatures.

When the young first make their appearance in the world, they are minute six-footed creatures, with antennæ, and two long hairs attached to the tail. The sexes resemble each other until the time comes for their change into the pupal state, when the males assume a quiescent condition, and are covered by a cocoon-like envelope. At the proper time they leave the cocoon, and are then seen to be little flying creatures, with only two active wings, the hinder pair being merely indicated by a couple of small projections.

Two hair-like appendages are attached to the tail; and in some species, such as that which is selected for illustration, the creature is further decorated with a tuft of downy and snowy white hairs. Before the male has changed into the pupal form, both he and his future mate have affixed themselves to the tree by means of their beaks, and have remained motionless for several months.

The change undergone by the female is, perhaps, the most singular in the annals of entomology. In her preliminary stages, she was a brisk, active, six-legged little creature, with antennæ at her head and two long setæ at her tail. But when she approaches the state which we must call " perfect," though in truth the creature seems to lose perfection as it advances in development, she fastens herself tightly to a branch, loses all vestige of limbs, head, or the least sign of her insect condition, the edges of her body enlarge, and are pressed closely to the bark, and her back swells enormously, until she looks like one of the galls that are often found on bark, and bears no small resemblance to a limpet with rounded back. In some species, the body is enveloped in a white cottony secretion, and in others a kind of wax is poured forth. The only advantage which she seems to possess over the male is, that she has a three-jointed beak, which she can drive into the tree, and by means of which she can obtain the nourishment necessary for this enormous growth. The male, on the other hand, has no mouth at all, nor does he need one.

Within the shell-like body the young Cocci are hatched, amid an abundant supply of white substance, something like flour. The mother by this time has died, but her shelly skin still remains, and forms a house wherein her children live until they are strong enough to enter the world. They are usually hatched towards the end of June, and the young escape at the end of July.

The COCHINEAL INSECT (*Coccus Cacti*) belongs to the same genus. This species is a native of Mexico, and lives upon a kind of cactus, called, from its insect guest, the *Cactus cochinellifer*. The wonderful amount of rich colouring matter which these insects contain is well known. The beautiful colours carmine and lake are obtained from this insect, and the best scarlets are likewise produced from the Cochineal, the difference of hue being due to a mixture of chloride of tin. The trade in the Cochineal is very great; and as the substance is very costly, and permits a parcel of great value to be compressed into a small bulk, it is often used in lieu of cash in mercantile transactions, and a package will go travelling backwards and forwards for a long time before it reaches its final destination.

Several other species, such as the POLISH SCARLET GRAIN (*Coccus Polonicus*), and the KERMES (*Coccus Ilicis*), are also valuable to colourists, and impart a very fine scarlet to substances treated with them, although the hues are not equal to those obtained from the cochineal. The latter of these insects was known both to the Greeks and Romans, and was used by them for the purpose of obtaining the purple dyes which were so much worn by the higher classes.

The LAC INSECT (*Coccus lacca*) is another member of this most useful genus. This species resides in India and the hotter parts of Asia. It is found attached to the twigs of trees, and is then called stick-lac, the shell-lac being the waxen secretion purified and shaped into thin shell-like plates.

HETEROPTERA.

THE insects belonging to the large and important order which is placed next to the Homoloptera, are readily known by several conspicuous characteristics. The wings are four in number, and the front pair are very peculiar in their structure, the basal portion being horny, like the elytra of beetles, and the remaining portion membranous like the hinder wings of the same insects. In some species, however, the wings are wanting, as in the common Bed-bug (*Cimex lectularius*). The body is always much flattened, the mouth is beak-like, and in the pupal stage the creature is active and resembling the perfect insect, except in its want of wings.

Many of the species are aquatic, and are always of dull black or brown tint, while the terrestrial species, and especially those of tropical climates, glow with all the colours of the rainbow. The greater number of the Heteroptera feed on vegetable substances; but there are many which are predaceous, feeding not only upon other insects, but killing those of their own kind. England possesses many examples of both the terrestrial and the aquatic Heteroptera, and it is remarkable that almost every terrestrial species is capable of emitting a very powerful odour, in most cases extremely unpleasant, but in others rather agreeable. This odour is usually evolved when the creature is touched or otherwise alarmed, and is evidently under the control of the insect.

As the space which can be devoted to the remaining insects is extremely limited, it will be merely possible to give a brief sketch of the different families.

In the first section of the Heteroptera, the species are aquatic and predaceous, some being very quick and active in the water, and others slow, but yet fierce and voracious, and obtaining by craft the prey which the others win by fair chase.

The first family of the Notonectidæ, or Back-swimmers, is represented in England by many species, of which the common Water Boatman (*Notonecta glaucus*) is the commonest. These insects are in the habit of lying on the back and propelling themselves by means of the hind legs, which are very long, and with the extremities expanded and fringed with stiff hairs so as to resemble and do the duty of oars. When lying on the back, the insect is wonderfully boat-like, the general shape much resembling an Indian canoe. These creatures breathe atmospheric air, for which they are forced occasionally to come to the surface, receiving it under their wings, just as is the case with the water-beetles already described. When the air has been taken in it is prevented from escaping by means of the stiff hairs with which the segments are furnished. Any one who has watched a pond in the summer-time must have noticed the Water Boatmen coming to the surface, poking their tails out of the water, and then descending. The beak of all this family is very sharp and strong, especially in the genus Corixa; and a heedless captor will often find a sudden pang shoot through his fingers, caused by the determined thrust of the insect's sharp beak, which is armed with a pair of spear-like points, the edges of which are deeply jagged. The wings of the Water Boatman are large and handsome, and the insects are powerful on the wing. They fly by night.

THE family of the Nepidæ is represented in England by the common WATER SCORPION, a very flat and leaf-like insect, which is found abundantly in slow running streams, ditches, and ponds. It derives its popular name from its scorpion-like aspect, the two slender filaments appended to the abdomen representing the sting-tipped tail, and the raptorial fore-legs resembling the claws. It is with these legs that the Water Scorpion catches its prey, which, when once grasped in that hooked extremity, is never able to make its escape. The beak is short, but very strong and sharp, and is not bent under the thorax as is the case with that of the water boatman.

The two filamentous appendages of the abdomen are used for the purpose of respiration, the air being drawn through them to the spiracles, which are situated near their base. The eggs of the Nepa are very remarkable, each egg being furnished with a crown of diverging filaments, which form a kind of cup, into which the succeeding egg is received, so that the joined eggs have much the appearance of the common mare's-tail or equisetum. The wings of the Nepa are large, and the insect is able to fly well. There is another tolerably common species of this family found in England, the Naucoris, a smaller insect than the Nepa, as good a flier, and a better swimmer, and a terrible one to use its sharp beak with painful effect. Some of the tropical species are really formidable, being three inches in length.

WATER-SCORPION.—*Nepa cinerea.*

Diactor bilineátus   *Hammatócerus conspillaris.*   *Dalader acuticosta.*

The next section of the Heteroptera includes insects which are mostly terrestrial, though some are fond of haunting the surface of water. The Hydrometridæ are well-known examples of the latter insects, and are popularly known by the name of Water-fleas. The common Gerris skims over the surface with wonderful rapidity, wheeling and turning as easily as a skater performing his manœuvres on the smooth ice. But the Hydrometra, a very slender creature, hardly thicker than a needle, and bearing a great resemblance to the walking-stick insect already described, glides slowly over the surface, mostly keeping among the aquatic plants at the margin, and passing silently as a shadow over the water. Silent and quiet as it looks, it is a terrible foe to the smaller insects, even making short excursions ashore for the purpose of seeking prey. I have seen one of these insects making its way back to the water, holding aloft in its fore-legs an unhappy fly which it had captured, and which it was bearing off in triumph. These insects are able to fly, and I have found specimens of the Gerris concealed under stones at the top of hills far distant from water.

The family of the Cimicidæ contains only one insect, the too common BED-BUG, a creature which is supposed to have been imported into England from America. This odoriferous, flat-bodied, rust-coloured insect has derived its very appropriate name from the old English word bugge, signifying a nocturnal spectre, and used in that sense by the old writers. These creatures are enabled, by means of their flat bodies, to creep into the smallest crevices; and when they have once taken possession of a room, can be with difficulty extirpated. A very effectual method of destroying them is to close all apertures in the room, and place in it a brazier with lighted charcoal and a quantity of sulphur. The living insects will be killed by the fumes, and though the eggs will escape, yet a second fumigation will mostly destroy the young that issue from them. All places where they have been should be carefully washed, and all crevices filled up with cement or any substance which will exclude the air.

The Reduviidæ comprise a great number of terrestrial insects, mostly exotic, but a few being natives of our country. Some of them are very large, and one species, the WHEEL-BUG (*Arilus serratus*), is said to possess electric powers. Its popular name is derived from the curious shape of the prothorax, which is elevated and notched, so as to resemble a portion of a cog-wheel. One species (*Reduvius personatus*) inhabits houses, and is said to feed upon the bed-bug. The larva and pupa of this insect are difficult to discover, on account of their habit of enveloping themselves in a coating of dust. The HAMMATOCERUS belongs to this family. The insect is remarkable for the curious structure of the second joint of the antennæ, which consists of numerous small articulations. The generic title is derived from two Greek words, signifying Link-horned, and is given to the insect in allusion to this peculiarity.

THE remarkable insect on the right hand of the illustration belongs to the Mictidæ, and is notable for two peculiarities, namely, the flattened expansion of the third joint of the antennæ, and the singular width and flatness of the abdomen, which is so very leaf-like as to remind the observer of the leaf insect. As may be seen by reference to the illustration, the abdomen is greatly expanded, and extends on either side far beyond the wings, which, when opened, exhibit the curious formation of the parts below. The thorax is also expanded and developed into a semi-lunar shape, the points being directed forward. The third joint of the antennæ is seen also to be expanded, flattened, and formed into a kind of battledore shape. In colour, this insect is very inconspicuous, being altogether of a dull brown, like a withered oak-leaf. It is a native of Northern India.

ON the left hand of the illustration may be seen another insect, remarkable for the curious formation of the hinder pair of legs. In the genus of which this species is an example, the tibiæ are expanded like the blade of a South Sea Islander's paddle, being very flat, and not so thick as the paper on which this account is printed. In the present species, the colouring is very splendid, the whole of the thorax being rich emerald green, with a peculiar lustre, as if incrusted with the minutest of gems, and diversified with two longitudinal streaks of light red. The latter colour, however, always fades in process of time, as is generally the case with all the shades of red in insects. The elytra are also green, but not so sparkling in effect. It is a native of Brazil, and the family in which it is placed is, according to the catalogue of the British Museum, the Anisoscelidæ. The broad leafy expansions of the legs are chestnut brown, spotted with a paler hue.

In the Tingidæ, the body is flat and broad, and the back short, three-jointed, and folded into a groove under the head. These creatures are mostly rapacious, their fore-legs being specially constructed for the capture of living prey. Some, however, are vegetable feeders, and attack flowers and leaves with such voracity, that in France, where they are common, the gardeners call them Tigers.

The Capsidæ are also found in gardens, and, unfortunately, are very fond of ripe fruit, especially raspberries, which they suck by means of their sharp beak, and render the rest of the fruit useless, imparting to it the horrid odour which emanates from their bodies, and which is as piercing and durable as the scent of the musk-rat. The genus Miris is very common in England, and is a pretty little grass-green insect, but its odour is unbearable, and clings to the fingers with such tenacity, that the repeated application of hot water, scented soap, and a brush, will hardly eradicate it.

The Lygæidæ are mostly small insects, and are found plentifully throughout Europe. Mr. Westwood mentions that in some years these creatures are seen in countless numbers, some islands off Torquay, and a marine rock of Teignmouth, being absolutely red with the bodies of these little insects. They are mostly vegetable feeders, but on occasions become carnivorous in their habits, and have no scruples in sucking the juices of their own kind. From recent observations, however, it seems that they will not attack living insects, but only feed upon them when dead.

Among the Coreidæ may be found some of the largest insects belonging to the order, particularly those that inhabit tropical climates. They are mostly vegetable feeders, sucking the juices of plants and trees through their beaks. Their wings are large and strong, and most of the species can fly well, preferring the hottest time of day for flight. Many species of Coreidæ are plentiful in England.

The last family of the Heteroptera is known by the name of Scutelleridæ, on account of the great size of the scutellum, viz. the triangular piece of horny substance just at the end of the thorax and between the base of the wings. It is a large family, and contains many British species, some of which are popularly called Bishop's Mitres, on account of their shape, which bears some resemblance to the conventional form of the episcopal head-dress.

The beak of these insects is very long, and is employed in extracting the juices from leaves, plants, or even from living creatures, some six or eight of these insects having completely cleared a room of the bed-bugs in the space of several weeks. They have also been observed sucking the life-juices of caterpillars. One of these insects has been known

Scutellaria grandis.   Catacanthus nigripes   Edessa curtus.

to take care of her young, which followed her about, some forty in number, and were watched over by maternal care. The reader will remember that the earwig displays similar parental affection. Many of the exotic species of the Scutelleridæ are remarkable for the gorgeousness of their colouring.

APHANIPTERA.

WE are now come to another order, deriving its name from the invariable absence of wings, the name being derived from two Greek words, the former signifying invisible, and the latter a wing. There are not many species belonging to this order, and they are all known by the popular name of Fleas. A magnified representation of the common FLEA is given in the extreme left of the illustration on page 555.

These insects are notable for their extreme agility and the hard shelly substance of their integuments, two characteristics which are very useful in defending them from foes, for in the first place they leap about so quickly that they are not easily caught, and in the second place they are so hard and polished, that even when seized they are apt to slip through the fingers before they can be immolated to the just wrath of the captor. The mouth of these insects is very complex in its structure, and is a veritable surgeon's case of lancets, saws, and probes. Although eager for blood to a proverbial extent, Fleas can endure a very long fast without much inconvenience. I have known a room to be unused for years. and yet, when I became its unfortunate first occupant, being rendered helpless by a broken leg and dislocated ankle, the Fleas came swarming in positive armies to their long-delayed feast, like the locust hosts descending upon a cornfield, and caused unspeakable miseries until they were routed by continual slaughter. What food these insects may have found in an empty room is not easy to say, as, though the larvæ might, perhaps, have continued to subsist on the feathers of the pillows, the perfect insects could not eat such juiceless substances, and must either have gone altogether without food, or drawn their subsistence from some unknown source.

Another species of Flea, the CHIGOE (*Pulex penetrans*), sometimes corrupted into JIGGER, is a terrible pest in tropical countries, attacking human beings, and by its peculiar habits causing severe injuries, unless they are checked at once. They mostly attack the feet, generally preferring the bare spot just between the toe and the nail. When they have made their way fairly under the skin, they swell to a very great size, the body becoming about the size and shape of a sweet pea, and being filled with a vast number of eggs. Generally, those who live in the Chigoe-infected regions are careful to have their feet examined every day, and the offending insects dislodged with the point of a needle. Sometimes, however, one may escape observation until it has obtained its full development, when its only external sign is a slight swelling, with a bluish colour. To extract one of these swollen insects is a matter of no small difficulty, for if the body be burst, and a single egg suffered to remain, the creature will be hatched in the wound, and the result will be a painful festering sore. If such an event should take place, the best plan is to pour a drop of spirit of turpentine into the wound, a process sufficiently painful, but yet preferable to the risk of the future sores.

The young negroes are very subject to the Chigoe, and every evening a chorus of outcries is usually heard, being sounds of lamentation from the children, whose toes are undergoing maternal inspection. The little creatures, with the shortsighted cunning of childhood, always try to hide the Chigoe bite, in hopes of escaping the resulting needle. But their cunning only meets its due reward, as when the Chigoe has made her burrow, the sharp eye of the negress is sure to discover it, and then the whole nest has to be excavated, and rendered untenable by red pepper, rubbed well into the hollow. Indeed, if it were not for the terror inspired by the red pepper, the children would hardly have a sound foot among them.

It may seem curious that the insect should be able to burrow under the skin without being discovered, but the fact is, that it sets about its work so quietly, and insinuates itself so gently, that the only perceptible sensation is a slight but not unpleasant irritation.

DIPTERA.

WE now pass to the DIPTERA, or Two-winged Insects, which may be known not only by the single pair of wings, but by the little appendages at their base, called halteres or balancers, and which are the only vestiges of the hinder pair of wings. Moreover, the wings are not capable of being folded. This order is of vast extent, and includes a whole host of species, many being extremely minute, and many others displaying so many uncertainties of form and habit, that the arrangement of this order is one of the greatest difficulties with which systematic entomologists have to contend. On the engraving a few examples are given of this order, for the purpose of illustrating some of the principal families.

A male specimen of the COMMON GNAT may be seen at the upper left hand of the engraving, and is an example of the family Culicidæ. The mouth of this pretty and graceful but very annoying insect, is fully as complicated as that of the flea, and under the microscope is a truly beautiful object. The male Gnat, which is easily known by the

COMMON GNAT.—*Oulex pipiens.*   GREAT CRANE-FLY.—*Tipula gigantea.*
FLEA.—*Pulex irritans.*   STALK-EYED FLY.—*Diopsis thorácicus.*
TSETSE.—*Glossina mórsitans.*
    BANDED HORNET-FLY.—*Asilius fasciátus.*   GAD-FLY.—*Tábanus bovinus.*   BOT-FLY.—*Œstrus equi.*

plumed antennæ, is not to be feared, not being a bloodsucker, that characteristic belonging solely to the female.

The eggs of the Gnat are laid in, or rather upon, water, and are built, as fast as laid, into a boat-like shape, which possesses such powers of flotation, that, even if water be poured upon it, the mimic vessel turns out the water, and rights itself as well as any life-boat. When hatched, the larvæ fall into the water, and begin at once to make themselves very conspicuous by their continual twisting and jerking themselves about. They are long-tailed, large-headed insects, and when they are at rest, they hang with their heads downwards, the whorl of hairs at the tip of the branched tail serving as a float. Through this tail the respiration is carried on, the little creature requiring to breathe atmospheric air. In process of time the larva changes into an active pupa, and lastly, when the perfect insect is about to make its appearance, it rises to the surface, the pupal skin splits along the back, and forms a kind of raft, on which the Gnat stands until its wings have attained sufficient strength for flight.

The Tipulidæ are very familiar to us through the well-known insects called DADDY LONG-LEGS, or CRANE-FLIES, one of which is represented at the upper right hand of the engraving. In their perfect state, these insects are perfectly harmless, although ignorant people are afraid to touch them. But in their larval condition they are fearful pests, living just below the surface of the ground, and feeding on the roots of grasses. Whole acres of grass have been destroyed by these larvæ; and two or three years ago Blackheath Park was so infested with them, that the turf was much injured, and in the beginning of autumn the ground was covered thickly with the empty pupa cases of the escaped insects.

THE central lower figure in the illustration represents the common BREEZE-FLY, a well-known British example of the Tabanidæ. It is also known by the popular names of GAD-FLY and CLEG. As in the gnats, the females are the only bloodsuckers, but they exert their sanguinary ability with terrible force. While staying in the New Forest, I suffered greatly from the bites of the Gad-flies, and was at last obliged to keep a little naphtha in a bottle, and rub it occasionally over my face and hands, for the purpose of repelling these blood-thirsty insects, which selected me for their victim, leaving my companions untouched. I have found the whole of the unprotected space round the neck covered with their bites, and my ears thickly stained with blood from the effects of their weapons.

To this family belongs the terrible TSETSE, the curse of Southern Africa, which destroys horses, dogs, and cows by thousands, though it causes no harm to man or to any wild animal. Fortunately, it is a very local insect, its boundaries being as sharply defined as if drawn on a map, one side of a stream being infested with this active insect, while the other is perfectly free.

The following account of its habits and the effects of its bite are given by Dr. Livingstone:—" In the ox the bite produces no more immediate effects than in man. It does not startle him as the gad-fly does; but a few days afterwards the following symptoms supervene: the eyes and the nose begin to run; the coat stares, as if the animal were cold; a swelling appears under the jaw, and sometimes at the navel; and, though the animal continues to graze, emaciation commences, accompanied with a peculiar flaccidity of the muscles; and this continues unchecked until, perhaps months afterwards, purging comes on, and the animal, no longer able to graze, perishes in a state of extreme exhaustion. Those which are in good condition often perish soon after the bite is inflicted, with staggering and blindness, as if the brain were affected by it. Sudden changes of the temperature produced by falls of rain seem to hasten the progress of the complaint; but in general the emaciation goes on uninterruptedly for months, and do what we will, the poor animals perish miserably.

When opened, the cellular tissue on the surface of the body beneath the skin is seen to be injected with air, as if a quantity of soap-bubbles were scattered over it, or a dishonest, awkward butcher had been trying to make it look fat. The fat is of a greenish-yellow colour, and of an oily consistence. All the muscles are so flabby, and the heart often so soft, that the fingers may be made to meet through it, and the lungs and liver partake of the disease. The stomach and bowels are pale and empty, and the gall-bladder is distended with bile."

The insect which occasions these terrible results is hardly larger than a house-fly, the figure being drawn of the natural size. It is curious that, although horses perish under its bite, mules, asses, and goats escape injury, and it seems that the bite of a single fly is sufficient to cause death. Another curious symptom is, that the blood loses its redness, and hardly stains the hands of the person who dissects the smitten animal. The source of all this mischief is to be found in a little poison-gland at the base of the mouth, not larger than a mustard-seed, and yet infinitely more deadly than the venom of the rattle-snake. The colour of the Tsetse is brown, with a few yellow bars across the abdomen. When it bites a man, the pain which it causes is very slight, and the worst results are a trifling irritation not more severe than that caused by the bite of a gnat.

The large insect on the right hand of the Tsetse is an example of the Asilidæ, or Hornet-flies, among which are found the most gigantic specimens of the order. The body of these insects is long, and clothed with stiff hairs. They are fierce and voracious, feeding mostly upon other insects which they catch on the wing, and out of which they suck the vital fluids through their powerful proboscis. One species of this family has been known to capture and carry off a hive-bee, a remarkable instance of a stingless insect attacking and overcoming a creature so formidably armed as the bee. Some of them are said to attack cattle after the manner of the Tabanidæ. As with the preceding family, the larvæ of the Asili reside under ground, and feed upon the roots of plants.

The family of the Syrphidæ, or Hoverer-flies, is rather large, and contains many interesting insects. Among them may be mentioned the Volucella flies, which feed, while in the larval state, on the larvæ of bees and wasps, and, as if to aid them in gaining admission into the nests of those formidable creatures, are shaped and coloured so like the insects which they invade, that at a little distance it is almost impossible to distinguish between them.

The DRONE-FLY (*Eristalis tenax*) belongs to this family. This insect bears a wonderful resemblance to the hive-bee, and has a habit of moving the abdomen in a manner that leads an unaccustomed observer to fancy that it possesses a sting. The larva of this insect is popularly known by the name of Rat-tail Maggot, on account of its peculiar construction. This larva resides in mud, with the head downwards. In order to enable it to breathe, the respiratory tubes are carried into a long and telescopic appendage attached to the tail, the end of which is furnished with a brush of hairs something like that on the tail of the gnat larva. The extremity of this curious organ is always held out of the muddy water, and it is most curious to see the grubs elongate their tails as the depth of water is increased.

All the vast family of Muscidæ, or Flies, are members of this order, and as at least eight hundred British species of this one family are known, it may be imagined that no description of them can be attempted.

On the right hand of the engraving may be seen a large and bold-looking fly. This belongs to the family of the Œstridæ, and is popularly known by the name of BOT-FLY. All these insects are parasitic in or upon animals. The larva of this Bot-fly resides in the interior of horses, and is conveyed there in a very curious manner. The parent fly deposits her eggs upon the hairs near the shoulders of the horse, where the animal is sure to lick them, in order to rid itself of the unpleasant feeling caused by agglutinated hairs. The eggs are thus conveyed to the stomach, to the coats of which organ the larvæ cling, and there remain until they have attained their full growth. They then loosen their hold, are carried, together with the food, through the interior of the animal, fall to the ground, and immediately begin to burrow. They remain underground until they have undergone their metamorphoses, and then emerge in the shape of the perfect insect. They do not seem to inflict any damage upon the animal from whose bodies they have drawn their nourishment, and some veterinary surgeons believe that they are rather beneficial than injurious.

Another kind of Bot-fly (*Œstrus bovis*) resides in the cow, but instead of being taken into the stomach, it burrows into the skin, and there forms large tubercles, that are popularly called worbles or wurbles. An aperture is always left on the top of the tubercle, and the larva breathes by means of keeping the two principal spiracles opposite to the orifice. When full grown, they push themselves out of the aperture, fall upon the ground, and there burrow and undergo their transformations.

The spiracles, to which allusion has been often made, are the apertures through which air is admitted to the system. Insects breathe in a very remarkable manner, the air being conducted through curiously-constructed vessels to every part of the body, even to the extremities of the feet and antennæ. It will be seen that the structure of these vessels must be very remarkable, on account of the opposite duties they have to perform. As they penetrate the entire insect, it is needful that they should be flexible, in order to permit the creature to move about at will, as if they were stiff-walled the joints would be rendered useless, and the insect would be unable to move a limb. Another characteristic, however, is required. They must be always kept sufficiently open for the free passage of air, and it is not easy to see how these qualities should be united, as a flexible tube will mostly, if abruptly bent, as is continually the case with the air-tubes of the limbs, lose its roundness at the angle, and shut off the communication. An india-rubber gas-tube is a familiar instance of this property of flexible tubes.

The difficulty is, however, surmounted by a simple and yet most effectual plan. The tubes are double, one within another, and in the interspace a fine but very strong hair-like thread is closely wound in a spiral. It will be seen that, by means of this structure, the tube can be bent in any direction without losing its roundness. The long flexible tubes of Turkish pipes are made in a similar manner, a spiral wire forming the basis, upon which is sewn the leather and silken outer tube—one of the many instances where the art of man has been anticipated in the animal creation. A third species (*Œstrus ovis*) is parasitic in the sheep, inhabiting the frontal sinus, *i.e.* the open space between the bones on the forehead and between the eyes.

The Hippoboscidæ, or Forest-flies, deserve a short notice. These are round-bodied insects with legs that can cling with wonderful force, and are capable of moving backwards, forwards, and sideways with equal swiftness, so that they are not easily captured, even when they do not take to wing. These insects are mostly found in or near forests, and are very annoying to horses and cattle. Their integuments are covered with hair, and are remarkably tough and leathery. The various species of Ticks belong to this family, and are closely allied to the Forest-fly.

Although not included in Mr. Westwood's list of insects, the Lice are mostly considered as belonging to this class, under the name of APTERA. There are very many species of these obnoxious creatures infesting different animals and tribes, and at least three species are found upon the human subject.

ON account of an unavoidable delay in the engraving, the accompanying illustration was omitted from its proper place, on page 519, and is now supplied. It must be inserted after the conclusion of the first parapraph, just before the Erycinidæ.

The two beautiful insects which are represented on the engraving are examples of the family Satyridæ, which has no British representatives. Both these creatures, unlike as they appear to be, belong to the same genus. The upper species is the *Hetæra Piera*, an insect which bears a wonderful resemblance to the transparent heliconia already described. Its wings are delicately transparent, and with the slightest imaginable tinge of yellow. On the lower wings there is a blush of orange-red, and the marks are darkish brown.

Hetæra Piera.
Hetæra dracontis.

The lower figure represents the *Hetæra dracontis*, a delicately marked, though not brilliant insect. The upper wings are very soft brown, traversed by a band of a greyish hue, and with a very slight tinge of chocolate. The lower wings are also brown, but with a faint wash of blue, and the light marks are azure. On the under side it is wholly brown, with two round spots of black edged with buff, and two or three whitish blotches.

CRUSTACEA.

HAVING now completed our brief survey of the insects, we proceed to the CRUSTACEA, a very large class, in which are included the lobsters, crabs, shrimps, water-fleas, and a host of other familiar beings. Even the Cirrhipeds, popularly known under the name of Barnacles, are members of this large class, and a number of curious animals, which have until lately been classed with the spiders, are now ascertained to belong to the Crustacea.

These beings can be easily separated from the insects on account of their general structure, the head and throat being fused into one mass, called technically the cephalothorax; the number of limbs exceeding the six legs of the insects; and the mode of breathing, which is by gills, and not by air-tubes. As a necessary consequence of the last-mentioned structure, the Crustaceans possess no spiracles, such as are found in all the stages of insect life, from the larva to the imago. They undergo a well-marked metamorphosis, and in those creatures whose development is best known, the change of shape is so entire as to have led the earlier zoologists to consider the undeveloped Crustacea as separate species. They may be also distinguished from the spiders by the presence of a series of feet, or rather of locomotive organs arranged under the abdomen, as well as by the metamorphosis of their earlier stages, a phenomenon which is not known to take place among the spiders.

The name of Crustacea is sufficiently appropriate, and is given to these creatures on account of the hard shelly crust with which their bodies and limbs are covered, a covering which, in some cases, is of such flinty hardness as to be used for the purpose of sharpening knives, and in others, attains a glossy polish which reminds the observer of glazed porcelain.

As our space is rapidly diminishing, we must proceed at once to the different families and genera, simply noting the more important characteristics as we proceed through the class.

The first section of these creatures are called the Podopthalmata, or Stalk-eyed Crustaceans, because their eyes are set upon footstalks. The first order is that of the Ten-legged Crustaceans, so called on account of the five pairs of legs that are set in each side. These are exclusive of the complicated apparatus of the mouth, and the jaw-feet which guard its entrance. The Crabs are placed first in the list of Crustaceans, and are technically called Brachyura, or Short-tailed Crustaceans, because their tails are of comparatively small size, and are tucked under the large shielded body. In the preliminary stages, however, the Crabs have tails as proportionally long as those of a lobster or a cray fish.

IN the accompanying illustration may be seen two examples of the first family of these animals, being a group of Crustaceans distinguished chiefly by the singular form of the carapax or upper shell, which is wide and abrupt at the base, but is prolonged in front, so as to form a long and pointed beak. In all these creatures the legs are long in comparison to the body, but in the LEPTOPODIA, which is seen in the lower part of the engraving, they are of such inordinate length, as to remind the observer of the round-bodied, long-legged harvest spider, which scuttles over the ground so rapidly when disturbed. On account of this great length of limb and small size of body, these crabs are often called Sea Spiders. The eyes of the Leptopodia are rather large, and not retractile. It may here be mentioned that the eyes of Crustacea bear some resemblance to those of insects, being compound organs, with a large number of facets, some square and some hexagonal. The eyes of the common shore crab or the shrimp afford excellent examples of this structure.

The generic name of the Leptopodia is very appropriate, being composed of two Greek words signifying "slender legs." It is a native of the West Indies.

The upper figure in the illustration represents one of the British Crabs. In the Stenorhynchus, the projecting beak is proportionately shorter than in the preceding genus, is cleft at the tips, and very sharp. The fore limbs, which are furnished with large claws, are stout and strong.

Stenorhynchus Phalángium.
Leptopodia sagittata.

As the shelly armour of the Crustaceans is, in most cases, so hard, strong, and unyielding, the mode of growth might be considered a problem not very easy of solution. For with the Crustaceans the growth continues during nearly the whole of life, or at all events for several years after they have passed through the various changes to which they are subjected in their imperfect stages of existence. Their increase of size and weight is marvellously rapid, and how it can be accomplished without subjecting the Crustaceans to the lot of the starveling mouse, who crawled into a jar of corn, but could not crawl out again after feasting on its contents, seems to partake of the character of an animated puzzle.

The answer to the problem is simply that the creature sheds its armour annually, expands rapidly while yet covered only by a soft skin, and is soon protected by a freshly-deposited coat of shelly substance. Even this answer contains a second problem little less difficult than that which it solves. How can a Crustacean, say a crab or a lobster, shed its skin? It is true that the cast shells are found, showing that the creature has escaped from its old and contracted tenement by a slit in some part of the body, such as the top of the carapace, and has left its shell in so perfect a state that it might easily be mistaken for the living animal.

But how did it manage about the claws? We all know what large muscular masses they are, how very small is the aperture in which the joint works, and how stiff and firm is the broad tendinous plate which is found in their interior. Examination shows that there is no opening on the claws through which the creature might have drawn the imprisoned limb, and it is also evident that the only method by which these members can be extricated, is by pulling them fairly through the joints. As a preliminary step, the hard, firm, muscular fibres which fill the claw and give it the well-known pinching power, become soft, flaccid,

Camposcia retusa.

and watery, and can thus be drawn through the comparatively small openings through which the tendons pass from one joint to another. The sharp and knife-like edges of the plates cut deeply through the muscle, which, however, is little injured, on account of its soft consistency, and heals with great rapidity as soon as the animal recovers its strength, and is gifted with a new shell. In the common edible crab, the flesh is quite unfit for consumption during this process, as any one can attest who has attempted to dress and eat a "watery" crab. Yet, in some of the exotic crustaceans, these conditions are exactly reversed, and the crabs are never so fit for the table as while they are soft and shell-less, after the old suit of armour has been thrown off, and before the new integument has received its hardening. It may here be observed, that the basis of the crustaceous armour is composed of chitine, the remarkable substance of which the elytra and other portions of the insect skeleton are composed.

The name of Leptopodia signifies slender feet, and is appropriately given to the long, slight-limbed creature which is represented in the illustration. When full grown, this species will mostly attain twice the dimensions of the figure. The Stenorhynchus is shown of its natural size.

THE two figures on the illustration, however dissimilar they appear to be, represent the same species; the upper figure showing the Camposcia as it appears when its exterior is free from extraneous substances, and the lower figure being a faithful image of the same individual as it appears while burdened with sponges and other marine growths.

As will be seen by reference to the engraving, the hairy limbs, as well as the whole of the body, are encrusted so thickly that their true shape is quite undistinguishable, and the animal seems to masquerade under a domino of living disguises. Even the joints

3.   O O

Duclèa calcitrapa.

can barely be ascertained, and, but for the continual movements to which they are subjected, it is very probable that the sponges would increase with such rapidity, that in a short time the limbs would be rendered immovable. These growths are so constant and rapid that the creature can only free itself at the time when it changes its skin; and it is likely that the crab may feel as comparatively light and disburdened after throwing off all this encumbrance of heavy voluminous substances, as does a thick-woolled sheep after the shears have removed the heavy fleece, and enabled the lightened animal to skip about the field astonished at its own activity and the sudden coolness shed over its body.

The reader will observe, on reference to the upper figure, that the Camposcia possesses all the characteristics already mentioned as appropriate to the family in which it is classed, and that the snout—if we may be allowed to employ the word—is elongated and very deeply cleft, so as to form a forked protuberance. The body is rounded at the base, and small in proportion to the limbs, though the apparent disproportion is not so marked as in the Leptopodia. This species is a native of the Philippines. The genus to which this creature belongs does not seem to be very large, only three or perhaps four species being known.

STILL keeping to the same family, we come to a curious genus containing some very remarkable creatures, among which the species represented in the illustration is one of the most notable.

In this genus the beak is comparatively small, but still contains the cleft tip, although the notch is not nearly so deep as in other members of this family. The claws are of considerable size and power, and the legs are long and furnished with an array of stiff bristles. The chief peculiarity, however, which most strikes the sight, not to say the

GOUTY CRAB.—*Lissa chiragra.*   *Chorinus.*   THORNBACK CHORINUS.—*Chorinus acanthonotus*

touch, is the formidable display of long and pointed spikes, which radiate from the body like the spines of a hedgehog. The eyes are not very prominent, being set on rather short footstalks, and nearly concealed by the projection from the shelly covering. This genus seems to be widely spread over the hotter portions of the globe, specimens having been taken off the Mauritius, in China, India, and the Philippines, of which latter locality the present species is a native.

All the crabs of this family are marine, and prefer the deeper parts of the sea, where they lurk among the waving masses of seaweeds, or crawl upon the oyster banks. As might be imagined from the length and slenderness of their limbs, they are but slow of progress, and seem to tumble over the ground in a very unsteady manner. Still, their long limbs are admirably calculated for the peculiar substances on which they pass their lives, and they are enabled to stride, as it were, over obstacles which would seriously encumber a creature with shorter legs. Their food consists almost wholly of small molluscs and other marine animals.

It is believed that in the winter these crabs hibernate in the mud, as when they have been captured in the early spring, the hollows in their shells have been found full of the same muddy substances which are obtained from the bed of the sea. Probably, it is at this period of their existence that their shells receive the germs of the seaweeds, zoophytes, and various growths with which they are often wholly obscured and overshadowed.

WE now come to another family wherein many of the same characteristics are preserved, but the legs are of moderate size. These creatures are popularly known by the name of Spider-crabs, and scientifically are termed Maiadæ.

Three examples of this genus are given in the illustration. The GOUTY CRAB has been gifted with its very appropriate name on account of the knobby and swollen limbs, which give it an aspect as if it were suffering from the painful but unpitied disease from which it derives its name. The specific title of Chiragra is of Greek origin, and bears a similar signification. The Gouty Crab is known to be an inhabitant of the Mediterranean, and is thought also to be a native of the West Indian seas. The specimen which is figured was captured in the former locality.

On the right-hand lower corner of the engraving is shown a creature which looks as if it had been made almost at random out of a thistle-bud and a handful of thorns. This is the THORNBACK CHORINUS, a species which can hardly be mistaken for any other, on account of its altogether eccentric shape, and the branching spines which spring on every side from its body and the joints of its limbs. It is but feebly provided with claws, these members being little larger than the ordinary limbs; and the eyes stand out on tolerably long footstalks.

The small specimen on the upper part of the engraving belongs to the same genus, but at the date of publication has not received the specific name by which it will ever afterwards be known. It has been selected for illustration on account of the substances with which its surface is partially covered, two nummulites being the most conspicuous of these objects. These animals are natives of the Eastern seas.

Nearly related to these species, we find two moderately common British crustacean which are interesting in their habits, though not particularly pleasing in their aspect. One of these is the FOUR-HORNED SPIDER-CRAB (*Arctopsis tetraodon*), a rather long-legged creature that seems to be very local in its habits, being rarely or never seen in some localities, while in others it is found in vast numbers. This crab generally hides itself under the overhanging masses of fuci which cover the submerged rocks, and thence is fond of descending into the lobster and crab pots, and so is made captive by the fishermen. The colour of this species is yellow, and the body and greater part of the limbs are densely covered with thick hair. The male is larger than the female.

The second species is the HARPER-CRAB, or GREAT SPIDER-CRAB, or SEA-TOAD (*Hyas araneus*), as it is sometimes called. This is commonly found on nearly all the British coasts, and prefers to range among the weeds just about the zone beneath low-water mark. It is one of the day-feeders, and will often leave the waves for the purpose of feeding upon the fish and other animal substances that have been flung upon the shore by the tide. In some places it haunts the stake-nets, and there makes a regal feast before it is disturbed by the proprietors.

Two figures of a very common and a very useful British species are seen in the illustration, one being represented on the left, and the other in the upper right-hand corner, sitting on the rock. The common THORNBACK SPIDER-CRAB, or SQUINADO, is plentiful upon our coasts, and, as may be seen by reference to the figure, is not a very prepossessing creature in external appearance, its body being one mass of sharp and not very short spines, and its whole frame possessing a weird-like and uncomely aspect.

Ugly though it may be in an artistic point of view, it is one of the most useful inhabitants of the sea, acting as a scavenger for the removal of the decaying animal matter that is ever found in the seas. More especially along the shore, where the refuse of mankind, such as unsaleable fish and crustaceans, are continually being cast into the waves, the Squinado is found to perform the necessary office of removing all such substances. It is a voracious creature, and being gifted with an acute sense of smell, is sure to discover without delay any substance on which it can feed, and to make its way thereto without delay.

Supposing, for example, that a fisherman has been employed, according to the wont of the race, in dressing-up, skinning, and trimming fish of various kinds, and has carelessly flung the offal into the sea, or on the shore where the waves will soon cover it, hardly a trace of the offensive substance will be found on the succeding tide, though bones and other hard portions may be seen in plenty. For a myriad nostrils—if the organs which correspond to our sense of smell can be called by that name—take cognisance of

COMMON THORNBACK-CRAB. (Female.)

COMMON THORNBACK-CRAB.—*Maia Squinado.*
THREE-SPINED SPIDER-CRAB.—*Pericera trispinósa.*

the putrefying substances, which to them are sweeter than the choicest perfumes vended at Cologne or elsewhere, and their owners crowd around on all sides, eager to enjoy the coming feast.

The Squinado, together with other crabs, sets to work boldly: with one claw holds tightly to the banquet, and with the other tears off morsels and deftly feeds himself therewith, putting them into his comical mouth with the regularity of clockwork, and with a rapidity that reminds the observer of a Chinese flinging rice into his mouth with his chopsticks. The strength and sharpness of the claws are such, that the toughest muscle cannot long withstand their power, and the flesh is torn from the bones as perfectly as if scraped away by a knife.

But this tearing and pulling is but rough work, and leaves a vast number of tiny atoms floating in the waves, that, in the aggregate, would be extremely injurious to the purity of the water. They are far too small for the large and coarse pincers of the crab to seize, but are exactly adapted for another set of scavengers which accompany the Squinado wherever it goes. If the reader will again refer to the illustration, he will see in the upper right-hand corner a small specimen of this crab, seated on a rock, and having a large tuft of some substance on either side. These tufts are composed of corallines and zoophytes, such as sertularia, coryne, campanularia, bugula, and the like. In this example, the extraneous growths have capriciously settled on the limbs, but in many instances they wholly cover the entire body and limbs, so that the crab moves along like Macduff's army under its leafy shade. These creatures live on the minute floating atoms of animal matter, just as the Squinado feeds on the larger pieces, and it is one of the many wondrous

RAM'S-HORN CRAB.—*Criocarcinus superciliosus.*   THORN-CLAW CRAB.—*Acánthonyx zébrida.*

examples of the care with which every creature is fed and every atom is utilised, that some twenty or thirty thousand tiny mouths should always accompany the crab for the purpose of feeding on the little morsels that are dispersed vaguely into the water as the crab tugs and pulls at the abundant feast.

There is something inexpressibly ludicrous in the action of this animal as it feeds itself. Although every one who is acquainted with the anatomy of the crustaceans knows well the position of the mouth, it is impossible to get over the feelings of ridicule which are raised by the sight of the animal feeding itself by pushing food into its chest. The steady business-like manner of its proceeding adds to the absurdity, and the gravest observer cannot repress a hearty laugh when he sees a Spider-crab busy at its meal. The ancients were well acquainted with the form of this crab, and held it in great respect, though from erroneous ideas, supposing it to be a peculiarly wise creature, and susceptible of the charms of music.

THE curious creature which is represented at the lower right-hand corner of the illustration is the THREE-SPINED SPIDER-CRAB, so called from the peculiar shape of the body, which, on account of the projecting beak and the strange modification of the carapace, has a kind of three-cornered aspect. Several species of this genus are known, and are found in the West Indian seas and off the Philippines The present species is one of the most common, and is found in the West Indies.

SOME remarkable examples of this family are to be found in the two species which are figured in the illustration. The larger specimen is the CRIOCARCINUS, a species which can be easily distinguished by the two long horn-like projections from the snout. The generic name, which is of Greek origin, and signifies Ram-crab, is given to the animal on account

HERALD-CRAB.—*Huénia heráldica.*   MICIPPA.—*Micíppa philyra.*
LONG-SNOUTED CRAB.—*Huénia elongáta.*

of this structure. The body is thorny, though not so wholly beset with spikes as in the spider-crab, and the eyes are placed on moderately long footstalks. The specific term, superciliosus, refers to a Latin word signifying an eyebrow, and is given to this crab on account of the overhanging plates under which the eyes are hidden when the footstalks are laid close to the head, as is the custom of the creature when alarmed. The present species is found in the New Hebrides.

ON the right hand and in the upper corner may be seen a curious-looking little crab, especially notable for the large and boldly hooked extremities of the limbs. The name of ACANTHONYX, or THORN-CLAWED, is given to the genus on account of this structure. At first sight, the Acanthonyx hardly seems to belong to the same family as the preceding species, the shape of the body being apparently the reverse to that which is characteristic of the Maiadæ. But on a closer examination, it is found that this difference is more apparent than real, and that though the body seems to be wider across the head, or rather, the cephalo-thorax, to speak accurately, the width is owing to mere projections and not to any increase of the actual body. The THORN-CLAW CRAB is found in many European seas, and is tolerably common in the Mediterranean.

OUR last examples of the Maiadæ are the three crabs which are represented in the accompanying illustration.

In these three species the reader will not fail to observe the curious variation of form that takes place in animals that belong to the same family and even to the same genus. The body of the MICIPPA is very large in proportion to the limbs, rounded, and covered with numerous protuberances of various sizes, mostly small tubercles, but sometimes being developed into bold spikes. The claw legs are remarkably small in proportion, and the claws themselves are even more feeble than might be inferred from the dimensions of the entire limb. Several species of Micippa are known, all of which are obtained from one or other of the Philippine Islands.

On the left hand of the engraving are seen two crabs very dissimilar in external appearance, and yet belonging to the same genus. The upper figure represents the HERALDIC CRAB, so called because the shape of its carapace presents a fanciful resemblance to the shield and mantle employed by heraldic painters in depicting coat armour. The sides of the carapace are developed into four singular projections, flat, and looking very much as if pinched out of the shell while its material was plastic. The snout is tolerably long, and very sharply pointed, and the eyes only iust project from under the protecting shell.

The lower figure represents the LONG-SNOUTED CRAB, a creature in which the carapace, instead of being wide, flattened, and formed with ring-like projections at the side, is drawn out to a wonderful length, and possesses two angular projections towards the base. Both these crabs are natives of Japan.

DOMED CRAB.—*Cryptopodia rotaicata.*

In the family of crabs which is known by the name of Parthenopidæ, we have a very different form, the carapace being more or less triangular, the beak or snout small and not notched, and the eyes very retractile. The claw-legs are generally large in proportion to the other limbs, which are often very short.

The DOMED CRAB is a very remarkable example of this family, and in addition to certain generic peculiarities, well displays the characteristics of the family. As may be observed by reference to the illustration, the claw-legs are very large throughout their entire structure, and are furnished at their extremities with short but powerful nippers. The carapace of this creature is extremely wide, but the width is due, not so much to the body as to the shell, which is expanded in such a manner as to conceal the legs under its shelter.

SPINOSE PARTHENOPE.—*Parthenope horrida.*

The generic name Cryptopodia is derived from two Greek words signifying Hidden-legs, and is an extremely appropriate title. Even the large claw-legs can be folded up and tucked away so neatly under the carapace, that when the creature lies still on the ground, no vestige of limbs can be seen, and it might easily be mistaken for a stone thrown casually on the shore. In fact, the whole contour of this crab, whether when moving or quiescent, irresistibly reminds the observer of the tortoise tribe, and bears a special analogy to the box-tortoise, which has already been described and figured. The eyes of this genus are very small, and, like the limbs, can be wholly retracted and hidden under the shell. The Domed Crab is a native of Japan.

The very singular and unprepossessing crab which is here represented, belongs to the typical genus of the family. At present this genus seems to be very small, the number of known species being decidedly limited. Owing to the marine residence of these creatures, and the extreme difficulty, not to say impossibility, of watching them in their watery homes, the habits of these Parthenopidæ are but little known, and in most cases can only be conjectured from the bodily form, just as the fossil animals are known to be carnivorous or herbivorous by the structure of their teeth and jaws, to be swimming creatures because they possess fins and paddles, or to be capable of flight because they are furnished with wings.

In the Spinose Parthenope, the carapace approaches to a five-sided figure, rather wider than its length, moulded into a series of the oddest imaginable protuberances, and covered with knobs, tubercles, and spines. The beak is sharp, short, pointed, and has a strong tooth just between the antennæ. The claw-legs are very large, armed with powerful forceps at their extremities, and covered thickly with such a multitude of knobs, spikes, and protuberances, that they really seem as if they were subject to disease and had thrown out a crop of unhealthy growths. The hinder limbs are comparatively small, but yet are strongly made, and armed with a whole array of thorny spines, so that, what with the claws and what with the spines, the creature is a truly formidable being, and one that may not be grasped with impunity by a careless hand.

This species inhabits some of the hotter parts of the world, and the individual from which the sketch was taken was procured from the Mauritius. It may here be mentioned that the illustrations of the crustacea have been taken from actual specimens, and that the originals may in almost every instance be found in the British Museum.

SPINE-ARMED LAMBRUS.—*Lambrus spinimanus.*   STRAWBERRY CRAB.—*Eurynome aspera.*

THE little STRAWBERRY-CRAB is very appropriately named, as its colour is of a pleasing red, and its surface studded with numerous tubercles, so as to bear some resemblance to the fruit whence it derives its popular name. It is a British species, and is generally found in deep water, so that the dredge is the instrument usually employed in its capture.

Mr. Gosse, who kept some of these pretty little crabs in a living state, makes the following observations on their habits: "The Strawberry-crab is a *climber.* If it were a terrestrial animal, I should say its habits were *arboreal.* True, it now and then wanders over the bottom of its abode, with slow and painful march, the hind feet held up at an angle above the level of the back, but generally it seeks an elevated position. We usually see it in the morning perched on the summit of some one of the more bushy weeds in the aquarium, as the *Chondrus*, or *Phyllopora rubens*, where it has taken its station during the night, the season of its chief activity, as of most other crustacea. It interested us much to see it climb: seizing the twigs above it, by stretching out its long arms alternately, it dragged up its body from branch to branch, mounting to the top of the plant deliberately, but with great ease. While watching it I was strongly reminded of the orang-outan at the Zoological Gardens; the manner in which each of these very dissimilar animals performed the same feat was so closely alike as to create an agreeable feeling of surprise."

The same writer then proceeds to draw analogies with other members of the animal kingdom, such as the spider-monkeys, the sloths, the long-legged beetles, and the spiders, for the purpose of showing that elongated limbs are mostly given to creatures that may be reckoned among the climbers.

The tubercles with which the body of this crab is so profusely studded are of a pink colour, and are terminated by a little knob either of red or white, but always highly polished. The ground colour is white, so that the creature presents a very pleasing contrast of colours. It is generally reckoned among the rarer British crustaceans, though perhaps without sufficient reason, as the depth of sea in which it resides renders it no very easy capture, and it is also accustomed to conceal its beautiful carapace under a coating of mud and small seaweeds, so that even when brought up in the dredge it might escape observation.

ON the left hand of the same illustration is seen a figure of the SPINE-ARMED LAMBRUS, a member of a moderately large genus, inhabiting the Mediterranean and the warmer seas of the world in general. The specimen from which the drawing is taken was captured in the Mauritius. In many respects the genus Lambrus resembles the parthenope, but is distinguishable by having one plate fewer in the abdomen, and in the manner in which the antennæ are jointed. The eyes of this Spine-armed Lambrus are retractile and placed on footstalks of an elaborate and curious construction.

THE large family of the Canceridæ now comes before us, and is familiarly known through the medium of the common EDIBLE CRAB.

This is a very common species, being plentiful around our rocky coasts, and generally remaining in the zone just under low-water mark. The fishermen catch it in various ways; but the most usual method, and that by which the greatest number of these crustaceans are captured, is by means of certain baskets, called crab-pots, cruives, or creels, according to the locality. These baskets are round, and in shape something like a flattened apple, and have an aperture at the top through which the crab gains access to the interior. When once within the basket, it cannot escape, because the opening is guarded by an inverted cone of osiers, like the entrance to a common wire mouse-trap, so that the elastic sticks yield to the expected prey while passing downwards, but effectually prevent all upward movement.

Supplied with a number of these creels, a corresponding amount of rope, floats, stones, and bait, the fisherman rows towards the best grounds, which are always where the bed of the sea is rocky, and the depth from three to twenty fathoms. The bait, consisting of haddock, skate, and other fish, is placed in the basket, together with a few stones, which serve to sink it, a line is attached, and the creel lowered out of the boat. A buoy is attached to the line and marked with the owner's name, so as to avoid mistakes as to the proprietorship of the creel. The fisherman then rows to a little distance, and sinks another baited creel, taking the precaution to place them so far asunder that the lines cannot be entangled in each other. It is worthy of notice that the bait must be perfectly fresh, or the crab will have nothing to do with it, while that for the lobster may be quite putrid. The fisherman, therefore, thriftily saves the overplus bait, and when it begins to be tainted, and will entice no crabs into the creels, uses it to catch lobsters.

Boys often employ their idle afternoons in crab hunting, always going among the rocks at low-water, and looking out for those rock masses that are covered with heavy seaweeds. They are armed with a kind of lance consisting of an iron hook fastened to a long stick, and with this they poke about in the crevices under the rocks, and twist out the crabs that have concealed themselves. These crabs, however, seldom attain any great size, the larger specimens remaining in the deeper water. The boys call them "pungers" to distinguish them from the green crab, which is hardly eatable.

It is fortunate for these creatures, that though they are terribly carnivorous in their habits, fierce and voracious as the tiger and cannibals withal, the sense of pain is but slightly developed. This is no mere assumption, based on dissection and examination of nerves, but is a well-known fact to those who have had the opportunity of studying them in their native state. A very amusing instance of this strange indifference to injuries which we should think productive of unendurable agony, is related by Mr. Rymer Jones, in his " Aquarian Naturalist."

After mentioning that he had committed the error of placing a number of crabs in an aquarium, he proceeds as follows : " We had no idea, when we consigned them to the same vessel, what a set of remorseless villains we had to deal with, and much question whether our reformatory prisoners would show such an example of appropriate classifica- tion. They were of different sizes; their dimensions progressively increasing in a regular ratio, the biggest being about as large as an ordinary saucer, and seemed each of them to be aware of the tender mercies he might expect from his companions, though such a presentiment, if it existed, apparently interfered not a bit with his premeditated designs upon the rest.

The game was not long in beginning : the first that ventured out was seized upon at once by the next in size, who, laying hold of his victim as though he had been a biscuit, with one pair of pincers, proceeded deliberately to break up his shell with the other, helping himself to the flesh by means of his finger and thumb, with as much deliberation as if he had been taking snuff from a snuff-box, and apparently caring little for the hungry eyes that seemed to glare with savage delight upon the atrocious spectacle.

The crab had, however, not very long enjoyed his cannibal feast, before his proceedings were, as we thought, very unceremoniously interrupted by the onslaught of a stronger foe, which, seizing him exactly as he had done the first, proceeded to break him up in a similar manner, helping himself with the utmost *sang-froid* to the flesh of his already well-fed victim ; while the latter, strange to say, by no means desisted from his meal upon the crab first slaughtered, until quite disabled by his ruthless assailant ; affording a remarkable illustration of the absence of suffering during the infliction of the well- deserved penalty."

The same writer then proceeds to observe that by the next morning all the crabs had undergone mutual consumption, with the exception of two specimens, which were of nearly equal dimensions, and fearing to engage in combat, betook themselves to opposite sides of the aquarium. At last, however, they did meet, and the weaker was killed and eaten by his more powerful relative.

As has already been mentioned, the crab is unfit for food when about to cast off its shell, and immediately after that operation, the firm muscle and fat having given place to a watery fluid. The simplest plan of ascertaining whether a crab is fit for food, is to shake it close to the ear, and listen for the splashing of liquid. If the least sound of a fluid be heard, the crab should be rejected. Weight, too, is a good criterion, as a watery crab is very light in proportion to its size. Male crabs, too, are preferred to the female, and may at once be known by the narrowness of the tail, that of the female being broad and flattish with a slight convexity.

When fairly caught, the crab is apt to feign death, and then becomes a dangerous and treacherous creature, ready to seize any object with its powerful claws, and holding with the gripe of a bull-dog. Should the reader ever be unfortunate enough to be caught in a crab's claw, let me advise him at once to twist off the entire joint, and then to pull the claw open. For, unlike the green crab, which gives a very severe bite, but soon loosens its hold, the Edible Crab will grasp as long as it lives, and when the enormous size is considered, to which it sometimes attains, and the mass of firm muscle which moves the claw, it will be seen that this species is really a formidable creature. In order, therefore, to pre- vent it from playing such tricks, it is better to put it *hors de combat* as soon as it is caught, and while it is feigning death. The best method of attaining this object is by administering a smart stab under the tail. The tail, by the way, affords a good indication of the crab's condition, and if it bulges well out, the creature is sure to be fat and full of firm meat.

The shell of this crab is seldom found entirely clean, being generally encrusted with acorn-barnacles, and various marine creatures. Sometimes, when the crab is a very old and large one, has ceased growing for several years, and consequently has needed no change of shell, it becomes absolutely loaded with all sorts of extraneous growths, and in

EDIBLE CRAB.—*Cancer pagurus.*

many cases is almost invisible under its load. The upper figure in the illustration was sketched from a specimen in the British Museum, where a number of oysters had affixed themselves to the shell, and consequently had been borne about with the crab in all its peregrinations.

Here I mention, that the inland naturalist will find himself greatly aided by occasional visits to the fishmongers' stalls, whither a variety of marine animals are sure to make their way. The general structure and anatomy of the crustacean can be admirably studied by means of the crabs, lobsters, shrimps, prawns, and the hermit-crabs which so often inhabit the periwinkle shells. Adhering to oysters, scallops, and the shells of old crustaceans, are numbers of molluscoids and zoophytes, while the molluscs themselves can be well studied by examining the oysters, scallops, mussels, whelks, and periwinkles. Indeed, no one need deplore the lack of specimens as long as he has access to a poulterer's or a fish-monger's, for he will be able to supply himself with subjects that will occupy his full time in their investigation, and will fill up every hour of the longest life.

Concerning some of the habits of this crab, Mr. Gosse has made the following interesting observations : " At the water's edge, at the outer base of the Capstone, at low-water spring-tide, I was looking about for actiniæ, when, peering into a hole, I saw a fine crab, not of the very largest, but still of very nice table dimensions. I poked in my arm and took hold of him, and though he made vigorous efforts to hold fast the angles and notches of his cave with his sharp toes, I pulled him out and carried him home. I noticed that there came out with him the claw of a crab of a similar size, but quite soft, which I suppose might have been either carried in there by my gentleman to eat, or accidentally washed in.

Æthra scruposa.

After I had got him out—for it was a male—I looked in and saw another at the bottom of the hole, which appeared to me considerably smaller. I debated whether I should essay this one also, but reflected that I could only eat one at a supper, and that moderation in luxuries is becoming. 'So,' said I, 'friend crab, I may find you here again on some other auspicious morning.' When I arrived at home, however, I discovered that I had left my pocket-knife at the mouth of the crab-hole. I felt loth to part with my old knife, and therefore at once put on my hat, running hard, for fear the tide, which had already turned, might be too high. I got to the place, however, just in time, found my knife, and then took another peep at the crab. It had not moved, and thinking that if I could not eat it myself, I might ask my neighbour's acceptance of it, I drew it out with my fingers, as I had done with the former.

But lo! it was a soft crab, the shell being of the consistency of wet parchment, and the colours, all except those of the carapace, being pale. It was a female, too, without any sign of spawn, and had lost one claw. Strange that I had not thought of connecting the soft claw that I had drawn out before with this crab that I saw at the bottom; but I carefully put the helpless creature into the hole again, and saw that it had settled its legs and body comfortably in its old quarters; and there I left it. . . .

What, then, are we to infer from this association? Do the common crabs live in pairs, and does one keep guard at the mouth of their cavern while its consort is undergoing its change of skin? If this is the case, it is a pretty trait of cancerine sagacity, and one not unworthy of their acute instinct and sagacity in other respects. The male displayed no appearance of the moult, its coat being of a shelly hardness. I have no doubt that the claw of its mate was unintentionally torn off in its efforts to grasp some hold when resisting my tugs in dragging him out."

Sometimes this species attains a very great size, a large male weighing fully twelve pounds. The female, however, rarely exceeds five or six pounds in weight. Crabs possess one excellent trait of character. They have no definite time of moulting, so that some specimens may always be found in proper condition for the table. Moreover, during the winter they merely withdraw into deeper water, and can be taken, though not very numerously, by laying the creels in twenty fathoms of water. Unlike the lobsters, which change from a deep bluish black to a bright scarlet when they are boiled, heated, or even immersed in spirits, the crab retains much the same hues during its life, and after it has been placed in a cooked state on the table. The cause of this change of colour is as yet very obscure, molecular action being generally thought to be the means whereby the change is effected.

In many external points the ÆTHRA resembles the domed crab, which has already been described on page 568. Like that being, the carapace is very wide, flat, and expanded at the edges, as may be seen by referring to the upper figure in the illustration, which represents the Æthra as it appears when seen directly in front. The limbs, too, are comparatively short, and can be concealed under the shell, which, from its hilly surface, covered with tubercles, and the irregular, notched, and ridged carapace, has but little of the cancerine aspect. Zoologists of the present day, however, have placed it in the same family with the edible crab. The claw-feet, with their forceps, are very like those of the parthenope, but are not so proportionately large, and their surfaces are concave, so as to fit into the trunk. The eyes are very small, and their orbits nearly circular.

All the species of this genus inhabit the East Indian and African seas, and the particular specimen was taken off the Mauritius. The figure is drawn of the average size, though large specimens attain a length of three, and a width of four and a half inches.

In the illustration on the next page we have three curious examples of this large family, each being notable for some peculiarity of form or habit.

MONTAGU'S CRAB belongs to a genus which finds several British representatives. It is a flat-bodied and strongly made creature, very restless in disposition, and with a curious fondness for getting under stones, and turning them over; probably for the sake of obtaining a meal from the smaller marine animals that are accustomed to shelter themselves in such localities. The shelly covering of this crab is remarkably strong and flinty, and the muscular power of the claws is gigantic, when the small size of the creature is taken into consideration. It is a tolerably common species on several of our coasts, appearing to be peculiarly plentiful on the southern side of the island.

The general colour of the carapace is reddish brown, deepening nearly into black at the sides. The "fingers," if they may be so called, of the claws, are jetty black. The figure of the Montagu's Crab occupies the left hand of the illustration, and is drawn of the dimensions to which the ordinary specimens attain.

On the extreme right of the same engraving is seen a curiously marked crab, the carapace being divided into a number of partitions, in which is a certain, though not very definite regularity. This is the RED-SPOTTED ÆGLE, an example of a genus which, as far as is at present known, has no British representative, but inhabits the warmer seas. The Mauritius and the Philippines are favoured haunts of the Ægle, and the specimen whose portrait appears in the illustration was taken in the former locality. The general form is well given in the figure, and the colour is red and whitish spotted.

THE upper figure represents the TOOTHED PERIMELA, another British example of this family.

In all these crabs the width and length of the carapace are nearly equal, and there are some peculiarities connected with the jaw-feet that will be mentioned at the end of the volume. This creature is not very frequently captured, although it is far more common than might be supposed from its rarity in cabinets. There seem to be few parts of the British coasts where the Toothed Perimela may not be found by those who know where to look for it. The zone just above low-water mark appears to be its favoured residence, and it is discovered lurking very craftily among the seaweeds that fringe the rocks just at the low-water line. It is a little species, and on account of its retiring propensities can easily escape observation. Even on our more northern shores this creature has been captured, and is to be found on the Irish as well as on the English coasts.

TOOTHED PERIMELA.—*Perimela denticulàta.*
MONTAGU'S CRAB.—*Xantho flórida.*   RED-SPOTTED ÆGLE.—*Ægle rufo-punctàta.*

The name of Toothed Perimela is given to this species in allusion to the shape of the carapace, which has the front edge rather flattened, and cut into a series of four or five strong teeth like those of a saw. The surface of the carapace is smooth, and is swollen into several decided projections, something like those softly rounded hills called by the French "*mamelons.*" Over the region of the liver, the carapace is concave.

NEARLY allied to the preceding species is the HAIRY CRAB (*Pilumnus hirtellus*), a creature which derives its popular name from the curious hairy covering with which it is decorated. The convex carapace is studded more or less thickly with longish hairs, and the four hinder pairs of legs are also protected in the same manner. This crab is not a very common one, and is mostly found on the southern and western coasts of England, as well as in some parts of Ireland. It seems to prefer moderately deep water, fifteen fathoms being the usual depth at which it is captured.

It may be easily known by the following characteristics. On the front edge of the carapace are arranged four spines set in the same line, and the front is divided by a deep notch down the middle. The claw-legs of this species are always unequal in size, and the first joint of the outer pair of antennæ is short. The ground-colour of the Hairy Crab is chestnut brown, with a reddish tinge of greater or less intensity according to the

BRASSY CRAB.—*Zozymus Æneus.*

individual. The legs are dusky red, relieved by a series of bands of a greyish yellow colour.

In many characteristics, the genus Zozymus resembles the genus Ægle so closely, that the two genera have been blended together by several systematic zoologists.

As a general fact, it is much to be wished that the modern fashion of breaking up the old and established genera into a host of new ones, many of which contain but a single species, had not proved so fascinating to the authors. In many cases, the characteristics employed as generic differences are so very trifling, that they are barely of sufficient importance for the establishment of a species. New families also have been invented with reckless profusion, and in many instances, known to every naturalist, the characteristics on which the family is founded serve equally for the genus and the species. Needful as is some definite system of nomenclature, and admirable as is the system which Linnæus founded, and which has since formed the basis of all arrangement, it can be pushed too far, and, as is well known to be the case, is so widely abused, that merely to learn the multitude of sesquipedalian titles with which the study of zoology is now encumbered, requires a greater exercise of memory than to study the habits and peculiarities of structures which alone form the true objects of zoological science.

The general shape of the Brassy Crab can be seen by reference to the engraving, as well as the curious protuberances into which the carapace is moulded, and which cover the claws and legs. The sketch is taken of the natural size of an ordinary specimen, and the original was captured off the Mauritius. Only three or four species are ranked under this genus and all of them are natives of the Mauritius or the Philippines.

SPOTTED CRAB.—*Carpilius maculatus.*

THE genus to which the SPOTTED CRAB is assigned is rather more comprehensive than the preceding, and contains six or eight species, all being remarkable for their round and smooth bodies, the peculiar notches and projections on the edge of the carapace, and the huge claws that terminate the first pair of legs.

None of the members of this genus seem to belong to the British seas, requiring the greater warmth of the tropical sunbeams to thrive and prosper. Some species are found about Jamaica and the West Indies in general, others inhabit the Philippines, and others, again, are natives of the Mauritius. Except from the peculiar spots with which the carapace is decorated, this crab might easily escape detection while lying with its limbs withdrawn, and its frame in a state of quietude, for it is so round and so smooth that it looks very like a large pebble that has received a partial polish from the action of the waves. Many specimens are covered more or less with vegetable and animal growths, such as corallines, algæ, barnacles, and zoophytes, and are therefore almost undistinguishable while they are quiet.

The spots which are seen upon the carapace are bright red. This creature is drawn of its natural dimensions, and is a native of the Mauritius.

WANT of space compels us to pass by several other genera of crabs belonging to this extensive family, many of which are of eccentric shapes and not unpleasing colours. They are all marine, and though they can endure removal from the water for a considerable time, are sure to die when the moisture of their gills has dried up. For the crustacea do not breathe by lungs, as in the mammalia, or by tubes, as in the insects, but by gills, or branchiæ, as they are technically called ; and it is absolutely necessary that these organs

TUBERCLED GALENE.—*Galéne dorsális.* SMOOTH GALENE.—*Galéne ochtódes.*

should be kept in a moist state. Even in the land crabs, which pass nearly the whole of their time upon the dry ground, the gills need moisture, and are constantly maintained in their wet condition by an internal chamber which is filled with water and supplies moisture to the branchiæ. This structure is analogous to that of the breathing organs in the climbing perch and other land-visiting fish, which have already been described in an earlier part of this volume. In order to retain a full supply of water, the land crabs are forced to visit a stream or pond daily, but a single dip is sufficient for the twenty-four hours. It is very remarkable that these particular crustaceans cannot endure total immersion in water, and that if they are submerged for any considerable time they will be drowned, though more slowly, yet as certainly, as a human being under similar circumstances, the gills being made for aërial respiration. In other land crabs, there are masses of membranous folds, and in some is found a spongy mass in which the needful water is stored. The drying of the branchial membranes produces a double effect and is in two modes hurtful to the respiratory functions. In the first place, all such membranes fail to discharge their functions when they become dry, a familiar example being found in the delicate membrane that lines the lips ; and, in the second place, as soon as the folds of the branchiæ begin to dry, they collapse, and so diminish the extent of surface which is presented to the air.

The gills may be easily examined on opening a crab or a lobster, in which animals they are popularly called "lady's fingers." Even to the unassisted eye they exhibit great beauty of structure, but when injected with coloured gelatine and placed under a good microscope, such a wonderful apparatus of interlacing vessels is observed as to baffle description, and almost to elude the most delicate pencil. Some of the lower crustaceans have a very curious system of respiration, of which the organs of locomotion are essential portions, and in allusion to this peculiarity, the creatures are called branchiopoda, or gill-footed crustaceans.

On the accompanying illustration are two figures representing two dissimilar species of the same genus. The left-hand figure is the TUBERCLED GALENE, so called from the profuse warty excrescences which grow upon the claw-feet and the pincers. The right-hand figure is the SMOOTH GALENE, in which the claws are very much smaller in proportion, and destitute of the tubercles which are so characteristic in the former species. Both these crabs are natives of the East Indies.

WE now arrive at the family of the Portunidæ, or Swimming Crabs, which may be recognised by the construction of the last pair of feet, which are flattened sideways, and have the last joint dilated into a thin oblique plate, which answers as an oar or a fin, and enables the creature to propel itself through the water. The first example of this family is the GREEN, or SHORE-CRAB, so familiar to every one who has passed even an hour on the coast between the time of high and low water. Although one of the commonest of our native crustaceans, it is at the same time one of the most interesting, and, owing to its diurnal habits, its fearless nature, and its love for the shallow waters, it is very easily observed. I have spent many a pleasant hour in watching the habits of this little creature, and could hardly have imagined the activity, the piercing sight, and the cleverness with which this crab is endowed.

It is a fierce and even voracious animal, chasing and fairly running down living prey, and actually leaping upon its victim with a spring like that of the hunting spider. I have seen the Green Crabs run after and catch even the active sandhoppers, calculating with nice precision the spot on which they alighted, and pouncing on them before they could get themselves into position for a second leap. If the prey should be of tolerable size, the crab does not leap at it, but darts out one of its claws with a stroke so sharp and quick, that the eye can scarcely follow it, and as true of aim as the serpent's dart, draws back the victim, seizes it immediately with the other claw, and begins to pull it to pieces before it can recover from the shock.

Any living thing that can be caught becomes prey to the Green Crab. I have seen it actively pursuing the flies that settle in great numbers on the decaying sea-weed, and once observed it to dash at, seize, and eat a burrowing wasp which had settled on a stone. When catching a small and active creature like a sandhopper or a fly, the crab moves quickly towards the spot, leaps at its victim, flinging its body over it, and enclosing it within a kind of cage formed by its legs. It then pokes one claw into the living cage, picks out the enclosed victim, pulls it to pieces, and tucks the morsels into its mouth. As for flies, they are eaten at once without being pulled to pieces; but the sandhoppers and larger creatures always undergo that preliminary process. It is a cruel cannibal, feeding on its own kind without the least hesitation, and displaying equal ingenuity in chasing a Green Crab that is smaller than itself, or escaping from one that is larger.

An amusing instance of its fox-like craft is related by M. de Quatrefages:—" One day I threw a large arenicola (i.e. lug-worm) into a pool of several feet in extent. A troop of little shrimps, who were sedately enjoying themselves in the clear element, dispersed in alarm, startled by the noise made by the fall of this strange body, but, recovering themselves in a moment, they rallied, and, while the annelid was endeavouring to bury itself in the sand, one of the youngest, and consequently the most venturesome of the party seized the creature by the middle of its body.

Emboldened by this example, the others lost no time in imitating it, and the poor arenicola was pulled about in all directions, until a full-grown shrimp, darting from behind a tuft of corallines, dispersed his feebler comrades, and appropriated the booty to himself. I soon saw, however, that he would be compelled to divide the spoil, for, at that very instant, there poured forth from the moving sand some scores of small turbos and buccinums, who, conscious that a victim was at hand, wished to participate in the feast. Without any sign of uncertainty or hesitation, they moved straight forward towards the arenicola, whose body was covered in the twinkling of an eye with those voracious molluscs.

I thought his fate definitively settled, when a small Shore-crab issued from beneath a stone, put to flight the shrimp, and, by dragging off the arenicola, very nearly upset all the turbos, who forthwith hurried back to their sandy haunts. Then, however, a large edible crab appeared upon the scene, and the poor little Mænas was obliged in his turn to beat a retreat, in order to escape out of reach of the formidable pincers of his stronger kinsman. But he still kept a watchful eye over the dainty morsel which he had tasted,

GREEN CRAB.—*Carcinus mœnas.*

and taking advantage of a moment when the larger crab was withdrawing from the field from some temporary emotion of alarm, he rapidly seized the long-disputed arenicola, and carried it for safety to some distance from the water's edge, where he might devour it at his ease on dry land."

No one who has not seen these animals while searching after food, can have any idea of the energy and fiery activity which they display. My own observations are slightly opposed to those which have just been quoted, for, instead of finding the young crabs to be the most venturesome, I have always noticed that they are the most timorous, as indeed they have good reason to be, scuttling off in all haste as soon as a larger relative makes his appearance, and hiding themselves in the first convenient cranny. Once I saw a very exciting race between a large and a small Green Crab, the latter succeeding in making good its escape by the simple plan of getting behind some sea-weed, and then darting under a dry stalk that was lying on the sand. The large crabs are the most audacious, I suppose, because they have so long been accustomed to rule, that they cannot conceive themselves in an inferior position. No sooner does a foe present himself, than the crab turns defiantly on him, elevates itself on its tail, and holds its claws threateningly in the air, like a boxer in pugilistic attitude. Nor does the creature put forward any groundless menace, for it will bite with terrible severity, and if a hand be placed near it, will strike with astonishing fury. To capture such a crab with the bare hand is no easy matter, and requires some little address.

This species has conferred a great benefit on science, as affording the specimens which set at rest the theory of its development, and proved not only that the crustaceans pass through a series of metamorphoses, but that certain supposed species of the lower crustaceans were nothing more than the undeveloped young of the Green Crab. When first hatched from the eggs the little crabs are hardly larger than the comma employed in the present pages, and are much more like shrimps than crabs. They have a large rounded shield by way of carapace, from which projects a large and long horn; and the tail is long, slender, and jointed, something like that of a scorpion, and with a large and forked tip beset with bristly hairs. The eyes are not set upon footstalks as in the perfect state, but are placed directly upon the shell, just like those of the lower order of these creatures, called Sessile-eyed Crustaceans. The colour of the animal, which goes by the name of Zoea, is very light green, and the whole shell is semi-transparent. Altogether, the Zoea, in general shape, and in its manner of progression, has no small resemblance to the well-known larvæ of the gnats which are so common in our water-butts throughout the summer.

After it has passed some time in this form it casts its shell, and is transformed into a being of a much higher nature, with a well marked crab-like aspect. The long horn on the carapace disappears, the claw-legs become boldly developed and are armed with stout pincers, the eyes are set on short and stout footstalks, and the tail is not unlike that of the lobster. While in this state its size is about equal to the capital letter T. Sometimes the crab-like shape is rendered more conspicuous by the bending of the tail under the body.

Another transformation now takes place, and the little creature emerges from the shell an unmistakeable crab, though yet not in the least resembling the form which it afterwards assumes. This perfect form is only attained after a considerable interval and through several more changes of skin; the base of the carapace becoming narrower at each change, the tail pressed more firmly against the body, the claw-legs attaining more strength and solidity, and the hinder legs developing into the flattened oar-like shape which is characteristic of the family.

It may here be observed that the cast skins of the developing Zoea afford most beautiful objects for the microscope, more especially in the horny coverings of the eyes, which exhibit all the facets in a most admirable manner, and are ready to be mounted on a glass slide without the trouble of preparing them. These cast skins are, however, so delicate in their structure, and so transparent, that they are not easily perceived, except by a practised eye, while their filmy substance and the minuteness of their dimensions render them so fragile, that they are not removable without much care on the part of the operator. The cast skins of the prawn or shrimp are equally useful for the microscope, and are beautifully delicate objects as they float about in the water.

THE little crustacean which is called by the name of the VELVET FIDDLER-CRAB, derives its popular and appropriate title from the movements which it makes while swimming through the water.

As may be seen by reference to the illustration, the last joints of the hinder feet are extremely flat, and it is by their movements that the crab is enabled to swim. Their motions are very like those of an oar when used in "sculling" a boat, and are popularly thought to resemble the movements of a fiddler's arm while playing a lively tune. The word "velvet" is affixed to the name, because the entire shell of a perfect specimen is thickly covered with shining hairs, short, silken, and soft, something like the pile of velvet or fine plush. It is seldom, however, that a really perfect specimen is seen, as the soft velvety pile is easily rubbed off, and in almost every instance has sustained some damage, so that the blackish shell is seen, with its polished surface. The edges are very seldom clothed with their normal coating of hair. When tolerably perfect, a full-grown specimen is a really handsome creature, with its coat of velvet pile, its striped feet and legs, its scarlet and blue claws, and its vermilion eyes set in their jetty sockets.

This species is not one whit less voracious or cruel than the edible or the green crab, and as it enjoys all their activity, with the additional privilege of swimming through the water, it is even a more formidable animal, chasing and killing every creature that it can overcome. Even the hermit-crab, that lies so snugly in its shelly cell, with the large fighting-claw guarding the entrance, and its body withdrawn into the inmost recesses of the shell, is frequently captured and killed by this doughty warrior. Every one who has tried to pull a hermit-crab out of its house, knows the difficulty of the task. The creature has the art of retreating into its dwelling so far, and pressing its claw and legs so firmly against the inner mouth of the shell, that there is nothing by which the animal can be grasped, except, perhaps, the antennæ; and the crab will allow itself to be pulled to pieces rather than loosen its hold. Yet the Fiddler-crab makes little account of the hermit, but pokes his claw into the shell, pinches the poor hermit across the thorax, and drags him out of his cell. It then pulls off and eats the soft abdomen, tears up the body and limbs, and flings them away in fragments, as if for sheer wantonness of destruction.

There are many species belonging to this genus, which are scattered all over the world, especially where the seas are warm or temperate. The Velvet Fiddler is tolerably

VELVET FIDDLER-CRAB.—*Portunus puber.*

common on our own coasts, more especially those of the southern parts of the island, and England can also boast of several other species. Of these the MARBLED FIDDLER (*Portunus marmoreus*) is perhaps the handsomest, on account of the regular patterns of buff, brown of various shades, and red, which are seen upon the body. The shape of the patterns is variable, but their arrangement is always symmetrical. These colours are, however, very fugitive, and can only be preserved by removing the whole of the soft parts, and dyeing the carapace with great care. All the species seem to be decidedly local, so that in the space of two or three miles of coast as many species of Portunus may be found, each in its own particular locality.

PERHAPS the very best swimmer in the family is the OCEANIC SWIMMING CRAB, a creature to which the generic name of Neptunus has been given on account of its wonderful mastery over the waves.

A glance at the illustration will show that this crab is made for speed, its flattened limbs and body being calculated to offer the least possible resistance to the dense fluid

OCEANIC SWIMMING CRAB.—*Neptunus pelagicus.*

through which it has to pass. The Oceanic Swimming Crab is among crustaceans what the albatross is among birds, being able to sustain itself for days together without needing rest, and whenever it does seek a brief repose, needing nothing but the floating algæ as a temporary resting-place. The movements of this species are achieved with an easy grace and freedom that remind the observer of the swallow's flight, as the crab flies swiftly through the water, its claws ready to seize their prey, and its limbs held in such an attitude that they offer scarcely any resistance to the element in which the creature lives.

This species has a very wide range, and is found throughout the warmer seas. It is common around India, Australia, and the Philippines, and from its bird-like fleetness and activity, has never failed to attract attention. Like others of its family, it feeds upon living prey, and chases its victims through the water with a speed as rapid, an aim as certain, and a voracity as unfailing as are exhibited by the shark itself. The species is notable for the shape of the carapace, and the sharp spine into which each side is developed.

THE strange-looking creature which is represented in the illustration, is nearly as good a swimmer as the oceanic crab, and has many of the same habits. Like that crustacean, the FORCEPS-CRAB roams the ocean as freely as the bird roams the air, shooting through the waves with arrowy swiftness in chase of prey, gliding easily along just below the surface, hanging suspended in the water while reposing, or occasionally lying across some floating seaweed. The chief peculiarity of the Forceps-crab is the structure from which its name is derived, the wonderful length of the first pair of limbs, and the attenuated forceps with which they are armed. Though not possessing the formidable

FORCEPS SWIMMING CRAB. — *Lupa forceps.*

power with which some crabs are armed, the Forceps-crab is yet as terrible an enemy to the inhabitants of the sea, for it can dart out these long claws with a quick rapidity that almost eludes the eye, and grasp its prey with unerring aim.

No one who has not watched the crabs in their full vigour and while enjoying their freedom, can form any conception of the many uses to which the claws are put and the wonderful address with which they are used. Their bony armour, with its powerful joints, appears to preclude all delicacy of touch or range of distinction, and yet, the claws are to the crab, what the proboscis is to the elephant. With these apparently inadequate members the crab can pick up the smallest object with perfect precision, can tear in pieces the toughest animal substances, or crack the shell of other crustaceans as a parrot cracks a nut in its beak. It can direct them to almost every part of its body, can snap with them like the quick sharp bite of a wolf, or can strike with their edges as a boxer strikes with his fists.

As may be seen by reference to the engraving, the paddle-legs are broad and well developed, so as to ensure speed, the front of the carapace is sharply and deeply serrated, and the sides are drawn out into long pointed spines. It is a native of the West Indian seas, and is represented about the dimensions of an ordinary specimen.

The NIPPER-CRAB (*Polybius Henslowii*) is a better swimmer than the fiddler-crab, being able, according to Mr. Couch's account, to ascend to the surface of the sea, and to pursue its prey through the waters. So well does this creature swim, and so voracious is its appetite, that it captures and eats even the swiftest sea-fish, having been known to pounce upon the mackerel and the pollack. Its method of proceeding seems to be to

NIPPER-CRAB.—*Polybius Henslowii.*   SENTINEL-CRAB.—*Podopthalmus vigil.*

dart upon its prey, grasp it firmly with its sharply pointed and powerful claws, and retain its hold until the unfortunate victim is quite fatigued and falls an easy prey. It is not so handsome as the velvet fiddler, having none of the beautiful scarlet and azure tints which decorate that species, and being mostly coloured with different shades of brown.

OUR last example of this interesting family is the SENTINEL-CRAB, so called from its extreme watchfulness and the wonderful manner in which its eyes are arranged so as to explore objects in every direction, without needing to move, or even to raise itself from its flat and crouching attitude. The generic name of this creature is of Greek origin, being composed of two words, the former signifying a foot, and the latter an eye, and is given to it on account of the strangely long footstalks on which the eyes are set. The figure represents the animal as it appears when alarmed, or when it desires to look around it, the eyes being lifted so as to command an extensive view. When the creature is at rest, the footstalks lie horizontally upon the body, and are received into two channels or grooves, where they lie hidden and safe from danger. These grooves may be seen on reference to the illustration.

A somewhat similar disposition is found in some of the land-crabs, but differing in the arrangement of the footstalks. Each of these curious organs consists of two pieces, and in the Sentinel-crab the first is long and the second very short, while in the land-crab exactly the reverse takes place, the length of the footstalk depending on the second joint. Only one species of Sentinel-crab is at present known, and is a native of the Indian Ocean. It never attains very great size, its length varying from two to four inches.

WE now leave the swimming and marine crabs, and turn to those which are able to spend a great part of their existence out of the water. The FLATTENED MUD-CRAB belongs to a tolerably numerous genus of crabs, which live along the banks of rivers or in damp forests, and are evidently a link between the aquatic and the true land-crabs. The THELPHUSA lives in burrows which it excavates in the mud to a considerable depth, and gives the fisherman no small trouble before it can be dug out.

One species of this genus, the GRANCIO of the Italians, is very common around Rome and is largely captured for sale in the markets, as its flesh is very delicate, and in great request on the fast days of the church. It is dug out of the mud and kept alive for sale, as it can endure removal from the water for a very long time, sometimes living a month upon dry land, the only precaution needful being that it should be kept in a damp spot, such as a cellar. It is a most useful species, as it can be eaten throughout the entire year, but is thought to be in best condition during and immediately after the moult. There are many ways of dressing this delicacy, some persons killing it by long immersion

FLATTENED MUD-CRAB.—*Thelphusa depressa.*

in milk, and others asserting that its flesh has more flavour if eaten raw, like that of the oyster. In the market these crabs are tied to strings, but always at such lengths that they cannot reach each other, or if they should do so they would of a certainty attack and maim their nearest neighbours.

The Lake of Albano is a very favourite resort of these crabs, which absolutely swarm in its soft muddy bed. On the first view, the Mud-crab looks very like the common green crab of the sea-shore, but can be distinguished by its colour, which is of a whitish or livid hue. It runs about with great speed, and when it fears the approach of an enemy, hurries into the water, burrows under the mud, or hides itself beneath a friendly stone. Should, however, its retreat be cut off, it proves that it can fight as well as run, and grips with such force, that it makes the blood flow before it can be shaken off. During the winter it dives deeply into the mud, and there remains hidden, until the warmth of spring induces it to leave its retreat.

ANOTHER family of land-crabs is well represented by the TOULOUROU BLACK CRAB, or VIOLET CRAB of Jamaica (*Gecárcinus ruricola*).

This singular creature is found in vast numbers, and for the most part lives in burrows at least a mile from the shore, and sometimes at a distance of two or even three miles, seldom indeed visiting the sea but for the purpose of depositing its eggs. About the months of December and January the eggs begin to form, and the crab is then fat, delicate, and in good condition for the table. In May, however, it is quite poor and without flavour, and does not recover its proper condition until it has visited the sea, deposited the eggs, and returned to its home. About July or August the Violet Crab is again fat and in full flesh, having in fact laid in a stock of fat which will afford it sufficient nourishment through the time in which it remains in a torpid state. It retires to the bottom of its burrow, into which it has previously conveyed a large amount of grass, leaves, and similar materials, closes the entrance, and there remains until the next year.

It is a very quick and active creature, scuttling off to its hole with astonishing rapidity, and is not to be captured without the exercise of considerable skill and quickness. Nor must it be handled without caution, for as it runs, it holds up its claws ready to bite, and if it succeeds in grasping its foe, it quickly throws off the limb—which continues to gripe and pinch as sharply as if still attached to its former owner—and makes good its escape while the claw is being detached. For the table, this crab is esteemed as one of the greatest delicacies, and is treated in various modes, sometimes stewed, but mostly cooked in its own shell.

PEA-CRAB.—*Pinnótheres pisum.*   LONG-ARMED MYCTIRIS.—*Myctiris longicarpus.*

THE figures in the accompanying illustration are examples of two families of crustaceans.

The two smaller figures represent the PEA-CRAB, a curious little crustacea that is found within the shells of the horse-mussel and one or two other bivalves. That this crab was a frequent inhabitant of the pinna was a fact well known to the ancient naturalists, who put forward a number of ingenious but rather fabulous theories to account for the singular alliance. By some writers it was said that the Pea-crab supplied the place of eyes to the blind pinna, and that its especial task was to warn it of the approach of the polypus or cuttle-fish, receiving board and lodging as a reward of its labours. Some thought that the Pea-crab performed the office which ancient tradition attributed to the jackal, and was sent out by the mollusc for the purpose of obtaining food, the host and guest dividing the spoil.

What may be the real reason for this strange habit is not quite clear, for though the Pea-crab will live in the same shell without inflicting any apparent injury to its host, it is yet very fond of mussel-flesh, and will eat it with much eagerness. Indeed, several specimens have been kept alive for more than a year by being fed upon that diet. Perhaps it may feed upon the juice and less important parts of the mollusc, just as the ichneumon larva feeds on the juices of the caterpillar. Sometimes two and even three specimens are found within a single shell, and on examining the mussels taken from an old bank where they have been permitted to rest quietly, almost every shell will contain one specimen of the Pea-crab.

The colour of the Pea-crab is reddish cream colour, and the dimensions are small, the specimens being shown of their natural size. The average diameter is half an inch. It is a very timid creature, as might be inferred from the remarkably retired spot in which it passes its life; and when it is alarmed, it contracts its limbs and pretends to be dead, remaining motionless for a very long space of time, and not moving until it feels sure that its enemy is out of the way.

On the right hand of the illustration is seen the LONG-ARMED MYCTIRIS, an example of a moderately large family of crustaceans, all of which inhabit the warm seas, and are most plentiful under the tropics. In this genus the carapace is very delicate, convex, and somewhat circular in form; the limbs are long and slender. In the present species the carapace is curiously divided by two longitudinal furrows into three convex protuberances, and projects slightly in front. The claw-feet are long and armed with pincers that are very powerful in proportion to the dimensions of the animal. It is a native of the Australian seas, and the specimen from which the illustration is sketched, was captured off the coast of New Holland. It is represented of the natural size.

WE now arrive at another family, called the Ocypodidæ, or Swift-footed Crabs, from their extraordinary speed, which equals or even exceeds that of a man.

The upper figure in the engraving represents the FIGHTING CRAB, a creature whose name is well deserved. As the reader may observe, one of its claws is enormously large in proportion to the body, being indeed, nearly equal in dimensions to the whole carapace, while the other claw is quite small and feeble. It is remarkable that sometimes the right and sometimes the left claw is thus developed. This animal is a most determined fighter, and has the art of disposing its limbs like the arms of a boxer, so as to be equally ready for attack or defence. The figure shows the crab in its posture of defence.

The Fighting Crab lives on the sea-shore or on the border of salt marshes, and burrows deeply in the earth, the holes being tolerably cylindrical and rather oblique in direction. In some places these holes are so close together that the earth is quite honeycombed with them, and the place looks like a rabbit-warren. Each burrow is tenanted by a pair of crabs, the male always remaining in the post of danger at the mouth of the tunnel, and keeping guard with his great claw at the entrance.

While running, it has a habit of holding the large claw aloft, and moving it as if beckoning to some one a habit which has caused one of the species to be named the Calling Crab. This action has in it something very ludicrous, and those who have watched the proceedings of a crab-warren say that there are few scenes more ridiculous than that which is presented by the crustaceans when they are alarmed and go scuttling over the ground to their homes, holding up their claws and beckoning in all directions. The generic name is derived from a Greek word signifying laughter, and is given to the crabs because no one can look at them without laughing. These crustaceans possess very long footstalks on which their eyes are placed, but, as has already been mentioned, the second joint of the footstalk is long and the first is short.

CLOSELY allied to these creatures is the RACING CRAB (*Ocypode cursor*), sometimes called the Sand-crab, from its habit of burrowing in the sand. Sir J. Emerson Tennent, in his "Natural History of Ceylon," writes as follows of this crab: "In the same localities, or a little inland, the Ocypode burrows in the dry soil, making deep excavations, bringing up literally armfuls of sand, which, with a spring in the air, and employing its other limbs, it jerks far from its burrows, distributing it in a circle to the distance of

FIGHTING CRAB.—*Gelásimus bellátor*
ANGULAR CRAB.—*Gónoplax anguláta*

many feet. So inconvenient are the operations of these industrious pests, that men are kept regularly employed at Colombo in filling up the holes formed by them on the surface of the Galle Pace. This, the only equestrian promenade of the capital, is so infested by these active little creatures, that accidents often occur through horses stumbling in their troublesome excavations."

These crabs run with surprising swiftness, and it is by no means easy to catch them before they escape into their burrows. Sometimes they are made to afford a few hours' amusement to military officers and other persons who have too much time on their hands, the struggle between man and crab being as exciting as the battle between an eagle and a salmon. One device is ingenious, simple, and often successful. Long strings are attached to flat pieces of slate or stone, which are carefully laid near a burrow, and some tempting food laid outside. The crabs crawl out to feed on the bait, and while they are engaged, the slates are quietly drawn over the entrance of the burrows. A sharp rush is then made, the crabs scuttle away to their homes, and one or two are generally captured before they have recovered their presence of mind sufficiently to leave their barricaded doors and ask for admission into another habitation. Another amusement is to chase the crabs on horseback, trying to ride them down by main speed, and to kill them with a gun. They mostly take an oblique line when running, so that a pursuer who is acquainted with their habits is more likely to succeed in his endeavours than one who employs nothing but main speed in the chase.

None of these crabs care much about the water, being quite satisfied if they can obtain sufficient moisture to keep their gills in working order. As is the case with most of their kindred, they seek the ocean when the time for laying their eggs has arrived. Even then, they remain but a very short time in the water. It is, however, conjectured that the first stages of existence must be passed either in the water or underground, as a very small Racing Crab never seems to be found. On account of the great speed of these creatures, the Greeks were accustomed to designate them by a name which signifies a horseman or knight. Opinion appears to be divided with respect to the value of their flesh, some species being highly esteemed, while others are totally rejected, and even decried as poisonous. It may be, however, that locality has some influence in these opposite opinions, and that in some places the crabs may feed on wholesome food and therefore be eaten with impunity, while in others they may perforce mix with their diet certain substances injurious to human health, and so become in some degree poisonous. The reader will doubtlessly remember that the common edible mussel is at one time perfectly harmless, and at another is so injurious as to cause serious effects upon the health of those who eat it. life itself having been threatened by the mysterious influence.

PAINTED CRAB —*Grapsus pictus.*

The ANGULAR CRAB is one of the British species. and in many respects bears some resemblance to the preceding species. The eyestalks of this crab are also long and movable, the carapace is wider than long, and the legs of the male are nearly five times the length of the carapace; in the female they are only twice the length. The Angular Crab is taken off the southern coasts of our island, and is either dredged out of rather deep water or found within the stomachs of fishes. It is a burrower, forming excavations in hardened mud, and always having each extremity of the habitation open. A Mediterranean variety of the same species prefers to live among rocks, and is a good swimmer, frequently coming to the surface of the water, but not being known to frequent the land. The claw legs are of great length, and the claws themselves are large and powerful. Its name of Angular Crab is given to it in allusion to the shape of the carapace.

IN the next family, of which the PAINTED CRAB is a good example, the eyestalks are very short, and the carapace is squared. The members of this family are found in nearly all warm parts of the globe, not, however, being natives of the British coasts. Now and then a FLOATING CRAB (*Planes Linneana*) is swept into our seas together with masses of the well-known Gulf-weed; but its presence is purely accidental and cannot entitle it to rank among the British species.

The Painted Crab is a native of the Antilles, and is a very active as well as beautiful species, haunting the sea-shore and running about nimbly in the spray. It is a good climber, and can ascend or descend nearly perpendicular rocks provided that they are washed by the waves. Some species of this genus prefer the mouths of tidal rivers, and remain mostly at the edge of the water. They seem to rejoice in the hottest rays of the tropical sun, and run about nimbly hither and thither, with the sunbeams flashing on their wet bodies. They are all wary and timid beings, betaking themselves to the water on the least alarm, and flinging themselves into the waves with such force that their flat bodies skim for some little distance over the surface, much as the flying squirrel skims through the air between two trees. While running along, they strike their claws against each other as if for the purpose of menacing their pursuer, and when a number of these crabs are startled in one locality, the clatter which they make is surprising. The colour of the Painted Crab is reddish, covered with spots and variegations of yellow. It is not at all a large species, the carapace being seldom more than two inches in length.

Two remarkable species of crabs are shown in the accompanying illustration.

The lower left-hand figure represents the CRESTED CRAB, a curious animal that inhabits Japan. Most of the species of this genus are found in the hot parts of the world, such as the Sandwich Islands, the Mauritius, and the West Indies. The whole shape of this creature is strange in the extreme, its carapace being covered with all kinds of tubercles and spines, and edged with saw-like teeth. Even the claws are covered with unexpected spikes and tubercles, and when folded in front of the body, assume a very crest-like aspect. The creature instinctively makes use of its extraordinary shape for the purpose of concealment, and when it is alarmed, it tucks its legs away under the broad carapace, folds its claws over its front, and remains perfectly motionless in spite of all annoyances. A sailor has been known to find one of these crabs on the seashore, to take it for a curious stone, and so to put it in his pocket. Some time afterwards, when he had laid down the supposed stone, he was not a little surprised to see it put forth a number of legs, and run away at best speed.

The name of Crested Crab is given to this species in allusion to the form of the closed claws. In the illustration, one claw is shown raised, for the purpose of exhibiting its remarkable shape, and the other is represented as it appears when folded, so as to bring before the reader's eye the inanimate aspect of the creature when it has closed all its limbs.

The ARMED CRAB is also a native of Japan and China, and belongs to the same family as the preceding animal. The chief peculiarity in this creature are the four sharp spines with which the carapace is armed, those at the side being of very great dimensions, each measuring half the length of the body. The claws are sharp and powerful, and are formed in a manner somewhat resembling the same members in the crested crab.

Allusion has more than once been made to the power of voluntarily throwing off a limb, a faculty which is inherent in all the crustacea, but in some species is prevalent to a wonderful degree. The land-crabs, for example, will always sacrifice their best claw as a means of purchasing safety, and seem able to part with almost any number of legs without feeling the loss. If, for example, a land-crab or mud-crab be taken up by the legs, it suddenly shakes itself loose, leaving in the captor's hands the limbs which he has grasped, and making off with the remainder. The animal always throws off its limb at one of the joints, seeming to achieve the feat by a sudden muscular contraction, like the movement which shakes off a blind-worm's tail, snaps away the wings from a flying ant, or breaks up the whole anatomy of a brittle-star into fragments. If the limbs be cut or severed between two of the joints, there is a flow of blood and the creature seems to feel the injury acutely. It soon, however, heals itself by shaking off the injured portion

CRESTED CRAB.—*Calappa philargus.*   ARMED CRAB.—*Thealia acanthóphora.*

at the joint immediately below the wound, and then seems to recover itself from the shock. This faculty is very needful to creatures who depend upon their claws for obtaining food, and who are so quarrelsome in disposition. As has already been mentioned, the crustaceans fight terribly, and in those cases where the combat is not *à l'outrance*, both parties have usually to deplore a limb or two crushed in the nippers of the opponent. Were no means provided for replacing the injured members, the poor creatures would die of starvation, as would an elephant if deprived of his proboscis, or a lion whose feet had been cut off and teeth drawn.

Every injured limb therefore is at once discarded at some joint, no bleeding takes place, and the stump heals almost immediately. After a short time, a little button seems to be protruding from the joint, and before many days have passed, a very small but perfect claw is seen to protrude. This new member grows regularly though slowly, and so in process of time the creature is re-supplied with its full complement of limbs. Every one has noticed the frequent inequality in the size of lobsters' claws, how one side is armed with a huge weapon nearly as large as a man's hand, while the other can only boast of a puny, soft-shelled claw an inch or so in length. This inequality is the result of some injury that has been inflicted on the limb from which the little claw has sprouted, and in almost every instance the original claw has been lost in battle. After the moult, and the induing of a fresh suit of armour, the growth of the new claw proceeds more rapidly.

It must be noticed that this power of reproduction of a lost or injured member always denotes that the creature possessing this capability is not very highly organized. Very few of the vertebrates, and those mostly belonging to the reptiles, are able to reproduce a lost member, and even in these few instances, the restorative power is very limited. A

TORTOISE CRAB.—*Camara Calappa.*

very few examples have been recorded where a limb has been lost and replaced, but such phenomena are extremely rare, and can only be looked upon as variations from the usual system.

THE faculty of avoiding danger by closing all the joints of the limbs and merging them as far as possible under the carapace, is carried to a wonderful extent in the TORTOISE-CRAB, a crustacean that derives its popular title from its general similitude to the reptile from which it derives its name. There are, indeed, many of the tortoise tribe which are not able to enclose themselves nearly so perfectly as does this crab, and excepting the box-tortoise, there is perhaps none that exceeds it in the very perfect concealment of all vestige of their limbs. The carapace is wide, flattened at the edges, and dome-like in shape, so as to afford a perfect cover to the limbs. Owing to the manner in which these crabs conceal their limbs, Cuvier called them by the appropriate name of Cryptopods, or Hidden Feet. In all of these creatures the carapace is domed or vaulted, so as to form a shelter for the legs, while in the typical genus the claw-feet are very large and compressed, with a decided upper edge which is notched or toothed so as to form a crest. The French know these crabs by various names, such as Migranes, Coqs de mer, and Crabes honteux.

The claws are broad, flattish, notched at the edge, and scooped in a peculiar fashion, so that when folded over the body they exactly fit to the shell, as if they were part of the same piece. Two sides of this species are given in order to show the crab in its upper and under aspects. The Tortoise-crab is a native of the Mauritius.

IN the family of the Leucosiidæ, the carapace is more or less rounded, and projects somewhat in front.

The larger figure on the accompanying illustration represents the URANIA-CRAB, an example of the typical genus. It has a smooth carapace with rounded edges, and the claw limbs are very large and powerful in proportion to the size of the body. On their edges they are covered with rounded tubercles, and one or two of these projections are scattered upon the surface of several joints. In all these crabs, the apertures through which water passes to cover the gills are in the form of canals. One curious characteristic of these crustaceans is that the external antennæ are very small indeed, and are inserted

SPOTTED LEUCOSIA.—*Leucósia hæmatóstica.*
URANIA-CRAB.—*Leucósia Uránia.*   KEELED CRAB.—*Myra carináta.*

in a narrow but deep notch near the eyes. In common with nearly all this genus, it is a native of the Philippines.

The upper figure represents a much smaller species of the same genus, in which are observable the same characteristics which have already been mentioned, namely, the very small and narrow snout, the round and flask-like body, the strong claw-feet, and the very small size of the external antennæ, which cannot be seen from the upper surface.

On the extreme right hand of the illustration is shown a strange, long-armed crustacean which is called the KEELED CRAB on account of the form into which its carapace is moulded, being pinched, as it were, into a kind of keel throughout its length.

In this crab the chief points of interest are the apparently shapeless carapace, which is moulded as if squeezed out of clay by a single grasp of the hand, and the very long claw-feet. In consequence of this latter structure, the Japanese call one of the species of this genus, the Tenkô-gani, or Long-handed Crab. In this creature, as in the last, there is no appearance of external antennæ when viewed from the upper surface, the claws are feeble in proportion to the long and somewhat powerful limbs to which they are attached, and the end of the carapace is drawn out into a long and sharp point. This species inhabits the Philippines.

THE Nut-crabs are members of the same family, and are found off the British shores. PENNANT'S NUT-CRAB is generally to be caught in about fifteen fathoms of water. It is rather a sluggish and inactive species, burying itself in the sand or mud at the bed of the sea, much after the fashion of the toad, and only leaving its eyes and claws at liberty to act. Thus it sits and waits for prey, behaving much like the ant-lion in its pitfall of sand. It seldom moves about except at night, and even when it does travel, its motions are very slow and deliberate. It is rather a pretty little crab, being of a tolerably bright yellow, with a red patch on the snout.

This, and other species of the same genus, are often found in the stomachs of marine fish, and as the shell is very hard, specimens are discovered in a good state of preservation.

Iphis septemspinosa.　　　　*Ixa cylindrus.*　　　　*Nursia plicata.*

In the course of the preceding pages we have seen many instances of curious structures which seem to be wholly supplementary and of no use whatever to the creature. The very fact of their existence is a proof that there is some use for them, although their office is so obscure as to elude all present researches. Such an example is given in the present illustration, of which it occupies the uppermost place, and is represented as climbing over a large flat stone. On each side of the body is a large cylindrical projection, so that the extreme measurement from side to side is nearly equal to three times the length of the body; the claws are long and feeble, all the force seeming to be thrown into the two projections. This crab is a native of several parts of Asia, and is found both in India and off the Philippines.

On the right hand of the illustration is seen another odd-looking crab, having its carapace scooped and grooved in a wonderful manner. This is an example of the genus Nursia, and is here given in order to show the strange variety of form which is assumed by the carapace. It is found in the Indian Ocean.

The last figure in the illustration represents the SEVEN-SPINED CRAB, so called on account of the seven sharp points that project from the carapace. There are several large species of this genus, one of which has nine spines instead of seven. In this curious creature the arms are longer and more slender than in the preceding species, and the claws at their extremities are exceedingly delicate and feeble. This crab is a native of the Eastern seas, and is generally captured off the coasts of India.

THE two larger figures in the next illustration represent the same species in two different attitudes, that on the left being intended to show its general aspect, and that on the right being given for the purpose of exhibiting the position which it assumes while watching for prey.

As will be seen by reference to the engraving, the MASK-CRAB buries itself in the sand or muddy bed of the sea, and only permits its snout to project, with the long antennæ, so as to feel (or, as some suppose, to listen) for approaching prey, and the eyes to look in all directions for any eatable creature that may haplessly wander within reach, and the claws, in order to seize the prey when it passes within the grasp of their long and formidable hands.

MASK-CRAB.—*Corystes Cassivelaunus.*   POLISHED CRAB.—*Thia polita.*

The antennæ are apt to become clogged with mud, and the crab is therefore gifted with an apparatus whereby they can be perfectly cleansed. In order to effect this object, the crab bends each antenna sideways, until it rests on the hairy base of its companion; it then draws it completely through the stiff bristly hairs, until every particle of extraneous matter is brushed away. Sometimes the Mask-crab buries itself so deeply, that it only leaves the tip of the antennæ above the sand. The name of Mask-crab is given to this crustacean, because the carapace is so formed that its two waved grooves mould the surface into an obscure likeness of the human face. This resemblance may be seen by reference to the left-hand figure.

The smaller figure on the right hand is the POLISHED CRAB, so called from the smooth, shining surface of the carapace. In this genus the carapace is somewhat heart-shaped, and very narrow behind; the claw-feet are small and short. Although this is one of the British species, it is not very commonly found, probably on account of its habit of burying itself rather deeply in the sand, so that the eye cannot perceive it, and the dredge passes over its sunken body without sweeping it into the net. It is rather a pretty little crab, though unfortunately its beauty is only skin-deep, and perishes after death. When living and in good health, the carapace is of a soft rose colour, and has a very pleasing appearance; but when the shell is emptied of its contents, or even after the death of the inhabitant, the pink hue rapidly fades into the dull greys so prevalent among dead crustaceans. In the Mediterranean the Polished Crab is very plentiful, and it is possible that if the British coasts were thoroughly searched, the Polished Crab would be found in tolerable plenty.

WE now come to another family, of which the WOOLLY CRAB is an excellent type.

This creature derives its name from the coating of thick short hair with which its body is covered. All the species of this genus possess several peculiarities; at each side of the shell, and just at the base of the claw-legs, is an aperture that looks as if it had been cut for a button-hole, and partly closed with a membrane. These apertures are in fact the openings through which the water passes for the purpose of supplying the gills with the needful moisture, and allowing it to escape when it has performed that office.

The limbs are very remarkable, both as to their shape and their disposition; they are unequal in size, and the two last pairs are elevated on the back in a very curious fashion. At their extremities is a large hooked nail, which is jointed to the limb, and can be folded back so as to take a firmer grasp. The Woolly Crab seldom approaches the shore, but prefers the deeper waters, and is only to be caught by a dredge with a very long line. On account of its locality very little is known of its habits though much is conjectured. It is mostly found in the Mediterranean.

SCALLOP CRAB.—*Caphyra pectenicola.*   WOOLLY CRAB.—*Dorippe lanata.*

On the left hand of the illustration may be seen a small scallop shell, with a very little crab lying inside it. This is the Scallop-crab, so called on account of its habits. The general shapes of this crab are not at all unlike those of the pea-crab, which has already been described; and the curious analogy that exists between form and habits, cannot but strike everyone who compares the two creatures.

In the sub-order which now comes before our notice, is seen a modification of structure which evidently forms one of the connecting links between the crabs and the lobsters, or, to speak more accurately, between the short-tailed and long-tailed crustacea. The two large divisions of the body bear scarcely any ordinary proportion to each other, the abdomen being exceedingly small, and the "cephalo-thorax" enormously large. Some of these creatures extend the abdomen from the body like the lobsters, while others bend it under them, like the crabs. In some species, of which the common hermit-crab is a familiar example, the last pair of legs are totally useless for walking, and are modified into a pair of appendages, by means of which the animal is enabled to grasp with a hold so firm, that it may often be torn asunder rather than be forced to loosen its gripe.

The right-hand figure of the next illustration represents a curious species belonging to the typical genus of the first family. In all these crustaceans the body is rather globular, and the carapace is bent downwards in front. The eyes are short.

The Hairy Crab is found in the hotter seas, and has been captured off the Cape of Good Hope. The reader will observe that the two hinder pairs of legs are very small in proportion to those limbs which are evidently intended for progression, and that they are furnished at their tips with a hooked claw. These modified and apparently stunted limbs are however extremely useful, their office being ascertained by studying the economy of the animal. With the claws at the end of these limbs the crab seizes pieces of sponge, shells, and other marine substances, so as to conceal its form under their shelter, thereby exhibiting a curious analogy to the well-known habits of the tortoise beetle while in its larval state. Some crabs in the British Museum are admirable examples of this peculiarity, one specimen of *Dromia lator* being nearly hidden under the mass of sponges under which it lies concealed, the sponge being nearly as large as a man's fist, while the crab is about the size of half an ordinary walnut.

BEARDED CRAB.— *Homola barbata.*　　　HAIRY CRAB.—*Dromia hirsutissima.*

One species of this genus, called, from the shape and mouldings of the carapace, the DEATH'S-HEAD CRAB, is found, though very rarely, on the British coasts, but is taken in fine condition among the Channel Islands. The scientific name of this crab is *Dromia vulgaris.* Its colour is deep brown changing to pink upon the claws, the carapace is strongly knobbed above, and the edge is notched so as to form four broad teeth. Some species of this genus are thought to be poisonous, but without any apparent reason. The hairy covering is not so extensive in other species, for the carapace of the common Death's-head Crab is quite smooth and polished, the hairs being restricted to the limbs, where they afford an excellent basis for sponges, corallines, and zoophytes.

THE BEARDED CRAB is an example of another family, in which the carapace is formed into a kind of beak, and is almost always covered with sharp spines. The fifth pair of legs are comparatively short, and are not employed in walking. In the Bearded Crab the eyes are very large and round, and the carapace is covered with short but sharp spines. The antennæ are long and the claws powerful, and are well suited for detecting and securing prey. The Bearded Crab is found in the Mediterranean.

THE strange and weird-like creature which is called by the appropriate name of the PORCUPINE-CRAB, is a native of Japan. In this species the characteristics of the family seem to be carried to the very utmost. The last pair of legs are extremely small; so diminutive, in fact, that they are folded under the body and not visible when the creature is viewed from its upper surface. The carapace is triangular and thickly covered with spines; and even the limbs bristle with thorny points set as closely as the horny bayonets

PORCUPINE-CRAB.—*Lithodes Hystrix.*

of the hedgehog. It appears to be rather dull and sluggish in its movements, crawling along the bed of the sea with slow monotonous action.

One species of this genus, the NORTHERN STONE-CRAB (*Lithodes Maia*), is found off the coasts of England, and is plentiful on many of the Scottish shores. It is covered with short thick spines which extend over its legs and claws, and in its general shape bears some resemblance to the spider-crabs already described. In spite, however, of its thorn-studded surface, it is much eaten by fishes, and is not unfrequently found entire in the stomachs of the fish that are taken off our coasts.

The Northern Stone-crab may be known by its very long beak furnished at the end with two short and rather diverging teeth, and by its bright scarlet colour when it is first taken from the water. The Porcupine-crab, which has the characteristics of its genus almost exaggerated, is found in Japan.

THE last of the Homolidæ is the strange and uncouth-looking creature which is shown in the accompanying illustration, and called the NODULED CRAB on account of its singular conformation.

This crustacean, instead of being covered with thorny points as in the stone-crabs, has its entire carapace, limbs, and claws so studded with tubercles, that it can scarcely be recognised as a living creature, and looks more like a rough stone encrusted with marine growths. The carapace is rather triangular in form, but its sides are so scooped into hollows and projections, its surface so moulded into elevations and depressions, and its shell so covered with tubercles of various shapes and sizes, that its true proportions are not easily distinguished.

NODULED CRAB.—*Echidnócerus cibárius.*

The claw-limbs are large and powerful, and are even more obscure in shape than the body, for the substance of the shell is thrown out into such a forest of tubercles that at first sight it seems to be covered with a very fertile crop of fungi, algæ, or the thick and fleshy molluscoids which spread so rapidly when once they have obtained a resting-place. Even the antennæ of this strange animal are furnished with long projecting points, and look something like the beautiful comb-shaped antennæ of the larger moths.

The specimen which is shown in the illustration is given for the purpose of exhibiting the remarkable aspect assumed by this crab while at rest, the limbs being tucked away under the body, and the large arms thrown forward so as to protect the eyes and front, and to be ready for seizing prey. The Noduled Crab is found in the Columbia river.

THE next family is a very small one, and is called Raninidæ, from the fancied resemblance which its members bear to the shape of a frog. In these crabs the carapace is something like the half of a jargonel pear, from which about half an inch has been cut at each end. The broader end is towards the front, and is scooped so as to form a number of tooth-like projections. The abdomen of these creatures is extremely small, and may be represented by about three-quarters of an inch of a French bean stuck on the small end of the pear which answers to the carapace.

The limbs are moderately large, and the crab is said to leave the water and travel on land. Some persons say that it climbs to the tops of houses, but without mentioning the height of the houses or the materials of which they are composed. The claws are rather large, flattened, something of a triangular shape, deeply toothed, and with the pincers bent inwards at almost a right angle. All the legs are very close together at their bases,

SYMNISTA.—*Albunea symnista.*   TOOTHED FROG-CRAB.—*Ranina dentata.*

and the last two pairs ascend upon the back. All the Ranidæ inhabit hot countries, and are found chiefly in the Mauritius, the Philippines, and India. The TOOTHED FROG-CRAB (*Ranina serrata*) is a good and tolerably common example of this family. It is a native of the Mauritius and Japan.

It is altogether a curious-looking creature, with a broad flattish carapace edged, in front with the most formidable-looking teeth, that hardly seem to belong to the shell, but to have been taken from the mouth of a shark and fastened artificially upon the front edge of the carapace. The legs of this creature seem quite insufficient to carry the great broad carapace, and the abdomen is almost absurdly small. The colour of the shell is very pale pink, and the spines which cover its surface are of a whiter hue, looking almost as if they had been pricked into the carapace by human means. All the points are directed forward, and have a very rough effect when the hand is drawn from front to rear. When full grown, the Toothed Frog-crab is about as large as a man's fist.

THE smaller figure on the same illustration represents a rather pretty-looking and decidedly curious crab, which is an example of another family, termed the Hippidæ. In this family the carapace is long, rounded, and rather thimble-shaped, in most cases slightly flattened above. The abdomen is very small, and from the upper view of the body is hardly visible. In one species of the typical genus, the ASIATIC HIPPA (*Hippa asiatica*), the carapace is very round, elongated, and altogether egg-shaped, so that it would hardly be taken for a part of a crustacean. Even its colour is a hue rarely seen among the shelly race, being a soft pale yellow very like the chrysalis case of the oak egger-moth, which indeed it also resembles in shape.

The colour of the SYMNISTA is very pale yellow. Its claws are suddenly broad, rather sharp, and bent over at a right angle. The antennæ are long, and beautifully fringed with hairs. It is a small species, only two or three inches in length.

THE general shape of the OAR-FOOT CRAB points it out as allied to the preceding species, although the antennæ are not so long, and their fringe not so conspicuous.

The name of Oar-foot is given to this species on account of the curious modification by which the false feet are developed into oar-like appendages with flat blades, which serve for swimming like the hinder feet of the swimming-crabs. The last ring of the abdomen is changed into a flattened and pointed paddle. The carapace is convex and of a tolerably regular oval. This species is a native of New Holland, and never attains to any great size. An illustration of the Oar-foot Crab will be found in one of the succeeding pages.

WE now come to a singular group of crabs which are remarkable for their soft and shell-less tails, and the mode employed to protect them. From their solitary habits they are called Hermit-crabs, and from their extreme combativeness they have earned the title of Soldier-crabs.

The best known of tnese crustacea is the common HERMIT-CRAB of England (*Pagurus Bernhardus*), which is to be found plentifully on our shores. Like all its race, the Hermit-crab inhabits the shell of some mollusc, in which it can bury its unprotected tail, and into which it can retreat when threatened with danger. The Hermit-crab usurps the deserted home of various molluscs, according to its size, so that, when young and small, it is found in the shells of the tops, periwinkles, and other small molluscs ; and when it reaches full age, it takes possession of the whelk-shell and entirely fills its cavity.

Any one may find these odd crabs by watching a rock pool after the tide has gone down. There are always plenty of shells in such places, and if the observer will remain very quiet, he will see one of the apparently empty shells suddenly turn over, and begin to run along at a great pace, much faster than if it had been inhabited by its usual occupant. On the least movement of the spectator, the shell stops as abruptly as it had started into action, and rolls over as before, seemingly dead and empty.

On picking it up, the mystery will be revealed, for within the shell will be seen an odd little crab, with a body curved so as to fit exactly round the shell, with one claw small and one very large. If it be touched it retreats still further into the shell, and defies any attempt to pull it out. Even if a claw be grasped, the creature cannot easily be withdrawn, and clings so tightly to its home that in most cases it may be torn asunder rather than loosen its hold. It is enabled to hold thus firmly by means of a pair of pincers situated at the end of the tail, and which are indeed the last pair of legs modified for that purpose. Sometimes the creature can be coaxed, as it were, out of its shell by a long and steady pull, but, as a general rule, to get a Hermit-crab uninjured out of its abode is a very difficult task.

I have often accomplished it by putting the shell upon an open actinia. The crab, feeling the tentacles of the actinia gradually surrounding its limbs, and not liking the aspect of the living gulf into which it is rapidly descending, makes the best of its way out of the shell, and can be snatched up before it has found time to recover its presence of mind. Sometimes a Hermit-crab may be captured while the inhabitant is three parts outside its shelly house, but, on the least alarm, the creature flies back to the farthest recesses of its home as if worked by a spring.

This crustacean is wonderfully combative, and will fight on scant provocation. Anything will serve for a cause of war, such as a piece of meat, a smaller crab, or a shell to which another individual happens to take a fancy. If two Hermits be removed from their houses, and put into a rock pool with only one shell in it, the combats which take place for the possession of that solitary shell are as fierce and determined as any that have taken place in the tourney or the field of battle. As with most of the crustaceans, the victor always eats his fallen foe ; and even though he be bereft of a few legs, he seems to care nothing for the loss, but eats away with perfect appetite.

Even when the crabs are suited with homes, their combats are fierce, deadly, and active, in spite of the heavy shell which they drag behind them, and which seems to

CRAFTY HERMIT-CRAB.—*Pagurus callidus.*

incommode them no more than the hundredweight of steel inconvenienced an ancient knight. They spar with great address, guarding the only vulnerable point with the large claw, and threatening the adversary at the same time with that weapon. At last one of them makes a dash. the pair grapple, the weaker is gradually overcome, the stronger pushes his claw into the failing adversary's shell, crushes his unprotected breast, draws him dying out of his shell, picks him to pieces and eats him.

To see a Hermit-crab fitting itself with a new shell is a very ludicrous sight. The creature takes the shell among its feet, twirls it about with wonderful rapidity, balances it as if to try its weight, probes it with the long antennæ, and perhaps throws it away. Sometimes, however, when the preliminary investigations have proved satisfactory, it twists the shell round until the tail falls into the opening, and then parades up and down for a little while. Perhaps it may be satisfied, and after twirling the shell about several times, whisks into it with such speed that the eye can scarcely follow its movements. Indeed it seems rather to be shot into the shell from some engine of propulsion than to move voluntarily into the new habitation. When the number of empty shells is great, the Hermit is very fastidious, and will spend many hours in settling into a new house.

A Hermit-crab when deprived of its shell presents a most absurd appearance. It is dreadfully frightened, crawls about with a terrified kind of air like that which is put on by a beaten dog, and will put up with anything by way of a house. I have seen a very large whelk-shell inhabited by a very little crab, so small and weak that it could not drag its huge home about, and was tumbled backwards and forwards as the waves washed over the shell. It was much too small to fix itself in the mouth of the whelk-shell as is the usual custom of Hermit-crabs, and had been forced to content itself with a hole that had been broken near the point.

This crab may be kept in an aquarium, as it is hardy, and can be fed with perfect ease. It is, moreover, less liable to fight with and kill its companions than the other crabs, probably on account of the shell, which protects the body, and renders a battle a very laborious undertaking. So that if two or three Hermits of similar dimensions are put into an aquarium, they will live on terms of armed neutrality, and if care be taken to feed them separately, they will survive for a long time. It is rather remarkable that when they become sickly, they are sure to leave their shells and lie listlessly on the stones or sea-weed. As soon as one of them is seen to act in this manner, it should be at once removed.

There is a curious notion prevalent respecting the Hermit-crabs. All the fishermen, and the sea-side population in general, firmly believe that the Hermit-crab is the young of the lobster, and that when it becomes large enough to protect itself, it leaves the shell, gets a hard tail, and changes into a real lobster. Any one who wishes to study the structure of the Hermit-crab can do so by visiting a fishmonger's shop, and looking over the stock of periwinkles, many of which are sure to be occupied by a Hermit-crab instead of a mollusc.

There are very many species of Hermit-crabs, those of the tropics being the largest and handsomest. Some of these larger species inhabit the trumpet-shell, some are found in the large turbos, and are handsome creatures, richly mottled with black and brown, and there are one or two species which live in the cone shells. These curious Hermits are shaped so as to suit the shell in which they reside, their bodies being quite flat and almost leaf-like, so as to enable them to pass freely into and out of the long narrow mouth of the shell.

The CRAFTY HERMIT-CRAB is found in the Mediterranean, and among other shells which it inhabits, the variegated triton is known to be a favourite. In the illustration the Crabs are supposed to have fought for the shell, and the vanquished is seen on the ledge above, whither it has been flung by the conqueror.

IN our next illustration we have a figure of the DIOGENES HERMIT-CRAB, a handsome and rather large species that inhabits Brazil and the West India Islands.

It occupies the shells of various molluscs, mostly, however, giving the preference to some large species of turbo ; and Mr. Bennett mentions that he possesses an unique shell which he found on the branch of a tree, having been taken from the sea by one of these crabs. While living, the Diogenes gives out a very unpleasant odour ; and as the crabs are in the habit of assembling in great numbers, the aggregate effect is rather overpowering. They gather together at the foot of trees or under bushes or brushwood, and even contrive to clamber up the branches or the trunks of trees, drawing themselves up by their powerful claws and limbs, and caring little for the heavy burden which they bear on their backs. They have the power of producing a curious noise, something like the croak of a frog alternating with sounds as if of drawing water through the lips.

They are active and voracious beings, and feed with equal avidity on animal and vegetable substances. They prefer animal food, such as fish, crabs, and, indeed, almost any kind of flesh, but they will greedily eat yams, cocoa-nut, and other fruits. They are timid creatures, croaking when disturbed, and seldom attempting to fight, but draw themselves smartly within their homes as soon as they feel alarmed. They are mostly nocturnal in their habits; and as they bring a large supply of shells to land, and are very fastidious about their accommodation, a great heap of empty shells is to be seen upon the shore, and there is a continual rattle during the night as the creatures knock the shells about in their movements.

DIOGENES HERMIT-CRAB.—*Cenobita Diogenes.*   OAR-FOOT CRAB.—*Remipes testudinarius.*

In all these creatures, the larger claw is very much developed; so that when the crab has withdrawn into the shell, the claw lies over the entrance and closes it like a living door, which has the further advantage of being used as an offensive weapon. The footstalks on which the eyes are set, are moderately long, stout, and jointed, and enable their possessor to see in all directions. The colour of this species is reddish brown, spotted thickly with black.

CLOSELY allied to the hermit-crabs, we find a very remarkable creature, called, from its habits, the ROBBER-CRAB. It is also known by the name of PURSE-CRAB. The habits of this creature, which will presently be described, are most remarkable, and there are several singular peculiarities of structure. The abdomen, for example, is no longer soft as in the hermits, but is covered above with strong plates, which overlap each other like those of the lobster's tail. The under-surface of the abdomen is soft and membranous. This is one of the crustacea that can endure a long absence from water, and is fitted with a peculiar addition to the breathing apparatus. There are twenty-eight gills, fourteen at each side of the body, and enclosed in a large hollow, which they do not nearly fill. Even when the footstalks are considered, on which the gills rest, they hardly occupy the tenth part of the hollow.

The Robber-crab is found in several parts of the Indian Ocean, is very common in Amboyna, and has been taken off the Mauritius. Mr. Darwin gives the following interesting account of this crab :—" It would at first be thought impossible for a crab to open a strong cocoa-nut covered with the husk, but Mr. Liesk assures me he has repeatedly seen the operation effected. The crab begins by tearing the husk, fibre by fibre, and always at that end under which the three eye-holes are situated. When this is accomplished, the animal commences hammering with its heavy claws on one of these holes till an opening is made ; then, turning round its body, by the aid of its posterior and narrow pair of pincers, it extracts the white albuminous substance of the nut.

I think this is as curious a case of instinct as ever was heard of, and likewise of adaptation of structure between objects apparently so remote from each other in the scheme of nature as a crab and a cocoa-nut tree. This crab is diurnal in its habits, but every night it is said to pay a visit to the sea, no doubt for the purpose of moistening its branchiæ. The young are likewise hatched and live for some time on the coast. These crabs inhabit deep burrows, which they excavate beneath the roots of trees, and here they accumulate surprising quantities of the picked fibres of the cocoa-nut husk, on which they rest as on a bed. The Malays sometimes take advantage of their labour by collecting the coarse fibrous substance, and using it as junk."

In the missionary voyage of Messrs. Tyerman and Bennett, a very spirited account is given of these crabs, and one or two interesting details are mentioned. For example, when the crab walks it raises itself well off the ground, standing nearly a foot in height, and gets along quickly, though with a clumsy and stiff gait. The antennæ are very sensitive, and it is said that if they are touched with oil, the creature immediately dies. Another mode of opening the shell is employed by these crabs besides that which is mentioned by Mr. Darwin, for, according to Messrs. Tyerman and Bennett, the crab, after tearing off the husk, insinuates the smaller joint of the claw into one of the holes at the end of the nut, and then beats the fruit against a stone until the shell is broken.

This Crab is by no means handsome, but is a very large and remarkably shaped creature. A fine specimen, when stretched out at length, will measure between two and three feet in length, and as it is stout in proportion to its length, it may rank with some of the largest of the crustaceans. The abdomen is of a curious form, and is evidently one of the structures intermediate between the crabs and the lobsters. Its general colour is pale yellowish brown, and its limbs are covered with little projections of a nearly black hue.

During the day the Robber-crab mostly hides in the fissures of rocks, or in holes at the foot of the trees, and in the evening issues from its concealment to prey upon the cocoa-nut. Its wonderful skill and power in opening this huge fruit have already been mentioned, but some writers give it credit for more extensive qualities, and say that it is in the habit of climbing up the palm trees for the purpose of obtaining the fruit. The particular palm which it is said to climb is the *Pandanus odoratissimus*.

It appears to be fierce in proportion to its strength, and Mr. Cuming found that if intercepted in its passage, it at first tried to intimidate its disturber by holding up the claws and clattering them loudly; and that even when it found itself obliged to give ground, it retreated with its face to the enemy, still maintaining a threatening attitude. The eyes of the Robber-crab stand on rather long but stout footstalks.

WE now come to the Porcelain-crabs, so called because their shells are smooth and polished as if made of porcelain, and have much of the peculiar semi-transparent gloss of that manufacture.

Several of these crabs are natives of the British seas, among which we may mention two species. The first of these is the common BROAD-CLAW PORCELAIN-CRAB, so called from the singular width and flatness of the claws, each of which is nearly as large as the whole body. Altogether this is a flat crab, and, like all flattened beings, is formed for a life under stones or in narrow crevices. It may be found easily enough by going to the very verge of low water-mark, and quickly turning over the loose stones which lie piled upon each other by the wave. Under these stones lies the Broad-claw, flat and quite at its ease, its great claws fitting beautifully into its shell, much like the same members in the domed crabs.

The food of the Broad-claw consists mostly of animalcules, which it catches by making regular casts with its hair-covered jaw feet, and sweeping its prey into its mouth by the action. The mode in which the hairs or bristles are set upon the foot is very beautiful, and is exactly calculated to act as a net, which will sweep up every object that crosses its path. Several of the terminal joints of these jaw-legs are edged with long and slightly curved hairs set nearly at right angles with the joint. It follows, then, that when the limb is flung out nearly in a straight line, these hairs diverge; but that as the limb is

BROAD-CLAW PORCELAIN-CRAB.—*Porcellana platychéles.*　　COMMON PLATED LOBSTER.—*Galathéa strigósa.*

bent while being withdrawn, the hairs become nearly parallel to each other, some of them cross, and form a very complete network of stiff bristles that sweep everything before it. Moreover, each separate bristle has a double row of still smaller hairs, projecting from each side, something like the vanes of a feather, and nearly touching those of the next hair on either side. This structure is evidently intended to ensure the capture of the very minute animalcules, which might be able to escape through the comparatively large meshes formed by the bristles.

Though the Broad-claw loves to hide in this manner, and remains so quiescent, it is fully able to move about, and can dart through the water with astonishing celerity, flinging out the abdomen, and giving a series of sharp flaps that urge it along just in the manner adopted by the lobster and its kin. Still, it does not attempt to swim, but merely darts towards some spot where it can find a hiding-place, and whence it will not stir for weeks together, finding in its narrow home all that it needs in the way of food.

ANOTHER British species is the LONG-HORNED PORCELAIN-CRAB, a little creature that is common in similar localities. It is much eaten by various fishes, and the codfish makes great havoc among its ranks. Both these crabs can pinch smartly with their flat but powerful claws, and, in spite of their insignificant appearance, cannot be handled with impunity. In them the last pair of legs undergo a curious modification, being very small, nearly hidden in the abdomen when not in use, and apparently objectless. They are, however, very useful limbs, being employed as brushes, and used for the purpose of cleaning the abdomen and part of the carapace from adhering substances.

WE now come to the second great division of the Crustaceæ, namely, those which have long and powerful tails. The lobsters and shrimps are examples of these creatures. In swimming rapidly through the water, the tail is the organ of propulsion which is employed, and a glance at its form will soon explain its use. This powerful mass of solid muscle is first stretched out to its utmost, and the fan-like appendage at the extremity is spread to its widest; the creature then closes its tail smartly under the body, so as to assume the attitude in which lobsters and shrimps are mostly brought to table.

The effect of this sudden contraction is, that the creature shoots swiftly through the water. Of course, the animal darts backwards, but so sharp are its eyes, and so true is its aim, that it can fling itself into a crevice barely large enough to contain it. Any one who wishes to see this manœuvre practised in all its force, may do so by watching the little sea-side pools wherein the shrimps and prawns are accustomed to disport themselves as long as the water lasts, and where, when it dries up, they bury themselves in the sand to await the coming tide.

This shooting mode of progression is not their only means of movement. By the ordinary use of their legs, nearly all the species can crawl among the sea-weed, or upon the rocks and bed of the sea, just as an insect crawls on the ground. And, when they are balancing themselves in mid-water, and are only desirous of moving gently about, they can do so by means of the numerous false legs under the body, which may be seen moving with great rapidity. Those who are fortunate enough to possess a marine aquarium, and can keep a prawn or a shrimp in the miniature ocean, will have many opportunities of watching the easy and graceful movements of these elegant crustaceans.

IN the preceding illustration is seen an example of the first family, called the Galatheidæ, in honour of the beautiful and unfortunate nymph vainly beloved by Polyphemus. Several species of this family are found on the British coasts, one of which is the common PLATED LOBSTER. This is a handsome little creature, the general ground colour being red, upon which are drawn a number of blue spots and streaks. Its activity does not correspond with its beauty, for, according to all accounts, it is a dull, sluggish creature, and from Mr. Couch's observations, is "incapable of any motion but backward, and rarely rises above the bottom; where, by a laborious motion of its tail, it contrives to retreat from its enemies; but its usual progress is by creeping, and by the legs only." Yet, although it is thus tardy while crawling, it can dart backward with all the agility of its race; and if alarmed, flashes through the water with arrowy speed, and can hardly be captured or its exact direction ascertained.

The beak of this species is triangular, and armed with seven strong teeth. By these characteristics it is distinguished from another species, MONTAGUE'S PLATED LOBSTER (*Galathea squamifera*), which has a short and wide beak, cut into nine spine-like teeth. The colour of this creature is greenish brown, tinged with red. It is to be found on most of our coasts, under stones at low-water mark. Several other species are found in England.

THE small but important family of the Scyllaridæ is easily recognised by the wide, flat carapace, the large and leaf-like outer antennæ, and the partly flexible tail-fan, by which the creatures drive themselves through the water. In consequence of their shape, they go by the popular name of FLAT or BROAD LOBSTERS. The habits of these crustaceans seem to be much alike. They live in moderately shallow water, where the bed of the sea is soft and

3. R R

JAPANESE SCYLLARUS.—*Scyllarus Sieboldii.*

muddy. Into this substance they burrow rather deeply, so as to be entirely concealed, and only issue from their retreat for the purpose of seeking food. In all the members of

SPOTTED IBACUS.—*Ibacus ciliatus.*

this genus, the carapace is longer than wide, and the sides parallel to each other. The common BROAD LOBSTER is found in various parts of Britain, and is exceedingly plentiful

SPINY LOBSTER.—*Palinurus vulgaris.*

in Greenland, where it forms the chief food of the Arctic auk (*Alca arctica*). The beak-like projection of its carapace is very wide, but does not project. The carapace is covered with little tubercles, and along the central line runs a series of spines. The outer antennæ are large and deeply toothed. The colour of this species is brownish, covered with red marks, dispersed in a simple, but very pretty pattern, which would serve as a model for embroidery, and would be particularly suitable for the heavy metallic ornamentation upon uniform coats. It is but a small species, measuring only three inches in length.

Some species of this family are eatable, and in Japan are considered as delicacies.

ONE species of these creatures, known by the name of the SPOTTED IBACUS, is a great favourite with the Japanese. In this genus the carapace is extremely wide, and is expanded in such a manner as to hide the feet, so as to remind the spectator of the domed crab, already described on page 594. The colour of the Japanese Scyllarus is red, covered with blue points.

THE well-known SPINY LOBSTER (*Palinúrus vulgáris*) belongs to the next family of crustaceans. In all this family the outer antennæ are very long and stout, and their basal joint is large.

The Spiny Lobster is also called the SEA CRAY-FISH, or the RED CRAB. Its claws are very small, and by no means formidable. It is mostly found on our western and southern coasts, and is caught in crab pots, like the common lobster. Its flesh is good and well flavoured, though rather tougher and coarser than that of the lobster; moreover, the want of the claws is a drawback to its excellence; so that it is not esteemed nearly so much as

SLOW SHRIMP. — *Arius stirhynchus.*　　　MUD BURROWER. — *Callianassa subterranin.*

the true lobster. Sometimes it is found entangled in the nets, and even upon the fisher-men's lines.

The average length of this species is eighteen inches, and its weight about five pounds, when adult. Its colour is purple-brown, with some irregular white spots, and its legs are reddish-white, banded longitudinally with brown. One species of this genus, *Palinurus ornatus*, sometimes attains to an enormous size, measuring from the end of the antennæ to the tail rather more than four feet.

On the preceding illustration are exhibited two more specimens of the British crustacea. Both of them belong to a family called the Thalassinidæ, in which the abdomen is long, its integuments rather soft, and the carapace small and compressed on the sides.

The MUD-BURROWER is found on our own coasts, but is not very often seen, as it lives in a burrow some two feet under the surface of the mud. It forces itself beneath the mud by means of the third pair of legs, and there passes the greater portion of its time. The shell of this species is very thin, and but for the enormous claw with which it is furnished, it would seem quite a helpless creature.

One species of this genus, the GREAT BURROWING CRAB (*Callianassa major*), inhabiting Florida and other parts of America, forms a very remarkable burrow; Mr. T. Say, who found this creature by digging in the sand, giving the following account of its habits. "It had formed a tubular domicile, which penetrated the sand in a perpendicular direction to a considerable depth; the sides were of a more compact consistence than the surrounding sand, projecting above the surface about half an inch or more, resembling a small chimney, and rather suddenly contracted at top into a small orifice. The deserted tubes of the Callianassa are in many places very numerous, particularly where the sand is indurated, by iron into the incipient state of sandstone; they are always filled up, but may be readily distinguished by the indurated walls and summit often projecting a little above the general surface."

The British Mud-burrower is rather a pretty little creature, being of a soft pink hue, sometimes changing to yellow on the sides. Very soon after death these colours fade, and change into dull grey. The haddock seems to feed largely on this species, as fragments are mostly found in the stomach of the fish.

THE second species represented in the engraving is the SLOW SHRIMP, a rather curious looking creature, much resembling the common shrimp, except that it possesses a pair of large and stout claws. Its popular name is derived from the sluggishness of its movements, as it has scarcely any idea of running or swimming away if alarmed, but only attempts to escape by burrowing in the mud. If, therefore, it should be intercepted upon some harder ground, where it is not able to burrow, it exhausts all its strength in unavailing efforts, and is easily taken prisoner. The best way of obtaining this creature is to dig it out of the sand. It is but a small species, measuring about three inches in length. Several other burrowers are inhabitants of the British shores. One of them is the MUD-BORER (*Gébia stelláta*), a small species, measuring about two inches in length. This creature often takes possession of the burrows which have been made and forsaken by the razor-shell, but it is doubtlessly able to bore holes for itself. It is rather a pretty little crustacean, being of a pale yellowish white, covered with very little star-shaped orange spots. On the front of the carapace are multitudes of little spines, arranged in longitudinal rows.

ANOTHER species, the DELTURA (*Gebia deltura*), inhabits the same and similar localities as the mud-borer. In many points it much resembles that crustacean, and has been thought by some persons to be the female of the same species. It furnishes abundance of food to various fishes, especially those belonging to the ray family, and its remains are found abundantly in their stomachs. It is much larger than the mud-borer. All the members of this genus have the carapace formed into a triangular beak, and the outer pair of fore-feet formed for walking. One more remarkable species of burrowing crustacea is the CALOCARIS (*Calocaris Macandrei*), which resides at a very great depth, having been ascertained to live at the bottom of the sea, more than a thousand feet from the surface. Here, like the rest of these creatures, it burrows in the mud, passing a kind of sub-marine mole-like existence. As, at this great depth, and under the mud, the ordinary visual powers would be of no avail, the creature has but the rudiments of eyes, which are small and quite without colouring. The Calocaris is mostly to be obtained from the stomachs of haddocks, rays, and flat-fishes.

The colour of this curious species is delicate pale rose while living, but, as is usual with this fleeting tint, it soon fades after death. The shell of the Calocaris is very delicate and thin, and the whole of the feet are covered with hairs.

WE now come to the family of the Astacidæ, which includes two well-known and very similar creatures, the fresh-water cray-fish, and the salt-water LOBSTER. The general shape and appearance of the Lobster is too well known to need any description. The vast numbers of Lobsters which are annually brought to the London markets are largely supplied from Norway, although there are many parts of our own coasts where these creatures can be taken plentifully. The Lobster is not much of a rover, seldom straying far from the spot on which it was hatched. It is rather remarkable that Lobsters are liable to permanent varieties, according to the locality in which they reside, and a good judge will be able to determine at a glance from what part of the country any given Lobster has been taken.

Sometimes a green specimen is brought to market, and the salesmen have a theory that it has obtained this change of colour by living in some spot where the ores of copper impregnate the earth. They consequently believe it to be poisonous. Both ideas, however, seem to be groundless.

Lobsters are always sold by number and not by weight, and their value is necessarily

LOBSTER.—*Hómarus vulgaris.*

dependent on the accurate eye of the dealer. The Lobsters are caught in creels or pots, like the crabs, but with greater ease and economy, as they are very fond of meat, be it fresh or tainted, and even if it should be putrefying will be attracted to it. Bright and shining objects seem quite to fascinate the Lobster, which will enter a "pot" even though the bait be nothing more than a number of empty oyster-shells placed so as to exhibit the shining white of the interior. A few years ago a curious bait was employed with great success. It was very simple, consisting of nothing more than a common phial bottle, silvered on the inside. This was hung in the lobster-pots, and served to attract the creatures to the bait. It has been suggested that the potency of this strange allurement may be attributed to its resemblance to the phosphorescent shining of putrid animal substances. But it is quite as probable that the glittering object may serve simply to attract the Lobster's attention, and that when it has approached in order to satisfy its curiosity, it perceives the bait, and immediately enters the trap. It is found that both bait and bottle are required, as if the latter is used alone, the Lobsters discover their mistake and quit a spot where they find no food.

Like many other crustaceans, the Lobster is a most combative animal, quarrelling on the slightest pretext, and fighting most furiously. In these combats it mostly loses a claw or a leg, being obliged to discard entirely a wounded member. A fresh leg or claw sprouts from the scar, and it is to this circumstance that the frequently unequal size of Lobster-claws is owing. Lobsters indeed part with these valuable members with strange indifference, and will sometimes shake them off on hearing a sudden noise. It is said that the commanders of certain preventive sloops were accustomed to levy a tax upon the Lobster-fishermen, threatening that unless a certain number of Lobsters were furnished to them they would fire cannon over the Lobster-grounds and make the creatures shake off their claws

If the fishermen find that they have wounded a Lobster, they have recourse to a very strange but perfectly efficacious remedy. Supposing one of the claws to be wounded, the creature would soon bleed to death unless some means were taken whereby the flow of blood may be stopped. The method adopted by the fishermen consists in twisting off the entire claw.   A membrane immediately forms over the wound, and the bleeding is stopped. The new limb that is to supply the place of that which was lost, always sprouts from the centre of the scar.

THE common CRAY-FISH or CRAW-FISH of our rivers (*Astacus fluviatilis*) has an almost exact resemblance to the marine lobster, which it resembles in many of its habits and qualities.   Like that creature, it hides itself in some crevice, and does not issue from its concealment except for the purpose of obtaining food.   It is equally quarrelsome and also displays many tokens of its combats in the shape of lost or minute members. It is quite a rare thing to find a large Cray-fish with both its claws of the same size.

This creature mostly hides under stones or holes in the bank, sometimes partially scooped out by the inhabitant, but mostly being the deserted tenement of a water-vole. Herein the creature sits, with its head towards the orifice, and its claws thoroughly protecting its home.   Even the sharp spikes of the head form no inconsiderable protection, for, if the hand be thrust into a hole tenanted by a Cray-fish, a sensation is perceived as if the fingers had been pushed against a quantity of needle-points.   From these dens it issues in search of prey, which consists of dead fish and any similar substances.

Cray-fish can be caught in various ways.   There are large "pots" or "creels," made of wicker-work, into which the creature is enticed by a bait, but out of which it cannot . escape.   There are Cray-fish nets, by which many hundreds can be caught in an afternoon.   These are simple circular nets fastened inside an iron hoop and having a piece of meat tied on the centre by way of bait.   A long string is attached to each net, and a forked stick, something like a clothes-prop, used for laying or taking them up.   The fisherman always has several dozen of these nets, which he disposes along the river-bank in the spots which he thinks best suited to Cray-fish.   By the time he has laid his last net, he must visit the first, which he pulls up quickly, and in which he mostly finds three or four Cray-fish eagerly eating the bait.   The net is then replaced, and he proceeds to the second.   On an average, each net produces three Cray-fish every round.

The flesh of the Cray-fish is something like that of the lobster, but far more delicate and without the indigestible qualities of the larger crustacean.   It is only in season for a comparatively short time, and in the other months of the year the flesh is soft, watery, and flavourless.

THE next family includes the true Shrimps, and contains but one genus. The Shrimp, which is so familiar on our tables, and which, until the marine aquaria became so common, was equally unknown in its living state, inhabits our shores, where it is produced in countless myriads.   In every little pool that is left by the retiring tide, the Shrimps may be seen in profusion, betraying their presence by their quick darting movements as they dash about in the water and ever and anon settle upon some spot, flinging up a cloud of sand as they scuffle below its surface, their backs being just level with the surrounding sand.   In consequence of this manœuvre, the fishermen call them "sand-raisers."   The small prawns are often confounded with the Shrimps and popularly called by the same title.   They can, however, be easily distinguished from each other, the beak of the prawn being long, and deeply saw-edged, while that of the Shrimp is quite short.

While living, the Shrimp wears tints so exactly like those of the sand, that when it is lying motionless, it harmonises exactly with the tawny bed of the sea, and cannot be discerned except by a practised eye.   When boiled, it does not change to so bright a red as is usually the case with eatable crustacea, but assumes a duller and more opaque hue. During life the Shrimp is a most beautiful creature, nearly translucent in many points, and when seen against the light seeming to possess some inward illumination.   Its habits are interesting, and can be successfully watched by means of an aquarium, though it is necessary to bestow some care on the creature, and keep it properly supplied with food, as it is, though so delicate, a very voracious animal and requires much feeding.

Shrimps are caught for sale in a peculiar wide and purse-like net set crosswise upon a pole, and pushed along the sand at the depth of about two feet or a little more.   By this method of procedure great numbers of Shrimps are gathered into the net as they dash along the sand, and together with them are various other inhabitants of the sea, quite useless to the shrimper, but very valuable to the sea-side naturalist. Any one who is studying the habits of the marine animals will do well to pay a shrimper for the right of examining the net and retaining whatever is useful or interesting. The method of burying itself in the sand is by using the hinder legs as scoops, settling into the small hollow made by them, and then flinging the sand over its back with its antennæ.

There are several species of true Shrimps inhabiting our shores, all good for food. and, although comparatively scarce, taken together with the common Shrimp. There is the BANDED SHRIMP (*Crangon fasciátus*), known by the narrow and rounded abdomen and the brown band that crosses the fourth ring. It is about an inch in length. It seems to be rather a rare species. Another Shrimp is called the SPINY SHRIMP, on account of five rows of teeth-like points upon the carapace. It is of a rather light brown colour, banded and striped above with greyish white, and spotted below with crimson. Another species, BELL'S SHRIMP (*Crángon sculptus*), is very small, being little more than three-quarters of an inch in length. There are several raised lines on the carapace, each with a few small

COMMON SHRIMP.— *Crangon vulgáris.*   SHORT-BEAKED RED SHRIMP.—*Alphéus brevirostratus.*

teeth. In colour it is extremely variable, but is mostly drab, with little black spots and chestnut specks, and is adorned with patterns of pale brown edged here and there with blue.

THE second figure in the preceding illustration represents the SHORT-BEAKED RED SHRIMP, a creature belonging to the family of the Alpheidæ. This is a Japanese species, but at least two species of the genus Alpheus are found on the English coasts. Their carapace projects over the eyes in a hood-like shape, and the beak is very small, sometimes indeed being altogether absent. The first pair of legs are always very stout and strong, and one claw is much larger and more powerful than the other. The greater number of the species belonging to the genus Alpheus live in the tropical seas, and those that have been found within British waters have clearly resided at a considerable distance from land. One species, for example, EDWARDS' RED SHRIMP (*Alphéus ruber*), has been found in the stomach of cod-fishes, mostly in fragments, but very rarely entire. Another species, the SCARLET SHRIMP (*Alphéus affinis*), is of a deep scarlet except the claws, which are marked with yellow. This Shrimp has been taken in the Channel Islands.

Another species, MONTAGUE'S SHRIMP (*Athanas nitescens*), inhabits the southern coasts of England, and is popularly thought by the fishermen to be the young of the lobster, its deep green colour and large pincers giving it a great resemblance to that crustacean. It is a sociable little being, congregating in some favoured spot, and assembling in considerable numbers. In fact, it is seldom found alone; and in clearing out a little sand-pool, six or seven may often be found in close companionship.

SWORD-SHRIMP.—*Penæus ensis.*   RING-HORNED PRAWN.—*Pándalus annulicornis.*

WE now arrive at the PRAWNS, a family which is easily known by the long and saw-edged beak that projects from the carapace. This family is very rich in species, many of which are most lovely creatures, resplendent in scarlet, azure, green, purple, and orange, and of a beautiful transparency, which gives double effect to the colours with which they are adorned.

The smaller figure represents the RING-HORNED PRAWN, in which the beak is extremely long, and slightly turned upwards. It is taken in large quantities on many parts of our coasts, but as it lives at some distance from the shore, it cannot be captured in the ordinary shrimp nets. The fishermen call it the Red Shrimp. The spines, or teeth in the upper edge of the long beak, do not spring at once from the substance of the beak, but are simply jointed to it, so that they can be moved slightly by pressure. A large number of species belonging to the genus Hippolyte are also found in England. In these creatures the beak is very large and strong. Several of the Æsop Prawns belong to this genus. They derive their popular name from the hump-like manner in which the abdomen is raised towards the centre and then bent downwards. COUCH'S ÆSOP (*Hippolyte Couchii*) is perhaps the most common of these beautiful little creatures, and has the characteristic hump strongly defined. It may be found plentifully in the shore-pools, flitting about the water with a movement much like the flirting and fluttering of a robin in a garden, and displaying its beautiful colours to the best advantage. It is a lovely little being, very variable in colour, but always marked with bright and peculiarly pure hues, mostly white, purple, and scarlet. Many of these Æsop Prawns are charming inhabitants of an aquarium, their pellucid bodies and beautiful colours making them fit inhabitants of the drawing-room or the conservatory. One species, WHITE'S ÆSOP PRAWN (*Hippolyte Whitei*), is an especially beautiful creature, being green with a white streak running along the back, and having a number of azure specks scattered over the body. Even the large EDIBLE PRAWN (*Palæmon serratus*) is a beautiful inhabitant of an aquarium. No one who has only seen Prawns on the table, red, opaque, and with their tails folded under them, can form the least conception of their wonderful beauty while living. As they swim gracefully through the water, the light passes through their translucent bodies and their beautifully streaked integuments, rich with transparent browns, pinks, and greys of various depths. Their delicate and slender limbs are ringed with orange and purple, and stained with pale blue.

At night, when a lamp is brought into the room, the effect produced by the Prawn is really surprising. The large globular eyes glow as if illuminated by some powerful light within ; and as the creature comes out of the darkness its eyes alone are visible, as they shine like two globes of living fire.

It is very interesting to watch the habits of this beautiful creature. It is extremely voracious, and seems always to be ready for food. I used to feed my own Prawns with the bodies of shrimps, hermit-crabs, and other marine crustacea that had died in the aquaria. All that was needful was to drop the dead animal into the water so that it should pass the spot where the Prawn had made its home. As soon as it approached, the Prawn used to dart out like a tiger from its den, its long antennæ waving in great excitement, and its forceps open and extended so as to be in readiness. The claws appear to be very feeble, but they are stronger than they seem, and are perfectly adequate to the task which they are called upon to perform. The creature would quickly grasp its prey with one claw, carry it off to its home, and there leisurely pick it to pieces, displaying considerable discrimination in choosing the most delicate morsels, and abandoning the remainder to its smaller companions who still lived in the same tank, and preserved their lives by hiding themselves in little nooks and crevices, wherein they were safe from their giant kinsman. The air of utter contempt with which the Prawn would twist off and fling aside the legs and antennæ of a shrimp or a hermit-crab was very amusing. Its greatest dainty, for which it would leave almost every other kind of food, was the soft abdomen of the hermit-crab.

The forceps employed for this purpose are those at the extremity of the second pair of feet, those of the first pair being used for a different purpose. Mr. Gosse has given the following account of those limbs and their use. After mentioning that they are covered with hairs set at right angles to the limb, like the bristles of a bottle brush, he proceeds as follows :—" These are the Prawn's washing brushes, especially applied to the cleansing of the under surface of the thorax and abdomen. When engaged in this operation, the animal commonly throws in the tail under the body, in that manner which we see assumed in the finest specimens that are brought to table, which is not, however, the ordinary position of life, the body being nearly straight. Then he brings his fore-feet to bear on the belly, thrusting the bottle brushes to and fro into every angle and hollow with zealous industry, withdrawing them now and then, and clearing them of dirt by passing them between the foot-jaws.

The reason of the inbending of the tail is manifest. The brushes could not else reach the hinder joints of the body, and still less the swimming-plates, but by this means every part is brought within easy reach. Sometimes the brushes are inserted between the edge of the carapace and the body, and are thrust to and fro, penetrating to an astonishing distance, as may be distinctly seen through the transparent integument. Ever and anon the tiny forceps of the hand are employed to seize and pull off any fragment of extraneous matter which clings to the skin too firmly to be removed by brushing ; it is plucked off and thrown away clear of the body and limbs. The long antennæ and all the other limbs are cleaned by means of the foot-jaws principally."

THE larger figure in the preceding illustration represents the SWORD-SHRIMP, a native of Japan, belonging to another family, termed the Penæidæ. All these creatures have a very long and much compressed abdomen, and the beak very small or absent. Very few species of this family are found in England. One of the British examples is the Grooved Shrimp (*Penæus sulcátus*), a common species in the Mediterranean, but rare in the British seas. It has three grooves on the carapace, two long and one shorter in the middle. It is a large species, sometimes attaining the length of seven inches.

Another species is the Sivado, sometimes called the Sword-shrimp, or the White Shrimp, the last-mentioned term, however, being applied very loosely by the fishermen. It is a very beautiful little creature, being of a translucent white colour, dashed and spotted with rich crimson. It is said that this species cannot endure exposure to the air, and that it dies immediately on being removed from the water.

Another order of crustaceans now comes before us, called the Stomapoda, or Mouth-footed Crustaceans, so called because their legs mostly issue from the neighbourhood of the

GLASSY ERICTHUS.—*Ericthus vitreus.*
ARMED ERICTHUS.—*Ericthus armátus.*
CHAMELEON SHRIMP.—*Mysis chameleon.*

CLUB-HORNED PHYLLOSOME.—*Phyllosóma clavicorne.*

mouth. The gills are external, and are formed in a most curious manner of a series of tiny cylinders. The greater number of Stomapods live in the hotter seas, but a few are inhabitants of our own coasts.

Our first example of these odd-looking creatures is the Chameleon-shrimp, perhaps the most common of its kind. This species is abundant on our coasts, and derives its popular name from the extreme variability of its colouring. It seems to alter according to the locality in which it resides. Those, for example, which live upon a sandy coast are of a grey hue, those which are found among the large dark sea-weeds are brown, and those that prefer the ulva and zostera beds are green, like the vegetation among which they live. These creatures are sometimes called Opossum-shrimps, from a curious modification of their structure. The last two feet are furnished with an appendage that forms a sort of pouch. In the male this pouch is small, but in the female it is large, and capable of containing a large number of eggs, which are carried about by the crustacean just as the opossum carries its young.

In the Northern seas these opossum-shrimps exist in vast multitudes, and form much of the food on which the great whale of those seas depends for its subsistence. Several species are thus eaten, and one of them, *Mysis flexuosus*, is largely eaten by the enormous shoals of salmon that visit these regions in the months of July and August, thereby aiding in giving to the fish that fineness of condition and fulness of flesh which ought to be possessed by a well-nurtured salmon. These creatures are fond of congregating at the mouths of rivers, probably because they find plenty of food in such localities, and during the winter, haunt the whole line of coast.

Many species of Opossum-shrimps are found upon our shores, and can be captured by the simple plan of hauling up masses of sea-weed, and seizing the little crustacea before they can escape. The Chameleon-shrimp is the lowest figure in the illustration.

ANOTHER example of these beings is the CLUB-HORNED PHYLLOSOME, a member of another and a very remarkable family. All the members of this family have the body exceedingly flat and leaf-like, formed by the carapace and part of the thorax. The abdomen is extremely small in proportion to the enormous size of the cuirass, and the limbs are so formed that they can be spread from the body so as to present a large radiating outline. These crustaceans are in the habit of floating on the surface of the water, extending their legs, and there lying quite at their ease. The body is beautifully transparent, and it would be almost impossible to see the Phyllosoma were it not that the eyes are of a most beautiful blue, and serve as indications of their owner's presence. This species is a native of the Atlantic Ocean. The Club-horned Phyllosome is seen on the right side of the engraving. The name Phyllosoma is derived from the Greek, and signifies Leaf-bodied. One or two examples of this creature have been found floating near the Channel Islands.

Two figures of a remarkable genus of Stomapod Crustaceans are given in the same illustration. In these creatures the upper part of the body is defended by a single and large cuirass, covering much of the head, being wide and free behind. The members of the genus Ericthus have the cuirass enormously developed, prolonged in front into a kind of beak, which projects over the head, and having behind several strong and rather long spines. These creatures have smaller claws than is found to be the case with the generality of the family, and all the limbs are of only moderate dimensions. The last segment of the abdomen is developed into a wide and flat fan-like blade. The eyes are large, round, and set on stout footstalks.

The GLASSY ERICTHUS derives its name from the translucency of its integuments, and the ARMED ERICTHUS is so called in consequence of the sharp spines that defend its shield. Both these species are inhabitants of the Atlantic. The latter is the central figure on the left side, and the former occupies the upper portion of the engraving. Both the figures are enlarged.

BEFORE passing to the next family, we must cast a brief glance at a very strange-looking crustacean, called the TRANSPARENT ALIMA (*Alima hyalina*). This remarkable animal looks much as if an Ericthus had been drawn out like wire to a considerable extent, retaining all the characteristics of the family, and some which belong to the genus. The abdomen is extremely long, something like the tail of a scorpion, and terminated by a flat paddle. The cuirass is so large and so loose that it hardly seems to belong to the creature, but to have been taken from some larger crustacean, and dropped upon its back. The eyes are large and globular, and stand on slender curved footstalks, bearing no small resemblance to a dumb-bell with a long and rather curved handle, each eye answering for the heads of the bell, and their united footstalks for its handle. The claw feet are long, slender and can be used with much quickness.

These creatures are natives of the warmer seas, such as the tropical portions of the Atlantic, the South Seas, and New Guinea. They all live at some distance from the shore.

WE now come to a curious family, called the Squillidæ. In these creatures the body

GOUTY SHRIMP.—*Squilla chirágra.*
MANTIS SHRIMP.—*Squilla mantis.*

is long and mostly flattened, and the first pair of legs are very large, and used for seizing prey, the last joint folding over serves to answer the purpose of a claw. The carapace is divided into three lobes. The best known of these crustaceans is the MANTIS-SHRIMP, so called from its great resemblance to the insect from which it takes its title. The carapace of the genus Squilla is small but long, and shields the mouth, the antennæ, and their appendages. The abdomen is very long and boldly jointed, and the appendages at its extremity are made in a manner that much resembles the fan-like tail of the lobster.

All the Squillæ are voracious, fierce, and active beings, and can strike as sharply with their long claw-feet as can the mantis with the corresponding limbs. From all appearance it seems as if the creatures were in the habit of hiding themselves in dark crevices, and from their dens striking quickly at passing prey.

This theory is much strengthened by the observations of Dr. Lukis, who kept a Mantis-shrimp alive for a short time. "It sported about, and after a first approach exhibited a boldness rather unexpected. When first alarmed, it sprang backwards with great velocity, after which it placed itself in a menacing attitude which would rather have excited the fear of exposing the hand to it. The prominent appearance of the eyes, their brilliancy and attentive watching, the feeling power of the long antennæ, evinced quick apprehension and instinct. I brought a silver teaspoon near them, which was struck out of my hand with a suddenness and force comparable to an electric shock. This blow was effected by the large arms, which were closed and projected in an instant with the quickness of lightning."

The Squillæ are seldom seen near land, specimens being mostly taken nearly six miles at sea, where the bed of the ocean is known to be of a sandy nature. They are good swimmers, darting quickly through the water by the action of the paddle at the end of the tail. The GOUTY SQUILLA derives its name from the largely tuberculated limbs, which look as if the animal were badly attacked with the gout. It is taken off the Mauritius.

OUR attention is now drawn to the second great group of crustaceans, called the Sessile-eyed Crustacea, because their eyes, instead of being placed on footstalks, are seated directly upon the shell. The body is divided with tolerable distinctness into three parts, for which the ordinary titles of head, thorax, and abdomen are retained as being more convenient and intelligible than the ingenious and more correct, though rather repulsive, titles that have lately been affixed to these divisions of the body.

They have no carapace, like the stalk-eyed crustaceans, nor do they breathe with gills, but by means of a curious adaptation of some of their limbs. None of the Sessile-eyed Crustacea obtain any large size, an inch and a half being nearly their utmost limit in point of length. Most of these animals reside along the sea-shores, where they are of very great use in clearing away the mass of dead animal and vegetable matter which is constantly found in the sea.

The first order of the Sessile-eyed Crustaceans is termed the Amphipoda, a word derived from the Greek, and signifying, "both kinds of feet," because they are furnished with limbs for walking and swimming; whereas in the Isopoda, or similar-footed crustaceans, the feet are all of the same character. The females are in the habit of carrying their eggs under the thorax, mostly between certain flattened appendages attached to the base of the legs.

In the next illustration we have several examples of the first family, called by the name of Orchestidæ or Jumpers, because they possess the power of leaping upon dry ground. The upper right-hand figure represents the most familiar of these little crustaceans, shown in the act of leaping. This is the well-known SAND-HOPPER, or SAND-SKIPPER, seen in such myriads along all our sandy shores, leaping about vigorously just before the advancing or behind the retiring tide, and looking like a low mist edging the sea, so countless are their numbers. Paley has a well-known passage respecting this phenomenon, too familiar for quotation.

The leap of the Sand-hopper is produced by bending the body and then flinging it open with a sudden jerk—in fact, the exact converse of the mode of progression adopted by the lobster and shrimp. The Sand-hopper feeds on almost anything that is soft and capable of decay, and seems to care little whether the food be of an animal or vegetable nature. Decaying sea-weed is a favourite article of food, and wherever a bunch of blackened and rotting sea-weed lies on the sand, there may be found the Sand-hoppers congregated beneath it, and literally boiling out when the sea-weed is plucked up.

Wherever there is sand, the Sand-hopper is to be found, even though no traces may be perceptible; and an experienced shore-hunter will seldom fail in obtaining as many as he wishes in the space of a few minutes. Even where the sand is extremely dry and level, and seems unfit to nourish Sand-hoppers, these little creatures are often snugly ensconced beneath, having burrowed deeper and deeper as the sand became dry. If a smart stamp of the foot be given, a vast number of little holes will make their appearance as if by magic. These are the burrows of the Sand-hoppers, which have been made while the sand was still wet, and over which a film of moist sand had formed itself. The shock caused by the stamp of the foot breaks the dried films, and the hole is at once made apparent.

To catch the Sand-hopper in fair chase is no easy task, but it can be captured without any difficulty by simply digging up the sand and throwing it aside. The Sand-hoppers seem so bewildered with their sudden change, that they merely sprawl about listlessly and can be picked up at leisure.

The teeth of this creature are strong and sharp, as indeed is needful for the tasks imposed upon them. The Sand-hopper will eat anything; and on one occasion when a lady had allowed a swarm of these little crustaceans to settle on her handkerchief, it was

C. PERNYS.—*Corophium longicorne.*
D SAND-SCREW.—*Sulcátor arenárius*

A. SAND-HOPPER.—*Talitrus saltator.*
B. KROYER'S SAND-SCREW.—*Króyera arenária.*

bitten to rags when she took it up. It is very fond of worms, will eat any kind of carrion, and sometimes, when pressed by hunger, has no scruple in eating its own kind. It has many enemies, as is sure to be the case when a little creature is produced in absolute clouds, when it is quite harmless, easily obtained, and excellent food. Sea-birds feed largely upon the Sand-hoppers, and many land-birds are in the habit of passing much of their time upon the shore, and eating their fill of these crustacea. The green crab is a terrible enemy to the Sand-hopper, even running it down in fair chase, as I have witnessed, and displaying wonderful ingenuity in pouncing upon the active little creature just as it descends from its leap. Even a little beetle, not a quarter its size, feeds upon the Sand-hopper, instinctively attacking it from below, where it is comparatively undefended by its shelly coat. Sometimes three or four beetles will unite in their attacks upon a single Sand-hopper. The technical name of this beetle is *Cillenum laterale.*

The SHORE-HOPPER (*Orchestia littorea*) is also plentiful on sandy coasts, preferring those where the sand is sprinkled with rocks. It may be known from the sand-hopper by its more compressed body, the partly clawed character of the two first pair of legs, and the comparatively small size of the first pair. Though it hops on the sand, like the preceding species, and has many similar habits, it is seldom found occupying the same locality, the sand-hopper taking to one part of the coast and the Shore-hopper to another.

ON reference to the illustration, a strange-looking creature will be seen occupying the lower left-hand corner. This is the common SAND-SCREW, an example of the next family. In these creatures the antennæ end in a lash-like point, called appropriately the flagellum, or little whip. The Sand-screw is so called from the odd movements which it makes when laid upon dry sand, wriggling along while lying on its side, and displaying an awkwardness in this respect which contrasts greatly with the wonderful power and freedom with which it can force its way through the wet sand. In the course of its burrowings, it makes many tortuous tracks in the sand, that are generally taken for the trace of some worm's passage.

There are many fossil remains said to be the relics of certain worms, but which are now thought by Mr. Albany Hancock to have been produced by some crustacean of similar habits to the Sand-screw. He has given a most interesting account of this discovery, and the following passages are extracted from his account:—"I went down to the beach, just as the tide was leaving the spot where the broad tracks were usually in great profusion. The sand was quite smooth, all irregularities having been obliterated by the action of the water. Here and there, however, the tracks had already made their appearance, but were as yet of very limited extent, and there was no longer any difficulty in taking the whole in in one view, and moreover, the extremities were perfectly distinct. It was only necessary to watch attentively to note the formation of the numerous and labyrinthine windings that had been so long a puzzle.

I had not long to wait before the sand at one of the extremities was observed to be gently agitated, and on this agitation ceasing, the track was found to have added nearly half an inch to its length. In the course of two or three minutes the sand was again put in motion, and the track once more a little prolonged. These movements were repeated over and over again, until it was quite clear that the track was formed by slow intermitting steps, and not, as might have been supposed, by one continuous gliding motion. Having satisfied myself of this, I took up the morsel of sand at the end of the track, just as it was again becoming agitated, and found that I had captured a small crustacean, the species of which was unknown to me, though in general appearance it was not altogether unlike the common sand-hopper, but not quite so long. I soon took in this way five or six specimens, all of the same species, and all forming tracks of precisely the same character; namely, broad, slightly elevated, flattened, and grooved.

While forming its track, the animal is never seen; it moves along a little beneath the surface of the sand, which it pushes upwards with its back, and the arch or tunnel thus formed partially subsides as the creature presses forward, and breaking along the centre, the median groove is produced."

An example of the tortuous track left by this animal is shown in the illustration.

IN the right-hand lower corner may be seen a more slender and delicate looking crustacean. This is KROYER'S SAND-SCREW, a creature which possesses some of the same habits as the last-mentioned species. It burrows horizontally beneath the sand, like the common Sand-screw, but differs in its mode of action, the back always appearing above the sand.

THE last figure in the illustration represents the LONG-HORNED COROPHIUM, a curious-looking and very interesting species, inhabiting the muddy parts of the sea-shore. This creature is common in the summer and early autumn, at which times it walks boldly upon the wet shore. During the later part of autumn and the winter it resides in holes which it burrows into the mud and clay, and in some places is so plentiful that the mud is quite honeycombed by its tunnels.

This species is very common on the French coasts, especially in the great mussel preserves near Rochelle. M. D'Orbigny, who observed their habits closely, has given a very animated account of their manner of feeding.

The whole of the muddy deposit along the shores is inhabited by myriads of marine worms, such as the nereis and lug-worm, and upon these the Corophium feeds. As the tide rises, the worms ascend to the mouths of their burrows for the purpose of eating the little animalcules that swarm on the shore. The Corophium wages continual war against these worms, darts at them with surprising speed, fastens on them, and eats them. Sometimes a great lug-worm will be surrounded by thirty or forty of these curious crustacea, all attacking it simultaneously, and forming a strange group as the worm writhes in its endeavours to escape, and carries with it the small but pertinacious foes under whose attack it is sinking.

Hundreds of the Corophium may be seen beating the mud rapidly with their enormous antennæ for the purpose of discovering their prey, and the energy of the movement and the evident excitement under which the creatures labour partake largely of the ludicrous. They do not restrict themselves to the worms, being equally ready to prey upon fishes, oysters, or indeed any animal substance that comes in their way. The fishermen, who know it by the name of Pernys, are very angry with this little creature, and declare that it robs them of their mussel harvest. They even assert that it climbs the posts of the complicated woodwork to which the mussels cling, cuts the silken threads by which these molluscs are attached, and having thus let them fall into the sea, eats them at leisure. As is the case with the sand-hopper, the Corophium is greatly persecuted by larger creatures, and is eaten in vast numbers by birds and many fishes.

All the members of this genus can be recognised by the enormous dimensions of their antennæ, which are extremely thick at the base, and look much more like a very large pair of legs than true antennæ.

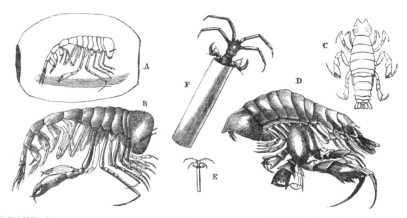

A, B FLEMING'S HERMIT-SCREW.—*Phrónima sedentária.*   C, D. *Dactylócera Nicæensis.*
E, F. CADDIS SHRIMP.—*Cérapus tubuláris.*

WE now come to some very curiously shaped crustacea, whose habits are fully as remarkable as their forms.

The two left-hand figures in the accompanying illustration represent the same species FLEMING'S HERMIT-SCREW, shown of its natural size above, and magnified below. It will be seen that the creature is enclosed in a nearly oval and transparent sac, which is found to be the body of one of the medusæ. In this case it is a beroë which has been chosen for this curious purpose. M. Risso tells us that, like the argonauts and carinariæ, these creatures may be seen in calm weather voyaging along in their glassy boats, and rising to the surface or sinking through the water at will. They live on animalculæ, and for the greater part of the year remain in the muddy depths of the ocean, ascending to the surface in the spring. How they enter their habitations, and their general economy, are subjects at present obscure.

There are several species of Phronima, all inhabiting similar dwellings. *Phronima sentinella,* for example, chooses the bodies of the æquoriæ and geroniæ for its home. These creatures are called by the name of Hermit-screws on account of the solitary life which they lead, each shut up in its cell or cocoon, as it may possibly be called. In all the Hermit-screws, the head is large and vertical, with two little antennæ, and the body is soft, nearly transparent, and ends in a number of bristle-like appendages. All the legs are long, slender, and apparently weak, except the fifth pair, both of which legs possess a large and powerful claw, and are directed backward.

A LITTLE crustacean belonging to an allied genus is not uncommon on our coasts. It has habits of a somewhat similar nature, dwelling in the chambers within several

3.

common medusæ. It will occasionally leave this curious residence, and return to it at will. It is about half an inch in length, has the two first pairs of feet shortest, tipped with a claw, and has the three last pairs of legs longer than the others. The name of this crustacean is *Metoecus medusarum.* Mr. Spence Bate separates all these parasitic animals into a distinct family, under the name of Phronimadæ. All the members of this family have the mandibles very large, some of the legs prehensile and oddly formed, and the head of enormous comparative size. Some of them attach themselves to fishes, and others to medusæ.

FIGS. C and D represent another strangely formed and closely allied crustacean, whose habits are however very imperfectly known, though it is presumed that they resemble those of the hermit-screw and its kin. In this genus, the head, though large, is not of such enormous comparative dimensions as in Phronima, and is rather squared in form. The strange and grasping legs may be seen by reference to the illustration, some of those members possessing great muscular development, and being armed at their extremities with formidable claws, the moveable joint bending over at right angles.

FIGS. E and F represent a small, but very remarkable crustacean, one of the few which really construct a home for themselves. The close resemblance between this creature and the well-known caddis-worm cannot but strike an observer. All the animals belonging to this genus inhabit a case which they are able to carry about with them. In spite of the awkwardness of such an appendage, the CADDIS-SHRIMP passes along at a brisk pace, moving by means of the two pairs of long antennæ, which not only look like feet, but are used for locomotion. The real feet are kept within the tube, with the exception of the two front pairs, which are almost wholly used for catching prey and feeding itself.

Some persons imagine that the tube of this creature is not of home manufacture, but is the deserted residence of some annelid. There is, however, no reason why a crustacean, which is much higher in the scale of creation, should not make as good a tube. The material of which these tubes are made resembles rough leather or papier-maché, and greyish brown in colour, and very tough. They are very small, in some species being not more than the sixteenth of an inch in length, and proportionately small in diameter. Sometimes the tubes are set so thickly upon the plant as to conceal its surface from view. They are set without the least order, and look as if they had been simply flung upon the sea-weed to which they adhere. The common carrageen (*Chondus crispus*), from which the well-known Irish moss is made, is the plant that is most favoured by their presence. When taken out of its cell, the little animal is not unlike a sand-hopper, except that the two pairs of antennæ are enormously developed, and the first few pairs of legs are furnished with small claws.

The generic name Cerapus is taken from the Greek, and is very appropriate, signifying "horn-footed." These strange antennæ are continually flung forward, grasping at everything that comes within their reach, and reminding the observer most forcibly of the peculiar actions of the cirrhipeds or barnacles. The Caddis-shrimp does not love the very shallow waters, and, except by use of the dredge, cannot be obtained but at the very low tides of March and September, those precious days so invaluable to the practical naturalist, where he finds laid out before him large tracts of the ocean-bed that, except for a few days, at intervals of six months, remain covered with water, and hide their treasures from all eyes.

THE central upper figure in the next illustration represents the common FRESH-WATER SHRIMP, or FRESH-WATER SCREW.

In common with the other Screws, this creature derives its name from its movements when taken from the water and laid upon the ground. Not being able to stand upright upon its feeble legs, it is forced to lie on its side, so that the perpetual kicking of its legs only forces it round in a screw-like fashion, similar to the conduct of the marine screw-shrimp when laid on the sand.

The Fresh-water Shrimp is extremely plentiful in every stream, and may be seen in great numbers even in the little rivulets that conduct the water from fields. They lurk in recesses in the bank or under stones that form the bed of the stream, occasionally darting out to seize some prey, and then making their way back again. Occasionally they push themselves a yard or two up the rivulet, but are sure to come floating back again before very long, allowing themselves to be passively swept along by the force of the water as if they were dead, but starting suddenly into active exertion as soon as they reach their former haunts.

In the water this crustacean moves by a series of jerks, and mostly lies on its side, though it often swims with its back uppermost, and frequently rotates as it passes along. It is a voracious creature, feeding upon dead fishes or any similar carrion. It is fond of the muddy parts of the stream, liking to conceal itself in the soft alluvium when fearful of danger. The eggs of the female are kept for some time under the abdomen, and the young remain in that situation until they have attained sufficient strength to shift for themselves.

The three remaining species are marine. At the lower right-hand corner of the engraving is seen a creature lying on its back. This is the WOOD-BORING SHRIMP, a

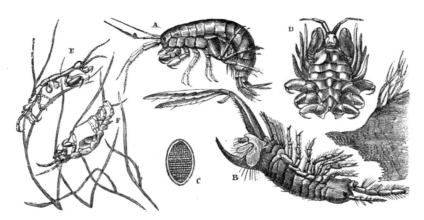

A. FRESH-WATER SHRIMP.—*Gammarus pulex.*   D. WHALE-LOUSE.—*Cyanu ováLis*

E, F. MANTIS SHRIMP.—*Caprella lineáris.*   B, C. WOOD-BORING SHRIMP.—*Chelúra térebrans.*

crustacean that nearly rivals the ship-worm itself in its destructive powers. The figure is represented in the act of attacking a piece of timber, which it has already much damaged. The Wood-boring Shrimp makes burrows into the wood, wherein it can conceal itself, and at the same time feast upon the fragments, as is proved by the presence of woody dust within its interior. Its tunnels are made in an oblique direction, not very deeply sunk below the surface, so that after a while the action of the waves washes away the thin shell and leaves a number of grooves on the surface. Below these, again, the creature bores a fresh set of tunnels, which in their turn are washed away, so that the timber is soon destroyed in successive grooved flakes.

According to Mr. Allman, its habits can be very easily watched, as if it is merely placed in a tumbler of sea-water, together with a piece of wood, it will forthwith proceed to work and gnaw its way into the wood. The apparatus with which it works this destruction is shown in fig. C, and is a kind of file or rasp which reduces the wood into minute fragments. The figures are all enlarged, and that of the rasp is greatly magnified.

In this creature the jaw feet are furnished with imperfect claws, and the tenth segment from the head is curiously prolonged into a large and long spine. The great flattened appendages near the tail seem to be merely used for the purpose of cleaning its burrow of wood dust which is not required for food. The creature always swims on its back, and when commencing its work of destruction, clings to the wood with the legs that proceed from the thorax. The Wood-boring Shrimp is one of the jumpers, and, like the sand-hopper, can leap to a considerable height when placed on dry land.

Another wood-boring shrimp will be figured and described in a succeeding page.

On the left hand of the illustration are seen two figures of a crustacean called appropriately the SKELETON-SCREW, or MANTIS-SHRIMP. There are several species belonging to this genus, and all possess similar habits. Their bodies are indeed skeleton-like in their bony lankness, but their appetites are by no means small in proportion to their size. Indeed, as is often the case with peculiarly meagre human beings, they are most voracious, preying incessantly on every small creature that comes in their way. They are furnished with terrible instruments of prehension, their first and second pairs of legs being devoted wholly to this purpose. The last joint but one is enormously large, and the last joint is thin, and shuts down like the blade of a clasp-knife into its haft, the groove being represented by a double row of spines between which the blade is received. The blade itself is finely notched along the edge. These claw-like terminations to the legs are used not only for seizing prey, but for grasping the branches and drawing the long attenuated body from one part to another.

Mr. Gosse, who has paid much attention to these curious beings, remarks that their movements among the marine vegetation are wonderfully like those of the spider monkeys among the branches, their long thin bodies adding to the resemblance. They run about with great agility, and are always to be found in the branches of the *Plumatella cristata*. The same writer has given a very interesting history of the Mantis-shrimp :—

" Their manners are excessively amusing. The middle part of their long body is destitute of limbs, having instead of legs two pairs of oval clear vesicles, but the hinder extremity is furnished with three pairs of legs armed with spines, and a terminal hooked blade like that already described. With these hindermost legs the animal takes a firm grasp of the twigs of the polypidom, and rears up into the free water its gaunt skeleton of a body, stretching wide its scythe-like arms, with which it keeps up a see-saw motion, swaying its whole body to and fro. Ever and anon the blade is shut forcibly upon the grooved haft, and woe be to the unfortunate infusorium, or mite, or rotifer that comes within that grasp ! The whole action, the posture, figure of the animal, and the structure of the limb, are so closely like those of the tropical genus *Mantis* among insects, which I have watched thus taking its prey in the Southern United States and the West Indies, that I have no doubt passing animals are caught by the crustacean also in this way, though I have not seen any actually secured.

The antennæ, too, at least the inferior pair, are certainly, I should think, accessory weapons of the animal's predatory warfare. They consist of four or five stout joints, each of which is armed on its inferior edge with two rows of long stiff curved spines, set as regularly as the teeth of a comb, the rows divaricating at a rather wide angle. From the sudden clutching of these organs, I have no doubt that they too are seizing prey ; and very effective implements they must be, for the joints bend down towards each other, and the long rows of spines interlacing must form a secure prison, like a wire cage, out of which the jaws probably take the victim, when the bending in of the antennæ has delivered it to the mouth.

But these well-furnished animals are not satisfied with fishing merely at one station. As I have said above, they climb nimbly and eagerly to and fro, insinuating themselves among the branches, and dragging themselves hither and thither by the twigs. On a straight surface, as when marching (the motion is too free and rapid to call it *crawling*) along the stem of a zoophyte, the creature proceeds by loops, catching hold with the fore limbs, and then bringing up the hinder ones close, the intermediate segments of the thin body forming an arch, exactly as the caterpillars of geometric moths, such as those for example that we see on gooseberry bushes do. But the action of the crustacean is much more energetic than that of the caterpillar. Indeed all its motions strike one as peculiarly full of vigour and energy.

I have seen the large red species swim, throwing its body into a double curve like the letter S, with the head bent down, and the hind limbs turned back, the body being in an upright position. It was a most awkward attempt, and though there was much effort, there was little effect."

THE last figure on the illustration, marked D, is the creature that is popularly known by the name of WHALE-LOUSE. All the species of this genus are parasitic, residing on the whale and dolphin. Their hooked and diverging legs, armed with their sharply curved claws, enable them to cling so tightly that not even the swift movement through the water, or the active exertions of the creature on which they reside, are sufficient to shake them from their hold. The different species of Whale-louse seem to prefer various parts of the body, one species clinging to the head, another to the side, and another to the fin. They all burrow rather deeply into the rough and thick skin of these marine mammalia.

Their bodies are flattened and rather oval; they have five pairs of legs, all prehensile; and on the second and third joints of the thorax, instead of legs there are long appendages for respiration, which usually are bent over the back.

IN this illustration are several of the Isopod crustacea, the signification of which word has already been given. In all the Isopods there is a great resemblance to the common woodlouse, and many of them might easily be mistaken for those common and destructive beings. The females have large horny plates on their legs, so formed as to produce a large pouch under the thorax, wherein the eggs are contained. In many species

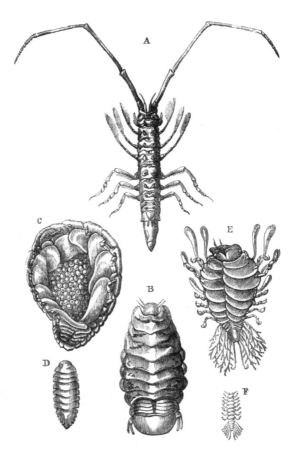

A. BAFFIN'S BAY ARCTURUS.—*Arctúrus Baffinii.*
C, D. SHRIMP-FIXER.—*Bopýrus orangorum.*   E, F. IONE.—*Ióné thorácicus*
B. FISH-LOUSE.—*Cymóthoë œstrum.*

some of the rings of the abdomen are connected so as to resemble a single joint.

The BAFFIN'S BAY ARCTURUS is shown at fig. A, and is one of the best developed of the whole order. In all the species belonging to this genus the body is long, and the first four pairs of legs are beautifully feathered at the ends. These cannot be used for walking, the three last pairs of legs being devoted to this purpose. The long antennæ are used as organs of prehension, and with them the creature captures its prey. The young are said to cling by their legs to the antennæ of the parent.

Several of these species take possession of corallines, each selecting a particular branch, and not permitting any other to intrude upon its premises, fighting with great valour against any assailant.

They resemble the fly-catchers in some of their habits, sitting patiently on their branch until they see some little creature passing within reach. They then dart at their prey, seize it, return with it to their resting-place and there eat it leisurely. They sit in a curious erect attitude, swaying the body about and occasionally cleaning the antennæ by drawing them through the tufted feet.

At fig. B is shown the common FISH-LOUSE, which is parasitic upon many species of fish, clinging tightly by means of their hooked legs. It is thought by many fishermen that the creature is by no means hurtful to the fish, but that it is absolutely beneficial, causing death if removed.

A RATHER curious and tolerably plentiful species of Isopod crustacean is seen at figs. C and D. This is the SHRIMP-FIXER, so called from its habit of affixing itself to shrimps and prawns, concealing itself under the side of the carapace. Any number of these curious parasites may be obtained from a fishmonger's shop, by the simple process of looking over his stock of prawns, and picking out those which have a swelling at the side of the carapace. The fishermen, who have the oddest ideas about marine objects, and know as little about shrimps as a ploughman about worms, generally fancy that these parasites are young soles! probably on account of the general shape of the male.

Fig. C represents the female of this crustacean much magnified, and showing the under side, for the purpose of exhibiting the mass of eggs which are congregated beneath the body, and kept in their places by the pouch formed by the plates attached to the legs. Owing to the pressure caused by the carapace of the prawn, the sides of the Shrimp-fixer are dissimilar, and distorted individuals, like that of the figure, are very common.

In all the members of this genus the male is much smaller than the female, being barely one-sixth the size of his mate, and is narrow and elongated, whereas she is wide, pear-shaped, and ending in a point. The false legs are ten in number, five on each side, and modified into triangular membranous plates, forming a pouch for the reception of the eggs.

The colour of this species is greenish, with a slight lustre above, and dark at the edges of the plates.

THE members of the genus Ione may be known by the appendages of the abdomen, which are thread-like and arranged round the body. Figs. E and F represent the two sexes of these curious crustaceans, the larger being the female. Both, however, are magnified, and the male is of minute dimensions when compared with the other sex.

This creature is also a parasite like the preceding, but makes its home within the thoracic plate of the burrowing crab (*Callianassa subterranea*), which has already been described on page 612. It forms a tumour on the side, and can be removed in a living state. It seems that both sexes are to be found under the same shell, the tiny male holding firmly to the appendages of his mate like a little child holding to its mother's dress. The colour of this species is orange-yellow, and the appendages are white.

Mr. Tuffen West has favoured me with the following remarks upon an allied crustacean :—" Some years ago, I assisted in the dissection and made drawings of the male and female of a remarkable crustacean taken from the gills of a hermit-crab. It was thought that a new genus would have to be constituted for it. In the female of this species there was a distinct space left between the plates covering the ova, for the accommodation of the male, and it is thought doubtful whether he ever takes any food. The males are model husbands : having once selected a mate, they never leave her."

Fig. A on the next illustration represents another wood-boring crustacean, called the TIMBER-BORING SHRIMP, or GRIBBLE.

Though belonging to another family, this creature is as destructive as that which has already been described, but makes its tunnels in a different manner, burrowing deeply into the wood instead of driving oblique passages. It proceeds in a very methodical manner, the tunnels being quite straight unless they happen to meet a knot, when they pass round the obstacle and resume their former direction. Small as is this crustacean, hardly larger indeed than a grain of rice, it is a sad pest wherever submarine timber is employed, for it works with great energy, and its vast numbers quite compensate for the small size of each individual. It appears to attack equally any kind of wood, though its progress is slower in oak and other harder woods than in deal. Sometimes it is found attacking the same timber as the chelura.

As with most of these creatures, the male is smaller than the female, being about one-third her size. The female may be distinguished by the pouch in which the eggs and afterwards the young are carried. About six or seven young are generally found in the pouch.

The Gribble is ashen grey in colour, with darker eyes. The timber into which these creatures have been boring looks very like old worm-eaten furniture. The creature is able to roll itself into a nearly spherical form, like the well-known pill-woodlouse. The tail is composed of many segments, and the antennæ are in pairs, set above each other.

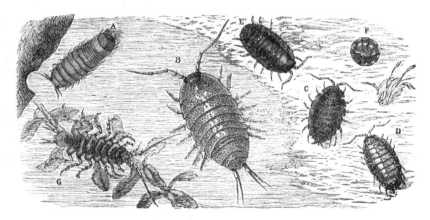

B. GREAT SEA-SLATER.—*Ligia oceánica.*

A. GRIBBLE.—*Limnória térebrans.*   E, F. ARMADILLO WOODLOUSE.—*Armadillo vulgáris*
G. WATER HOG-LOUSE.—*Asellus aquáticus.*   C, D. WOODLOUSE.—*Porcellio scaber.*

At fig. B is seen a creature much resembling the common woodlouse. This is the GREAT SEA-SLATER or SEA-WOODLOUSE, a species which, though extremely plentiful, is not seen as often as it might be imagined, owing to its extremely retiring habits and hatred of light. The Sea-slater lives on the stone and rocks of the sea-shore, and hides itself carefully during the day in the crevices, its flattened body enabling it to crawl into very small chinks. At early morning, however, and in the evening, these creatures may be found by thousands, and any one who will take the trouble to search the rocks by the aid of a "bull's-eye" lantern will find himself repaid by the vast number of nocturnal animals that have ventured out of their dens.

The female carries her young in a kind of pouch formed by the development of a number of horizontal plates along the abdomen. They remain in this natural cradle for some time, and even after they are able to run about, may be seen clinging to their parent. Mr. Tuffen West tells me that on one occasion he picked up a very large Sea-slater, but nearly let it fall again, startled by seeing four or five little ones run from the body. More and more followed, until twenty had made their appearance. Thinking that he had taken up a dead specimen, he put it down again, and was hardly less surprised to see it run off quite briskly.

The substance of the Sea-slater is rather softer than that of the common woodlouse. It appears to feed either on animal or vegetable substances, and is itself much preyed upon by birds and other enemies. The fish are very fond of these creatures, and some species have been known to hover about rocks during a storm for the purpose of preying on the Sea-slaters that are washed into the water. The colour is very variable, but is mostly some shade of brown or grey. This, as well as the succeeding species, belongs to the family of Oniscidæ.

FIG. G in this illustration represents the WATER HOG-LOUSE, a creature that is the aquatic representative of the sea-slater just described. This species is plentiful in fresh-water, whether still or running, and in general walks very leisurely, though when alarmed it can run swiftly. In this genus the proportion of the sexes is reversed, the male being larger than the female. Its average length is about half an inch.

AT figs. C and D are shown two views of the common WOODLOUSE, the latter figure being given in order to show the equal development of the legs. This creature is very plentiful in all damp places, and especially exult in getting under logs of wood or decaying timber. In cellars and outhouses they are common, and are generally to be found in dark and damp localities. Fowls are very fond of them, and there is no surer way of extirpating these sharp-toothed creatures than by allowing some fowls to scrape and peck about in the places where they have taken up their residence. Under the bark of dead and decaying trees is a very favourite residence with the Woodlouse, and in such localities their dead skeletons may often be found, bleached to a porcelain-like whiteness.

The colour of the Woodlouse is a darkish leaden hue, sometimes spotted with white.

An allied species, the LAND-SLATER (*Oniscus asellus*), is equally plentiful. This species may be distinguished by the two rows of yellow spots and the same number of white spots that run along the back. There are also eight joints in the outer antennæ, whereas there are only seven in the same members of the woodlouse.

THE well-known PILL-WOODLOUSE, or PILL-ARMADILLO, is seen at figs. E and F, the former representing the creature as it appears while walking, and the other shows it when rolled up into a globular shape. In this attitude, it bears a strong analogy to the common hedgehog, and a still stronger to the manis, as in the latter case the creature is defended by horny scales that protect it just as the external skeleton protects the armadillo. While rolled up this creature has been often mistaken for a bead or a berry from some tree, and in one instance a girl, new to the country, actually threaded a number of these unfortunate crustaceans before she discovered that they were not beads.

As they bear such a resemblance to pills, they have often had to pay the penalty of their likeness ; for in the earlier days of medicine, and even up to the present time, they have been employed in the pharmacopœia. Even now, though no modern physician would prescribe them for the cure of any disease, the Pill-woodlice may be seen in the recesses of druggists' shops. I have often seen a drawer half-filled with these creatures, and used to convert them into marbles, bullets for a toy cross-bow, and various other purposes in which they were quite as useful as if they had been employed according to the original design. The colour of the Pill-woodlouse is dark greyish brown, with a slight polish.

WE now enter upon a large subdivision of the crustacea, called scientifically the Entomostraca, a term derived from two Greek words, the former signifying an insect, and the latter a shell. All these strange creatures are aquatic, and their bodies are protected by a shell of horny or leathery consistence, sometimes in one single piece and sometimes formed of several portions. The gills are attached to the feet, or the jaws and the feet are jointed and fringed with hairs.

In the first section of these creatures the gills are attached to the feet and they are therefore termed Branchiopoda, or gill-footed. They all swim freely in the water. The first order, the Phyllopoda, or Leaf-footed Entomostraca, have the joints of the feet flat, leafy, and gill-like, and are fitted for respiration. Sometimes the body is naked, and at best, only the head and thorax are covered with the carapace. The first family of these creatures is the Apodidæ, or Footless Entomostraca, so called because all the feet are formed into breathing organs. There are no less than sixty pairs of these feet, all with many joints, and indeed, the number of joints which are required to form one of these apparently insignificant creatures, is almost incredible. With the wonderful patience of the German nation, Schæffer counted the joints, and found that they fell very little short of two millions.

One species, the CRAB SHIELD-SHRIMP (*Apus cancriformis*), is found in England, and has been taken in ponds in different parts of the country. It swims either on its back or in the usual attitude, and uses the branchial feet in its progression. Its food seems to consist of the smaller Entomostraca. The mandibles of this creature are very powerful, and capable of breaking up the shells of the creatures on which it feeds. Its colour is brownish yellow clouded with brown. The example in the illustration is a species discovered by Mr. Tuffen West, and represents a rather small specimen.

A. WEST'S SHIELD-SHRIMP.—*Apus.* B, C, D. BRINE-SHRIMP.—*Artémia salina.*
E. F. FAIRY SHRIMP.—*Chirocephalus diáphanus.*

AT figs. B, C, and D, are represented the BRINE-SHRIMP, fig. C showing its natural size, B the under surface, and D its young. This very remarkable being loves to reside in water so strongly charged with salt that every other creature dies in so saturated a solution, about a quarter of a pound of salt being contained in one pint of water. These animals may be seen by thousands in the salt-pans at Lymington, Hants, where the workmen call them Brine-worms.

They congregate thickly in the strongest brine, while in the ordinary sea-water they do not trouble themselves to venture. The workmen believe that the continual movements of these creatures have the effect of clearing the brine, and if they find that their own salt-pan is without the Brine-shrimp, they always fetch some from another pan.

The movements of this little creature are most graceful. It mostly swims on its back, its feet being in constant motion, and its course directed by means of its long tail. It revolves in the water, bends itself into varied curves, turns fairly over, wheels to the right or left, and seems thoroughly to enjoy the very fact of existence. Its colour is mostly red, and in some of the pans the Brine-shrimps congregate in such multitudes near the surface that the water looks quite pink with their bodies.

The figure of the young Brine-shrimp, which is greatly magnified, is introduced for the purpose of showing the great dissimilarity of form in the young and the adult. Indeed, through the whole of the Entomostraca, the young bear a great general resemblance to each other, but are quite unlike their parents.

THE two larger figures represent the FAIRY-SHRIMP, appropriately named, as a more fairy-like creature can hardly be conceived. It is to be found in several parts of England, and there is a pond on Blackheath where it used to be tolerably common. In spite of its comparatively large size, measuring more than an inch in length, it may easily escape observation, as its body is of glassy transparency, and scarcely visible in the water except by the red and blue tints of its tail, branchiæ, and feet. It always inhabits stagnant water, and may even be found in the half putrid mass of mud and water that lies at the bottom of casting nets.

In his valuable work on the British Entomostraca Dr. Baird gives the following account of this beautiful creature:—"They swim upon their back, and in fine, warm weather, when the sun is not too strong, they may be seen balancing themselves as it were, near the surface, by means of their branchial feet, which are in constant motion. On the least disturbance, however, they strike the water rapidly with their tail from right to left, and dart away like a fish, and hasten to conceal themselves by diving into the soft mud, or amongst the weeds at the bottom of the pool.

It is certainly the most beautiful and elegant of all the Entomostraca. The male is especially beautiful. The uninterrupted undulatory waving motion of its graceful branchial feet, slightly tinged as they are with a light reddish hue; the brilliant mixture of transparent bluish green and bright red of its prehensile antennæ, and its bright red tail, with the beautiful plumose setæ springing from it, render it exceedingly attractive to the view.

The undulatory motion of its branchial feet serves another purpose in addition to that of keeping the animal suspended in the water. The thorax or body of the animal has been described, when floating on its back, as like the cavity of a little boat, the feet representing the oars. When these are in motion, they cause the water contained in this cavity to be compressed, and to mount up as along a canal, carrying in the current the particles destined for its food towards the mouth. It seems to be constantly, when in this position, employed in swallowing and digesting its food, its masticatory organs being in perpetual motion."

In the illustration, the upper figure represents the male, and the lower the female.

WE now come to some of the minute species of these curious animals, and, in consequence, all the figures are much magnified. The creatures belonging to this order are termed the Cladócera or Branch-horns, because their antennæ are forked and branched. With the exception of the head, the body is wholly enclosed within two shells like the half shells of a walnut. There is only one eye, of very large comparative size; there are two pairs of antennæ, and the lower pair are used in swimming. In the first family, the Daphniadæ, the lower antennæ are very large and upper very small. There are five or six pairs of feet enclosed within the shell, and the intestine is straight.

The common WATER-FLEA is a good example of the typical genus. Eight or nine species of this genus are now known, and the habits are much the same in all. They are to be found in fresh water, whether still or running, but seeming to prefer the former. Even the horse-ponds, which are so trodden about by cattle that they seem to consist of equal parts of bad pea-soup and duckweed, are favoured residences of the Water-fleas, and several species may sometimes be found in one little pond.

The common Water-flea has five pairs of feet, the lower antennæ branched and branchleted, the lower pair being very large and powerful. The head is lengthened downwards into a prominent beak. These little creatures exist in vast numbers, and sometimes assemble in such heavy masses that they quite change the colour of the water. Dr. Baird has remarked that they will sometimes assemble so as to form a belt of a foot or so in breadth, and ten or twelve yards in length, and that the whole belt will pass round the pond ; thus obeying some strange instinctive command that, though inaudible to human ears, is conveyed to the myriads whom it concerns, and marshals their hosts with the same wonderful discipline that governs a flock of starlings. If a shadow should fall upon this belt of Daphnias, they all suddenly vanish, and appear again as soon as the darkness has passed away.

The eggs of the Daphnia are placed within the space between the shell and the body, and are there kept until they are hatched, and the mother permits them to escape from under her shelly covering. The young are not in the least like their parent, having no shell, no abdomen, and not nearly all their limbs. The reader will take notice that each of the antennæ is divided into two branches, and each of these branches is subdivided at its extremity into three branchlets. The specimen represented is a female, and is given for the purpose of showing the manner in which the eggs are lodged between the shell and the body of the mother.

Below the Water-flea, and at fig. B, is drawn another of these wonderful little beings, called the MOINA, a genus of which very few species are at present known. The colour

C. *Sida crystallina.*   F, G, H. *Cyclops quadricornis.*   A WATER-FLEA.—*Daphnia pulex.*
E. *Polyphemus pediculus.*   B. *Moina branchiata.*   D.—*Chydorus sphæricus.*

of this species is olive green, its head is round and blunt, and the lower antennæ are peculiarly large and muscular at the base. The carapace is much smaller behind, and at the end of the abdomen are eight very short spines and two long and stout claws. This species is to be found in the ponds on Blackheath, in which localities the microscopic naturalist will detect a vast number of beautiful and even rare creatures that are seldom seen in other places.

On the back of the Moina is seen a dark mass, which, on account of its shape, is called the ephippium, or saddle. This remarkable appendage appears upon the back of the female, particularly in summer, and is divided into two capsules, each portion containing one egg. The eggs which are placed in this singular resting-place are found to remain unhatched through the winter, and are probably defended by this living saddle from the effects of cold, just as pistols are sheltered from the inclemency of the weather in their holsters. Probably, also, the saddle may serve to protect the eggs from the effects of drought, when the pools are dried up by a hot and rainless autumn. When the mother moults, the saddle and its contents are thrown off together with the shell, and it is not uncommon to find specimens of the young swimming about with a portion of the saddle adhering to them, and looking like young beans just sprouting from the earth, and carrying with them the two lobes of the seed from which they sprang.

At fig. C is shown a very pretty little Entomostracan, belonging to a small sub-family called the Sidinæ. In all these beings there are six pairs of feet, the lower antennæ have two branches, and a row of sharp and rather strong filaments springs from the edge of the larger branch. In this genus, one branch of the lower antennæ has three and the other two joints. Though it is occasionally very active, passing through the water with great rapidity, it is mostly dull and stationary, having a curious habit of pressing the back of its head against some object, and there remaining for a considerable period without moving. It derives its specific name of crystalline from its beautifully transparent aspect.

IN the Lynceidæ there are two pairs of antennæ, the upper being very short, and the branches of the lower having three joints. They have five pairs of legs, and one eye, with a black spot a front of it. The abdomen is jointed. All the species are rapid swimmers, and their food consists of both vegetable and animal matter.

In the Chydorus, of which one or two British species are known, the body is nearly spherical, the lower antennæ are very short, and the beak is very long, sharp, and curved downwards. The colour is olive in the present species, and has a smooth shining exterior. It may be found in ponds and ditches throughout the year.

THE curious globular looking creature seen at fig. D is an example of another family, called the Polyphemidæ, having only four pairs of feet, which are not included in the shield. Their single eye is very large, and has given rise to the name of Polyphemus, which belonged to the one-eyed giant overcome by Ulysses and his companions. The lower antennæ have two branches, one with four joints and the other with three. In the lower part of the carapace there is a large empty space for the accommodation of the eggs and young.

AN example of the typical genus is given at fig. E, which represents the common POLYPHEMUS of our ditches and ponds. In this creature the abdomen is long and projects from the shell, and in the adult the eye is enormously large, seeming to occupy the whole head. There is a deep notch or groove in the Polyphemus, seeming to separate the body from the head. It appears always to swim upon its back, and uses both the antennæ and legs to drive it through the water.

Fig. F will be described presently.

FOLLOWING the arrangement of Dr. Baird, we now come to another section, termed the Lophyropoda, or Plume-footed Entomostraca. In all these animals the mouth is well supplied with teeth, and the body is enclosed in a cuirass, either covering the head and thorax, or shaped like a bivalve shell, and enclosing the whole animal. There are not more than five pairs of feet, only one eye, the gills are few and attached to the mouth, and there are two pairs of antennæ, one pair being used for locomotion.

In the first order, called Ostracoda, a term derived from a Greek word, signifying a shell, the cuirass is in two parts, and encloses the animal like a bivalve shell. The hind jaws are furnished with gills. In the family of the Cypridæ, the upper pair of antennæ are long, have numerous joints and a pencil of long filaments ; the lower pair are short, thick, and used as feet. There are two pairs of real feet. The CYPRIS, which is represented at fig. A, belongs to a genus which has many British species, and may be found in almost every pond or ditch. The body is enclosed thoroughly in its valved cuirass, something like a walnut in its shell, the fringed antennæ and legs protruding from between the valves and permitting the creature to move. It is a most elegant little being, the shell being gracefully curved, and the antennæ being fine and transparent as if they were threads of glass. Dr. Baird tells us that the valves are very brittle, and that on their exterior they are washed with a kind of varnish which protects them from the action of the water. Owing to this varnish, these creatures cannot venture even to rise to the surface; for as soon as the shell is exposed to the air, it becomes quite dry, and so buoyant that no exertion of the Cypris can sink it again.

These tiny animals will often live through a hot summer which dries up the pond in which they reside, and at the first rain will make their appearance again, swimming merrily about as if nothing had happened. As soon as they feel themselves being deserted by the water, they bury themselves deeply in the mud, and even their eggs retain their vitality, though the mud should be baked quite hard. When the Cypris changes its skin, it throws off the whole shell, the internal parts of the body, the beautiful comb-like gills, and the tiny hairs which clothe the bristles of the antennæ.

Two more examples of this pretty genus are given in this illustration, fig. B representing *Cypris claváta*, and fig. C, *Cypris vídua*, seen from above, in order to show the two valves of the shell and their line of junction. Seventeen or eighteen British species of Cypris are known.

In the family of Cytheridæ, the upper pair of antennæ have no long filaments.

A. *Cypris tristriáta.*   B. *Cypris claváta.*   C. *Cypris vídua.*   D. *Cythére minna.*
E. *Cythére inopináta.*   F. *Cythereis Jonesii.*   G. *Cythére impressa.*   H. *Cypridina Macándrei.*
I. *Canthocamptus minútus.*   K. *Cetochilus septentrionális.*

The members of the typical genus Cythere are mostly marine, and may be found in the little rock-pools at the sea-side, darting about among the branches of sea-weeds and zoophytes that live so plentifully in such situations. Safe in these sheltered spots, they care nothing for wind and waves, and the storm which flings the huge whale on the shore will fail to injure these tiny beings, whose very minuteness is their safety. Three species of Cythere are here given. *Cythere minna*, fig. D, is remarkable for being the largest British species seen by Dr. Baird. Its valves are white. It was found in deep water off Shetland, and taken in a dredge. *Cythere inopinata*, fig. E, derives its specific name of *inopinata* or unexpected, from the fact that the creature was found where no one would have expected its presence, namely, in a small pond in Middlesex. It is a very small species, and always remains at the bottom. Its colour is white, and there is a little orange coloured mark on the upper edge. An oblique view of this species has been chosen, in order to show the curious rounded projections upon the middle of each valve.

The *Cythere impressa*, fig. G, was found in sand at Torquay. The shell is dull black in colour, and is covered with little punctures impressed upon its surface, whence is derived its specific name.

A closely allied genus is represented at fig. F, and is remarkable for the manner in which the valves are ridged, irregular, covered with tubercles, and having their edges boldly toothed. This species was taken in the Isle of Skye.

IN the family of the Cypridinadæ there are two eyes, set as footstalks, and two pairs of feet, one pair being always within the shell. There is only one genus of these creatures, and all the species are marine. The shell is oval, sharply pointed at each end, and the front edge is deeply notched. The pair of feet that are retained within the shell are modified into one organ, which seems to be intended for the purpose of supporting the eggs. At fig. H is seen this curious organ. The present species was procured in seventy fathoms of water, near the Isle of Skye. Some of the exotic species are luminous.

The next order is termed the Copepodæ, or Oar-footed Entomostracans, because the five pairs of feet are mostly used for swimming. The body is divided into several rings, the cuirass covers both the head and thorax, and the mouth is furnished with foot-jaws.

In the family of the Cyclopidæ the head and body are merged together with the first ring of the thorax. There are two pairs of foot-jaws, and the fifth pair of legs are very minute.

The species which is represented at fig. F on the illustration at page 635 is very common in every pond and ditch, and the female may at once be recognised by the little egg-bags which she bears on the sides of the abdomen, like John Gilpin's wine-bottles at his belt. The colour of this species is exceedingly variable, differing according to the locality where the creature happens to reside. It is mostly white, but some individuals are brown, others greenish, while a few are red. Both salt and fresh water are inhabited by the CYCLOPS, and some of the marine species are so highly luminous, that they add in no slight degree to the phosphorescence of the ocean. At figs. G and H are shown specimens of the young in two stages of their growth.

TURNING again to the engraving on page 637, and referring to fig. I, we find a little creature with a long abdomen, which it is able to turn over its back, something after the fashion of the earwig or the cocktail beetles. In this Canthocamptus the thorax and abdomen are merged into each other, and gradually diminish in size to the extremity. All the species belonging to this genus have very small and simple foot-jaws. It inhabits ponds and ditches of fresh water. Mr. Tuffen West tells me that in February, 1861, he was examining some of the slime that had gathered upon the roof of the Cramlington Pit, at a vast depth from the surface, and that he found in the slime some of these minute crustaceans quite brisk and lively, whisking their tails up and down smartly, as shown in the figure. These creatures must have been washed down the pit while still unhatched, and have been thus carried down from the open air into the bowels of the earth.

Fig. K represents a creature, which though very small, not more than the sixth or seventh of an inch in length, is of exceeding importance to commerce, as it affords food to the herring, several whales, and other valuable beings. In the seas where this little creature lives, whole tracts are reddened with the multitude of their hosts, which swarm near the surface, and congregate in such vast numbers, that the wind has been known to catch up a whole bank of them, like a wave, and fling it into the vessel, covering the deck and the sailors with their bodies. The codfish feeds largely and luxuriously upon these abundant creatures, needing not to take any pains about them, but swimming lazily through their masses and opening its mouth, into which they pass without the least trouble.

The long antennæ are used as oars, being thrown backward at every stroke until their tips touch each other. This attitude, however, is only assumed while the creature is in haste, as it is often seen to pass gently through the water, with its antennæ at right angles to the body, as shown in the engraving. Dr. Sutherland, in his "Voyage to Baffin's Bay," writes of these elegant little beings :—" They are always on the alert to elude and escape from their pursuers. When the water is but slightly agitated, they dive from the surface, and in a few minutes, when it becomes still, they can be seen ascending slowly, but rarely using the antennæ. I could only obtain specimens by including them in a large quantity of water taken up suddenly, from which they could be separated subsequently by straining through a calico bag. A bucketful (two gallons) of water often produced twenty to thirty individuals, and sometimes twice that number. They never survived a single night, even though kept in their native element in a vessel. From their constant darting from side to side of the vessel, perhaps it is a safe inference that the fear of danger in their new situation may be one of the chief causes of the early extinction of life."

The colour of this species is light red, and the body is nearly translucent.

Another curious species deserves a word of mention. This is the *Notodelphys ascidicola,* which is found swimming in the bronchial sac of the ascidia.

IN this illustration we have examples of a group of Entomostraca which are parasitic upon fish and other inhabitants of the waters. They belong to Dr. Baird's third legion, called the Pœcilopoda, a term derived from two Greek words, signifying various-footed. They are so named because they are partly formed for walking or seizing prey, and partly for swimming and breathing. In the first order, the SIPHONO-STOMA, or tube-mouthed Entomostraca, the mouth is furnished with a tube containing sharp spike-like mandibles. The foot-jaws are well formed. The object of the tube and its sharp mandibles is evidently for the purpose of piercing the skin and sucking the juices of the beings upon which they cling, and the strong foot-jaws enable them to hold so firmly that they cannot be shaken off. The first tribe is called Peltocephala, or buckler-headed, because the head is shaped something like an ancient buckler; the head is also furnished with plates in front, and small antennæ of two joints. The first family of these creatures is called Argulidæ, and may be known by the circular shaped head shield, and the manner in which the second pair of foot-jaws are modified into a pair of powerful suckers.

The FISH-ARGULUS may be seen upon many of the ordinary river-fishes of England, the stickle-back being its favourite. I have seen it on the roach, and even upon the golden carp. It is not very small, being about the diameter of a small sweet pea, and may easily be watched if placed in an aquarium in which any fish are swimming. The little creature at once makes for the fish,

A *Argulus foliaceus.* B. *Caligus Mülleri.*
C. *Nicóthoë ástaci.* (Natural size.)
D. *Nicóthoë ástaci.* (Female, magnified.)
E. *Nicóthoë ástaci.* (Male, magnified.)
F. *Dichelestium sturiónis.*
G. *Chondracanthus zei.* H. *Lernæodiscus.*
I. *Jacculina.*

darting along with considerable speed, and fixes itself to the side just under the pectoral fins. It does not, however, remain fixed to the fish, but occasionally leaves it, and starts off on little voyages of discovery, always, however, returning at short intervals, as if for the purpose of assuring itself of a meal. It is wonderfully flat, looking very like the shed seed-vessel of some plant, and the resemblance is increased by its pale green colour.

The female is considerably larger than the male, and may at once be known by the black spot on each side of the abdomen.

The CALIGUS, which is shown at fig. B, is referred to another family.

This creature is mostly found upon the codfish and brill, and clings with great firmness. Mr. Tuffen West tells me that he has examined the Caligus carefully with the microscope, and assured himself that the suckers are present. "They are hemispherical, shallow in front, where their margin thins off to a translucent membrane ; and deep behind, where their concavity is bounded by a strong transversely striated membrane." The figure is magnified about two diameters.

Figs. C, D, and E represent a remarkable parasite, adherent to the gills of the lobster. This creature belongs to a different tribe, which may be known by the small and mostly blunt head, and the long and well-jointed antennæ. The family Ergasilidæ have the head rounded, the body oval, the abdomen well developed, and the feet small and branched.

The LOBSTER-LOUSE is sometimes found in considerable numbers fixed to the gills of the lobster, from which the female never moves after she has once taken a firm hold, though the male is more erratic in his habits, and swims about as he chooses. At fig. C it is shown of its natural size among the lobster-gills, which are popularly termed the "lady's fingers." At fig. D is seen a magnified figure of the female, and at E a figure of the male, also enlarged. During her early youth, the female is not much larger than the male, but as soon as she attaches herself to her new home, a pair of strange projections are seen to grow from the side, and by degrees become so large that they seem to constitute the entire creature. Below these projections the egg-sacs are developed.

At fig. F is seen a curious parasite that infests the sturgeon.

The *Dichelestium* is rather more than half an inch in length and the twelfth of an inch in breadth. This creature insinuates itself deeply into the skin, making its way to the bony arches upon which the gills are supported, but not appearing to touch the membranous gills themselves. Sometimes as many as ten or twelve are taken from a single fish. They can grasp very firmly by means of their forceps, and are able to turn round whenever they please. This curious creature belongs to the order of the Lerneadæ, in which the mouth is formed for suction, and the limbs scarcely visible. All these beings are parasitic upon fishes, and are often so deeply buried in the tissues that the whole body is concealed and only the egg-bearing tubes suffered to appear. As is the case with many creatures, especially those that occupy a low place in the scale of creation, the young enjoy a wider range than the parent, being able to roam about at will, and not settling down to a motionless existence until they have attained maturity. The figure represents the female Dichelestium.

There seems to be no bound to the wondrous forms which these parasites assume. Fig. G represents the *Chondracanthus*, so called because its body is covered with cartilaginous spines or tubercles. The name is derived from two Greek words, the former signifying cartilage, and the second a thorn. The broad and flattened egg-tubes are seen below. This strange being is found upon the gills of the John Dory.

At figs. H and I are seen two most extraordinary beings, which were discovered under the abdomen of a lobster. In both these creatures (which certainly seem to belong to the Lerneans), the whole of the head becomes modified into a set of branching fibres, much resembling the roots of a tree. There is no mouth whatever, all nourishment being transmitted through these fibres. They are quite recent discoveries, and after the engraving had been prepared, several of the figures were removed in order to make way for forms so novel, so interesting, and so unique.

MORE fish parasites are given in this illustration, many of them possessing the most bizarre and unexpected forms. As our space is rapidly diminishing, these creatures must be very slightly treated.

Fig. A represents the PERCH-SUCKER, and fig. B the male of the same species. The reader will notice the strange formation of the female, and the great dissimilarity between herself and her small mate. Fig. C exhibits an example of a curious genus termed Anchorella, in which the arm-like appendages are very short, and united from the base so as to look like a single organ. At fig. C is seen the female *Anchorella uncinata*. In this species the body is white, and the short arms end in a rounded knob. This creature is rather more than half an inch in length. At fig. D is shown the male of the same species, which would hardly be recognised as having any connexion with the long bodied creature that has just been described. The length of the male is about the forty-eighth of an inch in length. At fig. E is another species of the same genus, *Anchorella rugosa*, so called because the body is notched at the side. This creature is about the seventh of an inch in length. All these creatures infest the cod, haddock, and similar fishes.

A wonderful example of a parasitic crustacean in several stages of development is seen at figs. H, I, K. This is the *Tracheliastes*, with its long egg-bags and strangely developed upper extremity. At fig. I is seen the same creature when very young, and at fig. K it is again shown, though of larger size.

In the next tribe of Entomostraca the head is kept buried in the tissues of the animal to which the parasites cling, and are there held firmly by some horn-like processes that spring from the back part of the head. They are, in fact, living spears, the barbed heads being sunk into their prey. Two curious forms are shown at figs. L and M. The former represents the female of *Lamproglena pulchella*, and is given in order to show the under surface of the body and the small pairs of feet. These feet show that it belongs to the family of the Penelladæ. Fig. M represents the *Lernentoma asellina*. This is also a female.

A. *Acthéres percárum.* (Female.)
B. *Acthéres percárum.* (Male.)
C. *Anchorella uncináta.* (Female.)
D. *Anchorella uncináta.* (Male.)
E. *Anchorella rugósa.* (Female.)
F. *Lernéntoma cornúta.* (Female.)
G. *Lernéntoma cornúta.* (Male.)
H. *Tracheliastes polycolpus.* (Female.)
I, K. *Tracheliastes polycolpus.* (Young.)
L. *Lamprogléna pulchélla.* (Female.)
M. *Lernéntoma asellína.* (Female.)

TURNING to the next engraving, we shall find some more examples of these strange beings. Not the least strange is that which is shown at fig. A, inasmuch as it is parasitical upon a parasite. This is the *Chalimus scombri*, which is found adhering to a caligus. At fig. A a much magnified drawing is given, showing the long tube and sucker whereby it adheres to its prey, and at fig. B the creature is seen hanging to the lower part of the caligus like a fish at the end of a line. This is one of the many instances that prove the truth of that quaint and far-seeing old saying, namely—

"Big fleas and little fleas
Have lesser fleas to bite 'em :
The lesser fleas have smaller fleas,
And so, *ad infinitum*."

3.

Fig. C shows the *Cecrops*, a creature that is found upon the sunfish, and adheres to the gills. It is not always fixed to this habitation, but floats about by thousands in the Mediterranean, where it is preyed upon by many fishes.

Fig. D shows the female of the SHARK-SUCKER, a species that is found adherent to the eyes of the Arctic shark, and appears to blind it. The sharks to which this unpleasant appendage was attached seemed to be quite destitute of sight, and did not flinch in the least when a blow with a lance was aimed at them. The arm-like appendages of this creature are inserted into the corner of the eye for nearly one-fourth of their length. This parasite attains to the length of three inches. Fig. E represents the male of the same species. Fig. F is the male of an allied species found on one of the common dog-fishes known by the name of tope, and described on page 208 of this volume. Fig. G is the female of the same species. At fig. H is seen a strange elongated creature called *Penella filosa* from its extreme length. This species is found to penetrate into the flesh of the sword-fish, the tunny, and the mole-fish, all of which have been described in this volume. It is said to cause them considerable pain. At fig. I is seen a parasite of even stranger form, but belonging to the same genus. This is *Penella sagittata*. At fig. K is shown the head of the sprat, having one of these parasites, called the SPRAT-SUCKER, affixed to its eye. Sometimes this creature is tolerably common, many specimens being obtainable at a single fishmonger's shop, while for several years hardly one will be seen. The colour of this parasite is pale sea-green with a slight bluish cast. The eggs are very green. A magnified figure of this creature is seen at fig. L, and is given in order to show the barb-like head by which it adheres to the eye of its victim.

A. *Chalimus scombri*.   B. *Chalimus scombri*.   Attached to a Caligus.)
      C. *Cecrops Latreillii*.   D. *Lernæópoda elongáta*.   (Female.)
E. *Lernæópoda elongáta*.   (Male.)   F. *Lernæópoda gálei*.   (Male.)
      G. *Lernæópoda gálei*.   (Female.)   H. *Penella filosa*.
I. *Penella sagittáta*.   K. *Lernéntoma spratti*.   (Natural size.)
L. *Lernéntoma spratti*.   (Magnified.)   M. *Lernæócera cyprinácea*.
      N. *Lernæa*.   (From gill of codfish.)   O. *Lernæa gobina*.
                P. *Lernæa radiáta*.

At fig. M is shown a strange and seemingly shapeless parasite that is found to affix itself to the carp tribe, and at N a creature of somewhat similar form is notable for being found upon the gills of the codfish. This creature belongs to the typical genus. At figs. O and P are given two more forms of this genus, which are evidently of a much less perfect type than the preceding species, and, according to the language of naturalists, are said to be degraded forms.

IT is hardly possible to imagine any forms that are so strange, any habits so astonishing as those which are found in the crustaceans figured in this illustration. Although they have been known for some time, their proper place in the scale of creation has long been a disputed point among systematic naturalists, some considering them to belong to the crustaceans and others to the spiders. As, however, they undergo a true metamorphosis, which is not the case with any spider, they are now admitted to be real, but unique crustacea. Even such naturalists as Siebold and Milne-Edwards differed about them, the former placing them among the spiders and the latter ranking them with the crustacea.

Such strange creatures as these are not easily described, especially when the space that can be granted to them is so limited, for their whole economy is so thoroughly unique that they require a volume rather than a page. They are found upon the British coasts, and their history is briefly as follows.

At figs. B and C of the accompanying illustration are seen two strange-looking creatures with wonderfully small bodies and enormous legs, jointed and arranged in such a manner as almost to preclude the idea of their real character. Indeed, it seems passing strange how the tiny abdomen can absorb sufficient nutriment for the supply of those marvellous limbs.

Their economy is as strange as their form.

Some specimens of a well-known zoophyte (*Coryne eximium*) are seen growing as they appear while attached to the rocks or sea bed, and upon the branches are shown numerous pear-shaped objects, attached in a manner that reminds the observer of the tropical gourds. On opening these nodules, each is found to contain a specimen of the *Phoxichilidium*, which, in spite of the long limbs, is packed away in a very complete manner, the limbs being rolled round the body so as to form the creature into a kind of ball. At fig. I one of these beings is shown as it appears when removed from the envelope. The lower part of the engraving

A. *Pycnogonum littorale.*
B. *Phoxichilidium coccinium.* (Male.)
C. *Phoxichilidium coccinium.* (Female.)
D, E, F. *Phoxichilidium coccinium.*
(Young in different stages.)
G. *Phoxichilidium coccinium.*
(Enclosed in swellings of *Córyne eximium.*)
H. *Phoxichilidium coccinium.* (Young nearly complete)
I. *Phoxichilidium coccinium.* (Removed from coryne.)
K. *Phoxichilidium coccinium.* (Just before maturity.)

shows the young Phoxichilidium at different stages of its growth. Sometimes it possesses the rudiments of limbs, with long filamentous appendages; sometimes it throws them off, and contents itself with a pair of stout claws, and then again grows a fresh set of limbs and a pair of small and feeble claws.

LONG-TAILED MOLUCCA CRAB.—*Limulus longispinus.*

Strange as are these habits, there is still a kind of analogy with other modes of animal life. On page 625 is mentioned the curious little crustacean which resides within the body of a beroë, and in the present instance there is an evident analogy with the various galls and their inhabitants, the cells of the Phoxichilidium being in fact the galls of the coryne.

THE crustacea abound in strange forms. The LONG-TAILED MOLUCCA CRAB belongs to a separate order, called by the name of Xiphosúra, or Sword-tailed Crustacea, in allusion to the long and sharp spine which projects from the shell. These creatures, of which several species are known, can easily be recognised by their general shape. The body and limbs are covered by a curious shield composed of two parts, the junction taking place across the centre of the body, as may be seen by reference to the engraving. Though perfectly harmless, these creatures can be made very offensive, for the natives of Molucca are accustomed to use the long sharp tail spine as the head for an arrow or lance, and thus make a most formidable weapon. Many of these crustacea attain the length of two feet, so that the spike is nearly a foot in length, and is capable of inflicting a deadly wound.

The edges of the hinder portion of the shield are deeply toothed, and the space between each tooth is occupied by a rather long and sharply pointed spine, which is not fixed, but is moveable on its basis. The feet are mostly furnished with tolerably strong claws.

The Molucca Crabs often leave the sea and crawl upon the sand, where they may be taken without much difficulty. They cannot endure the heat of the sun's rays, and are in the habit of burrowing into the sand when the sunbeams beat too fiercely on their shells. Sometimes they do not bury themselves very deeply, and then they are discovered by the projecting tail spike, which shows itself above the level of the sand, and betrays the position of the animal. As they pass over the sand they present a very curious appearance, as their large shield-like shell entirely covers the limbs, and the creatures seem to be carried along by some external agency rather than to be propelled on their own limbs. Owing to the shortness of the legs, and the large rounded shell, the Molucca Crabs are almost helpless if laid on their backs, being obliged to wait until some friendly wave may strike them and enable them to resume their proper attitude. These crustaceans occur largely in certain strata, and are found in a fossil state, many species attaining to a very great size. One living species (*Limulus cyclops*) is a native of the East Indies, and goes by the popular name of PAN-FISH, or SAUCEPAN-CRAB, because the shell, when the limbs and body have been removed and the tail spine permitted to retain its place, has some resemblance to the useful culinary article from which it derives its name. It is often used as a ladle for dipping water out of a vessel.

WE now come to the last members of the crustacea, creatures which were for a long time placed among the molluscs, and whose true position has only been discovered in comparatively later years. Popularly they are called Barnacles, but are known to naturalists under the general term cirrhipedes, on account of the cirrhi, or bristles, with which their strangely transformed feet are fringed.

When adult, all the cirrhipedes are affixed to some substance, being either set directly upon it, as the common acorn barnacle, so plentiful on our coasts ; placed upon a footstalk of variable length, as in the ordinary goose-mussel ; or even sunk into the supporting substance, as is the case with the whale barnacles. When young, the cirrhipedes are free and able to swim about, and are of a shape so totally different to that which they afterwards assume, that they would not be recognised except by a practised eye. More will be said on this subject.

Along the under surface are set six pairs of limbs not furnished with claws, but being developed at their extremities into two long filaments, jointed and covered with hairs. By means of these modified limbs the cirrhipedes obtain their food. The common acorn-barnacle of our coasts affords a familiar and beautiful example of the mode by which this structure is made subservient to procuring a supply of food. The closed valves at the upper part of the shell are seen to open slightly, a kind of fairy-like hand is thrust out, the fingers expanded, a grasp made at the water, and the closed member then withdrawn into the shell.

This hand-like object is in fact the aggregated mass of legs with their filaments. As the limbs are thrust forward, they spread so as to form a kind of casting net ; and as they return to the shell, they bring with them all the minute organisms which were swimming in the water. This movement continues without cessation, as long as the Barnacles are covered with water, and appears to be as mechanically performed as the action of breathing as performed by the higher animals.

We will now cast a hasty glance at the transformations through which these creatures pass before attaining their perfect state. It has already been mentioned that the young cirrhipedes are free and able to wander about at will ; and as is generally the case in such instances, they are apparently of a higher organization when young than when adult. For example, the young Barnacle can swim freely with certain limbs. When adult, it loses those limbs. When it is young, it possesses eyes ; but when it attains maturity, it loses those valuable organs, which, although indispensable to a wanderer, are needless for a being which is fixed to one spot and needs not to move in order to obtain subsistence.

When first set free from the parent, the Barnacle is extremely minute, and has a striking resemblance to the young of one of the Entomostraca already described. It has three pairs of legs, with imperfect joints and ending in bristle-like appendages. By the vigorous flapping of these limbs the young Barnacle is driven quickly through the water, with a sharp but uncertain movement. In fact, a microscope of low power, when applied to the water wherein a number of these tiny creatures are swimming, discloses a swarm of merry little beings playing about just like the clouds of gnats over water, or the dancing motes in the sunbeam.

Just in the middle of the part of the body which by courtesy we will call the forehead, a single eye is placed, black, round, and shining as if it were a little jet bead inserted into the body. There are also two very large antennæ, which serve two useful purposes, for they aid the free and imperfect Barnacles to proceed through the water, while they are the means whereby the creature fixes itself to the rock when about to undergo its last change.

It then passes through a series of changes, casting off its skin at every change, and exhibiting the curious phenomenon that, in its form and general appearance, it presents a strange resemblance to the young of several Entomostraca, such as cypris and cyclops. When it is ready for the final change, the young Barnacle seeks some rock or other resting-place, and begins operations by pressing the large antennæ against the supporting substance. A curious cement or glue is then poured from their bases: as this cement is not soluble in water, it fixes the creature firmly to the rock. Almost as soon as it is fairly settled, the Barnacle again casts its skin, parting with the bivalve shell which guarded its body, casting away the eye which has hitherto directed its course, and assumes, though still of very minute form, the shape of the adult.

After describing these wonderful changes, Mr. Gosse makes the following apt remark. " Marvellous indeed are these facts. If such changes as these, or anything like them, took place in the history of some familiar domestic animal—if the horse, for instance, were invariably born under the shape of a fish, passed through several modifications of this form, imitating the shape of the perch, then the pike, then the eel, by successive casting off its skin ; then by another shift appeared as a bird, and then, gluing itself by its forehead to some stone, with its feet in the air, threw off its covering once more and became a foal, which then gradually grew into a horse ;—or if some veracious traveller, some Livingstone or Barth, were to tell us that such processes were the invariable conditions under which some beast of burden largely used in the centre of Africa passed ; should we not think them very wonderful ? Yet they would not be a whit more wonderful in this supposed case than in the case of the Barnacle, in whose history they are constantly exhibited in millions of individuals and have been for ages—and even in creatures so common that we cannot take a walk beneath our sea cliffs without treading on them by hundreds."

Having thus glanced cursorily at the general structure and habits of the cirrhipedes, we will proceed to the individual specimens which are figured in the illustration.

In the upper left-hand corner of the illustration is seen a group of the common Goose-mussel or Duck-barnacle, so called on account of the absurd idea that was once so widely entertained, that this species of barnacle was the preliminary state of the barnacle-goose, the cirrhi representing the plumage, and the valves doing duty for the wings.

This Barnacle is tolerably universal in its tastes. It clings to anything, whether still or moving, and is the pest of ships on account of the pertinacity with which it adheres to their planks. Its growth is marvellously rapid, and in a very short time a vessel will have the whole of the submerged surface coated so thickly with these cirrhipedes that her rate of speed is sadly diminished by the friction of their loose bodies against the water

When once the Goose-mussel has affixed itself to any object, the rapidity of its growth is positively startling. The minute young are poured from its shells in such multitudes that they look like cloudy currents in the water ; and after they have enjoyed their brief period of freedom, they settle down, attain maturity, and in their turn become the origin of a countless posterity.

I have seen a large log of timber, about fourteen feet in length by one foot square, so thickly covered with these Barnacles that the wood on which they rested was not visible. The same log, which had evidently formed part of the cargo of a timber ship,

Lepas anatifera.
Pollicipes mitella.
Corónula diadéma.

Conchoderma aurita.
Balanus crenátus.

Lepas fasciculáris.
Pyrgóma grande.
Bálanus tintinnábulum.
Tubinicella tracheális.

had been attacked by the ship-worm as well as the Barnacle, and had been tunnelled from end to end by that insatiable devourer. The log was so entirely covered by the Barnacle and the shipworm, that the wood of which the beam was composed was quite invisible, and could not be seen until the heavy masses of Barnacles were lifted up by the hand.

The old boatman who had picked up the log while fishing, and had ingeniously built a trough to receive the log, a tank of sea water to supply the trough, and a kind of tent composed of sails to hold the trough and the log together, was very full of a discovery that he had made. He was fully persuaded that the ship-worm and the Barnacle were identical, and that when the ship-worm was tired of boring into wood, it came to the surface, and was immediately changed into a Barnacle. He was quite impervious to reason, and always went into a passion whenever the facts seemed to contradict his theory.

If the objects were enumerated to which the Barnacle will cling, a volume would hardly be sufficient for the mere catalogue. It has been found on ships, boats, floating timber, shells, turtles, whales, and marine snakes. A moment is sufficient to give them a firm hold of any object, and when once they have fixed their antennæ, the fiercest storm cannot shake them off. Even after death, the force with which they cling is as great as during life, and they seem almost to form part of the substance to which it adheres. The length of the foot-stalk is extremely variable, in some measuring three or four times the length that it does in others. This species is found in nearly all temperate and warm seas.

A second but smaller group of Stalked Barnacles is seen in the corresponding right-hand corner. This is the FASCINE-BARNACLE, a larger and finer species, which, as may be seen by reference to the illustration, can be distinguished by the number and shape of its shelly valves. These valves, indeed, afford most important indications of the genus to which any species belongs, and in the arrangements of some zoologists they play the principal part in the formation of the system.

I may here mention that the whole of the figures upon the preceding illustration are drawn from specimens in the splendid collection of H. Cuming, Esq., who kindly lent them for the purpose, as were several of the more curious molluscous shells, such as the Chinese Pearl-mussel, shown on page 428. The same specimens were employed by Mr. Darwin in his elaborate work upon the cirrhipedes. The Fascine-barnacle is found in the Indian Ocean.

A rather singular form of Barnacle is seen just below the Goose-mussel, resting on short footstalks, and having somewhat triangular valves. This is the MITELLA-BARNACLE, which may be known by the stoutly shaped footstalks and the rough, shagreen-like character of their surface. At the base of the shell are seen a number of smaller accessory pieces, all pointed and marked with slight transverse lines. This species comes from China, the Philippines, &c.

Exactly in the centre of the illustration is placed a specimen of the EARED BARNACLE, which derives its appropriate name from the curious tubular projections which stand out boldly from either side, like the ears of a quadruped from the head. This species lives in the warmer seas.

A group of Eared Barnacles have been found attached to another genus of Barnacle, which lives on, or rather in, the skins of cetacea, and to which we shall presently allude. Indeed, these beings seem to care little about the substance to which they adhere, one species of Stalked Barnacle having actually been taken upon the delicate surface of a living Medusa.

We now leave the stalked barnacles, and proceed to those species which are placed directly upon the substances to which they adhere. A little to the right of the eared barnacle is seen a small group of upright shells, surrounded by buttress-like and pointed projections. This is the BELL-BARNACLE, found off the coast of Madeira, Africa, and other hot parts of the ocean. It sometimes attains a very considerable size, and is eaten by the Chinese, who think that it resembles the lobster in flavour.

Other species are also eaten, such as the PARROT'S-BEAK BARNACLE, a creature deriving its name from a curved projection something like the bill of a parrot. This enormous Barnacle is sometimes found measuring between five and six inches in height, and between three and four inches in diameter. It is found in large bunches, sometimes consisting of a hundred individuals, some adhering to the rocks and others to the shells of their companions. The bunches of Parrot-beaked Barnacles bear a decided resemblance to the strange cacti whose leaves are set so oddly upon each other.

This Barnacle is gathered, or rather hewn from the rocks in large quantities, and exported to Valparaiso and other places, where it is held in high estimation as a delicacy for the table. It is generally boiled, and eaten cold, like the common crab, and is said to resemble that crustacean in general flavour. It is a South American species, and is found most plentifully and of the largest size at Concepcion de Chile, and the best specimens are taken from a little island called Quiquirina, which lies across the mouth of the bay.

The two little specimens on the rock, just below the foot of the eared barnacle, are examples of the common ACORN-BARNACLE, so plentiful on our coasts. The surface of every stone and rock that is washed by the sea, the exterior of every pile of masonry that is lashed by the waves, is covered with the shells of this curious little creature, which is extremely valuable to the naturalist, as its habits are easily studied, and from its exceeding plenty any number of specimens can be obtained. They are very pretty inhabitants of an aquarium, but they require peculiar conditions to keep them in health, and if they die, are sure to corrupt the surrounding water to such an extent, that nearly every other inhabitant of the aquarium will share their fate. Spots over which the tide only runs for a few hours are thickly studded with these Barnacles, and it is interesting to see how quickly they open their valves and fling out their arms as soon as the water covers them at each returning tide. When the sea withdraws, they close their shells firmly, and retain within their interior a sufficiency of water wherewith to carry on the business of respiration until the next tide brings a fresh supply. Total submersion seems to be hurtful to them.

They are very awkward to the shore bather who does not know the coast, as the edges of their shells are exceedingly sharp and knife-like, and inflict very painful scratches when brought into collision with the unprotected skin. Even to those who are searching on the rocks for marine curiosities the Barnacles are very annoying, as they are constantly scratching the hands when an incautious searcher happens to stumble and tries to save himself by grasping at the rocks.

At the right-hand centre of the illustration are seen two little white dome-like objects perched upon a mass of coral, and having an opening above. These are examples of a rather curious cirrhipede, the CORAL-BARNACLE, which, as is evident from the material on which it is supported, is only to be found in those seas which are warm enough to produce the corals. Sometimes the growth of the coral is too rapid for the Barnacle, which is gradually covered by the increasing stony deposit, and at last is actually buried deeply in the mass, where it dies from starvation. The reader may remark that one of the molluscs described on page 379, is also in the habit of making its residence upon coral, and were it not for a peculiar adaptation of structure, would perish for the same reason. But the Magilus is able to extend its shell as fast as the coral deposits fresh substance, and therefore always contrives to keep itself within reach of the water. In the Pyrgoma the cone is composed of a single piece, very thick, rather compressed, and open above.

Some very strange forms of cirrhipedes are now presented to us, and are figured on the lower part of the illustration. On the left hand is seen a small group of the CORONET-BARNACLE, so called on account of the coronet-like shape of the body. This species is always found upon the skins of the cetacea which inhabit the Arctic Seas, such as the Greenland whale and the long-armed whale (Balænoptera longimana).

The specimens exhibited in the illustration are represented of their natural size, but Mr. Sowerby informs me that in a piece of whale skin only four inches in length, no less than six specimens of this creature are attached, all larger than those shown in the engraving. The cirrhiped does not merely adhere to the skin, but in process of time actually buries itself deeply into the tissues, and would seem to cause much annoyance to the creature on which it was parasitic.

A still stranger example of these curious parasites is the BURROWING BARNACLE, two figures of which are shown deeply sunk into the skin of a whale. This species plants itself in the skin of the whales belonging to the Southern seas. This pest of the cetaceans is nearly cylindrical in shape, and remarkable for a series of raised rings, which surround it like the hoops upon a barrel. As the creature increases in age, it also increases in length, and adds ring after ring, in proportion to the depth of its imbedment in the skin. The Burrower-barnacle is found in great numbers, and actually studs the whale's skin with its shells. Not only does the skin suffer from their presence, but the blubber is also infested by them, as they often pass completely through the skin, and sink deeply into the fatty tissues beneath. I have seen several fine examples of these sunken cirrhipedes, and could not but admire the wonderful adaptation of their structure to their mode of living.

CLASS ARACHNIDA.

ANOTHER class of animated beings now comes before us, which, under the general term of Arachnida, comprises the Spiders, Scorpions, and Mites.

These beings breathe atmospheric air, they have no antennæ, and they have four pairs of legs attached to the fore parts of the body.

In some of the higher Arachnida, there is a bold division into thorax and abdomen, and the former portion of the body is clearly divided into separate segments. By the earlier naturalists, the Arachnida were placed among the insects, but may readily be distinguished by several peculiarities. In the first place, they have more than the normal number of six legs, which alone would be sufficient to separate them from insects. They have no separate head; the head and thorax being fused, as it were, into one mass, called the cephalothorax. In many of the lower species there is not even a division between the thorax and abdomen; and the body, thorax, and abdomen are merged into one uniform mass, without even a mark to show their several boundaries. They undergo no metamorphosis, like that of the insects, for, although the young Spiders change their skins several times, there is no change of form.

Beginning with the true Spiders, we find that their palpi (i.e. the jointed antennæ-like organs that project from the cephalothorax) are more or less thread-like, and in the males are swollen at the extremity into a remarkable structure, as indicative of the sex as is the beard of man, the curled tail-feathers of the drake, and the gorgeous train of the peacock. In the different genera, these palpi are differently formed, and afford valuable indications for systematic zoologists.

In the accompanying illustrations, several examples of these palpi are given, belonging to Spiders which will be described in the course of the following pages. The reader will not fail to remark their exceeding diversity of form, and the readiness with which they can be distinguished from each other. They are all much magnified, as the largest specimen is hardly equal to the head of a minikin pin. Still, their structure is not very difficult to be comprehended, and a moderately good magnifying glass will mostly be sufficiently powerful to answer the purpose. The Spiders all breathe by means of certain lung-like organs, called the pulmonary sacs, though some species are also furnished with air-tubes. These sacs communicate with the external air by means of small apertures called "stigmata," which are analogous to the spiracles of insects. There are seldom more than two of these stigmata, and never more than four.

Acrosoma arcuata.

Hypoplatea celer.

Eripus heterogaster.

Gasteracantha cancriformis.

Acrosoma. (Another species.)

Cteniza nidulans.

In these strange creatures, the mandibles are furnished with a curved claw, perforated at the extremity, something like the poison-fang of a venomous snake, and used for a similar purpose. A gland furnishes a secretion which is forced through these organs, and is injected into any object that may be wounded by the sharp claw. The fluid which is secreted for the service of the fangs is nearly colourless, and is found to possess most of the properties that exist in the venom of the rattle-snake or viper. The very existence of this fluid is denied by some writers, and its poisonous nature by others. I can, however, state from personal experience, that the bite of an angry spider inflicts a really painful injury, not very dissimilar to the sting of a wasp. I have seen a lady's hand and arm swollen so as to be hardly recognisable as belonging to the human figure, in consequence of a bite inflicted by a large Spider on the back of her hand.

PALPI OF MALE SPIDERS.

A. *Lycósa.* D. *Linyphia.*
B. *Sálticus.* E. *'Atypus.*
C. *Clubióna.* F. *Epeíra.*

They all spin those remarkable nets which we popularly call "webs," and which differ wonderfully in the various species. These webs are, in very many instances, employed as traps wherein may be caught the prey on which the Spider feeds, but in other cases are only used as houses wherein the creature can reside. Some of the uses to which these wonderful productions are put, as well as some details of their structure, will presently be mentioned.

We now pass to the typical species of these curious animals.

The Spiders belonging to the family Mygalidæ may at once be known by the shape of their mandibles and the terrible claws which proceed from them. In the greater number of Spiders, the claws are set horizontally, but in the Mygalidæ they are bent downwards, and strike the prey much as a lion clutches at his victim with his curved talons. Several species of these Spiders are known, most of which attain to considerable dimensions, and some are so enormously large as to become really formidable creatures, which man himself does not like to attack except with a weapon of some kind, or, at all events, with a shod foot.

The GREAT CRAB-SPIDER belongs to the typical genus of this family, and is one of the formidable Arachnida that are said to prey upon young birds and other small vertebrates, instead of limiting themselves to the insects, and similar beings, which constitute the food of the generality of the Spider race. All Spiders are carnivorous, the dimensions of their prey varying with those of the destroyer, and it is by no means an illogical supposition that a Spider whose spread of limb equals that of a human hand, might suck the juices of some of the smaller and more helpless vertebrates.

In Madame Merian's well-known work on the insects of Surinam, there is a careful and forcible sketch of one of these great Spiders (*Mygale avicularia*) engaged in preying upon a humming-bird, which it seems to have taken out of its nest. She gives also a description of this Spider, mentioning that it chiefly feeds upon ants, but that when they fail, it climbs the trees and catches the humming-birds. For a time this account was believed, and the Spider received the specific name of *avicularia* in consequence of its bird-catching propensities. After a while, however, several persons ventured to discredit the story, and at last both the account and the illustration were set down as simple

CRAB-SPIDER OR MATOUTOU.—*Mygale cancerides.*

fabrications of the imagination. Experiments were also tried, dead humming-birds being put into the dens of these Spiders, without any result, and the whole of Madame Merian's account was boldly denounced as fabulous.

Yet there were many observers of nature who continued to think that so painstaking a naturalist as Madame Merian, who had spent many years of her life in constant investigations, was not likely to have given so circumstantial an account without some grounds for it. That she was quite correct in saying that the Spider fed generally on ants, was conceded even by her opponents, and it was just possible that she might not be wholly incorrect in the latter part of her statement.

Moreover, they thought that the experiments were by no means conclusive, and that the natural conditions were not fulfilled. It was true enough that when a dead humming-bird was pushed into the nest of a Mygale, the creature did not attempt to eat it, but retreated to the back of its den, or tried to get away. They thought that the Mygale could not be expected to act otherwise, and that there was a vast difference between a dead humming-bird pushed into a burrow in the daytime by a huge heavy-footed biped, and a living humming-bird, asleep at night in its nest upon a tree. An animal of any kind must be left undisturbed, if the observer wishes to gain an insight into its habits; and if he deliberately violates all the conditions, he can hardly expect favourable results. If a practical naturalist wishes to learn whether the Mygale, a nocturnal being, is in the habit of visiting the trees at night and robbing the nests of the humming-birds when it could not obtain its proper supply of ants, he would hardly set to work in so clumsy a manner as to poke a dead humming-bird into the creature's burrow by day.

Surely, the only method would be to ascertain, in the first place, that the Spiders could not obtain the ants on which they usually fed, and then to watch the nests of the humming-birds at night, to see if the Mygale paid them a visit. The experiments were simply futile. Humming-birds never think of getting into subterranean burrows, and if a Mygale saw such a bird making its way into his domicile, he would be justified in running away as fast as he could from so strange a phenomenon. Lately, however, the Mygale has been seen repeatedly to kill the young, not only of the humming-bird, but of other vertebrates, and thus Madame Merian's reputation for veracity remains intact. It is true that, in one or two places, she narrates circumstances which are not true; but then she always takes care to mention that such events were related to her by a third person; and whenever she speaks of any circumstance as having been witnessed by herself, her statements may be implicitly relied upon.

As a proof of her perfect veracity on this habit of the Mygale, I will quote a passage from M. Moreau de Jonnès, who spent many years in Martinique, and watched carefully the habits of these enormous Spiders:—

"It spins no web to serve it as a dwelling. It burrows and lies in ambush in the clefts of hollow ravines, in volcanic tufas, or in decomposed lava. It often travels to a considerable distance, and conceals itself under leaves to surprise its prey, or it climbs on the branches of trees to surprise the colibris (*i.e.* humming birds) and the *certhia flaveola* (a bird allied to our common tree-creeper). It usually takes advantage of the night to attack enemies, and it is commonly on its return towards its burrow that one may meet it in the morning and catch it, when the dew, with which the plants are charged, slackens its walk.

The muscular force of the Mygale is very great, and it is particularly difficult to make it let go the objects which it has seized, even when their surface affords no purchase either to the hooks with which its tarsi are armed, or to the claws which it employs to kill the birds and the anolis (a kind of tree-lizard). The obstinacy and bitterness which it exhibits in combat cease only with its life. I have seen some which, though pierced twenty times through and through the corslet, still continued to assail their adversaries, without showing the least desire of escaping them by flight.

In the moment of danger, this Spider usually seeks a support against which it can raise itself and mark its opportunity of casting itself upon its enemies. Its four posterior feet are then fixed upon the ground; but the others, half extended, are ready to seize the animal which it is about to attack. When it darts upon it, it fastens itself upon the body with all the double hooks that terminate its feet, and stretches to attain the superior base of the head, that it may sink its talons between the cranium and the first vertebra. In some of the American insects I have recognised the same instinct of destruction.

. . . . The Mygale carries its eggs enclosed in a cocoon of white silk of a very close tissue, forming two rounded pieces, united at their border. It supports this cocoon under its corslet by means of its antennulæ, and transports it along with itself. When very much pressed by its enemies, it abandons it for an instant, but returns to take it up as soon as the combat is concluded.

The little ones are disclosed in rapid succession. They are entirely white; the first change which they undergo is the appearance of a triangular and hairy spot which forms on the centre of the upper part of the abdomen.

I had preserved from 1,800 to 2,000 of these, all of which proceeded from the same cocoon. They were all devoured in the same night by some red ants, which, guided by an instinct that set at defiance all my cares, discovered the box in which I had inclosed the spiders, and insinuated themselves into it by means of an almost imperceptible aperture, through which myriads of them passed, one by one, in the space of a few hours. It is owing, in all probability, to the destructive war waged upon the avicularia by these insects that the number of these Arachnida is confined within such narrow limits, which by no means correspond with their prodigious capability of reproduction."

The talons of the spiders are scientifically called by the appropriate name of "falces," the word being Latin, and signifying a reaping-hook. By this name they will be called in the course of the following pages. The falces of the great Crab-spiders are of

SULZER'S ATYPUS.—*'Atypus Sulzeri.* TRAP-DOOR SPIDER.—*Cteniza nidulans.*

enormous size, and when removed from the creature and set in gold, they are used as tooth-picks, being thought to possess some occult virtue, which drives away the tooth-ache.

ON the left hand of the accompanying illustration may be seen two specimens of a British example of this family. This creature is shown rather magnified, in order to exhibit its peculiar structure. The reader will not fail to notice the manner in which the eyes are mounted on a kind of pedestal or watch-tower, so as to allow the creature to see objects in its front, which would otherwise be hidden by the enormous and elevated mandibles.

This spider is rare, but has been found in several places. It frequents damp situations, and makes a rather curiously shaped burrow, nearly horizontal at its commencement, but afterwards sloping downwards. The tunnel is lined with a kind of web of white silk, very strongly compacted, which serves to prevent the earth from falling into the burrow. Part of the tube projects outside the entrance, and acts as a protection. The female places her eggs in a little cocoon also composed of white silk, and keeps them at the bottom of the burrow until the young are hatched.

The length of this creature is nearly half an inch, and its colour is reddish brown, becoming paler and more ruddy on the limbs. The male is smaller and darker than his mate.

AT the right hand of the same illustration are seen two examples of the curious TRAP-DOOR SPIDER of Jamaica, erroneously called the Tarantula.

Like the preceding species, this spider digs a burrow in the earth and lines it with a silken web, but instead of merely protecting the entrance by a portion of the silken tube, it proves itself a more complete architect by making a trap-door with a hinge that permits it to be opened and closed with admirable accuracy. The door is beautifully circular, and is made of alternate layers of earth and web, and is hinged to the lining of the tube by a band of the same silken secretion. It exactly fits the entrance of the burrow, and when closed, so precisely corresponds with the surrounding earth that it can hardly be distinguished, even when its position is pointed out. It is a strange sight to see the earth open, a little lid raised, some hairy legs protrude, and gradually the whole form of the spider show itself.

The strength of the membrane is very considerable. One of the nests in my own collection has endured a large amount of rough handling, and has yet suffered but little damage. It will permit a man's finger to be slipped into its interior, and has a very soft and silken feel to the touch. The trap-door has, however, been somewhat injured, as most of the dry earth has been shaken off, and only the layers of web left in their places. I have also several of the spiders themselves preserved in spirits, and though they have lost their colours, as is the wont of most preserved Arachnida, their falces are very perfect, and the peculiar barbed mandibles are clearly perceptible.

The mode in which these Spiders procure food seems to be by hunting at night, and, in some cases, by catching the insects that are entangled in the threads that the creature spins by the side of its house. There are several species of Trap-door Spiders, and all seem to possess similar habits. In the daytime they are very chary of opening the door of their domicile, and if the trap be raised from the outside, they run to the spot, hitch the claws of their fore feet in the silken webbing of the door, and those of the hind feet in the lining of the burrow, and so resist with all their might. The strength of the Spider is wonderfully great in proportion to its size, and few persons would anticipate the force of its resistance.

Another figure of this species is given at the right-hand side of the illustration on page 651, in order to show the creature in the act of re-entering its curious home.

The Lycosidæ, or Wolf-spiders, are all ground livers, and take their prey in fair chase instead of catching it in nets. They are mostly found among herbage, low bushes, fallen leaves, and similar localities; and if they should happen to feel alarmed, they run for safety under stones, mosses, rocks, and into any accidental crevice in the earth. The family includes an immense number of species, which are found in almost every part of the world. They are fierce and determined hunters, chasing their prey wherever it may seek shelter. Some of them are semi-aquatic in their habits, and are not only able to run fearlessly upon the surface of water, but can descend along the aquatic plants until they are deeply immersed, breathing by means of the air which is entangled among the hairy clothing of their bodies.

The large central figure on the next engraving represents the celebrated TARANTULA-SPIDER, so called from the town of Tarentum, in Italy, where this Arachnid is very plentiful.

There was a deeply-rooted belief among the inhabitants of that town and its neighbourhood, that if any one were bitten by the Tarantula he would be instantly afflicted with a singular disease called tarantismus, which exhibited itself in one of two extremes, the one being a profound and silent melancholy, and the other a continual convulsive movement of the whole body. It was also thought that this disease could only be cured by music, and that a certain tune was needful in each particular case.

The disease undoubtedly existed, and might, not improbably, be cured by music; but its source was entirely unconnected with the Tarantula. It ran through towns and villages like wildfire, drawing into its vortex hundreds of persons of both sexes who came within the sphere of its influence. The patients would leap, and dance, and wave their arms, and shriek, and sing, as if the ancient Dionysia were being re-enacted in Christian times; and, indeed, it is by no means unlikely that the frenzied gesticulations of the ancient bacchanals were attributable to a similar cause. As soon as the music ceased to play, the patients ceased to dance, and fell back into the profound stupor from which the brisk sounds had aroused them. The disease was evidently a nervous affection, tending to propagate itself, like chorea and hysteria at the present day, and, in fact, seems to be little more or less than a rather aggravated form of the former of these maladies—if, indeed, they are not different developments of the same ailment.

That the tarantismus should be cured by music, and consequent dancing, is a natural result. The patient indulged in long and continuous exercise, fell into a violent perspiration, fell exhausted, slept calmly, and awoke cured. The Spider, upon whom the odium of

WOLF-SPIDERS.

Lycósa andrenivora.   *Lycósa tarántula.*   *Lycósa saccáta.* (Female).
                                                                      Lycósa saccáta. (Male).

this strange disease rested, is perfectly innocent, being as harmless to man as any other Spider, and only formidable to the insects on which it preys.

Another species of Lycosa, inhabiting the south of France, has sometimes been confounded with the true Tarantula of Italy. The habits of this species have been carefully studied by M. Olivier, and have afforded some interesting details respecting the economy of the creature. It frequents dry and uncultivated soils, and sinks therein a little pit, of a depth varying with its size and the length of its residence. The interior of this cell is strengthened with a web. At the entrance of this burrow it sits watching for its prey, and as soon as an unfortunate insect passes within range, it darts forward, seizes it in its talons, and bears the victim away to its den, where it feasts in peace and solitude.

The female is a kind parent, and extremely fond of her eggs and young. She envelopes the eggs in silk, and forms them into a globular ball, which she always carries about with her until the young are hatched. When the time comes for the little spiders to make their appearance in the world, the mother tears open the envelope, and so aids her young to escape. As soon as they are fairly out of the egg, they transfer themselves to the body of their parent, where they cling in such numbers, that she is hardly visible under her swarming brood. They remain with their mother through the winter, and in the following spring the bonds of mutual affection are loosened, and the young disperse to seek their own living.

About sixteen or seventeen British species of Wolf-spider are already known, and it is likely that they will be found to be still more numerous.

The left-hand figure derives its name from its habit of killing the smaller bees, such as the andrena and its kin. One of these bees may be seen half buried in a convolvulus flower, where it has just been espied by the spider. It is mostly found in old pastures and commons, and its colour is extremely variable, though brown and yellow are the prevailing tints. A very common species (*Lycosa campestris*) may be seen plentifully in all meadows and pasture-lands, or even on the lawns in our gardens. It is a brisk and

3.   U U

nimble creature, running quickly along, as, indeed, is needful for a being that depends on its agility for its living.

About June, the female has made up her little packet of eggs, inclosed in a snowy-white silken envelope, and carries this burden about wherever she goes. Nothing will separate her from her eggs. If the packet is forcibly removed, she remains on the spot, hunting in every direction, and evidently suffering great distress ; and if the white ball be laid near her, she soon spies it, darts at it almost fiercely, and carries it off. Her affection for her egg-ball is, however, quite instinctive ; like the feeling which induces a hen to sit upon a piece of white chalk, which she takes for an egg. If a little bit of white cotton-wool be rolled up so as to resemble the lost egg-packet, the spider will seize it and make off with it, not at all suspecting the imposition.

There are, on an average, about one hundred eggs in each packet. They are quite round, and very tiny, like the palest yellow translucent dust shot ; and their silken covering is drawn so tight, that their globular forms give it an appearance of being embossed. The colour of this species is greenish brown, with a few little dark spots, and the body is also banded and spotted with yellowish brown of various shades.

THE species on the right hand of the illustration is also common in England. It is rather a prettier species than that which has just been mentioned, being of a rich chestnut-brown, with a longitudinal bar of yellow along the body, and a number of yellow spots on each side of the bar, where it runs over the abdomen. The colours of the male are rather duller than in the female. The cocoon of this species is yellowish brown in colour, and contains about fifty eggs. A band of slighter texture and lighter hue surrounds the cocoon.

One species of this genus, the PIRATE-SPIDER (*Lycósa pirática*), deserves a brief notice. This creature is mostly found near water, or on marshy land. It is very quick and active, and can run on the surface of the water without sinking. If alarmed, it immediately takes refuge below the surface, crawling down the stems of aquatic plants, and can remain in that position for a long time. The egg-packet contains about one hundred eggs, generally rather less, and seems to be no impediment to the activity of the mother, who can run over the water even when thus incumbered.

The colour of the Pirate-spider is rather complicated. The cephalothorax is brownish black, edged on either side by a white band, and having a dull yellow streak along its centre. Along the upper part of the abdomen runs a chestnut patch, edged with white spots, and having an arch-like mark of pure white, the point of the arch being directed towards the tail of the spider. The rest of the abdomen is simple grey-brown. The male is smaller and duller coloured than his mate.

I may here remark, that a full account of these, and many other British Arachnida, may be found in Mr. Blackwall's splendid work on this subject.

THE two handsome spiders in the next illustration are inhabitants of Britain.

The pretty *Dolomedes mirabilis* is found in well-wooded districts. We learn from Mr. Blackwall's researches, that the cocoon of this species is of a dull yellow colour, smooth within and rough without, and containing more than two hundred yellow eggs, loosely tied up in the cocoon. She carries her yellow burden under the thorax, and supports it, not only by her limbs, but by some silken threads which serve to bind it to the body. When the young are about to leave the cocoon, the mother spins a rather large silken nest among grass or low bushes. This nest is of a dome-like shape, and under its shelter the young spiders are first set free. They immediately cluster upon the silken lines spun by themselves, and remain under the dome until they are strong enough to go out into the world on their own account.

The colour of this fine spider is yellowish brown, and at each side of the body runs a deep black band, having a narrow white line along its centre. When the female has laid her eggs, she loses these fine tints, and resumes a sober grey colour.

ON the left hand of the illustration may be seen another species belonging to the same genus, which is not only one of the handsomest, but one of the largest British Spiders, its

Dolomédes fimbriátus. *Dolomédes mirábilis.*

body alone being nearly an inch in length.   Like the Pirate-spider, it is found in the vicinity of water, and seems to be most common in the fens of Cambridgeshire.   It will often descend voluntarily below the surface of the water, its respiration being conducted by means of the air-globules which cling to the mass of hairs with which its body is covered.

The cocoon of the female is brown in colour, and of considerable size, containing more than two hundred eggs.   It is carried, like that of the preceding species, under the thorax. The colour of this spider is rich dark brown, with a broad band of yellowish buff down each side, and a double row of little white spots on the abdomen.   The legs are paler, and of a more ruddy hue.   Only three species are at present acknowledged as British.

WE now come to the beautiful Hunting Spiders, a family which is spread over the world, and which has several British representatives.   They are the very chetahs, or hunting leopards, of the Spider race, and have the mottled beauty as well as the active limbs of the mammalian leopards.   They can all run fast, and have also the power of leaping upon their prey to a considerable distance.   They are mostly found upon walls, among stones, or upon leaves.

The handsome little spider that is called by the name of *Eresus cinnabarinus* is by no means common in England, and may indeed be considered as one of our rarest species. It is not very large, being only one-third of an inch in length, but its colour is extremely beautiful, the cephalothorax being deep velvety black, edged towards its hinder margin with vermilion, and the whole upper part of the abdomen being coloured with the same brilliant hue.   On the upper part of the abdomen are six square black spots, the first four being large and the last two small.   Each of these spots is edged with pure white, and their effect against the rich scarlet of the abdomen is very fine.

AT fig. C is given a specimen of the common HUNTING SPIDER of England, sometimes known by the name of Zebra-spider from its boldly-striped markings.

This little spider is extremely common, and in the summer time may be seen on almost every wall and tree trunk, busily hunting for prey.   Even upon the window-sills the Hunting Spider pursues its chase ; and as it is very bold and allows itself to be approached quite closely, its proceedings are easily watched. When it sees a fly or other insect which it thinks suitable for food, it sidles quietly in the direction of its intended victim, keeping a most careful watch, and ever drawing nearer to its prey. As the fly moves, so moves the spider, until the two beings almost seem to be urged by a common instinct. Surely and gradually it makes its way towards the unsuspecting fly, and then, with a leap so quick that the eye can scarcely follow its movements, it springs upon its prey, rolls perhaps over and over in a short struggle, and in a few moments emerges victorious from the contest, its former antagonist dead or dying in its grasp. I have witnessed such a scene hundreds of times, as the garden in which I passed many years was furnished with long ranges of old walls full of crevices that were exactly suited to the purposes of the Hunting Spider.

Even on a perpendicular wall the spider will make these leaps. It is sure not to fall to the ground, because it always draws a silken cord behind as it moves, and so, whenever

A. *Sálticus Blackwállii.*   B. *'Eresus cinnabarínus.*   C. *Sálticus scénicus.*
D. *Myrmarachna melanocéphala.*   E. *Sálticus formicárius.*

it leaps upon its prey, it is saved by its self-woven ladder, and reascends, bearing its dead victim in its grasp.

While engaged in its search, the Hunting Spider is all full of animation. It traverses the wall with great speed and in a very jerky manner, first darting this way, then running that way, then diving into a crevice, then running out and looking around. Sometimes, when it wishes to extend its sphere of vision, it raises the whole front part of the body by simply straightening the fore-legs, and it is surprising what a knowing look it assumes when in that position.

This is a handsome species to examine when under a low power of the microscope, say about twenty-five diameters. Its colour is brown, banded obliquely with white. The female does not carry her eggs with her, but wraps them in either one or two cocoons, and hides them in some secure spot, such as the crevices in rocks, and under the bark of trees. Only fifteen or sixteen eggs are placed in each cocoon.

Figs. A and E represent two more British examples of this interesting genus. The former is a really large species, measuring one-third of an inch in length. Its colour is greyish black, spotted with a darker hue, and sundry short bands of the same colour are drawn diagonally over the cephalothorax and the edges of the abdomen. A band of dull yellow is drawn along either side of the abdomen. The latter species is extremely rare, and is remarkable for its ant-like shape. The great mandibles are rather dark brown, and the front half of the cephalothorax is nearly of the same hue, but with more black. The entire centre of the body is buff, and the latter half of the abdomen is black-brown, divided from the buff by a white band.

The last species shown on the preceding illustration, at fig. D, is an example of a very remarkable spider, even more ant-like than the preceding species. Its mandibles are of very great size, and the strange shape of its attenuated abdomen can be seen by reference to the illustration. It is a native of Bengal, and is wonderfully like the mutilla, that terrible ant which has already been described on page 499. It is notable for several reasons, among which may be the fact that its head seems to be nearly distinct from the thorax, a structure quite unlike that of the arachnida, from the mygale to the cheesemite. It is thought to eat ants as well as to resemble them. The head, if it may be so called, of this curious spider is black, and the remainder is red. It is about half an inch in length.

I may mention here, that spiders, like the crustacea, are apt to be terribly quarrelsome ; and the strangest part of their nature is, that they are most combative during the season of love. In many species, especially those where the male is of insignificant dimensions compared with those of the female, all courtship is conducted under the most unexpected difficulties. A male in love is equally a male in a fright, for if his addresses are not received favourably, he runs a great chance of being eaten on the spot. And even when he has not been repulsed, he still stands in great danger ; for many of these Arachnidan beauties are as cruelly deceitful as the enchantress of the "Arabian Nights," and kill their lovers ruthlessly as soon as they have granted their prayers. So, as Alphonse Karr well remarks, the stereotyped exclamation of "Love me, or I die !" is by no means a metaphor, but a simple enunciation of a fact.

When spiders of nearly equal powers fight with each other, the battle rages vehemently, and if the weaker can escape with life, it is sure to have lost several of its limbs. As with the crustaceans, however, the deprivation is only temporary, for the severed members are reproduced ; and though they hardly seem to attain the same dimensions as the original limbs, are yet to a degree serviceable.

THE spiders belonging to the genus Thomisus are, like the hunting spiders, dependent for their subsistence on their bodily powers and activity. Some, which are rather slow of limb, are in the habit of concealing themselves under leaves or in crevices, and thence pouncing suddenly on the insects that venture too near the treacherous precincts, but the generality are active creatures, running about swiftly, and much resembling the saltici in their movements. Sometimes these creatures are popularly called Crab-spiders, because they can move in any direction without needing to turn their bodies.

At fig. A is shown one of our common species of Thomisus. It is mostly found on the ground, or lurking among the foliage of old pasture-land. In its colour, and indeed in its whole appearance, it is singularly variable, and exhibits so many differences that the simple varieties have been treated by several zoologists as distinct species.

This is one of the many species which, when young, is accustomed to take aërial excursions, and to form that delicate substance popularly known as "gossamer." There is no gossamer spider, as is generally supposed, but many species are in the habit of spinning long loose threads and allowing themselves to be wafted into the air. Lycosæ are very fond of the same curious habit. Sometimes these gossamer webs, each with its minute aëronaut, may be seen floating by thousands in the air, glittering with iridescent light as the morning sunbeams fall on them, and covering the fields with their pearl-strung threads as far as the eye can reach.

The whole question of the spider's web is very curious and interesting ; and although our fast waning space will not permit of a full description, a few lines must still be granted to these beautiful structures.

The web is produced primarily from a fluid contained within the body of the spider, and secreted within certain glands, varying in number and dimensions according to the species. Like the thread of the silkworm, this substance becomes hard on exposure to the atmosphere, and is drawn out through tubes of exceeding minuteness. In the silkworm, these spinnerets, as they are called, are two in number, but in the spider they are almost innumerable, so that the apparently single thread of the tiniest spider, minute as it may seem, and really is in fact, is composed of many hundred finer threads all collected into one strand, like the fibres of hemp in a rope. The strength obtained by this form of structure is very great, and the line is not only strong, but elastic, capable of being drawn out like an India-rubber thread and resuming its original length when the extending force is removed.

As regarding the gossamer web, Mr. Blackwall makes the following observations: " Although spiders are not provided with wings, and consequently are incapable of flying, in the strict sense of the word, yet, by the aid of their silken filaments, numerous species, belonging to various genera, are enabled to accomplish distant journeys through the atmosphere. These aërial excursions, which appear to result from an instinctive desire to migrate, are undertaken when the weather is bright and serene, particularly in autumn, both by adult and immature individuals, and are effected in the following manner.

A. *Thómisus cristátus.*   C. **Thómisus Cambridgii.**   B. *Thómisus citreus.* (Female.)
E. *Arkys lancier.*   D. *Thómisus citreus.* (Male.)

After climbing to the summits of different objects, they raise themselves still higher by straightening the limbs ; then, elevating the abdomen, by bringing it from the usual horizontal position into one almost perpendicular, they emit from their spinners a small quantity of viscid fluid, which is drawn out into fine lines by the ascending current occasioned by the rarefaction of the air contiguous to the heated ground. Against these lines the current of rarefied air impinges, till the animals, feeling themselves acted upon with sufficient force, quit their hold of the objects on which they stand, and mount aloft.

The webs named gossamer are composed of lines spun by spiders, which, on being brought into contact by the mechanical action of gentle airs, adhere together till by continual additions they are accumulated into irregular white flakes and masses of considerable magnitude. Occasionally, spiders may be found on gossamer webs after an ascending current of rarefied air has separated them from the objects to which they were attached, and has raised them into the atmosphere ; but as they never make use of them intentionally in the performance of their aëronautic expeditions, it must always be regarded as a fortuitous circumstance."

The same writer also remarks that the various directions in which these gossamers are known to sail is in no way attributable to the will of the spider, but merely to the currents of air through which the webs float. He also reviews the opposite opinions regarding the production of the first lines of the web. Some writers say that the spider has the power of projecting its threads in any direction which it may choose, while others assert that it has no such power, and that the creature is forced to wait for a current of air which can bear the slender thread on its breath. After noticing the arguments and experiments on both sides of the question, he comes to the conclusion that the spider is indebted to the air and not to its own projectile capabilities.

Strong and elastic as these webs may be, they have never yet been put to any useful purpose, save to check the bleeding of a cut finger, or to form the cross-wires of an astronomer's telescope. The thread of many species is suitable enough for manufacture, but it cannot be supplied in sufficient quantities. Spiders cannot be kept in any number, as they would be always fighting and eating each other; and they are so voracious that they could not be properly furnished with food, flies being difficult to catch in many parts of the year, and in the cold months quite unattainable. As a proof that if the web could only be obtained in sufficient quantity it might be woven into various articles of apparel, there are now in existence several pairs of gloves, stockings, and other fabrics that have been made, though with very great difficulty, from this substance.

THE odd-looking spider called *Arkys lancier* is seen on the preceding illustration, at fig. E. It is a native of Brazil and the surrounding countries. The cephalothorax of the spider is orange-yellow, with a line drawn transversely over it, and changing to a brilliant red at the point on each side. The round spots on the abdomen are bright yellow; the hinder feet are covered only with short down, but those in front are furnished with strong spines.

To return to the illustration on page 662, and to our account of the genus Thomisus.

At figs. B and D are represented the two sexes of *Thomisus citreus*, for the purpose of showing the great difference in their dimensions and general shape, the female being twice as long as the male, and, as a necessary consequence, very much larger in cubic dimensions. This species is tolerably common, and is usually found on flowers, whether growing in gardens or in the field.

The female is a light citron-yellow, with some dark streaks on the cephalothorax, and a double row of round dark spots on the upper part of the abdomen. The yellow colour extends over all the limbs. The male, on the contrary, is light leafy green, with two black bands running down the abdomen, and a darker streak on each side of the cephalo-thorax approaching to brown. The first and second pairs of legs are dark chestnut-brown, while the others are green like the body, so that it is a very pretty-coloured creature, and so unlike the female that few persons would believe it to belong to the same species.

ON the next illustration we have more examples of this active family. At figs. A and C are given the two sexes of the same species, in order to show their dissimilarity, a phenomenon very frequent among spiders, and which renders the task of arranging them correctly a matter of no slight difficulty.

This is a very active spider, and can run swiftly even upon polished substances. It is found in well-wooded districts, and is remarkable for the speed with which it runs. The cocoon made by the female is rather large, being nearly a quarter of an inch in diameter, and containing about seventy pale yellow eggs laid loosely in a white cell. This cocoon is not carried about by the female, but is lodged in a larger cell of dull white silk; and this cell is generally placed within a leaf, the edges of which are drawn together by stout lines of the same silken fabric. A dead and already withered leaf is chosen for this purpose.

The colour is quite different in the two sexes. The female is rather prettily marked with brownish chocolate upon a ground colour of reddish yellow, while the male is deep black-brown, with a curious scribbled pattern of a paler hue along the back. The specific name of " dispar," or unlike, is given to the spider on account of this dissimilarity. It is worthy of notice, however, that in the immature state the colours are alike in both sexes. The reader will doubtlessly remember that this is the case with many birds, and that even when the adult male glows with all the hues of the rainbow and the adult female wears a mere dress of sober brown, black, and grey, the young birds are so similar in their plumage that it is hardly possible to distinguish one sex from another.

In the species represented at fig. D the two sexes are coloured in nearly the same manner, and the male is chiefly to be distinguished from his mate by the smaller extremities of the palpi.

Our last example of this genus, drawn at fig. E, is a small but rather pretty species, in which the male is rather smaller and slightly darker than the female. The cocoon of

A *Philódromus dispar.* (Male.)   B. *Sparassus smarágdulus.*   C. *Philódromus dispar.* (Female.)
D. *Philódromus oblongus.*   E. *Philódromus pállidus.*

this species is slightly made, and white in colour, and contains a large number of little spherical eggs, not adhering to each other. The colour is pale greyish brown, profusely speckled with tiny black dots, and marked in a very peculiar manner with dark chocolate-brown. On the upper part of the cephalothorax there is a large and nearly triangular patch of this colour, with a point directed towards the tail, and around it are arranged several short streaks all converging towards its point. At the end of the abdomen a number of similar stripes are drawn, but without the triangular patch.

At fig. B is given a specimen belonging to another genus, and being a certainly remarkable insect. The sexes are wonderfully dissimilar, but instead of one sex being brilliantly coloured, and the other only tinted with dull hues, as is mostly the case, both sexes are equally beautiful, though with boldly contrasting colours. This difference of hue is only in the adult spider, as, when immature, the male and female are coloured alike.

This spider is more than half an inch in length, and is found in tolerable plenty in the south of England, its beautiful colours rendering it very conspicuous. The adult female is pale green, with some darker stripes painted, as it were, upon the upper surface of the cephalothorax, and all drawn from the sides towards the centre; while along the middle of the abdomen runs a deep green streak, edged with greenish white. The male, who is smaller than his mate, has the whole front of the body coloured like that of the female. But the abdomen is totally different. The ground colour is pinky cream, speckled with brown, and three broad crimson bands are drawn longitudinally throughout its entire length, the central band having several protuberances at intervals.

THE family of the Drassidæ is spread over the greater part of the world, and has many representatives within the British Isles. They all have a rather remarkable habit of concealing themselves, not in holes or crevices, but in silken cells, spun by themselves among leaves, under stones, in chinks of walls, and, in fine, wherever their instinct leads them. They are active creatures, and catch their prey by fair chase, in one instance even pursuing the victim beneath the surface of the water.

A. *Clubióna holosericea.* (Male.)
B. *Drassus cúpreus.*
C. *Drassus cúpreus.* (Nest in dried leaf.)
D. *Clubióna holosericea.* (Female.
E. *Drassus micans.* (Eggs.)
F. *Drassus micans.*
G. *Clotho Durándti.*

The species shown at fig. B never attains to any great size, two-fifths of an inch seeming to be the ordinary length of a female, the male, of course, measuring rather less. It generally resides under stones in many parts of Great Britain, and but for that habit would be seen oftener than is now the case. The cocoon of the female is pure white, and contains rather more than one hundred eggs of a very pale yellow colour. The cocoon is then placed in a larger sac, also made of white silk, and placed in a hole in the ground. The mother spider generally includes herself in this second sac. The colour is alike in both sexes, being of a pale reddish brown.

At fig. F is shown another species of the same genus, and at fig. E is drawn a portion of a twig, showing the manner in which the female deposits her eggs. The reader will probably have seen these curious little egg pyramids on the branches of various trees. This species is very small, the male measuring barely the eighth of an inch in length. It is rather prettily coloured. The limbs are buff, with a large patch of chocolate-brown on the first two pairs. The cephalothorax is ruddy leaf-brown, with six white streaks, their points converging to a spot in the central line. The abdomen is deep black in the shade. but when the light shines upon it, various iridescent hues of purple, green, and copper are given out, rendering the creature a really beautiful species. The name of "micans," or glittering, is applied to the creature on account of its changing colours.

AT figs. A and D are shown the two sexes of a curious and prettily marked spider of moderate dimensions, the female measuring nearly half an inch in length.

This species is mostly found in well-wooded districts, living in a pretty white silken house, which it spins under the shelter of rough bark or shady leaves. The cocoon containing the eggs is placed in this cell, and affectionately tended by the parent. The cocoon is also made of white silk, and generally contains rather more than one hundred spherical eggs. These are very pale yellow in colour, and laid loosely in the cocoon. The cephalothorax of this species is pale dull green, and the abdomen is soft silken grey, with a peculiar velvety lustre, produced by the dense clothing of hair with which it is covered. The specific name, "holosericea," signifies silken, and is therefore very appropriately given to the species.

Another species of this genus is shown in the illustration on page 669. It is rather larger than the generality of the Clubionas, being nearly three-quarters of an inch in length. The reader must understand that the length is exclusive of the limbs, and is measured from the front of the cephalothorax to the end of the abdomen. This is a very rare creature, having only once been taken in England. This solitary specimen was found near Cheltenham, but it is likely that others must be in the country, inasmuch as its presence certainly indicates the existence of parents, and probably of relations. The only other supposition is, that a solitary specimen was by accident brought over from the Continent, found its way to Cheltenham, and then happened to fall into the hands of some one who was acquainted with its zoological value.

THE last example upon this illustration (fig. G) is a really remarkable creature, whose habits have been studied by M. Dufour. That careful naturalist found it in the Pyrenees, in Catalonia, and in the mountains of Narbonne. It has also been discovered in Egypt and Dalmatia.

Of this Arachnidan, M. Dufour gives a most interesting description, from which the following passage is extracted:—

"It makes at the inferior surface of large stones, and in the clefts of rocks, a cocoon, in the form of a cap, or little dish, a good inch in diameter. Its contour presents seven or eight emarginations, of which the angles alone are fixed upon the stone, by means of bundles of thread, while the edges are free. This singular tent is of an admirable texture: the exterior resembles the finest taffetas, composed, according to the age of the worker, of a greater or less number of doublings.

Thus, when the Uroctea (another name for the Clotho), as yet young, commences to establish its retreat, it only fabricates two webs, between which it remains in shelter. Subsequently, and, I believe, at each moulting, it adds a certain number of doubles. Finally, when the period marked for reproduction arrives, it weaves a cell for this very purpose, more downy and soft, where the sacs of eggs, and the young ones newly disclosed, are to be shut up. Although the external cap or pavilion is designedly, without doubt, more or less soiled by foreign bodies, which serve to conceal its presence, the apartments of the industrious fabrication are always scrupulously clean.

The pouches which inclose the eggs are four, five, or even six, for each habitation, which, nevertheless, forms but a single habitation. These pouches are of a lentacular form, and are more than four lines in diameter; they are formed of a kind of taffeta as white as snow, and furnished internally with a down of the finest kind. It is only at the end of December, or in the month of January, that the laying of the eggs takes place; it was therefore necessary, beforehand, to provide for the defence of their progeny against both the rigour of the season and hostile incursions. Everything of this kind has been carefully done. The receptacle of this precious deposit is separated from the web, immediately applied upon the stone, by a soft down, and from the external cap by the various stones of which I have spoken.

Among the emarginations which border the tent, some are altogether closed by the continuity of the stuff; others have their edges simply lapped over, so that the Uroctea, by raising them, may issue at will from the tent, and re-enter. When it quits its domicile to proceed to the chase, it has little cause to fear that its habitation should be invaded, for itself alone possesses the secret of the impenetrable emarginations, and the key to those by which it can introduce itself.

When the young ones are in a state to do without maternal cares, they take their departure and proceed to establish elsewhere their particular habitations, while the mother dies in her own tent. Thus the last is at once the cradle and the tomb of the Uroctea."

The colour of the Clotho is chestnut-brown on the cephalothorax and black on the abdomen, which is marked with five round, yellowish spots. It is rather less than half an inch in length.

THE curious and interesting WATER SPIDER is now far better known than was formerly the case, as the numerous aquaria that have been established over the kingdom have tended to familiarize us with this as well as with many other inhabitants of the water.

This creature leads a strange life. Though a really terrestrial being, and needing to respire atmospheric air, it passes nearly the whole of its life in the water, and, for the greater part of its time, is submerged below the surface. To a lesser degree, several other spiders lead a somewhat similar life, sustaining existence by means of the air which is entangled in the hairs which clothe the body. Their submerged existence is, however, only accidental, while in the Water Spider it forms the constant habit of its life.

Like the pirate-spider, this creature is purposely covered with hairs, which serve to entangle a large comparative amount of atmospheric air, but it has other means which are not possessed by the species already described. It has the power of diving below the surface, and carrying with it a very large bubble of air that is held in its place by the hind-legs; and in spite of this obstacle to its progress, it can pass through the water with tolerable speed.

The strangest part in the economy of this creature is, that it is actually hatched under water, and lies submerged for a considerable time before it ever sees the land. At some little depth the mother spider spins a kind of egg or dome-shaped cell, with the opening downwards. Having made this chamber, she ascends to the surface, and there charges her whole body with air, arranging her hind-legs in such a manner that the great bubble cannot escape. She then dives into the water, proceeds to her nest, and discharges the bubble into it. A quantity of water is thus displaced, and the upper part of the cell is filled with air.

WATER SPIDER.—*Argyroneta aquatica.*

She then returns for a second supply, and so proceeds until the nest is full of air.

In this curious domicile the spider lives, and is thus able to deposit and to hatch her eggs under the water without even wetting them. The reader will have noticed the exact analogy between this sub-aquatic residence and the diving-bell, now so generally employed. As to the spider itself, it is never wet; and though it may be seen swimming rapidly about in the water, yet the moment it emerges from the surface, its hairy body will be found as dry as that of any land spider. The reason for this phenomenon is, that the minute bubbles of air which always cling to the furred body repel the water and prevent it from moistening the skin.

The eggs of this spider are inclosed in a kind of cap-shaped cocoon, not unlike the cover of a circular vegetable dish. This cocoon may be seen in the illustration in the upper part of the cell. It usually contains about a hundred little spherical eggs, which are not glued together.

The Water Spider is a truly active creature, and its rapid movements can be watched by means of placing one of these Arachnida in a vessel nearly filled with water. If possible, some water plant, such as the vallisneria, or anacharis, should be also placed in the vessel. Here the spider will soon construct its web, and exhibit its curious habits. It must be well supplied with flies and other insects thrown into the water. It will pounce on them, carry them to its house, and there eat them.

It is a tolerably common species in England, being especially fond of inhabiting quiet and rather deep ditches, where it is well sheltered, and the stream is not rapid enough to endanger the security of its domicile. It is necessary that the water plants to which the nest is fixed should be sufficiently firm to prevent the nest from being swayed on one side, as, in that case, the air would escape, and the water make its entrance. Owing to the vast number of these spiders that have been sent to the London markets, it is not now nearly so plentiful as used to be the case, and indeed has been almost extirpated from several localities where it was formerly seen in great numbers. Near Oxford it was tolerably plentiful in the ditches that drained the fields into the Cherwell and Isis, and I have often watched its active movements through the water. Whenever it swims, it always keeps its head downwards, just as is the case with a human diver, and it urges itself through the water with quick smart strides of its hairy legs.

The limbs and cephalothorax of this species are brown, with a slight tinge of red; and the abdomen is brown, but washed with green. It is densely covered with hairs. On the middle of the upper surface of the abdomen are found round spots arranged in a square. The male is rather larger than the female, and his legs are larger in proportion. He may, however, be distinguished by the large mandibles and longer palpi.

OF the species represented in the next illustration, fig. A has already been described. The rest belong to the family of the Ciniflonidæ.

All these spiders are fond of residing in crevices in rocks, walls, and stones, or under leaves, or sheltered by old projecting bark; and near their hiding-place they weave nets of a most elaborate structure, not flat like those of the common garden-spider, but inclosing spaces of considerable size in comparison with the small dimensions of their architects. These webs are woven chiefly by means of a peculiar apparatus on the hinder legs, consisting of two rows of parallel and moveable spines. The web is most intricate in its arrangements, and connected with the hiding-place of the spider by means of a silken tunnel of variable length, through which the creature darts when it feels the vibration of an insect in its web, and to the bottom of which it retreats if it apprehends danger. Sometimes the spider makes more than one of these tubes.

Several species of Ciniflo are very plentiful in England, and may be found hidden in their dark silken caverns even in houses. Cellars often contain them, and they frequently swarm in the belfries of old churches. They are extremely ferocious, and mostly seem to be hungry, killing fly after fly with untiring assiduity.

The *Ciniflo ferox* is moderately plentiful, and may be found in old buildings, especially in the dark crevices behind the windows, and under stones. Its length is a little under half an inch. The cephalothorax is heart-shaped, of a pale yellowish brown, and clothed thinly with long black hairs. The abdomen is dark brown, and is variegated with buff markings of a shape not easy to describe, but which can be seen by reference to the illustration.

At figs. C and D are shown both sexes of a small, but interesting spider, termed *Ergatis benigna*. It is not unfrequent upon heaths and commons, and makes an irregular web at the tips of the gorse and heather. This web passes from one twig to several others, and is studded with the bodies of the captured prey.   Within the web the female spider places her cocoons, which are two or three in number, dish-shaped, and are fastened to the stems of the plants upon which the web is built.   There are about thirty eggs in each cocoon, and they may mostly be found about June.

The colour of the female is very dark brown, upon which is described a bold pattern of buff.   The male is smaller, darker, and the markings on the body are of a duller hue.   Fierce as is this little creature in its own way, it often falls a victim to the voracious asilidæ, or hornet-flies, which completely reverse the usual order of things, and instead of being devoured by the spider act the part of its destroyers.   The soft skin of this spider is easily pierced by the jaw-lancets of the harvest-fly, and, owing to this structure, the poor little spider learns practically the discomfort of being eaten.

A. *Clubióna nutrix.*   C. *'Ergatis benigna.* (Female.
B. *Ciniflo ferox.*   D. *'Ergatis benigna.* (Male.)

WE now come to some members of another family.

The pretty species of Agelena, which is shown in the next illustration at fig. C, is found plentifully on heaths and commons, and derives its specific name from the complicated nature of its web.   This is a very large structure when compared with the dimensions of its architect, and is spread almost horizontally over the tops of furze, heath, and the other plants which are found on commons.   It is a tolerably massive web, and well calculated for catching prey.   Unlike the garden-spider, which boldly sits in the middle of the web, trusting to the delicate meshes escaping the eyes of flying insects, the Agelena does not trust itself in sight, but sits in its dark cavern, which communicates with the web by means of a silken tunnel.

The female makes several dish-shaped cocoons, in which are contained a large, but variable number of eggs.   These cocoons are pure white, and are attached to the grasses, heath, or other herbage of the locality.   The eggs are spherical, of a pale yellow colour, and rather large in proportion to the dimensions of the spider.   As if in order to conceal the cocoon, and deprive its exterior of the snowy whiteness which might attract the eye of an enemy, the mother takes care to fix little bits of dead leaves, particles of earth, and other substances, upon the outermost layers.

This spider is prettily marked.   The ground colour is deep chestnut-brown, upon which is drawn a series of buff markings, with their edges as well defined, and their arrangement as carefully balanced, as if they were executed with a pencil.   The male is darker than the female, and the marks on the abdomen are of a lighter and brighter hue. Its average length is half an inch.

At figs. A and B are seen specimens of another genus, the two sexes being given.   This fine spider is mostly found in old houses, chiefly haunting the corners, and spinning a thick horizontal sheet of web, and forming a rather stout silken tube as a communication between the web and the den where the spider sits and watches for its prey.   In the old college buildings at Oxford, this species is very common; and the well-known Cardinal-spider, which is found at Hampton Court, and which frightens ignorant persons by its large size, is probably a variety of this species.   The web is always very completely supported by guy ropes, which are laid with the greatest care, and disposed as artistically as if arranged by a professional architect.

Like the preceding species, this spider makes several dish-shaped cocoons, surrounds them with a coarse envelope, and covers the white silken cases with bits of old mortar, fragments of wood, particles of whitewash, or any other substance that can be easily obtained.   These cocoons are to be seen in June and July.

The colour of this species is ruddy brown, dark brown, and buff, the first tint being drawn in a broad band along the whole body, the second being the ground tint, and the third being formed with rows of spots on each side of the central line.   The limbs are banded with reddish brown and black.   The male is smaller than the female, darker coloured, and has his legs proportionately longer.   The average length of the body is

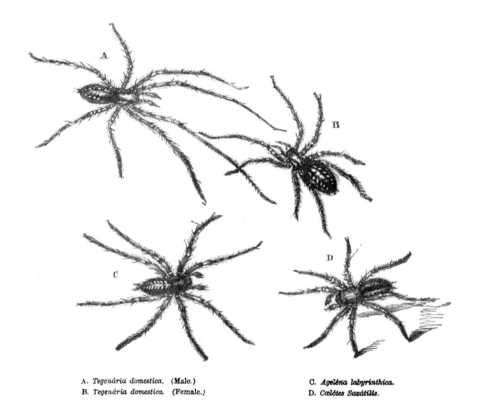

A. *Tegenária domestica.*   (Male.)
B. *Tegenária domestica.*   (Female.)

C. *Ageléna labyrínthica.*
D. *Cœlótes Saxátilis.*

nearly three-quarters of an inch, so that the spread of limb is, in a fine specimen, very considerable.   This species occasions dire tumults in the minds of housemaids, who sweep away the webs with ruthless broom, and give the spider no hope of a peaceful home.

There are several British species of Tegenaria, all having very similar habits.   In one species, it has been found that the spider changes its skin, or moults, nine times before they reach adult age, the first moult being achieved while in the cocoon, and the remaining eight after the young have left their silken cradle.   It has, moreover, been proved by actual experiment, that a limb may be removed at the joint and renewed many times in succession, the new limb appearing after the next moult.   The life of this species averages four years.

We now pass to the last figure in this illustration, which represents a member of another family.

This pretty spider is rather more than half an inch in length, but owing to the shortness of the limbs, it does not present so formidable an aspect as many of less real dimensions. This species was only known to be British in 1826, when it was discovered beneath rough fragments of rock upon Snowdon. Others, however, have been found in many parts of the kingdom; and although it is not a common species, it cannot lay claim to any great rarity. The female of this species makes a curious and rather large cocoon of a dish-like shape, measuring nearly half an inch in diameter, and attaches it firmly to the under surface of stones by means of a series of strong silken threads. The cocoon is found about May and June. The outer surface of the cocoon is rather profusely studded with patches of mud, in all probability to take off their too great brilliancy.

The colour of this species is simple, but pleasing. The cephalothorax and limbs are reddish brown, and the abdomen is yellow, over which is drawn a broad black streak, narrowing as it approaches the extremity of the abdomen.

A. *Theridion lineátum.* (Female.)   B. *Theridion lineátum.* (Nest.)   C. *Theridion lineátum.* (Male.)
D. *Theridion denticulátum.*   E. *Theridion nervósum.*   F. *Theridion lepidanórum.*

The very large genus, Theridion, of which several species are given in the accompanying illustration, is spread over the greater part of the world, and has many British examples. These spiders are mostly of small dimensions, some being extremely minute. Several of the tiny spiders, popularly called Money-spinners, belong to this genus; and fortunately for themselves, they are protected from destruction by the prevailing notion that it is unlucky to kill a money-spinner.

Their habits are extremely various. Some live under stones, others are found in houses, hiding themselves in shady corners, while the greater number prefer the open air, the fields, and commons, and spin their elaborate webs over the bushes, trees, or herbage.

One species (*Theridion benignum*), so called by M. Walckenaer, because the two sexes live peaceably with each other, contrary to the usual custom among Arachnida, is very common in gardens, and about autumn is especially conspicuous. It has a great liking for the grape-vine, and its webs may often be found covering the clusters of grapes. These webs should be sedulously protected, for, though slight in texture, they are an admirable protection to the grape, intercepting the many insects that love to prey upon that delicate fruit. In some places it is so plentiful that almost every bunch of grapes has its protecting network thrown over it.

A rather large species inhabiting Corsica, and known by the name of MARMIGNATTO, or MARMAGNATTO (*Theridion tredecim-guttatum*), seems to be rather a formidable creature, its bite causing much pain even to man, and, according to Rossi, inducing most serious symptoms, which are only removable by sharp treatment and copious perspiration. It lives in the open fields, and preys mostly upon insects of the grasshopper kind, stretching long threads across the furrows, which serve to entangle the feet of the active insect, and enable the slower Arachnida to make sure of its victims. When the spider finds a locust thus entangled, it further secures the struggling insect by fresh threads spun over its feet and legs; and when it has fairly bound all its limbs, it mounts upon its victim and inflicts a fatal wound at the junction of the head with the neck. As soon as the locust has received the bite, it is attacked with a violent convulsion through its whole frame, and dies almost instantaneously.

This action seems to be universal throughout the Theridia, wherever the spider attacks a large and powerful insect. In Webber's "Song Birds of America," there is an animated account of a battle between a large cockroach and a spider, which seems to belong to this genus. In this case, the cockroach struggled furiously, and was nearly escaping, had not the little spider bethought itself of a new manœuvre. "We had noticed him frequently attempting to bite through the sheath armour of the cockroach, but he seemed to have failed in piercing it. He now seemed determined to catch the two fore-legs that were free. After twenty trials at least, he noosed one of them, and soon had it under his control. This pair of legs was much more delicate than the others; he instantly bit through the captured one.

The poison was not sufficient to affect the large mass of the cockroach a great deal, but the leg seemed to give it much pain, and it bent its head forward to caress the wound with its jaws; and now the object of the cunning spider was apparent. He ran instantly to the old position he had been routed from on the back of the neck, and, while the cockroach was employed in soothing the smart of the bite, he succeeded in enveloping the head from the back in such a way as to prevent the cockroach from straightening it out again, and, in a little while more, had him bound in that position, and entirely surrounded by the web. A few more last agonies and the cockroach was dead, for the neck, bent forward in this way, exposed a vital part beneath the sheath; and we left the spider quietly luxuriating upon the fruit of his weary contest. This battle between·brute force and subtle sagacity lasted one hour and a half."

The colour of the Marmignatto is deep black, with thirteen round spots on the abdomen, one spot being blood-red.

Another Theridion has been seen to catch its prey in a somewhat similar manner, netting the insect in its silken toils, spinning thread after thread, and binding it tighter and tighter to the spot, and at last killing it when fairly tied down, and then carrying it off to its domicile. Another species of this genus is given in the ensuing illustration.

THE genus Linyphia has also many British representatives, and, as in the preceding genus, the generality of these spiders are of very small dimensions. One species (*Linyphia triangularis*) is very plentiful, and towards the end of summer or the beginning of autumn, its webs may be seen stretching across the branches. Though but a very little spider, not so large as a grain of rice, it makes webs of wide spread, laid horizontally, and carefully sustained by guy ropes attached to different objects around. Sometimes the guy ropes are so strong, and their elasticity so great, that they actually draw the net out of its flat horizontal direction, and make it swell into a very shallow dome.

The structure of the web is rather loose, and the fibres are necessarily very slender but is yet strong enough to arrest and detain tolerably large insects. The spider generally remains near the middle of and below the web, and as soon as a passing insect becomes entangled in the treacherous meshes the spider runs nimbly to the spot, wounds the insect through the web, and so kills it. The next move is to bite a hole in the web, pull the dead insect through, and then to suck the juices from its body.

When a number of these spiders are confined in the same box, they fight fiercely and kill each other unmercifully; but in contradiction to this combative disposition, the male is able to woo his intended mate without fearing death from her fangs. The cocoon of the spider is mostly of a rather loose structure and is placed near the web. The eggs are rather numerous, placed loosely in the cocoon, and of a pale reddish brown. They may be found at two periods of the year, June and September.

No small skill is required in placing the eggs within the cocoon, which in many species is as round as a cricket ball, and is so tightly drawn round the eggs that their outlines are visible through the web. This operation is conducted entirely by touch

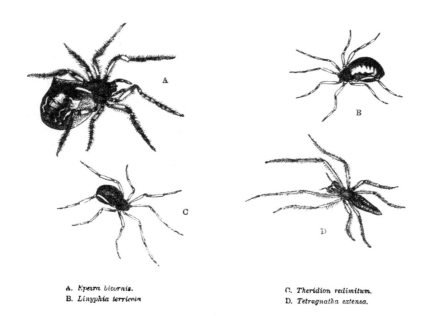

A. *Epeira bicornis.*
B. *Linyphia terricola*
C. *Theridion redimitum.*
D. *Tetragnatha extensa.*

without the aid of the eyes, and is achieved in the following manner :—The spider first spins a circular plate of silk, and then adds a raised border, so as to form a hemispherical cup. In this cup she then places the eggs, not only filling it, but piling them up with admirable regularity. She then binds them in their places by cross threads, and lastly spins a complete covering, thus forming the whole mass into a spherical ball.

THE curious spider represented at the lower right-hand corner of the illustration is called the *Tetragnatha.* In this spider the jaws are very large, long, widened towards their tips, and diverging from each other. The eyes are nearly of the same size, and are arranged in two regular lines, nearly parallel to each other. The web which this creature spins is vertical, like that of the garden-spider.

WE now arrive at the Epeiridæ, a family containing some of the strangest members of the spider race. The best known of this family is the common GARDEN-SPIDER, some-times called the CROSS-SPIDER, from the marks upon its abdomen. This is thought to be the best typical example of all the Arachnidæ. It is found in great numbers in our gardens, stretching its beautiful webs perpendicularly from branch to branch, and remaining in the centre with its head downwards, waiting for its prey. This attitude is tolerably universal among spiders; and it is rather curious that the Arachnidæ should reverse the usual order of things, and assume an inverted position when they desire to repose.

The web of this spider is composed of two different kinds of threads, the radiating and supporting threads being strong and of simple texture. But the fine spiral thread which divides the web into a series of steps, decreasing in breadth towards the centre, is studded with a vast amount of little globules, which give to the web its peculiar adhesiveness. These globules are too small to be perceptible to the unassisted eye, but by the aid of a microscope they may be examined without difficulty. In an ordinary web, such as is usually seen in gardens, there will be about eighty-seven thousand of these globules, and yet the web can be completed in less than three-quarters of an hour. The globules are loosely strung upon the lines, and when they are rubbed off, the thread is no longer adhesive.

Many interesting circumstances can be narrated of this spider, but our space will not permit of more than a brief description. Several species of Epeira are inhabitants of England and have different habits. The following account of an Epeira and its web is given by the Rev. D. Landsborough, in his "Excursions to Arran":—

"As he was rather a gigantic spider, his tent, instead of being on the ground, was elevated, like the house of a giant of whom in early life we have all read. It was built on the tops of the common grass, *Holcus lanatus*, more than a foot above the ground. Had he built his house on the top of one stalk of grass, the house and its inhabitant might have borne down a single slender stalk. But he had contrived to bring together several heads whose roots stood apart, and, with cordage which he could furnish at will, had bound them firmly together, so that his elevated habitation was anchored on all sides. From whatever *airt* the wind blew it had at once halser and stay. Not only did he bind the heads together, but he bent, doubled, and fastened them down as a thatch roof, under which his habitation was suspended.

As he was a larger spider than usual, his house was large; the more capacious apartment, which I believe was the nursery, being below; and the smaller one, which was his observatory or watch-tower, being above, from which he could pounce upon his prey, or, in case of hostile attack, could make his escape by a postern gate, so as to conceal himself among the grass.

During my visit in June last, I was anxious, as we returned from Whiting Bay, to ascertain whether this interesting colony of tent-makers was still in a thriving state, and not seeing any at first, I began to fear that a Highland clearance had taken place. When I at last discovered a few of them, I saw that, as there are times of low trade among our industrious two-footed artisans in towns, so are there occasionally hard times among our six-footed operatives in the country. The field in which they encamped had, I suppose, been overstocked. The stately *Holcus* had been eaten down; but these shifty children of the mist had availed themselves of the heather, doubling down the tops of some of the heath-sprigs, and under this thatched canopy forming their suspension-tabernacles. As yet, however, it was too early in the season. The house had only one apartment; the web of which it was formed was as yet thin, so that through it I could see the spider, which, being but half-grown, had not yet got in perfection its fine tiger-like markings. 'Go to the ant, thou sluggard;' go also to the spider. He who taught the one taught the other; and learning humility, let both teach thee."

IF the reader will refer to the large illustration on page 651, he will see several strange-looking creatures, having their bodies covered with points, knobs, and spines in a most formidable array. These curious spiders inhabit several of the hot parts of the earth, and are remarkable for the extreme hardness of their skin and the brilliancy of their colouring. The skin of these arachnids is as hard and firm as the shelly armour of the crustaceans, and really startling to the touch. There is, however, one spider, the

Sclerarachne, which even surpasses them in the hardness of its skin. This is a very small species, with six eyes, a native of Cuba, and evidently forms one of the links between the true spiders and the mites. The name Sclerarachne is of Greek origin, and literally signifies "hard-spider."

During their life-time these spiders literally glitter with resplendent hues, and gleam like living gems set in the deep verdure of the forests. Crimson, azure, emerald, and purple adorn these remarkable Arachnida, and in several species the skin looks exactly as if it were made of burnished gold and silver. After death these glaring colours vanish and change into dull browns and blacks, but in many cases a few relics of the former

GARDEN OR CROSS SPIDER.—*Epeira diadema.*

beauty are still discernible, especially in those specimens where the surface once glittered with metallic radiance.

A collection of spiders belonging to the genera Acrosoma and Gasteracantha presents a most extraordinary appearance. There seems to be no bound to the variety of spines and spikes with which the bodies of these creatures are armed; and had it not been for the lack of space, a few illustrations would have been wholly filled with their strange and weird-like forms. The object of these appendages is quite unknown. Some writers have suggested that they may be intended as defensive armour, and given for the purpose of deterring birds from eating them. But this opinion is quite untenable, as there is no reason why they should be thus guarded more than any other spiders. Indeed, this is another of the many mysteries of zoology, which will never be unveiled until we learn to look beneath the surface and to inquire not only the object of a colour or formation, but its meaning.

IN the illustration of the GARDEN-SPIDER, both sexes are given, in order to show their different shape and comparative dimensions. The female is one of the fiercest Amazons of the spider race; and in case she should object to the attentions of her intended spouse, he must needs flee for his life, a feat which he generally performs by flinging himself out of the web and lowering himself quickly to earth with his silken ladder.   This creature is called the Cross-spider, because the back of its abdomen is marked with a triple yellow cross upon a dark brown band that runs along the central line of the body.

IN this illustration we have examples of some remarkable spiders, one of which presents a very strange appearance.   This is the NOPS, an arachnid which has only two eyes instead of eight or six, but in which these organs are so enormously large that their dimensions compensate for their paucity of number.   On the front portion of the cephalo-thorax there is a black spot, and on this spot are seated the two eyes, round, globular, black, and brilliant.   It is one of the hard-skinned species, and appears to be allied both to Gasteracantha and Epeira, in spite of its two eyes.

It is mostly found under stones in woods, and in such localities is tolerably plentiful, but is very rare in houses, though it does sometimes make its appearance in the dwellings. The colouring of the Nops is very simple, the cephalothorax being ruddy brown and the

Nops Guanabacoæ.   Otiothops Walckenaeri.   Dysdera erythrina.

abdomen dark brown.   It is not a large species considering that it lives in a hot country, measuring rather less than half an inch in length.   It is believed to be the only known spider that possesses only two eyes.   This species is an inhabitant of Cuba.

On the same illustration is seen another remarkable arachnid, the OTIOTHOPS, which is especially notable from the fact that its two hinder eyes are united together.   This spider is a native of Cuba, and is generally found under stones in well-wooded places, and, like the last-mentioned species, has a hard and shelly skin.   In length it is rather under half an inch.

WITH the exception of the curious spiders just mentioned, the species which have been described bear eight eyes, or rather ocelli, very like the organs of the same name in insects, and arranged upon the cephalothorax in various patterns.   One well-known writer on the Arachnida has based his system entirely upon the number and arrangement of these ocelli; but the zoologists of the present day seem to think that such a system is insufficient for such a purpose, though very useful—and, indeed, palpably so—as a subordinate means of arrangement.   The next group of spiders are in reality separated by the fact that they possess only six eyes, and are therefore called Senoculata, or Six-eyed Arachnida, the preceding belonging to the group of Octonoculina, or Eight-eyed Arachnida.

This species of Dysdera has lately attracted much attention, for, although it is properly a native of Southern Germany and the surrounding countries, it has lately been discovered in England.   This solitary British specimen was found near Brighton, in a lump of clay upon a bank of the same substance ; and it is a remarkable fact that the entrance of the hole in which the creature lay, was stopped up with a dried carapace of the same species.

Should any reader wish to discover this rare spider in some other locality, he may easily identify it by its peculiar shape, which is given in the illustration, its straight jaws, its powerful falces, and its six eyes arranged in a form something like that of a horseshoe, two small ocelli in front and four larger behind.   It has altogether a reddish cast ; and its length is more than half an inch.

THE pretty SCYTODES shown in this illustration is found both in Europe and Africa, but only in the hotter parts of the former continent.   It may be known by its six eyes arranged in pairs, and its elegant colouring, which is pinky white, with two rows of black spots on the abdomen and black rings on the legs.   Its eyes are brilliant yellow.   The female always uses her jaws in carrying the cocoon, which is about the size of an ordinary pea.

The SEGESTRIUM also has six eyes, and is found in Europe.   It lives mostly in hollows of walls and rocks, spinning a silken tube in which it conceals itself, and holding in its feet the lines which communicate with the exterior. The tube is open at both ends, so

Phalangium lóngipes.

Segestrium senoculatum.   Scytodes thoracica.

that when the spider feels either of the lines shaken, it can dart out at once upon its prey.

The common HARVEST-SPIDER, or HARVEST-MAN, is a very common and well-known inhabitant of England, and, whether in gardens or in the open field, is to be found in very great numbers.

Sometimes the Harvest-spider is seen scrambling over the grass with wonderful speed, its little round body hardly discernible as it moves along, and its long straggling legs looking like animate hairs.   Sometimes it prefers to cling to a wall or fence, and there remains perfectly quiet, with its legs stretched out to their full extent, and occupying a wonderful spread of surface.   Sometimes again, especially on windy days, it seeks sheltered spots, such as crevices in old walls, or the rough bark on the leeward side of tree trunks.

One day this summer (1862), as I was bathing in the river Cray, just below a lasher, I happened to look under the cross-beam of the woodwork, and there saw something which I took for a mass of black horsehair.   Wondering how such a substance could get into such a situation, I went to examine it, and then found that the supposed horsehair was nothing more or less than a legion of Harvest-spiders, all gathered together, their little bodies nearly hidden by their bent legs.   There must have been some thousands of the creatures under the beam, all perfectly motionless.   An intelligent countryman, to whom I pointed out this curious assemblage, was quite as surprised as myself, never having seen anything of the kind before.

Like many other very long-limbed creatures, the Harvest-spider seems to set little store by its legs, and will throw off one or two of them on the slightest provocation. Indeed, it is not very easy to find a Harvest-spider with all its limbs complete; and if such a being should be captured, it is nearly certain to shed a leg or two during the process. It appears to be totally indifferent to legs, and will walk off quite briskly with only half its usual complement of limbs. I have even known this arachnid to be deprived of all its legs save one, and to edge itself along by this solitary member, in a manner sufficiently ludicrous. The cast legs contain much irritability, and even after they have been severed from the body continue to bend and straighten themselves for some little time.

A strange genus, termed Gonoleptes, is closely allied to the Phalangium. These curious spiders have the palpi very broad, very flat, and armed with thorns; and the body is flat, expanded behind, and covered with a hard shelly skin. The legs are extremely long, and the hinder pair are longer than the others. All the members of this genus are exotic.

THE formidable-looking arachnid which is called the GALEODES by no means belies its appearance, but, from many accounts, seems to be a really dangerous creature.

The bite of the Galeodes is much dreaded in the countries where the creature lives, and is said to produce very painful and even dangerous effects. Still, we may leave an ample margin for exaggeration; and when we consider the black catalogue of crimes that are attributed to the newt, the blindworm, and various other harmless creatures of our own land, we may well imagine that the popular opinion of the Galeodes is not likely to be very favourable.

The Galeodes is fond of warm, sandy situations, and like many of the Arachnida, is seldom seen except by night, when it comes from its hiding-place in search of prey. Under such circumstances, it is very likely to retaliate if injured by a bare hand or foot, and to inflict a wound causing considerable pain.

There are several species belonging to this genus.

Even the scorpion itself is hardly more formidable in aspect than the Galeodes, and to the generality of the insect tribe it is even a more dreadful foe. Armed with two pairs of powerful mandibles placed side by side, like the claws of lobsters without their jointed foot-stalks, the Galeodes sets off at night in search of prey. It runs with wonderful rapidity, more like a mouse than a spider, and, from the large size to which it sometimes attains, covers much more space than a mouse as it darts over the floor. One specimen in the British Museum is about two inches in length exclusive of the limbs, and measures exactly ten inches in total length as it stands ready for action, with its body on the ground and the limbs bent in the attitude represented in the engraving. If, therefore, the limbs were straightened, the length would be very greatly increased.

A large specimen of the Galeodes will attack any insect and almost any creature of small size. It has been known to leap upon a lizard, to cling to its back as the com-batants rolled about on the ground, to kill it by driving its fangs into the spinal cord at the junction at the head with the neck, and, finally, to eat it entirely with the exception of some of the larger bones. The lizard measured three inches in length, exclusive of the tail.

Much attention has been given to the Galeodes and its habits by Lieutenant-General Sir J. Hearsey, K.C.B. who has kindly communicated to me the following observations:—

When the Galeodes approaches any creature that it desires to attack, it thrusts out its long palpi, touches the body with the rounded tips of those members, and immediately raises them aloft, as if fearful lest they should be injured. The whole action is wonderfully like the manner in which an elephant flings its proboscis in the air after touching any-thing of which it is not quite sure. The tips of the palpi are rounded and soft, and when they are applied to any object a sort of phosphorent flame seems to be emitted from them. Having satisfied itself by the touch, the creature rushes in at once to the attack.

In order to ascertain whether the Galeodes would really attack and eat vertebrated animals, an ordinarily-sized specimen was captured and placed under a bell-glass. A very

GALEODES.—*Solpuga araneóides.*　　　BOOK SCORPION.—*Chélifer Wideri.*

young musk-rat was then inserted under the glass, the Galeodes being on the opposite side. As the creature traversed its transparent prison, it came suddenly on the young musk-rat, which was quite a baby and could not open its eyes. Without hesitation it sprang on the little animal, killed it, and in a very short time had eaten it.

The manner in which the Galeodes kills its prey is really remarkable. The double set of pincers are sharply hooked, like the beak of an eagle, and are capable of being separately opened and shut like lobster's claws, and of being used conjointly to secure prey between them ; and, moreover, the upper joint of each claw can be pushed far over the lower. When the creature seizes a large animal, such as the lizard above mentioned, it buries the pincers in the flesh, and deliberately shears its way onwards, each pair of pincers working alternately, one pair being engaged in holding the prey and the other in cutting.

The same Galeodes was then pitted against a little bat, about three or four inches across the wings. Though small, it was full-grown and lively. When placed under the glass shade it fluttered about, but was speedily arrested by the spider, which leaped upon it, proceeded to drive its fangs into the neck, and clung so tightly that it could not be shaken off. In vain did the bat try to beat off the enemy with its wings, or to rid itself of the foe by flying in the air. Nothing could shake off the Galeodes ; the long legs clung tightly to the victim, the cruel fangs were buried deeper and deeper in its flesh, the struggles gradually became weaker, until the point of a fang touched a vital spot, and the poor bat fell lifeless from the grasp of its destroyer.

The next antagonist of this redoubtable warrior was a scorpion, about four inches in length. The Galeodes seemed nothing daunted, seized the scorpion by the root of the tail, just where it could not be touched by the sting, sawed its way through the tail, severed that deadly weapon from the body, and then killed and ate the scorpion, together with its tail. There was, however, much uncertainty as to its mode of attack in this instance, for no one could exactly ascertain whether it was directed to the one point of safety by chance or instinct. Another similar scorpion was then procured and placed in the glass bell. The Galeodes darted as usual to the attack, but unfortunately seized its foe by the front. The scorpion immediately grasped the Galeodes in its nippers, quickly brought its tail over its back, and by a well-directed stroke succeeded in stinging its enemy. At the moment of receiving the stroke, the Galeodes started back, opened all its limbs, began to quiver throughout its whole frame, and rolled over quite dead.

The colour of the Galeodes is palish yellow, and the tips of the fangs are black. Their surface is very hard and polished, and when the light falls upon them they gleam as if covered with burnished gold. In a specimen now before me, the array of hairs with which the fangs are fringed glitter as if tinged with the rainbow.

One species of Galeodes inhabits the New World, being found in Havannah, but the greater number of them are inhabitants of the hotter portions of the Old World. In India the present species is plentiful, and is apt to be rather annoying, especially to a new-comer. On the same illustration with the Galeodes is seen a magnified representation of the curious CHELIFER, a little arachnid very much resembling a tiny scorpion without a tail. The body is flattened, and the palpi are much elongated and furnished with a regular claw at the end, like that of a true scorpion. The Chelifer is an active little being, running with much speed, and directing its course backward, forward, or sidewise, with equal ease. It lives in dark places in houses, between books in libraries, and similar localities, prefer-ring, however, those that are rather damp. It does no harm, however, to the books, but rather confers a favour on their owner, feeding on woodlice, mites, and other beings that work sad mischief in a library.

Its general colour is brownish red, and it is remarkable that the palpi are twice as long as the whole body. This, as well as an allied genus called Obisium, is found in England. The two genera can be easily distinguished by the cephalothorax, that of Chelifer being parted by a cross groove, and that of Obisium being entire.

PHRYNUS.—*Phrynus palmatus.*

WE are now approaching the true Scor-pions, and pause on the way to examine the remarkable arachnid represented in the illus-tration. In this, as well as the Scorpions, the abdomen is divided into segments, the palpi are very large and foot-like, and are furnished at their tips with claws like those of the crustacea. The cephalothorax is broad, semi-circular, and very slightly separated from the abdomen.

Of all the spider race, the Scorpions are most dreaded ; and justly so. These strange beings are at once recognised by their large claws and the armed tail. This member is composed of six joints, the last being modi-fied into an arched point, very sharp, and communicating with two poison glands in the base of the joint. With this weapon the Scorpion wounds its foes, striking smartly at them, and by the same movement driving some of the poison into the wound.

The effect of the poison varies much, according to the constitution of the person who is stung, and the size and health of the Scorpion. Should the creature be a large one, the sting is productive of serious consequences, and in some cases has been known to destroy life. Generally, however, there is little danger to life, though the pain is most severe and the health much injured for the time, the whole limb throbbing with shooting pangs, and the stomach oppressed with overpowering nausea. The poison seems to be of an acrid nature, and the pain can be relieved by the application of alkaline remedies, such as liquid ammonia, tobacco ashes, &c. Melted fat is also thought to do good service, and the nausea is relieved by small doses of ipecacuanha. Some of the poison can mostly

ROCK-SCORPION.—*Buthus afer.*

be brought to the surface by means of pressing a tube, such as a tolerably large key or the barrel of a small pistol, upon the spot, and the duration, if not the severity of the pain, is thereby mitigated. The great ROCK-SCORPION of Africa is much dreaded by the natives, whose only idea of cure is to tie a bandage firmly above the wound, and then make the patient lie down until the effects have gone off.

The Scorpions inhabit most warm countries, and everywhere are held in the greatest detestation. All kinds of precautions must be taken to guard against a sudden wound, for these creatures are very fond of warmth and afraid of light, and therefore crawl into houses, and conceal themselves in the warmest and darkest spots that can be found. They get into beds, creep under pillows, make their way into the toes of boots, crawl into clothes, hide themselves under cushions, and are, withal, so plentiful that no careful person thinks of thrusting his hand under a pillow or his foot into a shoe without ascertaining that no Scorpion has taken up its abode there.

They are fierce and rapid creatures, perfectly aware of the terrible weapons with which they are armed, and not unfrequently routing a foe only by the ferocity of their aspect. When threatened or alarmed, the Scorpion curls its tail over its body, flourishes the venomed weapon about in a most menacing style, and if it thinks that it cannot conveniently escape, it takes up the offensive, and boldly rushes to the attack, its claws and tail ready for the assault.

It is a rather remarkable fact, that the poison of the Scorpion gradually loses its effect upon a human being, and that a man suffers less and less each time that he is stung. One bold philosopher had the courage to follow out this principle to the furthest extent, and made Scorpions sting him repeatedly until he had become poison proof, and suffered but little inconvenience beyond the transient pain of the puncture.

The Scorpion, however repulsive in appearance and venomous in action, yet may excite some admiration for its attachment to its young. While they are yet small and feeble, they congregate upon the person of the mother, swarming over her back, her forceps, her limbs, and even clinging to her tail, and exist in such numbers that they quite conceal the outline of their parent. The little Scorpions remain upon the body of the mother until they are about a month old, when they separate, and are able to shift for themselves. It will be remembered that the young of several spiders behave in a similar manner.

In all these creatures the tail is composed of the six last joints of the abdomen, and the powerful limbs, with the lobster-like claws at the tips, are the modified palpi. The eyes of the Scorpions differ in number, some species having twelve, others eight, and others only six: these last constitute the genus Scorpio. On the lower surface of the Scorpions are seen two remarkable appendages, called the combs, the number of teeth differing in the various species. In the Rock-scorpion the teeth are thirteen in number, while in the red scorpion there are never less than twenty-eight teeth. A figure of this curious organ may be seen in the illustration on the preceding page. The Rock-scorpion is a large creature, measuring about six inches in length when fully grown.

FEMALE SCORPION AND FAMILY.

Like the other Arachnida, the Scorpion is carnivorous, and feeds upon various living creatures, such as insects and the smaller crustacea. They mostly seize their prey in their claws, and then wound it with the sting, before attempting to eat it. Even the hard-mailed coleoptera, such as the ground beetles, the weevils, &c. fall victims to this dread weapon, while the grasshoppers and locusts fall an easy prey before so terrible a foe.

WE will now turn our attention to the little, but annoying, creatures called Mites.

None of the Mites attain large dimensions, and the greater number of them are almost microscopic in their minuteness. Everywhere the Mites are found, in the earth, in trees, in houses, beneath the water, and parasitic upon animals. They haunt our cellars and swarm upon our provisions—cheese, ham, bacon, and biscuits are equally covered with these minute but potent destroyers; and even our flour stores are ravaged by the countless millions of Mites that assail the white treasures. Whether the cause or the effect of the malady, Mites are found in many forms of disease, both in man and beast, and will certainly propagate the infection if they are removed from the patient and transferred to a healthy person. They are even found deep within the structures of the vital organs, and Mites have been discovered in the very brain and eye of man.

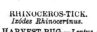

RHINOCEROS-TICK. HIPPOPOTAMUS-TICK.
Ixódes Rhinocerínus. Ixódes Hippopotamensis
HARVEST-BUG.—Leptus autumnális. Ixódes venustus.

Without dilating further upon their general habits, I now pass to our illustration of these tiny creatures.

At the lower left-hand corner of the illustration is shown a figure of a very common and most annoying species, the well-known HARVEST-BUG.

This little pest of our fields and gardens is very small, and of a dull red colour, looking exactly like a grain of cayenne pepper as it glides across a leaf. It is seldom seen until June or July, and is most common in the autumn, in some places swarming to such an extent that the leaves are actually reddened by their numbers. They are especially plentiful on the French bean; and I well remember that when I was a little boy I was horribly tortured by the Harvest-bugs, which came from the leaves of the French beans among which I was employed, and, crawling over my shoes, left a scarlet ring of intolerable irritation round my ankles.

While we are walking through the stubble-fields, the Harvest-bug is terribly apt to make successful attacks upon our ankles; and in the case of persons endowed with a very tender skin almost drives the sufferer to the verge of madness. Gilbert White, in his "Natural History of Selborne," tells us that warreners are "so much infested by them on chalky downs, where these insects swarm sometimes to so infinite a degree, as to discolour their nets and to give them a reddish cast, while the men are so bitten as to be thrown into fevers."

The Harvest-bug does not confine its attacks to human beings, but equally infests horses, dogs, sheep, and rabbits. It burrows under the skin in a very short space of time, and after a little while a red pustule arises, sometimes as large as a pea, occasioning great irritation at the time, and much pain if it be broken or wounded. On account of its red colour, the French call the Harvest-bug the ROUGET.

THE three other figures represent creatures belonging to another genus. The pretty species in the lower right-hand corner derives its name of "venustum," or beautiful, in consequence of the pretty colouring of its surface. The ground colour of this creature is deep black, upon which are set some patches of rich orange-red, edged with yellow. The little lines arranged round the body are also yellow, and its legs are red. It is moderately large, being about one-sixth of an inch in length.

The other two species are parasitic upon the rhinoceros and the hippopotamus, and derive their name from the creatures which they infest. The HIPPOPOTAMUS-MITE, or TICK, as it is sometimes wrongly called, is seen in the upper right-hand corner of the illustration. The general colour of this species is pale straw above, and deep liver-red below, the limbs being of the same colour as the upper surface, but rather paler. The lines and streaks upon the body are black. Its body is decidedly convex, and there is a very slight indication of a thorax. Its length is about a quarter of an inch.

The RHINOCEROS-MITE has also a convex body, the head and palpi are orange, and the blotches upon the body and the limbs are of the same rich hue. This creature is slightly larger than the preceding. It also belongs to Africa, being found on the Borele, sometimes called the Rhinaster (*Rhinoceros bicornis*).

Another species of Ixodes is seen in the succeeding illustration. All these creatures are furnished with suckers, through which they can draw the juices of the animals on which they are parasitic, and with a peculiar barbed modification of the parts of the mouth, which enables the parasite to anchor itself as it were with living grapnels. There is hardly any animal which is not subject to the attacks of these tiresome mites, and even the hard-shelled tortoise itself is not free from them. They fix themselves so firmly with their barbed grapnels that, if they are roughly torn from their hold, they either leave their heads in the wound, or carry away part of the flesh. Under the microscope the head of any Ixodes forms a beautiful object, and is easily prepared by means of Canada balsam and pressure.

These creatures often swarm in thick woods, and attach themselves for the nonce to the leaves of shrubs, at no great height, waiting for the time when some animal may wander near and become their victim. Sometimes they swarm upon an animal to such an extent that they have been known to kill even a horse or an ox from sheer exhaustion. The French call the Ixodes of the dog, the LOUVETTE, and in America all the mites belonging to this group are known by the name of PIQUES.

These "ticks," as they are popularly called, are extremely annoying in tropical countries, where they swarm in every forest, and infest every living creature that passes by, provided its skin be sufficiently soft to be penetrated by their beaks. They are small and flat when they first settle themselves on their victim, but they suck the blood with such vehemence and industry, that they speedily swell and redden, until at last, when fully gorged, they are as large as broad beans, and as easily crushed as ripe gooseberries.

In these countries, after a walk in a forest, every one is obliged to undergo a thorough inspection from head to foot in order to rid himself of the ticks. When found, they must by no means be pulled away, as their barbed heads would then remain in the wound, and cause a festering sore. The proper method of detaching them is to touch them with oil, when they immediately begin to work their way out of their holding places, and may then be removed and killed. Sometimes a tick is only to be found by the pain which it causes. A dull aching pang, for example, shoots at intervals up the arm, and the experienced forester at once begins to look for a tick somewhere about the roots of the fingers. The creature in such a case is usually very small, not very much larger than a cheese-mite, but it still has strength enough to make its presence felt.

Even in our own large forests, the ticks are numerous and unpleasant. In the New Forest, for example, they are far too plentiful to be agreeable; and after a day's walk in the wood I have often been obliged to serve numbers of ticks with an oily notice of ejectment.

FLOUR-MITE.
Acarus farinæ.
BEETLE-MITE.
Gamasus coleoptratorum.

SCARLET MITE.
Trombidium hirsutissimum.
Ixodes thoracicus.

THE figure in the upper left-hand corner of the illustration represents the FLOUR-MITE, a creature that is but too well known to millers and dealers in corn. Although it is a very tiny creature, it contrives to travel over the loose flour with considerable speed. The well-known cheese-mite is closely allied to the Flour-mite. In these creatures the body is covered with numerous stout hairs, which are capable of movement, so that each hair must have at least two muscles, together with their tendons. Despite, therefore, of the minute size of these mites, their structure is not a jot less complicated than that of many larger beings, and possesses a wonderful series of organs of which the higher animals are destitute.

The little vesicle at the end of the foot is a beautiful object in the micro-scope, especially if the mite can be kept alive while imprisoned under the field of the instrument. In these creatures the females are larger than their mates. The eggs of this mite are oval, very white, and covered with a sort of brown network.

Just below the Flour-mite is given a much magnified representation of the BEETLE-MITE. This genus is a very large one, containing a great number of species. Most persons who have been accustomed to see the common Watchman-beetle (*Geotrupes stercorarius*) in its wild state must have noticed the frequency with which the under part of the body is infested with certain pale yellow mites. This particular species is here represented. Sometimes the beetle is so covered with the mites that its whole body swarms with them; but, as a general fact, they confine themselves to the under surface. Many other insects are victims of mites belonging to the genus Gamasus, the humble-bee being very conspicuous in this respect.

Closely allied to the beetle-mite is the terrible RED MITE, so called by the bird-fanciers, in allusion to its colour when gorged with blood. When hungry it is of a light yellow colour, but when it has fed, the blood shows its ruddy hue through the transparent skin of the mite. It is a very small creature, and lives mostly in the crevices of the cage during the day, coming out to feed at night. I always used to destroy them by inserting insect-powder into the crevices of the cage, dusting the birds well with the same substance, and keeping a small camel's-hair brush charged with oil, with which any stray mite could be at once killed.

The dark-looking figure in the upper right-hand corner represents the well-known SCARLET MITE, so plentiful on banks and in gardens. It is covered with a soft down, which gives a very rich and pleasing depth to its colour. This species is not of large size, but in the East Indies a species is found which is three or four times larger than our own Scarlet Mite. It yields a bright red dye, and is therefore called *Trombidium tinctorium*.

THE two upper figures in this illustration represent creatures of a different family, the Pediculidæ. That on the left hand is the Hog-tick, which is found only upon swine, and not universally even upon those animals. It is of moderate dimensions, measuring about one line and a quarter in length. Its thorax is mostly brown, and its abdomen greyish yellow. Mr. Denny, in his "Monographia Anoplurorum," gives the following account of the Hog-tick :—

" *H. Suis.* This species is found in great abundance on swine, but it does not appear so generally spread as might be expected from the dirty habits of the animals. It most frequently occurs on those fresh imported from the Sister Isle. It was many months before I could obtain a single example. I had applied to both farmers and pig-butchers, neither of whom seemed to approve of the idea which I had conceived of *their* pigs being lousy, but referred me to those of the Emerald Isle as being sure to gratify my wishes, forgetting, I suspect, that the Irish pigs come to this market to meet English buyers.

I accordingly visited a colony just arrived, where I most certainly met with a ready supply. But here they were confined almost entirely to lean animals ; and wherever I found a pig fat and healthy, no game were to be seen. In walking, this species uses the claw and tibial tooth with great facility, which act as finger and thumb."

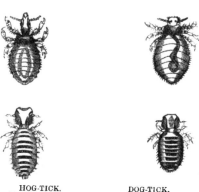

HOG-TICK.
Hæmatopinus suis.

DOG-TICK.
Hæmatopinus piliferus.

DEER-TICK.
Trichodectes longicornis.

HORSE-TICK.
Trichodectes equi.

The DOG-TICK is apt to be extremely troublesome, not only getting into the fur of the dog, but harbouring in their bedding, and almost defying all attempts at destruction. White precipitate seems the best solid substance for this purpose, and a very weak solution of nitric acid answers well as a liquid. But, in both cases, the dog must be muzzled to prevent it from nibbling at its fur, and thus imbibing some of the poison. Its colour is ashy flesh, with a slight chequering. The skin is so transparent that the intestine can plainly be seen, of a dull red colour. When gorged with blood, the creature becomes of a light scarlet. This species is also found on the ferret.

THE two lower figures in the same illustration refer to another family of these creatures. The HORSE-TICK is found both on the horse and ass, especially when fresh from pasture, and is very common under such circumstances. It is rather a pretty species, with a light chestnut head and thorax, and may be known by the squared thorax and the long club-like first joint of the antennæ.

The LONG-HORNED TICK, or DEER-TICK, seen in the left hand of the illustration, is also a common species, and is parasitical on the common fallow deer, assembling in great numbers on the inner side of the thigh. The general shape of the parasite can be seen by the engraving. The colour of the head and thorax is something like that of the last-mentioned species, except that there is more red in it. Moreover, it can be distinguished by the antennæ, which have the second joint the longest and the third acute. The eyes, too, are large and prominent.

THE four creatures represented in the accompanying illustration are parasitic on birds. The species shown in the upper left-hand corner is unpleasantly familiar to poultry-keepers, swarming among the feathers to such a degree that the hands are often covered with these parasites when the fowls are plucked or even lifted up. They cling very tightly, and are not easily brushed away, as their bodies are smoothly polished, and offer scarcely any resistance. The colour of this species is pale straw.

The upper right-hand figure represents a parasite found on the common swan and other aquatic birds, such as the bean goose and Bewick's swan. It is coloured after a rather peculiar fashion. The head, thorax, and legs are bright chestnut, smooth and shining; the abdomen is white, except the first segment, which is of the same colour as the thorax. There is also a chestnut spot on the third segment, and a row of short liver-coloured bands runs down each side.

At the lower left-hand corner is seen a parasite which is not very plentiful, but which is found on various birds, such as the rook, the raven, and blackbird. It has a hard shelly surface, and is marked with numerous dark bars. The last figure represents the SICKLE-HORNED TICK, so called from the shape of its antennæ, which are rather large, flattish, and curved. It is a pretty species, its squared head being of a light chestnut colour, and highly polished. The abdomen is broad, rather flat, and of a light tawny yellow, barred with deep red, and having the last segment of the same colour as the head. It is parasitical upon the common peacock, and may almost invariably be found after the death of the bird congregated in tolerable numbers about the base of the beak.

Ménopon pállidum.
Nirmus árgulus.

Docóphorus cygni.
Goniódes falcicornis.

IN the accompanying illustration are seen some details of the characteristic portions of many Arachnida, most of which have been casually mentioned in the preceding pages.

The two upper rows exhibit the manner in which various types of eyes are arranged, and are valuable as showing how important are the eyes in distinguishing one genus of spiders from another. The bold staring eyes on the right-hand side of the upper row are those of the genus Nops, and, as far as is known, are unique among spiders. They are large, black, glittering, and bear no slight resemblance to the glasses of "bull's-eye" lanterns. The central figure represents the eyes of the genus Salticus, or Hunting Spider, and on the extreme left are the eyes of Lycosa, or Wolf-spiders, which are set in a manner almost precisely the reverse of the preceding.

Passing to the right-hand figure of the second row, we come to the cephalothorax and eyes of the new British spider, *Dysdera erythina.* The central figure gives a profile view of the cephalothorax of a male *Walckenaera acuminata,* showing the long, slender, vertical process of its termination, and the manner in which the eyes are disposed upon it. On the left hand of the same row is shown the elevated front of the cephalothorax belonging to the genus Hersilia, showing the remarkable manner in which the eyes are arranged.

The third row contains some specimens of grasping organs. On the right hand is shown the under surface of the base of mandibles, belonging, I believe, to the genus Linyphia. The severe illness of the artist prevents me from ascertaining whether that particular genus afforded the subject, but the drawing corresponds closely with the mandibles of Linyphia. The central figure shows the maxillæ and falces of the Water-spider (*Argyronetra*), and the left-hand figure exhibits the same organs of the Hunting Spider.

The next three figures relate to the feet and legs. The right-hand sketch represents the foot and tooth-claws of a Tegenaria (probably *Tegenaria civilis*), show-ing the two large comb-like upper claws, and the single little claw below, toothed only at its base. The central figure represents the end of the foot of a Hunting Spider. It is worthy of notice that the feet of the genera Philo-dromus and Sparassus are formed in a like manner. The left-hand figure shows the foot of one of the Wolf Spiders (*Lycosa*).

The last object is the strange double comb of the scorpion. This wonderful organ is situated on the first ring of the abdomen, and no one knows its real use. It consists, as may be seen, of a pair of stout projections, jointed at their base to the abdomen, and furnished with a number of little tooth-like appendages. The con-sistence of this organ is rather tough, thick, and somewhat parch-ment-like; it can be raised but slightly from the abdomen, but plays easily backwards and for-wards by means of its joint. The number of teeth belonging to the combs is extremely variable in the different species, as it may be that even the same species varies according to the age of the in-dividual. In one species, for ex-ample, no more than thirteen teeth have been counted in each comb, while in others the teeth are about thirty in number, and in a few species even amount to fifty or sixty. The combs that are re-presented in the engraving were taken from the Rock-scorpion, described and figured on page 681.

Eyes, *Lycósa*.
Eyes, *Hersilia*.
Falces, *Sálticus*.
Foot, *Lycósa*.

Eyes, *Sálticus*.
Eyes, *Walckenaera*.
Falces, *Argyronetra*.
Foot, *Sálticus*.
Combs *Scorpion*.

Eyes, *Nops*.
Eyes, *Dysdera*.
Labium, *Linyphia*.
Foot, *Tegenária*.

MYRIAPODA.

IN accordance with the best systems of the present day, the MYRIAPODA are considered as a separate class.

Some writers have placed them at the end of the insects, on account of certain structural resemblances with certain insects in the larval state. There is also a strong resemblance to the Annelida, or Ringed Worms, which will be placed next in order; and indeed, when we come to examine the lower forms of animal life, we find ourselves quite bewildered with their many relationships, and uncertain as to their true position in the scale of nature. Van der Hoeven, after reviewing some of the difficulties of systematic zoologists, makes the following pertinent remarks :—"Thus is the entire animal kingdom a *net everywhere connected*, and every attempt to arrange animals in a single ascending series must necessarily fail of effect."

THE reader will remark that in the spiders the head and thorax are fused together into a single mass, the abdomen remaining separate. In the Myriapoda the reverse of this structure is seen, the head being perfectly distinct, while there is no outward mark to distinguish between the thorax and abdomen.

The Myriapoda are without even the rudiments of wings, and possess a great number of feet, not less than twelve pairs; and in some species there are more than forty pairs of legs. In allusion to their numerous feet, the Myriapoda are popularly called Hundred-legs, and their scientific title is even bolder, signifying ten thousand feet. To this class belong the well-known centipedes, so plentiful in our gardens, and the equally well known millipedes, found under decaying wood and in similar localities.

In England none of the Myriapods attain to great dimensions, but in hot countries, and especially under the tropics, they become so large as to be positively formidable as well as repulsive. Even the common centipede of the garden is by no means an attractive being, and there are few persons who can handle one of those creatures without some feeling of disgust.

In all the Myriapoda the feet are terminated by a single claw. Some species are totally blind, but those who possess visual organs have two masses or clusters of simple eyes, their number being variable according to species or in the different stages of development in the same individual.

The first order of the Myriapods, called by Mr. Newport the Chilopoda, may be known by several characteristics. The head is broad and somewhat prominent, and the segments of the body are unequal, each having a single pair of legs. The mandibles are long, sickle-shaped, sharp, and prominent. The first tribe of the Chilopods has antennæ of great length, longer indeed than the body, very slender, and composed of many joints. The tarsi are also many-jointed, unequal, and very long, as may be seen by reference to the illustration. The eyes are prominent and rather globular.

The family to which the NOBLE CERMATIA belongs is known by eight large bone-like plates or shields upon the back, looking very like the ridge tiles on the roof of an out-house. The members of the genus Cermatia, or Scutigera as it is sometimes called, are spread over the hotter parts of the world, and attain their greatest dimensions under the tropics. Specimens of these strange beings are found in the South of Europe, Madeira, many parts of Africa and Asia, Florida, New Holland, and Australia. The eyes of the Cermatia are unlike those of the generality of Myriapoda, the ocelli being crowded together so that the facets assume a hexagonal form like those of the insects and some of the crustacea.

All the Cermatiæ are exceedingly active, running about on their long legs with an action that reminds the observer of the common harvest-spider. Indeed, the whole creature has very much the look of being composed of a number of harvest-spider's legs attached to the body of a centipede. The Cermatia is carnivorous in its habits, feeding upon insects and having a great predilection for spiders. A full-grown Cermatia will attack even one of the large and formidable spiders of the tropics, and, safe in its shelly mail, succeed in killing and devouring its foe. In the struggle it will probably lose a few legs; but the creature is in no wise fastidious about its proper complement of limbs, and loses six or seven legs with perfect indifference, behaving in this respect like the harvest-spider, the crane-fly, and other "lang-leggit" creatures, whose affection for their limbs seems to be in inverse ratio to their length.

The colour of the Noble Cermatia is pale brown, with a yellowish line running down its centre. The limbs are strongly marked with yellowish brown, green, and rings of blue. It seldom exceeds two inches in length. This species is found in the East Indies and in the Mauritius.

Lithóbius rúbriceps.   Cermátia nóbilis.

THE second figure in the same illustration represents a myriapod belonging to another family, termed the Lithobiidæ. The members of this family may be known by the fifteen shields upon the back and their sharp, elongated angles. They are found in the open air, hiding under stones, a habit to which is due the title of the family and genus. The name Lithobius is composed of two Greek words, the former signifying a stone, and the latter to live.

In this species the head is large and squared, and of a deep red colour. There are fourteen eyes on each side, and they are small and very black. The long antennæ are yellow, and the mandibles are of the same colour, deepening into black at the points. The general colour of the body is olive-brown, the green tinge being more conspicuous in some individuals than in others, and the legs are yellow. This is a small species, rather less than an inch and a half in length. The figure is of the same size as the specimen from which it was drawn. This species inhabits the south of Spain.

More than twenty species of Lithobius are known, scattered over the greater part of the world, some being found even in comparatively cold countries. Several of them are natives of England, and one or two are very prettily coloured, such as the Variegated Lithobius (*Lithobius variegatus*), which has a double row of dark spots along its body, and the Black-eyed Lithobius (*Lithobius melanops*), which is of a yellowish green colour, with an orange head, one joint of each leg of the same bright hue, and twelve large black eyes on either side of the head. The Forked Lithobius (*Lithobius forficatus*) is very common

3.   Y Y

Scolopendra formósa. (Lower figure.)　　　*Scolopendra lútea.* (Upper figure.)

in this country, being found plentifully under stones and in similar localities. It is a quick, active creature, of a canary-red colour, and is of moderate dimensions, measuring from an inch to an inch and a half in length. It has fifteen pairs of legs.

WE now arrive at the true Scolopendræ, which, together with the allied genera, are popularly known by the name of Centipedes. The genus Scolopendra is a very large one, containing about sixty species, most of them inhabitants of the tropics, and many attaining a large size.

The great Scolopendræ are not only unpleasant and repulsive to the sight, but are really formidable creatures, being armed with fangs scarcely less terrible than the sting of the scorpion. These weapons are placed just below the mouth, and are formed from the second pair of feet, which are modified into a pair of strong claws, set horizontally in a manner resembling the falces of ordinary spiders, and terminated by a strong and sharp hook on each side. These hooks are perforated, and are traversed by a little channel leading from a poison gland, like that of the scorpion, so that the venomous secretion is forced into the wound by the very action of biting. These curious weapons cover the first pair of feet and the gnawing organs of the mouth. All the members of the order possess this remarkable modification of the feet, which has earned for them the title of Chilopoda, a term composed of two Greek words, the former signifying a beak, and the latter a foot.

Both the species of Scolopendra figured on the engraving are exotic, and of rather large dimensions. The *Scolopendra formosa* is well deserving of its specific title, which signifies beautiful, on account of the splendid colouring with which it is adorned while living, the feet being orange, with black teeth, and the edges of each segment being bright green. It is a native of the East Indies, and is about four inches in length. The second species is, as its name imports, of a yellowish colour, with a deep orange-coloured head and appendages. This is a native of the Caribbean Islands, and is of the same length as the preceding species.

GIANT CENTIPEDE.—*Scolopendra gigas.*

On the above illustration is given a reduced figure of the GIANT CENTIPEDE, a creature that well deserves its name, sometimes attaining a foot or rather more in length.

This truly formidable being is a native of Venezuela, and possesses a pair of such powerful venom-fangs that its bite is nearly, if not quite, as dangerous as that of the viper.

As an example of the effects of the poisoned wound inflicted by these large centipedes, I may cite a passage from Williamson's valuable work on "Oriental Field Sports:"—"Centipedes grow to nearly a foot long, and as thick as a man's little finger; their form is, indeed, flatter, or like tape. When young they are of a clay colour, but become darker with age. They bite by means of a pair of strong forceps placed horizontally at their mouth, nearly as large as the hooked thorns on a blackberry bramble, causing much pain and inflammation, and often occasioning fever. Being from their shape so peculiarly capable of secreting themselves, they sometimes occasion very ludicrous accidents. I once saw a friend apply a flute to his mouth to play on it, but scarcely had he begun, when a large centipede fastened to his under lip, causing him to change his note very abruptly. Several have been bitten while smoking their hookahs; and I was myself once made to smart in putting on my gloves, a centipede having taken possession of one of the fingers.

A very grave and respectable old gentleman, who was remarkably fond of starting an hypothesis and hunting it to death, and who would rather pay the piper than not have his dance out on all occasions, perceived a large centipede deliberately crawling up an old door at Bethsaron Gardens, near Chororinghee. The veteran assured the company that all venomous animals were in their nature inoffensive, and never wounded but when attacked. Experience having satisfied some present to the contrary, an argument arose, and the old gentleman, with much dignity, asserted that he would prove the validity of his position by placing his finger in the centipede's way. He did so, and received such a bite as occasioned a violent fever, from which a critical abscess under his arm-pit relieved him."

There are many of these fierce and venomous creatures scattered over the world, causing no small annoyance to the new comers, who cannot for a long time look with

Scolopendra angulata. (Lower figure.)   *Scolopendra variegata.* (Upper figure.)

indifference on a great centipede, some eight or ten inches long, running up the wall close to their heads, or traversing the floor within a short distance of their feet. Among military men the monotony of camp life is sometimes agreeably diversified by a centipede hunt, the creature being chased as eagerly as if it were a fox or a wolf, and neatly captured in a split bamboo, or between two sticks.

So extremely poisonous are the fangs of these myriapods, that they will even kill poultry without much difficulty, while the smaller creatures on which they prey die almost immediately under the bite. The force with which they can grasp is really terrible, the two hooked claws being driven into the flesh until they meet, and holding their position so firmly that the centipede will rather be torn asunder than loosen its grasp. The best way to assure oneself of the force and general structure of these fangs is to procure a specimen that has been preserved in spirits and dissect it, when the powerful muscles that work the poison-feet, the glands which supply the venom, and the perforated passage through which it is discharged into the wound, are easily made out.

The colour of the Giant Centipede is bright rusty red, with a deep green head and antennæ, and blackish or olive-coloured feet. A closely allied species of similar dimensions is found in Jamaica.

In the lower left-hand corner of the same illustration is seen a rounded object with an aperture on one side. This is the nest of one of these myriapods, and was found by Mr. Foxcroft while he was digging for beetles and other insects at Sierra Leone. The nest, if it may so be called, is formed in a reddish kind of earth, and many of these habitations were discovered in the same locality. In some of them the animal was resident when the nest was brought to light.

Two more examples of this large genus are given in the above illustration. The beautiful species which is placed uppermost in the drawing is a native of Demerara, and is generally about five inches in length when adult. Its colour is rather rich and striking. The general hue of the upper surface is deep chestnut, and the front edge of the head segment, the hinder edge of the dorsal segments, and the lower surface are light orange. The antennæ are olive-green, and the feet are orange banded with olive.

The second species is found in the Island of Trinidad, and in colour contrasts well with the preceding. Its length is not quite so great, measuring less than five inches. This creature is deep green, the lip and mandibles are reddish orange, and the feet are orange and green. It derives its specific name of *angulata* from the sharp angle on the sides of each segment.

WE now take our leave of the true Scolopendræ and pass to other genera.

Both the specimens represented in the next illustration are remarkable beings; the one for its noisy nature, and the other for its phosphorescent power.

If the reader will examine a specimen of the centipede that is figured in the lower part of the illustration, he will notice that the feet of the creature are modified into flat plate-like appendages. As the centipede moves along it makes a clattering noise with these plates, and derives from this curious habit both its scientific names. The Corybantes were an ancient tribe to whom the education of Jupiter was intrusted when he was sought by his father, who wanted to eat him, and who, in order to cover the sound of his

Eucorybas crótalus. (Lower figure.)   *Arthronómalus longicornis.* (Upper figure.)

cries, continually danced and played the castanets around the infant. The specific name, *crotalus*, signifies a rattle, and is, therefore, appropriate to the creature.

This centipede is a native of Southern Africa, and is found about Natal. Its colour is rusty brown, and the species may be easily recognised by the peculiar structure just described.

THE smaller figure represents a centipede which is found in England, and in some localities is very common. It is represented of its natural size, and, as may be seen, is in no way conspicuous for its dimensions. It is, however, remarkable on another account. It has the power of giving out a tolerably strong phosphorescent light, which is only visible after dark, but is then very conspicuous, and has often caused the centipede to be mistaken for a glow-worm. It is not unfrequently found within peaches, apricots, plums, and similar fruits, when they are very ripe, and lies comfortably coiled up in the little space between the stone and the fruit, where the sweetest juices lie.

The colour of this centipede is yellow; its head is deep rust colour; its antennæ are very hairy and four times as long as the head segment. There are from fifty-one to fifty-five pairs of legs. Its length varies from two and a half inches to three inches.

THE left-hand figure in the next illustration is remarkable for the enormous number of rings of which the creature is composed, and the consequent number of legs which are needed to carry it over the ground. Although this species is only from four to five inches, it has no less than three hundred and twenty-two legs. It is a native of the Philippine Islands, whence so many wonderful forms are brought.

The rather harsh generic name of this creature is composed of two Greek words, the former signifying an angle, and the latter the top of the head, and is given to the animal because the front edge of the head is formed into an acute angle. The general colour of the species is ashen grey, and the mandibles are black at their tips.

The second figure introduces us to a new group of Myriapoda, where the creatures have the power of rolling themselves up, more or less completely, like the hedgehog and the pill-woodlouse. The species shown in the illustration is given of the natural size. It is a native of Borneo, and was found by Mr. A. R. Wallace. The general colour of the creature is rusty brown, inclining to red; the head is edged with blackish brown, and the front segment is also edged with the same colour. The surface is shining and

Gonibrégmatus Cumíngii.   *Zephrónia impressus.* (New species.)

polished, and all the segments, except that at the end of the body, are marked with deep longitudinal impressions. For this reason, I call the species "impressus." Its length is nearly two inches. The specimen from which the drawing was taken is in the collection of the British Museum.

Before leaving these creatures, it may be as well to state that, during the earlier stages of their existence, these animals are much less perfect than when they have reached adult age; they have not their full complement of segments or limbs, nor an equal number of eyes. The metamorphosis, therefore, is complete, and serves to show the relationship between the Myriapoda and the insects.

Two new examples of this remarkable genus are next given, in order to show the beautiful markings which exist in some species, and to exhibit the aspect which they present when rolled up for defence.

The central figure represents the Actæon Millepede, which is here shown of its natural dimensions. It is a native of Madagascar, and was noted by the celebrated female traveller Madame Ida Pfeiffer. The surface of its body is very polished and shining, and the general colour is a livid yellow. A number of tiny puncturations are

Zephrónia versicolor. (Rolled up.)   *Zephrónia Actǽon.*   *Zephrónia versicolor.* (Extended.)

scattered rather sparely at the back of the head and between the eyes, but in the front they are more numerous, and along the sides they are nearly as thick as the little depressions on the end of a thimble. The species may also be known by the shape of the eighth to the eleventh segments included, which are curiously pointed, looking as if they had been snipped off diagonally with scissors.

The two figures at either side of the illustration represent the same species in two postures. The body is very smooth and beautifully coloured, being of a yellowish ground tint, boldly variegated with stripes and spots of deep black, so as to render it extremely conspicuous. The front of the head, the eyes, legs, and antennæ are pale green in the preserved specimens, and are thought to be darker during life. The head is marked with distinct points. This beautiful species is further remarkable from the fact that no two specimens ever seem to be coloured exactly alike. On the upper ridge of the face there are from eight to ten little short spines. This species is a native of Ceylon.

These creatures are all natives of the hotter parts of the earth, but we have an allied example in almost every garden, and certainly in every field throughout the greater part of England. This is the Pill-millepede (*Glomeris margináta*). It is found among moss and under stones, and, as it rolls itself up in a manner very similar to that which is employed by the armadillo-woodlouse, is often mistaken for that being. It may, however, be readily distinguished from that crustacean by the simple fact that the legs have their origin on a single line traversing the middle of the under surface, and that when the creature is walking, their extremities do not project beyond the edges of the shelly covering.

Like the armadillo-woodlouse, the Pill-millepede was formerly used in medicine, probably because it looks somewhat like a pill, and may be found among the old stock of druggists' shops, mixed with the veritable armadillo. Both these beings feed on the same substances, namely, decaying animal and vegetable matter. It seems to be rather a gregarious creature, as it is generally found in tolerable numbers in some favoured locality.

Polydesmus spléndidus.   *Polydesmus granulátus.*   (New species.)

WE now come to another genus, termed Polydesmus. This term is composed of two Greek words, the former signifying "many," and the latter a "bundle," in allusion to the numerous groups of limbs arranged along the body. In all these creatures the body is covered with a hard skin, and the segments are flattened and lengthened at the sides. The upper figure represents a handsome species found in India, which mostly attains the length of two inches. The colour and general aspect of this species are rather striking. Independently of the very deep depression of the segments, which has a very strange effect, the colour is bold and striking, being deep and very rich reddish brown, diversified by an angular spot of bright yellow placed in the hinder angles of each segment. The body is smooth and slightly shining. When alarmed it is able to roll itself into a partial spiral, so as to present merely the hard shelly armour to the foe, and to shield the limbs within the coil.

THE lower figure represents a species belonging to this genus, but which has not yet been described. I propose to give it the specific name "granulatus," on account of the peculiar appearance of the body, which is thickly covered with very minute raised tubercles of a white colour, such tubercles being called granules in scientific nomenclature. Perhaps I can give a better notion of the idea expressed by the word "granulated," by mentioning that it could be rightly applied to such substances as shagreen.

The general colour of this species is rather dark drab, and it may be at once recognised by the peculiar form of the segments, which are flattened and elongated even more than usual in this genus, and are set at their extremities with three distinct teeth. The length of the specimen from which this description is taken is rather more than three inches. It is in the collection of the British Museum.

A species of this genus (*Polydesmus complanatus*) is found in England, and is not uncommon in gardens. It is about three-quarters of an inch in length, is very narrow, and has thirty-one pairs of feet. The genus is a very large one, and contains a great number of exotic species.

Before passing to the next large family of myriapods, we may mention the pretty little PENCIL-TAIL (*Polýxenus lagúrus*), a tiny creature which seldom attains a greater

Spirostreptes cinctatus.

Spirostreptes annulátipes.
Julus terrestris.

length than the twelfth of an inch. It is found under the bark of trees, in clefts of walls, and in moss, and may be known by the twelve pairs of feet, the bunches of little scales on the sides, and the white pencil at the end of the tail.

IN the above illustration are given three examples of the curious family Julidæ. These creatures are very like those which have just been described and figured, but may be known from them by the fact that the edges of the segments are not flattened and lengthened, but are continued in an unbroken circle. They feed mostly on decaying vegetable matters, but have been seen to eat dead earthworms and molluscs.

They all exhale a peculiar and rather unpleasant odour, which is caused by a fluid secretion in certain little sacs along the sides, two on each ring. The little apertures through which this scented fluid exudes may be seen on examining the creatures closely, and by some of the earlier writers they were mistaken for spiracles, the sacs themselves being thought to be the breathing apparatus. The real spiracles may be seen on the under sides of the animal, close to the insertion of the feet. Like the preceding creatures, they can roll themselves up, but, on account of the length of their body, they can only assume a spiral form, as is shown by the left-hand figure in the illustration.

The lower and smallest figure represents the common MILLEPEDE of our gardens, shown of its natural size. This little creature is very plentiful, and may be found under decaying wood, or below stones. Its movements are very curious. The little delicate feet, looking like white threads proceeding from below, move in a regularly graduated order, so that, as the creature glides along, a succession of waves seem to pass over its body. On being touched it immediately stops, and coils itself into a spiral form, lying necessarily on its side.

The development of the Julidæ is curious and interesting. In the early part of the spring, the female deposits sixty or seventy eggs in the earth, digging a hole expressly for their reception. Here they lie until they are hatched, which occurs in about three weeks' time, when the young Julidæ make their way into the world. They are then without any limbs, and retain the two halves of the egg-shell by means of a filament, which fastens them to the body. After a little while they gain three pairs of feet, and then are able to separate themselves from the egg-shell. At this period of their existence, they bear a great resemblance to the larvæ of some beetles. As they continue to grow, however, the number of segments and limbs increase, so that they gradually lose their resemblance to the beetle larvæ, and attain the shape and form of their parents.

THE *Spirostreptes cinctátus* is a native of India, and sometimes attains considerable dimensions, reaching a length of nine inches. It is of a rusty red colour, in some individuals inclining to yellowish clay, and has a drab ring round the middle of each segment. The legs also have a ring of the same colour round the middle of each joint.

The large straight figure on the right hand of the engraving represents the *Spirostreptes annulátipes*. This is also an Indian species, and somewhat resembles the preceding, except that its colours are much deeper; there is a narrow black ring round the middle of each segment, and each joint is broadly banded with the same colour. There are seventy-five segments in this species, when it has reached full age.

ANNULÁTA.

A NEW class of animals now comes before us. These creatures are technically called Annulata, or sometimes Annelida, on account of the rings, or annuli, of which their bodies are composed. They may be distinguished from the Julidæ by the absence of true feet, although in very many species the place of feet is supplied by bundles of bristles, set along the sides. The respiration is carried on either by means of external gills, internal sacs, or even through the skin itself.

In most of the Annulata the body is long and cylindrical, but in some it is flattened and oval. The number of rings is very variable, even in the same species; so variable, indeed, that in some specimens of *Phyllodoce laminsoa*, no less than five hundred rings have been counted, while others possess only three hundred.

The group of worms which come first on our list is remarkable for the architectural powers of its members. In order to protect their soft-skinned body and delicate gills, they build for themselves a residence into which they exactly fit. This residence is in the form of a tube, and in some cases, as in the Serpulæ, is of a very hard shelly substance, and in some, as the Terebella, is soft and covered with grains of sand and fragments of shells.

The beautiful SERPULA is now very familiar to us, through the medium of marine aquaria, its white shell, exquisite fan-like branchiæ, and brilliant operculum, having lived and died in many an inland town where a living inhabitant of the ocean had never before been seen.

The shell of the Serpula is tolerably cylindrical, very hard, white, and moderately smooth on the exterior, though it is ridged at intervals, marking the different stages of its formation. The size of the tube increases with the growth of its inmate and architect, so that a perfect specimen is always very small at its origin, and much larger at its mouth. The Serpula is able to travel up and down this tube by the bundles of bristles, which project from the rings along the sides, and is able to retract itself with marvellous rapidity. It has no eyes, and yet is sensible of light. For example, if a Serpula be fully protruded, with its gill-fans extended to their utmost, and blazing in all its scarlet and white splendour, a hand moved between it and the window will cause it to disappear into its tube with a movement so rapid, that the eye cannot follow it.

Sabella unispira.

Serpula contortuplicata.

The gills, whose exquisitely graceful form and delicate colouring have always attracted admiration, are affixed to the neck, as, if they were set at the opposite extremity of the body or along the sides, they would not obtain sufficient air from the small amount of water that could be contained in the tube.

The beautiful scarlet stopper ought also to be mentioned. Each set of gills is furnished with a tentacle-like appendage, one of which is small and thread-like, and the other expanded at its extremity into a conical operculum or stopper, marked with a number of ridges, which form a beautiful series of teeth around its circumference. The footstalk on which this stopper is mounted is a little longer than the gills, so that when the animal retreats into its tube the gills collapse and vanish, and the entrance of the tube is exactly closed by the conical stopper.

The Serpula is a lovely inhabitant of the aquarium, but has an inconvenient habit of dying, sometimes coming out of the tube for that purpose, and sometimes retreating to its furthest recesses, and there putrefying, to the great damage of the aquarium. There are several kinds of Serpula, some of which are only attached by the lower part of the tube, and hold the rest of that wonderful structure upright in the water ; some, like the present species, intertwine their tubes very much like a handful of boiled macaroni ; while others, such as the *Serpula triquetra*, form tubes which do not project at all, but are affixed to their supports throughout their entire length. This species makes a triangular tube. There are many interesting circumstances connected with the habits and structure of these lovely worms, but our failing space will not admit of a longer description.

We now come to another pretty tube-inhabiting annelid, which is called Sabella, because it lives in the sand and forms its tube of that substance. Several species of Sabella are found on the British coasts, the most common of which is the SHORE SABELLLA (*Sabella alveolaria*), a little creature seldom exceeding three-quarters of an inch in length. As is the case with many of these worms, it has a thin tail-like appendage at the extremity of its body, which is doubled up within the tube. The head is furnished with a great number of little thread-like tentacles, which are very flexible, and under a good microscope are seen to have a groove running along the centre, and a double row of teeth along the edges, something like the snout of a saw-fish.

This is a useful species to the naturalist on account of its plentiful occurrence, and readiness to work while in captivity. If a Sabella be watched while it is building up its curious tube, it will be seen to choose the particles of sand with the greatest care, selecting and seeming to balance them with the tentacles, and cementing each in its place with a glutinous secretion, which has the property of setting while under water. If the creature can be induced to build its case against the side of a glass vessel the possessor has cause to be gratified, for the creature does not waste material, and will often make the glass answer for one side of its tube, thereby permitting the observer to watch its entire economy.

The skin of these worms is very tough. I remember once having to dissect the digestive organs for a lecture, and losing hour after hour in my endeavours to make a successful preparation. Just as the lecturer's servant came for the dissection, I had begun a fresh subject, and quite lost patience. So I gave the worm an angry tug with the forceps, when the whole skin of one side stripped off, leaving the digestive organs exposed as beautifully as if they had been carefully dissected.

To give the Sabella a variety of building materials, and to note which it accepts, is always an interesting amusement; for the worm is very fastidious, not to say capricious, in its choice, and always likes to have a stock of materials from which it may make its selection.

WHILE wandering along any of our sandy coasts, we frequently come across some moderately large tubes projecting from the sand, and rather conspicuous in the little puddles left by the receding tide. Round their mouth is usually a set of forked filaments which, like the tube itself, are composed of fragments of sand agglutinated together. The substance of this tube is very soft, but very tough, and will endure a tolerably hard pull without breaking. If the inhabitant of these tubes be sought, it will not be found without much labour, for the TEREBELLA retreats to the further extremity at the least indication of danger; and as the tube is a foot or more in length, and is always conducted under stones or among rocks, it is not easily dislodged.

As in the case of the Sabella, this annelid performs its architectural labours by means of its tentacles, which are most wonderfully constructed, so as to be capable of extension or retraction, and at the same time can seize or throw away a particle of sand at any part of the tentacle. The method of working is very well given by Mr. T. Rymer Jones:—"If a specimen be dislodged from its tube, it swims by violent contortions in the water, after the manner of various marine annelids; the tentaculæ and the branchiæ are compressed and contracted about the head, like a brush; and as the animal is very soon exhausted by such unnatural exertions, it soon sinks to the bottom. Should a quantity of sand be now scattered from above, the tentaculæ, speedily relaxing, extend themselves in all directions to gather it up, sweeping the vessel quite clean, so that in a very short time not a particle is left behind that is within their reach, the whole having been collected to be employed in the construction of a new artificial dwelling, adapted to shelter the naked body of the architect.

We will suppose a tube to have been partially constructed into the side of the aquarium, wherein a specimen is about to take up its permanent abode. During the earlier part of the day, the animal is found lurking in its interior, with only the extremities of the tentaculæ protruding beyond the orifice, and so it will remain till towards noon.

But scarcely has the sun passed the meridian, than the creature begins to become restless; and towards four or five it will be seen to have risen upwards, the tentaculæ extending with the approach of evening, until after sunset, when they are in full activity. They are now spread out from the orifice of the tube like so many slender cords—each seizes on one or more grains of sand, and drags its burden to the summit of the tube, there to be employed according to the service required. Should any of the tentaculæ slip their hold, the same organs are again employed to search eagerly for the lost particle of sand, which is again seized and dragged towards its destination.

Such operations are protracted during several hours, though so gradually as to be apparently of little effect. Nevertheless, on resuming inspection next morning, a surprising elongation of the tube will be discovered; or, perhaps, instead of a simple accession to its walls, the orifice will be surrounded by forking threads of sandy particles agglutinated together."

There are many species of Terebella, and even on our own coasts we may be gratified with several beautiful forms of these interesting annelids. They have, to a considerable extent, the power of reproducing lost portions of the body; and it has been found that even the whole mass of plumy tentacles can be removed without much injury to the Terebella, which retreats to its tube, and after a while reproduces the whole of the missing organs.

SHELL-BINDER.—*Terebella conchilega.*

The SHELL-BINDER is very plentiful on some of our coasts, especially those where the shells of various molluscs are found in profusion. The tube of this species is built almost entirely of little fragments of shell, and is of very great length—so long, indeed, and going so deeply into the sand and among the stones, that to procure a perfect specimen is almost an impossibility, except by some rare good fortune. As this creature makes its dwelling about midway between high and low water mark, it may sometimes be procured by setting to work as soon as the tide has retreated, and, with crowbar, pick, and shovel, making the best use of the few hours that can be given to the task. I have never yet succeeded in extracting an entire tube, though I have often tried to do so.

PASSING from the tube-inhabiting worms, we now come to those which are free and able to move about at pleasure.

No one who has walked on any of our sandy coasts can have failed to notice the numerous worm-casts which appear in the sand, between high and low water, being most numerous where the sand is level, and becoming scarcer in proportion to the steepness of the slope. Sometimes, when a large, marshy flat makes its appearance, which is never entirely dry even at low water, these worm-casts become so numerous that the foot can hardly be placed between them; and even while the spectator is gazing on the wet sand, coil after coil of dark sand emerges from below, as if Michael Scott's familiars were trying to fulfil their task of making ropes from sea-sand.

These sandy coils are the casts of the LUG-WORM, so valuable to fishermen as a bait, and which, when well settled upon the hook, and tipped with a mussel, prove most attractive to the whiting pout, rock cod, plaice, dabs, and other shore-loving fishes. At every low tide the fishermen's boys may be seen busily digging for Lug-worms, or Logs, as they generally term these annelids, and in a populous spot they will fill their square wooden pails in a wonderfully short time.

As a number of Lug-worms lie in a box, covered with sand, mud, and slime, twisting and writhing about in continual movement, they have by no means an attractive aspect, and might even be thought repulsive. But if a single worm be taken from the mass, washed, and placed in a vessel of clear sea-water, it assumes quite a different aspect, and becomes a really beautiful and interesting creature. Its colour is very variable, but usually is dark green and carmine, some specimens being almost entirely of the latter hue. Others, again, are nearly brown, and some of a deep red.

GREAT EUNICE.—*Eunice gigantea.*
LUG-WORM.—*Arenicola piscatórum.*

Along the sides runs a double row of the wonderful bristles by means of which the creature is enabled to propel itself through the sand, and projecting from the back are thirteen pairs of light scarlet tufts, which, on examination, are found to be the gills of the worm. These are most beautiful organs, and when magnified are seen to be composed of many tufts, like the branches of a thick shrub.

The Lug-worm has some of the habits of the tube-making annelids, for, although it is perfectly free and able to move where it likes, it does not push its way through the sand at random, but forms a tunnel of moderate strength, through which it can pass and repass at pleasure. As it bores its way through the sand, it pours out a small quantity of the glutinous matter which has already been mentioned in the Terebella, and thus cements the sides of the tunnel together in a manner somewhat resembling the brickwork of a railway tunnel. Like that work of engineering skill, moreover, the tube of the Lug-worm cannot bear removal, breaking up when it is unsupported by the surrounding earth. It is, however, amply strong enough for its use, and will withstand the beatings of ordinary waves without yielding.

In the whole of the genus Arenicola there are no eyes nor jaws, and the head is not distinct. Several species of this genus are known.

THE second figure in the illustration represents another annelid closely allied to the Nereidæ. In this family the body is very long and composed of numerous segments. The proboscis has, at least, seven, and sometimes nine, pairs of horny jaws. The species which is given in the illustration will sometimes attain a length of more than four feet, and comprise upwards of four hundred segments in its body, each segment furnished with its paddles, some seventeen hundred or more in number.

When in a living state, this is a most lovely creature, winding along its serpentine course with easy grace, and gleaming with all the colours of the rainbow as the sunbeams fall on its polished surface and active propellers.

AN example of the beautiful genus Nereis is now given. The Nereidæ have both tentacles and eyes, and the proboscis is large, often being furnished with a single pair of horny jaws. In the typical genus the eyes are four, arranged in a sort of square, and the tentacles are four in number. The proboscis is thick, strong, and armed with two jaws.

The beautiful Nereids may be found plentifully on our coasts, mostly hiding under stones and rocks, or hiding in the sand. They are well worthy of examination under the microscope; and, perhaps, the best method of making out the structure of these beautiful creatures is by taking a single segment and noticing its construction. On the back are seen certain tufts of different shapes in the various species, but all agreeing in being composed of numerous blood-vessels ramifying in a most complicated manner. These are the gills, or branchiæ, of the Nereis.

On each side are seen the organs of locomotion, sometimes consisting of a single, but mostly of a double, row of oars. Each oar is formed of a strong muscular footstalk, from the extremity of which proceeds a bundle of stiff bristles and a variously formed flap, which is technically called the "cirrus." If the bristles be examined separately, their wonderful forms cannot fail to attract admiration. They no longer appear as the simple hairs which the naked eye would assume them to be, but are transformed, as it were, into

Phyllódoce Paretti.
Nèreis Margaritácea.

a very arsenal of destructive weapons, the barbed spear—the scimetar, the sabre, the sword-bayonet, and the cutlass, all being represented; while there is no lack of more peaceful instruments, such as the grapnel, the sickle, and the fish-hook.

The Nereids will live for a time in a shallow basin half filled with sea-water, and are, therefore, valuable to those who really desire to study for themselves the beautiful forms with which they are surrounded, and which, but for the microscope, would ever be hidden from our eyes. The observer should not fail to examine the formidable proboscis with its terrible jaws. While the worm is at rest, this proboscis is retracted like the finger of a glove, and the jaws appear to be situated in the neck, where, indeed, they were once taken for a gizzard. But either by dissection or applying pressure in the right direction, the jaws can be drawn out, and are then found to be destructive weapons at the end of the proboscis. Many years ago, while examining, for the first time, a Nereis which I had found on the sea-shore, I took this structure for a gizzard, and find, on reference to my note-book, that a sketch of these internal jaws is marked with the title, "Gizzard of the above."

Let me here recommend every reader to sketch, however rudely, each novelty that he finds; always to note the colours, and, if possible, to tint the sketch itself. Those who cannot draw may always get a correct outline by means of the camera lucida, but it is better to give the rudest imaginable sketch rather than none at all. The value of such notes is incalculable; and though at the time they may seem to be of little service, it continually happens that some casual remark is made, and some peculiarity of structure noted, which serve afterwards to settle the most important controversies.

There is scarcely any instrument more generally useful than the camera lucida as adapted to the microscope, for it can not only be used with that instrument, but may be removed, and, by a very simple arrangement, will enable the possessor to take a sketch of a landscape, a house, or an insect, even to copy an illustration from a book, without the least knowledge of drawing. An account of this invaluable instrument, and the method of arranging it to suit various purposes, will be found in my little work on the "Microscope," published by Messrs. Routledge and Co.

THERE are very many species of these interesting worms, no few of which are found on the British coasts, among which we may mention the GLOW-WORM NEREIS (*Nereis noctiluca*), a little species seldom more than an inch in length, but which is remarkable for its power of emitting phosphorescent light in a manner that reminds the observer of the luminous centipede already described. There is also the PEARLY NEREIS (*Nereis margaritacea*), so-called on account of its pearly white colour. This is a much larger and more handsome species, measuring eight or ten inches in length when fully grown. Another species, the SHINING or IRIDESCENT NEREIS (*Nereis fulgens*), is remarkable from the fact that it constructs a tube of very thin silken texture transparent in itself, but often being studded with particles of sand. It seldom exceeds seven inches in length, and is of a deep orange-red colour, with a blackish line running along the back. Like most of the Nereids, it is iridescent on the surface.

Another magnificent species, called the LAMINATED NEREIS (*Phyllodoce laminata*), deserves a passing notice. This fine species sometimes attains the length of two feet, and is certainly the finest example of the family that is to be found in the British seas. Its colour is shining iridescent green, having a bluish tinge on the back, and changing gradually to a more leaf-green hue on the sides. There are no less than four hundred segments in the body of a full-grown Phyllodoce, and, consequently, eight hundred paddles and sets of bristles, by means of which it can swim through the water or craw upon the sand with equal ease and grace.

As Mr. T. Rymer Jones well remarks: "The mechanism of this creature, its parts and their powers, are to be ranked among the more conspicuous and admirable works of creation, nor can they be contemplated without wonder. Issuing forth from its retreat, it swims by an undulating serpentine motion. Its unwieldy body, gradually withdrawn from its hiding-place, has its multiplied organs unfolded in regular order and arrangement, so that, whether intertwined or free, they never present any appearance of intricacy or confusion—each part performs its own proper functions, and the general effect is produced by the united exercise of the whole. When inactive, the lateral paddles are laid close over the back, but when in activity they spread widely out, acting like so many oars to aid the animal's course by their united impulse on the water.

It is a pleasant thing to see a well-manned boat glide over the smooth surface of the sea, or to watch the long array of oars as silently they simultaneously dip and rise again, all flashing in the evening sunshine. But such a sight is but a paltry spectacle compared with that afforded by these gorgeous worms; four hundred pairs of oars, instinct with life, harmoniously respond in play, so active that the eye can scarcely trace their movements, save by the hues of iridescent splendour, violet and blue, and green and gold, the very rainbow's tints that indicate their course."

It is a remarkable fact, that in the Nereids their young are often produced by the simple process of breaking off a piece from the end of the body. The last ring but one becomes swollen and lengthens, and by degrees assumes the appearance of a young Nereid, with its eyes and antennæ. When it is sufficiently strong it is broken off, and goes forth to seek an independent life. Sometimes it happens that a second and a third are thus formed before the first is separated, and M. Milne-Edwards has seen a row of six young Nereids thus attached to their parent.

The CIRRHATULUS derives its name from the numerous cirrhi, or thread-like appendages, which project from its sides, and which serve for legs as well as organs of respiration. These curious appendages arise from the alternate segments of the body, and are continued in two rows along the back almost to the very end of the body.

If placed under a good microscope, the transparent walls of these cirrhi permit the blood to be seen coursing through them. This is not, however, a very easy operation, as the creature is very timid, and when touched will contract the cirrhi into a shapeless bundle. When, however, the Cirrhatulus is quite at its ease, recumbent in its rocky home,

Syllis maculósa.   Cirrhátulus Lamárckii.

it permits the cirrhi to lie flat on the ground, where they surround it like a mass of red worms continually writhing and twining throughout their length.

It is one of the light-hating creatures, always seeking a retreat under some stone or in a cleft of a marine rock; and it is, moreover, protected by a mass of sand, mud, and slime, which it collects around its body, so as effectually to disguise its shape. The length of the Cirrhatulus is about four inches, and its colour is mostly red, with a tinge of brown.

THE members of the genus SYLLIS may be easily distinguished by the shape of the tentacles, which are jointed in such a manner as to resemble closely the beads of a necklace. The number of the tentacles is always uneven, and this fact serves to separate them from an allied genus, where their number is even.

THE strange hair-covered grub which is shown in the following illustration appears a very ordinary being when merely depicted in the plain black and white of the printer;

3.   Z Z

PORCUPINE SEA-MOUSE.—*Aphrodite hystricella.*   SEA-MOUSE.—*Aphrodite aculedta.*

but when seen in its living state, or even when preserved in spirits of wine, proves to be one of the most gorgeous inhabitants of earth.

The SEA-MOUSE, as the creature is called from its hairy coat, possesses beauties which never fail to strike even the unobservant eye of a casual passenger, as the wondrous hues of ruby, emerald, sapphire, and every imaginable gem, flash from the coat of this breathing rainbow. Each hair of the Sea-mouse is a living prism, and when held singly before the eyes is a most magnificent object in spite of its small dimensions, flinging out gleams of changing colours as it is moved in the fingers, or the direction of the light is changed. I have often thought that if Shakespeare had only known of the Aphrodite, he might have furnished Queen Mab with a still more fairy-like conveyance.

Yet the habits of the creature seem to be quite out of accordance with its exceeding beauty. When the sunlight falls on its surface, the many-hued hairs give forth a chromatic radiance which is almost painful to the eyes from its very intensity; and it would be but natural to conclude that the Aphrodite made its home in the sunniest spots, and welcomed the dawn with gladness. Such, however, is not the case; for this beautiful creature, which wears all the colours of the humming-bird and seems equally a child of the sun, passes its life under stones, shells, and similar localities at the muddy bottom of the sea.

The whole group of the Aphroditacea is separated from the rest of the order by reason of the curious mode in which its respiration is conducted. If the beautiful hairs be pushed aside, a series of scales will be seen upon the back, which are guarded by a covering of a loose felt-like substance, composed of interwoven hairs. This felt, if it may be so called, acts as a filter, which is very necessary, considering the muddy localities in which the creature lives, and permits the water to pass in a purified state to the breathing apparatus, which is set beneath the scales. These scales or plates move up and down something like the gills of a fish, and by their alternating movements have the power of admitting the water and then expelling it in regular pulsations. If a recent specimen be examined, a considerable quantity of mud is always to be found entangled in the felt-like covering of the scales.

Sometimes this beautiful annelid attains a considerable size, reaching the length of five or even six inches. Generally, however, from three to four inches is the measurement. It is a slow-going, but very voracious creature, feeding even upon its own kind, and using its powerful proboscis as a means of capture.

In some species of this genus, the spines which edge the body are most marvellously formed. They are set upon projecting footstalks, and when not required for use can be drawn back into the body. Their shape, however, would seem to render such a proceeding dangerous, inasmuch as they are formed just like the many-barbed spears used by certain savage tribes. In the *Aphrodite hispida*, for example, they are just like doubly-barbed harpoons, and would wound the soft tissues of the body most severely when withdrawn. In order, therefore, to prevent this result, each spine is furnished with a sheath composed of two blades, which close upon the barbs when the weapon is withdrawn, and open again to allow its exit when it is protruded.

THE second species represented in the above illustration is easily distinguished from the preceding creature by the peculiar structure of the back, which is devoid of the felt-like substance, and the scales are consequently bare.

CHÆTOPTERUS.—*Chætópterus pergamentaceus.*

ON account of the singular structure of the CHÆTOPTERUS, it has been placed in a family by itself, of which it constitutes the sole genus.

This remarkable worm is one of the tube-builders, and makes a dwelling of a tough parchment-like consistency, measuring eight or ten inches in length. It is found in the seas about the Antilles. In this curious being there is no distinct head, and no vestige of maxillæ, but the proboscis is furnished with a lip, to which are attached two small tentacles. "Then comes a disk with nine pairs of feet, then a pair of long silky bundles, like two wings. The gills, in the form of laminæ, are attached rather below than above, and predominate along the middle of the body."

THE next family, of which the common EARTH-WORM is a very familiar example, is distinguished by the ringed body without any gills or feet, but with bristles arranged upon the rings for the purpose of progression.

In the well-known Earth-worm, the bristles are short and very stiff, and are eight in number on each ring, two pairs being placed on each side ; so that, in fact, there are eight longitudinal rows of bristles on the body, four on the sides, and four below, which enable the creature to take a firm hold of the ground as it proceeds.

EARTH-WORM.—*Lumbricus terrestris.*

Except that the worm makes use of bristles, and the snake of the edges of its scales, the mode of progress is much the same in both cases. The whole body of the creature is very elastic, and capable of being extended or contracted to a wonderful degree. When it wishes to advance, it pushes forward its body, permits the bristles to hitch against the ground, and then, by contracting the rings together, brings itself forward, and is ready for another step. As in each full-grown Earth-worm there are at least one hundred and twenty rings, and each ring contains eight bristles, it may be imagined that the hold upon the ground is very strong.

As every one knows, the Earth-worm lives a very solitary life below ground, driving its little tunnels in all direction, and never seeing its friends, except at night, when it comes cautiously to the surface and searches for company. In the evening, if the observer be furnished with a "bull's-eye" lantern, and will examine the ground with a very gentle and cautious step, he will be sure to find many worms stretching themselves out of their holes, retaining for the most part their hold of the place of repose by a ring or two still left in the hole, and elongating themselves to an almost incredible extent. If, while thus employed, an Earth-worm be alarmed or touched, it springs back into its hole as if it had been a string of india-rubber that had been stretched and was suddenly released.

The worms have a curious habit of searching for various leaves and dragging them into their holes, the point downwards, and are always careful to select those particular leaves which they best like. As a general rule, they dislike evergreens; and the leaf which I have found to be most in favour is that of the primrose. I have often watched the worms engaged in this curious pursuit; and in the dusk of the evening it has a very strange effect to see a leaf moving over the ground as if by magic, the dull reddish brown of the worm being quite invisible in the imperfect light.

The food of the Earth-worm is wholly of a vegetable nature, and consists of the roots of various plants, of leaves, and decaying vegetable substances. Many persons cherish a rooted fear of the Earth-worm, fancying that it lives in churchyards and feeds upon the dead. These fears are but idle prejudice, for the worm cares no more for the coffined dead than does the tiger for the full manger, or the ox for the bleeding gazelle. The corpse when once laid in the ground sinks into its dust by natural corruption, untouched by the imagined devourer.

The so-called worms that feed upon decaying animal substances are the larvæ of various flies and beetles, which are hatched from eggs laid by the parent; so that if the maternal insect be excluded, there cannot be any possibility of the larvæ. Moreover, neither the fly nor beetle could live at the depth in which a coffin is deposited in the earth; and if perchance one or two should happen to fall into the grave, they would be dead in half an hour, from the deprivation of air and the weight of the super-incumbent soil.

Let, therefore, the poor Earth-worm be freed from causeless reproach; and though its form be not attractive, nor its touch agreeable, let it, at all events, be divested of the terrors with which it has hitherto been clothed.

The Earth-worm is a timid and retiring creature, living below the surface of the ground, and having a great objection to heat and light. Heat dries up the coat of mucus with which its body is covered and which enables it to slide through the ground without retaining a particle of soil upon its surface. A very moderate amount of heat soon kills an Earth-worm; and if one of these annelids be placed in a spot where it cannot hide itself from the sun's rays, it soon dies, and either melts into a kind of soft jelly, or hardens into a thin strip of horny parchment.

The vexed question of its use to agriculture is too wide a subject to be treated at length in these pages; but we may safely come to two conclusions—first, that unless it were of some use it would never have been made; and secondly, that it will be wiser to find out wherein its use lies than to kill it first and then perhaps discover that its presence was absolutely needful and its absence injurious.

The Earth-worm is of no direct use to mankind, except, perhaps, as bait for the angler; and for this purpose they are easily obtained by the simple process of driving a garden-fork into the ground and shaking it about vigorously. The timid worms are very much alarmed at the tremulous earth, and come to the surface for the purpose of escaping, when they can be easily seized and captured.

The COMMON LEECH is almost as familiar as the earth-worm, and is one of a genus which furnishes the blood-sucking creatures which are so largely used in surgery. It belongs to a large group of Annelida which have no projecting bristles to help them onward and are therefore forced to proceed in a different manner.

All these Leeches are wonderfully adapted for the purpose to which they are applied, their mouths being supplied with sharp teeth to cut the vessels, and with a sucker-like disc, so that the blood can be drawn from its natural channels; while their digestive organs are little more than a series of sacs in which an enormous quantity of blood can be received and retained.

Every one who has had practical experience of Leeches, whether personally a sufferer or from seeing them applied to others, must have noticed the curious triangular wound which is made by the teeth. If the mouth of a Leech be examined, it will be seen to have three sets of minute and saw-like teeth, mounted on as many projections, which are set in the form of a triangle. The wound made by this apparatus is rather painful at the time, and is apt to be troublesome in healing, especially in the case of very thin-skinned persons, requiring the application of strong pressure and even the use of some powerful caustic.

At one meal the Leech will imbibe so large a quantity of blood that it will need no more food for a year, being able to digest by very slow degrees the enormous meal which it has taken. It is a very remarkable fact, that the blood remains within the Leech in a perfectly unchanged state—as fresh, as red, and as liquid as when it was first drawn —and even after the lapse of many months is found to have undergone no alteration.

The very great difficulty in inducing a Leech to make a second meal is well known, and can be well accounted for by the fact that it has already taken enough food to support existence during one-sixth of its whole life. In England this is almost impossible, as the time occupied in reducing the Leech to the requisite state of hunger is so long that it more than counterbalances the value of the creature itself. " Use up, and buy more," is the plan that is now pursued by the surgeon.

The Leeches that are used in this country are mostly imported from Brittany, where they live in great numbers and constitute an important branch of commerce, being sold by millions annually. The Leech-gatherers take them in various ways. The simplest and most successful method is to wade into the water and pick off the Leeches as fast as they settle upon the bare legs. This plan, however, is by no means calculated to improve the health of the Leech-gatherer. who becomes thin, pale, and almost spectre-like, from the constant drain of blood, and seems to be a fit companion for the old worn-out horses and cattle that are occasionally driven into the leech-ponds in order to feed these bloodthirsty annelids.

Another plan is to entangle the Leeches in a mass of reeds and rushes, and a third method is, to substitute pieces of raw meat for the legs of the man, and take off the

COMMON LEECH. — *Hirudo medicinalis.*

Leeches as they gather round the spoil. This proceeding, however, is thought to injure the health of the Leeches, and is not held in much favour.

Those who keep Leeches, and desire that they should be preserved in a healthy state, will do well to line the sides of the vessel with clay, and to place a bundle of moss, equisetum, and similar materials, with the annelids. These creatures are invested with a coat of slime, and, as is often the case with such beings, is obliged frequently to change its skin. This operation is performed every four or five days, and is rather a troublesome one, unless the Leech be furnished with some such materials as have just been mentioned. Like the snake, when in the act of casting its slough, the Leech crawls among the stems of moss, and thus succeeds in rubbing off its cast garment.

The Leech lays its eggs in little masses, called cocoons, each of which contains, on the average, from six to sixteen eggs. These cocoons are placed in clay banks, and are of rather large size, being about three-quarters of an inch in length. In some parts of France, attempts are made to rear the Leeches ; and it is found that these fastidious annelids will not lay their cocoons in small tanks, but require large reservoirs lined with clay and edged with weeds and other aquatic plants.

I may perhaps mention that our British waters contain other species of blood-sucking Leeches, which are found mostly in still or stagnant waters, and invariably gather to a spot where the mud is thick, soft, and plentiful. Last summer, while bathing, I waded through some mud in order to pick some very fine dewberries that were overhanging the bank, and when I began to dress found that my feet were covered with Leeches of different sizes. I counted eighteen on one foot, and then found that their numbers were so great that I ceased to count them.

IN the accompanying illustration we have two more examples of this curious family.

The upper figure represents the common HORSE-LEECH, which is so plentiful in our ditches and more sluggish rivers. This annelid is distinguished from the preceding by the character of its teeth, which are not nearly so numerous as in the medicinal leech, and much more blunt. It is a carnivorous being, and feeds upon the common earth-worm,

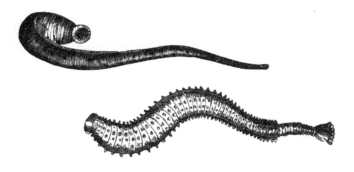

HORSE-LEECH.—*Hæmopis sanguisorba.*
SKATE-SUCKER.—*Albiöne muricáta.*

seizing it as it protrudes itself from the banks of the stream in which the Horse-leech resides. There is a popular prejudice against the Horse-leech, the wound which it makes being thought to be poisonous. This, however, is clearly erroneous, and the creature has evidently been confounded with another species, the BLACK LEECH (*Pseudobdella nigra*). The Horse-leech is much larger than the medicinal species, and may be known by its colour, which is greenish black; whereas that of the medicinal leech is green, with some longitudinal bands on the back, spotted with black at their edges and middle; the under surface yellowish green, edged, but not spotted with black.

THE lower figure in the same illustration represents the SKATE-SUCKER, so called because it is found adhering to several fishes, and is especially prevalent on the common skate and others of the ray tribe. Almost all the species of this genus are beset with the curious nodules upon the rings of the body, which give to the creatures so strange an aspect. In this genus, moreover, the portion containing the head is quite distinct and separated from the body by a sort of neck.

All these creatures have two modes of movement: they can crawl slowly along by means of moving their rings alternately, or they can proceed at a swifter pace by employing a similar mode of progress to that which is made use of by the larvæ of the geometrical moths. Being furnished with a sucker at either end, they first fix their hinder sucker against any object, and then extend the body well forwards. Having secured the sucker of the head, they loosen their hold of the posterior sucker, arch their bodies just like the looper-caterpillars, and so proceed. In the illustration of the common leech, one of the specimens is shown in this attitude.

Before bidding farewell to the Leeches, we must cast a casual glance at three remarkable members of this group.

The first is the BRANCHELLION or BRANCHIOBDELLA, a flattish and not very large creature, which is notable for being parasitic upon the torpedo, and retaining its hold in spite of the electric powers of the fish. Another species of the same genus is found on the lobster.

The second of these creatures is the wonderful NEMERTES, a leech-like being not furnished with sucker, and attaining the extraordinary length of thirty or forty feet. It is possessed of the most astonishing powers of extension and contraction, so that a specimen of twenty feet in length will suddenly contract to three or four feet. No one, indeed, seems to know the precise length to which a Nemertes can extend.

It is always writhing and coiling its long body into apparently inextricable knots, but never suffering any real entanglement, and has been known, without any apparent reason, to convert its whole body into a long and slender screw. I should much like to give a detailed account of this wondrous being, but am forced to pass to our next illustration.

The last of these beings is the LAND-LEECH of Ceylon (*Hirudo Zeylonica*), a terrible pest to those who travel through the forests, and often occurring in such vast numbers as even to endanger life. On every twig and grass-blade the Land-leeches sit, stretching out their long bodies towards the coming prey, and hastening after the traveller with direful speed. Instinctively they make for the unprotected parts. They crawl rapidly up his clothes, they insinuate themselves into his neck, his coat-sleeves, and his boots; and wherever a patch of bare skin can be found, there a small colony of leeches is sure to hang.

Lank and thread-like as they are when they first assail their prey, they soon swell to an enormous size with the blood which they imbibe; and in the cases of horses and cattle, hang in large clusters like bunches of black grapes. When travelling through the forest, the last of the train are always the most pestered by the leeches; for the blood-thirsty annelids have been aroused by the passage of the pioneers, and hasten up just in time to catch the rear-guard. Those who wish to gain a good idea of this leech may do so by consulting Sir J. Emerson Tennent's "Natural History of Ceylon."

WE now pass to the last members of this great class, the Entozoa, or Internal Worms, so called because they are all found in the systems of living animals. They have also, but wrongly, been termed Intestinal Worms, inasmuch as very many species inhabit the respiratory, or even the sensorial, organs, and are never found in the intestines. The Entozoa are very numerous, and are distributed throughout the world, inhabiting the interior of various living beings; and, indeed, their presence is so universal, that wherever an animal can live, there are Entozoa to be found within its structure.

To give an idea of the wide distribution of these strange beings, we will take one genus of Entozoa as exemplified by the specimens in the British Museum, and note the various animals in which the members of that single genus have been found.

The restricted genus Ascaris is the type of its family, and many specimens are in the collection of the British Museum. These have been taken from the following animals: man, mole, dog, fox, cat, seal, wood-mouse, sow, horse, grizzly bear, heron, tortoise (several). toad, frog, ruffe, blenny, fishing-frog, barbel, cod (several), turbot, flounder, eel, goshawk, barn-owl, lapwing, red-wing, cormorant, and grouse. These are taken according to the number of the specimen in the catalogue, without any attempt at arrangement according to their locality.

The history of these remarkable beings is, for the most part, shrouded in mystery, and we know but little of their true habits and the manner of obtaining entrance into the various beings on which they live. It is, however, ascertained that the young of the Entozoa have a very different shape from that of their parents, and that they may obtain entrance into their future homes under the disguise of various forms.

To this curious sub-class belong many remarkable creatures, among which the following may be briefly mentioned. The GUINEA WORM (*Filaria medinensis*) is one of the most developed of this group of animals. It is a strange-looking creature, eight or ten feet in length, and not thicker than ordinary sewing thread. It is found in many of the hotter parts of the world, especially in the country from which it takes its name. It is also found in America, though it does not appear to be plentiful except in the Island of Curaçao.

The Guinea Worm is much dreaded by the inhabitants of the countries where it resides, on account of the pain and inconvenience which it occasions, and the great difficulty in destroying it. It mostly takes up its residence in the leg, and there grows to

CAT-WORM.—*Tænia crassicollis.*
LARK-WORM.—*Tænia platycéphala.*

an inordinate length, causing much pain and swelling until the head of the worm makes its appearance. As soon as the sufferer perceives that the worm has made its appearance, he takes a small piece of wood or pasteboard, rolls the projecting end of the worm round it, and, after very cautiously drawing it out for a few inches, he winds up the loosened portion, and ties the stick close to the limb. Next day, he draws out a little more of the worm, and proceeds in a similar manner, until, in a fortnight or three weeks, the entire entozoon has been withdrawn and wound upon the roller.

This process is extremely simple, but demands the greatest care, as, if the worm should be broken, a most painful and even dangerous tumour is certain to arise. I have seen a moderately large specimen of the Guinea Worm extracted from an English sailor, into whose leg it had made an entrance, and who was quite incapable of work until his enemy had been destroyed.   It is rather flat, like some kinds of silken thread, and is of a very pale brown colour.

Passing by a few genera, such as Ascaris, Tricocephalus, and Strongylus, all of which are found in the human subject, we come to the large and important genus Tænia, which may be accepted as the type of all the Entozoa.   The well-known Tape-worm (*Tænia solium*) belongs to this genus.   It derives its name of *solium*, as also its French title of Ver Solitaire, from the supposed fact that only one individual can infest the same person.   It is, however, known that a few cases have occurred where the same individual has been afflicted with, at least, two specimens of Tape-worms.   Those which belong to this genus may be known by the long flat body, and the head with four suctorial spots, and almost invariably a circle of very small, but very sharp, hooks.   The whole structure of the Tænia is interesting to those who study comparative anatomy, but too purely scientific to be described in these pages.

The upper figure in the preceding illustration represents a species of Tape-worm that is found in the common cat ; and the lower figure affords an example of a Tape-worm that is found in several birds, such as the nightingale, blackcap, and the bird from which it has derived its popular name.   This figure is intended especially for the purpose of showing the peculiar head, with its four suctorial discs.   Another example of this genus is seen in the next illustration, exhibiting the curious head, and the extraordinary difference of size between the segments of the body.   This species is found in the black-backed gull.

There is another notable genus of Entozoa, commonly going by the name of Hydatids, on account of the large amount of liquid which is found within their cells.   Within each large cyst, or cell, myriads of smaller cysts may be found, some in an early stage of progress, and others being further advanced, and containing other cysts within themselves.

I have seen some enormous Hydatids taken from the interior of a female monkey. They were so large, and so full, that the owner of the animal thought that it had died from over-eating itself.   On opening the creature, however, the stomach was found to contain very little food, and almost the whole cavity of the chest and abdomen was filled with huge cysts, which had encroached upon all the vital organs, pushed the heart on one side, enveloped the whole of one lung ; and, in fact, had caused so strange a disturbance of the viscera, that the fact of the creature having supported life under such circumstances seemed almost incredible.   The large cysts contained a vast number of smaller cells, and these again were filled with cysts of still less dimensions.   A large quantity of fluid also existed, and floating in the liquid were found myriads of echinococci, very small, but with the characters exhibited perfectly well under the microscope.

We now come to our last example of the Entozoa.

The Ray-worm is, as its name imports, an inhabitant of fishes belonging to the ray tribe.   The genus to which this creature belongs is a rather remarkable one, the head affording some curious characteristics.   As may be seen, it is very large in comparison with the body, and has four deep clefts set opposite to each other.   Some of the species have only two of these clefts, which, in the Greek language, are called " bothria," but in all the genus either two or four of these clefts are to be found.   The generic title literally signifies " furrow-headed."

One species of this genus (*Bothriocephalus latus*) is the well-known Broad Tape-worm, which is so injurious to man whenever it takes up its abode within a human being. This creature will sometimes attain a length of twenty feet; and it may always be recognised by the peculiarly deep and opposite furrows on the head.   Like all its kind, this Tape-worm increases with great rapidity, multiplying its joints by division, and in that manner extending itself to the great length which has already been mentioned.

The whole history of these strange creatures is very obscure, on account of the impossibility of watching them in the spots wherein they take their residence.   They are all, as is generally the case with beings low in the scale of nature, exceedingly tenacious of life, and will survive treatment which would kill many beings apparently stronger and more capable of resistance than themselves. For example, several species of Entozoa have been found in a living state within meat after it has been cooked, especially in those spots where the heat had not penetrated sufficiently to destroy the natural ruddiness of its colour. It is known that "measly" pork derives its peculiar aspect from the presence of Entozoa, and that many of these Entozoa, or, at all events, their eggs, are swallowed by human beings, within whom they make their homes, and whom they condemn to infinite worry, pain, and weakness.

Even immersion in alcohol does not seem to inflict any serious damage upon these creatures. Rudolphi found a specimen of an Ascaris quite lively in the throat of a

GULL-WORM.—*Tænia porósa.*
RAY-WORM.—*Bothriocéphalus tumidulus.*

cormorant that had been steeped in spirits of wine for nearly a fortnight. Even the severe process of being thoroughly dried appears to be quite as ineffectual to destroy these beings. A number of Ascarides that had been removed from a fish and suffered to become quite dry, and were apparently nothing more than flat slips of membrane adhering to a board, began to revive as soon as they were wetted, and actually moved the moistened part of their bodies, while the remainder was still dry and adhering to the board.

RADIATA.

ECHINODÉRMATA.

WE now arrive at a vast and comprehensive division of living beings, which have no joints whatever, and no limbs, and are called Radiata, because all their parts radiate from a common centre. The structure is very evident in some of these beings, but in others the formation is so exceedingly obscure, that it is only by anatomical investigation that their real position is discovered.

The highest forms in this division have been gathered together in the class Echinodermata. This word signifies Urchin-skinned, and is given to the animals comprising it because their skins are more or less furnished with spines, resembling those of the hedgehog. In these animals the radiate form is very plainly shown, some of them assuming a perfectly star-like shape, of which the common star-fishes of our coasts are familiar examples. In some of the Radiates, such as the sea-urchin, the whole body is encrusted with a chalky coat, while in others it is as soft and easily torn as if it were composed of mere structureless gelatine.

The mode of walking, or rather creeping, which is practised by these beings, is very interesting, and may be easily seen by watching the proceedings of a common star-fish when placed in a vessel of sea-wate. At first it will be quite still, and lie as if dead, but by degrees the tips of the arms will be seen to curve slightly, and then the creature slides forward without any perceptible means of locomotion. If, however, it be suddenly taken from the water and reversed, the mystery is at once solved, and the walking apparatus is seen to consist of a vast number of tiny tentacles, each with a little round transparent head, and all moving slowly but continually from side to side, sometimes being thrust out to a considerable distance, and sometimes withdrawn almost wholly within the shell.

These are the " ambulacræ," or walking apparatus, and are among the most extraordinary means of progression in the animal kingdom. Each of these innumerable organs acts as a sucker, its soft head being applied to any hard substance, and adhering thereto with tolerable firmness, until the pressure is relaxed and the sucker released. The suckers continually move forward, seize upon the ground, draw the body gently along, and then search for a new hold. As there are nearly two thousand suckers continually at work, some being protruded, others relaxed, and others still feeling for a holding-place, the progress of the creature is very regular and gliding, and hardly seems to be produced by voluntary motion.

We will now proceed to our examples of these curious beings.

On the upper portion of the illustration is seen a specimen of the turritella-shell, from which protrudes a long, slender, worm-like being. This is the HERMIT SIPUNCULUS, a creature which is remarkable for the fact that it resides in the empty shells of molluscs, after the same fashion as is observed by the hermit crabs.

If taken out of the shell, the Sipunculus resembles a worm so closely, that it might easily be mistaken for an annelid ; and indeed, according to one of our best zoologists, it forms a link between these two great divisions, for in its person radiism sets and annulism begins. The end of the body, which is concealed within the shell, is capable of being enlarged into a bulb-like shape, which enables the creature to maintain a firm hold of its shelly retreat, and the other extremity is furnished with an external proboscis, at the end of which is a small circlet of tentacles.

Several species of this genus are eatable and held in great estimation by the Chinese, who catch them in a very ingenious manner. The EDIBLE SIPUNCULUS lives in holes in the sand, and always keeps the mouth of its burrow open. The Chinese fishermen arm themselves with a bundle of slender wooden rods, tapered to a point at one end, and

Priápulus caudátus.

SPOON-WORM.—*Thalassema Neptúni.*

Syrinx nudus.

Sipúnculus Bernhardus.
(Two figures.)

having a little round knob at the extremity of the point. They proceed to the sands at low water, and drop one of these rods into each burrow, where they leave it for twelve hours. During this time the Sipunculus is sure to swallow the button, and as the elastic tissues contract it is unable to release itself; and when the tide has again retreated, the rod, with the Sipunculus attached, is drawn out of the burrow by the fishermen.

The species which we are now examining is very careful of its own comfort, and in order to make the entrance of its shell exactly suitable to its own size it stops up the aperture with sand and similar substances. The individual which is represented in the illustration, and which was drawn from a living specimen, had used both sand and small pebbles in this work.

Lying by the side of the sipunculus is a rather curious creature, represented as lying on the ground, and notable for the long tuft-like appendage at its extremity. This is the Tailed Priapulus, a species which is found in the southern seas, and occasionally taken off our own coasts. The curious bundle of threads at its extremity is supposed to serve the purpose of respiratory organs. It has a retractile proboscis, but no tentacles round the mouth.

The Syrinx is distinguished from the Sipunculus by the proboscis, which in these creatures is rather short, and has an indented tentacular fold round the mouth. The figure shows an example with the proboscis protruded. The generic name, Syrinx, is derived from the resemblance of the creature to the reed from which the ancient pipes were made.

In the middle of the illustration are seen two smaller figures of different shapes. These, however, represent the same species, and are here given in order to show the wonderful capacity for changing its shape. The Spoon-worm is so called on account of the spoon-like appendage to the proboscis. Behind the proboscis are two shining hook-like bristles. All the members of this genus are remarkable for the wonderful power of contraction and expansion possessed by the skin, and the extraordinary manner in which

Psolinus brevis.   *Psolus phantapus.* (Two figures.)

they can alter their shape. In consequence of this extremely contractile structure, the whole of the water contained in the body is spirted out as soon as a wound is made, and intestines are seen to be forced out after the water. One species of Thalassema is used as bait by fishermen.

All the species on the preceding engraving belong to the family Sipunculidæ, and are found in British seas.

The specimens on the above and the next two illustrations belong to the family HOLOTHURIDÆ, and are popularly known by the name of SEA CUCUMBERS or SEA PUDDINGS. In these the body is mostly cylindrical, and is covered with a tough leathery skin, upon which are placed a number of scattered chalky particles. The mouth is surrounded with a set of retractile tentacles.

Some species of this family are eaten by the Chinese, and a large trade is carried on in these strange products of the sea; the annual merchandise being worth about two hundred thousand pounds. The price of the TREPANGS, as they are called, is very variable, according to the species, some kinds being comparatively cheap, and costing rather less than two pounds per hundredweight, while others will fetch thirty pounds for the same weight. There are, besides, "fancy prices" for some very scarce species of Trepang, which, however, are likely to be equalled in real value by the cheaper and commoner kinds. In the British Museum may be seen a valuable collection of the various kinds of Trepang, with their prices attached.

They seem to be very unattractive creatures, black, wrinkled, and looking much as if they had been made out of the upper-leather of old shoes. They are, however, convertible into a rich and palatable soup, and are also stewed in various ways, taking, in fact, the same rank among the Chinese that turtle does with us. The Trepang is prepared for the market by being carefully opened and cleansed, laid in lime, and then dried either in the sun or over wooden fires.

Crawling up the stem of the sea-weed may be seen an odd-looking little creature, called the PSOLINUS, remarkable for the great length of the ambulacræ, which lift it well above the object on which it walks. Owing to this fact, it has quite an intelligent aspect as it crawls along, with its beautiful crown of tentacles expanded, and waving in the water.

The two larger figures represent the same species in different aspects, the upper showing its under surface, and the lower exhibiting its appearance while walking. In these two curious genera, the ambulacræ are only distributed in the under surface, and in the present example are placed in three rows on a flattened disc, which occupies part of the under surface.

In the genus PENTACTE, the ambulacræ are placed in a series of parallel rows along the body, sometimes six, but mostly five in number. The two upper figures in the illustration on page 720 represent a member of this genus under two conditions.

It is a remarkable fact, that when one of the Holothuridæ is alarmed, or suffers from indigestion, or is affected in any way, it proceeds to an act which is the exact analogue of the Japanese custom of " happy despatch." Under any or either of these circumstances, it proceeds to disembowel itself, and does so with a completeness and promptitude that are almost incredible. It disgorges the whole of its interior, with all the complicated arrangements that render the Holothuridæ such singular beings to dissect, casts away all its viscera, its stomach, and even throws off the beautiful bell of tentacles.

Having done this, and reduced itself to the condition of an empty skin, which cannot eat because it has no mouth and no stomach, and will not walk, because it has no object for locomotion, it remains perfectly quiescent for some months. At the expiration of that period, a fresh set of tentacles begin to make their appearance ; they are followed by other portions ; and after a while, the animal is furnished with a completely new set of the important organs which it had cast away. It seems a singular cure for indigestion, but no one can deny its efficacy.

The uppermost figure was taken from a living specimen which had first flung off its tentacles, then turned its stomach inside out, then thrown it aside, and six months afterwards was still alive. The central figure represents the same species in its ordinary state, with its beautiful tentacles protruded.

The lowermost figure represents a fine species of SEA CUCUMBER, which has received its generic name from its great resemblance to that vegetable. The smaller species are appropriately named Sea Gherkins. The food of all these animals consists of marine molluscs and other small inhabitants of the sea. The complete but empty shell of several small molluscs have been found within the stomach of dissected specimens, proving that the creature must have swallowed the shell entire, and dissolved out its inhabitant by the process of digestion.

It may as well be mentioned that the only vestige of a skeleton in these creatures is a ring of chalky substance surrounding the beginning of the intestinal canal, and formed of ten pieces, five large and as many small. To this curious ring are attached the longitudinal muscles of the body, by which the creature can lengthen or shorten itself at will, the expansion and contraction of the body being due to a series of transverse muscular fibres. The longitudinal muscles are ten in number, and are arranged in five pairs.

THE uppermost figure in the illustration on page 721 is another example of the Cucumariæ, and is here given in order to show the beautiful mouth with its crown of tentacles.

The other figures represent a most singular being called the SYNAPTA, from a Greek word signifying to seize hold of anything. This name is given to it because, when the hand is drawn over its surface, the skin is slightly arrested by some invisible agency.

On taking off part of the skin of the Synapta and placing it under the microscope, a most wonderful sight is disclosed. The skin is furnished with a number of little tubercles

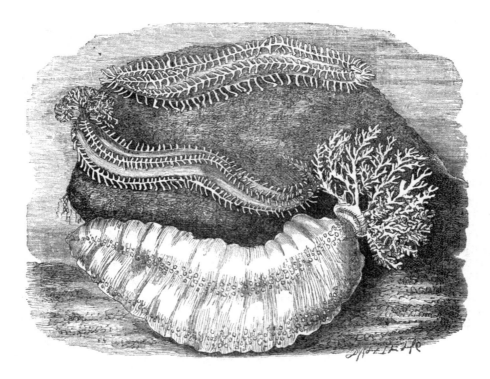

Pentacta pentactes.
Cucumária Drummondii.

on which are set numbers of tiny spicules which look as if they were anchors for a fairy fleet. They are of extremely minute dimensions, and are quite invisible without the aid of a microscope, but never fail to excite admiration when they are well exhibited. Perhaps the best method of bringing out their beautiful shapes is by using a parabolic condenser or a spotted lens, as then their translucent glassy forms shine out against a dark background.

These little objects are of exactly the same shape as the classic anchors of ancient times, and were it not for their extreme minuteness, the person who sees them for the first time is tempted to think that they have been manufactured by some ingenious impostor. But the hand of man is quite incapable of making these beautiful little objects, with their long shanks, their gracefully curved arms, and their sharply-pointed and regularly-serrated flukes.

Nor are the anchors the only wonders which so appropriately deck the skin of a marine animal. If the little prominences can be neatly placed under the microscope without being rubbed, each anchor is found to be affixed by the end of the shank to the end of a curiously-formed shield, made of the same translucent substance as the anchor itself, and pierced with a perfectly regular pattern like ladies' "cut-work" embroidery. These shields hold the anchor in such a way that, as the shield lies flatly upon the skin, the flukes of the anchor are held in the air. The object of this remarkable arrangement is not known.

There are several species of Synapta, all with the anchors and shields, but the pattern upon the shields is different in the various species, as is the shape of the anchor. These

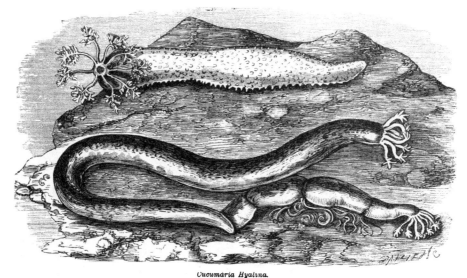

Cucumária Hyalína.

Synapta digitáta. (In health.)   Synapta digitáta. (Unwell.)

remarkable appendages have been compared by some authors to the little hooks on the calyx of the well-known burdock.

The central figure represents a Synapta in its usual state, and the lowest figure exhibits the same creature as it appears when it is annoyed and is explaining its feeling by the simple process of cutting itself to pieces. Insufficiency of food will induce the creature to perform this strange act, and impure water will have the same effect.

WE now come to a new and beautiful family of this order, called Echinidæ, because they are covered with spines like the quills of the hedgehog. Popularly, they are known by the name of SEA-URCHINS, or SEA-EGGS.

In all these curious beings the upper parts are protected by a kind of shell always more or less dome-shaped, but extremely variable in form, as will be seen in the following illustrations. The shell is one of the most marvellous structures in the animal kingdom, and the mechanical difficulties which are overcome in its formation are of no ordinary kind. In the case of the common SEA-EGG, which is shown on the following illustration, the shell is nearly globular. Now, this shell increases in size with the age of the animal; and how a hollow spherical shell can increase regularly in size, not materially altering its shape, is a problem of extreme difficulty. It is, however, solved in the following manner :—

The shell is composed of a vast number of separate pieces, whose junction is evident when the interior of the shell is examined, but is almost entirely hidden by the projections upon the outer surface. These pieces are of a hexagonal or pentagonal shape, with a slight curve, and having mostly two opposite sides much longer than the others. As the animal grows, fresh deposits of chalky matter are made upon the edges of each plate, so that the plates increase regularly in size, still keeping their shape, and in consequence the dimensions of the whole shell increase, while the globular shape is preserved.

If a fresh and perfect specimen be examined, the surface is seen to be covered with short sharp spines set so thickly that the substance of the shell can hardly be seen

COMMON SEA-URCHIN.—*Echinus sphœra*

through them. The structure of these spines is very remarkable, and under the microscope they present some most interesting details. Moreover, each spine is movable at the will of the owner, and works upon a true ball-and-socket joint, the ball being a round globular projection on the surface of the shell, and the socket sunk into the base of the spine. When the creature is dead and dried, the membrane which binds together the ball-and-socket joint becomes very fragile, so that at a slight touch the membrane is broken and the spines fall off. The beautiful specimen from which the drawing is taken is in the possession of Mr. Sowerby, and is remarkable for the fact that the spines are all perfect and in their place.

Other peculiarities of structure will be noted in connexion with the different species.

The Common Sea-urchin is edible, and in some places is extensively consumed, fully earning its title of Sea-egg, by being boiled and eaten in the same manner as the eggs of poultry.

The fishing for these creatures in the Bay of Naples is graphically and quaintly described by Mr. R. Jones:—"I had not swum very far from the beach before I found myself surrounded by some fifty or sixty human heads, the bodies belonging to which were invisible, and interspersed among these, perhaps, an equal number of pairs of feet sticking out of the water. As I approached the spot, the entire scene became sufficiently ludicrous and bewildering. . . . Down went a head, up came a pair of heels—down went a pair of heels, up came a head ; and as something like a hundred people were all diligently practising the same manœuvre, the strange vicissitude from heels to head and head to heels, going on simultaneously, was rather a puzzling spectacle."

After inquiry, it proved that these divers were engaged in fishing for Sea-urchins, which are especially valuable just before they deposit their eggs, the roe, as the aggregate egg masses are termed, being large and in as much repute as the "soft roe" of the herring.

These Sea-urchins are fond of burrowing into the sand, an operation which is conducted mostly by help of the movable spines. They will sink themselves entirely out of sight, but not without leaving a slight funnel-shaped depression in the sand, which is sufficient to guide a practised eye to their hiding-place.

THE species which is shown in the accompanying illustration is given, not so much for the sake of its own beauty or for any peculiarity of form, as to exhibit the structure and habits of the Echinidæ in general. The PURPLE-TIPPED SEA-URCHIN is here shown in

PURPLE-TIPPED SEA-URCHIN.—*Echinus miliáris.*

many attitudes. The specimens are shown as they appear in an aquarium, and the drawings were taken from life.

On the upper part of the illustration is seen a Sea-urchin crawling up the glass front of the aquarium, and exhibiting the beautiful suckers, or ambulacra, by means of which it is enabled to walk. The expanded tips of the suckers are especially to be noted. Two of these suckers are seen planted on the rock, for the purpose of exhibiting the beautiful construction of their extremities. These are, of course, magnified. Another specimen is seen on the right hand, climbing up a rock, and carrying with it a number of pebbles, which act as a kind of shield or protection. The specimen from which this drawing was taken was in the habit of climbing to the upper edge of a living modiola, and there remaining, with its stony umbrella held over its body. The pebbles are held by the ambulacra, which attach themselves just as if they were employed in moving the whole body. Sometimes the Sea-urchin will collect together such a number of stones and pebbles, that it is almost entirely hidden under the superincumbent mass. A specimen thus enveloped is seen at the lower part of the illustration.

Immediately under the first-mentioned specimen is a curious figure which demands some explanation. This is the internal skeleton of the Sea-urchin removed from the body, and comprises the complicated and beautiful machinery by which the Echinus is enabled to cut to pieces the hard substances on which it feeds. It will be seen, by reference to the illustration, that there are five long strips of bony substance, which, on examination, are found to bear a wonderful resemblance to the incisor teeth of the rodent animals, being of enamel-like hardness at their points, and soft and pulpy at their bases. Like the rodent teeth, moreover, they are continually worn away at their points, and as continually pushed forward by fresh growth from behind.

Indeed, a very good idea of the structure of these jaws may be obtained by taking five incisor teeth of the rabbit, and arranging them so that their points touch each other and their bases diverge. Let these be bound together by the muscles necessary for their peculiar movement, and they then will form a figure almost identical with the jaws of the Echinus. This curious apparatus was likened by Aristotle to a lantern, and has thence derived the popular name of Aristotle's lantern.

It will be seen that a sort of five-sided pyramid is formed by the entire apparatus, and that at its base are a number of stout levers which assist in the movements. This remarkable object ought to be examined while quite fresh, as the muscular bands are easily affected by drought, shrivel into thin thread-like bands, and quite fail to convey a correct idea of their real shape and duty.

Below the jaws is seen a magnified view of one of the large movable spines, showing the projecting ball on the surface of the shell and the socket in the base of the spine. The two curious objects on the rock, at the right hand of the illustration, represent the strange appendages to the shell, which are termed Pedicillariæ, and are given in a magnified form in order to show their structure. The Pedicillariæ are found scattered among the spines, and placed upon the shell, and are of very minute dimensions. They are of calcareous substance, and are furnished at their tips with a three-bladed head, the valves of which are continually opening and closing, and presenting the appearance which is exhibited in the illustration.

Pedicillariæ are found upon many other Echinodermata, and their object is very doubtful. The most rational theory seems to be that they are placed there in order, by their perpetual movements, to keep the shell clear from the various algæ that are sure to grow upon any marine substance that is not protected from their approaches, as also to perform the same office towards the young serpulæ, molluscoids, and zoophytes which fill the ocean with their multitudinous hosts, and settle upon every object that affords even a short period of quiet.

On the next illustration are seen two views of the same object, in order to show the great size to which the spines attain in some species.

The genus to which the PIPER-URCHIN belongs can always be recognised by the enormous comparative size of the tubercles sustaining the spines and the parallel rows of ambulacra. The tubercles are not perforated. In the illustration, one figure shows the Piper-urchin as it appears when entire, and the other exhibits its aspect when the spines are removed, so as to show the large size and bold protuberance of the spine-bearing tubercles.

The members of the genus Cidaris are mostly found in the hotter parts of the world, and are plentiful in the Indian Seas. The spines of several of the species have been made serviceable in the cause of education, being found to make excellent slate-pencils after being calcined. The missionaries have the credit of making this useful discovery.

The food of the Echini in general seems to consist of various substances, both of an animal and vegetable nature. Fragments of different sea-weeds have been found in the digestive cavity, as also certain portions of shells, which seem to prove that the Echinus had fed upon the molluscs, and broken their shells in pieces with its powerful jaws. The precise mode of feeding is not exactly ascertained; but it seems likely that the

Echinus can seize its prey with any of its ambulacra, no matter on what portion of the body they may be situated, and pass it from one to the other until it reaches the mouth,

PIPER-URCHIN.—*Cidaris papillata.*

which is placed in the centre of the open disk.   Both univalve and bivalve molluscs appear to be eaten by the Echinus.

THE creatures which are represented in the illustration on page 726 are appropriately named HEART-URCHINS, from their peculiar shape, and bear an evident resemblance to the heart-cockles already mentioned.   Many species of Heart-urchins are found in a fossil state, and are especially common in the chalk formations.   On the upper right-hand side of the illustration is seen the handsome PURPLE HEART-URCHIN in a perfect state, with all its array of slender armed spines ; while, on the other side, is a specimen of the COMMON HEART-URCHIN with its spines removed, showing well its peculiar shape.

The shell of this genus is slight and delicate, and is composed of very large plates, which, in consequence, are comparatively few in number.   There is always a furrow of greater or less depth at the upper end.   In the naked specimen the rows of pores through which the ambulacra pass are plainly perceptible, and even in the fossilized specimens, which have been buried in the earth for so many ages, these pores are still visible, and so plainly marked, that the genus and species of the dead shell can be made out with little less ease than if the animal were just taken out of the water.

The Heart-urchins are found in all parts of the world, and our own seas contain specimens of these curious beings.   In the Mediterranean they are extremely plentiful, and mostly appear to live below the sand.   They seem to feed on the animal substances that are mingled with the sand, for M. de Blainville found, on dissecting many specimens, that their digestive organs were always filled with fine sand.   The walls of the digestive cavities are exceedingly delicate, and have been compared to the spider's web.

In the lower left-hand corner may be seen another form of these remarkable creatures, where the shell is formed into two points.   This is the FIDDLE HEART-URCHIN, so called from the fiddle-shaped mark upon the shell.   At its right hand is another specimen denuded of its spines, for the purpose of showing the peculiar mark from which the species derives its name.   The Common Heart-urchin with all its spines is seen in the right-hand lower corner.   All these species are found in the British seas.

COMMON HEART-URCHIN. (Without spines.)   PURPLE HEART-URCHIN.—*Spatangus purpúreus.*
FIDDLE HEART-URCHIN.—*Brissus lȳrifer.*   COMMON HEART-URCHIN.—*Amphidótus cordátus.*
FIDDLE HEART-URCHIN. (Without spines.)

In some of the hotter parts of the world, such as the Indian seas, several species of Echinus are armed with sharp and slender spines, which are apt to pierce the bare foot of a bather, and to cause painful, and even dangerous, wounds. Most of these Echini live in the crevices of rocks, but sometimes crawl over the sand, and inflict much suffering upon those who unwittingly place a foot upon them. Mr. F. D. Bennett, in his account of a "Whaling Voyage," had practical experience of these sharp spines:—"On one occasion, when searching for fish in the crevice of a coral rock, I felt a severe pain in my hand, and, upon withdrawing it, found my fingers covered with slender spines, evidently those of an Echinus, and of a grey colour elegantly banded with black. They projected from my fingers like well-planted arrows from a target, and their points, being barbed, could not be removed, but remained for some weeks imbedded as black specks in the skin.

Its concealed situation did not permit me to examine this particular Echinus, but I subsequently noticed others of a similar nature fixed to the hollows in the rocks; they were equal in size to the *Echinus cidaris*, and their body was similarly depressed, but the spines were long, slender, and more vertically arranged, and their points finely serrated. Their colour was jet black. These animals adhered so firmly to the rocks, that they could not be detached without difficulty.

When closely approached, they gave an irritable shrug to their spines, similar to that displayed by the porcupine or hedgehog. It was difficult to say if the hand had been brought in perfect contact with this Echinus before it was wounded by its weapons. In some experiments, I approached the spines with so much caution, that had they been the finest pointed needles in a fixed state, no injury could have been received from them; yet their points were always struck into my hand, rapidly and severely. The natives are well aware of the offensive character of these animals, and caution the stranger against handling them."

The same author mentions that a species of Cidaris is largely eaten by the South Sea Islanders, and that in various places on the sea-shore there are large heaps of its shells and spines, showing that feasts have been lately held in that locality.

CAKE-URCHIN.—*Echinarachnius placenta.*

THE curiously-formed Echinus which is shown in this illustration is popularly called the CAKE-URCHIN, on account of its remarkably flattened form. It belongs to a family which are generally called Shield-urchins, from their flat, disk-like shapes. A second figure of this species is given as viewed from the side, in order to show the wonderfully flattened shell, and the manner in which it rapidly slopes from the centre to the circumference. The general shape and arrangement of the plates from which the shell is built may be seen in the larger figure. This remarkable creature is a British species, but is not very plentiful upon our coasts. The word "placenta" is Latin, signifying a flat cake, and is appropriately given to this species.

The development of the Echinus is so very remarkable, that it deserves a passing notice. This creature passes through a metamorphosis even more strange than that of the insect, and no one who was not acquainted with the animal could possibly recognise in the delicate framework of translucent spines the larval form of the globular Sea-urchin. At first, the little creatures are almost shapeless and globular, rolling about through the water in an uncertain kind of way. But by degrees they put forth a dome-like portion, from which proceed several slender calcareous rods, altogether making a figure that has been aptly compared to a skeleton French clock. In this state it was formerly known by the name of Pluteus. As if to carry out the comparison still further, the first indication of change to its more perfect form is the development of a circular

LOBED URCHIN.—*Encope grandis.*   KEYHOLE-URCHIN.—*Encope subclausa.*   WHEEL-URCHIN.—*Rotula Rumphii.*

disk which will represent the face of the clock, upon which are traced certain lines that answer to the hands and figures. By rapid degrees, the disk expands and covers the gelatinous substance of the animal, and puts on hour by hour more of the Echinus as it loses its former skeleton shape. The latter becomes rapidly covered by and absorbed into the former, and in due time the framework of long slender rods, which might also be well compared to an artist's easel, or the tripod stand of a theodolite, is converted into the well-known globular Echinus, with all its complicated apparatus of spines, pedicillariæ, and walking organs.

The reader may perhaps have noticed, that on inspecting a common Echinus, especially from the interior, it exhibits in a very distinct manner its close alliance with the well-known star-fishes. Take, for example, a common five-finger star-fish out of the water, lay it on its back, and then gather all the five points together. Now, supposing the creature to be dead, strip the skin from the rays, leaving it only adherent down the centre, join the edges of the strips, and there is a very good imitation of the Sea-urchin.

THE two curious specimens represented in this illustration also belong to the Shield-urchins.

The ROTULA, so called because of its wheel-like shape, is nearly as flat as a piece of money, and has a very slight elevation in the centre. It is remarkable for the very deep teeth into which one side of the disk is cut, giving the creature an aspect as if it were a cog-wheel in process of manufacture. The colour of this species is mostly greyish slate above, and dull white below. The under surface is veined over its whole extent, all the veinings radiating from the centre. The colour of this Rotula is, however, extremely variable.

The second species might be well called the KEYHOLE-URCHIN. This remarkable creature, instead of being toothed at the edge like the preceding species, has its disk pierced with oblong apertures of a shape much resembling a keyhole. These apertures are rather variable in their shape, sometimes being merely pierced through the disk of the Urchin, and sometimes extending fairly to the edge. When full-grown, this is rather a large species, much resembling an ordinary pancake both in shape and dimensions. There are many species of Encope, most of which are inhabitants of the hotter seas, some being found in Southern America. The colour of the Keyhole-urchin is dull grey. The whole family is a very remarkable one and affords numerous points of interest to the careful observer.

ASTERIAS. (With bivalve in mouth.)   BUTTHORN.— *Asterias aurantiaca.*

LEAVING now the Echini, we pass to the next large group of Echinodermata, called scientifically Asteriadæ, and popularly known as Star-fishes. These creatures exhibit in the strongest manner the radiate form of body, the various organs boldly radiating from a common centre.

Many of these creatures are exceedingly common upon our own coasts, so plentiful, indeed, as to be intensely hated by the fishermen. Of these, the common FIVE-FINGERS, or CROSS-FISH, is perhaps found in the greatest numbers. All Star-fishes are very wonderful beings, and well repay a close and lengthened examination of their habits, their development, and their anatomy. There are sufficient materials in a single Star-fish to fill a whole book as large as the present volume, and it is therefore necessary that our descriptions shall be but brief and compressed.

To begin with the ordinary habits of this creature.

Every one who has wandered by the sea-side has seen specimens of the common Five-fingers thrown on the beach, and perhaps may have passed it by as something too commonplace to deserve notice. If it be taken up, it dangles helplessly from the hand, and appears to be one of the most innocuous beings on the face of the earth. Yet, this very creature has in all probability killed and devoured great numbers of the edible molluscs, and has either entirely or partially excited the anger of many an industrious fisherman.

To begin with the former delinquency. It is found that the Star-fish is a terrible foe to molluscs, and although its body is so soft, and it is destitute of any jaws or levers, such as are employed by other mollusc-eating inhabitants of the sea, it can devour even the tightly shut bivalves, however firmly they may close their valves. On looking at a Star-fish, it will be seen that its mouth is in the very centre of the rays, and it is through that simple-looking mouth that the Star-fish is able to draw its sustenance.

Even if it should come upon a mollusc which, like the oyster, is firmly attached to some object, it is by no means disconcerted, but immediately proceeds to action. Its first process is to lie upon its prey, folding its arms over it so as to hold itself in the right position. It then applies the mouth closely to the victim, and deliberately begins to push out its stomach through the mouth, and wraps the mollusc in the folds of that organ. Some naturalists think that the Star-fish has the power of secreting some fluid which is applied to the shell, and causes the bivalve to unclose itself. But whether this be the case or not, patience will always do her work, and in time the hapless mollusc surrenders itself to the devourer. In the case of smaller prey, the creature is taken wholly into the mouth, and there digested.

A very remarkable example of the voracity of the Star-fish is shown in the engraving on page 729, sketched from a specimen in the British Museum. A Star-fish belonging to this genus had managed to swallow entire a bivalve mollusc, and had dissolved out all the soft parts from the shell. This it was unable to throw out, as is the usual custom of Star-fishes, and, in consequence, the empty shell of the bivalve became a fixture within the body of the Star-fish, producing the remarkable effect depicted in the engraving.

The second delinquency of the Star-fish is achieved as follows :—By some wonderful power the Star-fish is enabled to detect prey at some distance, even though no organs of sight, hearing, or scent can be absolutely defined. When, therefore, the fishermen lower their baits into the sea, the Star-fishes and crabs often seize the hook, and so give the fisherman all the trouble of pulling up his line for nothing, baiting the hook afresh, and losing his time.

The fishermen always kill the Star-fish in reprisal for its attack on their bait, and formerly were accustomed to tear it across and fling the pieces into the sea. This, however, is a very foolish plan of proceeding, for the Star-fish is wonderfully tenacious of life, and can bear the loss of one or all of its rays without seeming much inconvenienced. The two halves of the Asterias would simply heal the wound, put forth fresh rays, and, after a time, be transmuted into two perfect Star-fishes.

It often happens that the lounger on the sea-shore finds examples of this species with only four, or even three, rays, and, finding no vestige of a scar to mark the place whence the missing limb was torn, he is apt to fancy that he has found a new species which only possesses a small number of rays. The fact, however, is, that the interval is immediately filled up by the creature; the rays on each side of the injury close up together, and all mark of a wound is soon obliterated. I have seen these strange beings with only one ray, proceeding quietly along without appearing to suffer any inconvenience from their loss.

The movements of the Star-fish are extremely graceful, the creature gliding onward with a beautifully smooth and regular motion. It always manages to accommodate itself to the surface over which it is passing, never bridging over even a slight depression, but exactly following all the inequalities of the ground. It can also pass through a very narrow opening, and does so by pushing one ray in front, and then folding the others back so that they may afford no obstacle to the passage. It also has an odd habit of pressing the points of its rays upon the bottom of the sea, and raising itself in the middle, so as to resemble a five-legged stool. If the reader is desirous of keeping a few Star-fishes in an aquarium, the object may be easily accomplished by keeping them in a very cool place, as they are extremely impatient of heat, and soon die if the water becomes too warm. They also require that a supply of air be frequently pumped through the water in which they reside.

The bony apparatus, or skeleton, if it may be so called, of the Star-fish is a most beautiful and wondrous object. Without going into the tempting regions of anatomy, I may state that a few hours will be well bestowed in examining the structure of any of these beings. A very simple plan of doing so is to wash the creature well with fresh water, lest the salt should rust the scissors and scalpel, and then carefully look into the extraordinary array of tentacles, or ambulacra, on which the creature walks. Let it then be pinned to a flat piece of cork loaded with lead, and sunk about half an inch below the surface of clear fresh water. Slit up the skin along each ray, taking care to save a portion for the microscope, and turn the flaps aside.

In each ray will be seen the curious feathered and fern-like branches of the stomach, and under them lies the wondrous array of bone-like pieces of which the skeleton is made. Thousands upon thousands of pure white columns are ranked in double vistas, and are over-arched by an elaborate structure of the same white material on the pillars. I know nothing that can compare with this sight for delicacy and beauty. Imagine a cathedral aisle half a mile in length, which is supported by a double row of white marble columns, and whose roof is formed of the same beautiful material; then, let all the pillars be bowed towards each other in pairs, so that their capitals rest against each other, and a dim idea will be formed of the wonderful structure of the Star-fish.

The piece of skin must be preserved in order to examine, with the aid of the microscope, the pedicillariæ and minute spiracles that stud its surface. A tolerably stout pair of scissors are required for the purpose of cutting the skin, as its substance is tough; and it is besides furnished with such an array of hard stony appendages, that the edge of a more delicate instrument would certainly be turned, and its blade run some risk of fracture.

Before we pass to the remaining examples of this family, a few words must be given to the development of this wonderful creature.

The eggs of the Star-fish are numerous, almost beyond the power of arithmetic to calculate, and thus keep up the needful supply of these creatures whose enemies are so numerous, and powers of escape so trifling. When first excluded, the eggs are not allowed to pass freely into the sea, but are protected for a time in a kind of cage or chamber formed by the parent by raising itself on the tips of its rays, as has already been mentioned. When hatched, the young are round and almost shapeless, bearing a very close resemblance to an imprisoned animalcule. They by degrees put forth their rays, the feet issue from the rays, and, after a while, they are enabled to shift for themselves, and are dismissed from their parental home.

WE now proceed to the examination of some of the more conspicuous species of Asteriadæ.

The large upper figure on page 732 represents the common FIVE FINGERS, or CROSS-FISH, which needs no more description than has already been given. At its right-hand is seen another specimen, which affords a good example of the reproducing powers already mentioned. Having lost all its rays except one, it has contrived to live with the assistance of that solitary ray, and has begun to rectify the disaster by developing a fresh set of rays, whose points can be seen just budding around the disk.

The little flat figure at the left-hand lower corner of the illustration represents the GIBBOUS STARLET, a pretty little species, notable for the manner in which the rays are connected by a membrane as far as their tips. In the corresponding right-hand corner is a specimen of the KNOTTY-CUSHION STAR, so called on account of the thick rounded rays.

PASSING to the next illustration, we have examples of several other curious forms of Star-fishes, all being found on the British coasts. The BIRD'S-FOOT SEA-STAR derives its name from its singular shape, which is not at all unlike that of a duck's foot, with its spread toes and connecting membrane. This beautiful species is very thin of texture, and has a pentagonal form, caused by the five rays and the connecting membrane. If the surface of this Star-fish be examined with a good magnifier, it will be found to be covered with tufts of very tiny spines arranged in a regular series, and forming a kind of pattern.

The colours of the Bird's-foot Star are positively splendid. Each ray is marked with a double line of bright scarlet, a narrow belt of the same colour edges the connecting membrane, and the centre is also scarlet. The ground colour is light yellow, and the contrast of these two beautiful colours has a remarkably splendid effect. This species is seldom seen in the shallow waters or above low-water mark, and is, as a general rule, taken with the dredge.

The boldly-rayed species, which looks something like the front view of a sunflower, is a very common denizen of our coasts, and goes popularly by the appropriate name of SUN-STAR. It often attains to considerable dimensions, and is always a very conspicuous object from the glaring colours with which its surface is decorated, and the large amount of surface on which they can be displayed. The upper surface of this fine species is bright

CROSS-FISH.—*Uraster rubens.*
GIBBOUS STARLET.—*Asterina gibbósa.*

CROSS-FISH. (Reproducing lost rays.)
KNOTTY CUSHION-STAR.—*Goniaster equestris.*

vermilion, and as it sometimes is eight or nine inches in diameter, it is a very brilliant object as it lies upon the rocks.

Should any reader be desirous of preserving this or any other of the Star-fishes for a cabinet, he may do so without difficulty, by taking a few precautions. The first process is to wash the Star-fish in plenty of fresh water, and it will be better to follow up this step by removing the whole of the stomach and its appendages. This may be done from the under surface of the rays; and it will perhaps be useful if a little cotton wool be judiciously inserted, so as to prevent the skin from collapsing during the process of drying. Star-fishes may be easily dried, either before the fire or in the sun, but in either case they must be carefully washed in fresh water; and if a fire be employed, as must be the case in wet or dull weather, the board on which the Star-fish is should not be placed very near the fire, and should be occasionally watched, so that any tendency to warping may be corrected

In the EYED CRIBELLA, the rays are rather blunt at their extremities, and are cleft nearly to the centre, so that there is no definite disk. This species is rather stiffer to the touch than the others. It must, however, be remarked that the consistency of the Star-fishes is

SUN-STAR.—*Solaster pappósa.*

BIRD'S-FOOT SEA-STAR.—*Pálmipes membranáceus.*
EYED CRIBELLA.—*Cribella oculáta.*

extremely variable, even in the same species or the same individual. If, for example, a specimen of the common cross-fish be taken from the pool of water in which it is lying, a practised hand will at once know whether it is dead or alive. In the former case the creature is soft and flabby to the touch, yields readily to the impress of the fingers, and hangs down heavily like a mass of wet rag. If, on the contrary, any life should be left in the creature, the rays are tolerably firm and resisting to the touch, and when held by one ray it has altogether a firmer and more lively feeling about it. A simple, but effectual mode of ascertaining whether a Star-fish be alive or dead, is to turn it on its back in some sea-water. If it be dead there will of course be no movement, but if the least particle of life be still latent in that body from which it can hardly be expelled, the ambulacra, or feet, are seen to put themselves in motion, some being thrust out while others are being withdrawn.

On this illustration are shown some very curious species of Star-fish.

The BRITTLE-STARS, of which there are several species, are very appropriately named, inasmuch as they are able to break up their rays in the most extraordinary manner, a capability which they mostly exercise when they feel alarmed. The generic name, Ophiocoma, is derived from two Greek words, the former signifying a serpent, and the latter a lock of hair.

The whole of the Brittle-stars are curious and restless beings. They can never remain in the same attitude for the tenth part of a second, but are continually twining their long

COMMON BRITTLE-STAR.—*Ophiocoma rosula.*   WHITE SAND-STAR.—*Ophiurus albidus*

arms as if they were indeed the serpents with which Medusa's head was surrounded. The least impurity in the water will cause these strange beings to break themselves to pieces in this extraordinary manner, but they never seem to disintegrate themselves with such rapidity as when they are touched or otherwise alarmed.

The lamented Professor Forbes has left an admirably quaint description of this suicidal process. Having in vain attempted to secure a perfect specimen of a Brittle-star, he thought that he might achieve that object by having a pail of fresh water lowered into the sea, so that as soon as the dredge reached the surface of the sea it might be transferred to the bucket of fresh water, and all the inmates killed at once by the shock.

A fine specimen of the genus Luidia was then taken in the dredge. "As it does not generally break up before it is raised above the surface of the sea, cautiously and anxiously I sank my bucket to a level with the dredge's mouth, and proceeded, in the most gentle manner, to introduce Luidia to the purer element. Whether the cold element was too much for him, or the sight of the bucket too terrific, I know not; but in a moment he began to dissolve his corporation, and at every mesh of the dredge his fragments were seen escaping. In despair, I grasped the largest, and brought up the extremity of an arm with its terminating eye, the spinous eyelid of which opened and closed with something exceedingly like a wink of derision."

These Brittle-stars are, however, extremely capricious in their exercise of this curious power. It sometimes happens that, as in the instance so amusingly narrated, the creatures break themselves to pieces without any apparent provocation, while, in other cases, specimen after specimen may be taken, handled, killed, or wounded, without the loss of a ray. Even in the aquarium, they are equally uncertain in their habits, at one hour being entire and splendid specimens, and at the next being little but a solitary disk amid a ruined heap of broken arms.

THE word Ophiurus is of Greek origin, signifying snake-tail, and is therefore very appropriately given to these curious beings, whose slender arms twist and coil just like a handful of small serpents.

SHETLAND ARGUS.—*Astrophyton scutatum.*

The Ophiuri are quite as voracious as the ordinary Star-fishes, and are able by means of the long arms to convey food to the mouth, which is placed in the central disk. The young of these Echinodermata are quite as curious as those of the sea-urchins, to which indeed they bear some resemblance. They have long been known to naturalists under the title of Easel animalcules, on account of their peculiar shape, their real origin not being suspected until later years. It is totally unlike the form which it attains when mature, and the relationship between the adult Star-fish and the Pluteus, as the larva is termed, has been well compared to the relation of an embroidery frame to the pictured canvas within. Both the species shown in this illustration are inhabitants of the British seas.

THE wonderful creature which is called by the name of SHETLAND ARGUS is one of a most remarkable genus of Star-fishes, which are remarkable for the vast development of their arms.

Although the whole mass of arms is of so complicated a description, it will be found, on carefully examining the creature, to be formed by the simple process of twofold division. From the central disk spring five stout arms, each of which almost immediately divides into two smaller arms, and these again into two others; so that in a fine specimen the number of little arms or tendrils, if we may so call them, exceeds eighty thousand. All these organs are extremely flexible, and quite under the control of the animal, which is able to close or expand them at will. When the extremities of the arms are drawn together, it will be seen that the whole animal assumes the shape of a globular basket; and in consequence of this resemblance, the name of Basket-urchin, or Sea-basket, has been proposed for the creature.

ROSY FEATHER-STAR.—*Comátula rosácea.*

It takes its food by means of these wonderful arms, using itself, in fact, like a living casting net, surrounding the prey with the spread arms, and enclosing it within their multitudinous lines. It has been known to embrace in this manner a fisherman's bait, and to allow itself to be drawn to the surface without loosing its hold. It is one of the deep sea Star-fishes, and is very seldom taken except by means of the dredge or line.

The structure of the Shetland Argus is most marvellously complicated, inasmuch as each of the numerous arms is composed of an enormous number of small joints, each exactly in its place, and so beautifully connected together, that they are as flexible as silken cords, and yet as perfectly under the command of their owner as if they were restricted to the original five from which they take their origin.

THE elegant and graceful Star-fish which is appropriately named the FEATHER-STAR, is a native of our own coasts, and has always attracted the attention of sea-side observers. It is not very readily seen, being one of the deep water species, but it may be captured by means of the dredge, and will live for some time in the marine aquarium. It is a very active being, combining in its own person the accomplishments of many different Star-fishes. For example, it can crawl with tolerable speed over the ground, can swim through the water with sufficient power to direct its course, can float about at will, driven by the tide, and will sometimes clasp pieces of floating wood so as to be carried along by the waves without any fatigue.

Its habits while in the aquarium are very interesting, and have been well described by Mr. Gosse :—

"In captivity, the Feather-star sits upon the frond of a sea-weed or on a projecting angle of rock, which it grasps very firmly with its clawed filaments, so firmly that it is difficult to tear it from its hold. When violence is used, it catches hold of its support or any other object within reach with the tips of its arms, which it hooks down for the purpose, and with its pinnæ, so that it seems furnished with so many claws, the hard stony nature of which is revealed by the creaking scratching noise they make as they are forced from any hold, as if they were made of glass.

I was surprised to observe that several of the arms were unsymmetrically short; and on examining these with a lens, saw distinctly that each had been broken off, and was renewed; the new part agreeing in structure and colour with the rest, but the joints were much less in diameter; and this difference was strongly marked at the point of union, the first of the new joints being not more than one-third as wide as its predecessor. The appearance much reminded me of a lizard renewing its tail.

In sitting, the Feather-star bends its arms with a sigmoid curve, the tips bending upwards. It waves them now and then, but not much, and remains long without moving from its hold. Though I repeatedly took it out of water, removing it forcibly, it manifested no tendency to voluntary dislocation."

Perhaps, however, the strangest part of the Comatula's life is its early youth.

Every one who has the least smattering of geology is familiar with the fossils called Encrinites, and is well acquainted with them under the different popular names that they bear. They are, or rather were, Echinodermata set upon a long flexible stalk, and being constructed, like the Star-fishes, of an enormous number of joints. Popularly they are known by the name of Stone-lilies, or Screw-stones, and their disjointed members are very familiar under the title of St. Cuthbert's Beads. The number of joints in an adult Encrinite is almost incredible.

In the head only of one specimen, no less than one hundred and fifty thousand joints have been calculated to exist, exclusive of the numerous parts of which the stem is composed. These joints are frequently found separated from each other; and as they are perforated by a small hole through which a thread might be run, they were formerly strung together and used as rosaries. Encrinites were found very plentifully in many marbles, which, according to Dr. Buckland's energetic language, are as entirely made up of the petrified remains of Encrinites as a corn-rick is of straws. These wonderful beings could hardly be dissected out of the stone by any exertion of human labour, but it is found that water will achieve a task at once too laborious and too delicate for human hands to undertake. It often happens that the abrupt faces of marble cliffs exposed to the weather, so that the annual rains are driven forcibly upon them, and by their continual action wear away the soft surrounding substance of the stone, leaving the harder forms of the Encrinites as memorials of the time long passed away.

The Encrinites have long ago perished, but there are still some existing species of stalked Echinodermata, which are closely allied to them, and are still more nearly connected with the history of the Feather-star. These are termed Pentacrinites, because their joints are five-sided. Many fossil species of Pentacrinites are found, and are seen in positions which seem to prove that they must have been adherent by their bases to floating objects, and thus carried about from one place to another, like the barnacles, which have already been described and figured.

ONE living species of these strange creatures is still in existence, and is shown in the accompanying illustration. This strange being is appropriately called by the name of Medusa's Head, as the many arms that wave about its summit bear some resemblance to the fabled head of Medusa, with its burden of venomous serpents.

It is not a very large species when compared with some of its fossil relatives, for the largest specimens hitherto discovered are only a few feet in length, and have a stem about as large as a common drawing-pencil. Several fossil species, on the contrary, are at least eleven or twelve feet in length, and measure a full inch across the stem. The MEDUSA'S HEAD is the only species at present known, though it is probable that others may be yet discovered. In the illustration, two specimens of this creature are given, the one to exhibit the head as it appears when expanded, and the other to show its aspect when closed.

Pentacrinus Europæus.   i.e. Larva of Comatula.)   MEDUSA'S HEAD.--Pentacrinus Caput Medusæ.

The smaller figure represents a being discovered by Mr. Thompson, and called by him *Pentacrinus Europæus.* This little creature is, when full-grown, barely three-quarters of an inch in height, and with a stem no thicker than sewing silk. Without entering into the many and interesting details of structure, development. and the habits of this beautiful little creature, we need only observe that this being has been proved to be the young, or larval state of the Feather-star. During this stage of its existence, the young Comatula is affixed to its ever-lengthening stem, but when it has attained adult age, it leaves its footstalk and wanders freely through the ocean.

The reader will not fail to observe that herein the creature acts in precisely the opposite mode to that which is adopted by many beings which have already been described. In those marine animals of which the acorn-barnacle is a good example, the young enjoys freedom throughout its earlier stages, is furnished with certain organs which it afterwards loses, and does not settle down to one spot until it has attained adult age. In the case of the Comatula, the contrast between the two states of life is very strongly marked, the creature being of a more nomad nature than the rest of its kin, and, in swimming, presenting a curious resemblance to the Medusa, the arms contracting and expanding in a manner that strongly reminds the observer of the pulsating disk of the acaleph.

ACALEPHA.

WE now arrive at a large and important class of animals.

These beings are scientifically termed Acalepha, or Nettles, a word which may be freely rendered as Sea-nettles. The term is appropriate to many of the species which compose this large class, for a very great number of the Acalepha are possessed of certain poisoned weapons which pierce the skin, and irritate the nerves as if they were veritable stinging-nettles floating about in the sea. Popularly, they are known by the familiar term Jelly-fishes, because their structure is so gelatinous, mostly clear and transparent, but sometimes semi-opaque or coloured with most beautiful tints.

The whole history of these remarkable animals is curious and interesting in the extreme, for not only do they exhibit some of the most graceful shapes and pleasing hues that can add beauty to a living being, but they also afford examples of the earlier forms of organs and members which in the higher animals attain their fullest development.

When they have attained their adult condition, they roam the seas freely, though in their earlier stages they are fixed to one spot and assume a shape quite unlike that of their parent. Examples of this curious phenomenon will presently be given.

The function of nutrition is carried on in these animals in a method sufficiently simple. They are furnished with a cavity, corresponding to the stomach of higher animals, in which the food is placed, and from which a number of diverging vessels convey the nutritive fluid to the rest of the body.

For convenience sake, this class is divided into three groups or orders, the first of which is called the Siphonophora, and includes the best organized members of the class. In them the shape of the body is irregular, and there is no central cavity. They are furnished with sucking organs, and move by means of a certain cavity into which water is received gently and from which it is expelled forcibly, or sometimes by means of little sacs or vesicles charged with air.

Owing to the vast number of species contained in this class, it is manifestly impossible to mention all the curious and interesting animals which it includes. Care, however, has been taken to select those species which afford the best types of their orders, and it will be found that almost every group of importance will find its representation in the following pages.

ON the right hand of the accompanying illustration may be seen a remarkable creature, called by the popular name of SALLEE MAN, sometimes corrupted, in nautical fashion, into SALLYMAN.

In this curious animal the body is membranous, oval, and very flat, and may be at once recognised by the cartilaginous crest which rises obliquely from its upper surface, and the numerous tubercles which depend from its lower surface and surround the mouth. This cartilaginous substance marks out the Sallee Man as possessing a somewhat higher organization than its merely gelatinous relatives, and it is therefore placed at the head of its order.

The Velella is seldom seen on our own coasts, although it sometimes happens to be driven, by stress of wind and waves, to regions more chilly than those in which it entered the world. It is thought with justice that the upright cartilage can act the part of a sail, and, by means of its diagonal setting, drive the creature through the sea. The exact direction of its movements is in all probability decided by the numerous tentacles which hang from its lower surface, and which, by contraction or extension, can become living rudders.

The Velella is very widely distributed, and is found in every sea except those that are subject to the cold influences of the poles. It seldom approaches land, but may be met in vast numbers, sometimes being crowded together in large masses, and of various sizes.

There is an allied genus called RATARIA, in which the body is circular, and the row of tentacles round the mouth is single. The body is sustained by a flattened elevated cartilaginous plate, and possesses also a longitudinal crest above, muscular and moveable.

The internal cartilage of the Velelladæ are sometimes found strewn in great numbers on the surface of the water. Sailors believe that the delicate substance of the creature has been destroyed by the hot sunbeams, but naturalists have now ascertained that the true cause of their destruction is to be found in the sea-lizard (*Glaucus*), which feeds upon these

PORTUGUESE MAN-OF-WAR.—*Physalis pelágicus.*
VENUS' GIRDLE.—*Cestum Véneris.* SALLEE MAN.—*Velella vulgáris.*

curious inhabitants of the ocean, and devours the whole body with the exception of the firm cartilaginous plates.

THE upper figure represents the celebrated PORTUGUESE MAN-OF-WAR.

This beautiful but most formidable acaleph is found in all the tropical seas, and never fails to attract the attention of those who see it for the first time. The general shape of this remarkable being is a bubble-like envelope filled with air, upon which is a membranous crest, and which has a number of long tentacles hanging from one end.

These tentacles can be protruded or withdrawn at will, and sometimes reach a considerable length. They are of different shapes, some being short, and only measuring a few inches in length, while the seven or eight central tentacles will extend to a distance of several feet.

These long tentacles are most formidably armed with stinging tentacles, too minute to be seen with the naked eye, but possessing venomous powers even more noxious than those of the common nettle. " It is in these appendages alone," writes Mr. D. Bennett, " that the stinging property of the Physalis resides. Every other part of the mollusc may be touched with impunity, but the slightest contact of the hand with the cables produces a sensation as painful and protracted as the stinging of nettles; while, like the effect of that vegetable poison, the skin of the injured part often presents a white elevation or wheal.

Nor is the inconvenience confined to the hand; a dull aching pain usually proceeds up the arm and shoulder, and even extends to the muscles of the chest, producing an unpleasant feeling of anxiety and difficulty in respiration. Washing the injured part with water rather aggravates than relieves the pain, which is best remedied by friction with olive oil. The cables retain their urent property long after they have been detached from the animal, and their viscid secretion when received on a cloth retains the same virulent principle for many days, and communicates it to other objects."

It is most probable that these terrible appendages are employed for the purpose of procuring food, and that they serve to entangle and kill the creatures on which the Physalis lives. Several of these acalephs have been observed with the bodies of half-decomposed fishes entangled among the short tentacles.

The colours of the Physalis are always beautiful, and slightly variable, both in tint and intensity. The delicate pink crest can be elevated or depressed at will, and is beautifully transparent, grooved vertically throughout its length. The general hue of its body is blue, taking a very deep tint at the pointed end, and fading into softer hues towards the tentacles. A general iridescence, however, plays over the body, which seems in certain lights to be formed of topaz, sapphire, or aquamarine. The short fringes are beautifully coloured, the inner row being deep purple, and the outer row glowing crimson, as if formed of living carbuncle. The larger tentacles are nearly colourless, but are banded at very small intervals throughout their length, giving them the appearance of being jointed.

It is a common trick with sailors to induce a "green hand" to pick up a floating Physalis, and to make him buy a rather dear experience at the cost of several hours' smart.

The vesicular body seems to be permanently filled with air, the animal having no power of inflating or collapsing at will. Many of these beings may be found on the sea-shore, where they have been flung by a tempest, the tentacles all decayed, but the body still inflated with air.

OUR last example of this order is seen in the next illustration, where it occupies the lower part of the engraving. This is one of a tolerably large family termed Diphyidæ, or double animals, because they are formed, as it were, of two animals, one fitting inside the other. Their general form is bell-shaped. In the present genus, both animals are similar, and of a somewhat pyramidal shape, and have a few points round the aperture.

The connexion between the two portions of the DIPHYES seems to be very slight, inasmuch as the two halves are often found separated from each other. The progress of the animal is achieved simply by taking water slowly into the bells, and expelling it smartly, much after the fashion of the ordinary Medusæ.

Trailing from the interior of the bells may be seen a curiously-elongated appendage, studded with globules, which are, in fact, the offspring in different stages of development. A number of tiny discs set on footstalks are also distributed along this appendage, and save the power of adhesion to any object which they may happen to touch.

We now come to a fresh order named Ctenophora, or comb-bearers, because their bodies are furnished with rows of flattened cilia, set in rows above each other something like the teeth of a comb. There are many members of this beautiful order to be found in our seas, of which the common CYDIPPE is an excellent example.

This lovely creature may easily be captured by the simple process of towing a gauze net over the side of a sailing-boat. When removed from the water the net will be found studded with variously-sized knobs of transparent gelatine, not particularly attractive, and presenting no salient points whatever. Let, however, these apparently inanimate lumps of jelly be transferred to a vessel filled with sea-water, and then how different is their aspect!

Until the eye is accustomed to their shapes, they are not very easily seen, owing to their transparency and the similarity between their refractive powers and those of the water. I have often noticed persons looking at my glass jars without discovering that a single living creature was within them, though each jar was tenanted by two or three of these beautiful creatures.

By degrees, however, they become plainly visible, the chief points of attraction being the eight bands of ever-moving cilia that are drawn longitudinally over the body, and by means of which the creature performs its wonderful evolutions. The Cydippe is never still, but careers through the water with ceaseless movement, sometimes rising and falling in one spot, sometimes rolling over and over, sometimes spinning on its longer axis, but mostly pursuing a partly spiral course, turning slowly on itself as it proceeds through the water.

Sársia tubulosa.   *Willsia stelláta.*   *Cydíppe pileus.*
Diphyes campanulífera.

During these movements, a faint iridescence plays over the whole body of the Cydippe, but its chief glories are concentrated upon the bands of cilia which are drawn over the body. On these the colours are too brilliant, and yet evanescent, for description. Miniature rainbows seem to ripple along these living belts; and as the Cydippe glides gracefully along, it appears to be encircled with many diadems of self-illumined jewelry. If examined by the microscope, the ciliæ of which the locomotive bands are composed are seen to bear some resemblance to very narrow Venetian blinds, each lath closing or opening in regular succession.

Pendent from the body are further seen two long filaments, to which are attached a number of shorter and still finer threads, not unlike the hooks and snoods on a deep sea-line, and used, indeed, for a similar purpose. The Cydippe can protrude or retract these tentacles at will, and is continually throwing them out from the body or drawing them back again, so that they never seem to be exactly the same length, one being often three or four times as long as the other. The manner in which these tentacles trail after the creature is extremely graceful, and the observer cannot resist a feeling of wonder that they should avoid entanglement.

The tentacles are employed for the purpose of catching prey, the Cydippe having been observed in the very act of seizing and eating its food. The long threads arrested the object as soon as touched, and in a very short time they were drawn to the central mouth, and the prey safely lodged within. The smaller crustaceans appear to be the favourite food of the Cydippe. The vitality, or perhaps the irritability, of the cilia is very enduring, for they continue to act when the animal is cut into several parts, or even when a little piece is nipped off, and will carry the severed portions through the water quite merrily.

The development of the Cydippe is very interesting, the young being produced from minute vesicles, and passing through a series of stages before they assume their perfect form.

IF the reader will now refer to the illustration on page 740 he will see a long, flat, riband-like creature edged with a delicate fringe of cilia. This curious being is called VENUS' GIRDLE, and from its beauty fully deserves the name.

This lovely creature is found in the Mediterranean, where it attains to the extra-ordinary length of five feet, the breadth being only two inches. Rightly, the words breadth and length ought to be transposed, as the development is wholly lateral. The mouth of the Venus' Girdle may be seen in the centre of the body, occupying a very small space in proportion to the large dimensions of the creature to which it belongs. A very good idea of the appearance of the Venus' Girdle may be obtained by supposing a Cydippe two inches in length to be flattened and rolled out into a riband of five feet in length.

Owing to the great length and tenuity of this creature, it is seldom found quite entire, but it seems to care little for the loss of a foot or so of its substance.

WE now come to a very large order of acalephs, including all those beings which are so familiar under the title of JELLY FISHES, SLOBBERS, and similar euphonious names. They are all united under the name of Discophora, or disc-bearers, because they are furnished with a large umbrella-like disc, by means of which they are enabled to proceed through the water.

Each order is separated into several tribes, the first of which is termed Gymnophthal-mata, or Naked-eyed Medusæ, because the little ocelli, or eye-specks, are either uncovered or altogether absent. The edge is either simple or branched. The name of Medusæ is given to these creatures on account of the long trailing filaments which depend from them like the snaky locks of Medusa from her head. In the Naked-eyed Medusæ, the circulating vessels may be seen radiating to the edge either simple or branched.

The little figure in the upper left-hand corner affords an example of the family Sarsiadæ, which contains several genera found in British seas. All the Sarsiæ are pretty little creatures, and may be known by the four simple nutritive vessels and the egg-tubes placed in the footstalk. In this genus the umbrella is nearly hemispherical, and there are four tentacles set at the ends of the radiating vessels.

Though small, the Sarsiæ are interesting to the naturalist, on account of the curious method by which the young are produced, sprouting like buds from the footstalk, and presenting a very strange aspect as they project in different stages of development. In their first stage, the young Sarsiæ are nothing more than simple prominences upon the surface of the footstalk, and gradually increase in size, developing first one part and then another, until at last the little creatures are quite perfect, shake themselves free from the parent, and commence an independent existence.

There is a curious species of this genus, *Sarsia prolifera*, in which the base of every tentacle is supplied with a little bunch of young Medusæ, some just making their first appearance as mere lumps of gelatinous substance, some half-grown, and others nearly ready to free themselves from the parent stock.

The last figure in this illustration is an example of another family, known by the branching vessels. In this genus the umbrella is rather more globular than in either of the preceding cases, and the vessels are six in number, each being twice divided into two smaller vessels. The footstalk is short.

THE members of the next family are known by their flattened discs and the egg-tubes running linearly along the vessels. The left-hand figure represents a rather curious genus which is devoid of footstalks and appendages, and has a disc almost as flat as a biscuit. The simple radiating vessels are well shown in the illustration. The right-hand figure affords an example of the typical genus, several species of which are found in our own seas. In these pretty creatures the disc is rather more convex than in the preceding genus · the footstalk is very wide and expands into many lobes, with long and broad

Eulóra undulosa   Eqpóren cyánen.

fringes ; and the tentacles are very slender and variable in number. The present species inhabits the South seas.

In the accompanying illustration is given an example of a fine genus, several species of which are inhabitants of the British seas. This creature belongs to the next tribe of the order, wherein the eye-specks are covered by certain flaps, and the circulating vessels united into a kind of network. This tribe is further divided into two families, in the first of which, the true Medusæ, solid food is received into a mouth, and in the second, there is no mouth, but nourishment is absorbed through the ends of branching vessels.

The CHRYSAORA belongs to the first of these families, and may be recognised by the long unfringed but furbelowed arms. A fine species belonging to this genus, *Chrysaora cyclonota*, was kept for some time by Mr. Gosse, and has afforded many useful hints to the students of Natural History. Experiments were made for the purpose of ascertaining the method of obtaining food, and it was discovered that the furbelowed arms as well as the tentacles are used for catching prey. A dead whitebait was first given to the Medusa, and after having been caught by the tentacles and furbelows, was delivered to the former organs, the latter relinquishing their hold. Very gradually it was shifted towards the mouth of the footstalk, and there held for about an hour, when it was released and fell to the bottom of the vessel.

Thinking that the fish might have been too large a morsel for the Medusa, the experimenter next supplied the animal with a small piece of cooked meat. This was seized as the fish had been, and during the course of the night was conveyed into one of the four cavities of the footstalk. There it remained for about sixty hours, when it was rejected. On being examined it was found to be perfectly white, but not in the least decomposed, or having any putrescent smell.

A curious change then took place. "After I had kept this Chrysaora for about a week, its manners underwent a change. It no longer swam about freely in the water by means of its pumping contractions, nor was its appearance that of an umbrella. It began to turn itself inside out, and at length assumed this form permanently, its shape being that of an elegant vase or cup, with the rim turned over, and the tentacles depending loosely from it, the furbelows constituting a sort of foot.

The latter were now put to a new use; the animal began habitually to rest near the bottom of the vessel or upon the broad fronds of the *Iridæa*, which were growing in the water and preserving its purity, but occasionally it would rise midway to the surface and hang by one or two of the furbelows. A fold or two of the latter would come to the top of the water, and dilate upon the surface into a broad flat expansion, exactly like the foot of a swimming mollusc · from this the Medusa would hang suspended in an inverted position. All the other furbelows, and portions of this one that lay below the expansion, floated as usual through the water, except that on some occasions an accessory power was obtained by pressing a portion of another furbelow to the side of the glass and making it adhere just like the portion that was exposed to the surface of the air. The texture of the furbelows when thus stretched smooth was exquisitely delicate." This curious movement seemed to be a prelude to the production of eggs, which were seen in great numbers. As if its whole life powers were exhausted by this process, the creature soon became feeble and then died, its captive life having endured for almost three weeks.

Chysaóra lútea.

OUR last examples of the Medusæ are to be seen in the illustration on page 746.

On the right hand of the engraving may be seen an example of the typical genus of this family, which is a native of our own seas. This is a sufficiently common species, and may be found plentifully on our shores, together with its kindred. There are few more beautiful sights than to stand on a pier head or lie in the stern sheets of a boat, and watch the Medusæ passing in shoals through the clear water, pulsating as if the whole being were but a translucent heart, trailing behind them their delicate fringes of waving cilia, and rolling gently over as if in excess of happiness At night, the Medusæ put on new beauties, glowing with phosphorescent light like marine fireflies, and giving to the ocean an almost unearthly beauty that irresistibly recals to the mind the "sea of glass mingled with fire."

On the left hand, in the lower corner, is seen that scourge of the ocean, the VENOMOUS CYANÆA. This harmless-looking creature is, in truth, one of the few inhabitants of the sea that are to be feared by bathers on our favoured shores; but its presence is so much to be dreaded that no one who has once suffered from the lash of its envenomed filaments will venture to bathe without keeping a careful watch on the surrounding water. I have twice undergone the torment occasioned by the contact of this creature, and know by experience the severity of its stroke.

At its first infliction, the pain is not unlike that caused by the common stinging-nettle, but rather sharper, and with more of a tingling sensation. Presently, however, it increases in violence, and then seems to attack the whole nervous system, occasionally causing a severe pain to dart through the body as if a rifle-bullet had passed in at one side and out at the other. Both the heart and lungs suffer spasmodically, and the victim occasionally feels as if he could not survive for another minute.

Rhizostoma Cuvieri.

Cyanæa capillata.

Medusa aurita.

These symptoms last for ten or twelve hours before they fairly abate, and even after several days the very contact of the clothes is painful to the skin. The shooting pangs just mentioned are of longer duration, and I have felt them more than three months after the Cyanæa had stung me.

To the unaided eye the filaments which work such dread misery are most innocuous and feeble, being scarcely stronger than the gossamer floating in the air, and looking much as if the Medusa had broken away a spider's web, and were trailing the long threads behind it. The microscope, however, reveals a wondrous structure, which, though it cannot precisely compensate for the sufferings inflicted by these tentacles, can at all events endow them with an interest which would not otherwise be felt.

Lest any of my readers should become fellow-sufferers with myself, I advise them to be very careful when bathing after a strong south-west wind has prevailed, and if ever they see a tawny mass of membranes and fibres floating along, to retreat at once, and wait until it is at least a hundred yards away. Some may suppose that this advice is needlessly timid, but those who have once felt a single poison thread across their hand or foot will recognise that discretion is by far the wisest part to be played whenever there is the least danger of being stung by the Cyanæa.

The last family, of which the large upper figure affords an example, is easily known by the absence of a mouth. In the typical genus, Rhizostoma, the footstalk is deeply scooped into semi-lunar orifices, and the eight cartilaginous arms are without fringes. This beautiful genus is also seen on our coasts.

Before taking a final leave of these remarkable beings, it is needful that we should briefly notice the strange metamorphosis through which some of them pass before they assume their well-known form. Experiments were made on a species of Chrysaora, by Sir John Dalyell, with the following result:—When first sent into the world, the young Medusæ were little flat, worm-like creatures, too minute to be examined by any except the highest powers of the microscope. By degrees, these tiny beings settle down to one spot and affix themselves, the body lengthens, arms begin to be shown, and after a while the strange creature is developed into the being known as the *Hydra tuba*.

Satisfied apparently with its condition, the Hydra remains in the same spot for some time, and produces a number of young Hydras, which sprout like buds from its sides, and, when separated, resemble their parent. Here we might naturally imagine to be the end of its history, for with almost all animals, when a being is able to produce young it is considered as having attained the utmost development of which it is capable. The Hydra, however, has yet other phases through which to pass. Towards spring, its body becomes much lengthened and wrinkled, so as to form a number of folds, just as if a series of threads had been tied tightly round it, one below the other. The upper rings now rapidly expand and the folds deepen, until the animal resembles a number of saucers regularly increasing in size, laid upon each other. The edges of each saucer are developed into two-cleft rays, and in this condition the animal proves to be the beautiful zoophyte discovered by M. Sars, and called the Strobila.

These are indeed strange vicissitudes in life, changes more marvellous than even those wrought by water and magic words in the old days when Haroun Alraschid ruled the faithful. There is yet more to come. The uppermost and largest disc or saucer now lengthens its rays and assumes the form of an unmistakeable, though shallow-disked Medusa. Its arms rapidly gain strength, the attachment becomes hourly weaker, until at last the whole disc is broken away, and floats into the wide sea in its new form.

How wonderful is this phenomenon, and how full of interest is the study of animate nature! Here we have a being which first enters into active existence in a shape like that of the infusorial animalcules; then changing into a hydra, and while in this state becoming the parent of a numerous offspring; then developing into a Strobila; and lastly breaking up into a series of Medusæ.

ZOOPHYTES.

Quitting the Acalephæ, we come to the vast class of Zoophytes, or animal plants, so called, because, though really belonging to the animal kingdom, many of them bear a singularly close resemblance to vegetable forms. Their substance is always gelatinous and fleshy, and round the entrance to the stomach are set certain tentacles, used in catching prey and conveying it to the stomach. These tentacles are armed with myriads of offensive weapons contained in little capsules, and capable of being discharged with great force. Organs of sight, smell, taste, and hearing seem to be totally absent, though it is possible that an extended sense of touch may compensate the creature for these deficiencies.

Without entering further into the constitution of these singular beings, we will proceed to the examination of the various groups into which they have been divided.

In the family of the Lucernariadæ, the tentacles are arranged in detached groups, a peculiarity whereby the creatures may easily be recognised. These organs are placed upon the outer edge of the membranous and expanded disc, in the centre of which is the squared mouth. They are mostly found adherent, as may be seen by the illustration, by a stem to some object, but they can swim with tolerable rapidity, their bodies pulsating like those of the Medusæ. None of them attain any great size, the largest being about one inch in height. Pink is their usual colour.

OPELET.—'*Anthea céreus*
PUFFLET.—*Edwardsia vestita.*

Lucernária aurioula.
SCOTTISH PEARLET.—*Ilyanthus scóticus.*

Mr. Gosse, in his "British Sea-Anemones and Corals," remarks that the Lucernariadæ have closer affinities with the Medusæ than with the Actiniæ, on account of several structural peculiarities, among which may be mentioned the gelatinous texture, the expanded umbrella, the egg-sacs in the substance of the umbrella, and the squared mouth at the end of a free footstalk. I have, therefore, departed a little from the ordinary arrangement, and placed the Lucernariadæ immediately after the Acalephs, forming a kind of intermediate link between them.

THE highest form of true Zoophyte is, undoubtedly, that which is so familiar under the name of Sea-Anemone—a name singularly inappropriate, inasmuch as the resemblance to an anemone is very farfetched; while that to the chrysanthemum, daisy, or dandelion is very close. These creatures are called Actinoida, and are easily distinguished by having the stomach inclosed in a sac divided into compartments by radiating partitions. For convenience sake, this group is divided into two sub-orders, the first of which is the Actinaria, known by the number of tentacles (twelve or more), perforated above, and the radiating partitions sometimes depositing solid chalky plates, commonly called "coral." The tribe Astræacea is known by the imperfect series of tentacles, and the family Actiniadæ by their circular arrangement.

The beautiful OPELET may easily be recognised by the great length of its many tentacles, which wave, and twist, and twine, and curl like so many snakes. It has 'but little power of retracting the tentacles, and is, therefore, more conspicuous than many other species. It is tolerably hardy, enduring confinement well, but requiring food more often than is the case with the other British Actiniæ. Like all other members of this order, the Opelet is able to arrest passing objects by means of the tentacles, and does so by the aid of a wonderful array of weapons unexampled in the animal kingdom.

If a portion of a tentacle be examined under a moderately powerful microscope, it will be seen to be studded with tiny cells, in each of which lies coiled a dark thread. On applying pressure to the cell, it suddenly discharges the coiled thread, which proves on a closer examination to be a long, wiry dart, often of wondrously complex structure, and capable of penetrating into any soft substance with which it comes in contact. Elaborate accounts and drawings of these cells and their contained weapons may be found in Mr. Gosse's valuable "British Sea-Anemones and Corals," a work to which I gladly refer my readers for many interesting details respecting the beautiful creatures on which we are at present engaged.

Though the human skin be a tougher and harder substance than the prey generally brought into contact with the tentacles, it yet can feel the effects of the individually minute but collectively potent weapons with which these delicate tentacles are armed. A finger which is touched by a tentacle is instantly conscious of being seized, as it were, and forced to adhere to the soft waving membrane which it could crush with a single effort. On most persons this adherence has no particular effect; but those who possess delicate skins, and a sensitive nervous system, are much worried by blisters and pustules occasioned by the assaults of these microscopical weapons. A young eel, measuring six inches in length and half an inch in thickness, was killed in a few minutes by mere contact with the tentacles, and in a very short time was tucked quietly away in the creature's stomach. These weapons are most numerous at the tips of the tentacles, just where they are most needed.

In the right-hand lower corner of the illustration may be seen the SCOTTISH PEARLET, a member of a genus once thought very rare in England, but now necessarily expanded into a family, and found to contain a considerable number of species, even in our own seas. Most of the Pearlets are able to crawl over solid bodies; some inhabit tubes; others are found burrowing in the sand; while nearly all are able to puff out the hinder part of the column with water.

Little is known respecting the history of the Scottish Pearlet, save that it is a very rare species, and has only been found in deep water. All the tentacles are very slender, and marked with a dark line.

The PUFFLETS are so called because they possess the power of puffing out the hinder part of the column until it assumes a somewhat globular shape. A British species of this genus, the PAINTED PUFFLET (*Edwardsia callimorpha*), appears to be one of the burrowers, its body being hidden beneath the sand, and the beautiful tentacles just protruding from the surface. None of the Pufflets have many tentacles.

We may here briefly notice another example of the same family.

The VESTLET is one of those members of the family which inhabit tubes. All of them are remarkable from the fact that they possess no adherent base, but, as a compensation for this deficiency, are furnished with an adherent power upon the stem, enabling them to crawl freely over solid bodies. In this species, the tube is cylindrical, and very wide in comparison with the dimensions of the inhabitant: it is of tough, paper-like consistence, rather thick, and is composed of many layers of intertwining fibres, mixed with sand and mud. The ordinary length of the animal is six or seven inches, and the width of the flower-like plumes about an inch and a half. Mr. Gosse found that he was able to remove the creature from its opaque dwelling, and place it in a tube of glass, which the animal accepted as an useful substitute, without troubling itself to reconstruct another house.

The beautiful creature which is shown in the centre of the next engraving, under the name of PLUMOSE ANEMONE, is certainly the most magnificent of the known British species.

It may be at once recognised by its bold cylindrical stem, firm and sturdy as the oak trunk, standing out bravely from the object to which it is affixed, and crowned with its lovely tufted tentacles, fringed and cut like the petals of the pink. Its colour is extremely variable, being snowy white, olive, red, orange, cream, or pale pink; and of all

SNAKE-LOCKED ANEMONE. - *Sagártia viduata.* (Open.)   SNAKE-LOCKED ANEMONE.—*Sagártia viduata.* (Closed.)

PLUMOSE ANEMONE.—*Actinolóba diánthus.*

BEADLET.—*Actínia mesembryánthemum.*

GEM-PIMPLET.—*Bunódes gemmácea.* (Open.)   GEM-PIMPLET.—*Bunódes gemmácea.* (Closed.)

the varieties, the first is, in my eyes, the most beautiful. It is capable of much alteration in its general form, shrinking to a mere shapeless fleshy mass, and looking by no means a pleasing object; expanding itself to the fullest extent, or forming itself into many shapes, according to the caprice of the moment.

Fortunately for the owners of aquaria, the Plumose Anemone is hardy, and bears captivity well. It often separates itself into several parts, each of which becomes an independent being, and in some stages of this process looks as if two individuals had become fused together.

THE pretty SNAKE-LOCKED ANEMONE may be recognised by the long, slender stem, and the flexible, indistinctly-barred tentacles, with a dark line running down each side.

It is found in many parts of the British shore, seeming to be rather local, but tolerably plentiful in the spots which it chooses for its residence. Though not adorned with brilliant colours, it is a remarkably pretty species, with its crown of delicate tentacles waving "like a thin blue cloud" upon the summit of its elongated stem. One of these Anemones has been known to produce some curious changes in its tentacles, at one time thickening them into knobs, and at another throwing out branches.

The widely-spread Anemone, with the circlet of pearl-like beads at the base of its tentacles, is the well-known BEADLET, the most common of all this order on the British coast.

It is a singularly hardy species, living mostly on the rocks that lie between high and low-water mark, and in some places collecting in extraordinary numbers. I remember on one occasion, after meeting a party of unsuccessful anemone-hunters, I filled their baskets in a quarter of an hour, though night had set in, and the only method of discovering the creatures was by the touch. It is perhaps more variable in colour than any of the British Actiniæ, the body taking all imaginable hues, passing from bright scarlet to leaf-green, grading from scarlet to crimson, from crimson to orange, from orange to yellow, and from yellow to green. The spherical beads around its mouth are more persistent in colour than any other parts of the animal, being almost invariably a rich blue, just like a set of turquoises placed around the disc. These, however, are occasionally subject to change, and lose all colour, looking like pearls rather than turquoises. Even the same individual is subject to change of colour, being evidently influenced by various external conditions, such as light and shade, food, and the purity of the water in which it is placed.

In the aquarium it is wonderfully prolific, surrounding itself with many a brood of tiny young, whose minute forms are seen settled around their parent, opening their tentacles with a kind of competent air that has something of the ludicrous about it. The Beadlet is something of a wanderer, and will not only crawl slowly over the glassy sides of the aquarium, but, when it has reached the surface of the water, will invert itself so that the tentacles are downwards, make its base hollow, and float away, trusting itself freely to this shallow boat.

The GEM-PIMPLET may be recognised by the double series of large and small warty protuberances placed alternately on its body. There are six white bands on the stem, and the tentacles are thick, marked with white, oval spots. Like the preceding species, the Gem-Pimplet is not local, though gathering in considerable numbers in certain favoured spots. Even when closed, with all the tentacles withdrawn, it may at once be known by the six bands of white which radiate from the orifice, and the great resemblance which its body bears to an echinus stripped of its spines.

LEAVING the sea-anemones, we now proceed to the next tribe, the Caryophylliaceæ, in which there are many tentacles, in two or more series, and the cells many-rayed. Many of these beings deposit a corallum ; but out of our British species, more than one-third are without this chalky support.

The FUNGIA, or SEA-MUSHROOM, is so called from its great resemblance to a mushroom, the expanded disc and delicate lamellæ having a singularly fungine form. The hard corallum of this genus is not fixed, but the creature is protected from the violence of the waves by its habit of lying in clefts of rocks, or in the deep cavities of coral reefs, so that it enjoys free access of water, without the danger of being carried away by the currents or dashed ashore by the tempest.

When young, however, the Fungiæ are affixed for a time, sometimes on rocks, and sometimes on the stony remains of their own kinds, being attached to a stem which gradually vanishes as the creature increases in age. While in this state, they bear some resemblance to the genus Caryophyllia. Though all possessing the same general characteristics, they are not all circular, some being oval, and others bearing no small resemblance to slugs. The entire corallum is surrounded by the soft substance of the Fungia, which envelops it below as well as above.

The examples given in the illustration exhibit the aspect of the Fungia under several conditions, showing the fully expanded tentacles, the ordinary appearance when living, and the lifeless skeleton or corallum. Most of the Fungia are found in the Indian seas, especially among the coral-beds.

WE now pass to another group of these curious beings. The ENDIVE CORAL is so called from the resemblance which its corallum bears to the crumpled leaves of that vegetable. The animal has no tentacles, and the cells are small, conical, and rather oblique. The corallum is fixed, sharply edged, and expanded from the base to the tip—a peculiarity which has earned it the specific title of Pavonia, or Peacock's-tail Coral. All the living members of this pretty genus are to be found in the East and West Indian seas.

The three figures which occupy the left hand of the illustration represents one of

Fúngia agariciformis. (Living.)

Fúngia Paumotensis.   *Fungia agariciformis.* (Skeleton.)

our few native Corals, shown under three aspects. The large, rounded figure in the lower corner exhibits the DEVONSHIRE CUP CORAL as it appears when the tentacles are fully expanded ; that to the right shows the dead stony corallum of the same species, and the upper figure is given for the purpose of exhibiting the curious manner in which it multiplies itself by throwing off buds from its sides. It is not a very large, but it is a very pretty species, the colour of its corallum being generally pure translucent white, sometimes tinged with a delicate rosy hue, while that of the living animal is pearly white, variegated with rich chestnut, and the palest imaginable fawn.

It is mostly a deep-water species, not unfrequent on our southern coasts, and is seldom procured except by means of the dredge or grapnel. Sometimes, however, it is found near the coast, and at the equinoctial springtides may sometimes be procured from the rocks which are laid bare by the receding waters. Fortunately for the collectors, it is very gregarious in its habits ; and when one specimen is found, others may generally be secured within a very short distance. It is a pretty inhabitant of the aquarium, and, as a general rule, may be induced to expand its long tendrils to their fullest extent, by placing a morsel of food upon the orifice. When properly managed it is tolerably hardy, but it does not brook inattention—shrinking up daily, and at last perishing hopelessly. When new to the mysteries of aquarium-management, I never could keep a Cup Coral more than a month.

IN the family Oculinidæ, the corallum is branched and tree-like, and is here repre-sented by our only known British form, the TUFT CORAL. It is very rare, and but seldom taken in our seas. A remarkably fine specimen is figured by Mr. Gosse, who remarks that it was taken off Skye in the year 1852, entangled in the deep-sea-line of a fisherman

DEVONSHIRE CUP-CORAL.—*Caryophyllia Smithii.*   TUFT-CORAL.—*Lophophélia prolifera.*
    (With side buds.)   ENDIVE-CORAL —*Euphyllia pavónia*   TREE-CORAL·-*Dendrophyllia nigrescens*
DEVONSHIRE CUP-CORAL. (And skeleton.)

Another specimen, weighing six pounds, has been taken in a similar manner between the islands of Rum and Eig. As may be seen from the illustration, the corallium resembles a massive, thickly-branched tree. The individual corals are about half an inch in height and the same in diameter.

On the right hand of the illustration is seen a coral that has attained a singularly tree-like form, and, in consequence of this structure, has obtained the appropriate name of Dendrophyllia.

The regular branched form of this coral can be seen by reference to the illustration, together with the manner in which the individuals are set on their common stem. The cells are rather deep, and the animals possess tentacles which are cleft longitudinally. It is a native of the hotter seas.

On the accompanying illustration we have some examples of those beings which we call Madrepores.

In the genus MADREPORA the animals are rather short, with twelve simple tentacles. The cells are deep, irregularly arranged upon the surface, and are crowded together towards the tips of the corallium, though they are scattered rather widely at its base. The cells are nearly cylindrical in their general shape and project outwards from a centre, something like the grains on an ear of wheat.

The genus Echinopora is distinguished by the peculiar arrangement of the cells, which are set only upon the upper surface of the coral. They are boldly radiated and rather irregular. All the true Madreporæ inhabit the hot seas, and are most plentiful under the tropics.

In the lower left-hand corner is seen a specimen of the genus ASTRÆA, so called because the animals are sown over its surface like stars in the heavens. The cells are decidedly short, and the tentacles few in number. The genus is a very large one, including many recent and fossil specimens, many of which are familiar to us in the polished stones of which mantelpieces and other domestic ornaments are made. Owing

3.   3 c

Madrépora abrotanóides

Astréa pállida.

Madrépora cribifera.
Echinópora unduláta.
Mæandrina cerebriformis.

to the vast number of the animals, and the rapidity with which they increase, the groups of Astræa often assume enormous dimensions; and in the secondary and tertiary rocks they frequently occur in such huge masses that whole rocks are composed of their remains.

In the right-hand lower corner is seen a figure of that remarkable coral which is popularly called BRAIN-STONE or BRAIN-CORAL, because the convolutions into which the corallium is moulded much resemble those of the human brain.

The animals of this genus are always united together in long waved series, each having a distinct mouth and series of very short tentacles. The cells are very shallow, and the valleys formed by their union are separated from each other by distinct ridges. The shape of this coral alters greatly with age, somewhat resembling the top-shells when young, but becoming rounded above when adult. The Brain-coral is found in several of the hot seas.

AMONG the Asteriadæ, as these creatures are called, in consequence of the star-like appearance of the polype or animal, the ORGAN PIPE-CORAL is perhaps the most striking. It forms, as far as is yet known, the only example of the group to which it belongs, and which is called Tubuliporina, on account of the multiplied series of regular tubes from which it is formed. Most persons are familiar with this beautiful coral, with its sets of tubes arranged like the pipes of a church-organ, or the storied rows of basaltic columns of the Giant Causeway.

Gorgónia flabellum.
Tubipora syringa.

Corállium fecundum.
Corállium nóbile.

The colour, too, is very pleasing, being a delicate pink, so that even the empty and lifeless corallium forms a really beautiful object. When living, however, it may fairly lay claim to the title of magnificent, for each tube is clothed, formed, and vivified by a light green polype, whose colour contrasts beautifully with that of the structure which is raised by that soft and feeble body.

The branched objects on the right hand of the illustration are two species of true coral, such as are used so largely in the manufacture of ornaments.

This beautiful zoophyte seems to be found only in the Mediterranean, where regular fisheries are established and the coral dragged from its recesses. The appliances, however, are very rude; and it is likely that more elaborate machinery would reap a rich harvest, by permitting some selection to be made and by enabling the fishers to regulate the dimensions of the groups of coral branches. Although the stony centre is so thick and solid, the substance of the animal is quite delicate and membranous, enveloping the corallium like wetted gold-beater's skin.

The fan-like object on the left hand of the illustration is popularly called from its shape, the SEA FAN, and well deserves that title. In this genus the branching arms are united by a number of transparent pieces, which are, in fact, developments of the branches, are covered in a similar manner by the investing membrane, and bear the living polypes on their surface. The whole structure easily dries, and may be found in most curiosity shops, or in the dwelling-houses of mariners, who have brought home these remarkable objects as presents to their wives.

An allied species belonging to the same family (*Isis hippuris*) is formed in a very

SEA-PEN.—*Pennátula grísea*
SEA-RUSH.—*Virgulária mirábilis*

SEA-FINGER.—*Alcyonium digitátum*

strange fashion. Its branches are composed of a number of strong joints, united together by horny rings, so that a certain amount of flexibility pervades its structure. Owing to this formation, it is sometimes called the Horn-plant or Sea-shrub, titles surviving from the time when all the corals were thought to be vegetables, and the expanded polypes to be their flowers. They are always fixed by a base, and grow like trees, with their branches upwards. It is worthy of notice, that the Gorgoniæ are never bushy, and, for the most part, have their branches in the same plane.

In this illustration we have examples of some very interesting species, some of which are extremely beautiful.

The Sea-pen is so named because its whole form bears the most remarkable resemblance to a quill-feather, consisting of a central shaft, from which a double row of "pinnæ" is developed at right angles, bearing the polype on their upper margin. As may be seen by the illustration, the whole form of this curious being is remarkably graceful, and it really seems as if it had been modelled upon a quill-feather plucked from the wing of some bird.

The Sea-pen is never attached to solid substances, but remains quite free in the ocean. It does not, however, swim, but is a helpless sort of being, and only kept in its proper position by the base being thrust into the mud or sand at the bottom of the sea. Some species of Sea-Pinnæ are phosphorescent, and present a magnificent sight in the darkness. It was once thought that the creature was able to swim by means of the webs, or pinnæ, which flapped like the fins of a fish, but it is now ascertained that no such power resides in these organs. The stem is of a rather soft consistency, strengthened by a bony centre, which reaches nearly to its tip.

Close to the Sea-pen may be seen an object of somewhat similar form, but considerably elongated, and with the pinnæ proportionately shorter. This is the SEA-RUSH, an animal belonging to a genus that can easily be distinguished from the preceding by several peculiarities. The pinnæ are short, deeply scooped above, and, with their bases, partly surround the central stem. The polypes are set only upon the edges of the pinnæ. There is an allied species belonging to the same family, called by the name of *Pavonaria*. In this remarkable genus, the general shape of the lengthened mass is four-sided, and the polypes are arranged in a somewhat spiral form on the stem, but only one side of its latter half. In our own seas, the Sea-rushes do not grow to any great length; but under a tropical sun they reach great dimensions, some of them measuring more than a yard in length.

THE last figure represents a very curious inhabitant of our seas, which is in the habit of encrusting all kinds of marine bodies, such as shells, stones, and stems of the large algæ. Its general mass runs out into lobes, and is of a soft spongy consistence, pierced with little holes, from which the polypes make their appearance when in health. When closely examined, the little holes or pores are seen to be formed of eight rays, in a kind of star-like pattern, and corresponding to the tentacles of the polypes which inhabit them. These little cells are placed at the ends of canals, which permeate the whole mass, and serve to unite into one common body the vast number of polypes which are thus aggregated together. When examined by the microscope, the substance of the polypidon is found to be filled with tiny particles of chalky matter, which serve to give consistency to the fabric, and add to its elasticity.

WE now arrive at the Hydroida, which are known by the internal cavity being simple, and the creature increasing by buds thrown out from the sides.

The left-hand figure represents an example of the family Tubulariadæ, in which the buds grow from the base of the tentacles, and break off their attachment as soon as they have attained maturity. The buds, or young, are naked. The animals are sometimes naked, but are often inclosed in a horny, tubular covering, which we will term the polypidon. The first family is represented by its typical genus. The polypidon of this genus does not throw out branches, and the tentacles are delicate, thread-like, and arranged in two circles. The germs, or buds, are set on very short footstalks, and are gathered upon the bases of the lower tentacles. This being is represented of the natural size.

Before leaving this interesting family of zoophytes, we must pause awhile, to cast a cursory glance at one or two of the more prominent British examples.

The CLUB-ZOOPHYTE (*Clava multicornis*) has a large and rounded extremity, something like the head of a bludgeon, upon which are placed irregularly a number of thread-like tentacles.

The various species belonging to the genus Coryne are also worthy of notice. These conspicuous, though minute zoophytes are found on our own shores, and may be recognised by the globular tips of the tentacles. Sometimes the creatures are naked, and sometimes they are inclosed in a rude sort of tube. The word "Coryne" is Greek, and signifies a club. The head of each tentacle is most elaborately constructed, and adorned with very minute tentacles, each being furnished with a small bristle at its tip. These tentacles can be moved with tolerable rapidity, and are held in various attitudes, sometimes stretching out at right angles from the stem, but often bending upwards, with their heads directed towards a common centre, and have been happily compared to the bars of a turnstile or the weighted arms of a screw press.

Another genus is that which is appropriately named Eudendrium, from two Greek words, signifying a beautiful tree. As may be presumed from its name, it has a decidedly tree-like form, each twig terminating in a polype whose flower-like tentacles add in no slight degree to its beauty. It is found that, when in captivity, the Eudendrium is sadly apt to throw off all the lovely diadems with which it is crowned, but that it will in process of time supply the deficiency by new heads.   Its reproduction is quite as remarkable as that of any creature which has hitherto been mentioned, but of which our failing space will not permit a detailed account.

Sertulária filicula.
Sertulária rósea.   Campanulária volúbilis.   Plumulária pinnáta.
Sertulária filicula. (Magnified.)   Plumulária pinnáta. (Magnified.)

In the Sertulariadæ, the buds are inclosed in vesicles, and do not break away when adult.   They are placed in cup-like cells, which have no footstalks.

Any of the common Sertulariæ affords a good example of this family ; and as they are easily procured, they are very valuable aids to those who wish to study the structure of these beautiful beings.   Even the empty polypidon is not without its elegance, and is often made up into those flattened bouquets of so called sea-weeds, which are sold in such quantities at sea-side bathing towns.   But when the whole being is full of life and health, its multitudinous cells filled with the delicate polypes, each furnished with more than twenty tentacles all moving in the water, its beauty defies description.   These little polypes are wonderfully active and suspicious.   At the least alarm, they retreat into their cells as if withdrawn by springs, and when they again push out their tentacles, it is in a very wary and careful manner.

The reproduction of these beings is very curious, for it is known that they can be propagated by cuttings just like plants, as well as by cell vesicles, and that in the latter case the first stage of the young closely resembles that of the young medusæ already mentioned.   They also reproduce by offshoots ; and it is very likely that their capabilities in this respect are not limited even to these three methods.

The Campanulariæ, or Bell-zoophytes, may be distinguished from the last family by having the cells placed on footstalks.   Two figures of the Bell-zoophyte are given, in order to show it as it appears of the natural size, and as it looks when magnified.

The whole history of this creature is very interesting, but on account of failing space we must restrict ourselves to its chief peculiarities. Placed among the ordinary polype-cells may be seen, at certain times of the year, a few scattered egg-shaped objects, some eight or ten usually being found on a branch. Within these cells are seen a small number of very minute living beings, which gradually develop themselves. A restless movement prevails towards the upper part, some slender tentacles make their appearance at the end, and at last the whole of the tip breaks loose, displaying itself as a tiny medusa.

This change is indeed a wonderful one, perhaps even more marvellous than the mutual transformations of hydra tuba and medusæ, inasmuch as the Campanularia and the medusa belong absolutely to separate classes ; and that a medusa should spring from a zoophyte is hardly less surprising than that a perch should give birth to a human being.

These important discoveries were made simultaneously by Professor Van Beneden and Sir John Dalyell, and the former naturalist was able to observe a phenomenon which certainly seems to be the first step towards the return from the medusa into the zoophyte. Having isolated a specimen of the little medusæ, and made a careful drawing of it, he left it for about an hour, and on his return was surprised to find that the whole shape of the tiny being had altered. The convex disc had become concave, the tentacles were reversed, and the animal had changed the central footstalk of the medusa into the semblance of a zoophytic stem.

" My observations," remarks that accomplished naturalist, as quoted by Mr. T. R. Jones, " go no further ; but although I have not seen the medusa give origin to a polype stem, I observed it up to the moment when it was about to form a new colony ; and without fear of deceiving ourselves, we may form by analogy some idea of the changes which must necessarily occur. The Camp.nularia, in its medusa state, has only a single aperture, situated at the extremity of its central pedicle. We have already seen that its body becomes inverted like the finger of a glove, and that the marginal filaments become converted into true tentacles. The polype fixes itself by the extremity of its central appendage—that is, by what was previously its mouth ; the back of the umbrella becomes depressed at the same time that the tentacles change their direction ; and in the centre of the disc a new aperture is formed, which communicates with the central cavity, and becomes the permanent mouth, which is situated directly opposite to the original one.

" Being now fixed by its base, the body of the polype begins to grow ; and as its external sheath becomes hardened, buds sprout at regular intervals from its surface. In a word, the growth of the polype resembles that of the hydra, with this difference, that in the latter there is no polype stem, and their buds sprout from another part of the body."

The name of Campanularia is given to this zoophyte in consequence of the bell-like form of its cells, and is derived from the Latin word " campana," a bell.

THE delicate PLUMULARIA is so called on account of the feathery appearance of its polypidon. The cells are always small and the egg-vesicles are scattered. In some species the stem is composed of many parallel tubes, such as *Plumularia myriophyllum*, but in the present species it is quite simple. The egg-vesicles are rather widely scattered.

ROTIFERA.

ALTHOUGH the Rotifera, or Wheel Animalcules, are generally placed among the Infusoria, on account of their minute dimensions and aquatic habits, it is evident, from many peculiarities of their formation, that they deserve a much higher place, and in all probability constitute a class by themselves.

They are called Wheel Animalcules on account of a curious structure which is found upon many of their members, and which looks very like a pair of revolving wheels set upon the head. These so-called wheels are two disc-like lobes, the edges of which are fringed with cilia, which, when in movement, give to the creature an appearance as if it wore wheels on its head, like those of the fairy knight of ballad poetry. These wheels can be drawn into the body at will, or protruded to some little extent, and their object is evidently to procure food by causing currents of water to flow across the mouth. All, however, do not possess these appendages, but have a row of cilia, mostly broken into lobes, extending all round the upper portion of the body.

They have a well-defined muscular system, while their jaws are nearly, if not quite, as complicated as those of the echinus. Most of them can swim, some are able to attach themselves at will to any fixed objects, while others are fixed to one spot from which they do not stir.

Distinct sexes have been discovered in several genera of Rotifers; and in those cases where the male has not been found, it is generally thought that the very small size and eccentric shape of the opposite sex may be the reason why it has not been discovered. In those instances where his existence has been indubitably ascertained, he is always a strange being, very unlike the female, very small, and, what is even more strange, possessing neither jaws, throat, stomach, nor intestines. His life must therefore be very short, as is known to be the case with the male sex in many insects. It has been well suggested, that perhaps the males are only produced at certain times of the year, and are not therefore found so plentifully as their mates.

Fortunately for observers, the integuments of these animals are extremely transparent, so that it is possible to watch the whole of the vital processes, and to see the various functions carried on with as much ease as if the skin were of crystal. Their development is wonderfully rapid ; for although but a few eggs are produced at one time, they are so quickly hatched, and the animal is so rapid in its growth, that Professor Ehrenberg calculated that in the genus Hydatica, although only three or four eggs are produced at a time, a single individual will be the progenitrix of nearly seventeen million descendants within the space of twenty-four days.

In this class the arrangement is very perplexing to systematic naturalists, and nothing 's as yet settled about it.

These remarkable beings are mostly found in water that has become stagnant but is partially purified by the presence of the Infusorians, which always swarm in such localities. There is, however, one very strange residence of the common Rotifer, namely, within the leaf-cells of the common bog-moss (*Sphagnum*). These cells are very large in proportion to the size of the leaf, are kept open by spiral threads coiled in their interior, and their walls are pierced with large apertures, so as to form a general communication throughout the whole mass of cells. Within these curious chambers the Rotifer is found, and is able to pass freely from one cell to another. They probably gain their admission in the egg state, and find sufficient moisture in the cells for their seeds.

The typical genus of this class is known by the name of Rotifer, an example of which is seen in the illustration. In all the members of this genus the body is rather elongated, and furnished at the hinder end with a kind of telescopic tail, by means of which they can attach themselves at will to any object, and release themselves whenever they please. Sometimes they move their bodies gently about, while still grasping by the extremity of tail ; sometimes they are nearly motionless, while they frequently rock themselves backward and forwards so violently that they seem almost to be testing the strength of their hold.

These creatures can both swim and crawl, the former act of locomotion being achieved by the movement of the cilia, and the latter by creeping along after the fashion of the leech, the head and tail taking alternate hold of the object on which they are crawling.

The masticating apparatus is always conspicuous, whether the animal have the wheel protruded or withdrawn. It is situated behind the bases of the wheel-lobes, and looks, when the animal is at rest, something like a circular buckler with a cross composed of double lines drawn over its surface. Even in the very young and undeveloped animals which are seen within the body of the parent, these jaws form the most conspicuous portions of their structure, and enable them to be recognised long before they are able to go out into their watery world and shift for themselves.

All the Rotifers have a marvellous fund of vitality, and survive under circumstances where animals less tenacious of life would die a thousand deaths. They have been thoroughly dried by means of chemical acid, wetted and restored to life, dried again, wetted again, and subjected to this treatment through many successive alternations, without perishing.

THE right hand figure in the illustration represents a creature which is a good example of those Rotifers, which are fixed to one spot by the extremity of the body. At first sight, this animal bears a strong resemblance to several of the Molluscoids; but

Rôtifer citrinus.

Stephanóceros Eichhórnii.

a closer examination shows that the apparent tentacles are nothing more than extensions of the lobes on which the cilia are set, and the apparent cell is no cell at all, but a gelatinous secretion from the body. In one genus, however, a veritable tube is built up, composed of particles of solid matter, formed into little pellets by a special organ, and then deposited upon the edge of the tube. The organ which forms these pellets is set towards the front of the head, and on its under side, and looks like a little revolving disc.

RHIZOPODA.

THE whole arrangement of the beings which we are now about to examine is still very obscure, and the best zoologists of the present time have declared that any system which has been hitherto adopted can only be considered as provisional.

Some writers, for example, unite the Rhizopoda with the Infusoria, while others rank them among the Polyzoa; and others again consider them to be intermediate between the radiata and those simple forms of animal life which are appropriately named Protozoa. After taking into consideration the various systems that have been propounded by different authors, I have come to the conclusion that, at all events, as a provisional arrangement, the Rhizopoda ought to be ranked as a distinct class, and placed in the position which they here occupy.

The name Rhizopoda is of Greek origin, and literally signifies "root-footed." It is a very appropriate title, inasmuch as they put forth certain filamentous appendages from their bodies, which look very like the tender rootlets of plants, and serve a double purpose, namely, as organs of progression, and as instruments whereby they may catch their prey.

Some of these beings are quite unprotected, their soft, gelatinous bodies being devoid of any covering; others are inclosed in a horny case, pierced with openings, through

which the filaments can be projected; while the greater number of the known species
are furnished with shells very similar in form to those of the molluscs, and in some cases
wonderfully similar to the highly complicated dwelling of one of the highest molluscs, the
pearly nautilus.

These minute though beautiful beings exist in numbers that are only rivalled by the
sands of the sea for multitude; and the vast hosts of these creatures can be barely
estimated even when we know that many large cities are built wholly of the dead
skeletons of these microscopic beings, and that in a single ounce of sand from the
Carribean Sea nearly four millions of these shells have been discovered. The living
species are not nearly so numerous as the fossil, but even the known species of this class
form no inconsiderable addition to the British fauna. They can be captured in various
ways. If, for example, growing algæ be plucked, and placed in a glass vessel of sea-
water, the Rhizopods will leave the algæ, and settle on the sides of the vessel. If they
five in muddy substances, such as the "oyster-ooze," which is especially prolific in
Rhizopod forms, the upper layer of mud should be taken off and stirred up in a vessel of
clear sea-water, when the creatures will sink to the bottom of the vessel, and may easily
be separated.

These modes are adopted for living specimens, but if the dead skeletons only are
required, they can be procured in many ways. One of the simplest methods of finding
Rhizopod shells is, to shake the dust out of sponges, and to examine it when laid thinly
on black paper. An ordinary pocket magnifier is employed in the search, and the
shells are readily seen against the black background. For removing them I always
employ a single bristle, stuck into a handle—one taken from a shaving-brush is, perhaps,
the best adapted to the purpose—and take up the shells singly by wetting the tip
of the bristle.

There is also another method whereby the empty shells may be obtained in considerable numbers. The sand, mud, or other substance, in which they reside, should be well
dried, heated, and then stirred into water. As the chambered cells of the Rhizopods
will be filled with air, they will float on the surface of the water, and can be skimmed off
without much difficulty.

The first sub-class of these beings is the Foraminifera, so called on account of the
tiny openings, or *foramina*, with which the pretty shells are pierced. Sometimes,
however, this shell is wanting, and its place supplied by a cover composed of matted
sand-grains. The greater number of these creatures are formed by a succession of buds,
each bud remaining in connexion with that from which it sprung, and thus forming a
composite body, which sometimes is rather complicated in its structure. Sometimes,
when the buds are merely arranged in a line, the result is a straight, rod-like form,
divided into a series of joints, marking the spots where the buds have in their sequence
issued from each other. If, on the other hand, each bud grows a little on one side of its
predecessor, a spiral form is the result, and a nautilus-like shell is formed. The
resemblance to this mollusc is further increased when each bud becomes rather larger
than that from which it sprung.

The arrangements of the Foraminifera hitherto in use have mostly been founded
upon the mode of growth; but Dr. Carpenter has clearly shown that this character is so
extremely variable, that no reliance can be placed upon it. In a single genus, there is
every gradation between the straight and the spiral forms; and, in many instances, a
shell which commences in a spiral will end in a straight line.

As, therefore, the already existing systems have been shown to be based on false
principles, and the arrangement which is to supplant them has not been fully decided
upon, we will not occupy our space by insisting upon the characters by which the systems
are established, but merely proceed to a brief description of the localities in which the
various species of our illustration may be found.

The greater number of the species represented in this illustration are found in
England, and may be recognised by means of the figures, among which may be mentioned
the genera which are now known by the names of Dentalina, Polystomella, Rosalina, and
Quinqueloculina. The remaining species are to be found in Central America.

Another sub-class of Rhizopods is named Polycystina, and is notable for the singular structure of the shells, which are pierced in regular patterns, without orifices, and are often prolonged into curious spikes and projections that give them a most wonderful beauty when seen under a good microscope. They are, in general, smaller than the Foraminifera, and are found in the mud of various seas, especially those of the West Indian islands. The marvellous variety which is obtained by the carrying out of two principles, namely, the piercing of holes and the projection of spikes, is almost incredible ; and the delicate tracery of the patterns thus produced is so artistic as to have been happily compared to the hollow ivory balls carved by the patient hands of Chinese artists.

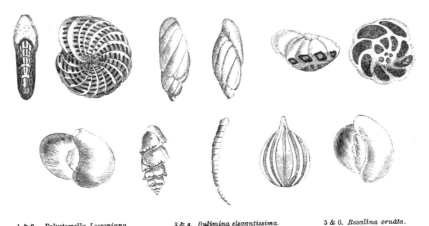

1 & 2. *Polystomella Lessoniana.*
7 & 11. *Quinqueloculina meridionalis.*

3 & 4. *Bulimina elegantissima.*
8. *Bulimina pulchella.*
10. *Oolina Isabelleana.*

5 & 6. *Rosalina ornata.*
9. *Dentalina acuta.*

THERE is one little creature, which is supposed by many physiologists to belong to the Rhizopoda, but whose position is very uncertain, and even its class not clearly ascertained. This is the NOCTILUCA, a tiny being, about as large as the head of a minikin pin, which swarms in our seas, and is remarkable for its phosphorescent power. If a vessel be filled with sea-water and brought into a dark room, the Noctiluca fills it with little sparklets of bluish light, which shine for an instant like stars in the firmament, and which can be induced to give out their momentary radiance by tapping the vessel, or even by a heavy footfall on the floor of the room.

Each of these little beings is furnished with a minute tail-like appendage, by means of which it is enabled to proceed through the water ; and on certain favourable occasions they fill the sea with their luminous hosts, and cause each wave to become a breaking mass of liquid fire. A ship passing through the sea leaves a fiery wake behind her keel, and when the boatmen lift their oars from the sea, they appear to drop flames from the blades as they are raised, all dripping, into the air at every stroke.

Although so small as to be microscopic in their dimensions, they are yet large enough to be discerned by the unaided eye, and can therefore be isolated without difficulty and placed in the field of the microscope.

ON the right hand of the accompanying illustration may be seen some odd-looking objects, which are considered as belonging to the Rhizopods, though not possessing any shell.

The Amœba is remarkable for the fact that it really has no outline and no shape, for its body is continually altering its figure; so that the rounded object which was seen in the microscope but a few minutes before, will, in that short space of time, have protruded a number of elongations that look like fingers of a glove or the rays of a star-fish.

It can elongate itself to almost any extent, can then throw out its strange protrusions so as to resemble a club with a spiked head, or it can gather itself into a rude globular mass, as if pinched out of dough by a single squeeze of the hand, allowing the soft substance to protrude between the fingers. It has no particular stomach, but extemporises that organ out of any part of its body with which its food happens to come in contact, literally pushing the food into its body and then digesting it without requiring any special apparatus for the purpose.

Some of the Lobose Rhizopods, as these creatures are called, are also furnished with a shelly or horny covering, such as the Arcellina, where the shield is cap-shaped, or the

Perinidium tripus.   Kerona polyporum.   Amœba princeps.
Perinidium michælis.   Amœba radiósa.

Difflugia, where it is pitcher-shaped, the animal protruding itself from that part which represents the mouth of the jug. Many physiologists suppose that the Amœba is not a perfect being, but it is merely the larval state of some animal with a higher development, such as the Arcella and other shell-bearing Rhizopods.

The genus Perinidium may be known by the furrow that runs transversely round the body, and is furnished with cilia. The integument of the body is membranous. The Tripos Perinidium is remarkable for its power of shining by night. It may be recognised by the shelly case, which is concave, smooth, and is developed into three horns, two being long and the other comparatively short. The longer horns are in front. Its length is about 150th of an inch. The Kerona also belongs to this order, and is found in fresh water, where it may often be seen in considerable numbers. Besides the usual cilia, it is furnished with instruments of progression that enable it to climb and creep, and are formed like bristles or hooklets. Its length is rather variable, but is about equal to that of the Perinidium.

INFUSORIA.

WE now come to the Infusoria, creatures which are all of very minute dimensions, and respecting which there is great uncertainty prevailing. As with the preceding class, no definite system has yet been invented by which they can be arranged; and in many cases physiologists are undecided whether the tiny beings are veritable species, or whether they are but the larval forms of higher beings; while, in some cases, it cannot be precisely ascertained whether they belong to the animal or vegetable kingdom.

Without, therefore, occupying our space with disquisitions which would require a volume for their full elucidation, we will proceed at once to some of the more remarkable forms among these curious beings.

Vorticella citrina.   *Stentor polymorphis.*

Two species of Infusoria are given in the accompanying illustration, both of which may be found in soft water that has been allowed to remain in the open air, and in which any vegetable matter has been permitted to decay. As may be seen, by reference to the illustration, both these creatures are affixed by footstalks to some object on which they make their residence, and both agree in having a bell-like mouth, edged with a fringe of cilia.

These organs are set upon the edge of the mouth, and their object is indirectly to draw food into the system by creating certain currents in the surrounding water. When the cilia are exposed to a good microscope, they appear to be formed like the cogs of a little wheel, which is rotating with great rapidity; and it is not until a close examination detects the real cause of this appearance that its illusory nature is discovered. As in the case of the cilia attached to the higher animals, of which a notice has already been given, each fibril bends in regular succession, so as to produce the effect of waves upon the eye.

When the Infusoria are free, the continual movement of the cilia causes them to move with greater or lesser swiftness through the water, each fibril acting as a minute paddle, and having a distinct feathering movement like that of an oar handled by a skilful rower. It is a most curious sight to observe the admirable manner in which they make their strokes, the flattened sides striking the water so as to give the greatest force to the blow, and the back stroke being made with the edge, so as to meet with the least possible resistance.

If the reader will refer to the figure of the VORTICELLA, he will find that the footstalks on which the bell-like cup is seated are of considerable length, and capable of being shortened by being coiled into a spiral form. This is by no means an uncommon Infusorian, and is very liberal in displaying this capability. It is usually found associating in groups, so that there is hardly a stage in its life of which some example cannot be discovered. Though devoid of apparent organs of sense, this creature is marvellously timid, shrinking in a moment if the water be shaken, and tightening its coils until they resemble the spiral rings of a vine's tendril. It soon, however, recovers itself, and by slow degrees permits the spires to uncoil, and waves its fringed head boldly in the narrow prison to which it has been consigned. Sometimes the Vorticella breaks away from its footstalks, and is then carried rapidly through the water by the action of its ever-waving fringe of cilia.

As is the case with many of its kindred, the Vorticella is able to increase its numbers by the simple process of splitting itself into two distinct beings, each of which is afterwards a complete and perfect being. At first, a single notch is seen upon the edge of the lip, but as time passes on the notch deepens, the cleft becomes more apparent, and in a wonderfully short space each half of the Vorticella is changed into a perfect individual, which in its turn is ready to divide and subdivide itself *ad infinitum*. It is a truly strange process this subdivision, and forms one of the links that bind animals of a higher type of organization with these lowly, but not imperfect beings. Thus, therefore, the Vorticella never need die of old age, for it renews its youth, as it were, by this voluntary division, just as if a man of sixty were to split himself down his spine, and thus become two young men of thirty, or, by further subdivision, four lads of fifteen.

THE second figure represents the STENTOR, so called because its general shape bears some resemblance to that of a speaking-trumpet. This is a comparatively large species, being visible to the naked eye, and readily distinguished by a practised observer. Sometimes it is found singly, either attached by its base or swimming boldly through the water; but in most instances it gathers itself round duck-weed, or floating sticks, and is produced in such numbers that its vast multitudes quite resemble a fringe of soft, filmy slime.

Like the preceding animal, the Stentor multiplies by self-division; but it is very likely that many other methods of increasing its numbers are employed. There is, for example, in these creatures, the remarkable phenomenon called "conjugation," which is almost identical with the same act as performed by some of the microscopic vegetables. If two free Infusoria of the same species—say, for example, the common Paramecium, that swarms so largely in stagnant waters—happen to meet at the proper season of the year, they adhere firmly to each other, as if they were magnets and iron, and go spinning about the water with no less speed than when each urged its single course. A vast number of very minute eggs are then produced by both of the individuals, but the further development of these eggs is not yet known. Sometimes, as in the Stentor, the Infusoria are fixed by their bases, and in such instances they bend their mouths towards each other, and so contrive to unite themselves in pairs.

PORIFERA.

WE now arrive at a large class of beings, which, if they really do belong to the animal kingdom, and are not to be ranked among vegetables, are by common consent allowed to form the very lowest link in the animal chain.

The name Porifera is given to them because the whole of their surface is pierced with holes of various dimensions, the greater number being extremely minute, while others are of considerable dimensions. The well-known Turkey Sponge, so useful for the toilet, will afford a good example of the porous structure.

Yet, no one can form an adequate idea of the living Sponge from the dry, dead skeleton which is sold under that name. Many of the species are decked with delicate colours, while all are truly beautiful creatures when viewed in full life and action. They are to be found widely distributed through the seas, and there is hardly a solid body on which a Sponge will not grow. Sponges are generally found hanging from the under sides of projecting rocks at some distance below the surface of the sea, or clinging to the roofs of submarine caverns. Some, however, are strong, sturdy, and branched, and stand boldly erect like the earth-plants which they so wonderfully resemble.

Even the living inhabitants of the sea are liable to become the resting-places of many a Sponge, and the crustacea are often forced to bear on their shells the additional burden of living Sponges and other zoophytes much more massive than their whole body.

The true living being which constitutes the Sponge is of a soft and almost gelatinous texture, to the unaided eye; and with the aid of the microscope is found to consist of an

Grantia coarctata.
(New species.)

Grantia compressa.

aggregation of separate bodies like those of the Amœbæ, some of which are furnished with long cilia. By the constant action of the cilia a current of water is kept up, causing the liquid to enter at the innumerable pores with which the surface is pierced, and to be expelled through the larger orifices. A Sponge in full action is a wonderful sight: the cilia drives the water in ceaseless torrents, whirling along all kinds of solid particles, arresting those which are useful for digestion, and rejecting those which it cannot assimilate.

The reader will at once see that a creature thus composed will stand in need of some solid framework on which the delicate fabric can be supported; and on examining a series of Sponges with the microscope, we find that it is mostly composed of a fibrous and rather horny network, strengthened with spiculæ of a hard mineral substance. The shape of the spiculæ is extremely variable, some being simple translucent bars, some looking much like rough flints rendered transparent, others star-shaped with several points, while the greater number resemble knotted clubs made of differently-coloured glass, and having a lovely effect under the microscope.

In the genus Grantia, which is well known to marine zoologists as having furnished valuable information respecting the nutriment and reproduction of the Sponges, no horny network can be found, but its place is supplied by the singular form of the spiculæ, which are composed of three long-pointed spines arranged so as to form a star of three rays. These rays, on account of their shape, form an entangled mass, and answer the purpose of the ordinary horny framework. The new species shown in the illustration inhabits the East Indies.

There are several British species of the genus Grantia. Some of them are hollow, and stand out with tolerable boldness from the objects on which they are set, while others are always found as whitish incrustations upon stones and other massive substances. Their structure is tolerably firm, and, on account of the absence of the horny framework, is not so elastic as are the generality of the Sponge tribe; and the texture is very close, but still porous. With a microscope of tolerably high power, magnifying from two to three hundred diameters, the layer of spiculæ can be readily made out, interlacing with

Halichóndria pulmata.

each other in wonderful profusion, and so completely intermixed that a single spicula is scarcely ever separable from the general mass.

The shape of the species belonging to this genus is extremely variable, but in all the structure is remarkably simple, the wall being extremely thin, so that the ramifying canals are not needed, and the water is merely absorbed through the minute pores of the wall and expelled through the large orifice which forms the mouth of the sac. If the spiculæ of this or other Sponges be wanted in a separate state, the animal matter can be removed by heat; but a better, though slower process, is to immerse the specimen in strong nitric acid or liquor potassæ, according to the flinty or chalky nature of the spicules. When separated they may be mounted in two ways, namely, as dry and opaque objects, or in Canada balsam.

We must now briefly examine a rather important genus of Sponges, which has many representatives in the British seas. It is a very extensive genus, and its members are variously shaped, all, however, agreeing in those salient points on which the group has been founded. They are all spongy, elastic, not slimy, and with a very porous surface.

The example which is given in the illustration is generally called the MERMAID'S GLOVE, because it is apt to spread into a form that bears a somewhat remote resemblance to a glove with extended fingers. It is certainly the largest of the British Sponges, sometimes attaining a height of two feet, and stretching out its branches boldly into the sea. The branches are rather flattened, and when full-grown are about an inch in width. They do not always remain separated throughout their whole extent, but are apt to coalesce in various parts, and sometimes to form rudely-shaped arches.

The colour of this Sponge is generally of a pale straw-yellow, and to the touch its exterior is decidedly rough, on account of the myriads of spiculæ which slighty project from the surface. These spiculæ are needle-like, sometimes slightly curved and sometimes straight. Mostly they are pointed at both ends, but as they are fragile and snap asunder with the least violence, they often look as if they were only pointed at one end. They lie nearly parallel to each other, and many are so placed that their points are presented outwards. This Sponge is found in deep water in many parts of the British seas.

LING-HOOD.—*Halichóndria ventilábrum.*
Halichondria fucórum.

FUNNEL-SPONGE.—*Halichóndria infundibulórum.*

ALL the figures in this illustration represent species of this large genus.

The FUNNEL-SPONGE may easily be distinguished by its shape, which closely resembles that of an ordinary funnel. Its structure is very finely porous, and it is rather a pretty and elegant species. The spiculæ which contain the softer parts of the FUNNEL-SPONGE are long, slender, and sometimes curved. In most cases they are pointed at each end, but in others only one end is sharp, while the other is rounded, so that the spicule resembles a needle without an eye. They are rather loose, and either lie in bundles or crossing one another. The width of an ordinary specimen is about three inches, and its length is equal to its width.

The LING-HOOD has a shape which, when it is young, reminds the observer of the preceding species. It may, however, be readily distinguished from that Sponge by the thick coating of hair-like spines with which its surface is covered. It always becomes shallower by age, and is therefore extremely variable in its form. The edge is seldom so smooth and regular as that of the Funnel-sponge, being mostly cut into notches and the intervals developed into lobes.

One remarkable characteristic of this species is the very brittle exterior, which can be broken away by the fingers, and is found to consist almost wholly of flinty spicules, cemented together by the glutinous substance of the animal. Sometimes it almost loses the cup-like form, and spreads out like a fan, deriving therefrom the popular title of SEA-FAN. As the term, however, is applied to many other marine beings, it is not thought so useful as the name which has already been mentioned.

OUR last example of the Halichondriæ is seen investing the upper part of a zoophyte. Many of these creatures, especially the sertularia and its kin, are very liable to the gentle but irresistible attacks of this Sponge, which, settling upon them, increases rapidly and more or less envelops them in its own mass. Its mode of growth is always variable. Sometimes it follows all the branches of the zoophyte on which it is parasitic, causing it to resemble a tree thickly covered with ivy: while at other times it spreads out so widely,

NEPTUNE'S CUP.—*Thaldssema Neptuni.*

that it gathers all the branches together, covers them with its own substance, and forms them into a shapeless spongy mass from which a few ends of the branches vaguely protrude.

This Sponge does not, however, confine itself to zoophytes as resting-places, but settles upon stones, sticks, shells, and other objects. Its colour is generally greyish white, but it sometimes deepens its tint and becomes of a yellowish brown. The texture of its substance is always rather coarse, but its softness differs greatly according to the object on which it has established itself and the locality in which it happens to be. The spiculæ are always short, rather curved, and are sharpened at one end and rounded at the other. About forty species of the genus Halichondria are found in our seas, several of which are remarkable for having the spiculæ knobbed at both ends.

THE extraordinary object which is called by the appropriate name of NEPTUNE'S CUP is one of the most magnificent, as well as one of the most notable, of the Sponge tribe. It hardly looks like a Sponge; and when a specimen is shown to persons who have no knowledge of the subject, they can hardly ever be made to believe that the exhibitor is not endeavouring to play a practical joke upon them.

The Neptune's Cup is of enormous dimensions, often measuring four feet in height and having a corresponding width. Its exterior is rough, gnarled, and knotted like the bark of some old tree ; and if a portion were removed from the side, it might almost be mistaken for a piece of cork-tree bark. Many persons have imagined that the strangely-shaped object was made of the skin of an elephant's leg, and I have even heard a teacher telling her pupils that it was an old Roman wine-jar.

It is hardly possible to disabuse strangers of the notion that it is not the result of human ingenuity until they are allowed to lift it, and test personally its exceeding lightness. It is hollow, and is not at all unlike an old font that by some misfortune has been deprived of its base. Its capacity is enormous, and it would not only form a cup for Neptune, but even Polyphemus himself might have filled its depths with the ruddy wine which he loved, and failed to empty the huge vessel at a draught.

The substance of this enormous sponge is porous, rather stiff, and without much elasticity. It yields but slightly to pressure, and almost feels to the touch as if it had been made from cork.

This is one of the exotic Sponges, being found only in the hotter seas. In general shape it has some resemblance to the Funnel-sponge already described, but is of much coarser texture, and, save for its gigantic size, is not nearly so attractive.

PUMICE-STONE SPONGE. — *Dactylóchalix pumicea.*

In the PUMICE-STONE SPONGE, we have an admirable example of the flinty structure developed to the utmost degree. The framework is wholly composed of flinty spiculæ, all fused together, and forming a highly porous mass, which at first sight resembles a madrepore rather than a true Sponge. It has not the least elasticity, but is as hard and as stiff as if it had been carved from stone. On account of its extreme porosity, it is very light, seeming to weigh not more than a piece of cork of the same size.

The whole surface, above and below, is plentifully sown with pores, which have a lovely effect under a magnifier, when the sunbeams fall on the glittering spiculæ of which the mass is composed. A number of the large apertures appear on both sides, and all converge towards the centre. The general shape of this remarkable Sponge is cup-like, but exceedingly shallow, and on the inside it is tolerably smooth, becoming rougher and deeply grooved on the outer surface. It has a peculiarly rough feel to the touch, almost exactly resembling the well-known rasping effect produced by rubbing pumice-stone upon the skin; and it is in consequence of this resemblance that it has gained its popular title.

The peculiarities of this very beautiful Sponge consist in the following distinctive characters, the most remarkable of which is its being formed entirely of silex, the reticulate structure of the mass being composed of transparent glassy tubes, the silex forming the mass itself, and not, as in other instances, arranged as spiculæ in the horny membranes ; consequently, it is perfectly rigid and sonorous when struck.

When viewed under a microscope of about seventy-five diameters, the net-like meshes are seen to be composed of beautiful glassy tubes, uniting one with the other in every direction, the external surface of the cylinders having a rugged aspect. The newest or last-formed portions appear to emanate from centres, and at certain distances from spherical knobs, from which straight tubes again arise, thus forming the net-like mass.

BEFORE taking leave of these interesting beings, we must glance rapidly at the method by which they distribute themselves so widely and increase with such marvellous rapidity.

It will be remembered that the soft animal matter of which the true Sponge is formed is composed of multitudinous bodies which closely resemble the Amœbæ, and many of which are furnished with thread-like cilia. In certain months of the year, which in our country are generally found to be October and November, a vast number of very minute yellowish particles are to be seen studding the body of the Sponge. They are not often seen near the surface, but are gathered plentifully within its multitudinous cells. Small as are these yellow particles, they are formed of many eggs, or " gemmules," as they are called, of the Sponge, which gradually increase in size, and at last are expelled from the larger orifices, and thrown at random into the wide sea.

There they are, flimsy, minute, shelterless, feeble, and apparently helpless. Small, however, as they may be, they still possess the power of transporting themselves through the water by means of the cilia with which their bodies are abundantly studded. Their shape is very like that of a pear ; and as they are wholly covered with cilia, except the narrow end, it is evident that their larger end must always be in front. They lead a free life for several days after their expulsion from the parental home ; and even in this early stage some indications of the future framework are to be seen.

After the lapse of some little time, these gemmules meet with some object which affords them a suitable resting-place, and accordingly affix themselves to the spot, from which they never afterwards can move. The rounded body soon becomes flattened, as it adheres with a close grasp, and spreads itself into a nearly circular film. The cilia still exist on the upper surface of this film, but the effect of their action is then not to propel the Sponge, but to create a current of water which can pass over it.

As time passes on, the distinctive spiculæ become visible, and, after three weeks or a month have passed away, the spiculæ have been gathered into little bundles, which by their arrangement tend to preserve the shape of the Sponge and to keep the orifices open. The little being now spreads rapidly, by a process which much resembles the subdivision of the Infusoria, and the whole mass of the Sponge is evidently composed of a vast number of the Amœba-like bodies which have already been described. Thousands upon thousands of these gemmules are passed out into the sea from every Sponge that inhabits its waters ; and the only wonder is, that, in consequence of such marvellously prolific properties, the Sponges do not swarm to such an extent as to fill the whole seas, and poison the entire earth with the odour of their decay.

CLASS.—REPTILES.

Animals with distinct bony skeleton—Skin clothed with horny plates or scales, covered with thin epidermis; often shed periodically—Respiration by lungs—Heart with two auricles; single ventricle divided into cells—Young produced from eggs; sometimes hatched before birth—The young resemble the parent, and undergo no transformation.

Section. SHIELDED REPTILES (CATA-PHRACTA).—Body covered with square imbedded shields. Bones of skull thick, hard, and united into solid mass. Bones of jaws united and solid. Young hatched after birth by egg.

Order. **CHELARIANS.**—Body enclosed in a case formed of the bones of the thorax, the limbs and head protruding. Jaws without teeth, and covered with horny coat.

Family. **TESTUDINIDÆ.** — Feet club-shaped, claws blunt. Shields at tail curved inwards and united together. Neck retractile.

Genus. TESTUDO.—Thorax convex, rather globular, and solid. Breastbone solid, with twelve shields; those of the throat separated. Five toes on fore feet, four on the hinder pair. Most warm parts of the world.

Genus. PYXIS.—Thorax solid. Front lobe of breastbone moveable, plates of breast twelve. Madagascar.

Family. **EMYDIDÆ.** — Head flattened, covered with skin. Neck retractile, vertebræ bent in perpendicular bow. Feet flat and broad, claws sharp. Tail conical and shielded beneath. Thorax solid, with horny plates. Egg oblong and white.

Genus. EMYS.—Head and tail moderate. Breastbone solid and deeply notched behind. Head with thin hard skin. Toes, five on fore limbs, four on hind pair, webbed to claws. Europe, Africa, Asia, and America.

Genus. MALACLEMYS.—Head with soft skin, toes as in preceding genus. America.

Genus. CISTUDA.—Breastbone divided by a single suture across the centre, and rounded in front. Southern Europe, America, and Asia.

Genus. KINOSTERNON.—Breastbone divided by two cross sutures. Chin bearded. Feet and tail nearly naked. America.

Genus. CHELYDRA.—Head very large, covered with hard wrinkled skin. Breastbone slender, sharp at each end, cross-like and leathery in the middle. Chin with two beards. Tail crested. America.

Family. **CHELIDIDÆ.**—Head flat, broad; nostrils long and tubular; jaws horny. Neck long, and bent under side of shell during repose. Feet webbed, and lobes between the claws; claws long and sharp. Shell rather flat. Pelvis large.

Genus. CHELODINA.—Head large and lengthened; neck very long, covered with thin skin. Claws, four on each foot. Breastbone broad, rounded in front, with a notch behind. Australia.

Genus. CHELYS.—Head very flat, and fringed with wart-like appendages; nose elongated at proboscis; jaws with soft lips. Neck thick, flat, long, and bearded at each side. Tail short. South America.

Family. **TRIONYCIDÆ.**—Head flat and oval, jaws with hanging fleshy lips, chin not bearded. Neck long and contractile. Toes, five on each foot; claws, three. Tail short. Shell covered with hard gristly skin; breastbone ring-shaped, covered with skin, and armed with sharp tubercles in the adult.

Genus. TRIONYX.—Head narrow in front. Breastbone narrow at each end, edge of shield flexible; eight pair of ribs. North America.

Genus. TYRSE.—Ribs eight pair; rest as in Trionyx. Europe, Asia, and Africa.

Genus. DOGANIA.—Head very large, broad behind, narrow in front. Ribs, eight pair, united into solid disc when adult. Asia.

Family. **CHELONIADÆ.** — Head rounded, shielded; jaws naked and horny. Feet very long, front pair largest, not retractile, flat, and pin-shaped; toes expanded at tips. Tail short and thick. Ribs free at end. Breastbone ring-like, united to thorax by gristly suture.

Genus. SPHARGIS.—Shell with leathery skin. Claws wanting. West Indies.

Genus. CAOUANA.—Shell with at least fifteen horny shields. Breastbone with a strong keel. Claws wanting. Atlantic and most warm seas.

Genus. CARETTA. — Shell with thirteen overlapping shields. Breastbone keeled on both sides. First or two first toes with claws. American and Indian seas.

Genus. CHELONIA.—Shell with thirteen smooth shields, not overlapping. First toe clawed. Hot seas.

Order. **EMYDOSAURI.** — Body elongated and covered with bony plates, arranged in longitudinal rows. Limbs four in number and free. Jaws with teeth, inserted in sockets and replaced by others formed below them. Back of neck with two sets of bony plates, the first called nuchal, second cervical. Toes, four on fore legs, five on hinder pair; claws on three inner toes.

Family. **CROCODILIDÆ.**—Lower canine teeth fitting into a notch in edge of upper jaw.

Genus. GAVIALIS.—Teeth all same length; jaws very long and slender, and rounded; nose of male inflated. Plates of back and cervicle set united. Asia.

Genus. MECISTOPS.—Teeth, ninth of upper jaw and eleventh of under larger than others; jaws oblong, slender, and rather flat; nose not swollen. Cervical and dorsal plates united. Africa. Hind feet webbed.

Genus. CROCODILUS.—Jaws oblong and flat. Nuchal, cervical, and dorsal plates separated by little scales. Africa, Asia, and America.

Family. **ALLIGATORIDÆ.**—Canine teeth of lower jaw fitting into a hollow in edge of upper jaw.

Genus. YACARE.—Jaws with ridge across face. Hind feet little webbed. Nostrils separated by division of gristly substance. Eyelids fleshy. New World.

Genus. ALLIGATOR.—Jaws with small longitudinal ridge between eyes. Hind feet fully webbed. Nostrils separated by division of bony substance. New World.

Order. **AMPHISBÆNIA.**—Body long and cylindrical; covered with squared plates imbedded in the skin, and arranged in cross rows. Limbs either wanting or very feeble. Jaws with teeth.

Family. **AMPHISBÆNIDÆ.**—Teeth on inner side of jaws. Limbs wanting. Body and chest with similar square shields. Pores before vent distinct.

Genus. AMPHISBÆNA.—Head flattened, muzzle round, rostral shield triangular and erect. Two shields of nose large, square, and extending across muzzle. Brazil and West Indies.

Family. **CHEIROTIDÆ.**—Teeth conical, curved backwards, and separate. Limbs two in front and weak, with five nearly equal toes. Pores before vent distinct.

Section. **SQUAMATA, OR SCALED REPTILES.**—Body covered with overlapping scales. Skull formed of separate bones; bone of ear external and jointed to skull. Tongue free, long, and nicked at the tip.

Order. **SAURA, OR LIZARDS.** — Mouth not dilatable, jaws toothed, bones of lower jaw united in front. Limbs four, and mostly distinct; toes clawed. Body long and rounded. Ribs distinct, and moveable. Tail long and tapering. Eggs covered with a hard skin.

Sub-order. LEPTOGLOSSÆ. — Tongue flat, long, and forked at tip

Tribe. CYCLOSAURA. — Scales of back and tail rounded, or granular; those of abdomen squared and set in cross bands. Eyes with two valvular lids. Feet strong, toes compressed.

Family. **MONITORIDÆ.**—Head with little, flat, many-sided shields. Tongue sheathed at base.

Genus. REGENIA.—Tail compressed, with keel of double row of scales. Nostrils large, oblique, and near the eyes. Toes short. Africa.

Genus. MONITOR.—Tail as above. Nostrils round, nearly half way between eyes and muzzle. Africa, Asia, and Australia.

Family. **TEIDÆ.**—Head with large, regular, many-sided shields; teeth solid; tongue flat and long, mostly free. Scales of back granular or keeled, and rhombic in shape. Sides flat. Throat scaly, with double collar. All the Teidæ inhabit the Western world.

Genus. TEIUS.—Throat with large hexagonal scales between folds of throat. Ventral shields small, long, and smooth. Five toes on each foot. Pores of thigh distinct. Brazils.

Genus. AMEIVA. — Teeth with three lobes, and compressed. Ventral shield broad and smooth. Tongue sheathed at base. Brazils and West Indies.

Genus. CNEMIDOPHORUS.—Tongue not sheathed at base. Teeth compressed longitudinally. Toes five on each foot.

Genus. CENTROPYX.—Throat with collar of large shields. Tail round. Scales of back large, those of sides granular. America.

Genus. ADA.—Collar and ventral shields smooth and long. Tail compressed. Scales of back unequal; throat with two folds.

Family. **LACERTINIDÆ.**—Head pyramidal, with regular many-sided shields. Throat scaly. Tongue long, flat, and deeply forked. Teeth hollow. Scales of body keeled and rhombic, those of sides small and granular. Sides flattened.

Genus. ZOOTOCA.—Nostrils erect in nasal shield. Toes simple and compressed. Collar distinct. Lower eyelid scaly and opaque. Europe, Africa, and Australia.

Genus. LACERTA.—Lower eyelid opaque. Chin fold distinct. Shields of abdomen narrow behind. Europe, Africa, Asia.

Genus. ACANTHODACTYLUS.—Toes keeled beneath, and fringed on the sides. Palate without teeth. Africa.

Genus. EREMIAS.—Nostril horizontal, between three scales. Eyelid distinct. Toes compressed, keeled beneath, not toothed on edge. Many ventral shields, in several series. Africa.

Genus. OPHIOPS.—Eyelid rudimentary, eye circular and exposed. Palate not toothed. Pores on thighs distinct. Shores of the Mediterranean.

Family. **ZONURIDÆ.**—Head pyramidal or flattened, with regular many-sided shields. Tongue flat and niched at tip. Scales of back and tail large and rhombic. A fold along the sides. Eyes with two lids.

Genus. ZONURUS.—Head flat. Scales of back and sides square and keeled. Lower eyelid opaque. Tail spined. Pores of thigh distinct and numerous. Africa.

Genus. PSEUDOCORDYLUS.—Head flat. Scales of back nearly oval, surrounded with granules. Tail spined. Africa.

Genus. GERRHOSAURUS.—Head pyramidal and squared. Tongue scaly. Five toes on each foot. Body spindle-shaped. Tail not spined. Africa.

Genus. SAUROPHIS.—Toes four on each foot, and very short. Body and tail very long. Tail not spined. Africa; perhaps Australia.

Genus. TACHYDROMUS. — Head pyramidal and long. Collar of keeled scales. Ventral scales keeled. Tail not spined. Asia.

Genus. PSEUDOPUS.—Fore pair of limbs wanting, hinder pair very small and undivided. Body snake-like. No pores on thigh. Africa, parts of Europe, and Asia.

Genus. OPHISAURUS.—Limbs wholly wanting. Two deep grooves along body. Palate toothed. North America.

Family. **CHALCIDÆ.**—Head with regular many-sided shields. Tongue scaly, palate without teeth. Limbs four, but very small and feeble. No pores on thigh. Body long and cylindrical.

Genus. CHALCIS.—Hinder feet not divided. Scales square. America.

Family. **ANADIADÆ.**—Lower eyelid transparent. Scales of back and sides six-sided, not overlapping; those of abdomen four-sided, and arranged in longitudinal series. Sides rounded. Limbs four, five toes on each foot; numerous pores on thigh.

Genus. ANADIAS.—Same characteristics. America.

Family. **CHEIROCOLIDÆ.**—Lower eyelid transparent. Tongue scaly. Collar double. Ears under skin. Body and tail long and nearly cylindrical. Scales of back, sides, and tail long, sharp, overlapping, and keeled, arranged in alternating rings. Legs four, and short; pores numerous; toes five on each foot, hinder toes very long.

Genus. HETERODACTYLUS.—Same characteristics. America.

Family. **CERCOSAURIDÆ.**—Upper eyelids very short. Ears distinct. Sides rounded with scales like those of back. Large keeled scales in series along back and upper part of tail. Under parts with flat squared scales. Limbs four, toes five on each foot, tail round.

Genus. CERCOSAURA.—Body long; scales of back oblong, with blunt keels. No pores on thighs. America.

Family. **CHAMÆSAURIDÆ.**—Palate without teeth. Lower eyelid scaly. Body and tail long, and covered with regular rings of keeled scales. Limbs rudimentary. Ears distinct and small. Sides without groove, and scaled like back.

Genus. CHAMÆSAURA.—Feet undivided, like fins. Back with six series of broad scales. Africa.

Tribe. GEISSOSAURA.—Scales of belly rounded, overlapping, and covered with thin epidermis. Sides rounded, scaled like the back. Tongue short, flat, narrow, slightly nicked at the end. Head with regular many-sided shields, neck nearly as wide as head.

Family. GYMNOPHTHALMIDÆ. — Eyelids rudimentary. Teeth of jaws conical, none on palate. Limbs four, weak. No pores on thigh.

Genus. CRYTOBLEPHARUS.—Head conical, scales smooth. Five toes on each foot. Ears exposed. Australia.

Family. PYGOPIDÆ.—Pupil of eye round or oblong. Shields of abdomen six-sided, in two or three series. A central series of larger shields on the tail. Fore limbs wanting, hinder pair very feeble.

Genus. PYGOPUS.—Scales of back keeled. Pores before vent numerous. Hinder limbs rather long. Pupil of eye circular. Australia.

Genus. DELMA.—Scales smooth. No pores before vent. Pupil oval and erect. Hind limbs short. Australia.

Family. APRASIADÆ.—Nostrils between two shields. Limbs wanting. Scales above and below nearly similar. Pores none.

Genus. APRASIAS.—Same characteristics. Australia.

Family. LIALISIDÆ.—Head-shields slightly overlapping; cheeks scaly. Nostrils in little scale on ridge of face. Fore limbs wanting; hinder undivided, short, flat, and scaly. Tail long. Series of pores before vent.

Genus. LIALIS.—Same characteristics. Australia.

Family. SCINCIDÆ.—Head conical. Eyelids distinct, and capable of closing over eye. Nostrils in a shield on nose. Limbs four, sometimes hidden under skin. No pores on thigh.

Genus. SCINCUS.—Scales thin, smooth. Rostral plate keeled in front. Tail round, tapering, without spines. Body spindle-shaped, and flat below. Toes five on each foot, flat, and fringed on sides. North Africa.

Genus. PLESTIODON.—Lower eyelid scaly. Limbs strong. Body spindle-shaped, and tail tapering. Scales large. Palate toothed. Africa, America, and Asia.

Genus. MABOUYA.—Lower eyelid transparent. Rest as in Plestiodon. West Indies and Asia.

Genus. ANGUIS.—Body and tail cylindrical, and very long. No external limbs. Head conical and blunt. Teeth rather long and sharp. Lower eyelid scaly and opaque. Europe, Asia, and Africa.

Genus. TRIBOLONOTUS.—Scales thick and wrinkled, those of back strong and spine-like. Body spindle-shaped. Head flat and plated, armed with spines behind. Lower eyelid scaly. Limbs four, and strong; toes five on each foot. No pores on thigh. Tail compressed, with five spiny keels above. New Guinea.

Genus. CELESTUS.—Body spindle-shaped; tail round, tapering, not keeled above. Scales grooved. Head flattened. Five toes on each foot, the claws small and sharp. West Indies.

Genus. DIPLOGLOSSUS.—Body and tail cylindrical and elongated. Rest as in preceding genus. Tropical America.

Family. OPHIOMORIDÆ.—Head shielded, rostral shield triangular and erect. Palate without teeth. Tongue flat, scaly, and with a shallow nick. Ears hidden under skin. Body and tail long and cylindrical; limbs none. Scaly, smooth, six-sided.

Genus. OPHIOMORUS.—Only genus. Northern Africa.

Family. SEPSIDÆ.—Rostral plate square. Palate toothless, and with a longitudinal groove. Limbs four or two, weak.

Genus. SPHÆNOPS.—Rostral plate long and notched. Head wedge-shaped ; lower eyelid transparent. Legs four. Northern Africa.

Genus. SEPS.—Rostral plate rounded. Head pyramidal. Body long and cylindrical. Lower eyelid with transparent disc. Limbs four ; toes three on each foot. Europe and Northern Africa.

Genus. GONGYLUS. — Body somewhat spindle-shaped. Limbs four ; toes five on each foot. Shores of Mediterranean.

Family. ACONTIADÆ. — Head small and shielded ; upper eyelid very small. Tongue scaly. No pores on thigh. Ears minute.

Genus. ACONTIAS.—Limbs none. Southern Africa.

Family. TYPHLINIDÆ.—Rostral shield large and cup-shaped. Body and tail cylindrical. Eyes and ears hidden under skin.

Genus. TYPHLINE.—Limbs none. Ventral shield large and single. Southern Africa.

Family. TYPHLOPSIDÆ.—Head flat, broad, rounded in front; rostral plate large, oblong, and erect. Mouth semilunar, under the muzzle. Eyes under scales or wanting. Body cylindrical. Tail cylindrical, suddenly contracted, and obliquely convex at tip.

Genus. TYPHLOPS.—Rostral plate very large, turned down in front. Tail short. Asia.

Genus. ONYCHOPHIS.—Rostral shield moderate ; oblong. Asia, Africa, Australia, and the Philippines.

Genus. ARGYROPHIS.—Head flat, rounded ; central shield oblong, arched, and erect. Eyes distinct. Asia and Tropical America.

Sub-order. PACHYGLOSSÆ.—Tongue thick and convex, slightly nicked at end.

Tribe. NYCTISAURA.—Scales of back and sides granular ; of abdomen, small and overlapping. Eyes nocturnal, lids circular. Body rather flat. Limbs strong.

Family. GECKOTIDÆ.—Only family.

Genus. PTYODACTYLUS.—Toes slender, dilated at tips, with two diverging series of plates beneath, not webbed. Body simple and tail round. Northern Africa.

Genus. GECKO.—Toes dilated throughout, with a series of transverse membranous plates beneath, last joint short, not webbed. Back with tubercles.

Genus. THECODACTYLUS.—Toes dilated, last joint short, bent under and hidden in notch ; two series of transverse plates beneath, half webbed. Tail granular. Tropical America.

Genus. PTYCHOZOÖN.—Head, body, and tail fringed with deeply-scooped membrane on each side. Toes webbed. Java.

Genus. TARENTOLA.—Toes dilated, with single series of plates beneath two claws on each foot. Rostral shield very long. Europe, Africa, America, and West Indian Islands.

Genus. SPHÆRODACTYLUS.—Toes slender, with single round disc at end. Claws none. Rest like Tarentola. America.

Genus. PHYLLURUS. —- Toes clawed, rather dilated at base, long, slender, and joints bent at angle. Tail very flat, tapering, and rounded at tip. New Holland.

Tribe. STROBILOSAURA.—All scales overlapping. Eyes diurnal, lids valvular, pupils round.

Family. IGUANIDÆ.—Teeth round at root, but leaf-like and toothed on edge, set on inner side of jaws ; young teeth growing at base of others and pushing them forward.

Genus. POLYCHRUS. — Body compressed, and covered with little squared scales arranged in rings. Pores on thigh. Toes slender, not fringed. Back not crested. Throat compressed and toothed in front. Eyes prominent. Brazil.

Genus. IGUANA.—Pores on thigh. Throat-pouch large and compressed ; toothed in front. Scales of back moderate. Tail compressed. Tropical America.

Genus. ALOPONOTUS. — Scales of back very m'nute. Throat-pouch not toothed. Rest like Iguana. St. Domingo.

Genus. METOPOCEROS.—Throat-pouch small, with a cross fold. Middle toes of hind feet toothed. Head-shields fl.t, a large horn-like tubercle on forehead, and two pair of large plates between nostrils. Pores on thigh in two rows. St. Domingo.

Genus. OREOCEPHALUS.—Head-shields conical and deeply wrinkled. Outer toes of hind feet very long. Rest like preceding genus. Galapagos.

Genus. BASILISCUS.—Head long, with crest behind ; a pin-like crest along the back and tail. No pores on thigh. Throat with cross fold. Outer toes of hind feet webbed at base, not fringed at sides. Tropical America.

Genus. XIPHOSURUS.—Throat-pouch large and dilatable. No pores on thigh. Toes dilated under last joint but one. Tail with bony crest. Central America (?).

Genus. ANOLIUS.—Back simple, or with little double crest of small scales. Rostral shield erect, nose rounded. Rest like Xiphosurus. North America.

Genus. PHRYNOSOMA. — Body flattened and covered with small scales. Sides toothed. Head armed with spines. Tubercular scales scattered over back and tail. California.

Family. AGAMIDÆ.—Teeth on edge of jaws ; tongue short, sometimes without a niche at the tip. Limbs strong, thumb of hind feet on same line as the toes.

Genus. DRACO.—Body compressed, scales overlapping ; ribs greatly elongated, supporting expansion of skin. Three pouches on throat. Ears not scaled ; nostrils below face-ridge. Java.

Genus. GONYOCEPHALUS.—Ribs simple, back with crest. Five toes on each foot. Tail with long-keeled scales beneath ; ears exposed. Java.

Genus. CHLAMYDOSAURUS.—Pores on thigh. Scales irregular and overlapping. Toes fringed ; throat with a fold ; back crested. Neck with a large frill-like expansion on each side. Australia.

Genus. GRAMMATOPHORA.—Back without crest, covered with cross rows of large scales. Pores of thigh numerous. Neck without frill. Australia.

Genus. STELLIO.—Body flattened, throat with a cross fold, back covered with overlapping scales. Several rows of ventral pores. Ears exposed. Tail covered with large spine-like scales in regular rings. Northern Africa and Asia.

Genus. AGAMA.—Single row of ventral pores. Scales keeled. Back not crested or toothed. Northern Africa.

Genus. MEGALOCHILUS.—No ventral pores. Ears hidden. Toes fringed on side. Edge of mouth fringed. Russia.

Genus. UROMASTYX.—No ventral pores, but some on the thighs. Ears exposed. Tail broad and flat, covered with regular rings of spiny scales. Northern Africa and Russia.

Genus. MOLOCH.—No pores. Ears exposed. Neck with a large convex tubercle on its upper surface. Body and limbs covered with large spiny tubercles. Australia.

Tribe. DENDROSAURA.—Body compressed. Scales granular and arranged in circular bands. Eyes projecting, globular, covered with circular lid pierced with a very small round hole. Tongue nearly cylindrical, long, extensile, and rather large at the tip. Ears hidden under skin. Toes five on each foot, arranged in two opposing groups. Tail long and prehensile. Only one genus.

Genus. CHAMELEO.—Characters of tribe. Old world.

Order. OPHIDIA.—Body elongated, true feet none. Mouth dilatable, bases of jaws inseparable. Eyelids and

external ears wanting. Tongue long, flat, deeply forked, and sheathed in lower jaw. Vertebræ fitting into each other with ball and socket joint, the socket being in front.

Sub-order. VIPERINÆ.—Jaws weak, upper toothless, but having large fangs in front ; lower jaw toothed. Shields of belly broad and band-like. Top of head flat, eyes at side, nostrils at sides of muzzle.

Family. CROTALIDÆ.—Face with a large pit on each side, between eye and nostril.

Genus. CRASPEDOCEPHALUS.—Head scaly, a single shield over eyebrow ; shields of hinder part of lips large ; scales lancet-shaped and keeled. Tail ending in a spine. Plates under tail in two rows. America and islands.

Genus. LACHESIS. — Plates under tail in four rows. America.

Genus. CENCHRIS.—Head shielded. Plates under tail in one row. Tail ending in a spine. America.

Genus. CROTALOPHORUS.—Head shielded. Plates under tail in one row. Tail ending in a rattle. Nine large shields behind eyes. America.

Genus. UROPSOPHUS.—Head with scales behind ; shields of temples and lips rather large. Tail ending in rattle. America.

Genus. CROTALUS.—Head with scales behind ; shields of temples and lips very small and rather convex. America.

Family. VIPERIDÆ.—Head without any pit between the eyes and nostrils.

Genus. DABOIA. — Head flat and long, with sharply-keeled scales. Nostrils large. Pupils of eyes round. India.

Genus. CLOTHO.—Pupil of eye erect and oblong. Head short and broad, the keeled scales ending in a spine. Rest like Daboia. Africa.

Genus. CERASTES.—Nostrils small and semilunar. Keel of scales not reaching to tip. Two rows of plates under tail. Eyelid of male horned. Northern Africa.

Genus. ECHIS.—Nostrils very small, between two little scales. Eyelids of male not horned. One row of plates under tail. Africa and India.

Genus. VIPERA.—Head with shields, flat, and high on sides ; nostrils in middle of a shield. Nose rather curved upwards. Europe.

Genus. PELIAS.—Nose blunt ; front of head with small shields. Europe.

Genus. ACANTHOPHIS.—Head flat and projecting over eyes ; nostrils between two shields. Tail ending in a hooked spine. Australia.

Sub-order. COLUBRINÆ.—Both jaws strong and toothed, with fangs in front or grooved teeth behind.

Family. HYDRIDÆ.—Ventral shield narrow. Eyes and nostrils above, nostrils with valve. Fangs moderate, and mixed with teeth of jaw. Pupil small and round.

Section I.—Abdomen keeled, and furnished with long, lancet-shaped shields or scales like those of back. Tail mostly compressed.

Genus. PELAMIS.—Head shielded, flat and long, gape wide ; nasal shield very large. Scales square or six-sided. Tail compressed. Pacific.

Genus. HYDROPHIS.—Head short ; nasal shield four-sided ; rostral shield five-sided and with three teeth below. Tail compressed. Asia.

Genus. ACROCHORDUS.—Head scaled, eyes surrounded by series of little scales. Body spindle-shaped, with three cuspid scales. Tail conical and tapering. Java.

Section II. — Abdomen rounded beneath, with broad shields. Tail conical and tapering.

Genus. ERPETON.—Scales keeled ; nostrils in a ring of small scales. Tail scaled beneath.

Genus. CERBERUS.—Head shielded, scaled above ; nostrils in centre of shield. Scales of body keeled and grooved. Asia.

Family. BOIDÆ.—Hinder limbs developed under skin, and protruding in form of a horny spur on each side of the base of the tail. Tail short.

Section I.—Tail strong and prehensile. Muzzle blunt, head distinct.

Genus. MORELIA.—Crown of head with scale-like plates. Deep pits on lip-shields. Pupil of eye elliptical and erect. Plates under tail two-rowed. Australia.

Genus. PYTHON —Crown of head shielded to behind eyes, rest scaled. Pits on lip-shields. Pupils of eyes linear and erect. Plates under tail two-rowed. India.

Genus. HORTULIA.—Head with irregular scales, shielded in front of forehead and on muzzle. Pits on lip and rost al shields. Plates under tail two-rowed. Africa.

Genus. EPICRATES.—Scales smooth, plates under tail one-rowed. Crown of head scaly. America.

Genus. XIPHOSOMA.—Lip-plates short and all pitted. Rest as in Epicrates. America.

Genus. BOA.—Scales smooth. Lip-shields short, not pitted. Crown of head scaled ; nostrils at sides, between two plates. America.

Genus. EUNECTES.—Crown with irregular shields ; nostrils between three plates. America.

Genus. CHILABOTHRUS.—Crown of head with regular shields ; nostrils between three scales. Jamaica.

Section II.—Tail very short, rarely prehensile. Head short and indistinct.

Genus. TORTRIX.—Seven shields on head. Pupil of eye circular and covered with a plate, transparent in centre. Scales smooth. One row of plates under tail. Spurs very small and sunk in pits. Tropical America.

Sub-order. COLUBRINÆ.—Abdomen with broad, band-like plates. Spurs none. Both jaws toothed. Tail conical and tapering.

Section I.—No grooved or perforated fang in front of the jaw.

Family. CALAMARIDÆ.—Body cylindrical. Head not distinct from neck, gape small, nostrils small. No keels on plates of abdomen.

Genus. CONOCEPHALUS.—Scales above keeled, in seventeen rows. Two rows of plates under tail. Head conical and narrow. Teeth equal and smooth. America.

Family. CORONELLIDÆ.—Belly flat, back broad. Head rather flat, shields of head regular. Front teeth shortest.

Genus. PSAMMOPHYLAX.—Rostral shield not protruding, rounded, and with a long process pointing backwards. South Africa.

Genus. CORONELLA.—Rostral shield moderate ; muzzle short, rounded ; and flat above. Last maxillary tooth smooth or grooved, and rather larger than the others. Most parts of world.

Family. NATRICIDÆ.—Body rather stout. Tail distinct. Head broad and distinct from neck, muzzle rounded. Gape wide. No keels on plates of belly.

. *Genus.* TROPIDONOTUS.—Scales of back keeled. Muzzle rather narrow. Last maxillary tooth longest. Frontal scales narrow. Most parts of world.

Genus. HETERODON.—Body short and thick, head short and flat ; rostral shield with a sharp ridge above. America and Madagascar.

Family. COLUBRINÆ.—Body moderate, tail tapering. Shields of head irregular. Scales oval, overlapping, and nearly equal. Two rows of scales under tail. Teeth not grooved.

Genus. COLUBER.—Rostral shield moderate and rounded. Body with rounded back. Teeth equal. Europe, Asia, and North America.

Genus. CORYPHODON.—Teeth of jaws increasing in size behind and smooth. Asia, Africa, and America.

Family. DRYADIDÆ.—Body very long ; tail long and gradually tapering. Head distinct from neck. Shields of head regular. Scales long and lancet-shaped ; two rows beneath the tail.

Genus. HERPETODRYAS.—Body slightly compressed. Two nasal shields. Teeth equal and smooth. America.

Genus. CYCLOPHIS.—Body rounded ; one nasal shield. Teeth equal and smooth. East Indies and North America.

Genus. PHILODRYAS.—Hinder jaw-teeth grooved. South America and Madagascar.

Genus. DROMICUS.—Hinder jaw-tooth separated from others by a space, longest and not grooved. West Indies and South America.

Family. RACHIODONTIDÆ.—Tail tapering, head short and high, snout short and blunt. One nasal shield, one anterior and two posterior oculars ; those on crown small and regular. Scales keeled, lancet-shaped. Throat-teeth formed by processes of neck vertebræ. Africa.

Genus. DASYPELTIS.—As family.

Family. DENDROPHIDÆ.—Body very long and slender, rather five-sided. Upper jaw longer than lower. Shields of head long and regular. Scales very narrow.

Genus. BUCEPHALUS.—No keels on plates of abdomen. Head stout, eyes very large. Africa.

Genus. AHÆTULLA.—Slight keels on plates of abdomen. Head long, crown flat. Africa and South America.

Family. DRYIOPHIDÆ.—Body very long and slender ; head long, narrow ; snout lengthened. Upper jaw much longer than lower ; rostral shield protruding. Scales narrow.

Genus. DRYIOPHIS.—Snout pointed, rostral solid. Africa, Asia, and South America.

Genus. PASSERITA.—Snout with short flexible appendage. India.

Genus. LANGAHA.—Snout with long and scaly appendage. Madagascar.

Family. DIPSADIDÆ.—Body slender and compressed. Head short and mostly triangular, sometimes squared. Shields of crown blunt.

Genus. EUDIPSAS.—Head flat and triangular. Scales large over vertebræ. Front teeth of jaws and palate longest. South America and East Indian Islands.

Family. ELAPIDÆ.—Head mostly squared, snout short, crown flat. No loreal shield. Scales smooth. Grooved fangs immoveable.

Genus. BUNGARUS. — Scales over vertebræ hexagonal. East Indies.

Genus. HAMADRYAS.—One or two smooth teeth behind fang. Three pairs of large shields round occipitals. East Indian Islands.

Genus. NAJA.—One or two smooth teeth behind fangs. Neck distensible into a hood. East Indies and Africa.

Genus. ELAPS.—No smooth teeth behind fang. Anal scale not cleft. East Indies, Africa, and America.

Family. DENDRASPIDÆ.—Long and slender, abdomen flat. Head squared and long, distinct from neck. Perforated fang in front. Front tooth of lower jaw longest.

Genus. DENDRASPIS.—As family. Africa.

Family. ATRACTASPIDIDÆ.—Body rounded, tail ending with sharp conical spine. Head short and flat, not distinct from neck. Perforated fangs in front of enormous length.

Genus. ATRACTASPIS.—As family. Africa.

Order. BATRACHIA.—Vertebrate animals, with naked skin (very rarely with a few imbedded scales). Neck not distinct, ribs none or very short. Young subject to a meta-morphosis. Breathing mostly by lungs, sometimes by gills.

Sub-order. BATRACHIA SALIENTIA. — Body short and flattened ; no tail.

A.—**Aglossa,** or Tongueless Batrachians.

Family. **DACTYLETHRIDÆ.** — Teeth in jaws ; toes webbed ; sacral vertebræ flattened ; no paratoids (*i.e.* glands on sides of head).

Genus. DACTYLETHRA.—As family. West and South Africa.

Family. **PIPIDÆ.**—No teeth in jaws ; rest as last family.

Genus. PIPA.—Head flat, triangular. Female with pouches on back. South America.

B. **Opisthoglossæ.**—Batrachia with tongue adhering to the front of the lower jaw, and free behind.

Series. With cylindrical or tapering fingers.

Section. With teeth in jaws ; well-developed ear.

Family. **RANIDÆ.**—Toes webbed ; processes of sacral vertebræ cylindrical. No paratoids.

Genus. PSEUDIS.—First finger opposite to three others. South America.

Genus. TOMOPTERNA.—Toes half-webbed. Two bony processes on lower jaw. Flattish spur on metatarsus. South Africa.

Genus. RANA.—Finger not opposite, tongue with deep notch behind. Teeth on vomer, *i.e.* central ridge of palate. Everywhere.

Genus. CERATOPHRYS.—Horn-like process over each eye. South America.

Family. **CYSTIGNATHIDÆ.**—Toes not webbed ; no paratoids.

Genus. CYSTIGNATHUS.—Two series of teeth on vomer. America and Australia.

Family. **DISCOGLOSSIDÆ.**—Toes webbed. Processes of sacral vertebræ wide. No paratoids.

Genus. DISCOGLOSSUS.—Teeth of vomer in straight row. Tympanum or drums of ear hidden. Shores of Mediterranean.

Family. **ALYTIDÆ.**—Toes webbed. Processes of sacral vertebræ wide. Paratoids.

Genus. ALYTES.—Toes slightly webbed. Europe.

Genus. SCAPHIOPUS.—Toes completely webbed. North America.

Section. No teeth in jaws. Ear perfectly developed.

Family. **BUFONIDÆ.**—Toes webbed. Processes of sacral vertebræ wide. Paratoids.

Genus. BUFO.—Crown of head nearly flat, muzzle round or blunt, skin smooth on sides. Paratoids large. Everywhere (except Australia).

Series. Fingers widened.

Section. Teeth in jaws. Ear perfect.

Family. **POLYPEDATIDÆ.**—Toes webbed. Processes of sacral vertebræ cylindrical. No paratoids.

Genus. ACRIS.—Disks of toes small, web broad, fingers not webbed. Tongue heart-shaped. North America.

Family. **HYLIDÆ.**—Processes of sacral vertebræ dilated. Rest as last family.

Genus. HYLA.—Head with soft skin, fingers not opposed, toes all webbed. No pouch on back of female. Everywhere.

Genus. NOTOTREMA. — Female with pouch on back. Central America.

Genus. TRACHYCEPHALUS. — Head bony above. No pouch. West Indies.

Family. **PELODRYADÆ.**—Toes webbed. Processes of sacral vertebræ dilated. Paratoids.

Genus. PELODRYAS.—As family.

Family. **PHYLLOMEDUSIDÆ.**—Toes not webbed. Rest as last family.

Genus. PHYLLOMEDUSA.—As family. Australia.

Section. No teeth in jaws. Ear perfect.

Family. **BRACHYMERIDÆ.**—Toes not webbed. Processes of sacral vertebræ dilated. No paratoids.

Genus. BRACHYMERUS.—As family. South Africa.

Family. **HYLAPLESIDÆ.**—Toes not webbed. Processes of sacral vertebræ cylindrical.

Genus. HYLAPLESIA.—As family. Tropical America.

C. **Proteroglossa,**—Tongue free in front.

Section. No teeth in jaw. Ear imperfect.

Family. **RHINOPHRYNIDÆ.** — Toes webbed. Processes of sacral vertebræ dilated. Paratoids. Muzzle truncated.

Genus. RHINOPHRYNUS.—As family. Mexico.

Sub-order. BATRACHIA GRADIENTIA.—Body long and tailed. Limbs four, small, weak, and equal. Tail seldom retained through life. Teeth in jaws, often in palate. Ears hidden. Ribs distinct or rudimentary. Eggs isolated and oval. Legs of larva developed nearly together.

Family. **SALAMANDRIDÆ.**—Teeth of palate in two longitudinal series. Tongue broad.

Genus. SALAMANDRA.—Back with series of perforated warts. Paratoids large and glandular.

Genus. TRITON.—Skin granular, a series of pores along part of sides.

Genus. LOPHINUS.—Skin smooth. No pores.

Family. **PLETHODONTIDÆ.**—Teeth of palate in short series across end of vomers.

Genus. ONYCHODACTYLUS.—Toes four or five, not webbed, furnished with claws. Hind legs dilated. Japan.

Genus. AMBYSTOMA.—Toes and fingers not clawed. Tongue oblong. Tail keeled above, round below. North America.

Order. **PSEUDOSAURIA.**—Body long, tailed, lizard-like ; legs four, and rudimentary. Gills internal and rudimentary, their apertures permanent.

Family. **PROTONOPSIDÆ.**—Legs strong. Teeth of palate set across in an arched series.

Genus. PROTONOPSIS.—As family. North America.

Family. **AMPHIUMIDÆ.**—Legs weak. Teeth of palate in double diverging series.

Genus. AMPHIUMA.—Two toes on each foot. North America.

Genus. MURÆNOPSIS.—Three toes on each foot. North America.

Order. **PSEUDOPHIDIA.**—Body snake-like. No legs. Skin smooth and wrinkled, with minute imbedded scales.

Genus. CÆCILIA. A pit under each nostril. Africa and South America.

Order. **PSEUDOICTHYAS.**—Body fish-like, covered with strong overlapping oval scales. Fur-like membranes, strengthened with soft jointed rays, above and below. Legs four, rudimentary. Gill external. Teeth of palate and lower jaw very large.

Genus. LEPIDOSIREN.—Fifty-five pairs of ribs. Brazil.

Genus. PROTOPTERUS.—Thirty-six pairs of ribs. Gambia.

Order. **MEANTIA.**—Body long, smooth, and naked. Legs two or four, weak. Gills external and permanent. Teeth on palate and lower jaw.

Family. **PROTEIDÆ.** — Legs four; tail compressed. Single series of teeth on front edge of vomer.

Genus. PROTEUS.—Head long. Three toes on fore feet, four on hinder. Carniola.

Genus. NECTURUS.—Head broad and compressed. Four toes on all the feet. Mississippi.

Family. **SIRENIDÆ.**—Limbs two in front. Body nearly cylindrical. Teeth in each jaw and on vomer in numerous cross series.

Genus. SIREN.—Four fingers. Gills with cilia. North America.

CLASS.—FISHES.

Vertebrated animals respiring through gills, living in water—Heart with one auricle and one ventricle; blood cold and red.

Order. **CHONDROPTERYGII.** — Skeleton cartilaginous. No sutures in skull.

Sub-order. ELEUTHEROPOMI.—Gills free.

Family. **ACIPENSERIDÆ.**—Head and skin covered at intervals with cartilaginous or bony shields.

Genus. ACIPENSER.—Nose conical and tapering; mouth under muzzle, with beards. Eyes and nostrils at side. Body with series of bony plates. Single dorsal fin, behind ventral and over anal. Short under lobe to tail fin. Many parts of world.

Genus. SCAPHIORHYNCHUS.—Nose flat, convex above, concave below. Mouth beneath, with four beards. Five series of scales on body. North America.

Family. **POLYODONTIDÆ.**—No tubercles or plates on body.

Genus. POLYODON.—As family. North America.

Family. **CHIMÆRIDÆ.**

Genus. CHIMÆRA.—Beak conical. Dorsal fins nearly contiguous. Northern seas.

Genus. CALLIORHYNCHUS.—Beak ending in leaf-like appendage. Dorsals far apart. Southern seas.

Sub-order. TREMATOPNEA.—Gills fixed by their edges to bars across the openings.

Sub-section. SQUALI.—No frontal cartilage on skull. Gill-openings at sides.

Family. **SCYLLIDÆ.**—Body long, nearly cylindrical. Pectoral fins moderate. Spout-holes. Gill-openings five, the last being over the root of the pectorals. Tail not forked; under lobe slight.

Genus. SCYLLIUM.—First dorsal between ventral and anal; second between anal and tail. Spiracles behind eyes.

Genus. PRISTIDURUS.—Snout long. Row of flattish prickles on upper rim of tail. Many seas.

Family. **SQUALIDÆ.**—Body tapering to tail. Dorsals one or two, first between pectoral and ventral.

Genus. SQUALUS.—Snout flat and rounded; nostrils in the middle, between the mouth and the point of the snout.

Genus. SPHYRNIAS. — Head hammer or heart-shaped, with the eyes at each side. Many seas.

Genus. GALEUS.—Head flat, spiracles small. Upper lobe of tail with one or two notches. Many seas.

Genus. MUSTELUS.—Spiracles large. Teeth mosaic-like. First dorsal nearly in centre, tail short. Many seas.

Genus. ISURUS.—Gill-openings large and before pectorals. Tail keeled on each side and crescent-shaped. Spiracles very small, far behind eyes. Teeth flat and three-sided. Many seas.

Genus. CARCHARODON.—Teeth three-sided, with notched edges. Many warm seas.

Genus. CETORHINUS.—Snout short and blunt. Gill-openings very large, spiracles small. Teeth minute, conical, points directed backward. Skin covered with little pointed scales. Northern seas.

Genus. ALOPIAS. — Snout short and conical. Gill-openings small, the last set over the pectorals. Teeth flat, three-sided, not notched. Upper lobe of tail very long, with a little pit at the root. Many seas.

Genus. ACANTHIAS.—A sharp spine before each dorsal. Head flat. Spiracles large. First dorsal between pectoral and ventral, and second between ventral and anal. Many seas.

Genus. DALATIAS.—No spines before dorsals. Head flat or compressed. Spiracles behind eyes. Gill-openings small. No pit at root of tail. Many seas.

Family. **SQUATINIDÆ.**—Body flat; eyes on upper part of head. Spout-holes very large, and behind eyes. Pectorals greatly expanded in front.

Genus. SQUATINUS.—As family. Many seas.

Sub-section. RAII.—Body flat; spiracles. Gill-apertures five, set before ventrals and under pectorals.

Family. **PRISTINIDÆ.**—Snout prolonged into a saw-like form, with teeth on each edge. Skin covered with mosaic scales.

Genus. PRISTIOPHORUS.—Beak with hooked spines of various lengths, prickles attached to skin. Long barbel under snout, and a keel on either side. Australasian seas.

Genus. PRISTIS.—Teeth of saw straight, similar, and somewhat lancet-shape. Teeth flat and mosaic. Tail-fin short. Many warm seas.

Family. **RAIDÆ.** — Body very flat and broad, with slender tail.

Genus. TORPEDO.—Disc round and blunt in front; back rounded and naked; tail-fin three-cornered. Teeth pointed and attached to a cartilage. A row of tooth-like projections on edge of spiracles. Warm and temperate seas.

Genus. RAIA.—Snout pointed and keeled, disc broad, tail very slender. Teeth mosaic-like; in males pointed during breeding season.

Genus. TRYGON.—Tail armed with one or more barbed spines, and as long or longer than the body. Teeth either pointed or wrinkled in the middle. Most seas.

Genus. MYLIOBATIS.—Teeth large, six-sided, mosaic. Tail very long, slender, and with a powerful spine. Nostrils close together in middle of snout. Warm seas.

Genus. CEPHALOPTERA.—Mouth at end of snout, and very broad; teeth small, flat, and leaf-like in lower jaw; none in the upper. Head with horn-like projections of the fins. Tail very long, and with a spine. Many seas.

Order. **ACANTHOPTERYGII.**—Part of the rays of the dorsal, ventral, and anal fins not jointed, and forming spines.

Family. **GASTEROSTEIDÆ.**—Body long and rather compressed ; teeth in both jaws, not on tongue or palate. Three branchiostegals (*i.e.* the slender curved bones by means of which the gill-aperture is closed). Bones of operculum (or gill-cover) not armed. Parts of skeleton forming external armour. Scales none, or in plates along side. Single spine before dorsal fin, and ventral armed with a strong spine, and sometimes a short ray.

Genus. GASTEROSTEUS.—As family. Rivers and brackish waters of cool latitudes.

Family. **BERYCIDÆ.** — Body rather deep and compressed ; eyes large and lateral. Slender teeth in both jaws, often on palate. Branchiostegals, four or eight. Bones of operculum armed. Scales, rarely bony, ctenoid (*i.e.* with comb-like edge), or wanting. Ventrals under breast.

Genus. MONOCENTRIS.—Scales forming a bony mail, a few isolated spines before dorsal. Seas of Japan.

Genus. HOPLOSTETHUS.—Abdomen notched. No teeth on vomer. One dorsal. Mediterranean.

Genus. MYRIPRISTIS.—Two dorsals. No præopercular spine (see page 197). Indian seas.

Genus. HOLOCENTRUM.—Two dorsals. Præopercular spine Upper jaw not projecting. Asiatic seas.

Family. **PERCIDÆ.**—Scales ctenoid, lateral line mostly continuous. Operculum armed. Branchiostegals, seven or eight. Teeth large, pointed, on jaws, palate, and vomer. Dorsal with spiny and soft portion ; ventrals under breast, with one spine and five soft rays.

Genus. PERCA.—Dorsal with thirteen to fifteen spines. Præoperculum and præorbital (*i.e.* the bone before the orbit of the eye) notched ; head naked above ; scales small. Many fresh waters.

Genus. LABRAX.—Dorsal, nine spines. Teeth on tongue ; præoperculum toothed beneath. Many seas.

Genus. ACERINA.—No bony ridge on operculum, no teeth on palatine bones. Opercula armed ; scales small. Many fresh waters.

Genus. LUCIOPERCA.—Two dorsals, two spines on anal. Dorsal spines twelve to fourteen. Operculum armed, præoperculum notched. European rivers and lakes.

Genus. CENTROPOMUS.—Body flattened, operculum not spined. North America.

Genus. CENTROPRISTIS.—One dorsal. Canine teeth very small ; teeth on vomer in angular or triangular patch. Operculum spined, præoperculum serrated. North America.

Genus. SERRANUS.—Scales small, teeth on palatine bones. Dorsal nine to eleven spines, anal with three. Operculum spined. Many warm seas.

Genus. POLYPRION.—One dorsal ; anal two or three spines. Operculum with strong bony ridge, and the bones toothed. Many seas.

Genus. OREOSOMA.—Anal spines wanting. Scales none, bony protuberances on body. Atlantic.

Genus. APOGON.—Lateral line commencing behind head. Teeth on palatines. Præoperculum with double serrated ridge. First dorsal with six—seven spines, anal with two ; operculum spined. Indian and Australian seas.

Family. **PRISTIPOMATIDÆ.**—Body compressed, scaled, teeth of scales sometimes wanting. Branchiostegals five, six, or seven. Barbels none. Dorsal one ; no molar or cutting teeth.

Genus. PRISTIPOMA.—Anal with three spines. Branchiostegals seven. Groove behind joint of lower jaw. Vertical fins not scaly, or only at base. Indian and Australian seas.

Genus. HÆMULON.—Soft parts of vertical fins scaly. Seas of Tropical America.

Genus. DIAGRAMMA.—Branchiostegals six -seven. Pores in lower jaw, no groove ; upper profile of head rounded. Indian seas.

Family. **MULLIDÆ.**—Body long, slightly compressed ; scales large. Two long barbels from under jaw. Branchiostegals four. Dorsal fins two, separate ; ventrals with one spine and five rays.

Genus. MULLUS.—No teeth in upper jaw. Many seas.

Genus. UPENEUS.—No teeth in palate, those of jaw in a single series.

Family. **SPARIDÆ.**—Body as last family ; tail not armed ; lateral line continuous. Branchiostegals five, six, or seven. Teeth, either cutting in front of jaws or lateral series of molars. One dorsal, anal three-spined ; ventrals under thorax, with one spine and five rays.

Genus. PAGRUS.—Teeth conical in front, double series of upper molars. Canines. Most seas.

Genus. PAGELLUS.—No canines ; rest as Pagrus. Most seas.

Genus. CHRYSOPHRYS.—Three or more series of upper molars. Canines. Mediterranean.

Family. **SQUAMIPENNES.**—Body much compressed and deep, scaled. Lateral line continuous. Branchiostegals six—seven. Teeth in broad or narrow bands, sometimes teeth on palate. Dorsal three or four, spined ; vertical fins covered with small scales. Lower rays of pectorals branched ; ventrals under thorax, with one spine and five soft rays. All tropical species.

Sub-family. Chætodontinæ.—No teeth on palate or vomer.

Genus. CHÆTODON.—No præopercular spine. Dorsal not notched. Muzzle short.

Genus. CHELMO.—Muzzle very long.

Genus. HENIOCHUS.—Fourth dorsal spine very long.

Genus. HOLOCANTHUS. — Præoperculum with strong spine, dorsal with twelve to fifteen spines.

Sub-family. Toxotina.—Teeth on palate. Dorsal fin on latter half of back.

Genus. TOXOTES.—Dorsal with five spines.

Family. **CIRRHITIDÆ.**—Body with cycloid (*i.e.* circular) scales ; lateral line continuous. Branchiostegals three, five, or six. Anal fin three-spined. Lower rays of pectorals simple. Ventrals under thorax, one spine and five rays. Tropical and warm seas.

Genus. CIRRHITES.—No teeth on palate ; dorsal spines ten ; intermaxillary bone not notched.

Family. **TRIGLIDÆ.** — Cleft of mouth extending to sides of muzzle, and partly vertical ; eyes directed upward. Præoperculum jointed to the sub-orbital ring. Two dorsals. Ventrals under thorax. Branchiostegals five to seven. Most seas, especially the Northern.

Group I. Body scaled or naked.

Section I. Spinous dorsal developed as much as or more than the soft and anal.

Sub-family. Scorpænina.—Anal fin short.

Genus. SCORPÆNA.—Body scaled. Occiput with a groove. First dorsal with eleven spines, second with one spine and nine or ten soft rays. Anal with three spines and five soft rays.

Genus. PTERÖIS.—Body scaled. Many spines and rays isolated, rays branched. No pectoral appendages. Indian seas.

Genus. APISTOS.—One pectoral appendage. Red Sea.

Genus. HEMITRIPTERUS.—Scales rudimentary. Spinous dorsal, with deep notch in front.

Genus. SYNANCEIA.—Scales none. Head grotesque. No teeth on vomer. One dorsal. Indian seas.

Genus. PELOR.—Scales none. Two pectoral appendages.

Sub-family. Cottinæ.—Spinous part of dorsal less developed than soft part and anal. Body scaled or naked, or plated.

Genus. BLEPSIAS.—Head and body compressed; head with tentacles, skin with prickles. Ventrals hardly visible.

Genus. COTTUS.—Head broad and flat. Scales none. Dorsals moderate. No palatine teeth.

Genus. PLATYCEPHALUS.—Head very broad. Ventrals under thorax, and behind root of pectorals. First dorsal spine short and separate.

Genus. TRIGLA.—Pectoral appendages three. No palatine teeth. Scales very small.

Sub-family. Cataphracti.—Body covered with bony plates.

Genus. AGONUS.—Two dorsals, pectorals moderate; no pectoral appendages.

Genus. PERISTETHUS.—Two pectoral appendages.

Genus. DACTYLOPTERUS.—Pectorals very large and used for flight.

Family. TRACHINIDÆ.—Body long, naked, or scaled. Teeth in bands, no molars or cutting teeth. Spinous portion of dorsal less developed than the soft portion; ventrals with one spine and five rays. Gill-opening under, branchiostegals five to seven.

Genus. URANOSCOPUS.—Eyes on upper part of head; lateral line continuous. Dorsals two. Mediterranean.

Genus. TRACHINUS.—Eyes lateral. Head not cuirassed. Lateral line continuous. Cleft of mouth oblique. Two dorsals, ventrals under throat. Teeth on palate.

Genus. SILLAGO.—No palatine teeth. Ventrals under thorax. Rest as Trachinus.

Genus. PERCOPHIS.—Cleft of mouth horizontal, lower jaw projecting. Dorsals two, ventrals under throat. Palatine teeth, canines strong. Brazil.

Family. SCIÆNIDÆ.—Body compressed and long. Scales ctenoid. Lateral line continuous. Branchiostegals seven. Teeth in bands, none on palate; no cutting teeth or molars. Two dorsals, the soft one more developed than the spinous or the anal; anal two-spined. Pectoral rays branched.

Genus. POGONIAS.—Chin with many small barbels. North America.

Genus. EQUES.—No barbels. Dorsal and tail fins scaly, first dorsal much elevated. Atlantic.

Genus. SCIÆNA.—No barbels. Dorsal not scaly. Upper jaw equal to or larger than lower. Anal spines weak. Many seas.

Genus. CORVINA.—No barbels. Anal spines strong. Rest as Sciæna. Mediterranean.

Genus. OTOLITHUS.—No barbels. Lower jaw longest. Canines conical. North America.

Family. POLYNEMIDÆ.—Lateral line continuous, extends along tail. Mouth under snout. Branchiostegals seven. Dorsals two; second dorsal, anal, and tail-fin covered with minute scales. Several thread-like appendages below pectorals. Ventrals under breast, one spine, five rays.

Genus. POLYNEMUS.—Teeth on vomer; anal fin equals soft dorsal. Tropical seas.

Family. SPHYRÆNIDÆ.—Scales small and cycloid. Mouth wide, teeth strong. Branchiostegals seven. Two dorsals; ventrals on abdomen.

Genus. SPHYRÆNA.—As family. Temperate and tropical seas.

Family. TRICHIURIDÆ.—Body long and compressed. Both portions of dorsal and anal very long and with many rays. Tail often with finlets.

Genus. LEPIDOPUS.—Ventrals wanting or very small. One dorsal; tail-fin distinct. Mediterranean, &c.

Genus. TRICHIURUS. — Tail tapering, without a fin. Atlantic.

Genus. THYRSITES.—Two to six finlets. Teeth in palate. South Africa. Australia.

Family. SCOMBERIDÆ.—Body naked or scaled. Spinous part of dorsal less developed than soft or anal, sometimes wanting. Soft dorsal and anal often divided into finlets.

Sub-family. Scomberina. — Spinous portion of dorsal separate. Body oblong, naked, or with very small scales.

Genus. SCOMBER.—Finlets five or six. Scales equal over body. Many seas.

Genus. THYNNUS.—Finlets six to nine. Scales of breast forming a corslet. Tail keeled on each side. Teeth on vomer and palate. Mediterranean, Atlantic, &c.

Genus. PELAMYS.—Finlets seven to nine. Teeth on palate, not on vomer. Mediterranean, Atlantic.

Genus. AUXIS.—Finlets seven to nine. No teeth on palate. Mediterranean, &c.

Genus. NAUCRATES.—Finlets none. First dorsal only a few free spines. Tail keeled. Many seas.

Genus. ECHENEIS. — Spinous dorsal transformed into sucking disc on head. Many warm seas.

Sub-family. Cyttina.—Dorsal divided, spinous part less developed than the soft. Body very deep, scaled or shielded.

Genus. ZEUS.—Bony plates along bases of dorsal and anal fins. Many seas.

Sub-family. Coryphænidæ.—One long dorsal. No teeth in gullet.

Genus. CORYPHÆNA.—Ventrals under thorax. Dorsal long, beginning on head. Scales very small. Warm seas.

Genus. PTERACLIS.—Ventrals under throat, with four rays. Dorsal beginning on forehead. Warmer seas.

Genus. LAMPRIS.—Ventral fins with many rays. Temperate seas.

Family. CARANGIDÆ.—Separated from the preceding family on account of the number of vertebræ, which are ten in the abdomen and fourteen in the tail, those of the Scomberidæ exceeding that number.

Sub-family. Carangina.—The soft dorsal fin and anal of nearly equal extent.

Genus. TRACHURUS.—Lateral line covered with shields. Many seas.

Genus. CARANX. — Lateral line partly smooth. Præoperculum toothed. Many seas.

Family. XIPHIIDÆ.—Body rather long. Teeth none, or minute. Upper jaw long and sword-shaped. Ventrals absent or rudimentary. Branchiostegals seven.

Genus. XIPHIAS.—Ventral fins wanting. Warm seas.

Genus. HISTIOPHORUS. — Ventral fins present. Warm seas.

Family. GOBIIDÆ.—Body long. Teeth mostly small. Spinous portion of dorsal flexible and less developed than soft part. Anal as soft dorsal. Ventrals with one spine and five rays, sometimes united into a disc.

Sub-family. Gobiina.—Ventrals united into a disc. Two dorsals.

Genus. GOBIUS.—Ventrals not adherent to abdomen. Body scaled. Teeth conical and fixed. Many seas.

Sub-family. Callionymina. — Ventrals separate. Two dorsals.

Genus. CALLIONYMUS.—One small gill opening on either side. Præoperculum armed. No single ventral ray.

Family. DISCOBOLI. — Body thick, surface naked or with tubercles. Teeth small. Ventrals with one spine and five small rays surrounded by fleshy disc.

Sub-family. **Cyclopterina.**—Two dorsals. Twelve abdominal, sixteen tail vertebræ.

Genus. CYCLOPTERUS.— As sub-family. Temperate seas.

Sub-family. **Liparidina.**—One dorsal. Twelve abdominal, thirty tail vertebræ.

Genus. LIPARIS.—As sub-family. Temperate seas.

Family. **BATRACHIDÆ.**—Body short and wide. Teeth conical and small. Spinous dorsal very short, soft, and anal long. Ventrals under throat.

Genus. BATRACHUS.—Head broad and flat. Spinous dorsal represented by three strong spines. Many seas.

Family. **PEDICULATI.**—Head and front of body very large. No scales. Carpal bones prolonged into a pectoral anus.

Genus. LOPHIUS.—Head flat. Six dorsal spines, three on the head and isolated. Temperate seas.

Genus. ANTENNARIUS.—Three isolated dorsal spines. Head compressed. Warm seas.

Family. **BLENNIIDÆ.**—Body long, low, cylindrical, naked or small scaled. Dorsals occupying nearly whole of back; anal long; ventrals under throat, composed of a few rays, and sometimes absent.

Genus. ANARRHICHAS.—Molar teeth; tail-fin separate. Temperate seas.

Genus. BLENNIUS.—No molars. Spinous and soft parts of dorsal fin nearly equal. Body naked. Single series of immoveable teeth in jaws. Many seas.

Genus. SALARIAS.—Series of small moveable teeth in gums. Tropical seas.

Genus. CENTRONOTUS.—Ventrals none or very small; scales small; no lateral line. No anal spine. Gill membranes united below throat. Temperate seas.

Genus. ZOARCES.—No tail-fin. Temperate seas.

Family. **TRACHYPTERIDÆ.**—Body long, much compressed, and naked. One dorsal occupying whole of back, and supported by flexible but non-jointed rays. No anal fin, and tail-fin not in line with axis of fish.

Genus. TRACHYPTERUS.—Ventrals large. Warm and temperate seas.

Genus. REGALECUS.—Ventrals reduced to long filaments.

Family. **ACANTHURIDÆ.**—Body compressed. Lateral line continuous. Mostly spines on sides of tail. Single series of flattish teeth in each jaw. One dorsal; anal with either two or three spines. All tropical fishes.

Genus. ACANTHURUS.—Scales minute; moveable spine each side of tail.

Genus. NASEUS.—Ventral with three soft rays. Immoveable plates on sides of tail.

Family. **LABYRINTHICI.**—Body rather compressed; scales moderate. Head also scaled. Lateral line absent or broken. Bony laminated organ in a cavity above the gills. Fresh waters of East Indies and South Africa.

Genus. ANABAS.—Teeth on vomer, none on palate.

Family. **ATHERINIDÆ.**—Body long, scaled; lateral line indistinct. Mouth moderate; teeth weak. Branchiostegals five—six. Two dorsals. Ventrals on abdomen; one spine, five rays. Vertebræ very numerous.

Genus. ATHERINA.—Scales cycloid. First dorsal separate from second. Snout blunt; cleft of mouth straight and oblique. Warm and temperate seas.

Family. **MUGILIDÆ.**—Body oblong, compressed; no lateral line; scales cycloid. Cleft of mouth narrow; teeth none, or feeble. Branchiostegals five—six. Dorsals two, short, first with four stiff spines. Ventrals on abdomen, one spine and five rays. Twenty-four vertebræ.

Genus. MUGILUS.—No true teeth in jaws Warm and temperate seas.

Family. **OPHIOCEPHALIDÆ.**—Body long; head flattened, and with shield-like scales above. Teeth in jaws and on palate. A cavity for water, opening on the gills. One long dorsal and anal, not spined. Many vertebræ.

Genus. OPHIOCEPHALUS. — With ventral fins. Fresh waters of East Indies.

Family. **CEPOLIDÆ.**—Body very long and compressed; scales small and cycloid. Dorsal and anal very long, with soft rays. Ventrals under thorax, one spine and five rays. Branchiostegals six.

Genus. CEPOLA.—As family. Many temperate seas.

Family. **CENTRISCIDÆ.**—Body deep, compressed, and bones of mouth produced into a long snout. Body shielded or cuirassed. Two dorsals, spinous, short, and with a single strong spine; ventrals small and on abdomen. Few vertebræ.

Genus. CENTRISCUS. — Body with scales or prickles. Many seas.

Family. **FISTULARIDÆ.**—Body much elongated; bones of head produced into a long tube, ending in a small mouth. Body shielded. Ventrals on abdomen, with six rays; no spine. Branchiostegals five. Many vertebræ.

Genus. FISTULARIA.—Body not scaled. Tropical seas.

Family. **MASTACEMBELIDÆ.**—Body long and eel-like, very small scales. Jaws very long, slightly moveable. No ventrals. Gill-opening only a small slit under side of head. Branchiostegals six.

Genus. MASTACAMBELUS. — Appendage of snout not striated below. Fresh waters of East Indies.

Family. **LABRIDÆ.** (See page 350).—Mouth with thick fleshy lips. One dorsal, spined in front, rayed behind. Ventrals under pectorals. Scales cycloid. Parchment-like teeth on pharyngeal bones.

Genus. HALICHÆRES.—Dorsal low and long, reaching from back of head nearly to tail. Anal rather long. Tail-fin rounded. Head rather long and snout pointed. Tropical seas.

Genus. EPIBULUS.—Jaw-teeth in single row, those of pharynx conical. Tail-fin scooped. Scales large, lateral line broken. Warm and tropical seas.

Genus. LABRUS.—Jaw-teeth conical in single row, or with lesser teeth behind front row; those of pharynx cylindrical, globular, or conical. Upper lip duplicate. Cheeks and operculum scaly. Lateral line not broken. Many seas.

Genus. SCARUS.—Head rather convex; jaws covered with overlapping teeth. Lips not duplicate. Lateral line broken and irregular. Tropical seas.

Order. **MALACOPTERYGII.** — Rays of the fins jointed, and mostly divided at the point.

Sub-order. SUB-BRACHII.—Ventral fins placed under or in front of pectorals.

Family. **PLEURONECTIDÆ.**—Body very much compressed and scaled. Both eyes on same side of head. Pectorals sometimes wanting.

Genus. SOLEA.—Jaws hidden under skin; teeth in several rows, on lower side only, cylindrical and minute. Many seas.

Genus. MONOCHIRUS.—Pectoral wanting on lower side. Temperate seas.

Genus. ACHIRUS.—Pectorals wanting. Temperate seas.

Genus. PLEURONECTES.— Jaws apparent, not scaled. Tail-fin distinct. Temperate seas.

Genus. PLATESSA.—Teeth in single row. Dorsal fin commencing over upper eye. Temperate seas.

Genus. HIPPOGLOSSUS.—Teeth sharp, long, and curved. Body rather narrow. Temperate seas.

Family. **GADIDÆ.**—Body long and scaled. Head symmetrical.

Genus. GADUS.—Head smooth. Teeth conical, in intermaxillaries, vomer, and lower jaw. Three dorsals. Branchiostegals seven. Ventrals under throat and pointed. A barbule on the chin. Colder seas. (Including Morrhua.)

Genus. MERLANGUS.—As Gadus, but no barbule. Many seas.

Genus. MERLUCIUS.—Two dorsals, one long anal. No barbule.

Sub-order. APODES. — No ventrals. Body long and snake-like; skin thick and soft; often with minute scales hidden in skin.

Family. OPHIDINI. — Swimming-bladder either closed or wanting.

Genus. AMMODYTES.—Body long, with minute scales. Lower jaw longer than upper. Teeth none. Branchiostegals seven. Dorsal and anal very long, tail-fin forked. Many seas.

Family. ANGUILLIDÆ.—Teeth comb-like. Gill-openings on sides. Pectoral fins conspicuous, dorsal and anal running round tail.

Genus. ANGUILLA.—Head and tail blunted. Seas and rivers.

Family. CONGERIDÆ.—Dorsal commencing close to head; skin without scales.

Genus. CONGER.—Front nostrils in short tubes at end of snout. Teeth of palate and vomer in straight, chisel-shaped rows. Many seas.

Family. MURÆNIDÆ.—No pectorals; gill-openings only narrow slits.

Genus. MURÆNA.—Single row of teeth on nasal bone. Many seas.

Family. GYMNOTIDÆ.—No dorsal; anal very long.

Genus. GYMNOTUS.—Scales none.

Family. LEPTOCEPHALIDÆ. — Small, scaleless, very much compressed, and transparent. Lateral line formed by intersection of muscles. Many seas.

Genus. HYOPROBUS.—Jaws long, straight; body deep behind head. Tail pointed.

Genus. TILURUS. — Membranous dorsal beginning at head. Anal none. Pointed teeth in both jaws. Tail very long.

Genus. LEPTOCEPHALUS.—Muscular bands bent regularly forward. Tail moderate.

Sub-order. ABDOMINALES.—Ventrals on abdomen and behind pectorals, sometimes absent. Swimming-bladder in most, and with an air-duct.

Family. HETEROPYGIDÆ.—Ventrals small. Dorsal set far back. Gill-apertures small. Body scaled. Teeth small and sharp in palate, none on vomer or tongue. Branchiostegals seven. Eyes none external.

Genus. AMBLYOPSIS.—As family. North America.

Family. CLUPEIDÆ. — Fatty dorsal wanting. Body with distinct and mostly large scales. Two inferior pharyngeal bones.

Genus. ENGRAULIS.—Head pointed, upper jaw longest, gape very deep. Ventrals before dorsal. Branchiostegals twelve. Many seas.

Genus. CLUPEA.—Body compressed, scales loose and large. Teeth minute or wanting. One dorsal. Abdomen keeled. Many seas.

Family. ESOCIDÆ.—Dorsal fin set far back, near tail-fin. Edge of upper jaw partly formed by intermaxillary bone.

Genus. EXOCÆTUS.—Pectorals large and very long, used for flight. Lower lobe of tail-fin elongated. No teeth on palate. Warm seas.

Genus. HEMIRAMPHUS.—Lower jaw long and projecting. Teeth small and placed in a belt round each jaw, none on palate. Body much elongated. Pacific.

Genus. BELONE. — Both jaws much lengthened and slender, lower longer than upper. Teeth small in both jaws, of unequal sizes. Many seas.

Genus. ESOX. — Body rather long; snout flattened, Mouth, jaws, and palate, crowded with sharp teeth. Fresh waters.

Family. SALMONIDÆ.—Second dorsal of a fatty consistence, without rays.

Genus. SALMO.—Head smooth. Teeth on vomer; palatine and maxillary bones. Branchiostegals often unequal on the two sides. Seas and fresh waters.

Genus. SCOPELUS.—Teeth sharp and conical in both jaws. Branchiostegals nine. Palatine teeth very small. Many seas.

Genus. SERRASALMUS.—Teeth in both jaws, triangular, flat, sharp-edged, mostly in a single line. Body rather compressed. Abdominal line serrated. Rivers of South America.

Family. CYPRINIDONTIDÆ.—Jaw-teeth very small. Throat crowded with teeth. Branchiostegals five—six. Whole of head scaled. No fatty fin.

Genus. ANABLEPS.—Teeth on palate. Cornea of eye divided transversely. Head flattened between eyes. Nostrils in short tube. Surinam.

Family. CYPRINIDÆ.—Body with rounded scales. One rayed dorsal. Mouth small and toothless. Inferior pharyngeal bones with large teeth. Branchiostegals three.

Genus. CYPRINUS.—Head naked. Branchiostegals flat and broad. Anal fin short. Fresh waters.

Genus. TINCA.—Dorsal and anal fins short. Barbules very small. Scales small. Ponds and rivers.

Genus. ABRAMIS.—No barbules. Body deep and compressed. Base of anal long. Lakes and rivers.

Genus. LEUCISCUS.—Dorsal and anal short. Barbules none. Fresh waters.

Family. SILURIDÆ.—Body naked or shielded. In most the first ray of dorsal and pectoral a spine with moveable joint.

Genus. SILURUS.—Body naked. Teeth very numerous. Dorsal small, single near head. Anal long. Barbules four to six, some very long. Seas and rivers.

Order. PECTOGNATHI.—Upper jaw-bones immovably joined to sides of intermaxillaries. Body with shields or spines and scales.

Family. SCLERODERMI.—Snout conical or pyramidal. Mouth toothed, in front of snout. Body with hard scales or cuirassed.

Genus. OSTRACION.—Teeth ten to twelve in each jaw. Body tri- or quadrangular, cuirassed, with holes for tail, fins, and mouth. Tropical seas.

Genus. BALISTES.—Body compressed. Eight teeth in each jaw. First dorsal formed of one or several spines. Second dorsal near tail, soft. Tropical seas.

Family. GYMNODONTES.—Jaws projecting from mouth, covered with solid tooth-ivory.

Genus. DIODON.—Jaws not divided. Warm seas.

Genus. TRIODON.—Upper jaw divided. Warm seas.

Genus. TETRODON.—Both jaws divided. Warm seas.

Genus. ORTHRAGORISCUS.—Body very short, deep, and compressed. Jaws not divided. Dorsal and anals connected with tail-fin. Many seas.

Order. LOPHOBRANCHIATA.—Gills composed of small round tufts, the operculum very large and the aperture very small. Body covered with bony rings.

Family. **PEGASIDÆ.**—Breast very wide; pectorals long, with nine or ten spiny rays. Mouth below snout (like that of the sturgeon). Breast-ring very broad, with two spines. Tail-fin with ten strong rays.

Genus. PEGASUS.—As family. Warm seas.

Family. **SYNGNATHIDÆ.**—Body long. No first dorsal or ventral. Anal very small or wanting. Males with egg-pouch. Gill-aperture small and circular.

Genus. HIPPOCAMPUS.—Tail prehensile at tip, without fin. Hind part of head elevated. Most seas.

Genus. SYNGNATHUS.—Snout long, cylindrical, without spines. Body seven-sided. Most seas.

Order. **GANOLEPIDOTI.**—Body covered with strong, bony, enamelled scales.

Genus. LEPIDOSTEUS.—Jaws long and narrow. Scales in oblique rows, tending backward. Dorsal small and near tail. Vertebræ extending to upper lobe of tail. North America.

Order. **CYCLOSTOMI.**—Body cylindrical. Branchiæ with six or seven spiracles on each side.

Family. **PETROMYZONIDÆ.** — Lip circular or horse shoe shape. Two dorsals.

Genus. LAMPETRA (including Petromyzon).—Mouth circular, with hard tooth-like processes. Branchial apertures seven. No pectorals or ventrals. Seas and rivers.

Genus. GEOTRIA.—Skin of throat dilated into pouch. Rivers, Australia.

Genus. AMMOCÆTES.—Lip of horseshoe shape. Mouth not toothed. Rivers.

Genus. MYXINE.—Dorsal very long and broad, running round tail. Lip surrounded with barbules. Mouth with a single hook-like tooth, tongue with two rows of teeth on each side. Branchial apertures two. Many seas.

Order. **LEPTOCARDII.**—Pulsating vessel instead of heart; blood pale. Dorsal cord instead of vertebræ.

Genus. AMPHIOXUS.—Body sharp at both ends. Dorsal running round body. Mouth below, a mere longitudinal slit. Many seas.

DIVISION.—INVERTEBRATA.

Animals without vertebræ, true heart, or brain.

CLASS.—MOLLUSCS.

Invertebrate animals with soft bodies, enclosed in a muscular skin—Mostly with an univalve or bivalve shell.

Sub-class. **CEPHALOPODA.** — Two sides symmetrical. Jaws powerful and vertical. Eyes large, on sides of head; cephalic ganglion (*i.e.* the knot of nervous matter which takes the place of the brain) protected by cartilage. Arms set round head.

Order. **DIBRANCHIATA.**—Eyes sessile and prominent; jaws horny; arms with suckers. Two gills. Ink-gland present.

Section. OCTOPODA.—Arms eight; eyes fixed.

Family. **ARGONAUTIDÆ.** — Dorsal arms of female webbed at tips, and secreting a delicate shell. Mantle supported in front by a single ridge on the funnel.

Genus. ARGONAUTA.—As family. Warm seas.

Family. **OCTOPODIDÆ.**—Arms not webbed, united at base by mantle. Two short styles in mantle instead of shell.

Genus. CIRRHOTEUTHIS.—Two fins on body. Arms united nearly to tip; suckers in single row. Northern seas.

Section. DECAPODA.—Arms ten. Tentacles two. Suckers on footstalks. Shell loose in mantle.

Family. **TEUTHIDÆ.**—Body long; fins short and broad. Shell horny.

Genus. SEPIOLA. — Body short; fins on back and rounded. Suckers in two rows or crowded on arms, four rows on tentacles. Many seas.

Family. **SEPIADÆ.**—Shell chalky. Tentacles long and wide at tips.

Genus. SEPIA.—Body oblong; fins on side, long as body. Four rows of suckers on arms. Many seas.

Order. **TETRABRANCHIATA.**—Animal creeping, shell external and chambered. Eyes on footstalks, jaws chalky, arms numerous. Branchiæ four.

Family. **NAUTILIDÆ.**—Body-chamber large. Siphuncle internal.

Genus. NAUTILUS.—Few whorls of shell. Warm and tropical seas.

Order. **GASTEROPODA.** — Sides not symmetrical Broad muscular foot under body.

Sub-order. PROSOBRANCHIATA. — Abdomen large and protected by a large shell. Branchiæ before the pulsating vessel.

Section. **SIPHONOSTOMATA.** — Shell spiral, aperture notched or produced. Operculum horny. Gills placed obliquely over back.

Family. **STROMBIDÆ.**—Shell with wide and deeply-notched lip. Operculum claw shaped and notched. Eyes large, on footstalks; foot small and feeble.

Genus. STROMBUS.—Spire short, aperture long, outer lip expanded. Warm seas.

Genus. ROSTELLARIA.—Spire long and many-whorled. Posterior canal running up spire. Warm seas.

Genus. PTEROCERAS.—Outer lip formed into several long claws. India and China.

Family. **MURICIDÆ.**—Canal straight in front, none behind. Eyes without footstalks.

Genus. MUREX.—Varices continuous, three or more rows. Aperture rounded, beak mostly very long, canal partly closed. Many seas.

Genus. TRITON.—Varices not continuous; lips toothed. Warm seas.

Genus. RANELLA.—Two rows of continuous varices. Many warm seas.

Genus. PYRULA.—Pear-shaped; spire short, outer lip thin; canal long and open. Warm seas.

Genus. FASCIOLARIA.—Shell long; canal open. Several oblique folds in inner lip. Warm seas.

Genus. FUSUS.—Shell spindle-shaped, spire short, outer lip thin; canal long and open. Warm seas.

Family. **BUCCINIDÆ.**—Shell with canal pent backwards, or notched.

Genus. BUCCINUM.—Few and rounded whorls, aperture large, canal reflected. Many seas.

Genus. CONCHOLEPAS.—Aperture very large, spire very short. Central America.

Genus. HARPA.—Aperture wide, and notched in front; spine short; ribs numerous and regular. Warm seas.

Genus. PURPURA.—Spine short, aperture moderate and notched in front; inner lip flattened. Shell streaked or tuberculated. Many seas.

Genus. MAGILUS.—Shell spiral and thin when young; prolonged when old into irregular tube, solid behind. Red Sea and Mauritius.

Genus. TEREBRA.—Long, pointed, many whorls; aperture small, canal short. Warm seas.

Genus. EBURNA.—When adult, inner lip spreads and covers umbilicus. Substance solid and smooth.

Genus. DOLIUM.—Shell rounded, spirally furrowed; spire small, aperture very large. Warm seas.

Genus. CASSIS.—Shell swollen, varices irregular; spine short, aperture long, outer lip toothed, canal recurved. Tropical seas.

Genus. OLIVA.—Shell cylindrical and polished, spire short; aperture long, narrow, and notched in front. Foot very large, lobes of mouth meeting over shell. Warm seas.

Genus. COLUMBELLA. — Aperture long and narrow, outer lip thick and toothed. Warm seas.

Family. **CONIDÆ.**—Shell forming double cone; aperture long and narrow; outer lip notched near or at suture.

Genus. CONUS.—Shell tapering regularly; spire short and many-whorled; notch of outer lip at suture. Tropical and warm seas.

Family. **VOLUTIDÆ.** — Shell turreted or convolute; aperture notched in front; columella plaited obliquely. Foot very large.

Genus. MITRA.—Spindle-shaped; spire sharp; aperture small. Many seas.

Genus. VOLUTA.—Shell swollen and thick; spire short; aperture large. Warm seas.

Genus. CYMBA.—Nucleus large and globular, whorls few. Rest like Voluta. Warm seas.

Genus. MARGINELLA. — Spire very short; aperture truncated in front; outer lip truncated when adult; columella plaited. Warm seas.

Family. **CYPRÆIDÆ.**—Shell convoluted and enamelled. Spire not visible; aperture narrow, with a channel at each end. Outer lip thickened.

Genus. CYPRÆA.—Shell swollen; canals short; outer lip ridged. Most seas.

Genus. OVULUM.—Inner lip smooth. Rest like Cypræa. Many seas.

Section. **HOLOSTOMATA (SEA SNAILS).** —Shell spiral; edge of aperture entire. Animal with gills like combs or feathers set across back on to neck. No siphon. Operculum hard and shelly on exterior.

Family. NATICIDÆ.—Shell globular; whorls few; spire small and blunt; aperture semilunar. Animal with long proboscis, and large foot and mantle.

Genus. NATICA.—Shell thick and smooth; umbilicus large; callus spiral. Animal without eyes; front of foot and sides with large lobes.

Genus. NATICELLA.—Aperture very wide.

Family. **NERITIDÆ.**—Shell globular and thick; spire small; aperture semilunar; outer lip sharp.

Genus. NERITA.—Shell smooth or spirally grooved; outer lip thickened; epidermis horny. Warm seas.

Genus. NERITINA.—Inner lip straight and toothed.

Family. **CERITHIDÆ.** — Shell long, spiral, many-whorled; aperture channeled in front.

Genus. CERITHIUM.—Shell turreted, faint varices; aperture small, with twisted canal in front; outer lip expanded; inner lip thickened. World-wide.

Genus. APORRHAIS. — Spire long, many-whorled, and tuberculated. Short canal in front of narrow aperture. Outer lip expanded, with finger-like projections. Many seas.

Family. **TURRITELLIDÆ.** — Shell spiral or tubular; upper part partitioned off.

Genus. TURRITELLA.—Shell long, many-whorled, striped spirally; aperture rounded with thin edge. Many seas.

Genus. VERMETUS.—Shell tubular and irregular; several partitions in tube; aperture round. Many seas.

Genus. SILIQUARIA. — Shell tubular, with longitudinal slit. Warm seas.

Genus. SCALARIA.—Shell turreted, many-whorled, with many transverse ribs; aperture round. Most seas.

Family. **LITORINIDÆ.**—Shell spiral, opaque; aperture rounded.

Genus. LITORINA.—Shell thick, few whorls; outer lip sharp. Most seas.

Genus. SOLARIUM.—Shell target-shaped, flattened; umbilicus very deep. Warm seas.

Genus. PHORUS. — Shell concave beneath; flattened margins to whorls, often with stones and shells attached; aperture oblique; outer lip very thin. Warm seas.

Genus. TRUNCATELLA.—Shell small, cylindrical; aperture oval and entire. Most seas.

Family. **PALUDINIDÆ.** — Shell globular or conical; epidermis thick; aperture round; operculum hard. Muzzle broad; tentacles long and slender; eyes on short footstalks.

Genus. AMPULLARIA.—Whorls round; operculum horny. Muzzle long; neck with lappet on each side, forming a siphon. Northern Hemisphere.

Family. **TURBINIDÆ.**—Shell spiral, pearly inside.

Genus. PHASIANELLA.—Shell long and polished; whorls convex; aperture oval and not pearly; outer lip thin. Many seas.

Genus. TROCHUS.—Shell pyramidal, base flat, many-whorled; aperture pearly inside; outer lip thin. World-wide.

Genus. DELPHINULA.—Shell shield-like, flattened, few-whorled, rough on exterior; aperture round and pearly. Warm seas.

Family. **HALIOTIDÆ.**—Shell spiral or ear-shaped; aperture very large and pearly; outer lip notched or perforated.

Genus. HALIOTIS.—Shell ear-shaped; spire small and flat; outer angle perforated with a row of holes. Most seas.

Genus. STOMATIA.—Outer lip furrowed, not pierced. Warm seas.

Genus. JANTHINA.—Shell spiral, thin, and translucent, few-whorled; aperture squared; lip very thin. Vesicular float attached to foot. Atlantic.

Family. **FISSURELLIDÆ.**—Shell conical and limpet-shaped, front margin notched or tip perforated.

Genus. FISSURELLA.—Shell oval, tip perforated, surface radiated. Most seas.

Genus. PARMOPHORUS. — Shell long and flattened, front edge arched. Warm seas.

Family. **CALYPTRÆIDÆ.** — Shell limpet-like, interior often with shelly process. Most seas.

Genus. CALYPTRÆA.—Shell conical, edge irregular, cup-shaped process in interior. Most seas.

Genus. PILEOPSIS.—Shell conical; apex **spirally** recurved; aperture rounded. Most seas.

Family. **PATELLIDÆ.**—Shell conical ; apex turned forwards.

Genus. PATELLA.—Shell oval, interior smooth ; apex nearly central. World-wide.

Family. **DENTALIDÆ.**—Shell tubular, curved, regular, open at each end, tapering from mouth ; aperture circular.

Genus. DENTALIUM.—As family. Most seas.

Family. **CHITONIDÆ.**—Shell composed of eight overlapping plates, lodged in a leathery mantle, each plate with a process from its front edge.

Genus. CHITON.—As family. World-wide.

Genus. CHITONELLUS.—Border velvety, exposed portions of the plate small. Hot seas.

Order. **PULMONIFERA.** — Foot broad. Breathing organ a simple chamber lined with network of respiratory vessels.

Section. **INOPERCULATA.**—Without operculum.

Family. **HELICIDÆ.**—Shell external. Animal with four retractile tentacles, upper pair largest and having _ye-specks at tips. Foot distinct ; body spiral. Respiratory aperture on right side, under edge of shell. Mouth with horny-toothed upper mandible.

Genus. HELIX.—Aperture of shell transverse, oblique, rounded, or semilunar ; margins distinct. Animal with long foot pointed behind. World-wide.

Genus. BULIMUS.—Shell oblong or turreted ; longitudinal edges of aperture unequal ; columella entire. Animal like Helix. Warm seas.

Genus. PUPA.—Shell oblong or cylindrical ; aperture rounded. Animal with short foot, pointed behind. World-wide.

Family. **LIMACIDÆ.** — Shell small or rudimentary, internal, or concealed by mantle, placed over respiratory cavity. Animal long ; foot not distinct.

Genus. LIMAX.—Shell internal, oblong, edge membranous. Foot very long ; mantle shield-shaped ; head and tentacles retractile. World-wide.

Genus. TESTACELLA.—Shell small, ear-shaped at end of body. Back of animal with two long furrows, from which others ramify. Many parts of world.

Genus. ARION.—Shell oval, or made of chalky granules. Tail rounded, with gland at tip. Most parts of world.

Family. **LIMNÆIDÆ.**—Shell large and thin, lip sharp. Muzzle short and wide ; eyes at base of tentacles ; mouth with upper mandible.

Genus. LIMNÆA.—Shell spiral, rather long, last whorl very wide ; aperture rounded in front. Tentacles triangular. Most parts of world.

Genus. PHYSA.—Spiral to left, and polished. Tentacles long and slender ; edge of mantle with a deep fringe.

Genus. PLANORBIS. — Shell flattened, many whorls ; aperture semilunar. Foot round and short ; eyes at inner base of tentacles.

Section. **OPERCULATA.**—With an operculum.

Order. **OPISTHOBRANCHIATA.**—Shell small or absent. Branchiæ towards end of body.

Sub-order. TECTI-BRANCHIATA.—Gills covered by shell or mantle.

Family. **BULLIDÆ.**—Shell rounded, thin ; spire small ; aperture long, rounded, and waved in front. Animal investing shell ; disc-like head ; eyes none, or in disc. One gill on right of back. Edge of mantle overlapping shell.

Genus. BULLA.—Shell oval, swollen, partly invested by animal ; aperture larger than shell. Head-disc two-lobed behind. Warm and hot seas.

Genus. CYLICHNA.—Shell cylindrical ; aperture narrow and rounded in front. Animal not investing shell. Many parts of world.

Genus. APLUSTRUM. — Shell oval, swollen ; aperture blunt in front ; outer lip sharp. Foot capable of investing the whole shell. Warm seas.

Genus. PHILINE.—Shell oblong ; aperture very wide in front. Head large and oblong ; eyes none ; foot short and broad. Warm and temperate seas.

Family. **APLYSIADÆ.**—Shell none, or small and covered by mantle. Animal like slug ; head distinct ; foot long, forming a tail behind. Very large lobes at sides.

Genus. APLYSIA.—Shell oblong, flexible, translucent. Animal oval ; neck long ; four tentacles. Sides with large lobes used as fins ; gill in centre of back. Most seas.

Genus. DOLABELLA.—Shell hatchet-shaped. Gill near end of body. Hot seas.

Genus. LOBIGER.—Shell oval, transparent ; on middle of back covered with epidermis. Animal slender ; eyes small, on sides of head. Two pairs of rounded lobes on sides. Warm seas.

Family. **PLEUROBRANCHIDÆ.**—Shell mostly limpet-like ; mantle covering back ; gill on sides.

Genus. UMBRELLA.—Shell limpet-like, flat, round, and concentric. Foot large, deeply notched in front. Mantle small. Gill on right side, forming a series of plumes. Warm and hot seas.

Sub-order. NUDIBRANCHIATA.—No shell in adult. Gills always external on back or sides of body.

Family. **DORIDÆ.**—Oblong ; gills in circle on middle of back. Tentacles two. Skin with spiculæ.

Genus. DORIS.—Oval, flattened ; mantle large. Tentacles retractile. All seas.

Family. **TRITONIADÆ.** Gills along sides of back ; tentacles retractile.

Genus. DENDRONOTUS. — Long ; tentacles laminated ; branched appendages in front of head. Single series of gills on each side of back. Northern seas.

Genus. DOTO.—Slender ; sheaths of tentacles trumpet-shaped ; tentacles slender, not laminated. Northern seas.

Family. **EOLIDÆ.**—Gills papillose, along sides of back ; tentacles not retractile. No distinct mantle.

Genus. EOLIS.—Rather oval. Gills numerous and overlapping ; mouth with horny upper jaw. World-wide.

Genus. GLAUCUS.—Long and slender. Gills on three pairs of lateral lobes. Atlantic and Pacific.

Order. **NUCLEOBRANCHIATA.**—Respiratory and digestive organs forming a kind of nucleus on posterior part of back.

Family. **FIROLIDÆ.**—Long, cylindrical, translucent ; ventral and tail fins. Mantle with circular lip.

Genus. FIROLA.—Head long and slender ; fin with small sucker. Tail keeled ; branchial processes numerous, conical, and slender. Tentacles four ; eyes black and distinct. Warm and hot seas.

Genus. CARINARIA.—Shell glassy, limpet-shaped, keeled. Animal large, translucent ; head thick and cylindrical. Warm and hot seas.

Order. **PTEROPODA.**—Sides towards the front furnished with swimming flaps.

Sub-order. THECASOMATA. — Animal with external shell ; head indistinct ; foot and tentacles scarcely visible.

Family. **HYALEIDÆ.**—Shell straight, curved, globular, or needle-shaped. Animal with two large fins. Body inclosed in mantle.

Genus. HYALEA.—Shell globular, translucent. Animal with long appendages to mantle, passing through slits in shell. Warm and hot seas.

Genus. CLEODORA. — Shell pyramidal, triangular, tip sharp. Eyes and tentacles scarcely visible. Warm and hot seas.

Genus. THECA.—Shell straight, conical, pointed; aperture triangular. Temperate and warm seas.

Genus. CYMBULIA.—Shell soft, slipper-shaped, sharp in front, blunt behind. Fins large and rounded. Warm and hot seas.

Order. **BRACHIOPODA.**—Bivalve molluscs, valves unequal; mostly toothed.

Family. **TEREBRATULIDÆ** — Shell covered with minute points; lower valve with large beak and two hinge-teeth. Animal attached either by footstalk or by ventral valve. Dorsal valve with hinge and loop.

Genus. TEREBRATULA. — Shell smooth, convex; beak perforated and blunt; loop short and attached to hinge-plate. Animal attached by footstalk. Mediterranean.

Family. **RHYNCHONELLIDÆ.** — Shell without the points, beaked; opening behind beak. Hinge-plate deeply divided.

Genus. RHYNCHONELLA.—Shell three-sided, beak sharp; dorsal valve depressed at sides; ventral valve hollowed on centre. Animal with long spiral arms directed inwards. Most temperate and cold seas.

Family. LINGULIDÆ.—Shell oblong; attached by footstalks passing between valves. Animal with fringed mantle.

Genus. LINGULA.—Shell compressed, gaping at both ends; dorsal valve rather short. Branchial veins arranged in series of loops. Warm and hot seas.

Order. **CONCHIFERA.**—Bivalves with body compressed, covered on both sides by a flap of the mantle. Gills at sides of body, under the mantle flaps. Valves connected by an elastic ligament at hinge.

Sub-order. ASIPHONIDÆ.—No respiratory siphons.

Family. **OSTREIDÆ.**—Valves unequal; ligament internal; adductor impression single. Animal marine; mantle open; two crescent-shaped gills on each side.

Genus. OSTREA.—Shell attached by left valve; hinge toothless; structure laminated. Margin of mantle double; lips plain; gills nearly equal. World-wide.

Genus. ANOMIA.—Shell thin, translucent, attached by a plug passing through a hole in notch in right valve. Short double fringe on mantle. World-wide.

Genus. PLACUNA.—Shell free, very flat, translucent, rounded, resting on right valve. Indian seas.

Genus. PECTEN.—Shell regular, rounded, mostly ribbed. Animal with a row of black ocelli and a fringe of delicate filaments. Gills crescentic and delicate. World-wide.

Genus. HINNITES.—Shell regular, and with byssus when young, cementing itself and becoming irregular when old. Europe.

Genus. LIMA.—Shell oblique, oval, front side straight and gaping. Edges of mantle separate, inner fringed with tentacles; ocelli very small. Europe.

Genus. SPONDYLUS.—Shell irregular, ribbed radiately, rather covered with spines or leaf-like projections. Lips of animal foliaceous. World-wide.

Family. **AVICULIDÆ.** — Valves unequal, oblique, attached by byssus; hinge-line straight and long, hinge smooth or very slightly toothed. Foot spinning a byssus. All in tropical and warm seas.

Genus. AVICULA.—Shell obliquely oval; cartilage pit single, oblique; one or two small teeth on hinge. Adductor impression large, near centre; pedal small and near umbo.

Genus. MALLEUS.—Shell when adult, T-shaped. China, Australia.

Genus. PINNA.—Valves equal, wedge-shaped, posterior on side gaping; hinge toothless. Edge of mantle with double fringe; byssus very large and powerful; gills long. Many seas.

Family. **MYTILIDÆ.**—Valves equal, oval, or long, close throughout; epidermis thick and dark. Animal spins a byssus; two gills on each side; foot cylindrical and grooved.

Genus. MYTILUS.—Shell pearl-lined, rounded behind, wedge-shaped; hinge with very small and shallow teeth. Edges of mantle slightly projecting, edges of gills fringed, byssus thick. World-wide.

Genus. MODIOLA.—Shell oblong, widened in front; hinge toothless; epidermis often deeply fringed. Most seas.

Genus. LITHODOMUS.—Shell cylindrical, wedge-shaped behind. Many seas.

Genus. DREISSENA.—Shell like Mytilus, but not pearl-lined; a blunt keel to each valve. Many seas.

Family. **ARCADÆ.**—Shell regular; epidermis thick; a long series of teeth on hinge. Mantle open; foot large and bent; gills very oblique.

Genus. ARCA.—Valves nearly equal, deeply-ribbed; hinge straight, with many transverse teeth. Foot long, pointed, and deeply grooved; ocelli on mantle; gills long and narrow. World-wide.

Family. **UNIONIDÆ.**—Valves equal, closed, pearly; epidermis thick. Front hinge-teeth thick and grooved. Edges of mantle united between openings of siphons; branchial orifice fringed. Foot very large; gills long.

Genus. UNIO.—Shell oval or long, thickening with age; hinge-teeth rather deep. Edges of mantle only united between orifices of siphons. World-wide.

Genus. DIPSAS.—Valves produced into a thin elastic wing. Warm rivers.

Genus. ANODON. Shell like Unio, but without teeth, oval and smooth. Northern Hemisphere.

Sub-order. SIPHONIDA. — With respiratory siphons; lobes of mantle united.

α. Siphons short; pallial line simple.

Family. **CHAMIDÆ.**—Shell thick, attached; two hinge-teeth in one valve, one in the other. Mantle closed; orifices of siphons, and foot, small.

Genus. CHAMA.—Valves foliated, upper the smaller; hinge-tooth of free valve between two of other. Two rows of filaments on edges of mantle; orifices of siphons wide apart; gills deeply plaited. Hot seas.

Family. **TRIDACNIDÆ.**—Shell blunt in front, very deeply ribbed; edges toothed. Animal with byssus, or free; orifices of siphons with thick border; two narrow deeply-plaited gills on each side.

Genus. TRIDACNA.—Shell very thick, three-sided, deeply ribbed and foliated; edges deeply toothed; byssal opening large. Hot seas.

Genus. HIPPOPUS. — Valves close, two hinge-teeth in each; byssus small. Coral seas, on reefs.

Family. **CARDIADÆ.**—Valves equal, free, heart-shaped, ribbed; hinge-teeth two. Mantle open in front; orifices of siphons fringed. Two oblique gills on each side; foot large and curved.

Genus. CARDIUM.—Shell swollen, sometimes gaping behind; edges slightly notched. Edges of mouth plaited; foot long. Northern Hemisphere.

Family. **CYPRINIDÆ.** — Shell regular; valves equal and close; epidermis thick; ligaments external. Hinge-teeth in both valves. Lobes of mantle joined behind, and pierced for the orifices of the siphons. Foot thick; two large gills on each side.

Genus. ISOCARDIA.—Shell heart-shaped; umbones spiral. Foot triangular; openings of siphons close together. Most parts of world.

β. Siphons long; pallial line waved.

Family. **VENERIDÆ.**—Shell regular, close; ligament external. Animal locomotive; siphons unequal, united; foot tongue-shaped; gills large and squared.

3 E 2

Genus. VENUS. — Shell thick, edges indented; three hinge-teeth in each valve. Edges of mantle fringed; siphons partly separated. World-wide.

Genus. CYTHEREA. — Shell like Venus, edges not indented. Edges of mantle plain; siphons half united. World-wide.

Family. **MACTRIDÆ.** — Shell three-sided, close or slightly gaping; ligament internal; epidermis thick; two diverging teeth on hinge. Mantle open in front; foot compressed.

Genus. MACTRA. — Shell nearly equal-sided; front hinge-tooth angular. Mantle open to siphons, edges fringed; outer gills shortest. World-wide.

Family. **TELLUIDÆ.** — Shell compressed; not more than two hinge-teeth; ligament on shorter side of shell. Mantle widely open in front; siphons separated, very long; gills unequal.

Genus. SCROBICULARIA. — Shell oval, thin; hinge-teeth small. Edges of mantle toothed; siphons slender; orifices simple. Many seas.

Family. **SOLENIDÆ.** — Shell long, gaping at both ends. Foot very large and powerful; siphons short and united; gills narrow, projecting into branchial siphon.

Genus. SOLEN. — Shell very long, nearly cylindrical, nearly or quite straight, edge parallel; two hinge-teeth in each valve; ligaments external, long. Mantle closed except at front end; foot blunt and cylindrical. All except frozen seas.

Family. **MYACIDÆ.** — Shell thick, gaping behind; epidermis wrinkled. Mantle nearly closed; foot small; siphons united, retractile; two long gills on each side.

Genus. MYA. — Shell oblong, left valves swollen, with a flattened gristly process. Foot straight; siphons fringed at orifices; an inner series of fringes in branchial orifice. Gills not projecting into siphon. Northern Hemisphere.

Family. **GASTROCHÆNIDÆ.** — Shell gaping; valves thin, toothless, sometimes cemented to a shelly tube. Siphons very long, contractile; orifices fringed. Edges of mantle very thick in front and united; gills narrow, projecting into branchial siphon.

Genus. GASTROCHÆNA. — Shell wedge-shaped, close behind, open in front; ligament external. Foot grooved; siphons united nearly to tips. Most seas.

Genus. ASPERGILLUM. — Shell smooth, cemented to end of shelly tube; only umbones visible; tube closed by perforated disc. Mantle fringed in front. Warm seas.

Family. **PHOLADIDÆ.** — Shell open at both ends, thin, hard, brittle, and marked like a file externally; hinge-plate reflected over umbones. Animal worm-like; foot short; mouth closed in front; siphons long, large, united nearly to tips; orifices fringed; gills narrow, projecting into exhalent siphon.

Genus. PHOLAS. — Shell nearly cylindrical, accessory valves. Foot large; body with fin-like termination.

Genus. TEREDO. — Shell globular, open at both ends; valves three-lobed, with a transverse furrow. Foot-opening very small; viscera within valves; gills long, passing into siphon. Siphons with two shelly styles, orifices fringed. Most seas.

Order. **TUNICATA.** — Head not distinct, inclosed in leathery or membranous sac, and finished with two apertures. Second sac within the former, also perforated. Gills internal.

Family. **ASCIDIADÆ.** — Animal solitary or gregarious, simple and fixed. Oviparous.

Genus. ASCIDIUM. — Body sessile; tests leathery or gelatinous; branchial orifice with eight lobes, a circle of filaments within; branchial sac simple. Northern Hemisphere.

Genus. CYNTHIA. — Test leathery, sessile; orifices with four lobes; branchial sac plaited, tipped with filaments. Northern Hemisphere.

Genus. PELONÆA. — Body long, cylindrical; orifices four-cleft; mantle adhering to test. Northern Hemisphere.

Genus. BOLTENIA. — Test leathery, globular, on foot-stalk; orifices at sides four-cleft; branchial sac plaited. Many seas.

Family. **CLAVELLINÆ.** — Animals compound and fixed, connected by tunic or gelatinous base.

Genus. CLAVELLINA. — Body long, erect, on footstalk; test transparent and smooth; orifices simple. Many seas.

Genus. SYNTETHYS. — Animals gelatinous, sessile; orifices simple. Northern seas.

Family. **BOTRYLLIADÆ.** — Animals compound; tests fused together. Oviparous and gemmiferous.

Sub-family. **Botryllina.** — Animals united in systems round a common cavity. Thorax and abdomen not distinct.

Genus. BOTRYLLUS. — Test gelatinous or gristly; systems round or star-shaped, individuals lying horizontally. Northern seas.

Family. **PYROSOMIDÆ.** Animal compound, free.

Genus. PYROSOMA. — Body not contractile, cylindrical, hollow, open at one end; animals grouped in whorls on exterior. Warm seas.

Family. **SALPIDÆ.** — Animals free, alternately solitary and aggregated.

Genus. SALPA. — Animal oblong, nearly cylindrical. Test thin, transparent; gill rudimentary. Alternate generations unlike each other. Northern Hemisphere.

Sub-class. **POLYZOA.** — Animals compound; digestive canal with two apertures, replicate, the posterior ascending by side of front portion. Mouth surrounded with tentacles. Front portion retractile by inversion.

Order. **INFUNDIBULATA.**

Sub-order. **CHEILOSTOMATA.** — Aperture of cell filled with thin membranous or chalky veil; mouth crescent-shaped; lip moveable.

Family. **CATENICELLIDÆ.** — Cells connected by flexible joints.

Genus. CATENICELLA. Cells facing same way; branches dichotomous; polyzoary erect; ovicells below opening of cell in front. Bass's Straits, New Zealand.

Genus. CALPIDIUM. — Avicularium on each side of cells; two or more apertures all facing same way; cells single at dichotomous fork. Bass's Straits.

Family. **SALICORNIARIDÆ.** — Cells round imaginary axis; branches cylindrical; polyzoary dichotomous, erect.

Genus. SALICORNARIA. — Front of cell depressed, surrounded by ridge, forming surface of polyzoary into hexagons or rhomboids. No apertures. Avicularia irregular. European and other seas.

Family. **CELLULARIDÆ.** — Cells in same plane; branches linear, dichotomous, erect.

Genus. CELLULARIA. — Cells bi-triserial, oblong or rhomboidal, contiguous, perforated behind; no avicularium, seldom vibraculum. Britain, &c.

Genus. MENIPEA. — Sessile avicularia on front of cell below aperture. Britain, &c.

Genus. SCRUPOCELLARIA. — Cells rhomboidal. Sessile avicularium on each cell. Aperture oval. Britain, &c.

Genus. CANDA. — No sessile avicularium; root like Scrupocellaria. Britain, &c.

Section. Polyzoary continuous.

Family. **SCRUPARIADÆ.** — Cells in single series. Joints not flexible.

Genus. SCRUPARIA. — Cells decumbent; aperture oblique near end. Branches from front of cell below aperture. Britain, &c.

Genus. ÆTEA.—Cells tubular, erect, from creeping fibre. Britain, &c.

Genus. BEANIA.—Polyzoary horny or chalky. Cells rising from slender tube and open in front; edges with hollow spines. Britain, &c.

Family. **FARCIMINARIADÆ.**—Cells round imaginary axis, alternate; polyzoary erect, dichotomous.

Genus. FARCIMINARIA.—Cells horny, flexible; edge raised; apertures large. Tasmania.

Family. **GEMELLARIADÆ.**—Cells opposite, in pairs.

Genus. GEMELLARIA.—Cells back to back, pairs facing same way; aperture oval, oblique. Britain, &c.

Genus. DIMETOPIA.—Cells back to back; pairs at right angles. Bass's Straits.

Genus. NOTAMIA.—Avicularia tobacco-pipe-shaped above each pair of cells. Britain.

Family. **CABEREADÆ.** — Polyzoary dichotomous; vibracula or sessile avicularia one to several cells.

Genus. CABEREA.—Vibracula large, on back of branches. Algoa Bay, &c.

Family. **BICELLARIADÆ.**—Vibracula none. Polyzoary dichotomous.

Genus. BICELLARIA.—Cells distinct; aperture upwards; several spines. Britain, &c.

Genus. BUGULA.—Cells elliptical, contiguous; aperture large. Britain, &c.

Family. **FLUSTRADÆ.**—Polyzoary flat, flexible, leafy, erect. Cells many.

Genus. FLUSTRA.—Cells contiguous, both sides of frond. Britain, &c.

Genus. CARBASEA.—Cells contiguous, one side of frond. Britain, &c.

Genus. DIACHORIS.—Cells separate, each connected with six others by tubular processes. Bass's Straits.

Family. **MEMBRANIPORIDÆ.**—Polyzoary flat, chalky, mostly encrusting. Cells horizontal.

Genus. MEMBRANIPORA.—Cells irregular or quincuncial, with raised margins; part of front membranous and flexible. Britain, &c.

Genus. LEPRALIA.—Polyzoary encrusting, spreading from centre. Britain, &c.

Family. **CELLEPORIDÆ.**—Cells vertical to axis or plane, heaped or irregular.

Genus. CELLEPORA.—Polyzoary chalky, rigid, heaped or quincuncial. Ascending beak or mouth, with avicularium.

Family. **ESCHARIDÆ.**—Polyzoary erect; cells quincuncial in same plane.

Genus. ESCHARA.—Cells on both surfaces, back to back, horizontal to plane of axis. Britain, &c.

Genus. RETEPORA. — Polyzoary leafy, chalky, net-like. Cells immersed, open on surface.

Family. **SELENARIADÆ.**—Polyzoary flat and rounded, plano-convex or concave. Vibracula large and strong.

Genus. CUPULARIA.—Each cell with vibraculum at apex. Warm seas.

Genus. LUNULITES.—Cells in series from middle, bifurcating; vibracula in alternate lines. Warm seas.

Genus. SELENARIA.—Certain cells with vibracula. Bass's Straits.

Sub-order. CYCLOSTOMATA.—Cell with round simple orifice.

Family. **CERISIADÆ.**—Polyzoary planet-like, jointed and branched. Cells tubular, in one or two series; orifices in opposite directions.

Genus. CERISIA.—Cells in two rows; orifices at tips.

Family. **TUBULIPORADÆ.** — Polyzoary solid, chalky, circular, lobed or two-branched. Cells tubular; orifice round and prominent.

Genus. PUSTULOPORA. — Erect, cylindrical; cells half sunken; orifice prominent.

Genus. IDMONEA.—Erect, two-branched. Cells on one side, in rows, divided by a longitudinal line.

Genus. DISCOPORA.—Base thin, sessile and cup-shaped.

Genus. ALECTO.—Creeping, branched irregularly. Cells horizontal, in one or more rows; ends free.

Genus. TUBULIPORA.—Base defined, branch-like. Cells nearly erect, aggregate or in imperfect rows.

Family. **VESICULARIADÆ.**—Polyzoary plant-like, horny, tubular; cells free, exteriorly flexible.

Genus. BOWERBANKIA.—Mottled and creeping, or erect and branched irregularly. Cells thickly clustered. Animals eight to ten tentacles.

Genus. SERIALARIA.—Cells aggregated in sets, on a stalk and on same side; animals with eight tentacles.

Genus. BUSKIA.—Cells nearly separate, on footstalks, and flask-shaped, with short processes from walls.

Family. **ALCYONIDIADÆ.** — Polyzoary spongy, irregular; orifice of cells contractile.

Genus. ALCYONIDIUM.—Polyzoary erect, lobed or simple. Cells sunken and five-angled.

Family. **PEDICELLINADÆ.** — Polyzoary plant-like, creeping, adherent, from which spring polypes on footstalks.

Genus. PEDICELLINA.—Animal globular. Tentacles curled inwardly, not retractile.

Order. **PHYLACTOLÆMATA.**—Lophophore bilateral; mouth with epistome.

Sub-order. LOPHOPEA.—Arms of lophophore free.

Family. **CRISTATELLIDÆ.**—Polyzoary free and locomotive.

Genus. CRISTATELLA.—Only genus.

Family. **PLUMATELLIDÆ.**—Polyzoary rooted.

Sub-family.—Lophophore with two long arms.

Genus. LOPHOPUS. — Polyzoary bag-shaped; ectocyst gelatinous; statoblasts oval; no spines. Britain, &c.

Genus. PECTINATELLA. — Polyzoary massive; ectocyst gelatinous; statoblasts disc-shaped, with spines. Britain, &c.

Genus. ALCYONELLA.—Polyzoary tubular, tubes united; ectocyst membranous. Britain, &c.

Genus. PLUMATELLA.—Polyzoary tubular, tubes distinct. Britain, &c.

Sub-family.—Arms of lophophore nearly funnel-shaped.

Genus. FREDERICELLA.—Only genus.

CLASS.—INSECTA.

N.B.—If no country is mentioned, the insect is British.

Ringed animals, breathing by tracheæ—Head distinct—Legs six, antennæ two ; subject to metamorphosis in early stages, mostly developing wings.

Order. **COLEOPTERA.**—Front wings horny or leathery, concealing the second pair of wings when at rest, horizontal, united down back by straight suture. Hinder of wings membranous, folding cross-wise and longitudinally. Mouth with transversely moveable jaws. Pupa incomplete.

Section. **PENTAMERA.**—Tarsi with five joints, fourth of ordinary size.

Sub-section. **ADEPHAGA.**—Outer lobe of maxillæ distinct and jointed ; inner edge of lower jaws with strong spines, tip hooked.

Stirps. GEODEPHAGA.—Legs long, formed for running ; jaws long, horny, and curved ; eyes prominent.

Family. **CICINDELIDÆ.**—Terminal hook of maxillæ articulated at base ; head wider than thorax ; mandible long, sharp, and toothed.

Genus. CICINDELA.—Front tarsi of male with three first joints dilated. Wings large. Many lands.

Genus. MANTICORA.—Elytra soldered together. Tarsi not different in sexes. Mandibles projecting and long ; head large. Africa.

Genus. TRICONDYLA.—Wings none. Body long and narrow. Thorax globular. Hot climates.

Family. **CARABIDÆ.**—Maxillæ not clawed. Labial palpi three-jointed, set on footstalk. Head not wider than thorax. World-wide.

Genus. CARABUS.—Labrum wavy in middle. Thorax squared, edges rounded or slightly heart-shaped. Abdomen oval. Wings not fit for flight.

Genus. BRACHINUS.—Labrum broad and short. Mandibles projecting, with two basal teeth. Joints of tarsi entire. Thorax short ; elytra broader than thorax and truncated.

Genus. ANTHIA.—Labrum rounded, covering base of large mandibles. Thorax narrow behind. Eyes prominent. Wings absent.

Genus. LEBIA.—Claws of tarsi toothed beneath. Mentum with two lobes ; no tooth. Last joint of labial palpi hatchet-shaped.

Genus. MORMOLYCE.—Antennæ very long. Head very long, narrow, especially behind. Elytra short, with lobes wider than abdomen. Java.

Genus. HARPALUS.—Tarsi of four front feet with three or four first joints wide, and heart-shaped.

Stirps. HYDRADEPHAGA. — Legs for swimming, two latter pairs flat and fringed with hairs. Body oval and mostly flat. Mandibles short and strong. Thorax broader than long.

Family. **DYTICIDÆ.**—Front legs short ; antennæ long.

Genus. DYTICUS.—Tarsi of males wide and flat. Eyes at edge of head, which is drawn into the thorax as far as eyes. Last joint of maxillary palpus oval, and larger than others.

Genus. ILYBIUS.—Elytra smooth. Body very convex, narrow between ; second and third joints of labial palpi nearly equal.

Family. **GYRINIDÆ.**—Front legs long ; antennæ short.

Genus. GYRINUS.—Antennæ shorter than head, nine-jointed. Eyes divided by lateral edge of head. Elytra not covering tip of abdomen.

Sub-section. **RYPOPHAGA.**—Palpi four ; antennæ more or less complex ; basal joints of male tarsi dilated.

Stirps. BRACHELYTRA.—Elytra very short ; legs for swimming ; antennæ increasing gradually to tip.

Family. **STAPHYLINIDÆ.**—Head very distinct ; neck narrow. Antennæ set between or in front of eyes. Mandibles long, having on the inside a fringed membranous apparatus.

Genus. STAPHYLINUS (including OCYPUS and CREOPHILUS).—Antennæ distant at base.

Stirps. NECROPHAGA.—Elytra mostly covering abdomen. Antennæ often knobbed at tip.

Family. **SILPHIDÆ.** — Body flat ; head bent rather downward. Maxillæ two-lobed. Antennæ thickened at tips. Legs not contractile.

Genus. SILPHA.—Antennæ longer than head, club pear-shaped ; a hook inside maxillæ. Head small ; thorax broad.

Genus. NECROPHORUS.—Antennæ short, club globular. No hook or maxillæ.

Section. **CHILOGNATHOMORPHA.** — Larvæ with long and cylindrical body ; end of abdomen curved under breast.

Stirps. CLAVICORNES.—Antennæ club-shaped.

Family. **HISTERIDÆ.**—Body squared, shining. Antennæ short, elbowed ; club three-jointed. Elytra short ; legs toothed.

Genus. HISTER.—Head small. Elytra shorter than abdomen, blunt. Head retractile.

Stirps. LAMELLICORNES.—Club of antennæ formed of teeth or plates.

Family. **LUCANIDÆ.**—Club-toothed.

Genus. LUCANUS. — Mandibles of male very large, like stag's horns ; those of female, short, sharp, stout, and curved. Club of antennæ four-jointed.

Family. **MELOLONTHIDÆ.**—Labrum two-lobed ; mandibles strong. Maxillæ with several teeth at tip. Club of antennæ formed of flattened plates. Elytra not quite reaching extremity of abdomen.

Genus. MELOLONTHA.—Club with six or seven joints ; claws equal, with basal tooth.

Family. **GEOTRUPIDÆ.**—Maxillary lobes membranous ; mandibles projecting. Elytra rounded behind, and covering abdomen. Club large, globular. Thorax broader than long.

Genus. GEOTRUPES. — Antennæ eleven-jointed, third joint larger than second ; neither thorax nor head armed. Body rounded.

Family. **SCARABEIDÆ.** — Clypeus large, concealing mandibles, which have a fringed membranous plate at their tips. Antennæ eight or nine-jointed. Body rather flattened.

Genus. SCARABEUS.—Head similar in sexes, not horned. Thorax of male often horned or tubercled. Hot countries.

Genus. CHALCOSOMA.—Head, one horn reflected, toothed behind. Maxillary palpi four-jointed ; labial three-jointed. Thorax three-horned, central very short. Hot countries.

Family. **RUTELIDÆ.**—Antennæ ten-jointed ; club three-jointed. Mandibles short, horny, notched on outer edge near tip. Maxillæ short, with several teeth. Elytra not reaching to end of abdomen.

Genus. CHRYSOPHORA.—Hind legs of male with broad thighs, and tibiæ armed with long triangular spur, turned inwards. Peru.

Family. **DYNASTIDÆ.** — Head or thorax, or both, horned or tuberculated in males. Clypeus small and triangular, pointed or toothed in front. Jaws strong, horny, toothed. Antennæ ten-jointed, club three-jointed.

Genus. CHELORHINA.—Thorax not horned. Head three-horned, the centre horn projecting, curved upwards, and two-cleft at tip. Tropics.

Section. **PRIOCERATA.**—Antennæ serrated.

Stirps. MACROSTERNI.—Thorax large and long.

Family. **BUPRESTIDÆ.**—Body long ; tips of elytra often toothed. Head very short ; antennæ serrated. Tarsi short.

Genus. CHRYSOCHROA.—Thorax very broad in middle, narrowing to neck. Eyes large. Elytra narrowing behind. Colours always brilliant. India.

Family. **ELATERIDÆ.**—Body hard ; antennæ short and serrated ; head sunk in thorax as far as eyes. Thorax with hinder angles lengthened at spines. Limbs very short.

Genus. PYROPHORUS.—Antennæ rather stout ; elytra sharply pointed. Light-emitting spot near each hinder angle of thorax, and two more under elytra. Brazil.

Stirps. APROSTERNI. — Thorax moderate ; body mostly soft.

Family. **LAMPYRIDÆ.** — Body long and flat ; head small ; thorax extending over head ; antennæ often toothed or serrated.

Genus. LAMPYRIS.—No beak to head, which is covered by thorax. Female wingless.

Family. **TELEPHORIDÆ.**—Both sexes winged ; head projecting. Last joint but one of the tarsi with two lobes.

Family. **PTINIDÆ.** — Body oval or cylindric. Head sunk into thorax as far as eyes ; antennæ longish ; thorax broad in front.

Genus. MEZIUM.—Elytra globular ; thorax furrowed ; wings none.

Genus. PTILINUS.—Body cylindric. Antennæ of male comb-like.

Genus. ANOBIUM. — Antennæ with three last joints largest, last joint egg-shaped. Body longish and egg-shaped.

Section. **HETEROMERA.**—Four front tarsi five-jointed, last pair four-jointed.

Family. **PYROCHROIDÆ.**—Body flattened ; neck distinct ; front part of body narrow. Abdomen broader than thorax, rounded at tip. Antennæ of males serrated.

Genus. PYROCHROA.—Antennæ larger than head and thorax ; comb-like in male.

Family. **MORDELLIDÆ.**—Body elevated and arched ; head very low.

Genus. RIPIPHORUS.—Scutellum concealed ; antennæ of male comb-like.

Family. **CANTHARIDÆ.**—Head wide behind eyes, then narrowed to short neck. Supplementary claws to tarsi. Abdomen soft ; thorax narrow behind.

Genus. CANTHARIS.—Narrow, two wings ; elytra long.

Genus. SITARIS. — Elytra narrowed, not overlapping ; wings two. Antennæ simple.

Genus. MELOE.—Elytra short, overlapping. Wings none. Antennæ various.

Section. **ATRACHELIA.**—Hinder part of head very wide, sunk into thorax as far as eyes. Elytra firm ; joints of tarsi not widened.

Family. **TENEBRIONIDÆ.** — Elytra free, wings for flight. Thorax as wide behind as base of elytra.

Genus. TENEBRIO. — Narrow, long ; thorax squared ; antennæ thread-like.

Section. **PSEUDOTETRAMERA.**—Fourth joint of tarsi very minute.

Stirps. RHYNCHOPORA.—Head produced into long beak. Antennæ short and mostly elbowed.

Family. **BRUCHIDÆ.**—Beak short, broad, flat.

Genus. BRUCHUS. — Antennæ often toothed, slender, sometimes thicker towards tips. Elytra squared and oblong.

Family. **ATTELABIDÆ.**—Antennæ not elbowed ; beak mostly curved and cylindrical.

Genus. RHYNCHITES.—Antennæ straight, eleven-jointed. Neck thick ; beak wide at tip.

Family. **CURCULIONIDÆ.**—Antennæ elbowed ; basal joint long, club at tip. Jaws at end of beak.

Genus. BALANINUS.—Body triangular. Beak nearly or quite as long as body ; antennæ set behind middle of beak.

Genus. CALANDRA.—Antennæ nine-jointed, funiculus six-jointed.

Genus. EUPHOLUS. — Antennæ moderate. Beak short, thick, nearly straight. Eyes round. Elytra nearly cylindrical. Tarsi large. New Guinea.

Family. **SCOLYTIDÆ.**—Beak short, third joint of tarsus two-lobed. Body cylindric or oblong, rounded above. Antennæ with basal-joint long, end-joints forming a kind of solid mass.

Genus. SCOLYTUS.—Body truncated behind. Funiculus of antennæ six-jointed.

Section. **LONGICORNES.**—Antennæ very long, sometimes serrated or tufted. Body long, rather flattened ; elytra broader than thorax. Head short, not beaked. Legs long.

Family. **PRIONIDÆ.**—Head not narrowed into neck. Mandibles very large and strong. Antennæ inserted between base of mandibles and eyes.

Genus. XENOCERUS.—Antennæ enormously long. Beak short, wide and bent. Eyes scooped. Thorax rather long, or transverse ; stripe before base. Elytra rather cylindrical.

Family. **CERAMBYCIDÆ.**—Eyes semilunar or kidney-shaped. Antennæ inserted close to eyes, not serrated. Body long.

Genus. CERAMBYX.—Body rather rounded, no spines on antennæ. Cross wrinkles on thorax.

Genus. CLYTUS.—Thorax globular, not armed. Antennæ horned ; elytra entire at tip.

Genus. DISAULAX.—Thorax rounded, wide in middle. Antennæ densely tufted in middle.

Section. **PHYTOPHAGA.**—Body thick, oval, round, or squared. Head short, sunk to eyes in abdomen. Antennæ short.

Family. **CASSIDIDÆ.** — Body flattened and broad. Mouth under head ; body oval or round.

Genus. CASSIDA.—Rounded, flat beneath ; mouth concealed by part of thorax ; head covered by thorax.

Genus. ASPIDOMORPHA.—Thorax wide and shield-like. Elytra rounded over body, then spread flatly. Heart-shaped when united.

Family. **CHRYSOMELIDÆ.**—Antennæ set widely apart. Body hemispherical ; legs equal-sized. Mandibles notched at tips.

Genus. CHRYSOMELA.—Winged ; palpi hatchet-shaped.

Genus. TIMARCHA.—Wingless ; elytra soldered.

Section. **PSEUDOTRIMERA.** — Third joint of tarsi minute, recurved between the two lobes of second joint. Elytra covering abdomen.

Family. **COCCINELLIDÆ.**—Body convex and hemispherical. Thorax short and transverse. Second joint of tarsi large.

Genus. COCCINELLA.—Hinder angles of thorax sharp.

Order. **DERMAPTERA.**—Front wings leathery, small ; suture straight, partly covering large hinder wings. Forceps at end of tail. Pupa resembling perfect insect, but without wings.

Family.—As Order.

Genus. FORFICULA.—Antennæ fourteen joints ; elytra and wings perfect.

Order. **ORTHOPTERA.**—Front wings on elytra parchment-like, large, heavily veined, overlapping at tip. Hinder wings large, folded longitudinally. Tail often spiked.

Section. **CURSORIA.**—Legs all for running.

Family. **BLATTIDÆ.**—Body oval and flat. Head small, often hidden under thorax. Antennæ very long, hair-like ; legs spurred. Two short-jointed appendages to abdomen.

Genus. BLATTA.—As family.

Section. **RAPTORIA.**—Body long, narrow, nearly cylindrical. Abdomen broader than front of body. Head exposed. Upper wings horizontal. Fore-legs enormous, used for capture.

Family. **MANTIDÆ.**—As Section.

Section. **AMBULATORIA.**—Body long and slender. All legs alike. Front wings comparatively small, hinder wings large.

Family. **PHASMIDÆ.**—As Section.

Genus. PHYLLIUM.—Body flat, broad ; front wings like leaves ; legs provided with leaf-like appendages. Antennæ of male half length of body, those of female still shorter. East Indies.

Genus. BACTERIA.—Wingless. Antennæ haired, larger than head. East Indies.

Section. **SALTATORIA.**—Hind-legs for leaping.

Family. **ACHETIDÆ.**—Antennæ very long and slender. Wings large, forming, when filled, a pair of projecting spines ; wing-covers of male with talc-like spot at base. Both legs flat on body. Jaws strong, with several sharp teeth.

Genus. GRYLLOTALPA.—Fore-legs broad and palmed. No external ovipositor.

Genus. ACHETA. — Fore-legs not palmed. Ovipositor projecting.

Family. **GRYLLIDÆ.**—Wings and covers forming a slanting roof. Ovipositor mostly flat and sabre-like.

Genus. PHASGONURA.—Wings and covers long. Ovipositor long and straight. Four short spines on end of abdomen.

Family. **LOCUSTIDÆ.**—No projecting ovipositor. Antennæ short.

Genus. LOCUSTA.—Antennæ simple. Wings and covers large.

Order. **THYSANOPTERA.** — Wings four, similar, long, narrow, not folded, lying along back. Mandibles like bristles. Tarsi two-jointed.

Order. **NEUROPTERA.**—Wings four, large, equal. Mandibles powerful. Ovipositor with several valves.

Section. **BIOMORPHOTICA.**—Pupa active.

Family. **TERMITIDÆ.**—Both sexes winged. Neuters not winged. Body oblong, flat. Antennæ short ; mandibles flattened, toothed, triangular. Wings nearly twice as long as body.

Genus. TERMES.—Labrum four-cleft ; palpi three-jointed. Tarsi four-jointed. Hot countries.

Family. **PERLIDÆ.**—Hinder wings very large and folded. Tarsi three-jointed.

Family. **EPHEMERIDÆ.** — Front wings very large. Tarsi five-jointed.

Genus. EPHEMERA.—Wings four ; three appendages to abdomen. Head of larva horned.

Family. **LIBELLULIDÆ.**—Head large, globular ; abdomen long ; thorax deep. Antennæ very short. Mandibles very powerful, and masked by lips. Appendages to abdomen not jointed.

Genus. LIBELLULA (including ICTINUS).—Hinder angle of second pair of wings rounded. Ocelli in a triangle.

Genus. CALEPTERYX.—Head transverse, hammer-shaped. Cells of wings very numerous, dark spot hardly visible.

Section. **SUBNECROMORPHOTICA.**—Pupa quiescent, but becoming active just before last transformation.

Family. **PANORPIDÆ.**—Head lengthened into a beak. Body long and slender. Legs long. Tarsi five-jointed ; two spurs on tibiæ.

Family. **HEMEROBIIDÆ.**—Body short and delicate. Wings very broad and membranous. Head small ; eyes prominent ; antennæ long and delicate.

Family. **MYRMELEONIDÆ.**—Body long and slender. Head small ; no ocelli. Wings large and netted, with a strong forked vein. Abdomen long.

Genus. MYRMELEON.—Wings equal size. Hot countries.

Genus. NEMOPTERA.—Hind-wings very long and narrow. Hot countries.

Order. **TRICHOPTERA.** — Wings four ; front pair mostly hairy, second pair large and folded when at rest. Long spines at end of tibiæ. Mouth not fitted for eating. Larva mostly resident in sub-aquatic case. Pupa mostly inactive.

Order. **HYMENOPTERA.**—Wings four, transparent, membranous ; veins few, hinder pair smallest. Mandibles powerful. Abdomen of female either with a saw or a sting. Larva footless ; pupa inactive.

Section. **TEREBRANTIA.**—Abdomen of females furnished with a boring instrument. No poison gland.

Sub-section. PHYTIPHAGA.—No footstalk to abdomen.

Tribe. SERRIFERA.—Boring instrument saw-like.

Family. **TENTHREDINIDÆ.**—As Tribe.

Genus. CIMBEX. — Club of antennæ composed of two united joints ; five joints preceding club.

Genus. ATHALIA.—Antennæ nine or ten-jointed, third joint long ; club indistinct.

Genus. NEMATUS.—Antennæ nine-jointed. Tarsi simple.

Family. **UROCERIDÆ.**—Boring instrument awl-like.

Genus. UROCERUS.—Ovipositor long. Front of thorax not forming a neck. Maxillary palpi very short.

Sub-section. ENTOMOPHAGA.—Abdomen with footstalk.

Tribe. SPICULIFERA.—Ovipositor needle-like. Abdomen not tubular at tip.

Family. **CYNIPIDÆ.**—Ovipositor-interior, nearly spiral. Antennæ straight.

Genus. CYNIPS.—Abdomen nearly egg-shaped. Antennæ of male fifteen-jointed, of female fourteen-jointed. Wings large.

Family. **ICHNEUMONIDÆ.**—Abdomen attached to end of thorax. Ovipositor and antennæ straight, latter mostly more than sixteen joints.

Genus. IOHNEUMON.—Abdomen long and egg-shaped, first segment rough. Ovipositor not protruding. Antennæ simple. Outer cell of front wings complete.

Genus. RHYSSA.—Abdomen long, convex, and shining. Faint wrinkles across thorax. Ovipositor very long.

Genus. MICROGASTER. — Abdomen very small. Eyes hairy. Antennæ eighteen-jointed.

Tribe. **TUBULIFERA**.—Abdomen telescopic and re-tractile at tip. Antennæ elbowed.

Family. **CHRYSIDIDÆ**.—As Tribe.

Section. **ACULEATA**. — Ovipositor modified into sting, with poison-glands. Sting defended by two-bladed sheath.

Sub-section. **INSECTIVORA**.—Front wings not folded. Larva solitary.

Family. **CRABRONIDÆ**.—Head long and squared. Man-dibles narrow. No elongated beak. Front tibiæ with a combed spur.

Genus. CRABRO.—Face very wide. Antennæ strongly elbowed. Mandibles two, cleft at tip. Maxillary palpi with six joints, labial with four.

Genus. PHILANTHUS.—Abdomen long and rather egg-shaped. Cell on edge of front wings very narrow and pointed.

Family. **BEMBECIDÆ**.—Head transverse, eyes extend-ing to hinder edge. Mandibles sharp, with one tooth inside. End lobe of maxillæ long. Tongue mostly long.

Genus. MONEDULA.—Body long, pointed at tip. An-tennæ set in middle of face. Tarsi with short spines. Males with one or two spines in tubercles, on under surface of abdomen.

Family. **SPHEGIDÆ**.—Collar large, extending to base of wings. Legs very long, and formed for digging. An-tennæ slender, often curled in females.

Genus. POMPILUS.—Thorax oblong. Antennæ inserted near mouth. Sting not protruding.

Genus. CHLORION.—Head as wide as thorax. Mandibles toothed inside. Abdomen on footstalk. Third ring of thorax transversely streaked. Hot countries.

Family. **SCOLIIDÆ**. — Collar extending to base of wings. Legs short and stout. Antennæ mostly short and thick. Both sexes winged.

Genus. SCOLIA.—Antennæ short and curled. Legs with thick spines. Labium composed of three parts set in a line. Hot countries.

Family. **MUTILLIDÆ**.—Females wingless. First and third joints of antennæ long. Mandibles notched. Sting very long.

Genus. MUTILLA.—Antennæ set above front edge of shield. Thorax not divided. Hot countries.

Sub-section. **SODALES**.—Front wings not folded. Larva social.

Family. **FORMICIDÆ**.—Females much larger than males. Both sexes winged; neuters wingless. Basal joint of antennæ long; elbow sharp. Eyes not reaching to hinder edge of head.

Genus. FORMICA.—Sting absent. Eyes distinct.

Genus. PONERA.—Sting large. Eyes of neuter wanting.

Genus. MYRMECIA. — Sting present. Basal joint of antennæ long. Footstalk of abdomen two-jointed.

Sub-section. **DIPLOPTERYGA**.—Front wings folded.

Family. **EUMENIDÆ**.—Solitary. No neuters. Mandibles rather long. End joints of antennæ often hooked in males. Thorax short. Abdomen sometimes on footstalk. Legs not spined.

Genus. EUMENES.—Basal segment of abdomen narrow, and forming a footstalk.

Genus. ODYNERUS.—Basal segment of abdomen short, rather bell-shaped. Abdomen rather conic.

Family. **VESPIDÆ**.—Shield of head nearly square. Mandibles short, toothed at the tips. Antennæ twelve-jointed in females and neuters; thirteen-jointed in males.

Genus. VESPA. — Abdomen egg-shaped. Mandibles oblique at tip.

Sub-section. **MELLIFERA**. — Basal joint of last tarsi formed into a spoon-like plate. Proboscis mostly present. One spur on each of two front pairs of tibiæ.

Family. **APIDÆ**.—Proboscis long. Basal joint of antennæ long.

Genus. NOMADA.—Body long. Marginal cell broad. Maxillary palpi six-jointed.

Genus. CHRYSANTHEDA.—Tongue long as body. No pollen-basket on hind-legs.

Genus. XYLOCOPA.—Body broad, hairy. Mandibles of female blunt.

Genus. BOMBUS. — Body thick, rounded, and hairy. Spurs on hinder tibiæ.

Genus. APIS.—Body moderately slender. No spurs on hinder tibiæ.

Order. **STREPSIPTERA**.—Front wings changed into two short twisted organs; hinder wings very large, folded like a fan. Mandibles slender.

Order. **LEPIDOPTERA**.— Wings four, membranous, wide, covered on both sides by overlapping scales. Nerves branching. Mouth formed for suction. Tibiæ spurred.

Section. **RHOPALOCERA**. — Antennæ mostly knobbed at tip. Proboscis long.

Family. **PAPILIONIDÆ**. — Legs for walking in both sexes.

Sub-family. **Papilionides**.—Hinder inner edge of second pair of wings concave or folded. Middle longitudinal nervure of front wings giving out four nervures. Central cell of wing closed. Club of antennæ elongated.

Genus. ORNITHOPTERA.—Hind-wings not tailed. Labial palpi moderately short. Abdomen of male deeply grooved on outer side. Indian Archipelago.

Genus. PAPILIO.—Hind-wings tailed. Labial palpi very short.

Genus. LEPTOCIRCUS.—Hinder wings folded longitudi-nally, very long, slightly curved at extremity. Abdomen short. Palpi very short. Java.

Genus. THAIS. — Inner edges of hinder wings deeply scooped. East Indies.

Sub-family. **Pierides**.—Hind-wings forming a gutter for the reception of the abdomen.

Genus. PIERIS.—Wings rounded, partly bare. Club of antennæ compressed.

Family. **HELICONIDÆ**. — Fore-legs small. Central cell of hind-wings closed. Palpi short, and set far apart. Club small.

Genus. HELICONIA.—Front wings long, narrow, entire; hinder wings narrow, receding from abdomen. All wings nearly transparent. Palpi rising above head. Front tarsus wide and slightly toothed. West Indies.

Genus. LEPTALIS.—Front wings curved and scooped, so as to form a hook.

Genus. MECHANITIS.—Wings very long, entire. Hinder wings narrow at base. Body slender. Hot countries.

Genus. ITHONIA. — Wings almost wholly transparent, except round edges, and a short bar on front edge. Central America.

Genus. EUPLÆA.—Front legs short and imperfect. An-tennæ set close together, club long and curved. Palpi short, thickly haired. Hot countries.

Genus. DANAIS.—Antennæ very long. Palpi set apart. Wings with triangular outline. Central cell of hind-wings closed. Hot countries.

Family. **NYMPHALIDÆ**.— Fore-legs rudimentary in both sexes, and thickly haired. Central cell of hinder wings either open or closed, with a very slender nerve.

Genus. CYRESTIS.—Hinder wings scooped from tip of abdomen. Outer edge of front wings nearly straight. Hot countries.

Genus. CETHOSIA.—Palpi set close, but diverging, and tipped with slender extremity. Lower wings embracing abdomen. Hot countries.

Genus. AGRAULIS.—Wings scooped round hinder edges. Hind-wings not tailed. Hot countries.

Genus. MARPESIA.—Front wings deeply scooped and rather hooked; hind-wings with two tails, outer three times as long as inner. Hot countries.

Genus. VANESSA.—Club of antennæ short. Eyes hairy. Front wings angular.

Genus. CATAGRAMMA.—Wings large, rounded on hinder edges. Under side of lower wings marked with rings, mostly so as to form the figure 8. New World.

Genus. NYMPHALIS.—Palpi close, erect, hairy. Club gradual. Hot countries.

Genus. ADOLIAS.—Thorax large. Abdomen small. Wings slightly scalloped along outer edges. Club gradual. Asia.

Genus. HETÆRA.—First pair of wings entire; second pair sometimes scooped. Wings wholly or partly transparent. Hot countries.

Genus. MORPHO.—Club very gradual. Central cell of hinder limbs open behind.

Family. **ERYCINIDÆ.**—Males with four walking feet; females with six. Claws of tarsi minute.

Genus. ERYCINA. — Upper wings rather triangular; lower much elongated and diverging. Wings nearly transparent. Hot countries.

Genus. ZEONIA.—Hinder wings with a very long, slender tail, starting at an obtuse angle with the wing.

Genus. HELICOPIS.—Club rather curved. Hinder edge of lower wings with six short tails. Hot countries.

Genus. CALYDNA. — Hind-wings not tailed. Central America.

Genus. EURYGONE.—Almost straight line along outside edge of both pair of wings. Abdomen short. Central America.

Family. **LYCÆNIDÆ.**—Small. Claws of tarsi minute; fore-legs fitted for walking. Last joint of palpi small and naked.

Genus. THECLA.—Club long. Eyes hairy; hind-wings mostly tailed.

Family. **HESPERIDÆ.**—Legs for walking. Antennæ wide apart, and often hooked at tip. Maxillæ very long. Body robust.

Section. **HETEROCERA.**—Antennæ not clubbed.

Family. **SPHINGIDÆ.**—Proboscis mostly long; antennæ prismatic, tipped by thread or feather. Hinder pair of wings small. Body long and sharp behind.

Genus. SMERINTHUS. — Wings angulated. Proboscis moderate.

Genus. ACHERONTIA. — Wings not angulated, sharp. Proboscis short. Antennæ hooked at tips.

Genus. SPHINX (including CHÆROCAMPA).—Spiral tongue very long. Antennæ not club-like. Wings sharp.

Genus. DEILEPHILA.—Tongue long; antennæ club-like.

Genus. MACROGLOSSA.—End of abdomen with large tuft; wings covered with scales.

Genus. SESIA.—End of abdomen tufted. Wings partially transparent.

Family. **ANTHROCERIDÆ.**—Antennæ not feathered at tip, sometimes simple in both sexes, sometimes comb-like in male. Abdomen never tufted; wings bent down in repose.

Genus. INO.—Antennæ straight, those of male doubly combed. No spur on hinder tibiæ.

Family. **ÆGERIIDÆ.**—Body long; wings naked and often transparent. Antennæ simple. Hinder legs spurred.

Genus. ÆGERIA.—Proboscis and antennæ short. Abdomen not tufted.

Genus. TROCHILIUM.—Proboscis and antennæ long. Large tuft on tip of abdomen.

Family. **URANIIDÆ.**—Wings with spring and socket. Antennæ long, never prismatic nor tufted. All exotic.

Genus. URANIA. — Antennæ slender, thread-like, gradually tapering to tip. Hinder wings tailed.

Genus. CASTNIA.—Antennæ club-like at tip. Proboscis long. Hinder wings not tailed.

Family. **HEPIALIDÆ.** — Antennæ short, gradually pointed. Proboscis wanting, or very short.

Genus. HEPIALUS.—Antennæ thread-like, much shorter than thorax.

Genus. COSSUS.—Antennæ long as thorax, toothed to tip in both sexes.

Genus. ZEUZERA.—Antennæ of male doubly combed to middle, those of female not combed.

Family. **BOMBYCIDÆ.**—Proboscis and palpi very small, or absent. Body thick and hairy. Wings large and broad.

Genus. ATTACUS.—Wings entire, eyed, sometimes with transparent spot. Antennæ deeply combed. Warm countries.

Genus. BOMBYX.—Antennæ deeply combed, longer than thorax. Abdomen extending beyond hind-wings. Asia.

Genus. OIKETICUS.—Female wingless. Male's body hairy. Antennæ combed at base, serrated at tip, no joints. Wings hairy. Abdomen more than half its length beyond hind-wings. Honduras.

Family. **ARCTIIDÆ.**—Antennæ of male deeply combed

or notched. Wings bent down in repose. Proboscis small, or wanting.

Genus. STAUROPUS.—Front wings narrow, tufted on disc. Antennæ of male combed, except at tip.

Genus. NOTODONTA.—Front wings slightly toothed at tip, tufted on inner edge. Thorax not crested. Antennæ of male combed.

Genus. HYPOGYMNA.—Proboscis wanting. Wings opaque. Antennæ combed in both sexes. Abdomen of male with a brush at tip.

Genus. HYPERCOMPA.—Proboscis much longer than head. Antennæ simple in both sexes.

Genus. DORATIFERA.—Proboscis not visible. Antennæ more than half the length of body, combed in male half length, rest serrated to tip. Abdomen as far as hind-wings.

Family. **LITHOSIIDÆ.**—Antennæ mostly slender, sometimes combed or feathered in male. Proboscis developed. Body slender.

Genus. PERICOPIS.—Body cylindrical. Palpi hairy in male. Antennæ slightly combed in male, serrated in female. Wings long, partly transparent. Central America.

Genus. HYELOSIA.—Palpi nearly as long as head, third joint very small. Antennæ more than half length of body and deeply combed in male; shorter and serrated in female.

Genus. CALLIMORPHA. — Fore-wings rather triangular. Antennæ of male slightly feathered. Palpi three-jointed.

Family. **NOCTUIDÆ.**—Body stout. Antennæ mostly simple. Wings moderate, strongly nerved. Proboscis developed.

Genus. CALOCAMPA.—Fore-wings long and narrow. Palpi short. Thorax slightly crested. Antennæ short and stout.

Genus. THYATIRA.—Wings large, blunt, broad, rather pointed at tip. Crest across thorax. Antennæ short and hairy.

Genus. CATOCALA.—Front wings forming triangle when bent down. Antennæ slender. Slight crest on thorax.

Family. **GEOMETRIDÆ.** — Body slender and limbs feeble. Proboscis short, weak, and membranous. Wings large. Thorax not crested.

Genus. BISTON.—Rather stout. Proboscis very short. Antennæ deeply double-combed. Wings thick, rather small, in both sexes.

Genus. OURAPTERYX.—Body moderate. Antennæ simple in both sexes. Front wings sharp at tip, hinder with small tails. Both sexes winged.

Genus. HALIA.—Wings short, triangular, powdery, hinder pair toothed. Antennæ of male doubly combed. Proboscis long. Both sexes winged.

Genus. MELANIPPE.—Proboscis long, antennæ simple. Both sexes winged; wings wide and rounded.

Genus. CHEIMATOBIA.—Proboscis very short. Antennæ of male hairy. Female nearly wingless, wings of male broad and thin.

Genus. PLATYPTERYX.—Front wings curved at tip, with hinder edges toothed. Proboscis scarcely visible. Antennæ doubly combed in both sexes.

Family. **TORTRICIDÆ.** — Small. Fore-wings broad, forming triangle with body when at rest. Body slender. Antennæ mostly simple. Labial palpi broad and flat. Maxillary palpi very small.

Genus. TORTRIX.—Antennæ very slightly haired in male. Fore-wings rounded at shoulder, hind-wings large.

Family. **YPONOMEUTIDÆ.** — Labial palpi long and slender, maxillary palpi generally wanting. Antennæ long and slender. Hinder wings fringed.

Genus. ADELA.—Palpi four. Antennæ very long. Eyes very large. Wings metallic; head thickly haired.

Genus. YPONOMEUTA.—Palpi two. Antennæ rather hairy. Wings large and curled, front pair long and round.

Family. **TINEIDÆ.**—Wings narrow. Labial palpi seldom seen; maxillary palpi very large. Antennæ moderate.

Genus. GALLERIA.—Thorax and abdomen stout. Front wings of male scooped at tip, those of female blunt.

Genus. GRACILLARIA.—Wings very narrow. Labial palpi very slender, long, and curved; maxillary palpi short, slender, and diverging. Fringe of hind-wing very long.

Genus. TINEA.—Head crested in front. Labial palpi longer than proboscis. Fore-wings rather long, rounded behind.

Family. **ALUCITIDÆ.**—Body very slender; wings very narrow, often deeply cleft and fringed. Proboscis long.

Genus. PTEROPHORUS.—Front-wings two-lobed, hinder three-lobed.

Genus. ALUCITA.—Both pairs of wings six-lobed.

Order. **HOMOPTERA** — Wings four, membranous; front pair larger than hinder, not overlapping. Tarsi not more than three joints. Body convex.

Section. **TRIMERA.**—Tarsi with three joints. Antennæ very small. Wings with definite cells.

Family. **CICADIDÆ.**—Head short and broad; eyes large and prominent. Three ocelli, placed in a triangle on back of head. Beak long and three-jointed. Legs not for leaping. Males with drums at base of abdomen. Females with ovipositor of two saw-edged blades.

Genus. CICADA.—Antennæ seven-jointed.

Family. **FULGORIDÆ.**—Antennæ three-jointed; ocelli two, placed beneath eyes. Legs mostly for leaping. Males without drums.

Genus. HOTINUS.—All wings variegated. Head elongated into a long process, slightly curved upwards. Exotic.

Genus. PŒCILOPTERA.—Hinder wings not variegated. Head not elongated. Ocelli wanting.

Family. **CERCOPIDÆ.** — Antennæ three-jointed, last joint forming a long bristle, set between eyes. Ocelli two. Legs for leaping. No drum in male.

Genus. CERCOPIS.—Head small. Ocelli apart, set on upper side of head.

Genus. APHROPHORA. — Head moderate. Ocelli close together, on crown of head.

Genus. BOCYDIUM.—Prothorax developed into a curious arrangement, consisting of a footstalk, with a branch backwards and a cross-bar in front, bearing four knobs, two central, on short footstalks. Brazil.

Genus.—TETTIGONIA.—Distance between ocelli and ocelli and eyes equal.

Section. **DIMERA.**—Tarsi two-jointed. Wings partly with cells. Antennæ moderate.

Family. **APHIDÆ.** — Antennæ seven-jointed. Wings numerously veined. Beak four-jointed. Ocelli three, in a triangle. Front wings largest.

Genus. APHIS.—Abdomen with two projections. Antennæ long, hair-like. Proboscis short.

Section. **MONOMERA.**—Tarsi one-jointed. Wings not celled. Antennæ from six to twenty-five joints.

Family. **COCCIDÆ.**—As Section.

Genus. COCCUS.—Male winged; abdomen with two long filaments at side and shorter appendage in middle. Females gall-like, fixed.

Order. **HETEROPTERA.** — Wings four, front pair largest, partly overlapping. Basal part leathery, rest membranous. Body flattened.

Section. **HYDROCORISA.** — Aquatic. Legs for swimming. Fore-legs short and folding, for prehension. Antennæ very short, hidden in furrows under eyes.

Family. **NOTONECTIDÆ.**—Hind-legs long, fringed, and oar-like. Body boat-shaped. Beak short, sharp, and thick.

Genus. NOTONECTA.—Body oval, cylindrical. Front tarsi three-jointed.

Family. **NEPIDÆ.**—Front legs strongly raptorial; tibia united to tarsus. Body very flat.

Genus. NEPA.—Body broad. Two bristles at end of abdomen.

Section. **AUROCORISA.**—Antennæ long, not concealed. Legs for running, not for swimming.

Family. **HYDROMETRIDÆ.**—Body long, narrow. Antennæ four-jointed.

Genus. GERRIS.—Basal joint of antennæ largest. Four hind-legs very long, two fore-legs distant from hind limbs.

Genus. DIACTOR.—Joints of antennæ of equal length. Hind pair of legs with large flattened plates. Brazil.

Family. **REDUVIIDÆ.**—Beak short, thick, and naked. Head narrowed behind into neck. Ocelli two. Tarsi three-jointed.

Genus. ARILUS.—Prothorax elevated and notched like a cog-wheel. West Indies.

Genus. HAMMATOCERUS.—Second joint of antennæ composed of great number of minute joints.

Genus. DALADER.—Abdomen flat, wide. Third joint of antennæ flattened and leaf-like. Thorax expanded. India.

Family. **CIMICIDÆ.**—Wingless. Antennæ four-jointed. Abdomen very flat and rounded. Tarsi three-jointed.

Genus. CIMEX.—As Family.

Family. **TINGIDÆ.**—Small. Body broad and flat. Beak three-jointed, received into groove on under side of head.

Family. **CAPSIDÆ.**—Body convex, soft; second joint of antennæ often thickened at tip. Females with long, sabre-shaped ovipositor.

Genus. MIRIS.—Long, narrow. Antennæ very long, second joint largest, first joint thickest. No ocelli.

Family. **LYGÆIDÆ.**—Small. Antennæ four-jointed. Tarsi three-jointed, with two cushions between claws.

Family. **COREIDÆ.**—Large. Antennæ four-jointed, end joint large, thick, or long. Ocelli near each other. Beak four-jointed, moderate.

Family. **SCUTELLARIDÆ.**—Beak four-jointed, long. Antennæ often five-jointed, last joint like others. Scutellum large.

Genus. EDESSA.—Thorax very wide, produced into a long sharp point on each side.

Genus. SCUTELLARIA.—Scutellum very long, reaching almost to extremity of abdomen.

Order. **APHANIPTERA.**—Wings four, very minute, like scaly plates on sides of thorax. Antennæ minute. Mouth for suction; tarsi five-jointed.

Family. **PULICIDÆ.**—As Order.

Genus. PULEX.—Labial palpi jointed.

Order. **DIPTERA.** — Wings two, membranous, not folded. A pair of winglets at base, second pair of wings replaced by balancers. Mouth for sucking. Tarsi five-jointed.

Section. **CEPHALOTA.**—Head distinct from thorax. Claws of tarsi not toothed.

Stirps. *NEMOCERA.*—Antennæ more than six joints; palpi with four or five joints.

Family. **CULICIDÆ.**—Very long beak, rather thick at tip, composed of seven pieces. Head small. Antennæ plumed in male.

Genus. CULEX.—Palpi of male larger than thorax, very short in female.

Family. **TIPULIDÆ.** — Proboscis very short. Palpi larger than proboscis, four-jointed.

Genus. TIPULA.—Antennæ larger than head, not plumed, thirteen-jointed. Last joint of palpi moderately long. Seven cells at end of wings.

Section. **BRACHOCERA.** — Antennæ very short. Palpi with one or two joints.

Stirps. *NOTACANTHA.*—Antennæ apparently three-jointed, last joint articulated.

Family. **TABINIDÆ.**—Mouth with six lancets in female, four in male. Maxillary palpi two-jointed. Eyes very large.

Genus. TABANUS.—Antennæ seven-jointed, not longer than head.

Genus. GLOSSINA. — Mouth with exserted proboscis, furnished with poison-gland at base. Thorax much longer than wide. South Africa.

Stirps. *ATHERICERA.*—Antennæ two or three joints, bristle at tip.

Family. **SYRPHIDÆ.**—Proboscis with four internal pieces.

Genus. ERISTALIS.—Body stout and hairy; fourth longitudinal nervure of wing much curved before tip. Eyes of male touching each other. Bristles of antennæ naked or haired.

Family. **MUSCIDÆ.** — Proboscis enclosing two pieces, short, thick, and membranous. Antennæ three-jointed, third largest. Nervures of wings few.

Family. **ŒSTRIDÆ.**—No visible mouth, or only minute tubercles. Antennæ very short.

Genus. ŒSTRUS.—Nervures not extending to hinder edge of wings.

Stirps. *PUPIPARA.*—Head sunk in thorax. Claws toothed.

Family. **HIPPOBOSCIDÆ.**—Head distinct. Legs short, stout, fifth joint of tarsi largest.

Genus. HIPPOBOSCA.—Wings ample. Nervures not extending to tip.

CLASS.—CRUSTACEA.

Ringed animals, with horny or calcareous exterior; limbs articulated—Respiration by gills or skin; sexes separate.

Sub-class. **PODOPTHALMATA.**—Eyes on moveable footstalks.

Order. **DECAPODA.**—Walking legs ten. Mouth-limbs twelve. Gills on side of carapace.

Sub-order. BRACHYURA.—Abdomen very small, folded under the body.

Tribe. *OXYRHYNCHITA.*—Carapace narrow in front, and forming a beak. Mouth apparatus squared.

Family. **MACROPODIADÆ.**—Legs long and slender, second or third pair larger than first.

Genus. STENORHYNCHUS. — Carapace triangular, prolonged into two-forked beak. Eyes not retractile. First joint of outer antennæ narrow, third larger than second. Abdomen six joints. Many seas.

Genus. LEPTOPODIA.—Claw-legs well developed, short. Antennæ feathered. West Indies and warm seas.

Genus. CAMPOSCIA.—Carapace smooth; claw-legs small; limbs with stout bristles. Antennæ short and partly feathered. Mauritius.

Genus. DOCLEA.—Beak small; claw-limbs large and powerful; carapace with many long and stout spines. Hotter seas.

Family. **MAIADÆ.**—Legs moderately long, first pair often larger than following; base of outer antennæ large.

Genus. LISSA.—Beak wide, notch shallow. Claw-legs moderate. Carapace and limbs noduled. Warm and hot seas.

Genus. CHORINUS.—Body large and rounded; limbs rather short; beak long. Limbs, and sometimes body, noduled or spined. West Indies.

Genus. ARCTOPSIS.—Carapace triangular, front with four horns, the two central forming a beak. Eyes on short footstalks. Abdomen seven joints. Many seas.

Genus. HYAS.—First joint at base of outer antennæ flattened and widened on outer side. Carapace wide and flat. Many seas.

Genus. MAIA.—Moveable stem of outer antennæ within corner of orbit. Carapace oval, beak stout. First pair of legs scarcely larger than others. Abdomen with seven joints. Most seas.

Genus. MICIPPA.—First joint of outer antennæ bent and widened at upper end. Carapace truncated in front, and terminated by a toothed bill. Philippines.

Genus. PERICERA.—Lateral antennæ inserted under muzzle. Seven segments in abdomen. West Indian seas.

Genus. CRIOCARCINUS. — Beak bifid, very long, and diverging. Eyes on long footstalks. Claw-limbs large; carapace spinose. New Hebrides.

Genus. ACANTHONYX.—Beak short, deeply cleft; claw-limbs small, all the toes sharply hooked. Six segments in abdomen of male. Mediterranean.

Genus. HUENIA.—Lower edge of forehead sharp. Hinder feet slightly keeled on front edge. Japan.

Family. **PARTHENOPIDÆ.**—First pair of legs much larger than others, large, and stout. Carapace triangular.

Genus. PARTHENOPE.—First pair of feet elbowed at claws. Beak short. Abdomen with seven joints. Warm seas.

Genus. LAMBRUS.—Abdomen of male with five joints; rest like Parthenope. Warm and hot seas.

Genus. CRYPTOPODIA.—Carapace wide; limbs capable of concealment under shell. Japan.

Genus. EURYNOME.—Beak with two triangular horns. Abdomen with seven joints. Eyes retractile. Carapace rough. First joint of outer antennæ ending at inner corner of orbit. Many seas.

Tribe. *CYCLOMETOPITA.*—Carapace wide, bow-like in front, narrower behind.

Family. **CANCERIDÆ.** — Carapace wider than long. Legs not adapted for swimming; hind-legs with pointed and claw-like toes.

Genus. CANCER.—Sides of carapace in front, many-lobed. Inner antennæ folded longitudinally. Many seas.

Genus. ÆTHRA.—Carapace much flattened; claw-feet incapable of being raised perpendicularly. Mauritius.

Genus. XANTHO.—Back rather flat. Abdomen with

seven joints in female, five in male. Inner antennæ transversely folded.

Genus. PERIMELA.—Carapace about as long as broad. Abdomen as in last genus. Outer jaw-feet with fourth joint inserted into third of its inner edge. Many seas.

Genus. PILUMNUS.—Carapace convex and hairy. First joint of outer antennæ small; four hinder pairs of legs hairy. Many seas.

Genus ÆGLE (including ZOZYMUS). — Carapace wide, toothed in front. Claw-limbs strong; legs broad and rather flattened. Indian seas.

Genus. CARPILIUS.—Carapace round, smooth, sparely toothed in front, wider than long. Claws very large and powerful; legs moderate. Mauritius.

Genus. GALENE. — Cephalothorax slightly bowed in middle; right pincers largest in both sexes; front legs squared. Hot seas.

Family. **PORTUNIDÆ.**—Last pair of feet flattened, the toe developed into a thin oval plate.

Genus. CARCINUS. — Carapace nearly as long as wide, front projecting. Tarsi of hinder legs narrow. Many seas.

Genus. PORTUNUS.—Carapace broader than long. Abdomen of male triangular. Second to fourth pairs of legs with long, narrow-grooved tarsus. Many seas.

Genus. NEPTUNUS.—Carapace much broader than long, sharply toothed in front, developed into spines on each side. Tarsi of hind-legs very wide. Warm seas.

Genus. LUPA. — Carapace wide and pointed at sides. Claw-legs very long, claw slender. Hinder feet flat and broad. West Indies.

Genus. POLYBIUS.—Carapace nearly circular. All legs flattened, tarsi of fifth pair oval. Many seas.

Genus. PODOPTHALMUS. — Carapace flat, very wide. Claw-feet large and long. Footstalks of eyes enormously long and moveable, resting in groove when not raised. Claws short. Mauritius and Philippines.

Tribe. *CATOMETOPITA.*—Carapace nearly square or oval, front nearly straight. Eyes on long footstalks.

Family. **THELPHUSIDÆ.**—Carapace more or less oval. Eye-footstalks short; fourth joint of jaw-feet not inserted into external angle of preceding.

Genus. THELPHUSA.—Carapace flat, smooth, broad, and heart-shaped. External antennæ very short, placed near footstalks of eyes. Warm seas.

Family. **GECARCINIDÆ.**—Carapace wide, thin, long, or rounded. Eye-footstalks long; fourth joint of jaw-feet inserted into external angle of preceding.

Genus. GECARCINUS.—Carapace heart-shaped, broad in front, stout, blunt behind. Footstalks of eyes short, lying in rounded furrows. South America.

Family. **PINNOTHERIDÆ.**—Carapace circular, front mostly narrow. Eye-footstalks very short; fourth joint of jaw-feet springing from top or outer angle of third joint.

Genus. PINNOTHERES.- Inner antennæ concealed under front of carapace. Body circular, rounded above. Abdomen of male very small, of female large and swollen. Many seas.

Family. **MYCTIRIDÆ.**—Carapace rather triangular. Body lamellated. Beak small, under which the eyes lie concealed. Sternal plate wider than long.

Genus. MYCTIRIS.—Carapace delicate, convex, rather circular or narrower in front. Limbs long and slender. Claw-feet long. Warm seas.

Family. **OCYPODIDÆ.**—Carapace rhomboidal, elevated in front. Eye-footstalks long. Claws very unequal in size.

Genus. OCYPODE.—Eyes occupying greater part of footstalks. Abdomen of male very narrow, last joint triangular; that of female is oval. West Indies.

Genus. GELASIMUS.—Last segment of abdomen in male semicircular, in female nearly round. Hot seas.

Family. **GONOPLACIDÆ.**—Carapace rhomboidal, front margin and sides forming continuous lines. Front very wide. Outer antennæ lodged below front and horizontal. Eye-footstalks very long.

Genus. GONOPLAX.—Carapace very wide in front. Eye-footstalks inserted near middle of front, very long, slender, reaching to front angles. Claws in males long and cylindrical. Abdomen of both sexes with seven joints. Temperate seas.

Family. **GRAPSIDÆ.**—Body compressed. Eye-footstalks short, thick, the eye occupying half their length. Front limbs short.

Genus. GRAPSUS.—Carapace not narrower in front than behind. Middle antennæ lodged under front. Hot seas.

Genus. PLANES.—Carapace longer than wide, front projecting. Tarsi thick and spined. Temperate and warm seas.

Tribe. OXYSTOMATA. — Outer jaw-feet when closed forming a triangular point. Outer antennæ very small.

Family. CALAPPIDÆ.—Last four pairs of feet concealed under carapace, first pair folded and coinciding with the general outline.

Genus. CALAPPA.—Antennæ like those of Cancer. Claws very large, compressed, and notched at upper edge, folded so as to cover mouth. Warm and hot seas.

Genus. THEALIA.—Carapace rounded, with projecting spikes at sides and back. Legs rather long. China.

Genus. CAMARA.—Carapace round, smooth, and wide, simple at edges. Claw-feet very flat. Mauritius.

Family. LEUCOSIIDÆ. — Outer jaw-feet when closed sharp-pointed and triangular in front. Carapace more or less circular. No branchial openings before front-legs.

Genus. LEUCOSIA.—Internal antennæ in little pits. Eyes small. Shell solid. Hot seas.

Genus. MYRA. — Carapace rather egg-shaped. Front limbs long. Indian seas.

Genus. IXA.—Carapace wider than long, sides prolonged into cylinder or cone. Philippines.

Genus. IPHIS.—Carapace rhomboidal. Conical spurry points on each side. Philippines.

Genus. EBALIA.—Carapace rhomboidal, knobbed. Abdomen seven-jointed, in male narrow, in female broad; last joint very small. Temperate and warm seas.

Genus. NURSIA.—Carapace rhomboidal, sides angulated or scooped. Indian Ocean.

Family. CORYSTIDÆ. — Outer antennæ very large. Sternal plate long and narrow.

Genus. CORYSTES.—Carapace longer than wide, with a few teeth, front forming a beak. Outer antennæ longer than carapace, with row of long hairs on upper and lower edges. Fore-legs of male very long. Temperate and warm seas.

Genus. THIA.—Carapace smooth, heart-shaped, narrow behind, front projecting. All legs small. Temperate and warm seas.

Family. DORIPPIDÆ.—Fourth and fifth pairs of legs elevated on back, not furnished with paddles. Eyes on simple footstalks.

Genus. DORIPPE.—Claws small, short, and equal ; third pair of feet largest, two last pairs having a hooked nail at end.

Family. CAPHYRIDÆ.—Carapace squared, rather convex, slightly wider than long. Eyes retractile. Sternal plate rounded.

Genus. CAPHYRA.—Carapace smooth. Pincers with spines on back. Front projecting. Philippines.

Sub-order. ANOMOURA.—Last, or two last, pairs of legs not used for walking, and are set above level of other legs.

Family. HOMOLIDÆ.—Carapace beaked and spinose. Inner antennæ long. Fifth pair of legs not used for walking.

Genus. HOMOLA.—A series of sharp spines set radiately on the carapace. Eye-footstalks short, eyes very large and round. Mediterranean.

Genus. LITHODES.—Carapace thickly spined. Beak very long. Fifth joint of outer jaw-feet oblong. Hind-legs received into branchial cavities. Warm seas.

Genus. ECHIDNOCERUS. — Carapace and limbs thickly noduled. Antennæ feathered. Columbia River.

Family. DROMIADÆ.—Body rather globular ; front of carapace bent down. Eyes short and looped in their orbits ; inner antennæ very short and lodged in groove.

Genus. DROMIA.—Carapace thickly covered with short hairs. Two hind pairs of legs small, elevated, ending in double claw. Many seas.

Family. RANINIDÆ.—Carapace large and pear-like, the broader end in front. Abdomen small, but partly projecting. All feet except first pair fin-like. Indian seas.

Genus. RANINA.—As Family.

Family. HIPPIDÆ.—All upper segments solid. Last pair of legs slender and folded. First segments of abdomen short and broad, last long and triangular. Antennæ plumed. Hot seas.

Genus. ALBUNEA.—Claws triangular, compressed. Last joint of next pairs of feet hooked. Carapace flat, squared, but rounded behind ; toothed in front.

Genus. REMIPES.—Claws rather long, conical, and feeble. Antennæ close together. Last joint of second and third pairs of feet flat and oar-like. Last segment of abdomen long and paddle-shaped.

Genus. HIPPA.—Claws egg-shaped and compressed, not fingered. Eye-footstalks long and thread-like. Last joint of second and third pairs of feet triangular, of fourth pair crescentic.

Family. PAGURIDÆ.—Abdomen soft, and nearly membranous. Two hind pairs of legs very short.

Genus. PAGURUS.—Abdomen turned on itself, a pair of unsymmetrical appendages at tip. Inner antennæ short. Most seas.

Genus. CENOBITA. — Antennæ advanced ; middle pair nearly as long as outer. Thorax egg-shaped, conical, long, narrow, and greatly compressed. Indian Ocean.

Genus. BIRGUS. — Abdomen rather solid, having two rows of flattened appendages beneath. Fourth pair of feet about half as large as third pair. Thorax heart-shaped. Indian Ocean.

Family. PORCELLANIDÆ.—Tail ending in swimming-plates. Fifth pair of legs small and folded. Sternal plate wide and circular.

Genus. PORCELLANA.—Carapace wide as long, flattened above. Outer antennæ very long ; outer jaw-feet larger. Claw-legs large and flat. Abdomen wide ; seven segments ; five swimming-plates.

Sub-order. MACROURA.—Abdomen long, plated, composed of seven segments and with five swimming-plates at tip. First five segments mostly with a pair of false feet tipped with flat fringed plates.

Family. GALATHEIDÆ.—No moveable plates to outer antennæ. Body flat. Fifth pair of legs slender, not used for walking, and folded. Carapace rather wide and flat, longer than broad, and beaked. Claws rather large. Abdomen larger than carapace.

Genus. GALATHEA.—Carapace covered with hairy grooves ; beak extended ; eyes large ; no orbit. Abdomen extended. Many seas.

Family. SCYLLARIDÆ.—Carapace wide. No moveable plate to outer antennæ. Fifth pair of legs not folded. All legs with one claw. Outer antennæ large and flat ; abdomen very wide ; swimming-plates partly flexible. In male, first pair of false feet very large, with two wide plates ; rest with one plate. Many seas.

Genus. SCYLLARUS.—Carapace larger than wide ; sides parallel. Many seas.

Genus. IBACUS.—Carapace wider than long, deeply cleft on either side, notched all round. Abdomen with ring-plates, flat, pointed, and wide. Indian seas.

Family. PALINURIDÆ. — Body cylindrical. Outer antennæ thick and long ; basal joint very large. Feet with single toe. Sternal plate wide.

Genus. PALINURUS.—Carapace spiny, with three horns in front. Many seas.

Family. THALASSINIDÆ.—Abdomen long, not very solid ; carapace small and compressed. First pair of legs large. Sternal plate long and narrow.

Genus. CALLIANASSA.—Carapace very small, not beaked. Eye-footstalks flat. Third pair of legs wide at end ; second pair with two toes. One claw much larger than the other. Many seas.

Genus. AXIUS.—Outer antennæ long ; small moveable spine on footstalk ; inner antennæ with two long bristles. Claws compressed, unequal ; last joint of abdomen long and triangular. Many seas.

Genus. GEBIA.—Beak triangular. Outer jaw-feet like walking feet. Abdomen wider in middle than at base ; first ring with a pair of little appendages. Many seas.

Genus. CALOCARIS.—Carapace large, beaked. Large triangular scale at base of outer antennæ. Claws long, flat, and slender. Eyes very small, without any pigment on cornea. Many seas.

Family. ASTACIDÆ. — Outer antennæ with moveable plate at base. Beak moderate, just hiding base of eye-footstalks. Claw-feet large ; second and third pairs of feet two-clawed ; fourth and fifth one-clawed. Side of each abdominal ring extended. Cross-joint in each swimming-plate.

Genus. HOMARUS.—Beak narrow, with several teeth. Eyes spherical. Basal plate of outer antennæ tooth-like. Many seas.

Genus. ASTACUS.—Beak wide at base, depressed ; one tooth on each side. Fresh running waters.

Tribe. CARIDITA.—Abdomen very large, horny. Outer antennæ with very large plate at base. Base of false feet covered by projections of segments.

Family. CRANGONIDÆ.—Both pairs of antennæ on same line. First pair of legs with small pincers.

Genus. CRANGON.—Carapace flattened ; claw also flattened. Second and third pairs of legs very slender ; fourth and fifth strongest. Many seas.

Family. ALPHEIDÆ.—Antennæ in two rows, nine above. Beak small and flat ; three last pairs with one claw.

Genus. ALPHEUS.—Carapace projecting like a hood ; beak small or absent. Claw-legs large ; unequal. First and second pairs of legs two-toed ; third, fourth, and fifth, one-toed. Many seas.

Genus. ATHANAS.—Beak small, not toothed. Three filaments on inner antennæ. First pair of legs large, powerful, and unequal ; pincers thick. Eyes not prominent, but projecting beyond carapace. Many seas.

Family. PALÆMONIDÆ.—Beak long, compressed, and toothed. Antennæ in two rows. Legs stout ; third, fourth, and fifth pairs one-toed.

Genus. PANDALUS.—Outer antennæ long, inner ending in two many-jointed filaments. First pair of legs one-toed. Beak turned up at tip, doubly toothed. Second pair of legs two-toed. Many seas.

Genus. PALÆMON.—Inner antennæ with three filaments. Second pair of legs strongest. Eyes large. Many seas.

Genus. HIPPOLYTE.—Beak large and compressed. Inner antennæ tipped with two many-jointed filaments. Many seas.

Family. PENÆIDÆ.—Antennæ in two rows, lower or both pairs long. Beak long and compressed ; abdomen also long and compressed.

Genus. PENÆUS.—Inner antennæ short ; first joint of upper antennæ very large and excavated above, forming a socket for the eyes. Carapace crested along the middle ; a groove on each side of the crest. Many seas.

Genus. PASIPHÆA.—Carapace narrow in front, beak short, not toothed. Basal joint of inner antennæ slender and with two many-jointed threads. Abdomen long ; false legs of first ring with a single plate, rest with two plates. Many seas.

Order. **STOMAPODA.** — Gills external, formed of cylinders. Seven or eight pairs of legs, mostly near the mouth.

Family. **MYSIDÆ.**—Carapace with folded edges, thorax thick and compressed at sides. Abdomen large ; six to eight pairs of legs on thorax, each with a large palpus, looking much like another leg.

Genus.—MYSIS.—No gills on thorax. False feet of abdomen very small, and without gill appendages. In females, branchial appendages to last two pairs of feet very large. Many seas.

Family. PHYLLOSOMIDÆ.—Carapace flat, membranous, transparent, formed of two bucklers. Feet small, mostly slender. No scales at base of outer antennæ. Abdomen small. One genus. Atlantic and Eastern seas.

Family. ERICHTHIDÆ.—Carapace membranous, transparent, folded at sides, prolonged in front into a point. Eyes thick and globular.

Genus. ERICHTHUS.—First joint of eye-footstalks shorter than second. Hinder angles of carapace sharply pointed.

Family. SQUILLIDÆ.—Body long. Carapace three-lobed. First pair of thoracic legs very large, last joint folding upon its predecessor. Second, third, and fourth pairs much smaller and furnished with pincers ; next three pairs slender and nearly cylindrical.

Genus. SQUILLA.—Last three pairs of legs with long pointed appendages, first joint of first pair of legs flat and deeply toothed on the inner edge. Many seas.

Sub-Class. EDRIOPHTHALMATA.—Eyes sessile ; body in three portions. Respiration through modified limbs.

Order. **AMPHIPODA.**—Head formed of a single segment, mandibles mostly with a palpus. Gill-vesicles under thorax ; first five pairs of abdominal limbs differ in form and used for locomotion.

Family. **ORCHESTIDÆ.**—Formed for leaping ; body

compressed, no palpus on mandibles. Two short appendages to tail. Upper antennæ shorter than basal joints of lower antennæ.

Genus. TALITRUS.—First pair of legs simple, no pincers on second pair. Lower antennæ long. Most seas.

Genus. ORCHESTIA.—First and second pairs of feet with slight pincers. Jaw-feet blunt at tip. First pair of legs smaller than second. Many seas.

Family. GAMMARIDÆ.—Formed for swimming, palpus on mandibles. Antennæ tipped with lash-like appendages. Legs long.

Genus. KROYERA.—Shell of head projecting like a hood and pointed. Jaw-feet with well-developed pincers. Many seas.

Genus. SULCATOR.—Upper antennæ half length of lower, forked, with second joint flattened. Second and third pairs of legs two-clawed. Most seas.

Genus. GAMMARUS.—Upper antennæ slender, small bristles at base of fourth joint. Tufts of spines on tail. First and second pairs of legs with slight pincers. Most seas.

Family. COROPHIDÆ.—Formed for walking. Body rather flat, abdomen straight, all segments free. Palpus on mandibles. Antennæ-like legs.

Genus. CERAPUS.—No lash on antennæ. Two of the toes with two joints ; animal inhabiting a tube. Many seas.

Genus. COROPHIUM.—Lower antennæ very large and without lash, first pair of legs with a lash. Body long, cylindrical, rather compressed. Many seas.

Family. CHELURIDÆ.—Two or three segments of abdomen united. Antennæ short and leg-like. Six points on tail.

Genus. CHELURA.—Upper antennæ shorter than lower, and with an appendage. Fourth to sixth segments fused together. Many seas.

Tribe. HYPERITA.—Head very large ; mandibles also large. First pair of jaws of three joints ; thorax six or seven-jointed. Some legs prehensile ; end of abdomen for swimming.

Family. PHRONIMADÆ.—Second pair of antennæ stiff and bristle-like ; some of true legs prehensile.

Genus. PHRONIMA.—Head vertical ; tail ending in thread-like appendages. Body soft, semi-transparent. Fifth pair of legs largest, and with pincers. Many seas.

Genus. METŒCUS.—Two first pairs of legs shortest, and with small pincers. Many seas.

Order. **LÆMODIPODA.**—Abdomen very small, hardly visible. Body cylindrical or depressed. Head very small.

Family. CYAMIDÆ.—Body depressed, oval. Eyes compound. Antennæ close at base. Five pairs of legs, all prehensile. Respiratory apparatus on second and third joints of thorax.

Genus. CYAMUS.—Head small, fused into first segment of thorax. Many seas.

Order. **ISOPODA.**—Body depressed, abdomen large. Mostly fourteen legs, and in the females the greater number have a large horny plate. Several segments of abdomen fused together.

Family. IDOTEIDÆ.—Appendages of last false legs very large and flat, covering all under side of abdomen. First segment very large and shield-like.

Genus. ARCTURUS.—Body long. Three last pairs of legs used for walking ; four first pairs flat at tips. Antennæ leg-like. Northern Seas.

Family. ASELLIDÆ.—Body long. First pair of antennæ very small. Abdomen jointed, last joint large and shield-like ; tip of abdomen with pointed appendages.

Genus. LIMNORIA.—Body capable of being curled up. Eyes distinct. Antennæ in pairs, one pair over the other. Tail of many segments. Many seas.

Genus. ASELLUS. Front legs largest. Body oblong, flattened, nine-jointed ; one joint for the head, seven for the thorax, and one for the abdomen. Tail appendages long. Fresh waters.

Family. ONISCIDÆ.—One pair of antennæ. Abdomen of six segments, last rather small. Legs slender.

Genus. LIGIA.—Outer antennæ with last joint compound. Four appendages to tail, two on each side.

Genus. ONISCUS.—Antennæ eight-jointed, base hidden by projecting front of head. Tail appendages four-jointed.

Family. PORCELLIO.—Antennæ with seven joints

Family. **ARMADILLIDÆ.**—Oval, capable of rolling up into a ball. Appendages of abdomen not beyond edge.

Genus. ARMADILLO.—Body convex and blunt. Outer antennæ seven-jointed. Last false feet with large basal joint.

Family. **CYMOTHOADÆ.**—Thorax seven-ringed, free. First five rings of abdomen mostly free. First three pairs of legs (or all) strongly clawed.

Genus. CYMOTHOA. — Antennæ nearly equal. Eyes scarcely visible. Last segment of tail in a transverse square. Northern seas.

Tribe. EPICARDITA. — Males permanently minute, females very large. Antennæ rudimentary. Abdomen slightly developed. Feet very short. Mouth adapted for suction and gnawing. In males, body thirteen or fourteen-jointed; one for head, seven for thorax, rest for abdomen. In females, all segments confused together.

Family. **IONIDÆ.**—Abdominal appendages fringe-like.

Genus. IONE.—Male, body narrow and long; abdomen six-jointed, each joint with a pair of membranous appendages. Two pairs of antennæ. Females, body pear-shaped, flat; appendages branched. Many seas.

Family. **BOPYRIDÆ.**—Abdominal appendages flat and hidden under abdomen.

Genus. BOPYRUS.—Male with very small antennæ, very short legs; abdomen with minute false legs. Female unequally pear-shaped, flat, abdomen wide at base. Five pairs of false legs, each with a triangular plate. Most seas.

Sub-class. ENTOMOSTRACA.—Shell horny or leathery, in one or more pieces. Branchiæ attached to feet or jaws. Feet jointed and fringed.

Legion. Branchiopoda.—Mouth formed for biting; branchiæ numerous, attached to feet. Body naked or bucklered. All feet with branchiæ. Two- or four-jointed antennæ. Eyes one to three.

Order. **PHYLLOPODA.**—Naked or partly bucklered. Feet eleven to sixty pairs, joints leaf-like, mostly respiratory. Two to three eyes.

Family. **APODIDÆ.**—One pair of short antennæ. Two eyes. Sixty pairs of feet. Body composed of many rings, almost wholly covered by shell.

Genus. APUS.—As Family. Ponds.

Family. **BRANCHIOPODIDÆ.**—Body naked; two pairs of antennæ (lower pair prehensile in male); eyes two, on footstalks. Eleven pairs of feet.

Genus. CRIOCEPHALUS. — Abdomen nine-jointed; tail with two plates. Lower antennæ of male cylindrical, with appendages at base. Ponds.

Genus. ARTEMIA.—Abdomen long, six-jointed. Lower antennæ of male flat and without appendages. Salt-pans.

Order. **CLADOCERA.**—Head projecting; body enclosed in double buckler, joined on back. Four to six pairs of feet, their joints cylindrical and leaf-like. Eye one, large. Antennæ two pairs; lower pair large, branched and used as fins. No palpi on mandibles.

Family. **DAPHNIADÆ.**—Upper antennæ small; five or six pairs of feet; enclosed in shell. Intestine straight.

Sub-family. Daphninæ.—Five pairs of feet. Lower antennæ two-branched; one branch three, the other four-jointed.

Genus. DAPHNIA. — Head produced downwards into a beak. Upper antennæ one-jointed, under the beak, small; lower large. Ponds.

Genus. MOINA.—Head round and blunt. Upper antennæ one-jointed, long; lower antennæ thick at base. Ponds.

Sub-family. Sidinæ.—Six pairs of feet. Lower antennæ two-branched; a row of spines on edge of large branch.

Genus. SIDA.—One branch of lower antennæ three- the other four-jointed. Ponds.

Family. **POLYPHEMIDÆ.** — Lower part of carapace empty. Four pairs of feet, not included in carapace. Eye very large. Lower antennæ two-branched; one branch four-the other three-jointed.

Genus. POLYPHEMUS.—Head distinct from body. Abdomen long, projecting from shell. Eye occupying nearly whole head. Ponds and ditches.

Family. **LYNCEIDÆ.**—Two pairs of antennæ; upper short, lower branched; each branch three-jointed. Five pairs of legs. One eye; abdomen jointed.

Genus. CHYDORUS.—Nearly spherical. Beak long, sharp, depressed. Lower antennæ very short. Ponds and ditches.

Legion. Lophyropoda.—Mouth formed for biting. Branchiæ few, attached to mouth. Body either with single or double shield. Not more than five pairs of feet; joints cylindrical. One eye. Two pairs of antennæ; one pair for locomotion.

Order. **OSTRACODA.**—Body in two-valved buckler; branchiæ on hind-jaw. No external ovary. Two or three pairs of feet.

Family. **CYPRIDÆ.**—Antennæ two pairs; upper with a pencil of long filaments, lower thick and foot-like. One eye. Two pairs of feet.

Genus. CYPRIS. — Lower antennæ with pencil of long hairs. Ponds.

Family. **CYTHERIDÆ.**—Antennæ two pairs; upper not with pencil of filaments. Three pairs of feet.

Genus. CYTHERE.—Feet not enclosed in shell. Abdomen short. Eye one. Valves in front and behind. Marine pools.

Genus. CYTHEREIS. — Valves of shell oblong; edges parallel above and below, and toothed or notched. Surface ridged and knobbed. In sand.

Family. **CYPRIDINADÆ.**—Antennæ two pairs, foot-like. Feet two pairs, one always within shell. Eyes two, on footstalks. Abdomen tipped with broad plate.

Genus. CYPRIDINA.—As Family.

Order. **COPEPODA.**—Body divided into wings; head and thorax bucklered. Mouth with foot-jaws. Feet five pairs, mostly for swimming. Ovary external.

Family. **CYCLOPIDÆ.**—Head and body fused together. Foot-jaws two pairs; fifth pair of legs very small. One eye.

Genus. CYCLOPS. — Foot-jaws large, strong, branched. Ovaries double. Ponds.

Genus. CANTHOCAMPTUS.—Foot-jaws small, not branched. Ovary single. Fresh and salt water.

Family. **CETOCHILIDÆ.** — Head distinguishable, but fused to first ring of thorax. Foot-jaws strong; three pairs. Eyes two.

Genus. CETOCHILUS.—Foot-jaws not branched. Head with two small pointed appendages. Thorax with six, abdomen with four segments. Last pair of legs like others.

Genus. NOTODELPHYS.—Thorax four, abdomen four, segments. Upper antennæ many-jointed; lower prehensile. Foot-jaws four pairs. Ovary large, within walls of body.

Legion. Pœcilopoda.—Mouth formed for sucking. Foot partly for walking or prehension, partly for swimming. Body mostly enclosed in buckler. All parasitical on fishes and other aquatic animals.

Order. **SIPHONOSTOMA.**—Mouth with tube which has pointed mandibles. Thorax ringed, with several pairs of feet. Foot-jaws large.

Tribe. PELTOCEPHALA.—Head shield-shaped, with frontal plates and two flattened antennæ.

Family. **ARGULIDÆ.**—Head like circular shield. Antennæ short, two-jointed. Second pair of foot-jaws modified into suckers.

Genus. ARGULUS.—Oval, flat, transparent, veined; shell notched on either side in front.

Family. **CALIGIDÆ.**—Feet four pairs, with long hairs. Antennæ flat and two-jointed. Second pair of foot-jaws, not sucking discs. Eyes two.

Genus. CALIGUS.—Thorax two joints. Small sucking disc on under surface of frontal plates. Fourth pair of feet one-branched, slender.

Genus. CHALIMUS.—Thorax four joints. No sucking disc on frontal plate.

Family. **CECROPIDÆ.**—Head like small buckler, with frontal plates. One plate-like appendage on upper part of thorax. Ovaries tube-like and twisted.

Genus. CECROPS.—Thoracic appendage small and rounded. All feet in male; first three pairs in female for walking.

Tribe. PACHYCEPHALA.—Head small, thick, and blunt. Antennæ long, five or six-jointed.

Family. **ERGASILIDÆ.**—Body oval or pear-shaped. Feet very small and branched.

Genus. NICOTHOE.—Antennæ slender, many-jointed. In female, thorax enlarged into two wing-shaped lobes. Feet four pairs.

Order. **LERNEADA.** — Mouth for suction. Thorax not jointed, all its organs very small. Males unlike females

Tribe. ANCHORASTOMACEÆ. — Females fixed to prey by strong and hooked foot-jaws. Antennæ one pair.

Fami'y. **CHONDRACANTHADÆ.** — Foot appendages large ; foot-jaws three pairs.

Genus. CHONDRACANTHUS.—Body covered with spines.

Genus. LERNENTOMA.—Egg-tubes long, sometimes stout, sometimes slender and twisting.

Tribe. ANCHORACARPACEÆ.—Two long append-ages from thorax, ending in sucking disc or knob.

Family. **LERNÆOPODADÆ.**—Appendages united at tip.

Genus. ACTHERES.—Body rounded ; head large ; egg-tubes pear-shaped.

Genus. LERNÆOPODA.—In female, body long and oval ; head short and thick. Foot-jaws two pairs. Male divided into two nearly equal portions.

Family. **ANCHORELLADÆ.**—Arm-shaped appendages short and united from base.

Genus. ANCHORELLA.—Female, head at end of very long neck. Foot-jaws three pairs ; first and second large, third very small. Egg-tubes moderate. Male very small.

Tribe. ANCHORACERACEÆ.—Females attached to prey by head, which is held by horns from hinder part.

Family. **PENELLADÆ.**—Several pairs of small feet on under surface of body, near head.

Genus. LERNÆONEMA.—Body long, slender, narrowed in front and ending in swollen head, with two or three curved appendages. Abdomen long ; ovaries long and slender.

Genus. PENELLA.—Body thick ; neck narrow ; egg-tubes thread-like.

Family. **LERNÆOCERADÆ.** — No feet or appendages on under side of body.

Genus. LERNÆOCERA.—Body long and straight. Head with horn-shaped appendages, simple, symmetrical. Ovaries straight.

Genus. LERNÆA.—Body twisted. Horn-shaped append-ages branched. Ovaries twisted into round masses.

Order. **PYCNOGONIDES.**—Trunk of four segments, a tubular joint at each end, that in front constituting the mouth. Feet of male four pairs, of female five pairs ; the last pair adapted for bearing the eggs. Marine.

Genus. PHOXICHILIDIUM.—Feet very long ; two antennæ. Pincers.

Gnus. PYCNOGONUM.—Feet hardly longer than body. No pincers.

Order. **XIPHOSURA.**—Shell thick, two-jointed ; the junction transverse. Thoracic feet round mouth, abdominal feet bearing branchiæ. A long sharp appendage to hinder shield, jointed to shield and covered with very thick skin.

Genus. LIMULUS.—Front shield rounded in front ; hinder angles sharp and pointed. Hinder shield irregularly hexa-gonal, with fixed teeth and moveable spines alternating on the sides. Eyes two, sessile, compound, on front shield. Hot seas.

CLASS.—ARACHNIDA.

Animals breathing atmospheric air—Antennæ none—Legs four pairs, attached to foreparts of body.

Order. **ARANEIDEA.**—Not more than four breathing holes. Abdomen covered with skin. Head and thorax united. Palpi male swollen at last joint.

Tribe. OCTONOCULINA.—Eyes eight.

Family. **MYGALIDÆ.**—Spinnerets four. Breathing apertures four. Falces bent downwards, curved.

Genus. MYGALE.—Eyes (:. .:). Feet hairy, nearly equal, first and fourth pairs largest. Exotic.

Genus. ATYPUS.—Eyes on a "watch-tower" in front of cephalothorax. Fourth pair of legs longest.

Genus. CTENIZA.—Mandible with a row of horny barbs on outer side, near claw. Jamaica.

*Family.—***LYCOSIDÆ.**—Spinnerets three pairs. Falces jointed vertically.

Genus. LYCOSA.—Eyes (:. .:). Third pair of legs shortest.

Genus. DOLOMEDES.—Eyes (:. .:). Fourth pair of legs usually longest, third pair shortest.

Family. **SALTICIDÆ.**—Three pairs of spinnerets. Falces jointed vertically on an inclined plane. Two breathing aper-tures in front of abdomen and on its lower surface.

Genus. SALTICUS. — Eyes (.**.). Legs stout, variable in length.

Genus. MYRMARACHNA.—Head nearly distinct. Skin hard. Abdomen swollen in middle. Bengal.

Genus. ERESUS.—Four larger eyes in a square (. ..). Legs moderate, nearly equal in length.

Family. **THOMISIDÆ.**—Three pairs of spinnerets. Tarsi mostly two-clawed.

Genus. THOMITSUS.—Eyes in two rows, one in crescent on part of a circle, with the convexity in front, nearly same size. First and second pairs of legs longest. Lip blunt at tip.

Genus. PHILODROMUS. - - Eyes in two transverse rows. Legs long, variable. Lip triangular or oval.

Genus. SPARASSUS.—Eyes in two transverse rows, second longer than first. Lip semicircular or oval.

Family. **DRASSIDÆ.**—Three pairs of spinnerets. Tarsi with two claws and numerous papillæ, or three claws and no papilla.

Genus. DRASSUS.—Eyes in two transverse rows, alike in size. Lip long, oval, rounded at tip.

Genus. CLUBIONA.—Eyes as before, slightly different in size. Lip rather oval or squared.

Genus. CLOTHO.—Two upper spinnerets long. Falces small and not toothed. Tongue triangular. North Africa.

Genus. ARGYRONRETA. — Eyes (. . . .). Inter-mediate of front row smallest. Lip long, tri-angular, wide at base, rounded at point. First pair of legs longest.

Family. **CINIFLONIDÆ.** — Falces jointed vertically. Spinnerets four pairs, lowest pair shortest. Three claws on tarsi.

Genus. CINIFLO. — Eyes (.·.·.). Intermediate of both rows sessile, the pairs on sides placed on tubercle. Lip wide in middle. First pair of legs longest, third shortest.

Genus. ERGATIS.—Eyes (. . . .) on a tubercle. Lip large, triangular. First pair of legs longest, then second pair, third shortest.

Family. **AGELENIDÆ.**—Falces jointed vertically on an inclined plane. Spinnerets three pairs.

Genus. AGELENA.—Eyes in two transverse curved rows, with convexity directed backwards. Lip squared or oval. Fourth pair of legs longest.

Genus. TEGENARIA.—Eyes in two transverse rows, front row straight, hinder curved, convexity backwards. Lip squared, notched or hollowed at tip. Legs slender, either first or fourth pair longest.

Genus. CÆLOTES.—Eyes (.·**·.) on a tubercle. Lip curved on sides, blunt at tip. Legs strong, fourth pair longest.

Genus. THERIDION. — Eyes (.·:·.). Sometimes two side pairs close together. First and second pairs of legs longest.

Genus. LINYPHIA.—Eyes, four middle placed in two rows, hinder pair more remote, two side pairs near each other. First pair of legs longest.

Genus. TETRAGNATHA.—Eyes in two nearly parallel straight rows, nearly equal in size. Legs long and slender.

Family. **EPEIRIDÆ.**—Jaw widened from base, forming rounded flattish plate.

Genus. EPEIRA.—Eyes (.·:·.). Skin soft.

Genus. GASTEROCANTHA. — Skin hard and bony, covered with spines, tubercles, and prominences of various shape. Tropical countries.

Genus. OTIOTHOPS.—Hinder pair of eyes united. South America.

Tribe. *BINOCULINA.*—Eyes two.
Genus. NOPS—As tribe. Central America.
Tribe. *SENOCULINA.*—Eyes six.
Genus. DYSDERA. Eyes near each other, in two rows; first row with two eyes, second with four. First pair of legs longest.
Genus. SEGESTRIUM.—Eyes in two rows; front row with four eyes, hinder with two. Europe.
Genus. SCYTODES. — Eyes equal, in pairs, forming a triangle. Africa.

Order. **PHALANGITA.**—Cephalothorax united with abdomen, which is ringed or transversely folded. Palpi thread-like.
Family. PHALANGIDÆ.—Legs long, feet with one claw.
Genus. PHALANGIUM.—Tarsi with ten or fifteen joints. Second and last pair of legs longest and nearly equal.

Order. **SOLIFUGÆ.** — Cephalothorax distinct from abdomen, which is ringed. Palpi as long and large as legs.
Genus. SOLPHUGA. — Mandibles double-clawed, lower finger moveable. Palpi longer than first pair of feet. All feet but first pair with two claws. Eyes two, on a tubercle. Hot countries.

Order. **PSEUDOSCORPIONES.** — Cephalothorax united to abdomen. Abdomen ringed. Palpi large, with pincers at tip.
Genus. CHELIFER. — Eyes two. Cephalothorax divided by transverse furrow.
Genus. OBISIUM. — Eyes four. Cephalothorax not divided.

Order. **PEDIPALPI.**—Palpi foot-like, with pincers at tip. Abdomen divided by segments. Breathing orifices eight or four.
Family. PHRYNIDÆ. Slight groove between abdomen and cephalothorax. Palpi spined. Eyes eight. Two front feet long, without pincers.
Genus. PHRYNUS.—Palpi long as body, with horny claws at tip. First pair of legs very long and slender. Body rather flat.
Family. SCORPIONIDÆ.—Abdomen sessile. Palpi with pincers, larger than legs. Six last segments of abdomen narrowed, last segment with curved sting. Two comb-like appendages on lower part of body, behind fourth pair of feet.
Genus. BUTHUS.—Eyes eight, three lateral on each side of cephalothorax.
Genus. SCORPIO.—Eyes six. Two lateral on each side.

Order. **ACARINA.** — Cephalothorax and abdomen united, not divided into segments.
Family. IXODIDÆ. — Body with leathery covering. Beak blunt, toothed at tip, barbed at middle. Last joint of feet two-clawed, with vesicle.
Genus. IXODES.—As Family.
Family. GAMASIDÆ.—Palpi free, thread-like. Mandibles with pincers. Eyes none, or indistinct.
Genus. LEPTUS.—Palpi tolerably defined, also sucker. Body egg-shaped, soft.
Genus. GAMASUS.—Mandibles with pincers and toothed. Labium three-cleft. Leathery shield on back.
Family. ACARIDÆ.—Head terminated in front by a scooped lip, or a single two-cleft projection. Palpi indistinct. Mandibles with pincers. Eyes none distinct.
Genus. ACARUS.—Body divided by a transverse furrow between second and third pairs of feet. Four of the legs close at their bases.
Family. PEDICULIDÆ.—Mouth composed of small retractile beak, attached at base. Tarsi with a single arched claw.
Genus. HÆMATOPINUS. — Antennæ eight-jointed, first joint large. Thorax narrower than abdomen. Abdomen flat, legs thick.
Family. PHILOPTERIDÆ.—Antennæ thread-like, three or five-jointed. Mandibles strong. No maxillary palpi.
Genus. TRICHODECTES.—Antennæ three-jointed. Tarsi one-clawed.
Genus. MENOPON.—Head broad. Antennæ shorter than head, club-shaped. Abdomen with ten segments.

Genus. NIRMUS.—Body narrow, head moderate, temporal edges rounded. Antennæ not branched.
Genus. DOCOPHORUS. — Head large, nearly triangular, broad at base. Abdomen broad oval.
Genus. GONIODES.—Head large, body rounded. Antennæ sickle-shaped; in male forming a claw.

CLASS.—MYRIAPODA.

Head distinct—Thorax and abdomen united—Not less than twelve pairs of feet—No wings.

Order. **CHILOPODA.**—Head wide and prominent; mandibles sharp and curved. A pair of feet on each segment.
Tribe. *SCHIZOTARSIA.* — Antennæ many-jointed, slender, larger than body. Legs long, unequal. Eyes compound and globular.
Family. CERMATIIDÆ. — Eight dorsal shields, each covering two ventral segments.
Genus. CERMATIA.—Eyes prominent. Head transverse. Dorsal shields scooped. Hot countries.
Tribe. *HOLOTARSIA.*—Tarsi mostly three-jointed, rarely twelve-jointed. Head of two segments. Antennæ not larger than body. Eyes aggregate, simple, or wanting.
Family. LITHOBIIDÆ.—Fifteen dorsal shields, squared, with long, sharp angles.
Genus. LITHOBIUS.—Eyes numerous.
Family. SCOLOPENDRIDÆ.—Foot-bearing segments, twenty-one or twenty-three. Hinder feet thickened, first or second joint spined.
Genus. SCOLOPENDRA.—Head heart-shaped, overlapping. Four pairs of eyes.
Genus. ENCORYBAS.—Antennæ nineteen-jointed; twenty-one pairs of feet, last largest, and developed into leaping plates. South Africa.
Family. GEOPHILIDÆ. — Segments nearly equal, numerous, each formed of two unequal segments. Last segment with short, spiny feet. Antennæ fourteen joints.
Genus. ARTHRONOMALUS.—Head squared.
Genus. GONIBREGMATUS.—Head heart-shaped. Philippines.

Order. **CHILOGNATHA.**—Head vertical, rounded. Mandibles thick and strong. Numerous segments, unequal. Feet jointed to abdomen.
Tribe. *PENTAZONIA.* — Body oval, contracted at globe, mostly convex above and flat below. Feet fixed to free moveable plates.
Family. GLOMERIDÆ.—Body smooth. Eyes distinct.
Genus. ZEPHRONIA.—Eyes numerous, aggregated. Antennæ club-shaped, six or seven-jointed. Twenty-one pairs of feet. Tropical countries.
Genus. GLOMERIS.—Eyes eight, in a curved line. Thirteen segments, seventeen pairs of feet.
Tribe. *MONOZONIA.* — Body long and worm-like. First half of each segment cylindric, second half with widened sides. On each segment a double plate carrying two pairs of feet.
Family. POLYXENIDÆ.—Head arched and prominent; body wide. Long tuft at end of abdomen.
Genus. POLYXENUS.—Body short, covered with minute pencils of scales. Thirteen pairs of feet.
Family. POLYDESMIDÆ.—Eyes none, or obscure.
Genus. POLYDEMUS.—Body flattish, rather convex, with horizontal plates at sides.
Tribe. *BIZONIA.*—Body cylindric, no plates on sides. Antennæ club-shaped, seven-jointed. Many segments, each formed of two fused segments, and bearing two pairs of legs.
Family. JULIDÆ.—Feet fixed to immoveable plates.
Genus. JULUS.—Head convex. Antennæ long. Sides of prothorax triangular.
Genus. SPIROSTREPTES.—Antennæ short, with funnel-shaped joints. Eyes transverse. Sides of prothorax lengthened or widened. Asia.

3.   3 F

CLASS.—ANNELIDA.

Elongated — No jointed feet, often with bristles or tubercles for progression—Respiration by gills, sacs, or skin.

Order. **SETIGERA.**—Body with bristles or bristle-bearing rudiments of feet.

Family. **AMPHITRITÆ.**—Head not distinct. Eyes none. Body mostly in a tube.

Genus. SABELLA.—Mouth transverse, across gills. Gills two, feathery; funnel comb-shaped, spiral, and large. Cylindrical stopper. Front tubercles with hooks and bristles. Tube gelatinous, covered with sand.

Genus. SERPULA.—Gills two, combed, fan-shaped. Tube not erect, calcareous, twisted.

Genus. TEREBELLA.—Upper lip surrounded by long tentacles. Tubercles with double row of hooks. Gills branching. Tube composed of sand and shells.

Family. **NEREIDÆ.**—Body long, slender; head distinct, with tentacles and eyes. Gills like lobes of feet. Tubercles throughout all body. Proboscis large, often with two horny jaws. No tube.

Genus. PHYLLODOCE.—Head small. Eyes two; tentacles four or five, the last very minute. Jaws none; proboscis thick and tentacled. Two bristles at end of body.

Genus. NEREIS.—Eyes four; tentacles four, short. Jaws powerful; proboscis thick and cylindrical.

Family. **EUNICEÆ.**—Body long, many segments. Proboscis with seven to nine horny jaws.

Genus. EUNICE.—Head distinct, tentacles five. Comb-like gills in front of body. Jaws seven.

Family. **ARICIÆ.**—Head scarcely distinct. Body small at each end. Proboscis short; jaws none.

Genus. CIRRHATULUS.—Body long, round, a few bristles on back and belly; cirrhi long. Gills like very long cirrhi in front of body.

Family. **ARENICOLIDÆ.** — Gills tufted in middle of body. Double row of bristles along sides. Head not distinct; eyes and jaws wanting.

Genus. ARENICOLA.—Segments divided by folds, thick in front, smaller behind; no bristles behind last pair of gills. Mouth with retractile proboscis.

Family. **CHÆTOPTERIDÆ.**—Front and hinder tubercles with a bundle of bristles, but without hooks. Those of middle with a large membranous appendage.

Genus. CHÆTOPTERUS.—Long, enclosed in leathery tube.

Family. **APHRODITIDÆ.**—Body oval. Head with tentacles; eyes two or four. Flaps distinct, with spine and bristles.

Genus. APHRODITE. — Head hidden under scales or bristles. Eyes two; tentacles three. Jaws very small or wanting. Body rather flattened.

Family. **LUMBRICIDÆ.**—No gills. Body with bristles, but no tubercles.

Genus. LUMBRICUS.—Nearly cylindric, narrow at both ends, blunt behind. Mouth nearly at tip. Bristles not retractile. Saddle composed of swollen rings in front of body.

Order. **SUCTORIA.**—Body without bristles, with a sucking disc at one or both ends. No external gills.

Genus. HIRUDO.—Body oblong, rather flat. Jaws three, with double row of crowded teeth. Eyes ten.

Genus. HÆMOPIS.—Double row of few teeth. Eyes ten.

Genus. BRANCHIOBDELLA. — Body with large and few rings. Eyes none. Jaws two, horny.

Genus. ALBIONE. — Body narrowed in front, rings numerous. Mouth small, narrowed at bottom of front sucker. Hinder sucker at end of body.

Order. **TURBELLARIA.**—Body cylindrical or flattened, mostly not jointed, or wrinkled throughout, covered with cilia.

Family. **NEMERTIDÆ.**—Body very long, very contractile, indistinctly ringed.

Genus. NEMERTES.—Several eyes; two pits at side of head, surrounded by cilia.

Sub-class. **ENTOZOA.** — Parasitical. Mostly elongated. No special organs of breathing. Found inside living animals.

Order. **STERELMINTHA.**—Intestine absent in some, vascular in others.

Family. **CESTOIDEÆ.**—Body long, flattened, soft. No mouth, head mostly with sucking discs.

Genus. TÆNIA.—Body long, jointed. Head with four sucking discs, mostly with beak surrounded with hooklets.

Genus. BOTHRIOCEPHALUS.—Head swollen, with two or four opposite furrows.

Genus. ECHINOCOCCUS---Vesicles single or enclosed in cryst. Many minute Entozoa in interior surface.

Order. **CÆLELMINTHA.**—Intestine in a distinct cavity.

Tribe. *NEMATOIDEA.*—Body round, elastic.

Family. **STRONGYLIDIDÆ.**—Mouth at or nearly at end of body, not surrounded with retractile hooks. Body round, long, and elastic.

Genus. STRONGYLUS.—Body very long, sharp in front. Mouth round or triangular.

Genus. ASCARIS.—Body sharp at both ends. Head three-valved.

Genus. TRICOCEPHALUS.—Body thread-like in front, suddenly thickening behind.

Genus. FILARIA.—Body very long, thread-like, nearly equal throughout. Mouth rounded.

CLASS.—ECHINODERMATA.

Radiate, skin more or less leathery, having variously-shaped chalky pieces imbedded.

Sub-class. **DITREMATA.**—Digestive canal distinct, with two openings; outline of body entire.

Order. **HOLATHUROIDEA.**—Shell none; no chalky pieces. Body cylindrical, moving by alternate extension and contraction.

Family. **THALASSEMADÆ.**—Proboscis with long, fleshy appendage. No tentacles. Vent at end of body.

Genus. THALASSEMA.—Body cylindrical, rounded, and smooth behind. Proboscis short.

Family. **PRIAPULIDÆ.**—Proboscis retractile. Tentacles none. Vents at end of long tail.

Genus. PRIAPULUS.—Body blunt behind. Tail branched and pointed.

Family. **SIPUNCULIDÆ**--Proboscis retractile, tentacled at tip. Vent at base.

Genus. SYRINX.—Proboscis shorter than body. Tentacles fingered.

Genus. SIPUNCULUS.—Proboscis long as body. Tentacles not fingered.

Order. **HOLOTHUROIDEA.**—Shell none. Skin with chalky netted pieces; a ring of same substance round gullet. Locomotion mostly by suckers.

Family. **PSOLIDÆ.**—Suckers three rows on oblong disc; two rows rudimentary.

Genus. PSOLUS.—Body irregular, slug-like. Tentacles ten.

Family. **PENTACTADÆ.** Suckers in five regular rows. Body irregular.

Genus. PENTACTA (includes CUCUMARIA). — Suckers alternate in each row, close. Tooth-plates nearly square. Tentacles ten.

Genus. PSOLINUS.—Body curved. Suckers distinct from each other. Under suckers bent. Teeth short. Tentacles ten

Family. **SYNAPTADÆ.**—Suckers none.
Genus. SYNAPTA.—Body cylindrical, slender, long. Tentacles long, and fingered at tips.

Order. **ECHINOIDEA.**—Shell chalky, either spherical or disc-like, formed of immoveable plates. Locomotion by spines and suckers.
Family. **CIDARIDÆ.** — Body globular. Mouth above and vent below, in centre of body. Five unbroken avenues. Teeth complicated.
Genus. ECHINUS.—Spines of same form, the knobs of shell not perforated.
Genus. CIDARIS. - Spines of several forms ; knobs of shell perforated.
Family. **SPATANGIDÆ.**—Mouth below, not in centre. Vent at one side. Avenues not continuous. Teeth wanting.
Genus. SPATANGUS. — Body heart-shaped, flattened. Avenues above and below. No dorsal impression.
Genus. AMPHIDOTUS.—Body heart-shaped, convex. An impression in four of the dorsal avenues.
Genus. BRISSUS. — Body heart-shaped, convex , dorsal impression enclosing avenues.
Family. **CLYPEASTERIDÆ.** — Mouth and vent both below ; mouth in centre, toothed. Avenues not continuous.
Genus. ECHINARACHNIUS. — Body nearly circular, very flat, e lge nearly entire. Avenues short. Vent between mou'a and hinder edge.
Genus. ENCOPE.—Body flat, nearly circular, lobed and pierced with keylike-shaped holes. South America.
Genus. ROTULA.—Body flat, front of disc cut into deep teeth ; hinder edge not lobed, nearly entire. South America.

Sub-class. **HYPOSTOMATA.**—Stomach sac-like, only one aperture. Body lobed or rayed.

Order. **ASTEROIDEA** — Shell composed of many moveable pieces. Skin sometimes leathery, sometimes calcareous. Digestive canal symmetrical.
Family. **ASTERIADÆ.** — Rays developed from body. Locomotion by suckers under rays.
Genus. ASTERIAS.—Body star-shaped. Rays flattish and short, edged with plates. Avenues edged with three sets of spines, two rows of suckers in each avenue.
Genus. URASTER.—Body rather conical, rays rounded, two rows of suckers in each avenue, and three sets of spines.
Genus. GONIASTER.—Body conical, edged with spiny plates. Avenues edged with cross rows of spines.
Genus. ASTERINÆ.—Body covered with short spines on both surfaces. Avenues edged with a single row of spines.
Genus. PALMIPES.—Body flat, thin, with small tufts of spines. Avenues edged by longitudinal tufts of spines.
Genus. SOLASTER. — Many rays, covered with tufts of spines. Avenues edged with three sets of spines, and furnished with two rows of suckers.
Genus. CRIBELLA.—Rays few, covered with spiny tubercles. Avenues edged with two sets of spines.
Family. **OPHIURIDÆ.**—Rays long, affixed to round disc like the body of an Echinois. Locomotion by suckers.
Sub-family. Ophiurina.—Rays simple.
Genus. OPHIURUS. — Rays scaly, prolonged at disc. Large shield-like plates at origin. Threads between spines simple.
Genus. OPHIOCOMA. — Rays scaly, not prolonged at disc. Small pentagonal plates at origin. Threads between spines pinnate.
Sub-family. Euryalina.—Rays much branched.
Genus. ASTROPHYTON.—Rays repeatedly two-branched.

Order. **CRINOIDEA.** — Digestive canal not symmetrical. Chalky shell formed of many moveable pieces.
Genus. COMATULA.—Rays five, pinnate. When adult, free ; when young, seated on stern.
Genus. PENTACRINUS.—Permanently on long footstalk, which is pentagonal and formed of numerous joints. Filaments placed in whorls on footstalk, which does not alter in diameter. Warm seas.

CLASS.—ACALEPHA.

Radiate, gelatinous, free—A digestive cavity with radiating vessels—Poison-threads—Metamorphosis.

Order. **SIPHONOPHORA.** — Body irregular. No central digestive cavity. With sucking organs. Locomotion by contractile cavity or sacs filled with air.
Family. **VELELLADÆ.**—Animal stretched over gristly plate.
Genus. VELELLA.—Body flat, oval. Oblique crest above ; tubular mouth below, surrounded by short tentacles.
Family. **PHYSALIADÆ.**—Animal a large air-filled sac, with many tentacles.
Genus. PHYSALIA.—Air-sac large, oblong, crest-wrinkled. Tentacles near one end, long, variously-shaped.
Family. **DIPHYIDÆ.** — Animals bell-shaped, double, fitting into each other.
Genus. DIPHYES. — Animals pyramidal, a few points round aperture.

Order. **CTENOPHORA.**—Body symmetrical, not disc-shaped. Large central digestive cavity. Locomotion by parallel rows of cilia.
Family. **CALLIANIRIDÆ.**—Two long, contractile filaments in chambers, one on each side.
Genus. CYDIPPE.—Body globular, no lobes nor tentacles.
Family. **BEROIDÆ.** — No contractile filaments nor chambers.
Genus. CESTUM.—Body very flat, wide, and short, edged with cilia. Hot seas.

Order. **DISCOPHORA.**—Body disc-shaped, oval, or less convex. Large central digestive cavity. Locomotion by expanding and contracting disc.
Tribe. *GYMNOPTHALMATA.*—Eye-specks uncovered or wanting. Circulating vessels running to margin, either simple or branched.
Family. **SARSIADÆ.**—Vessels simple, four. Ovaries in substance of footstalk.
Genus. SARSIA.—Hemispherical ; four tentacles at edge of vessels. Eye-specks four. Stomach in extensile footstalk.
Family. **WILLSIADÆ.**—Vessels branched.
Genus. WILLSIA.—Globular. Ovaries six. Vessels six. Eye and tentacle opposite each branch. Footstalk short.
Family. **ÆQUOREADÆ.** — Vessels more than eight. Disc flattish. Ovaries in a line along vessels.
Genus. ÆQUOREA.—Disc rather convex. Vessels very numerous. Footstalk wide, cleft into many broad lobes. Tentacles slender.
Genus. EUDORA.—Disc very flat. No tentacles nor footstalk. Hot seas.
Tribe. *STEGANOPTHALMATA.*—Eye-specks covered by flaps. Circulating vessels forming a network.
Family. **MEDUSADÆ.**—Stomach with a mouth, through which solid food is taken.
Genus. CHRYSAORA.—Disc hemispherical, lobed. Many tentacles on edge. Four very long furbelowed arms, not fringed.
Genus. MEDUSA.—Disc hemispherical. Many tentacles on edge. Eyes eight. Ovaries, chambers, and fringed arms four ; one opening in centre, four at sides.
Genus. CYANEA.—Disc flattened, rather scolloped ; four fringed arms. Tentacles of edge very numerous, set beneath disc.
Family. **RHIZOSTOMADÆ.** — No central orifice to stomach. Food received through ends of branching canals.
Genus. RHIZOSTOMA.—Ovaries four. Footstalk with four or eight semilunar apertures.

CLASS.—ZOOPHYTA.

Radiate, gelatinous, or fleshy, with tentacles round mouth —Stinging weapons.

Order. **ACTINOIDA.** — Internal cavity enclosing stomach, and divided into compartments by radiated walls. Germs thrown out of cavity at sea.

Sub-order. ACTINARIA. — Tentacles twelve or more. Seldom tubercled ; radiating partitions often depositing solid chalky plates.

Family. ANTHEADÆ.—Base adherent at pleasure. Column smooth, not pierced ; edge simple.

Genus. ANTHEA.—Base slightly adherent, broader than column. Summit of column expanding. Tentacles numerous, in several rows, very long, hardly retractile.

Family. ILYANTHIDÆ. — Base not adherent, lower extremity rounded and simple.

Genus. ILYANTHUS.—Tentacles of one kind, on edge. Column thick and pear-shaped. Mouth simple.

Genus. EDWARDSIA.—Column slender, with epidermis.

Family. SAGARTIADÆ. — Base adherent. Column pierced.

Genus. ACTINOLOBÆ. — Tentacles moderate. Disc retractile. Column with suckers.

Genus. SAGARTIA.—Column with suckers.

Family. BUNODIDÆ.—Column warty. Base adherent.

Genus. BUNODES.—Warts bold, arranged in vertical lines.

Family. ACTINIADÆ.—Edge headed.

Genus. ACTINIA.—Acontia wanting. Base much wider than column. Tentacles wholly retractile.

Family. LUCERNARIADÆ.—No coral. Tentacles in remote groups. (N.B.—Mr. Gosse, in his "British Sea-Anemones and Corals," excludes this family, believing it to be properly placed with the Medusæ.)

Genus. LUCERNARIA. — Body bell-shaped, adherent or free at will. Tentacles knobbed, in groups on angles of disc.

Family. FUNGIDÆ.—Polypary free, rounded, or oval ; convex above, concave below. Plates radiating from centre to circumference.

Genus. FUNGIA.—Round, the plates with very fine teeth. Hot seas.

Family. CARYOPHYLLEADÆ. — Tentacles crowded ; mouth capable of protrusion.

Genus. CARYOPHYLLEA.—Corallum simple, or fixed by adhesion. Plates converging in three series ; sponge-like mass of twisted plates at bottom of cup. Animal with mouth ribbed ; tips of tentacles knobbed.

Genus. LOPHOPHELIA.—Corallum branched ; stem long, solid, and smooth.

Family. GYROSA.—Cells united into waving furrows. Plates on both sides.

Genus. MEANDRINA.—Polypary stony and hemispherical. On convex side stars winding and lamellated. Hot seas.

Genus. EUPHYLLÆA.—Polypary leaf-like ; stars on both sides. Hot seas.

Family. MADREPORIDÆ. — Tentacles twelve, short ; polypary stony.

Genus. MADREPORA.—Cells lamellated, isolated ; interstices prominent. Warm seas.

Family. TUBULIPORIDÆ. Polypary composed of parallel tubes, fixed by transverse partitions.

Genus. TUBULIPORA.—As Family. Indian seas.

Family. CORTICATA.—Polypary fixed, branching ; axis solid and stony. Polypes retractile.

Genus. CORALLIUM.—Shaft uniform, rigid, with five longitudinal furrows. Mediterranean.

Genus. ISIS —Jointed ; the junctions horny. Hot seas.

Sub-order. ALCYONARIA.—Tentacles warty or plumed ; branchlets perforated. Corallum, when present, mostly chalky, sometimes horny, rarely flinty.

Family. GORGONIADÆ.—Fixed, plant-like, forming secretions from epidermis.

Genus. GORGONIA.—Axis horny ; not jointed. Crust soft and fleshy.

Family. PENNATULADÆ.—Free, and feather-shaped.

Genus. PENNATULA.—Shaft nearly cylindrical. Pinnæ two-ranked and spreading, having polypes on upper edges.

Genus. VIRGULARIA.—Polypary long and slender ; polypes on a series of curved projections.

Family. ALCYONIIDÆ.—Polypary attached, fleshy, with numerous chalky spicules ; irregular.

Genus. ALCYONIUM.—Polypary lobed ; pores star-shaped, from which the polypes emerge.

Order. HYDROIDA.—Internal cavity simple, increasing by buds from sides.

Family. TUBULARIADÆ.—Buds not permanent, near base of tentacles. Animals naked, or in horny tubular envelope.

Genus. TUBULARIA.—Inclosed. Not branched. Polypes with two circles of tentacles ; egg-germs on footstalks.

Family. SERTULARIADÆ.—Egg-germs in vesicles. Buds permanent, in horny envelope. Polypes in sessile cells.

Genus. SERTULARIA.—Plant-like ; stem tubular, branched, and jointed. Cells alternate, or in pairs on opposite sides.

Genus. PLUMULARIA.—Simple or branched, feathery ; cells mostly in angle of branches. Egg-vesicles scattered.

Family. CAMPANULARIADÆ.—Cells on footstalks, rest as in preceding family.

Genus. CAMPANULARIA.—Creeping or erect ; cells at ends of long-ringed stalks or in whorls ; embryos medusa-shaped.

CLASS.—ROTIFERA.

Microscopic, contractile, with cilia at front of body, resembling wheels when in motion—Intestines distinct ; apertures two, one at each extremity of body.

System as yet uncertain.

CLASS.—RHIZOPODA.

Form mutable, no external organs or cilia—Moving by processes protruded or withdrawn at pleasure.

System as yet uncertain.

CLASS.—INFUSORIA.

Microscopic—Form often mutable ; no apparent vessels or nerves—Mostly with internal cavities.

System as yet uncertain.

CLASS.—PORIFERA.

Form various, fixed, composed of horny fibres or flinty or calcareous spiracles.

System as yet uncertain.

INDEX.